An anthology
of articles on
Sound Reinforcement
from the pages of the
Journal of the
Audio Engineering
Society
Vol. 15–Vol. 44
(1967–1996)

Copyright © 1996 Audio
Engineering Society, Inc.
Library of Congress Catalog
Card No. 78-61478.
ISBN No. 0-937803-30-8.
Selections from Volumes 15 through 44,
1967–1996. First printing 1996 August
(all rights reserved). Except for brief
passages to be used in review
or as citation of authority,
no part of this book may be
reproduced without prior
permission from:
Audio Engineering Society, Inc.,
60 East 42nd Street, New York,
New York, 10165-2520, USA.
Telephone: 212 661 8528.
FAX: 212 682 0477.

preface

Since the first *Sound Reinforcement* anthology of papers published in the *Journal of the Audio Engineering Society* was issued in 1978, advances in the state of the art in this technology have included: 1) improvements in the ability to control loudspeaker directional characteristics; 2) greater understanding of loudspeaker interaction and coverage effects for more than one loudspeaker in a cluster; 3) more accurate and easier prediction and evaluation of intelligibility; 4) greater application of digital processing and control; and 5) introduction and widespread application of computer sound system adjustment and control.

These advances are reflected in the papers chosen from the *Journal* for inclusion in this anthology.

John M. Eargle and Mark R. Gander made sufficient contribution to the selection process to be considered co-editors. I am sincerely grateful to them.

David Lloyd Klepper 1996 January
White Plains, New York, USA
and
Jerusalem, Israel

contents

A. ELECTRONICS AND ELECTRONICS–TRANSDUCER INTERACTION

Computer-Controlled Sound Systems. Bob Moses (Vol. 42, pp. 938–955 [1994 November]) 10

Theory and Design of a Digital Audio Signal Processor for Home Use. David Griesinger (Vol. 37, pp. 40–50 [1989 January/February]) 22

Multiple-Point Equalization in a Room Using Adaptive Digital Filters. S. J. Elliott and P. A. Nelson (Vol. 37, pp. 899–907 [1989 November]) 33

Effects of Cable, Loudspeaker, and Amplifier Interactions. Fred E. Davis (Vol. 39, pp. 461–468 [1991 June]) 42

Considerations in Grounding and Shielding Audio Devices. Stephen R. Macatee (Vol. 43, pp. 472–483 [1995 June]) 50

Fundamentals of Grounding, Shielding, and Interconnection. Kenneth R. Fause (Vol. 43, pp. 498–516 [1995 June]) 62

B. LOUDSPEAKER PERFORMANCE AND THE LOUDSPEAKER–ROOM INTERFACE

Acoustical Measurements by Time Delay Spectrometry. Richard C. Heyser (Vol. 15, pp. 370–382 [1967 October]) 82

Determination of Loudspeaker Signal Arrival Times: Part 1. Richard C. Heyser (Vol. 19, pp. 734–743 [1971 October]) 95

Determination of Loudspeaker Signal Arrival Times: Part 2. Richard C. Heyser (Vol. 19, pp. 829–834 [1971 November]) 105

Determination of Loudspeaker Signal Arrival Times: Part 3. Richard C. Heyser (Vol. 19, pp. 902–905 [1971 December]) 111
Editor's note. David Lloyd Klepper 115

Acoustical Impulse Response of Interior Spaces. Richard G. Cann and Richard H. Lyon (Vol. 27, pp. 960–964 [1979 December])...............................116
Comments. Don Davis (Vol. 28, p. 817 [1980 November]) 120
Authors' Reply. Richard H. Lyon and Richard G. Cann (Vol. 28, p. 817 [1980 November]) 120

Central Cluster Design Technique for Large Multipurpose Auditoria. E. T. Patronis, Jr. and Catharina Donders (Vol. 30, pp. 407–411 [1982 June]) 121

Loudspeaker Coverage by Architectural Mapping. Ted Uzzle (Vol. 30, pp. 412–424 [1982 June]) 126
Editor's note. David Lloyd Klepper 138

The Design of Distributed Sound Systems from Uniformity of Coverage and Other Sound-Field Considerations. Rex Sinclair (Vol. 30, pp. 871–881 [1982 December]).....139
Corrections. (Vol. 31, p. 256 [1983 April]) 149

An Accurate and Easily Implemented Method of Modeling Loudspeaker Array Coverage. John R. Prohs and David E. Harris (Vol. 32, pp. 204–217 [1984 April]) 150

Computer Simulation of Loudspeaker Directivity. David G. Meyer (Vol. 32, pp. 294–315 [1984 May]) 164

Digital Control of Loudspeaker Array Directivity. David G. Meyer (Vol. 32, pp. 747–754 [1984 October]) 186

A Graphic Method for Choosing and Aiming Loudspeakers for Reinforcement. Peter W. Tappan (Vol. 34, pp. 269–277 [1986 April]) 194
Comments. Peter W. Tappan (Vol. 34, p. 664 [1986 September]) 203

Horn Layout Simplified. Bob Thurmond (Vol. 35, pp. 976–983 [1987 December]) 204

Cluster Suitability Predictions Simplified. Bob Thurmond (Vol. 36, pp. 337–341 [1988 May]) 212

Introduction to Special Issue on Loudspeaker Arrays. Neil A. Shaw (Vol. 38, p. 203 [1990 April]) 217

Measurement and Estimation of Large Loudspeaker Array Performance. Mark R. Gander and John M. Eargle (Vol. 38, pp. 204–220 [1990 April]) 218

Multiple-Beam, Electronically Steered Line-Source Arrays for Sound-Reinforcement Applications. David G. Meyer (Vol. 38, pp. 237–249 [1990 April]) 235

Prediction of the Full-Space Directivity Characteristics of Loudspeaker Arrays. Kenneth D. Jacob and Thomas K. Birkle (Vol. 38, pp. 250–259 [1990 April]) 248

An Array Filtering Implementation of a Constant-Beam-Width Acoustic Source. Jefferson A. Harrell and Elmer L. Hixson (Vol. 38, pp. 221–230 [1990 April]) 258

Low-Frequency Sound Reproduction. Mark E. Engebretson (Vol. 32, pp. 340–346 [1984 May]) 268

Near-Field and Far-Field Performance of Large Woofer Arrays. G. L. Augspurger (Vol. 38, pp. 231–236 [1990 April]) 275

Large Arrays: Measured Free-Field Polar Patterns Compared to a Theoretical Model of a Curved Surface Source. John Meyer and Felicity Seidel (Vol. 38, pp. 260–270 [1990 April]) 281

Comparative Performance of Three Types of Directional Devices Used as Concert-Sound Loudspeaker Array Elements. Paul F. Fidlin and David E. Carlson (Vol. 38, pp. 271–295 [1990 April]) 292

Effective Performance of Bessel Arrays. D. (Don) B. Keele, Jr. (Vol. 38, pp. 723–748 [1990 October]) 317

Microalignment of Drivers via Digital Technology. John A. Murray (Vol. 42, pp. 254–264 [1994 April]) 343

C. RECONCILING SPEECH AND MUSIC

Sound Reinforcement Systems in Early Danish Churches. Dan Popescu (Vol. 28, pp. 713–717 [1980 October]) 356

The Acoustical Design of a 4000-Seat Church. A. H. Marshall, C. W. Day, and L. J. Elliott (Vol. 35, pp. 897–906 [1987 November]) 361

A Holographic Approach to Acoustic Control. A. J. Berkhout (Vol. 36, pp. 977–995 [1988 December]) 371

Acoustical Negative Feedback for Gain Control. David Klepper (Vol. 39, pp. 64–65 [1991 January/February]) 390

The Acoustics of St. Thomas Church, Fifth Avenue. David Lloyd Klepper (Vol. 43, pp. 599–601 [1995 July/August]) 392

D. INTELLIGIBILITY EVALUATION

Acoustic Feedback—Its Influence on Speech Intelligibility. Sverre Stensby,
Asbjørn Krokstad, and Svein Sørsdal (Vol. 27, pp. 887–890 [1979 November]) 396

Those Early Late Arrivals! Mr. Haas, What Would You Do? Cecil Cable and
R. Curtis Enerson (Vol. 28, pp. 40–45 [1980 January/February]) 400

The Practical Application of Time-Delay Spectrometry in the Field. Cecil R. Cable
and John K. Hilliard (Vol. 28, pp. 302–309 [1980 May]) 406
Comments. F. J. Fahy (Vol. 29, p. 262 [1981 April]) 414
Authors' Reply. C. R. Cable and J. K. Hilliard (Vol. 29, pp. 262–264
[1981 April]) 414

The Modified Hopkins–Stryker Equation. Don Davis (Vol. 32, pp. 862–867 [1984
November]) 417

Subjective and Predictive Measures of Speech Intelligibility—
The Role of Loudspeaker Directivity. Kenneth D. Jacob (Vol. 33, pp. 950–955
[1985 December]) 423
Comments. Farrel M. Becker (Vol. 34, pp. 560–561 [1986 July/August]) 429
Additional comments. Don Davis (Vol. 34, pp. 561–562 [1986 July/August]) 430
Author's reply. Kenneth D. Jacob (Vol. 34, pp. 562–563 [1986 July/August]) 431

Measurement of %AL$_{cons}$. Carolyn P. Davis (Vol. 34, pp. 905–909
[1986 November]) 433

Application of Speech Intelligibility to Sound Reinforcement. Don Davis and
Carolyn Davis (Vol. 37, pp. 1002–1019 [1989 December]) 438

Correlation of Speech Intelligibility Tests in Reverberant Rooms with Three
Predictive Algorithms. Kenneth D. Jacob (Vol. 37, pp. 1020–1030
[1989 December]) 456
Editor's note. David Lloyd Klepper 466

Speech Intelligibility in Some German Sports Stadiums. K. Rijk, F. Breuer, and
V. M.A. Peutz (Vol. 39, pp. 39–46 [1991 January/February]) 467
Correction. (Vol. 39, p. 266 [1991 April]) 475

Use of the L. G. Marshall–Crown–Techron ELR Program for Adjusting Digital
Units in Sound Reinforcement Systems. David Lloyd Klepper (Vol. 43,
pp. 942–945 [1995 November]) 476

An Analysis Procedure for Room Acoustics and Sound Amplification Systems
Based on the Early-to-Late Sound Energy Ratio. L. Gerald Marshall
(Vol 44, pp. 373–381 [1996 May]) 480

E. SPECIAL SYSTEMS

Conference Systems Using Infrared-Light Techniques. H.-J. Griese (Vol. 27,
pp. 503–506 [1979 June]) 490

A.

electronics and electronics–transducer interaction

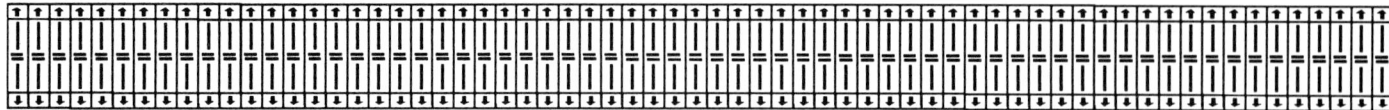

Computer-Controlled Sound Systems
Transformation of the Audio Industry in the Information Age

Bob Moses
AES Member

Rane Corporation
Mukilteo, Washington, USA

Computer control has become a dominant issue in our industry, coinciding with the public's fervor over the information superhighway and the commercialization of digital media technologies. The audio industry is facing a major transformation from traditional analog-based systems, to systems that integrate powerful new digital information technologies. Foremost in everyone's mind are two questions: how will computer control change sound systems, and when will this technology be ready for widespread application? Unfortunately, there are no simple answers to these questions. How and when computer control proliferates depends on a number of random and unpredictable factors. The transformation will occur naturally once the technology demonstrates enough value to overcome its cost and complexity. This article explores how computer control might change the implementation and operation of future sound systems, and how we can influence the process of this transformation.

FUTURE SYSTEMS...

It's 6:30 pm and the band still hasn't arrived for sound check. The show is scheduled to begin in just one hour. Several years ago the sound engineer would be in a panic. But tonight she is confident the show will go smoothly with or without a sound check because she has already recalled the entire system configuration from memory, stored after the group's last performance in this club several months ago. She's thankful she stored everything: the house mix and EQ, monitor mixes and EQ, the settings of the effects processors, etc. Even the light cues. When the band arrives, they should be able to plug in and play. Let's hope they show up...

Meanwhile, across town, evening services are about to begin at a small church. A wedding was held in the church earlier in the day, and the DJ left the sound system configured for playing dance music after the reception. Several years ago the preacher would have struggled with a plethora of switches and dials as he attempted to reconfigure the sound system to reproduce his lavalier microphone, at a proper level, with no feedback. But this church has a new computer-controlled sound system that provides a single knob on the lectern selecting between "minister," "organ+minister," "choir," "reception," and so on. When the minister makes a selection, all the audio equipment in the system recalls a preset configuration preprogrammed by a consultant. The DJ's oversight is not a problem tonight. The minister twists the knob, selects "organ+minister," and cues the organist.

A couple of months ago the minister complained to the consultant that the volume was too low during "choir" mode. The consultant was 2000 miles away and did not wish to make a house call to the church, so he dialed into the system from his office via modem and modified the preset remotely. Both people were very pleased that it was so easy and inexpensive to calibrate the system.

The consultant runs a prosperous business maintaining dozens of these computer-controlled systems around the country. He implements them using a multi-vendor computer-controlled system. Most systems are based on the same basic set of generic hardware devices, and are customized by creating an application-specific software control panel. A block diagram of the system is provided in Fig. 1. The reader will recognize that the audio portion of this system is similar to most traditional systems, except for the addition of a computer-controlled network.

When setting up a system, the consultant hires an industrial engineer with expertise in ergonomics to design the control panels, and configures the audio equipment himself. The application in this church was designed to allow the minister (who knows nothing about operating a sound system) to control the system with a single knob mounted on his lectern (Fig. 2). The evening service is organized as a sequence of scenes, such as: "minister," "choir," "minister over CD," and so on. Each scene is a snapshot of the settings on every device in the system, and is selected by twisting the knob. Several examples of scenes are provided in Table 1.

During each scene, the mutes and faders on the mixer are set for proper source selection and mix. Overall volume is set via the mixer and amplifier level controls. (To maintain optimal gain structure, gain is turned up at the mixer, and turned down at the amplifier.) The CD and tape players are either stopped or playing. The equalizer is set with one of two preset curves. The first EQ preset is used when a person's microphone is open, and has a boosted midrange (to give presence to the voice) and sev- ➥

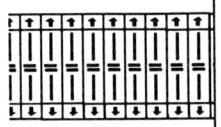

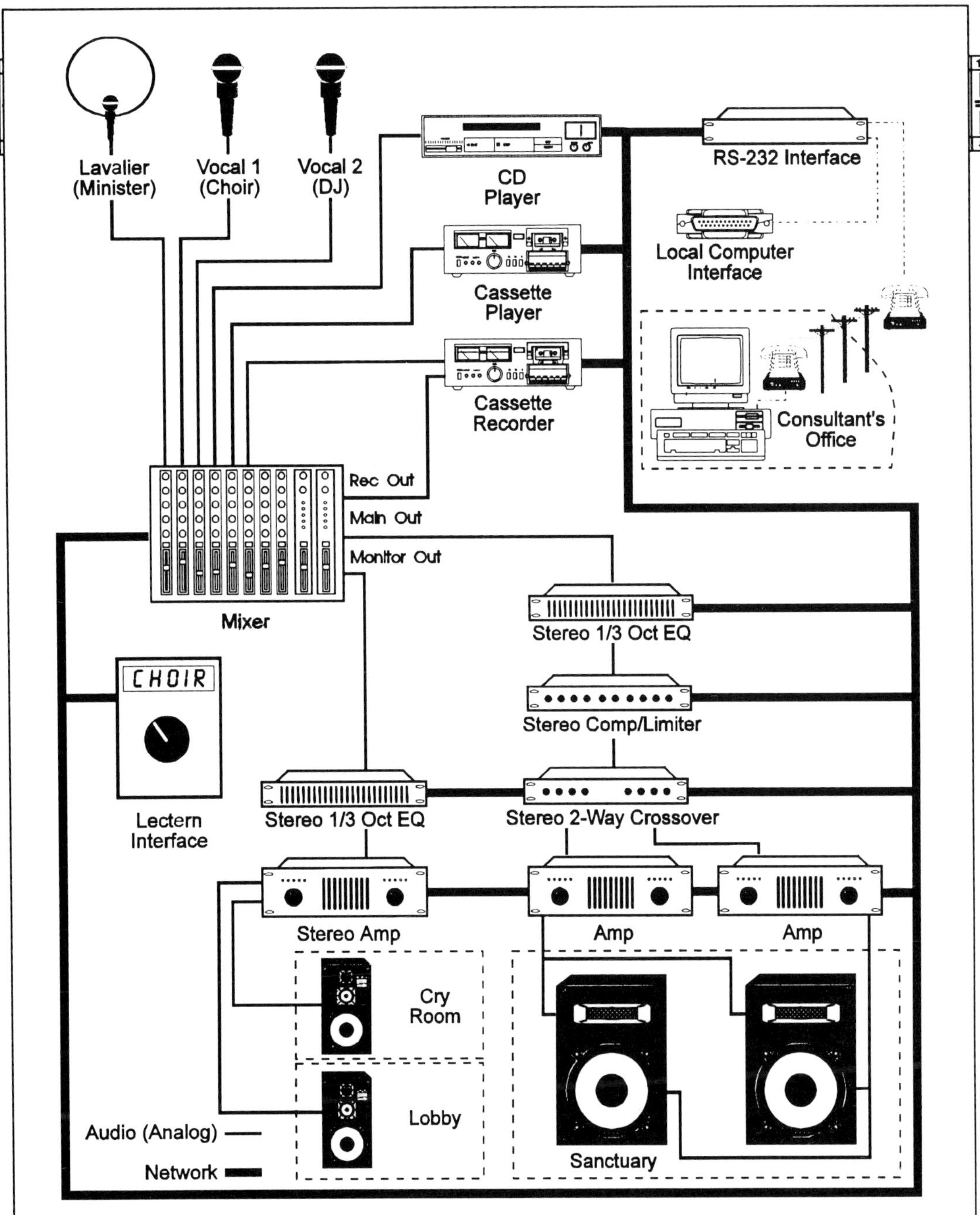

Fig. 1. Computer-controlled church sound system

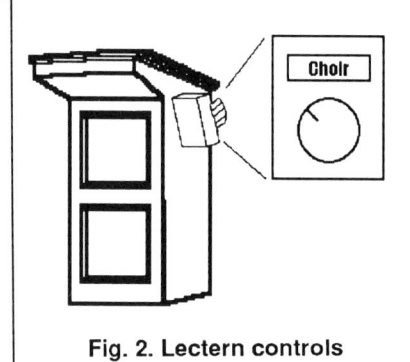

Fig. 2. Lectern controls

Table 1. Typical scenes for a church service							
Scene	Source(s)	Amp (volume)	CD Player	Tape Deck	EQ	Compressor
Minister-sermon	minister mic	medium	stop	stop	voice	on
Minister-meditation	minister mic	low	stop	stop	voice	on
Choir	choir mic	medium	stop	stop	music	off
Minister over CD	minister mic, CD player	medium	play	stop	voice	on
Music-CD	CD player	medium	play	stop	music	off
Music-tape	tape player	medium	stop	play	music	off
Reception	CD player	loud	play	stop	music	off

Computer-Controlled Sound Systems

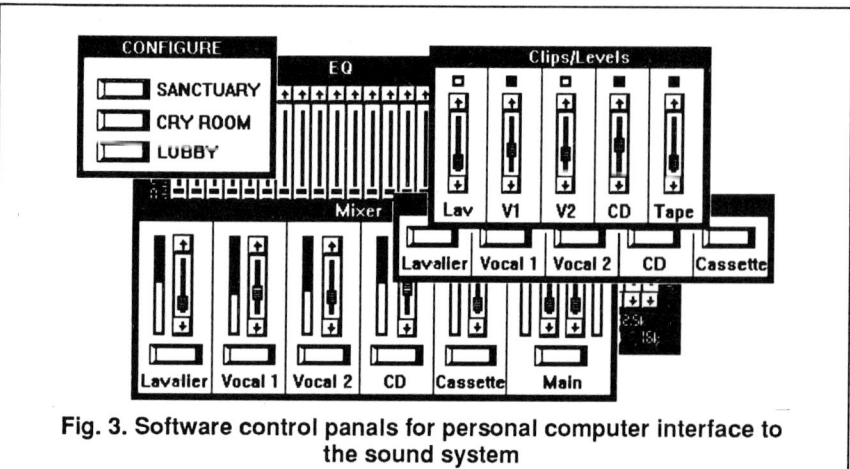

Fig. 3. Software control panels for personal computer interface to the sound system

eral notches for combating feedback. The second EQ preset is the common "smile" curve with boosted high and low frequencies. This curve is used while the system is playing music. The compressor is also preset for each scene. Compression ratio is increased while the minister speaks, since his voice varies from quiet meditations to charismatic outbursts. Compression is disabled during music passages to preserve the dynamics of the music.

The equipment in this system is interfaced with an industry standard local area network. The human interface to the network is provided via a single selector knob (mounted on the lectern) which recalls network presets. The network also supports an RS-232 port for a personal computer (PC), available to knowledgeable operators who wish to fine tune the system. The user interface software running on this PC is composed of control panels designed especially for this church (Fig. 3), and allows an operator to control and monitor every parameter on every device on the network. Special control panels have been created with logically grouped controls, such as all the source mutes, all the clip lights and corresponding level controls, all the equalizer filters, and so on. A modem is also connected to the PC, allowing someone to log on and operate the system remotely.

Meanwhile, the band just arrived at the nightclub mentioned earlier. The show is scheduled to begin in 10 minutes, and the sound engineer is scurrying to connect their instruments to the sound system. A diagram of this system is provided in Fig. 4. This system was designed and installed by the same consultant who did the church across town. Many of the hardware components in the two systems are the same. The major differences between the two systems are the number of channels and the user interface. The one-knob interface to the church's sound system would never suffice for this system. The club's user interface is organized into a series of hardware and software control panels optimized for various modes of operation. There is a system setup control panel that provides access to signal routing, gain structure controls, room equalization, and so on. There are also a number of control panels that provide run-time front ends to the mixer, effects processors, amplifiers, and so on. These control panels are displayed on a large touch screen monitor in the mix island. A hardware console is also provided with generic switches, faders, meters, etc. The controls on this console can be assigned to device parameters anywhere in the system, and familiar devices such as a graphic equalizer or a mixer can be emulated.

Of course, the manager of this club is not well-versed at operating sound systems, and prefers an interface consisting of a minimum number of simple controls. When there is no live music, the sound system is used to play CDs and cassettes. The only controls the consultant dared trust the manager with were source and volume. Source is restricted to CD or cassette tape player, and volume is restricted to a safe operating range. These two user options are adjusted via two wall-mounted knobs (Fig. 5). One knob, a 2-position rotary switch, selects between the CD and cassette player. It is mapped to the two channel mutes corresponding to these sources on the mixer. The other knob is mapped to the amplifier volume controls, with a restricted range. The manager is shielded from the sound equipment and its complicated user interfaces. As far as he is concerned, this system is composed of just the CD and cassette players. The lower level complexities are hidden from him, as they should be.

THE AUDIO INDUSTRY MEETS THE INFORMATION AGE

It is intriguing to imagine future systems such as these. Indeed, digital technologies, especially information technologies, have captivated the public's attention. Today, it is impossible to hide from terms such as "information superhighway," "interactive multimedia," "virtual reality," "global village," "cyberspace," and so on. These fads have certainly found their way into the audio industry. Audio is one of the most important threads in the tapestry of converging digital media. As its practitioners, we are faced with the opportunity (and responsibility) of guiding audio technology into the information age. As with most radical new technological developments, this responsibility initially rests on the shoulders of early visionaries (often considered to be heretics). But as the vision catches on, these new technologies will touch us all. Naysayers should be careful not to ignore history. When the Bell Telephone Company (known as AT&T today) began service in the late 1800s, most people dismissed the telephone as a novelty with limited commercial value. Today, the company is the largest in the world, employing over 1,000,000 people. Similar experiences were shared by the inventors of the personal computer and other ubiquitous technologies.

Today's technological visionaries are predicting an age when high-speed local area networks will carry information between devices in a system, facilitating interoperability between devices, and freeing the human operator from a never-ending struggle. There are strong indicators that such a transformation may occur very soon. Dozens of profes-

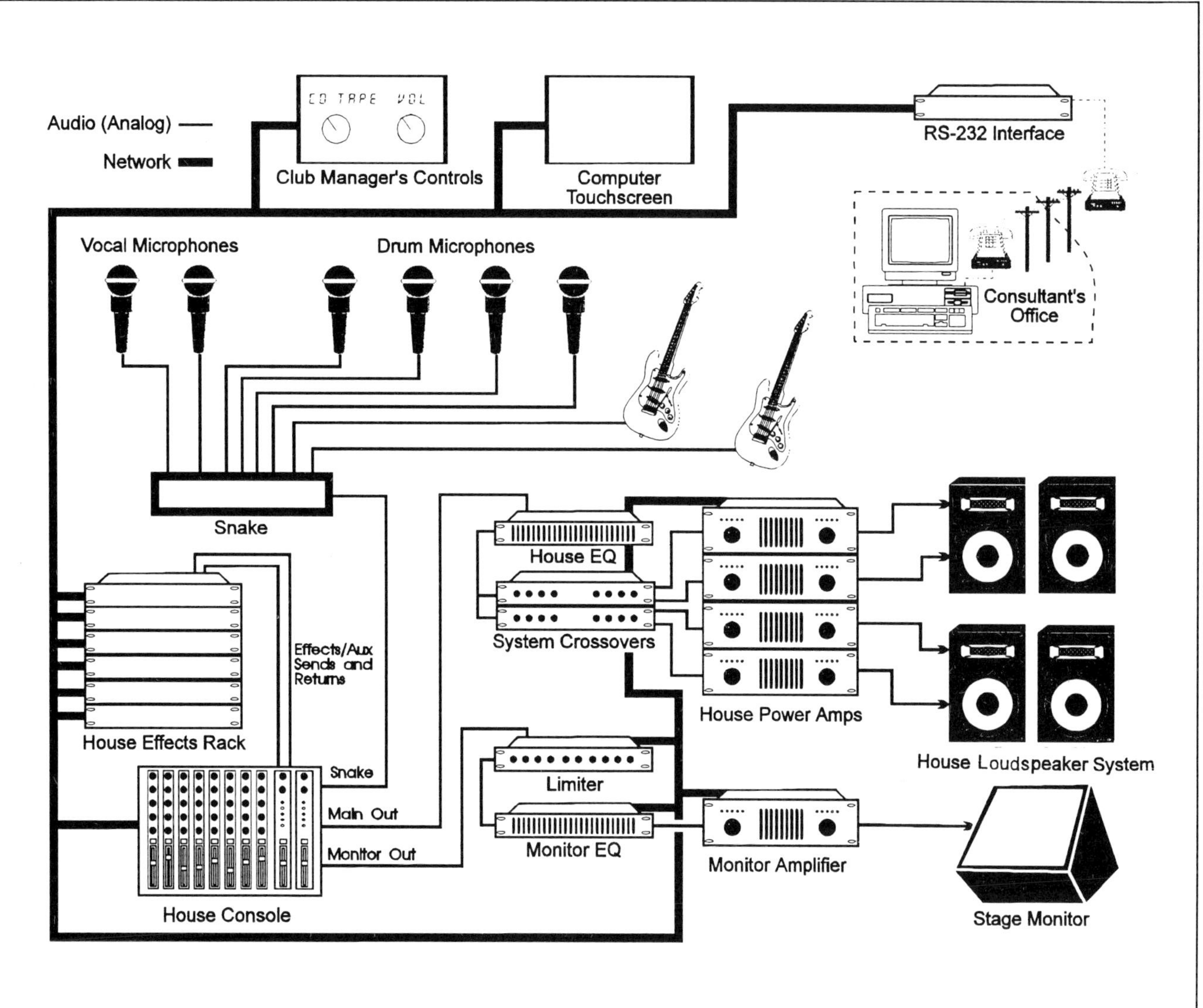

Computer-Controlled Sound Systems

Fig. 4. Computer-controlled nightclub sound system

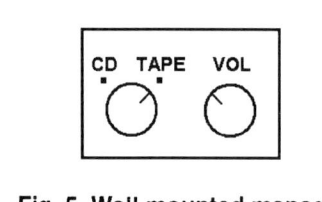

Fig. 5. Wall-mounted manager controls

sional audio manufacturers have announced their intention to produce equipment conforming to one computer control system or another. The trade press is laden with feature articles about this technology. And the Audio Engineering Society Standards Committee SC-10 Subcommittee on Sound System Control has attracted over 100 participants from all corners of the industry in its effort to create local area networking standards for audio systems.

But no amount of hype or technical wizardry will ever replace the market economy. If computer control is expensive, unreliable, too complex, or otherwise inappropriate, it will fail. But if computer control provides great value, it will flourish. Therefore, to gain insight into when computer control will finally become ubiquitous, and what form it will take, we must examine how it provides value to sound systems.

THE VALUE OF COMPUTER CONTROL

Computer control offers designers and users of sound systems a number of benefits. Some of these are:
- Centralized control
- Repeatable, accurate settings
- Interoperability and automation
- Enhanced human interface
- Multimedia synchronization
- Economy.

Centralized Control

A computer-controlled system often provides a central point of control so that all functions can be operated and monitored from a single position. The ideal system is controllable from a number of arbitrary positions, such as the front-of-house sound booth, the monitor mixer position, backstage, behind the bar, or even remote via modem. This gives the human complete freedom to operate the system wherever it is most appropriate and convenient.

It is important to distinguish between centralized control and master driven systems. By centralized control, we mean everything in the system can be controlled from one position. This position is arbitrary and there may be multiple positions. A master-driven

Computer-Controlled Sound Systems

system can only be controlled by a single master. This limitation often prevents such a system from being usable in venues where several people may need access to the controls.

Aside from operator convenience, a centralized system topology offers a more important advantage: the elimination of conflicts in the system. To understand how, we must consider an important aspect of systems theory called complexity. Complexity varies from a low level system, in which all components of the system are autonomous with no external control, to a high level system, in which a master has absolute control over all the components. Low level systems are generally better at responding to local needs, are simpler and cheaper, and are more flexible. On the other hand, low level systems are prone to conflicts as autonomous entities compete over limited resources. A high level system reduces conflicts (since all control is concentrated in one place), but centralizing control leads to increased information transfer, therefore increasing complexity and cost, and reducing efficiency. The rule of thumb for systems complexity states: Make every decision at the lowest possible level, but be prepared to shift control of the system to a higher level if conflicts occur.

Perhaps an example will shed more light on the above concepts. Consider the situation in medieval England in which a common pasture was shared by all the members of a community to graze their livestock. Each livestock owner believed that the more cows he owned, the more food he could provide to his family and sell to his neighbors. Since grazing was free, it was to his benefit to increase the size of his herd as fast as possible. Unfortunately, as each farmer increased his herd, the cattle ate the grass faster than it could grow. The grass disappeared, the cows starved, people fought over limited food supplies as they starved, and the system collapsed. This illustrates the conflicts that can occur in an extremely low level system.

The solution to this problem is for each farmer to give up enough autonomy to coordinate controlled use of the land. This represents a migration upward from an extremely low level system to a higher level system. One might be tempted to take this to the extreme and implement a system at the highest level, with complete central control. Unfortunately, a completely centralized system also suffers. How can the cows and land be divided equally? How does the bureaucracy survive without levying burdensome (and counter productive) taxes (wasting system resources)? How can resources be shared among the system entities without the master becoming a bottleneck and slowing efficiency? If the master breaks down, the entire system is unable to do anything. Indeed, both extremes are problematic, and a rule of thumb must be employed to provide control of the system at the optimal level. In the above example this could be accomplished by limiting the amount of property available to each farmer, giving him the incentive to make the best of it.

But what about audio systems? This theory seems to contradict the assertion that centralized control is advantageous in a computer-controlled sound system. It would seem most appropriate to provide control of the system at a very low level, for example, at the front panel of each device, instead of a single computer screen. This is exactly the situation we have had for many years with systems composed of racks of analog devices with manual controls. And as the complexity theory predicts, these low level systems have experienced many conflicts.

Consider the church example, before a computer-controlled system was installed. Conflicts between the human operator and equipment were common. This church (as with most) could not afford to hire a professional sound engineer, so it relied on the ability of laypeople to operate the sound system. Before the advent of computer control, these laypeople were confronted with a sea of mysterious knobs on the mixer and signal processors. Mistakes were common as the operators struggled with improper mix levels, feedback, misadjusted compressor/limiter ratios and thresholds, and so on. The operator, preoccupied with being part of the actual event, would often miss a cue to turn the tape deck on, or to mute a microphone when a speaker made his or her exit. Complicating the problem was the physical layout of the system. Some equipment was located in a sound booth in the church's sanctuary, while the rest of the equipment was located in a backroom closet. It was not possible to control or monitor all equipment from either location. Indeed, the system was too complicated and intimidating to many operators, which is why the church hired the consultant to install a computer-controlled system.

Conflicts in this system were not limited to human–machine interaction. Because the system was composed of autonomous signal processors, it was prone to conflicts between competing devices. A prime example was the interaction between the anti-feedback equalizer, sweetening equalizer, and gain controls. The feedback equalizer notched out several frequencies in the vocal range in its attempts to keep feedback under control. The sweetening equalizer was adjusted to obtain the best possible sound, and therefore attempted to undo the work of the feedback equalizer by generating a curve that boosted the vocal range. Raising the gain of the system only made matters worse. The result was chaos, two angry equalizers, a frustrated sound engineer, and annoyed listeners.

Indeed, conflicts were commonplace in this low level system. Computer control added enough centralization, that is, raised the level of control, to eliminate most of them. By programming presets into the system and providing a single-user interface (the knob on the lectern), conflicts between the human operator and the system were eliminated. The one-knob interface is not acceptable for a skilled operator, however, consistent with the warning in complexity theory not to get too carried away with an extremely high level system. The RS-232 interface to the system provides access to these low level controls, providing the balance between high and low level control. Equipment conflicts are also reduced by creating presets (that is, coordinated use of the resources in the system) with proper gain structure, equalization, etc., programmed by an expert.

What would happen if control was at too high a level? Complexity theory predicts that the efficiency of the ➡

system would suffer, as would complexity and cost. A system with a single master that controls all the slaves would represent this extreme situation. It should be obvious to the reader that this system would exhibit the predicted problems of lowered efficiency (the master becomes a bottleneck) and a disastrous failure mode (if the master dies, the system halts). It is commonly accepted in networking that peer-to-peer systems provide the optimal level of complexity. In a peer-to-peer system, all entities are able to communicate directly with each other (therefore eliminating the communication bottlenecks associated with a master–slave system), and are therefore more efficient and flexible.

In conclusion, the optimal level of system control allows devices to communicate freely (as in a peer-to-peer network), but provides enough centralization to eliminate the conflicts that occur when devices are unaware of their environment. The winning computer control system will be the one that maintains this delicate balance.

Repeatable, Accurate Settings

Computers are much better at accurately adjusting parameters on equipment than people are. When you instruct a computer to set a control to 10.0 dB, you get 10.0 dB, not 10.5. Computers can easily memorize settings, thereby allowing entire system configurations to be saved and recalled later. This can be extremely useful. For example, a live sound system is often shared by two or more bands in any one night. During sound check, the sound engineer painstakingly adjusts the system to tune out feedback in the monitors, equalize the room, get a good mix, and so on. But if his or her band is not first on the bill, these adjustments will be altered during the earlier performance(s). A computer-controlled sound system allows the sound engineer to store the system configuration after sound check so it can be recalled later. In fact, this system setup can be recalled next time the band returns to the club, months or even years later. Each time the system is used, the configuration can be adjusted to get incrementally better performance from it. The sound engineer does not have to start from scratch every time the band returns to the venue.

Interoperability and Automation

A computer-controlled system functions by transmitting digital information between equipment. There is no reason this digital information has to originate from human control. A computer may generate information and share it with devices in the system without the human ever being involved. This allows powerful automation to be programmed into the system. Interoperability allows system entities to communicate among themselves, without human intervention. Getting the human out of the loop is often the best thing that can be done for the system (and the human).

Interoperability is possible when control objects can send and receive messages to and from each other. Automation is possible when a process algorithm can be placed in the message flow path. Consider two objects in a system, connected through an algorithm. One object is the source object, the other is the destination object. The source object outputs a message that is altered by the algorithm. The output of the algorithm is received by the destination object. The algorithm could be a signal processing function, a security clearance, or virtually any other function. The algorithm can be preprogrammed by the system installer or it can be learned dynamically by an expert system (artificial intelligence). Through this interoperability model, objects can control other objects, limited only by the complexity of the algorithm.

As an example, consider the church system's computer-controlled graphic equalizer. Some of its filters are set to full cut to remove frequencies susceptible to feedback. The equalizer is also used to tame the room response and sweeten the sound. Clearly, feedback suppression has priority over the other functions. The equalizer must not be allowed to increase the gain in the trouble spots. A system with interoperability could place an algorithm between the user's equalizer controls (objects on a virtual control panel on a computer screen) and the actual hardware equalizer filters (Fig. 6). This algorithm disallows control in the troublesome frequency bands. In addition, another algorithm could interface the system's main volume controls to the depth of the notch filters so that the notch filters deepen as gain is increased, maintaining a safe gain-before-feedback margin (Fig. 7). An advantage of the latter algorithm is that it also relaxes the depth of the notches when system gain is low, allowing the settings for sweetening to become active. Another algorithm could bind the setting of the main volume control to the equalizer, allowing a dynamic "loudness" contour to follow the gain of the system. As gain is turn down, the high and low frequencies are raised to maintain the perception of loudness. As gain is turned up, the loud-

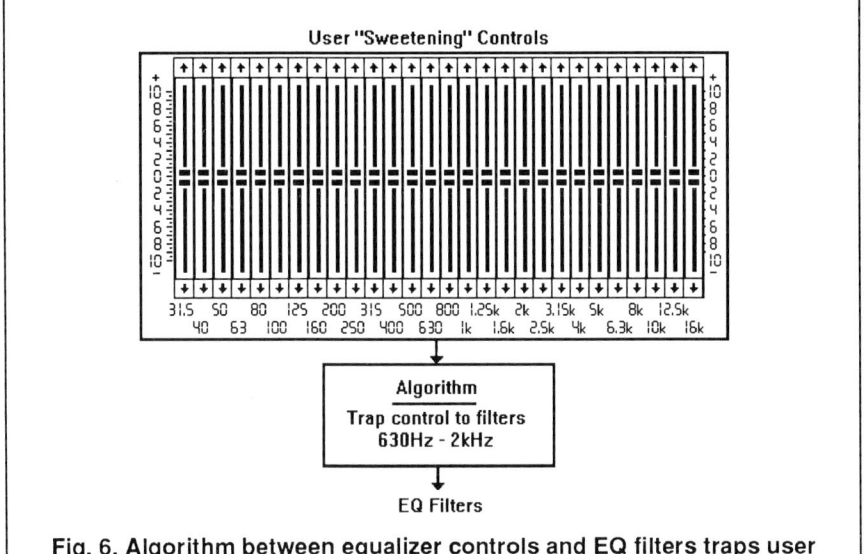

Fig. 6. Algorithm between equalizer controls and EQ filters traps user attempts to alter critical frequency bands

ness contour relaxes.

Other powerful examples of interoperability are:

• Air conditioning system controls are fed through an algorithm that increases system gain when air conditioning is turned on (to overcome the high ambient noise of air conditioning).

• Clipping indicators are fed through an algorithm that lowers gain as faults are detected.

• Output of a real-time analyzer is fed through an algorithm that controls an equalizer and delay to maintain a specified room response, or too seek and destroy feedback.

• An algorithm could average the response of a VU meter, and adjust a compressor/limiter to maintain a constant signal level within a specified range.

Ultimately, future systems may incorporate artificial intelligence and expert systems. An expert system observes how a human operator reacts to certain conditions in the system and learns how to make high level decisions automatically. For example, an expert system could learn that every time an audio channel clips on the mixer, the corresponding level control is turned down a few dB. Or, whenever the lavalier microphone is in the mix, the compressor is adjusted to maintain a constant signal level as the speaker turns his head from side to side.

Indeed, interoperability allows resources to work together as meta-resources, thereby realizing the benefits of a higher order control system (see Centralized Control above). Ultimately, the computer control system that provides the most efficient and flexible means for interoperability will succeed in the marketplace.

Enhanced Human Interface

People use a combination of visual, auditory, and kinesthetic means to interact with the world. Of course, everyone uses a different combination of these styles. Unnatural, or unfamiliar, interfaces are inherently more difficult to operate than natural/familiar ones. The ultimate goal of a good human interface is to interact with the person in the person's world, instead of the computer's world, by creating sounds, images, and experiences that people are accustomed to. For example, reaching out a hand and twisting a virtual knob with a data glove (or a real one on a hardware control panel) is much more natural than poking at a tiny icon of the knob on a flat screen with a mouse. Whether the virtual knob, or a real knob, is more desirable for controlling a particular function depends on the application and the operator. Sometimes the real knob is the best interface, but other times a customizable virtual control surface is better. A good computer–human interface adapts to the person using it.

Consider the familiar graphical user interface (GUI). A GUI allows custom control panels to be designed for a particular person, and a particular task. GUI panels can display meaningful graphics, such as photographs of equipment and bird's eye views of a building. Alternatively, generic physical control panels, with a number of assignable manual switches, faders, knobs, lights, etc., are very useful for people who prefer to reach out and grab a real control. Graphical user interfaces and physical control panels are excellent for concentrating large numbers of controls into a small space. This can be very beneficial. For example, sophisticated mixing consoles are very large because they have to provide many controls. A computer-controlled mixing console can economize on size (and therefore cost and performance) by implementing a customizable control surface with "soft" controls. Functions that are rarely used (or shouldn't be) can be hidden. Groups can be logically arranged on the control surface.

A significant benefit of centralized, customizable controls is that they allow very complicated systems to be controlled by laypersons. Consider the lectern knob and LCD human interface in the church example (Fig. 2). When the knob is turned, preset names appear in the LCD. The operator selects a system configuration by simply dialing the knob to the desired preset name. All the details of how the mixer, signal processors, amplifiers, and so on, are adjusted are hidden from the operator. This user interface can also be implemented on a touch screen ➡

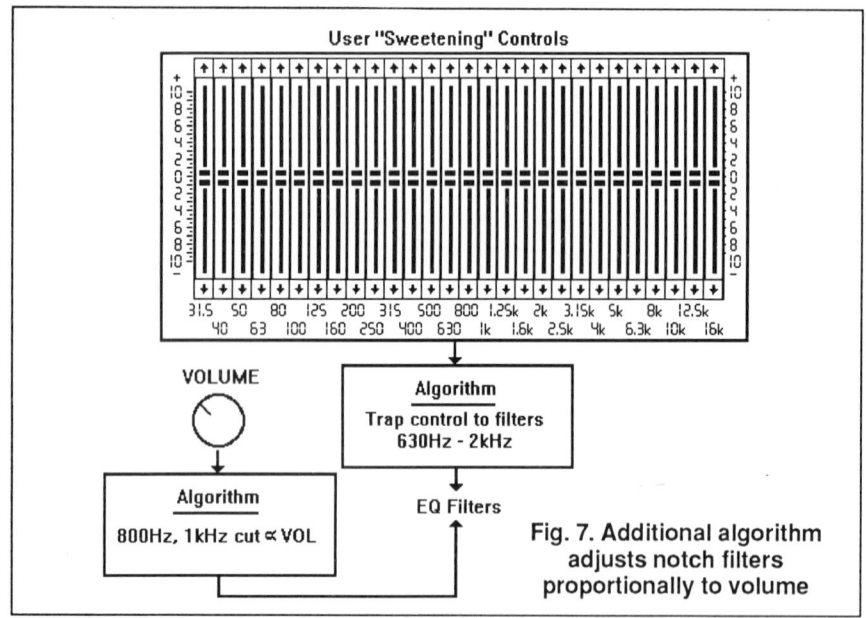

Fig. 7. Additional algorithm adjusts notch filters proportionally to volume

Fig. 8. Example touch screen interface with icons representing system presets

display, with an assortment of icons representing the various presets (Fig. 8). Again, all the details of how to adjust the equipment for each preset are handled automatically by the system. The human is concerned only with what he or she wants from the system (music, speech, etc.). The system is concerned with how to fulfill these needs, hiding the details. This minimizes the complexity of making scene changes, prevents human errors (unless the human chooses the wrong preset), and provides accurate, repeatable configurations every time.

A number of new human interface technologies are being developed by human factors and virtual reality researchers. Examples are tactile and force feedback devices, head mounted displays, gesture recognition, eye tracking, spatial tracking, and voice recognition. These technologies promise to revolutionize the way people interact with computer-based systems. The head mounted display (HMD) allows the human operator to see information no matter where he or she is, and where he or she is looking. For example, an HMD could display important signal level metering and other indicators as the sound engineer watches performers for visual cues. Eye tracking could be employed with a generic control surface to map the controls to what the operator is looking at. Gesture recognition could be used to map performers' motions to changes to the system configuration. The possibilities are endless, and very exciting. With the commercialization of many of these technologies, professional applications will become economical and practical. A danger is to get too carried away with this gadgetry and lose the naturalness of the interface. In the end, the technologies that allow the human to interact with systems naturally will find welcome homes in sound systems.

Multimedia Synchronization

Multimedia is probably the most overhyped information technology today. I'll try not to write very much about it, but a lot of people are missing its importance. We interact with the natural world using a combination of our senses (sight, touch, hearing, and so on). But art and technology are generally optimized to interact with only one or two of these senses at a time. Multimedia art combines visual, aural, and possibly other media, resulting in a richer, more natural experience. Multimedia systems of the future will dissolve the boundaries between audio, video, lighting, and other media. These sytems will allow artists to create more profound experiences for their audiences, and possibly provide more natural human interfaces to their operators. Unfortunately, people are preoccupied with applying multimedia for immediate gratification today. But as the technology and its practitioners mature, multimedia will become a powerful paradigm.

Let us consider the multimedia applications supported by the computer-controlled system in the nightclub example. The system provides centralized control over the music, sound, lighting, video, smoke machines, and other equipment. Some of this equipment operates on sub-nets, such as DMX 512 for the lighting equipment, MIDI for the music equipment, and AES-24 for the sound equipment. These sub-nets are interfaced together into one cohesive multimedia meta-system. This meta-system has a fully customizable human interface that supports a number of graphical and physical controls for simultaneously controlling music, lights, pyrotechnics, and so forth. The club is famous in town for having the most stunning performances. Artists and audiences come in droves. The multimedia computer-controlled system is a definite boon to the club's business.

Ultimately, any system that cannot integrate a variety of media will be short lived in the coming age of multimedia. The computer-controlled system that provides this flexibility has the best chance of succeeding.

Economy

There is no denying that computer control adds cost to a system. But it also adds value to the system, which offsets this increase in cost. Indeed, computer control even lowers the cost of some devices. For example, one of the most expensive devices in a traditional audio system is the mixing console. The high cost is largely due to the number of physical controls on the console, rather than the internal electronic circuitry that does the actual mixing. A computer-controlled mixer does not require all the physical controls, and can therefore be smaller and more cost effective.

Through centralization, much of the user interface hardware can be transferred to a centralized controller. Rather than duplicating expensive controls (an LCD, keypad, rotary encoder, and so on) on every device in the system, often only one set of controls is necessary at the system controller. The cost of this intelligence is paid once and amortized over the entire system. Moreover, manufacturers can concentrate on perfecting generic modular building blocks. One programmable filter device (for example, a digital signal processor) can act as an equalizer, notch filter, crossover element, side chain filter, etc. This increases production volumes on a fewer number of end products, which reduces costs for the manufacturer and ultimately the consumer.

Finally, by removing all the LEDs and LCD backlights from our front panels, we will save a lot of electricity. This will ease our energy drain on the planet, saving the ozone, salmon, and humans. If none of the other benefits sell you on computer control, at least do it for the salmon.

Conclusion

These examples highlight just a few of the benefits computer control will bring to sound systems in the future. The ramifications of evolving from purely analog systems to computer-controlled digital systems will not be subtle. From enhanced performance, to operator convenience, computer control will undoubtedly revolutionize sound systems. Indeed, we are on the brink of a significant transformation as digital information technologies sweep through our industry, causing a convergence of audio, video, computer, and other heretofore separate systems. This transformation affects everyone from manufacturer to end-user, student to seasoned professional. It is natural for people to feel anxious about the future. What will happen? How can a person prepare? Who will be ➥

Computer-Controlled Sound Systems

the winners and losers? Is this all just snake oil and rhetoric, or is there reality behind the hype? When will we really get to use these new technologies?

There are no easy answers to these questions. The author has been involved in computer control standardization efforts for several years and has witnessed a disturbing situation. Many people are demanding short term solutions to long term problems that cannot be controlled. There is disagreement over what the goals of the industry should be, and who should institute them. This chaos disturbs people, and causes even more problems.

The process of transformation is complex, chaotic, and intimidating, and therefore begs close examination. By understanding the transformation process, we have a better chance of making wise business and engineering decisions, and facilitating industry-wide cooperation in the creation of a multi-vendor computer control system. In the following sections, it is shown that chaos and uncertainty are necessary and exciting elements of our evolution into the information age. For those who wish to influence the outcome, there is hope. Read on...

TRANSFORMATION OF THE AUDIO INDUSTRY

Perspective

The audio industry was born late in the 19th century as Edison, Bell, Marconi, Poulsen and other inventors discovered that sound could be captured, manipulated, transmitted, and reproduced. Early audio devices such as the phonograph, telephone, radio, and recorder were based on *analog* technology, in which an acoustic waveform was replicated by an *analogous* waveform in another medium using mechanical, electrical, or magnetic techniques. Analog technologies for recording, processing, and reproducing audio signals have evolved for over 100 years. Indeed, many people have (wrongly) developed a world view, or paradigm, that manipulating audio is necessarily an analog process.

In the late 1950s scientists began experimenting with *digital* techniques to manipulate audio signals. As opposed to a continuous-time analog waveform, a digital waveform is a sequence of discrete-time samples, each representing the magnitude of the original audio signal at an instant of time. Scientists discovered techniques to capture, manipulate, and reproduce sequences of numbers using computers that were more accurate and robust than their analog counterparts. As computers evolved in the 1960s and 1970s, digital technology was successfully applied to more and more audio applications. The seemingly magical properties of digital technology paved the way for its commercial success. Our faith in the analog paradigm began to erode and digital audio equipment proliferated.

In the 1980s digital technologies flourished in audio systems. The digital compact disc (CD) was introduced to the consumer marketplace, replacing the analog vinyl LP almost overnight. The musical instrument digital interface (MIDI) was created and revolutionized the way artists produce music. The evolution of digital signal processing (DSP) brought forth extremely powerful, cost effective, tools for manipulating audio signals. Digital audio tape (DAT) recorders became a commodity, offering artists the ability to create pristine recordings on any budget. Personal computers found applications everywhere.

As discussed in the preceding sections, we are on the verge of a tremendous transformation to computer-controlled systems implementing the latest technologies for multimedia, information processing, virtual reality, and so on. This transformation will test our patience and flexibility.

The Transformation Process

The process of transformation has been studied in many different fields. As an illustrative example, Table 2 presents transformation models from three very dissimilar branches of science and philosophy. Ilya Prigogine's "Dissipative Structures" model describes how chemical and physical systems transform as equilibrium is disturbed. Thomas Kuhn's "Scientific Revolutions" model describes how scientific thought evolves through the ages and transforms society in its wake. Finally, Karl Marx's work concentrates on the transformation of social systems. These three examples are fundamentally different, yet the models described for the transformation process are strikingly similar.

Consider, Prigogine's model. Systems are always in a state of fluctuation. Systems in equilibrium experience small fluctuations. As these fluctuations increase, systems become unstable. At some time, for one reason or another, one or more of these fluctuations experience positive feedback within the system and cross a threshold —shattering the existing organization. It is impossible to predict which fluctuation will cross this threshold, since the system is chaotic. This transforma- ➡

	Table 2. Sampling of transformation models		
	Ilya Prigogine *Dissipative Structures*	Thomas Kuhn *Scientific Revolutions*	Karl Marx *Historical Determinism*
Initial state	Fluctuations within defined boundaries	Normal science	Steady state
↓	Fluctuations past a threshold	Growth of anomalies	Growing dissatisfaction
			Conflicts
	Crisis	Crisis	Crisis
		Revolution	Revolution
New state	Jump to a higher order	Normal science within new paradigm	New order

tive moment, called the bifurcation point, marks the time when the system either advances to a higher order structure or disintegrates into further crisis. Thus, according to Prigogine, the system passes from an initial equilibrium, through a state of high entropy and chaos, to a higher structured equilibrium or disintegration. This model agrees well with the other two examples in Table 2.

A Generalized Transformation Model

Thus, a general model for transformation (Table 3) begins with a system in

Table 3. Generalized transformation model	
1	Equilibrium
2	Growth of Anomalies
3	Chaos
4	Crisis
5	Bifurcation & Radical Change
6	New Equilibrium

equilibrium, emergence of disruptive anomalies, rapid acceleration of these disturbances into crisis until a radical change to the system occurs, and subsequent transformation to a new equilibrium state. This new state is not necessarily a better state—the system could also transform into a state of further crisis (as witnessed in the recent transformation of the former Soviet Union). Because there is a period of chaos during the process, it is not possible to control or predict the outcome. However, by reinforcing possible sources of positive feedback in the system, we can influence the transformation process.

An Example: The MIDI Revolution

Lest the reader believe that this transformation model does not apply to the audio industry, let us examine a recent example. The music industry witnessed a significant transformation in the 1980s as MIDI became a universal standard for interoperability. Early electronic music equipment was predominantly analog, and enjoyed a relatively stable marketplace as new technological developments (usually of an analog nature) were in pace with the obsolescence of old ones. The music industry, as an analog system, was in equilibrium. By the late 1970s, digital technologies were cropping up in the form of digital synthesis, digitally controlled equipment, low-speed communication links, and so on. These digital technologies introduced new capabilities (in Prigoginian terms, fluctuations) that disturbed the entrenched analog status quo. Consumers were faced with difficult purchasing decisions as they weighed the performance advantages of digital over the cost advantages of analog. Manufacturers struggled to keep up with the accelerating pace of technical advancements. By 1980 the marketplace reached a state of chaos as manufacturers introduced proprietary systems, and consumers grappled with their incompatibilities. Pressure mounted until Dave Smith and Chet Wood proposed the Universal Synthesizer Interface at the AES 70th Convention in 1981. This radical idea experienced positive feedback as manufacturers and consumers immediately realized the benefit of a common communications standard. The bifurcation occurred, transforming the industry into a state of higher order: MIDI. The industry transformed, breeding many new products, businesses, musical aesthetics, and so on.

The Audio Industry Today

Although MIDI completely transformed the music industry, and to a significant degree, the professional audio industry, audio systems are still based largely on analog technology. Most audio signal interfacing between equipment, in particular, is still implemented in the analog domain. Human interfaces to audio systems are based largely on analog controls (mechanical switches, potentiometers, and so forth). However, this 100-year-old analog system paradigm is rapidly giving way to a new system paradigm based on digital information technologies. Today we have dozens of different computer control protocols, digital audio interfacing formats, computer systems, software interfaces, etc., and most of them are incompatible with each other. These incompatibilities are causing significant distress in the marketplace, not unlike the experience in the electronic music marketplace prior to MIDI. We are certainly in the stages of chaos, preceding a significant transformation to predominantly digital systems.

The reaction from many audio engineering professionals has been to join together and create standards to combat these incompatibilities. The Audio Engineering Society has been central in these efforts. One example of a successful standard is the AES3 (or so-called AES/EBU) digital audio interface, which is found on most professional digital audio equipment. The creation of this standard allowed manufacturers to build equipment that could exchange digital audio data streams with equipment built by other manufacturers, and consequently allowed multi-vendor digital audio systems to grow.

Another AES effort to standardize digital information technologies is being undertaken by the SC-10 Subcommittee on Sound System Control. SC-10 was established in the fall of 1992, with the mission of creating standards for local area networking in audio systems. But, as many industries (computer hardware, software, audio, video, and so on) converge in proverbial cyberspace, it is impossible to predict exactly what will transpire. This complicates the efforts of the committee, as it is difficult to keep up with and commit to rapidly evolving technologies. This raises important questions such as: what can such a committee be realistically expected to accomplish? How firmly do we wish to commit ourselves to today's technologies, in light of what we know is coming in the near future? Should a committee legislate new technologies, or should we wait for innovative technologies to emerge out of a competitive marketplace? How can we maintain control, or at least survive, in this unpredictable process of transformation?

Communication is Our Main Leverage Point

As the generalized transformation model predicts, there is little we can do to control a significant transformation. This lack of control often intimidates people, and spawns feelings of fear and discouragement. However, as the generalized transformation model predicts, the current state of chaos is a natural part of the evolution of audio systems. We cannot enjoy transformation to a higher order system without passing through this necessary stage. Rather than feeling discouraged by the escalating chaos, we should look forward to the upcoming stages of revolution and ascendancy to a higher structure. Of course, there is no guarantee that the system will transform to a higher order structure, it could bifurcate into deeper crisis! Therefore, the primary mission is to identify desirable sources of positive feedback in the transformation process, and guide the industry through them.

One source of positive feedback, which we can control, is communication. Communication spreads information, and therefore promotes a unified effort to evolve a new, universal, paradigm. Communication also breaks a serious "catch 22" which hinders the commercial success of a radical new technology. Manufacturers often struggle to sell a new technology until they can reassure consumers that it is stable and reliable. But, consumers are not reassured until the technology has been thoroughly tested and debugged by a large number of other people. Similarly, the initial cost of a new technology can be extremely high until volume drives it down. But volume remains low when the cost is high. Breaking these catch 22 situations usually requires the manufacturer and/or consumer to accept risk. Communication can soften the risk by reassuring both parties that a sincere effort will be undertaken to work in earnest to apply the new technology effectively.

To ensure a healthy business environment, competing manufacturers must communicate to create a multi-vendor system. As witnessed by the MIDI example, a multi-vendor system is beneficial to virtually everyone. Manufacturers can selectively concentrate on a small portion of the overall system, thereby perfecting their craft. Small start-up companies have a base onto which they can offer innovative new technologies. The consumer is given much more choice within a single multi-vendor system, than between several closed proprietary systems. A multi-vendor system allows everyone to share the risk and benefit by the advances. Competition, necessary for a healthy market economy, flourishes within a multi-vendor system that supports thousands of companies.

Finally, students and seasoned professionals must communicate. Students leave academia with knowledge of the latest advances in science, but lack the experience and wisdom of the professional to apply them appropriately. The professional, preoccupied with running a successful business, often does not have the time to keep up with the latest advances in science. Therefore, it is important for students and professionals to communicate about the opportunities afforded by new technologies, and their practical application. These discussions lead to innovative new products, and the growth of our industry.

Indeed, communication provides potent sources of positive feedback for the evolution of useful new technologies. As predicted by transformation theory, this positive feedback is necessary to lift us out of chaos. Therefore, to positively influence the upcoming transformation, the audio industry must plan a vision for future systems and work in unity to realize it. According to the generalized transformation model, we cannot predict which "fluctuation" will cause the bifurcation. This means we do not know which of the available computer control systems will "win" in the marketplace. Critical mass of support is necessary for the industry to transform to any one (or several) of them. The system that provides the most value, makes the most sense to everyone, and is the most open to universal adoption, will flourish.

Communication in Action

The importance of communication during this volatile period of transformation cannot be understated. To facilitate this process, the Audio Engineering Society is holding the 13th International Conference on Computer-Controlled Sound Systems in Dallas, Texas, December 1–4. The conference program will cover all aspects of this subject, from designing equipment to installing systems. Everyone is advised to participate in this conference, as it is an excellent opportunity to share information and contribute to the healthy evolution of computer control technology in our industry.

SUMMARY

The audio industry is on the brink of a major transformation from analog-based systems to digital-based systems. Digital technology carries with it powerful new capabilities which, if fully exploited, yield enhanced human interfaces, interoperability, automation, and integration of many media.

The industry is currently in a state of chaos and uncertainty as it transforms to this higher order. This state of heightened entropy is normal for a system in transformation, and should not discourage us from working toward the new digital paradigm. The best (and only practical) option we have to influence the transformation is to build a shared vision of this higher order system by communicating our ideas and needs. When this vision reaches critical mass, the system will naturally transform to it. We cannot legislate transformational change. Concentrating on what we want in the end, rather than how we get there, is our best hope of leading the industry through a transformation to a desirable end.

And ultimately, computer control will flourish when and where it provides value.

REFERENCES

[1] I. Prigogine, *Order Out Of Chaos*, (Bantam Books, New York, 1984).

[2] Amir Levy, "Second-Order Planned Change: Definition and Conceptualization," Organizational Dynamics, vol. 15, p. 15 (1986 Summer).

[3] P. Senge, *The Fifth Discipline* (Doubleday, New York, 1990), chap. 11.

[4] D. Kauffman, *Systems One: An Introduction to Systems Thinking* (S.A. Carlton, Minneapolis, 1980) pp. 32–35.

[5] R. Moses, "Report of AESSC Subgroup WG-10-1 on Communication Formats," *J. Audio Eng. Soc.*, vol. 40, p. 52 (1992 Jan./Feb.).

ACKNOWLEDGMENTS

I would like to thank those people who helped me prepare this article. Anne Cotter taught me everything I know about systems and transformational change theory. Steve Turnidge drew some of the figures. Many of the ideas in this paper were the result of discussions with Mark Lacas, David Scheirman, Bob Rogers, Jeff Berryman, and numerous participants in the SC-10 committee.

Bob Moses is a senior digital audio engineer at Rane Corporation, where he has spent the last seven years developing digital audio, MIDI, and computer-controlled audio products. He holds three patents, with more pending, in the areas of digital audio and data communications systems.

He got his start in audio electronics at the age of six when his father helped him build a Heathkit radio. He received his Bachelor's degree in electrical engineering from McGill University in 1987. While at McGill, he researched compact disc technology and invented a 72-MIPS time domain digital oversampling processor. His student thesis detailing this technology won the Life Member award from the IEEE and first place honors in a national student research competition. He patented this technology and cofounded Wadia Digital Corporation (with his father, Donald Moses) to bring it to the marketplace. Immediately after helping found Wadia Digital Corporation,

Bob Moses

Bob followed his heart to Seattle, where he married his college sweetheart, and joined Rane Corporation.

In 1992 he helped form an arts cooperative called Northwest Cyber-Artists, which brings artists and technologists from different walks of life together to collaborate on projects. In 1993 he coproduced the group's production "Synesthetics," a three night event that featured a number of unique technologies and aesthetics resulting from the convergence of music, dance, video, computer animation, virtual reality, kinetic sculpture, chaos algorithms, and audience participation. In October 1994, Bob coproduced the "Beyond Fast Forward" festival at the Seattle Center (site of the Space Needle), which featured installations such as MIDI-controlled water fountains, wired jugglers, virtual reality installations, and workshops on multimedia production. His next project is a series of interactive multimedia bird feeders (the birds do the interacting) for the Audobon Society. Bob is also involved in efforts to create a cafe in Seattle where people can interact with people in other parts of the world using video conferencing, wide-area networking, and other information technologies.

Bob has been very active in the AES, and is currently serving as chairman of the AES Standards Committee (AESSC) SC-10-3 working group on information. He has been a longtime participant in the AES SC-10 efforts, and has chaired working group activities since 1991. He is also serving as the Program Cochairman of the upcoming AES 13th International Conference on Computer-Controlled Sound Systems, in Dallas, Texas.

Bob is a prolific writer on the subject of computer-controlled sound systems, multimedia and interactive arts, and digital audio, for the AES and a number of trade publications. He recently coauthored a book with Craig Anderton and Greg Bartlett entitled: *Digital Projects for Musicians,* which teaches people how MIDI equipment works, and how to build custom devices for themselves.

In his free time, Bob enjoys the serenity of the Cascade Mountains, skydiving, sailing, and exploring the culinary treasures of the Pacific Northwest.

Theory and Design of a Digital Audio Signal Processor for Home Use*

DAVID GRIESINGER

Lexicon, Waltham, MA 02154, USA

Studies of concert-hall acoustics have shown that energy from lateral reflections is particularly important in listener satisfaction and involvement. A digital signal processor has been developed for home playback which dramatically increases the lateral sound energy from recorded music and films, both with and without multiple loudspeakers.

0 INTRODUCTION

Bringing the processor described in this report to life was a great deal of fun and hard work for many people. The author's major task was to decide what such a processor could do to *really* change the way recorded sound was heard in the home, and develop solutions to these problems in software. It is on these questions and solutions that this engineering report will dwell.

To complete the task we had to draw on our own research and that of many others. We needed to find out just what is really involved in listening to recorded sound, identify the characteristics of sound and spatial hearing which maximize listening pleasure, and then use our digital technology to bring these characteristics to the home. The solutions turned out to be multifaceted. There is no one technique that works in all situations—but there are many, both old and new, that can make a significant contribution to home listening. All the techniques have a common thread—they raise the lateral sound energy in the listening room.

1 CONCERT HALLS

Over the last 30 years there has been a great deal of research on the importance of lateral reflections in concert-hall design. The work of Marshall, Barron, Schroeder, Ando, Blauert, and others has shown that

---

* Presented at the 85th Convention of the Audio Engineering Society, Los Angeles, 1988 November 3–6; revised 1988 December 5.

listeners prefer halls with significant sideways reflected energy. These researchers used different methods to study the effect of lateral reflections on listener preference. Barron used a laboratory approach in which a listener was surrounded by loudspeakers in an anechoic chamber. Various delayed signals were played from different directions, and the listeners' responses were analyzed. This is also the method employed by Blauert (1986). Both researchers found that lateral reflections were vital to creating a sense of spaciousness or envelopment, which Barron named spatial impression (SI). Barron and Blauert have analyzed the effects of direction, delay, and frequency content on the amount of SI.

Schroeder used interaural crosstalk elimination to play binaural recordings made in actual concert halls. This technique allowed different halls to be compared directly. He found that (among other parameters, such as reverberation time) listeners preferred halls where the interaural correlation between the two ears was as low as possible. In a hall interaural correlation is reduced only by lateral sound energy.

Our object was to design a processor that would substantially improve the listening environment at home, and increasing the lateral energy seemed the best way. The author's previous research indicated that the problem is frequency dependent. At low frequencies lateral room modes can be excited by out-of-phase stereo signals, and these can contribute greatly to listener satisfaction. This is the primary mechanism for the creation of spaciousness in an ordinary playback room. Higher frequency-reflected sound also contributes to SI in the home. You need only listen to stereo in an anechoic chamber or outdoors to realize how much the room contributes. Unfortunately room reflections are usually not similar to the lateral reflections in concert halls. The time delays are too short, and the directions are wrong. We wanted to re-create as much as possible the sound fields that existed in the original concert hall. If there was no hall, we wanted to generate sound fields that would draw the listeners into the music as if they were in a good hall.

2 CONCERT-HALL SIMULATION

The most obvious approach is concert-hall simulation, in which the listener is surrounded by extra loudspeakers in a relatively reflection-free room. The loudspeakers provide appropriately delayed sound from the proper directions. The idea is not new. The group at Göttingen, including Meyer, Burgtorf, and Damaske, may have been the first to try it, but Barron, Blauert, Horral, and Ando have all worked on the problem.

Fig. 1 shows the 12 loudspeakers used by Horral and part of the carefully deadened room. Fig. 2 shows the 16-tap delay line and the reverberation chamber used to make the signals to feed the loudspeakers. A practical home system needs to have fewer loudspeakers, but they can be driven by much more complex electronics. How many loudspeakers do we need, where do we put them, and how reflection-free does the room have to be?

To answer these questions the work of Barron, Blauert, and Horral is helpful. Borish's work on a concert-hall simulator was a major help in the question of how to program the electronics.

A home concert-hall simulator is illustrated in Fig. 3. Hardware and cost considerations limited us to a digital processor with two inputs and four outputs. We decided to design the product to work with two to eight loudspeakers in concert-hall simulation mode. The main stereo pair is placed conventionally, and music from the Compact Disc (CD) or other source is sent directly to it without passing through the digital processor. When only two loudspeakers are available, additional signals are added from the digital processor, which uses interaural crosstalk elimination to simulate side loudspeakers. When more loudspeakers can be used, the first ones we add are the sides, followed by two behind the listener. These are driven by four independent outputs from the digital processor.

2.1 Size and Shape

The algorithm for the digital processor was developed with a ray-tracing program (similar to the one described in Borish, but using two source positions), which determines the direction and loudness of reflected energy in concert halls (Fig. 4). The program ran on our VAX computer and automatically loaded the results into a standard digital processor (Lexicon 480). This system allowed us to investigate a great number of hall shapes and positions for the loudspeakers. Two shapes, rectangular and fan, were ultimately chosen as being the most characteristic and useful. Within each shape we found the most important parameter was the size of the hall, and we settled on a selection of three basic sizes. Size selection is vital. Ando and Imamura have shown that the optimal time delays for reflected energy depend critically on the correlation time of the music being played. Slow symphonic music needs a long time or a large hall, while fast-moving jazz or modern music requires a hall or room of much smaller size. Our work indicated that the best hall shapes sounded good in both large and small sizes, and the different sizes had to be matched to the music.

2.1.2 Stereo Processing

The final algorithm was developed by the author. It

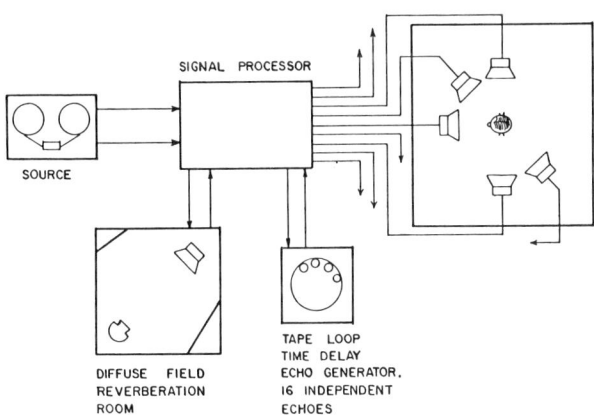

Fig. 2. Horrall's concert-hall simulator electronics.

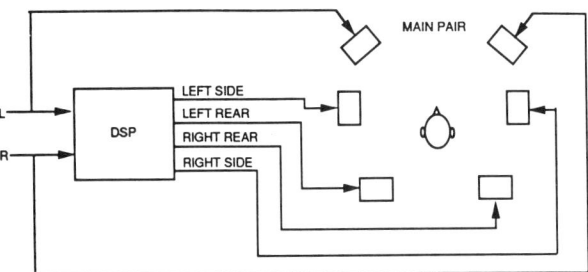

Fig. 3. Electronics for Lexicon CP-1 concert-hall simulator.

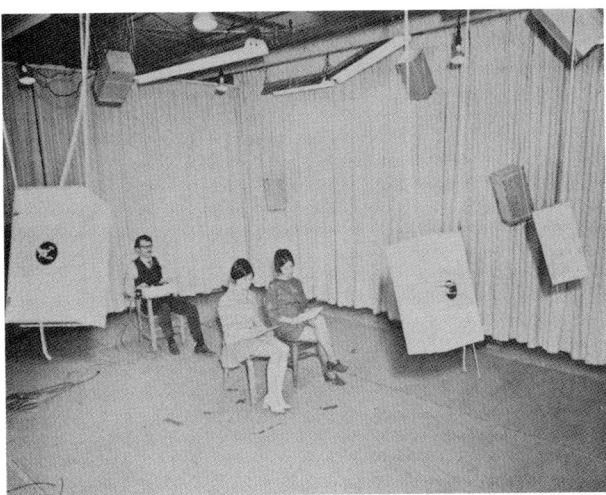

Fig. 1. Horrall's concert-hall simulator.

combines the most important features of the shape and direction information from the computer synthesis with sufficient reverberation to make a convincing acoustic result. One of the important features of the algorithm is that it processes the signals in stereo. The input to the ray-tracing program was not a single source but a pair of sources, asymmetrically placed on the stage of the hall. The listener was also not exactly in the middle. When a stereo signal is played into the simulator, the two input channels are processed differently, and significant separation can be heard in the synthesized reflections. This adds additional depth and spaciousness to the synthesis, as well as significantly improving the timbre.

2.2 Side Loudspeakers

In the absence of interaural crosstalk elimination (see Sec. 4) you need at least two extra loudspeakers

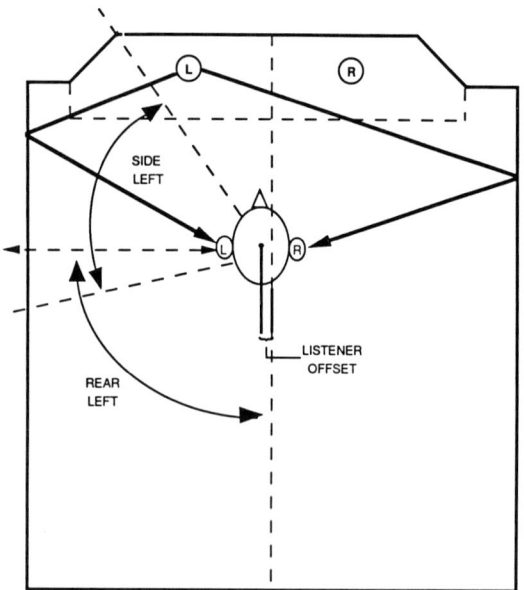

Fig. 4. Room model for CP-1 concert-hall simulator. Ray tracing of a three-dimensional hall model. Left and right sources are traced separately; reflections are assigned to side or rear, depending on side and rear cutoff angles chosen.

placed nearly at the sides ($\pm 85°$) of the listener. In Fig. 5 these are loudspeakers 2.

2.2.1 Evidence from Interaural Correlation

This result is both obvious and controversial. If you think that your goal is to produce maximum lateral sound energy, putting loudspeakers at the sides of the listener is an obvious choice. However, if you think you are trying to minimize interaural correlation, this is not the case. Ando's attempts to find the minimum interaural correlation (in the absence of sound from the front) suggested that the best position for lateral loudspeakers was about $\pm 55°$ from the front (loudspeakers 1 in Fig. 5). He recommends that in concert-hall design lateral reflections should be concentrated at these angles.

We investigated this and found that the correlation between the ears of a person or a dummy-head microphone is a simple function of direction and frequency. At low frequencies there is always positive correlation between the two ears. As the frequency goes up, at some angle the time delay between the two ears is sufficient to give more than 90° of phase shift, and the correlation becomes negative. At still higher frequencies the correlation becomes positive again. The angle of minimum correlation for a broadband source may indeed be at about $\pm 50°$, depending on the actual bandwidth used. Blauert has used earphones to study the apparent localization of music and noise with different correlations, and finds that there may be an optimum positive value of correlation.

2.2.3 Experimental Results

We feel that the results from interaural correlation are misleading. Our experiments suggest that it is the hall reflections themselves which cause listener preference, and not the low correlation they induce. In real halls most of the energy is from the front, ceiling, or rear, and negative correlations rarely occur. It is well known that at least for isolated hearing events the hearing mechanism uses correlation information only to determine the direction of the sound source, and our

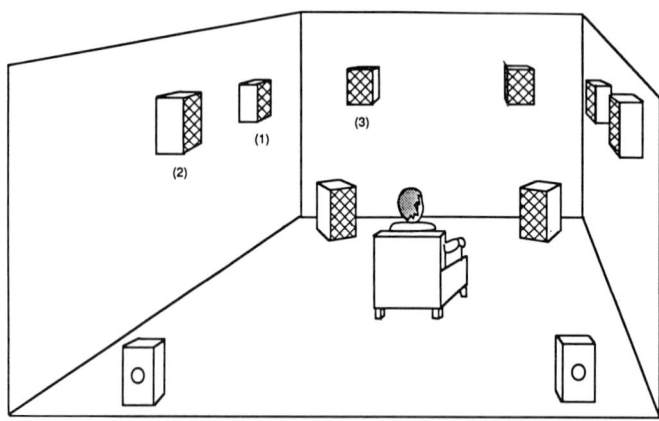

Fig. 5. Close listening position. Side position 2 is better than 1. Positions 2 and 1 together is best. Position 3 makes little additional improvement.

experiments indicate that at least at some level of the brain people are able to determine the direction of a reflection, even in the presence of a loud sound from the front. This result is consistent with the model of hearing proposed by Theile, where spatial information is extracted first.

2.3 Direction and Frequency Effects

We set up an experiment to investigate the effect of loudspeaker angle on spatial impression. We played somewhat dry stereo music through main loudspeakers at ±30°, while adding delayed reflected sound through one of two additional pairs of loudspeakers at two different angles around the listener (for example, positions 1 and 2 in Fig. 5). The listener was asked to decide which of the two extra pairs made the sound more spacious. We found that spatial impression increases as the angle between the side loudspeakers is increased, up to a maximum at 90° (loudspeakers fully to the sides). We also studied the effects of frequency by band-limiting the signals to the extra loudspeakers, and found (consistent with Blauert and Lindemann, 1986) that spatial impression is enhanced by all frequencies. Previous work by Barron (quoted extensively by the author in a previous paper on "Spaciousness and Localization") that low frequencies were particularly important was not confirmed. Low frequencies are indeed important, but no more so than others. The spatial impression from the band-limited loudspeakers varied with the angle in the same way as the non-band-limited loudspeakers, with the maximum at 90° for all bands. This confirms many of the results of Barron and Blauert. This observation appears to contradict a recent paper by Morimoto and Maekawa, who found bass frequencies to be more important than upper frequencies. This is possibly because in their experiments, acoustic loudness was held constant, whereas we kept the broadband input to the bandpass filter constant as the bandwidth was varied. We feel that our method is more helpful in understanding concert halls.

Note that our strong recommendation of a side loudspeaker pair differs from the approach recommended by some other manufacturers. We found it was not possible to generate much SI with loudspeakers in front and behind the listener (position 3 in Fig. 5). The conventional "quad" loudspeaker array is simply not an effective generator of SI unless it is aided by room reflections.

2.3.1 Implications for Concert-Hall Design

This work confirms the results of Marshall and Barron that SI is strongest for reflections that come from the sides of the listener, with a maximum at ±90° from the front. All frequencies appear to be equally important. More frontal directions also improve the hall sound, but contribute less strongly to SI as defined by Barron. The environment processor we developed, like the concert-hall synthesizer of Horral, is a useful tool for studying concert-hall design.

2.4 Additional Loudspeakers

Rear loudspeakers are next in importance after the side loudspeakers, especially when longer reverberation times are desired. However, we did notice that while spatial impression was greatest when the side loudspeakers were at ±90°, there was a better blend of the side and main sound when the loudspeakers were at lesser angles. The sound was really best with two pairs of auxiliary loudspeakers in front, one pair at ±85° and one at about ±45°. Although our product literature does not emphasize this point, the product has several ways of driving this extra pair. If an additional loudspeaker pair is placed in the front of the room and is wired in parallel with the rear loudspeakers, a convincing reverberant field results, along with some improvement to the front blend. This extra front pair could also be driven by the side loudspeaker signal. Alternately the processor can be configured as if side loudspeakers were absent, in which case the main loudspeakers with the help of crosstalk elimination will fill in both reflections and reverberant energy between the main and side loudspeakers.

The recommendation that side loudspeakers be placed at the sides of the listener should also not be slavishly followed when the listening setup differs substantially from the ones we were testing. Fig. 6 shows an ar-

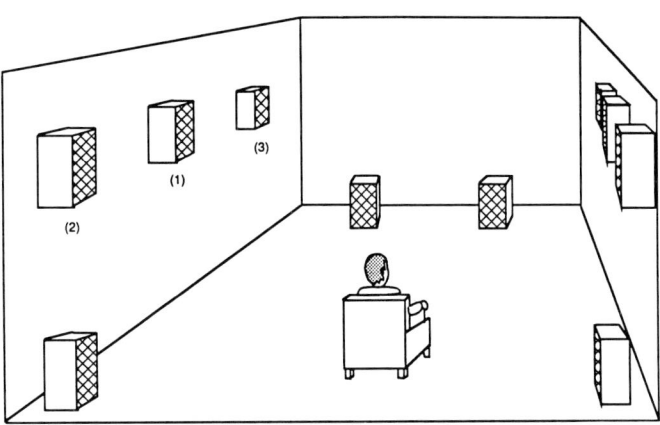

Fig. 6. Distant listening position. Side position 1 is better than 2. Positions 1 and 3 together is good. Position 2 is helpful, but must be softer than 1.

rangement devised by a customer, where the main loudspeakers were at one end of a long room and the listening position was at the other. Placing the side loudspeakers in the rear half of the room (position 2 in Fig. 6) sounded decidedly unnatural. There was too large a gap between the main loudspeakers and the sides. A position closer to the main pair (position 1 in Fig. 6) was much more effective. This customer liked adding an additional pair at position 3, driven in parallel with the rear loudspeakers.

2.5 Height

We make no specific recommendation for the height of the extra loudspeakers. In real halls much of the reflected sound comes from areas slightly above the listener, but mounting loudspeakers high can be a lot of work. We tried putting the rear loudspeakers near the ceiling of the room, and the sound was very good. Putting the side loudspeakers too high will probably adversely affect SI, since the side loudspeakers will then be closer to the top of the head than they are to the sides.

2.6 Playback Room Acoustics

The more absorptive the room is, the better it will sound. The environment processor can improve the sound of quite live rooms, but it does not realize its full potential. With multiple loudspeakers the room walls should not be reflective—the lateral sound is supplied by lateral loudspeakers. Reflected energy from the room walls has insufficient delay and will only increase the loudness and the muddiness of the direct sound. Although diffusors can help somewhat, it is better to absorb as much of the reflected sound as possible. We have found that fabric-covered 2-in (50-mm) fiberglass panels are quite effective when placed on the walls of a room, especially if they can be augmented by carpet and drapes. It does not appear possible to make the room too absorptive. The author was fortunate to be able to try the environment processor in a special room where the walls and the ceiling were paneled with fiberglass separated from the wall by about 18-in (0.46 m). This room had a broadband absorption of 0.8 all around. Ordinary stereo in this room was poor, but with side and rear loudspeakers and hall synthesis the sound was wonderful, even on fast-moving popular music and jazz.

3 REVERBERATION

The simulation algorithm was optimized for reproducing the spatial properties of halls with rather short reverberation times. To allow the creation of other spaces a scaled-down version of a standard Lexicon reverberation algorithm was added. This algorithm also has a stereo input, with four outputs intended for the sides and rear of the listener. This program allows very large spaces to be constructed, as well as offering very high diffusion. There is less of a sense of a particular size and shape, however.

4 CROSSTALK ELIMINATION—PANORAMA

When side loudspeakers are not possible, all is not lost. Crosstalk elimination can be used to simulate side loudspeakers, and this technique is used in the concert-hall synthesis program (Ambience) and the reverberation program. Crosstalk elimination is also very interesting for playing ordinary stereo and binaural recordings without ambience generation. We call this technique ambience extraction.

Schroeder's original studies of concert halls used crosstalk elimination, developed earlier by Atal and Schroeder. Two loudspeakers were placed in front of a listener in an anechoic chamber. Digital signal processing (nonreal time) was used to alter the loudspeaker signals such that the listener's left ear would hear only the left input signal, and the right ear only the right. Damaske and Mellert developed an acoustic technique that did the same thing. Both methods allowed playback of binaural recordings without the in-the-head localization associated with earphones, and required the listener to be in a precise spot. The author knew of the existence of these systems through Schroeder (1974) and Blauert (1981).

One of our first experiments for the proposed environment processor was to develop and experiment with such a technique. We did it entirely in the digital domain, of course, and in real time. The algorithm turned out in retrospect to be very close to that of Atal and Schroeder. Like that system, ours not only cancels the signal that diffracts around the head of the listener (the first-order correction); it also cancels the signal used for the cancellation (a higher order correction). It uses a simple spherical model for the head, augmented by experimental data on head diffraction and pinnae effects (Fig. 7).

We found that good results could be obtained in an ordinary room as long as it was not too small or reverberant. The listening area was also larger than expected. Ideal crosstalk elimination at all frequencies demands very precise placement, within an inch of the exact centerline between the loudspeakers. With some training it turns out to be easy to hear when your head is in the right place, and it is easy to maintain the position. For lower frequencies the positional requirements are much more relaxed. Indeed, at low frequencies the crosstalk elimination is equivalent to a phase-compensated left minus right bass boost, such as has been described by the author in previous papers. Such a boost increases SI throughout the room and can be enjoyed by all listeners.

Coloration off axis turned out to be minimal, greatly increasing the usefulness of the technique. This equivalence at low frequencies to a left minus right "shuffler" has also been noted by Cooper. It has an interesting practical consequence.

When binaural recordings are played through the crosstalk eliminator, the poor low-frequency separation in these recordings is greatly improved, and they sound very good. However, if ordinary stereo recordings are played, excessive out-of-phase bass energy will result—at least when the canceling circuit is used in an ordinary room. For this reason we modified the spherical-head model to avoid excessive bass with normal recordings. The "normal" or "wide" settings on the CP-1 are for this purpose.

Binaural recordings that have already been corrected for this effect (see Griesinger, this issue) should be played with the "normal" setting, not the "binaural" setting.

The advantage of crosstalk elimination, which we call Panorama, is that the ear of the listener can be fooled into thinking that there is a side loudspeaker present, even when there is none. For a listener in the right place, two loudspeakers in the front *can* generate significant SI, and not just at low frequencies. Ordinary stereo is significantly improved, and listening to binaural recordings with Panorama can be exceedingly realistic. With a good recording, Panorama through two loudspeakers brings the listener close to being present at the original recording. The loudspeakers disappear, the front image is wide, and the hall surrounds the listener realistically.

4.1 Adjustment

Ideally the Panorama program requires loudspeakers with good imaging, placed away from the walls (or inset into the walls) of a large absorptive room. To help the user adjust the system, we included a digital noise source, which is supplied to either the left input alone, the right input alone, or both inputs uncorrelated. When the listener finds the right place to sit, the left-only noise appears to be heard only in the left ear, with a sharp null appearing in the right ear as the head is moved laterally. The sound appears to come almost entirely from the left of the listener, with pinna effects pulling some of the high frequencies toward the real loudspeaker location. Many listeners find that the noise appears much farther to the side in one ear than in the other. This frequently indicates a high-frequency hearing loss in the opposite ear.

4.2 Binaural Recordings

When the system is adjusted correctly, the performance on binaural recordings can be amazing. There is a radio play CD in German where voices appear to float

Panorama can be used with music, films, or from within the Reverb and Ambience programs to simulate side speakers if the listener is inside the effective area between loudspeakers.

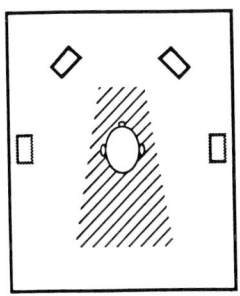

Imagine a click in the left speaker...

Sound from speaker L travels to the left ear and also to the right ear, a time Δ t later.

If we supply a negative delayed signal to the right speaker, this crosstalk can be canceled.

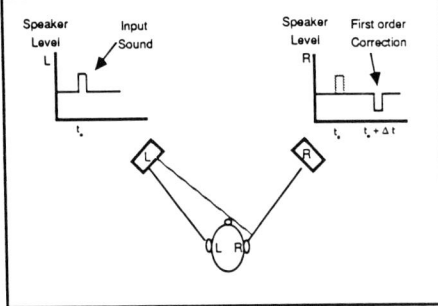

First order correction travels to left ear, where it will be heard unless canceled by an additional correction. When these higher order corrections are supplied, accurate cancellation is possible.

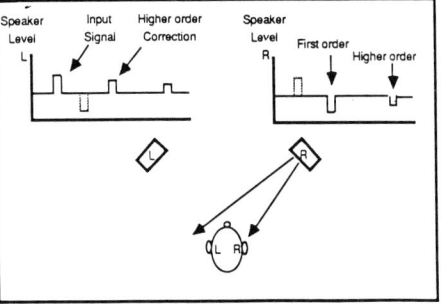

Fig. 7. Crosstalk elimination—Panorama.

magically all around the listener. The loudspeakers disappear entirely. On another CD of binaural recordings Dr. Theile walks around the listener. The first time the author heard this he was not able to localize speaking behind the dummy head, but after several listenings this became easy to do, and now it happens every time. It seems possible to train yourself to take advantage of whatever cues are being offered by this technique, just as it is possible to learn to enjoy ordinary stereo.

4.3 Spatial Equalization

In our version of Panorama there is provision for supplying delayed left minus right information to the rear loudspeakers if they are present, and this can be very pleasant. A phase-compensated spatial equalizer has been included, allowing the listener to adjust the width, and thus the SI, of low frequencies. A "normal" setting is provided for stereo with significant out-of-phase bass energy, and a "wide" position for pop and rock. There is also a setting for true binaural, which uses a true spherical-head model. This model results in a phase-compensated left minus right bass boost similar to the one recommended by the author for making binaural recordings. If the recording has already been corrected in this way, the normal setting should be used. The program also allows the listener position to be adjusted, both for the closeness of the listener to the loudspeakers and for small lateral offsets. When the angle between the loudspeakers is small, the listening area becomes larger, and the effect of Panorama is more dramatic, but less precise (Fig. 8).

4.4 Phasiness

For some professionals who have listened for years to ordinary stereo Panorama may be hard to accept at first. Panorama works by a cancellation technique. It increases the out-of-phase component of the loudspeaker signals, so the leakage around the head will be eliminated. Ideally this requires an anechoic environment and very precise localization of the listener's head. In more ordinary environments cancellation is less than perfect, and these out-of-phase components of the sound can be heard. Cries of "I hear phasiness!" are often uttered, as if this were the end of the world. For less trained mortals the program can be quite exciting.

5 FILMS—DOLBY PRO LOGIC

Film producers are beginning to realize the potential of lateral sound to draw a viewer into the film. Most modern films are mixed and released in Dolby Surround, a matrix-encoded surround system (see Julstrom). The decoder used during the mix is a professional unit developed by Dolby Labs. Playing these films in the home through a logic-enhanced surround decoder with similar characteristics is a real pleasure. To accommodate these films we developed a full-logic decoder for the environment processor, the first to work entirely in the digital domain. The results exceeded our expectations. The initial performance of the decoder was very good and close to Dolby specifications. We then worked closely with Dolby Labs on the selection of time constants and features. The joint effort was extremely productive, with many suggestions from both sides continually improving the result. In the end we had a decoder that was both of very high quality and fully compatible with their system.

We took advantage of the digital implementation to do some things that are difficult in analog. We added an input delay of about 20 ms, which allows the logic circuits to decode the direction of a signal before it actually appears. This makes for more accurate steering and fewer artifacts. The level detectors have a wide range, so steering is effective at very low levels. The major advantage of the digital implementation, however, is in using some leftover processing time to correct common problems in the source material. We do this by continuously monitoring and correcting both the azimuth and the balance of the film.

Dolby Surround, like all matrix surround systems, relies on level and phase differences to steer sound energy to different loudspeakers. In a film the strongest signal is usually the dialog, which is matrixed equally into the left and right channels. Stereo music and effects are panned between left and right as usual, and surround information is mixed into both channels, but out of

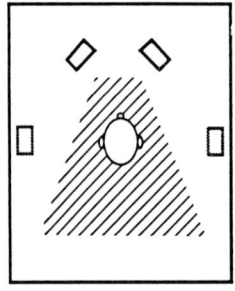

When the front speakers are close together, the Panorama Effect is less precise but more dramatic, and it works over a larger area.

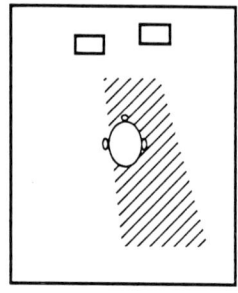

Speaker alignment is important. A 6" difference in the distance to the rear wall can greatly change the effective area, unless compensated for by the LISTENER POS parameter.

Fig. 8. Panorama effect.

phase. The simplest surround decoders direct the left channel to the left loudspeaker, the right to the right loudspeaker, the sum $(L + R)$ to the center, and the difference $(L - R)$ to the rear. With this form of decoding, separation is low. Any signal will be reproduced by at least three loudspeakers (Fig. 9).

A logic decoder should steer the dialog to the center loudspeaker and remove it from the left and right. The rear or surround loudspeaker will contain no dialog even without logic, since it plays the $L - R$ signal.

There is a simple way to remove dialog from the left and right loudspeakers, and that is to simply turn them off in the presence of dialog. This is the form of enhancement used by early quad decoders and some film decoders. Unfortunately when music is also present, this will reduce the apparent separation of the music to monaural whenever dialog is present. A better way, and the one adopted here, is to subtract the dialog from the left and right channels, replacing it by additional energy from the $L - R$ channel. When this is done correctly, the total energy of all the signals remains constant, and very little pumping or decrease in separation can be heard. A properly designed decoder will do a similar subtraction in all directions, effectively removing sound from channels where it is not wanted, while preserving the total loudness of all signals.

Unfortunately all bets are off if the source material has phase or amplitude errors, and these are very common in films. We have seen many tapes and disks with azimuth errors of more than 50 μs and some with over 100 μs. Azimuth errors cause leakage of dialog, not only into the L and R channels, but in the $L - R$ or surround channels as well. Earlier decoders used sharp filtering to remove leakage from the rear, and simply turning off the side channels is perhaps less distracting on bad material than constant energy subtraction. Thus we have the difficulty that on some currently available material a simple nonstandard decoder might well outperform a standard Dolby decoder.

Fortunately we can escape this problem by correcting the incoming material. We monitor the signal for phase and balance, and correct it for best results in the decoder. Correction of phase to an accuracy of a few microseconds and balance to a fraction of a decibel usually only takes a few seconds, and from then on the decoding is excellent.

5.1 Loudspeaker Placement

The miminum loudspeaker placement for film is illustrated in Fig. 10. The presence of a first-rate logic decoder in an environment processor that is also designed for music leads to some conflicts and opportunities. For best results surround films require at least four loudspeakers, and the locations of these loudspeakers are different than for standard music stereo. First of all, there must be a center loudspeaker—otherwise there is very little point in having logic decoding

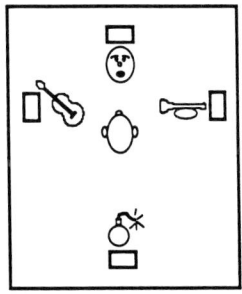

Films originally have four channels: one for dialog and three for music and effects. To make a Dolby Stereo film, these are combined to two.

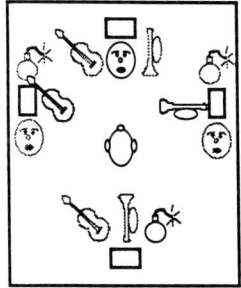

With conventional surround any sound comes from at least three directions.

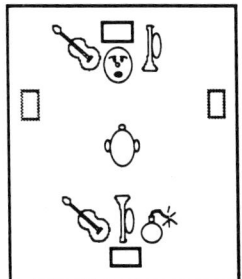

Simple logic decoders turn down the left and right speakers during dialog. This seriously affects music and effects.

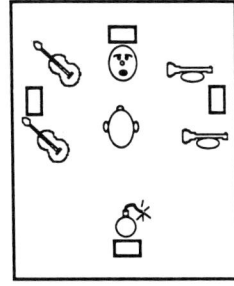

Pro Logic decoders remove dialog from the left and right channels, while maintaining stereo as much as possible.

Pro Logic requires phase accuracy. Common azimuth errors cause ghost dialog in all channels unless the azimuth error is corrected.

Fig. 9. Film sound decoding.

at all. Because the center loudspeaker fills in the center region, the left and right loudspeakers can be placed unusually wide, thus directly increasing the SI felt by the listener. Adding at least one rear loudspeaker completes the arrangement. If you want to reproduce both music and films, five to seven loudspeakers are recommended (Fig. 11). With a subwoofer in a corner somewhere the total is eight.

Once again we did a series of experiments on the ideal location for the loudspeakers and ran into some strong feelings. For almost everyone, including the author, placing the main left and right loudspeakers as wide as possible was desirable. If side loudspeakers are present as well as main loudspeakers, the side loudspeakers should carry the main left and right channels with the dialog removed and are essentially wired in parallel with the main loudspeakers.

This loudspeaker setup maximizes SI. It also makes the sonic image—the sound-effect space within which the characters move—much wider than the screen. To most people this increased width is gripping and compelling. The viewer quickly accommodates to the extra width and is drawn into the action. In a standard theater arrangement the left and right loudspeakers are placed at the edges of the screen, and the side loudspeakers are wired in parallel with the rear loudspeakers, all reproducing the surround signal in mono. The surround signal is often soft in films, except when the mixer wanted a rearward effect, and so this loudspeaker arrangement usually adds little SI.

We did several experiments where we switched back and forth between these two setups. The people who liked the wide sound liked it very much—and they were the great majority. A small and very vocal minority (many of whom were professionally associated with filmmaking or criticism) much preferred the standard arrangement. This included a film sound professional who claimed the wide sound image (and consequent occasional out-of-phase material) made him distinctly queasy.

Our decoder is the first that makes the wide loudspeaker arrangement practical. On other decoders azimuth errors cause so much leakage of dialog into the left and right channels that the side loudspeakers are distracting if they carry this information. As professional decoders of similar quality become available, we expect that more theaters will use a wider presentation, and gradually films will be mixed to take advantage of it, as special theaters such as Omnimax currently do. We decided to make the wide presentation of films the norm in our product. Customers can decide to set their systems the other way if they desire.

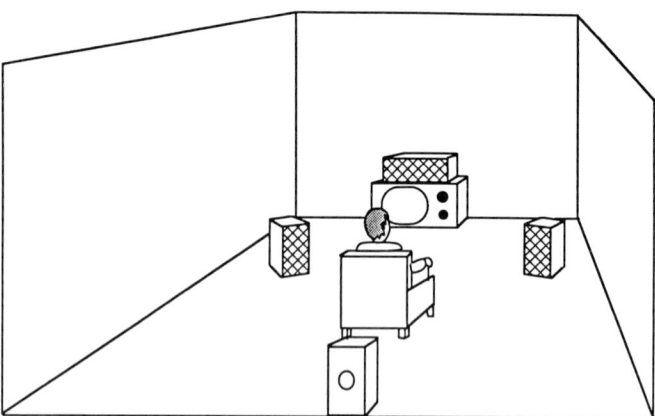

Fig. 10. Placement of minimum number of loudspeakers for film.

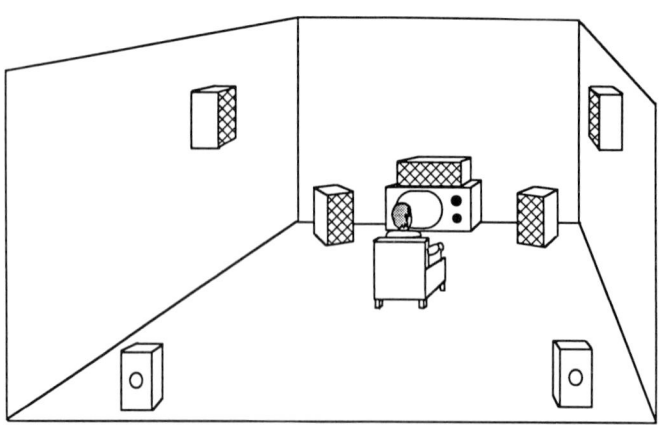

Fig. 11. Loudspeakers placement for film and music.

5.2 Stereo Logic

The film decoder also makes for a very interesting way to listen to music. Much popular music in particular is mixed in a way similar to films. Lead vocals tend to be centered, with instruments panned across the front. Reverberation and effects tend to be reproduced as surround. We included a position in our decoder designed to reproduce music in this way (it is similar to the film decoding, but the amount of directional enhancement is adjustable), as well as some obscure parameters such as the rear channel delay and rolloff. We call this setting Stereo Logic.

5.3 Mono Logic

When a mono film is played through a logic decoder, all the sound is directed to the center channel. Since many films are only available in mono, we designed a program that dynamically separates the dialog from music and effects such as background noise, wind, streets, or ocean. It presents the dialog to the center loudspeaker and directs the music and effects to the input of an acoustic simulation program set to resemble the size and shape of a small theater. The side and rear loudspeakers add the sound of being in a larger place, while leaving the dialog largely unaffected. We call this program Mono Logic.

6 LOUDSPEAKER SETUP PROCEDURE

The environment processor was designed to work with a great variety of loudspeaker placements. The customer sets the machine for the number of loudspeakers present. When only two loudspeakers are present, the Panorama program works normally and Ambience and Reverb are automatically routed through Panorama before being added into the main channels. As more loudspeakers are added, the machine can be configured to work with them, up to a maximum of eight—including two mains, two sides, two rears, a center, and a subwoofer. If an additional pair of loudspeakers are wired in parallel with the rears of sides and are placed in the front, a total of ten may be used. There is a digitally generated noise source which allows all the loudspeakers to be balanced, and another noise source in Panorama to adjust the program to the listener's favorite position.

7 CONCLUSIONS

The programs we call Panorama, Ambience, Reverb, and Stereo Logic increase musical realism. They each use different mechanisms to increase lateral sound. Users can easily choose which program works best for their room and their music. Crosstalk elimination (Panorama) in combination with true binaural recordings provides the most realistic reproduction of height, depth, and surround that the author has heard, and requires only two loudspeakers. With multiple loudspeakers the other programs can extend this performance through a large listening area. The film programs Pro Logic and Mono Logic draw the listener into the action through the dramatic increase in lateral sound that good surround can create.

8 BIBLIOGRAPHY

Y. Ando, *Concert Hall Acoustics* (Springer, NY, 1985).

——and M. Imamura, "Subjective Preference Tests for Sound Fields in Concert Halls Simulated by the Aid of a Computer," "*J. Sound Vibration*, vol. 65, pp. 229–239 (1979).

B. S. Atal and M. S. Schroeder, "Apparent Sound Source Translator," U.S. patent 3,236,949 (1966 Feb. 22).

M. Barron, "The Subjective Effects of First Reflections in Concert Halls—The Need for Lateral Reflections," *J. Sound Vibration*, vol. 15, pp. 475, 494 (1971).

——, "The Effects of Early Lateral Reflections on Subjective Acoustical Quality in Concert Halls," Ph.D. dissertation, University of Southampton, England (1974).

J. Blauert, *Spatial Hearing* (MIT Press, Cambridge, MA, 1983).

——and W. Lindemann, "Spatial Mapping of Intracranial Auditory Events for Various Degrees of Interaural Coherence," *J. Acoust. Soc. Am.*, vol. (1986 Mar.).

——and—— "Auditory Spaciousness: Some Further Psychoacoustic Analyses," *J. Acoust. Soc. Am.*, vol. 80 (1986 Aug.).

J. Borish, "An Auditorium Simulator for Domestic Use," *J. Audio Eng. Soc.*, pp. 330–341 (1985 May).

D. H. Cooper and J. L. Bauck, "Prospects for Transaural Recording," *J. Audio Eng. Soc.*, vol. 37, this issue, pp. 3–19.

P. Damaske, "Head-Related Two-Channel Stereophony with Loudspeaker Reproduction," *J. Acoust. Soc. Am.*, vol. 50, pt. 2, pp. 1109–1115 (1971 Oct.).

——and V. Mellert, "Ein Verfahren zur richtungstreuen Schallabbildung des oberen Halbraumes über zwei Lautsprecher," *Acustica*, vol. 22, pp. 153–162 (1969/70).

D. Griesinger, "Spaciousness and Localization in Listening Rooms and Their Effects on the Recording Technique," *J. Audio Eng. Soc.*, vol. 34, pp. 255–268 (1986 Apr.).

——, "Equalization and Spatial Equalization of Dummy-Head Recordings for Loudspeaker Reproduction," *J. Audio Eng. Soc.*, vol. 37, this issue, pp. 20–29.

T. R. Horral, "Auditorium Acoustic Simulator: Form and Uses," presented at the 39th Convention of the Audio Engineering Society, *J. Audio Eng. Soc. (Abstracts)*, vol. 18, p. 702 (1970 Dec.), preprint 761.

S. Julstrom, "A High-Performance Surround Sound Process for Home Video," *J. Audio Eng. Soc.*, vol. 35, pp. 536–549 (1987 July/Aug.).

A. H. Marshall, "A Note on the Importance of Room

Cross-Section in Concert Halls," *J. Sound Vibration*, pp. 100–112 (1967).

E. Meyer, W. Burgtorf, and P. Damaske, "Eine Apparatur zur elektroakustischen Nachbildung von Schallfeldern," *Acustica*, vol. 15, pp. 334–339 (1965). See also Blauert (1983), p. 284.

M. Morimoto and Z. Maekawa, "Effects of Low Frequency Components on Auditory Spaciousness," *Acustica*, vol. 66, pp. 190–196 (1988).

M. R. Schroeder, "Binaural Dissimilarity and Optimum Ceilings for Concert Halls: More Lateral Sound Diffusion," *J. Acoust. Soc. Am.*, vol. 65, pp. 958–963 (1979).

——, "Progress in Architectural Acoustics and Artificial Reverberation: Concert Hall Acoustics and Number Theory," *J. Audio Eng. Soc.*, vol. 32, pp. 194–203 (1984 Apr.).

——, D. Gottlob, and K. F. Siebrasse, "Comparative Study of European Concert Halls: Correlation of Subjective Preference with Geometric and Acoustic Parameters," *J. Acoust. Soc. Am.*, vol. 56, pp. 1195–1204 (1974).

Dr. Griesinger's biography appears on p. 29.

Multiple-Point Equalization in a Room Using Adaptive Digital Filters*

S. J. ELLIOTT AND P. A. NELSON

Institute of Sound and Vibration Research, University of Southampton, Southampton SO9 5NH, UK

A method is presented for designing an equalization filter for a sound-reproduction system by adjusting the filter coefficients to minimize the sum of the squares of the errors between the equalized responses at multiple points in the room and delayed versions of the original electrical signal. Such an equalization filter can give a more uniform frequency response over a greater volume of the enclosure than a filter designed to equalize at one point only. The results of computer simulations are presented for equalization in a "room" with dimensions and acoustic damping typical of a car interior, using various algorithms to adapt automatically the coefficients of a digital equalization filter.

0 INTRODUCTION

In sound-reproduction systems an equalization filter is sometimes used to modify the frequency spectrum of the original source signal before feeding it to the loudspeaker, in an attempt to compensate for unevenness in the frequency response of the loudspeaker and the listening room. Such an arrangement is illustrated in Fig. 1, in which a microphone, whose response is assumed to be flat, is substituted for the human observer in order to make the net response of the reproduction chain purely electrical. Such equalization filters can take many forms. One common form is a parallel combination of bandpass filters, the outputs of which have a manually adjustable gain and are added together to produce the output. Such filters can compensate for gross deficiencies in the frequency response of the sound-reproduction chain, which includes the electroacoustic response of the loudspeaker and the acoustic response of the listening room. The transient properties of narrow-bandwidth filters are, however, notoriously bad, and this can lead to a degradation in the impulse response of the equalized reproduction chain.

Another approach is to design an equalization filter by making the impulse response of the equalized sound-reproduction chain as close as possible to that desired. A net impulse response of a delta function, for example, would mean that the sound-reproduction chain had been equalized perfectly. In general it is, however, not possible to achieve such perfect inversion of the equalization chain since the acoustic path usually has delays and other nonminimum phase behavior associated with it [1]. The ability of the equalization filter to invert the response of the reproduction chain is much improved if the equalized output is compared with a delayed version of the original signal. Such a "modeling delay" is illustrated in Fig. 2. In Sec. 1 we formalize the design of such single-channel systems, and we extend the theory to the case of multiple microphones in Sec. 2.

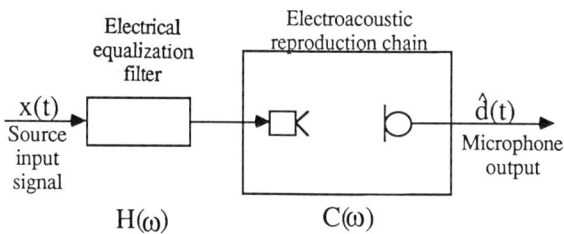

Fig. 1. Use of equalization filter in sound-reproduction system.

---

* Manuscript received 1989 July 13.

1 SINGLE-CHANNEL EQUALIZATION

We assume that the equalization filter to be designed is digital and has an all zero (FIR) structure with coefficients h_0 to h_{I-1}. We also assume that the response of the unequalized reproduction chain is modeled by a digital FIR filter with coefficients c_0 to c_{J-1}. If the sampled source signal is $x(n)$, the sampled signal fed to the loudspeaker is $y(n)$, and the sampled output from the microphone is $\hat{d}(n)$, then

$$y(n) = \sum_{i=0}^{I-1} h_i x(n - i), \qquad \hat{d}(n) = \sum_{j=0}^{J-1} c_j y(n - j)$$

so

$$\begin{aligned}\hat{d}(n) &= \sum_{j=0}^{J-1} c_j \sum_{i=0}^{I-1} h_i x(n - i - j) \\ &= \sum_{i=0}^{I-1} h_i r(n - i)\end{aligned} \qquad (1)$$

where

$$r(n) = \sum_{j=0}^{J-1} c_j x(n - j) \ . \qquad (2)$$

The summation of Eq. (1) can be written in vector form as

$$\hat{d}(n) = \boldsymbol{r}^T(n) \boldsymbol{h} \qquad (3)$$

where

$$\begin{aligned}\boldsymbol{r}^T(n) &= [r(n), r(n - 1), \ldots, r(n - I + 1)] \\ \boldsymbol{h}^T &= [h_0, h_1, \ldots, h_{I-1}].\end{aligned}$$

The most usual method of defining how $\hat{d}(n)$ is the "best" approximation to $d(n)$ is to minimize the mean-square difference between these two signals, that is, to adjust the coefficients of the equalization filter to minimize the performance index,

$$J = E\{e^2(n)\} \qquad (4)$$

where $e(n) = d(n) - \hat{d}(n)$ and E represents the expectation operator. It should be noted, however, that this performance index is not the only criterion that can be used to define the difference between desired and equalized signals [2]. One advantage of the mean-square performance index J is that it is a quadratic function of each of the coefficients in the equalization filter,

$$J = E\{d^2(n)\} - 2\boldsymbol{h}^T E\{\boldsymbol{r}(n)d(n)\} + \boldsymbol{h}^T E\{\boldsymbol{r}(n)\boldsymbol{r}^T(n)\}\boldsymbol{h} \qquad (5)$$

which has a globally minimum value for some unique set of filter coefficients [since the matrix $E\{\boldsymbol{r}(n)\boldsymbol{r}^T(n)\}$ is positive definite]. Using fairly standard optimization methods, this optimum set of filter coefficients can be shown to be given by

$$\boldsymbol{h}_{\text{opt}} = [E\{\boldsymbol{r}(n)\boldsymbol{r}^T(n)\}]^{-1} E\{\boldsymbol{r}(n)d(n)\} \ . \qquad (6)$$

In practice, adaptive algorithms can be used to adjust the coefficients of $\boldsymbol{h}$ automatically to be a close approximation to $\boldsymbol{h}_{\text{opt}}$, and these are discussed in a later section. We are concerned here with the physical consequences of designing an equalization filter according to this criterion.

Previous studies of such equalizing filters [3], [4] have demonstrated that it is possible to obtain good equalization at a single microphone position, but that the equalized response away from this point can be worse than the unequalized response. In order to illustrate this point, and to introduce the acoustic model that is used in later sections, we consider using this equalization strategy in an enclosure with dimensions and acoustic damping typical of a car interior.

The acoustic response from an acoustic source in one position in the enclosure to a microphone in another was modeled as the sum of the contributions of a finite number of acoustic modes in the enclosure [5]. The size of the enclosure was 1.9 m long by 1.1 m high by 1.0 m wide, and all modes with a natural frequency below 1200 Hz were included in the modal summation (about 500 modes), even though the response was only calculated for frequencies up to 512 Hz, the sample rate being 1024 Hz. The damping ratio of all the modes was set to 0.1. This purely acoustic response was then convolved with a filter that had a zero at direct current and a zero at half the sample rate, which represents the high-pass filtering action of a loudspeaker and the low-pass filtering action of the anti-aliasing and reconstruction filters that would be used in any practical system. In the coordinate system used for the computer simulation the origin was in the front bottom right-hand corner of the enclosure (as seen from the interior), and the coordinates (x_1, x_2, x_3) represent the distance back, across, and up, respectively, from this origin. The loudspeaker was represented by a monopole acoustic source at (0.0, 0.9, 0.7), which is approximately the position of a front dashboard loudspeaker on the left-hand side of a real car. The frequency response was calculated from this loudspeaker to a microphone at (0.1, 0.1, 0.9), which corresponds ap-

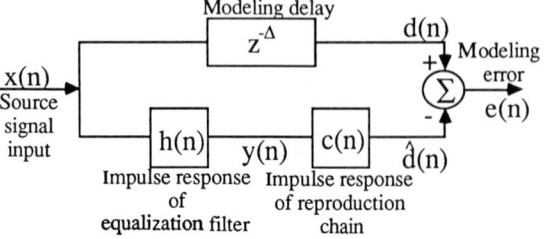

Fig. 2. Block diagram of single-point equalization problem with sampled signals.

proximately to the position of the driver's right-hand ear in a real car with right-hand drive. This frequency response, as well as a truncated version of the corresponding impulse response, obtained by inverse Fourier transformation from the frequency response, are shown in Fig. 3. These responses do have the gross characteristics of those actually measured in car interiors. Of particular note is the delay of about 3-4 ms before the dominant point in the impulse response, which corresponds to the acoustic delay time for the distance from the loudspeaker to the microphone (0.92 m).

A 50-coefficient FIR filter was used to equalize the frequency response of Fig. 3 with a modeling delay of 15 samples. This filter was adapted to minimize the mean-square modeling error using an algorithm discussed in Sec. 3. The frequency response and the impulse response of the equalization filter after convergence are shown in Fig. 4. It can be seen that the frequency response in Fig. 4 is a good approximation to the inverse of that in Fig. 3. This is further illustrated in Fig. 5, which shows the original (solid) and the equalized (dashed) responses at the equalization microphone (microphone 0), and also the original response and the effect of this equalization filter at three other microphones: microphone 1, at position (0.9, 0.9, 0.9), corresponding approximately to the front passenger's left-hand ear; microphone 2, at (1.9, 0.1, 0.9), corresponding approximately to the right-hand rear passenger's position; and microphone 3, at (1.9, 0.9, 0.9), corresponding approximately to the left-hand rear passenger's position.

It is clear that although the frequency response has been improved significantly at microphone 0 and somewhat at microphone 1, this equalization filter makes the responses more peaky at the rear microphone positions. This is largely due to the presence of the first longitudinal acoustic mode in the enclosure, with a natural frequency of about 90 Hz. It has little effect at microphones 0 and 1, since they are close to the nodal plane of this mode. These microphones have a relatively low response at about 90 Hz, which is boosted by the equalization filter. The microphones in the rear of the enclosure (2 and 3) pick up this mode strongly, however, even before boosting by the equalization filter. Thus the effect of the equalization filter is to produce a peak of some 15 dB above the average response at 90 Hz.

Fig. 5 illustrates the point made by Mourjopoulos [3], Farnsworth et al. [4], and Geddes and Blind [6] that equalization at one point can significantly disturb the response at other points in the enclosure. The results presented here are in a lower frequency range than those presented in [3] and [4], however, and may still represent a practical equalization strategy for, say, the front two seats if it were applied only to the low-frequency ("woofer") unit of an in-car entertainment system.

For completeness, Fig. 6 also shows the time domain response from the loudspeaker to each microphone before and after equalization. An extra delay equal to the modeling delay has been added to the unequalized responses for clarity. It is again clear that although the equalized impulse response at microphone 0 is very close to that desired (a delayed delta function), a considerable amount of low-frequency ringing has been added by the equalization filter to the responses in the rear of the car.

2 MULTIPLE-POINT EQUALIZATION

The failure of single-point equalization schemes to control the response at points away from the equalization microphone within the enclosure suggests that the problem of equalization at a number of points might

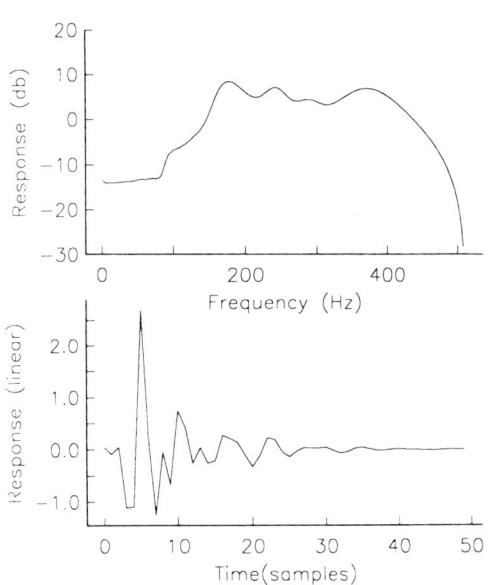

Fig. 3. Typical frequency and impulse responses in enclosure used for computer simulations.

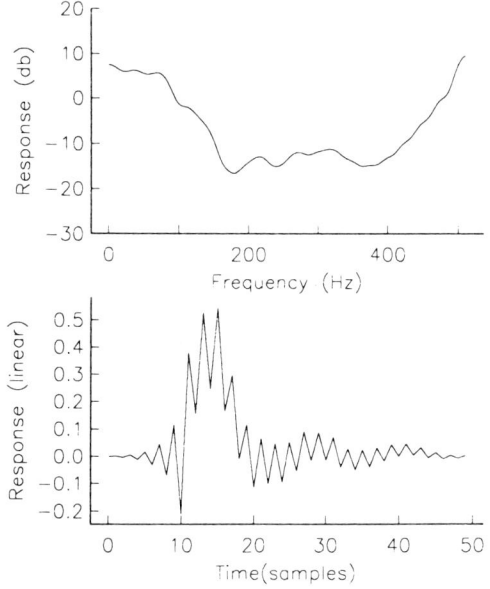

Fig. 4. Frequency and impulse responses of filter that equalizes enclosure response of Fig. 3 with a modeling delay of 15 samples.

be cast as a more general least-squares problem. This is illustrated in Fig. 7, in which the output of a single equalization filter is coupled to multiple microphones via multiple room impulse responses. Each microphone output is subtracted from a desired signal, formed by passing the source signal through an individual modeling delay (of Δ_l samples for the lth microphone), to obtain an error signal at each microphone.

The vector of output signals can now be represented [7], [8] as

$$e(n) = d(n) - R(n)h \qquad (7)$$

where

$$e^T(n) = [e_1(n), e_2(n), \ldots, e_L(n)]$$
$$d^T(n) = [d_1(n), d_2(n), \ldots, d_L(n)]$$
$$R^T(n) = [r_1(n), r_2(n), \ldots, r_L(n)].$$

$r_l(n)$ and h are defined similarly to the vectors in Sec. 1. The object of the equalizing filter is now to minimize the sum of the squares of each of the errors. This new performance index may be written as

$$J = E\{e^T(n)e(n)\}$$

so that

$$J = E\{d^T(n)d(n)\} - 2h^T E\{R^T(n)d(n)\}$$
$$+ h^T E\{R^T(n)R(n)\}h . \qquad (8)$$

The performance index again has a globally minimum value for a unique set of equalization filter coefficients given by

$$h_{\text{opt}} = [E\{R^T(n)R(n)\}]^{-1} E\{R^T(n)d(n)\} . \qquad (9)$$

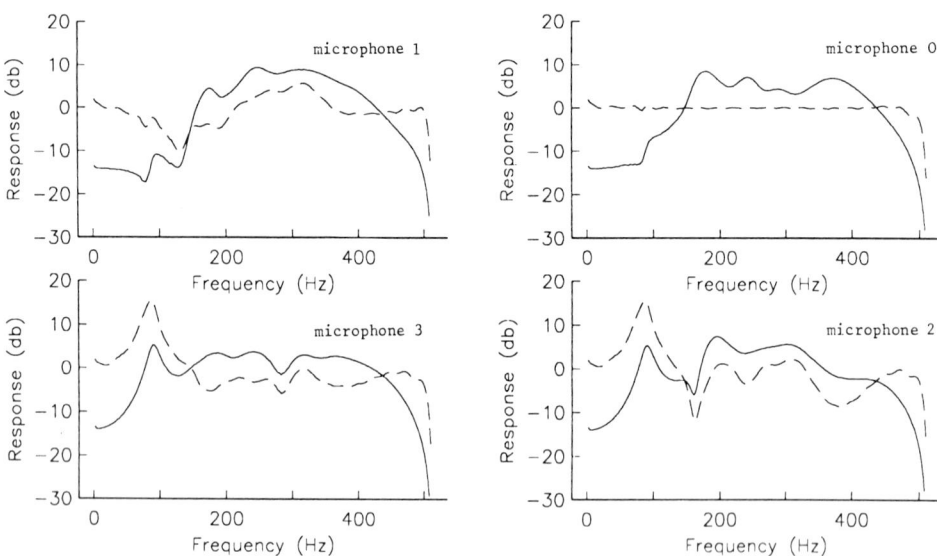

Fig. 5. Original frequency response from loudspeaker to four microphone positions in enclosure (solid lines) and response after introduction of an equalization filter designed to equalize response at microphone 0 (dashed lines).

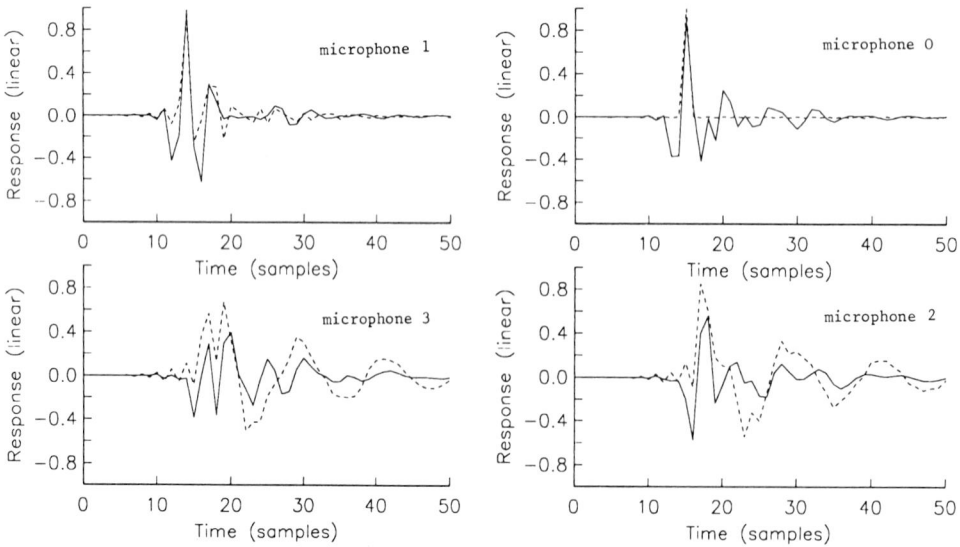

Fig. 6. Original impulse response from loudspeaker to four microphone positions in enclosure (solid lines) and net impulse responses after introduction of an equalization filter designed to equalize response at microphone 0 (dashed lines).

An adaptive algorithm is presented below for automatically adjusting the coefficients of **h** to be a close approximation to h_{opt}. This algorithm has been used to obtain an equalizing filter for the enclosure described in Sec. 1. This equalizing filter, however, now attempts to do the best job of equalizing at all four microphone positions by minimizing the sums of the squares of the differences between the microphone outputs and delayed source signals.

The frequency and impulse responses of this new equalization filter are shown in Fig. 8, and Fig. 9 compares the equalized responses at all microphone positions to the original responses. It is clear that the peaks common to all four microphone responses, such as those at about 200 Hz, have been largely removed. However, the equalizing filter has to cope with conflicting requirements at about 90 Hz, those of increasing the response in the front of the enclosure and those of suppressing the response in the rear. In fact the equalization filter does suppress the peak in the rear at the expense of creating a dip in the front at this frequency, since this strategy generates a smaller total residual error than boosting the response in the front and having the response in the back rise even further. Apart from the dips in the equalized responses at about 90 Hz in the front and at about 180 Hz at microphone 2, the filter is doing a reasonable job of equalizing at all points. The variation in the frequency response function from 2 to 500 Hz, averaged across the microphones, is about 15 dB when this equalization filter is used, compared to the original average variation in the frequency response function of some 28 dB over this frequency range. Fig. 10 shows the original and the equalized impulse responses at the four microphones, again with a shift equal to the individual modeling delays added to the original responses for clarity. The equalized responses at each of the microphones tend to be more "compact" in time than the unequalized responses, and a significant pair of "reflections" at about 25 and 30 ms have been largely removed from the responses at the front microphones.

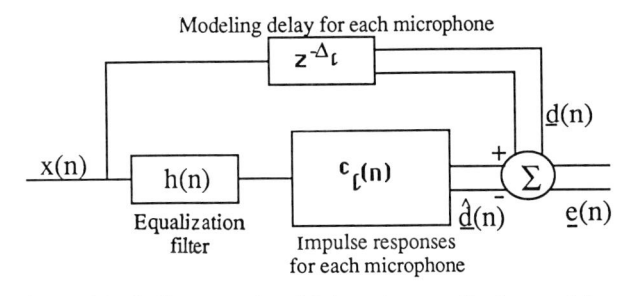

Fig. 7. Block diagram of multiple-point equalization problem with sampled signals.

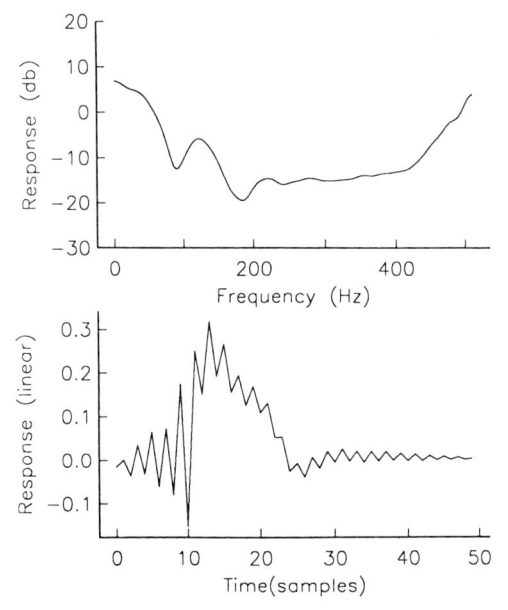

Fig. 8. Frequency and impulse responses of the equalization filter that minimizes the sum of the mean-squared modeling errors at all four microphones.

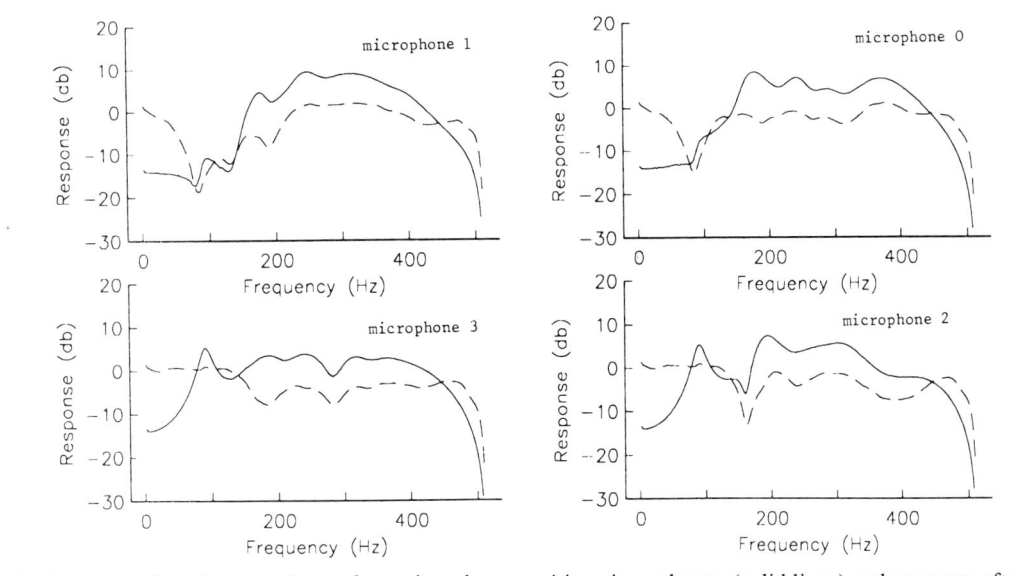

Fig. 9. Original response from loudspeaker to four microphone positions in enclosure (solid lines) and response after introduction of an equalization filter that minimizes the mean-square modeling errors at all four microphones (dashed lines).

The modeling delays used to generate the desired signals $d(n)$ at each microphone for these results were chosen for microphones 0, 1, 2, and 3 to be 15, 14, 18, and 17 samples, respectively. It was found that if all the modeling delays were set to be equal, a significantly poorer equalized response was obtained overall. It is interesting to note that the differences in the modeling delays used are approximately equal to the differences in the propagation times of a direct acoustic wave from the loudspeaker to each of the microphones. This suggests that the equalization filter can equalize the response at each microphone best by simulating a plane propagating wave in the enclosure.

3 ADAPTIVE ALGORITHMS

3.1 Single-Channel FIR Algorithms

All of the algorithms used here for adaptively adjusting the coefficients of the equalization filters are based on the instantaneous gradient descent methods introduced widely by Widrow [9]. The simplest and most common form of this algorithm is the least-mean-square (LMS) algorithm in which the single error signal $e(n)$ is given by

$$e_1(n) = d(n) - \mathbf{x}^T(n)\mathbf{h}(n) \tag{10}$$

where

$$\mathbf{x}^T(n) = [x(n), x(n-1), \ldots, x(n-I+1)]$$
$$\mathbf{h}^T(n) = [h_0(n), h_1(n), \ldots, h_{I-1}(n)] \ .$$

The LMS algorithm adjusts each of the filter coefficients in $\mathbf{h}(n)$ at each sample time by an amount proportional to the gradient of the instantaneous mean-square error [9]:

$$\mathbf{h}(n+1) = \mathbf{h}(n) - \alpha \frac{\partial e_1^2(n)}{\partial \mathbf{h}(n)} \ .$$

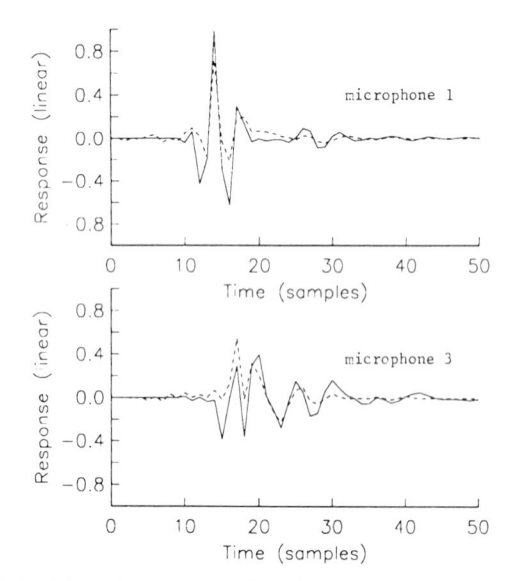

Therefore

$$\mathbf{h}(n+1) = \mathbf{h}(n) + \alpha \mathbf{x}(n) e_1(n) \tag{11}$$

where α is the convergence coefficient, which determines the speed of convergence of the algorithm. If α is too large, however, the algorithm will become unstable and an estimate of the largest stable value of α, α_{max}, has been given by Haykin [10], from a consideration of the second-order statistics of $h(n)$, as

$$\alpha_{max} \approx \frac{2}{\overline{x^2} I} \ . \tag{12}$$

An adaptive filter using the LMS algorithm for single-channel equalization is illustrated in Fig. 11(a). One disadvantage of the algorithm, however, is that the coefficients will be biased by any measurement noise at the input to the adaptive filter, that is, the output of the system to be inverted [9]. For this reason, Widrow introduced the "filtered x" LMS algorithm illustrated in Fig. 11(b), which also has the advantage that it may be used on line and was later found to be generalizable to multiple channels, as we shall see. The output error for this arrangement may be written, as in Sec. 1, as

$$e_2(n) \simeq d(n) - \mathbf{r}^T(n)\mathbf{h}(n) \ . \tag{13}$$

Applying the same philosophy of minimizing the instantaneous squared error leads to the algorithm

$$\mathbf{h}(n+1) = \mathbf{h}(n) + \alpha \mathbf{r}(n) e_2(n) \ . \tag{14}$$

This is called the filtered x algorithm because $r(n)$ consists of the reference signal $x(n)$ filtered by the impulse response of the system to be inverted, as in Eq. (2). In practice this impulse response can only be estimated imperfectly, and an approximation to $r(n)$

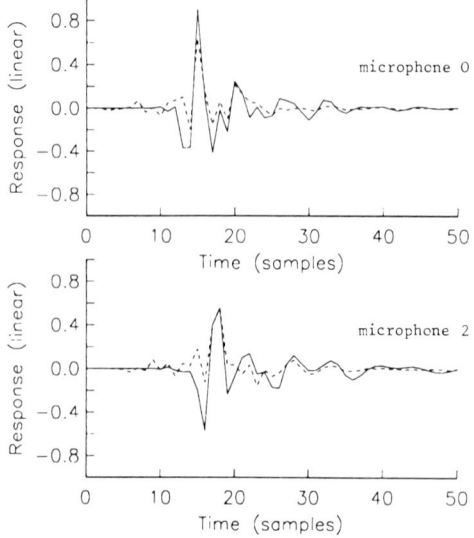

Fig. 10. Original impulse response from loudspeaker to four microphone positions in enclosure (solid lines) and net impulse response after introduction of an equalization filter that minimizes the mean-square modeling errors at all four microphones (dashed lines).

must be used in the algorithm, although it is found that the algorithm is very robust to errors made in the generation of this reference signal.

Once again α is a convergence coefficient, which has a certain maximum value before the algorithm becomes unstable. A consideration of the first-order statistics of $h(n)$ suggests that α_{max} is inversely proportional to $\overline{r^2}$ in this case rather to than $\overline{x^2}$. However, here the maximum convergence coefficient is found to depend not only on the length of the adaptive filter and the spectrum of the filtered reference signal, but also on the delays inherent in the error path [that is, in $c(n)$]. Computer simulations, using a white noise reference signal and a pure delay (of δ samples) in the error path, have been used to establish the maximum convergence coefficient under a variety of conditions. The results of these simulations are shown in Fig. 12. They suggest that α_{max} in this case is of the form

$$\alpha_{max} \approx \frac{2}{\overline{r^2}(I + \delta)} . \qquad (15)$$

If the filtered reference signal is not white, however, which was the case when this algorithm was used in the arrangement of Sec. 1, in which the white reference signal $x(n)$ was passed through a filter with a nonuniform frequency response to give $r(n)$, this simple formula no longer holds. In particular, $1/\alpha_{max}$ becomes approximately proportional to $1.2I$, rather than to $0.5I$, as in Eq. (15). This more complicated behavior is probably due to the correlation between the samples of the filtered reference signals, which was not present in the simulations used to obtain the results presented in Fig. 12. This may be similar to the "eigenvalue spread" problem discussed in [9] and [10]. Certainly the convergence of the squared error, in the simulations of the filtered x algorithm used to obtain the results in Sec. 1, shows evidence of several "modes" of convergence.

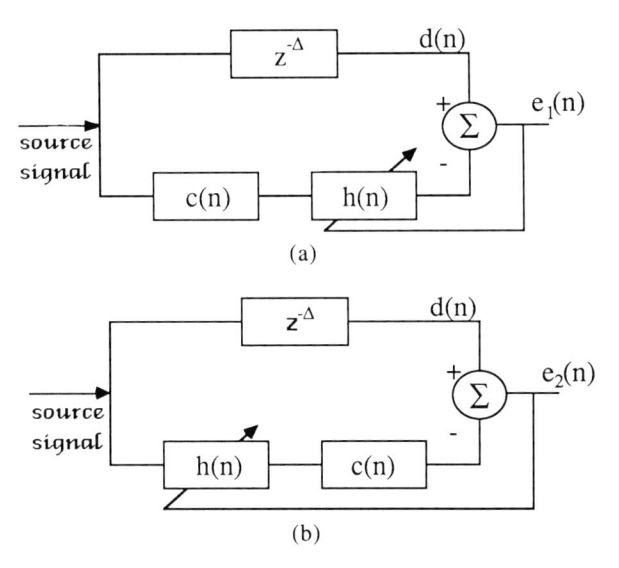

Fig. 11. Block diagrams showing difference between (a) conventional LMS algorithm used for equalization of $c(n)$, and (b) filtered x LMS algorithm.

3.2 Multiple-Channel FIR Algorithm

The generalization of the single-channel filtered x algorithm to multiple channels has been discussed in [7] and may be expressed, using the notation of Sec. 2, as

$$\boldsymbol{h}(n + 1) = \boldsymbol{h}(n) + \alpha \boldsymbol{R}^T(n)\boldsymbol{e}(n) . \qquad (16)$$

The expression for the maximum convergence coefficient in the single-channel case [Eq. (15)] suggests a generalization for α_{max}, for the case of L microphones, of the form

$$\alpha_{max} \approx \frac{2}{\sum_{l=1}^{L} \overline{r_l^2}(I + \delta_l)} \qquad (17)$$

where $\overline{r_l^2}$ is the mean-square value of the lth reference signal, and the lth error path has a delay of δ_l. In the simulations of the four-channel filtered x algorithm performed for Sec. 2 the delays in the error path were relatively small ($\delta_l \approx 4$ to 8 samples) so their effect was difficult to determine, but the variation of α_{max} with filter length I conformed reasonably well to the expression

$$\alpha_{max} \approx \frac{1}{1.2 \sum_{l=1}^{4} \overline{r_l^2} I} . \qquad (18)$$

3.3 Adaptive IIR Algorithm

One other algorithm was investigated for use in this application, which differed from the preceding ones in that the equalization filter had a recursive form, that is, its output $y(n)$ was related to its input $x(n)$ by

$$y(n) = \sum_{i=0}^{I/2} a_i x(n - i) + \sum_{j=1}^{I/2} b_j y(n - j) \qquad (19)$$

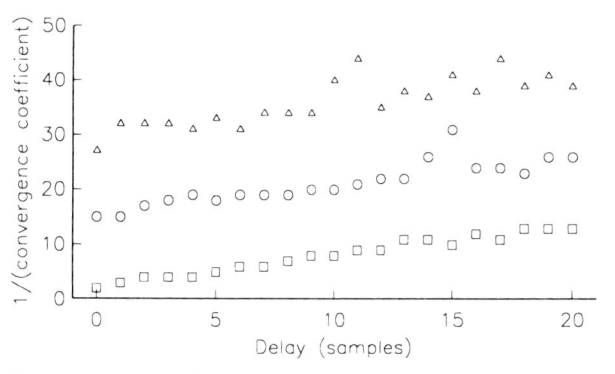

Fig. 12. Reciprocal of largest convergence coefficient before instability in simulations of the filtered x algorithm with a white noise reference signal, for various pure delays in error path and different filter lengths: □—1 coefficient; ○—25 coefficients; △—50 coefficients.

where the total number of filter coefficients a_i and b_j is now $I + 1$. The motivation for such recursive (IIR) filters is that they are potentially more efficient than nonrecursive (FIR) ones, that is, they can achieve the same performance with a smaller number of coefficients. Unfortunately the recursive nature of Eq. (19) potentially complicates any adaptive algorithm designed to adjust the coefficients. This is partly due to the fact that for some combinations of the coefficients b_j, the filter may be inherently unstable, and an adaptive algorithm may find it difficult to recover from such an instability. Another difficulty is that the mean-square error is now no longer a simple quadratic function of each of the filter coefficients, so gradient descent methods such as those used in the preceding are no longer guaranteed to converge.

Despite these potential disadvantages, single-channel adaptive recursive filters have been used by Eriksson et al. [11] in active noise control applications, which are formally very similar to the equalization problem considered here, and these authors reported that the algorithm they used was reliable and robust. A multiple-channel generalization of the Eriksson algorithm was reported in [12] and has been used in a simple comparison with the results of nonrecursive adaptive filters described in Sec. 2.

The results suggest that the algorithm reported in [12], when used to minimize the modeling error at four microphones in Sec. 2 with 26 nonrecursive and 25 recursive coefficients, does converge on a time scale comparable with that of the convergence of the FIR algorithm. The final residual error is also very close to that found with the 50-coefficient FIR equalization filter. If the responses at each of the microphones are calculated when equalized with the IIR filter, they are barely distinguishable from those using the FIR filter, reported in Fig. 9. The converged IIR filter coefficients have been used to calculate the first 100 samples of the (infinite) impulse response of this equalizer, which is shown in Fig. 13 together with the result from the 50 coefficients that comprise the complete impulse response of the converged FIR filter from Fig. 8. It is clear that the impulse responses of the two equalization filters are very similar, and although the recursive property of the IIR filter allows it to have some response beyond 50 samples, this response is small and has little effect on the frequency domain behavior.

This preliminary investigation suggests that although the multiple-channel IIR adaptive algorithm does converge, in the application considered here, it gives no better results than an FIR adaptive filter with a comparable number of coefficients.

4 CONCLUSIONS

The equalization of the acoustic response of rooms using adaptive digital filters has been considered. The problem of equalizing the response at one position only in the room is discussed first, and this is formulated in the time domain by minimizing the mean-square error between the equalized response and a delayed version of the original signal.

The acoustic response of a small enclosure, with dimensions and acoustic damping typical of a car interior, has been modeled, using a modal summation, up to a frequency of about 500 Hz. Using an adaptive FIR digital filter with 50 coefficients and a modeling delay of 15 samples, it is found that the response at one position can be equalized very effectively. The consequence of equalization, at this one point, on the response at other points in the enclosure has also been calculated. It is found that at some other points in the enclosure the response is degraded significantly by the equalization filter, as has previously been suggested by Mourjopoulos [3] and Farnsworth et al. [4].

An alternative equalization strategy has been presented in order to try to overcome this problem, namely, that of minimizing the sum of the mean-square errors between the equalized responses at several points in the enclosure and individually delayed versions of the original signal. Results from applying this approach at four rather widely spaced positions in the same enclosure have been presented, which demonstrate that some improvements can be effected, although there inevitably remain differences in the equalized responses at the four positions. The modeling delays used to form the error signals at the four equalization positions, which gave the most convincing overall equalization, suggest that the equalization filter does best by attempting to simulate a plane progressive wave in the enclosure.

Various algorithms have been presented for adapting the coefficients of an FIR equalization filter in practice to achieve a good approximation to the single- or multiple-point least-squares solution. The maximum convergence coefficients of these algorithms have been considered in relation to the delays in the path to be equalized and the number of coefficients in the adapting filter. An algorithm for adapting the coefficients for an IIR digital filter for multiple-point equalization has also been investigated. This filter converges to a solution very similar to that found by an adaptive FIR of comparable length, and there would appear to be no advantage to using such a filter over an FIR filter in this particular application.

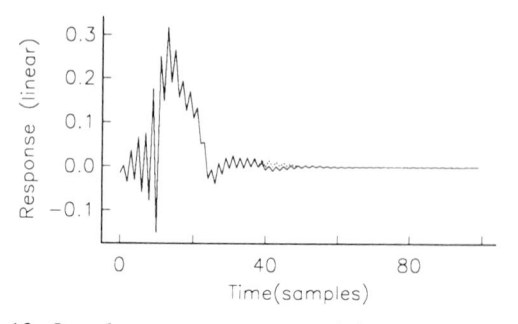

Fig. 13. Impulse response computed from coefficients of adapted IIR filter used for equalization at four microphones (solid line). Also shown are the coefficients of adapted FIR filter (dotted line), which extend only for the first 50 samples.

5 ACKNOWLEDGMENT

This work was supported in part by the National Aeronautics and Space Administration under Contract NAS1-18107 while one of the authors (S. J. Elliott) was visiting the Institute for Computer Applications in Science and Engineering (ICASE), NASA Langley Research Center, Hampton, VA 23665.

6 REFERENCES

[1] S. T. Neely and J. B. Allen, "Invertibility of Room Impulse Responses." *J. Acoust. Soc. Am.*, vol. 66, pp. 165–169 (1979).

[2] J. Mourjopoulos, P. M. Clarkson, and J. K. Hammond, "A Comparative Study of Least-Squares and Homomorphic Techniques for the Inversion of Mixed Phase Signals," *Proc. ICASSP* (1982).

[3] J. Mourjopoulos, "On the Variation and Invertibility of Room Impulse Response Functions, *J. Sound Vibration*, vol. 102, pp. 217–228 (1985).

[4] K. D. Farnsworth, P. A. Nelson, and S. J. Elliott, "Equalisation of Room Acoustic Responses over Spatially Distributed Regions," *Proc. Inst. of Acoustics, Autumn Conference: Reproduced Sound* (1985).

[5] A. J. Bullmore et al., "The Active Minimisation of Harmonic Enclosed Sound Fields; Part II: A Computer Simulation." *J. Sound Vibration*, vol. 117, pp. 15–33 (1987).

[6] E. Geddes and H. Blind, "The Localized Sound Power Method," *J. Audio Eng. Soc. (Engineering Reports)*, vol. 34, pp. 167–173 (1986 Mar.).

[7] S. J. Elliott, I. M. Stothers, and P. A. Nelson, "A Multiple Error LMS Algorithm and Its Application to the Active Control of Sound and Vibration," *IEEE Trans. Acoust., Speech, Signal Processing*, vol. ASSP-35, pp. 1423–1434 (1987).

[8] P. A. Nelson and S. J. Elliott, "Least Squares Approximations to Exact Multiple Point Sound Reproduction," ISVR Memo 683 (1988). Also submitted to *IEEE Trans. Acoust., Speech, Signal Processing* as "Adaptive Inverse Filters for Stereophonic Sound Reproduction."

[9] B. Widrow and S. Stearns, *Adaptive Signal Processing* (Prentice-Hall, Englewood Cliffs, NJ, 1985).

[10] S. Haykin, *Adaptive Filter Theory* (Prentice-Hall, Englewood Cliffs, NJ, 1986).

[11] L. R. Eriksson, M. C. Allie, and R. A. Greiner, "The Selection and Application of an IIR Adaptive Filter for Use in Active Sound Attenuation," *IEEE Trans. Acoust. Speech, Signal Processing*, vol. ASSP-35, pp. 433–437 (1986).

[12] S. J. Elliott and P. A. Nelson, "An Adaptive Algorithm for IIR Filters Used in Multichannel Sound Control Systems," ISVR Memorandum No. 681 (1988).

THE AUTHORS

S. Elliott

P. Nelson

Stephen Elliott graduated with joint honours in physics and electronics from the University of London in 1976, and in 1979 was awarded a Ph.D. from the University of Surrey for a thesis in musical acoustics.

After short periods as a research fellow working on acoustic intensity measurement and a temporary lecturer, he was appointed as a lecturer at the Institute of Sound and Vibration Research at the University of Southampton, U.K., in 1982, under the SERC Special Replacement Scheme. He was made senior lecturer at ISVR in 1988.

His research interests include the connections between the physical principles and signal processing aspects of speech analysis, and the active control of sound and vibration. It is partly the analogy between the problems of active noise control in enclosures, where the "desired signal" is silence, and sound reproduction, where the "desired signal" is the original source, which prompted the work presented in this paper.

●

Philip Nelson received the B.Sc. degree in mechanical engineering in 1974 from the University of Southampton, U.K. He went on to undertake research into aerodynamic sound production at the Institute of Sound and Vibration Research and received a Ph.D. degree in 1981.

From 1978 to 1982 he worked as a research, development, and consulting engineer at Sound Attenuators, Ltd., Colchester, U.K., where he first became interested in the active control of sound. In 1982 he was appointed lecturer at the Institute of Sound and Vibration Research under the SERC Special Replacement Scheme. Since then he has worked primarily on the active control of sound and vibration, and was appointed senior lecturer in 1988.

Effects of Cable, Loudspeaker, and Amplifier Interactions*

FRED E. DAVIS

Engineering Consultant, Hamden, CT 06517, USA

Loudspeaker cables are among the least understood yet mandatory components of an audio system. How cables work and interact with loudspeaker and amplifier is often based more on presumption and speculation than on fact. The literature on loudspeaker cable behavior and effects is minimal. Measurements were made with 12 cables covering a variety of geometries, gauges, and types. The measured data indicate distinct differences among the cables as frequency-dependent impedance, subtle response variations with loudspeakers, and reactance interactions between amplifier, cable, and loudspeaker. In some cases the effects of the amplifier overwhelm the cable's effects. Mathematical models that provide insight into the interaction mechanisms were constructed and compared to the measured data.

0 INTRODUCTION

A variety of specialty loudspeaker cables can be found advertised in almost any audio magazine from the last 10 years. All promise the same result—better sound—yet they span the gamut of electrical characteristics, geometries, and materials. How loudspeaker cables work is often based more on presumption and speculation than on fact. Few articles are published exploring the behavior of these mandatory components in journals [1] and popular magazines [2]–[6]. Debates continue on computer network newsgroups on audio [7]. "White papers" available from manufacturers (but otherwise unpublished) are frequently more marketing than science [8]–[11].

Using a simplistic view of how loudspeakers and cables work, conventional wisdom would suggest that since loudspeakers exhibit a low impedance (nominally 4–8 Ω), then the cable should have even lower resistance. As a result, "monster" cables were introduced. Then a more complex view of cables emerged, suggesting that loudspeaker cables would perform better with less capacitance or more inductance, or the skin effect, phase shift, and dispersion were veiling high frequencies, or they behaved like transmission lines. These factors are the essence of 'high-end' cables. Greiner addressed some of these issues in his papers [1]–[3]. In short, he proposes that loudspeaker cables are not transmission lines (audio frequency wavelengths are much too long compared to the length of the cables); phase shift and dispersion effects are too small to be audible (typically less than 0.3 deg/m at 20 kHz, and differences of less than 60 ns/m for most cables between 100 Hz and 10 kHz); and the skin effect has only a small effect on heavy conductors (skin depth in copper at 20 kHz is 0.5 mm).

It is no secret that loudspeakers offer a complex load to the amplifier [12]–[13]. While an isolated loudspeaker is predominantly inductive, the complex impedance of most loudspeaker systems with multiple drivers and passive crossover networks exhibits both negative and positive phase angles at given frequencies, indicating capacitive reactance as well as inductive reactance.[1] Otala and Huttunen [13] show that given complex waveforms, commercial loudspeakers require up to 6.6 times more current than an 8-Ω resistor for the same waveform, suggesting a dynamic impedance as low as 1.2 Ω.

The ideal loudspeaker cable should transfer all audio frequencies into any loudspeaker load with flat voltage response. Real cables will always show some loss due to resistance, but better cables will both minimize this loss and still transfer all frequencies unscathed. The acoustical result will depend on many factors, but the electrical interaction of loudspeaker, amplifier, and cable forms an essential foundation. This engineering

---

*Manuscript received 1990 August 21; revised 1990 December 24.

[1] Some issues of *Audio* that illustrate Nyquist plots of loudspeakers' complex impedances are vol. 74, p. 100 (1990 Nov.); vol. 74, p. 94 (1990 Aug.); vol. 73, p. 111 (1989 June); vol. 73, pp. 88, 108 (1989 Sept.).

report examines the mechanisms for this interaction and shows how it can affect the response of the system.

1 SAMPLE OF CABLES TESTED

The sample of cables gathered for this test represents a variety of commonly and uncommonly available wire. Most of the samples were 3.1 m in length. Some are *very* expensive (over $419 per meter), others cheap ($1.91 per meter), and some are not loudspeaker cables at all. This is not an exhaustive examination of every loudspeaker cable available. The following is a brief description of each type with sample numbers as they appear in Figs. 1–3. They are presented in order of ascending resistance per meter. When known, the organization of the strands is shown in parentheses as (quantity*gauge). Unspecified gauges were estimated from conductor diameter and resistance.

1) *Levinson HF10C.* Many very small copper strands in two parallel conductors (each about 6.4 mm in diameter) spaced about 12.7 mm apart (between centers of the conductors). Approximately 3 AWG. Extremely flexible for such a heavy conductor.

2) *Auto Jumper Cables.* Literally from the garage. Two thick parallel (9.5-mm diameter) conductors of approximately 7 AWG (19*20).

3) *Krell "The Path."* Independent wires of about 15.9-mm diameter, each of complex layer construction. The conductor is 4.8 mm in diameter, the remainder is insulation. It has several groups of tightly twisted very thin enameled wires wound in helices around heavier enameled wires. (This construction is similar to Music Interface Technologies' "Vari-Lay" and Monster Cable "Time Correct"). All conductors are soldered together at each end with heavy, crimped terminations. Approximately 5 AWG. They are labeled "transconductant speaker cable."

4) *AudioQuest Green "Litz."* Six conductors (approximately 10 AWG) of many small enameled copper wires, lightly twisted over a stranded plastic core, altogether about 12.7 mm in diameter. Equivalent to about 6 AWG.

5) *Kimber 16LPC.* These are 16 independent wires, woven together in a flat cable, Teflon insulation. Each individual wire is equivalent to 19 AWG, and is composed of seven strands of variable gauge from 31 AWG to 24 AWG. Equivalent to 7 AWG.

6) *Spectra-Strip 843-138-2601-064 Ribbon Cable.* Abbreviated 138-064. Made of 32 twisted pairs of 26 AWG wire (7*34), arranged in a flat ribbon. Intended for high-speed differential digital data transmission. Equivalent to about 8 AWG.

7) *Belden 9718.* Belden's 12 AWG (65*30) loudspeaker wire with clear PVC insulation and parallel construction, like "zip" cord (sample 12).

8) *Music Interface Technologies' CVT.* A large 18-mm diameter cable using MIT's Vari-Lay construction (multiple conductors of different gauge and length). The manufacturer claims this will permit "all frequencies to travel through a given length of MIT cable at exactly the same rate of speed," hence the name constant velocity transmission (CVT). Two groups of three Vari-Lay bundles form the two main conductors, with a coaxial cable connected in an unknown fashion (due to potting compound) inside a proprietary coupler at the amplifier end. At $419 per meter, the most expensive cable tested. Equivalent to 12 AWG.

9) *Kimber 8LPC.* Very similar to sample 5, except eight independent wires, woven in a flat cable, Teflon insulation. Each individual wire is equivalent to 19 AWG, and is composed of seven strands of variable gauge from 31 AWG to 24 AWG. Equivalent to 10 AWG.

10) *Kimber 4PR.* An unusual cable made from eight independent wires of 23 AWG (7*31) braided together, PVC insulation. Equivalent to 14 AWG.

11) *Spectra-Strip 843-191-2811-036 Ribbon Cable.* Abbreviated 191-036. Made of 36 wires of 28 AWG (7*36), arranged in a flat ribbon; intended for digital interconnections. The least expensive cable tested at $1.91 per meter. Equivalent to about 15 AWG.

12) *Belden 19123.* 18 AWG (41*34) "zip" (lamp) cord. Brown PVC insulation, parallel construction.

2 ELECTRICAL PARAMETERS OF CABLE SAMPLES

The standard electrical parameters of the cables were measured with an ESI model 252 impedance meter and normalized to 1 m. The results are shown in Figs. 1–3. It is resistance, capacitance, and inductance that will decide the performance of the cable, since exotic materials and layer geometries can only affect these fundamental characteristics.

Cable resistance in milliohms per meter is shown in Fig. 1 (remember that this includes the resistance of both conductors). Resistance is not a major factor in cables of reasonable length. Based on resistance alone, it would require about 23.4 m of 18 AWG cable to show −1 dBV drop with an 8-Ω load. 12 AWG seems more than adequate even for demanding systems, high

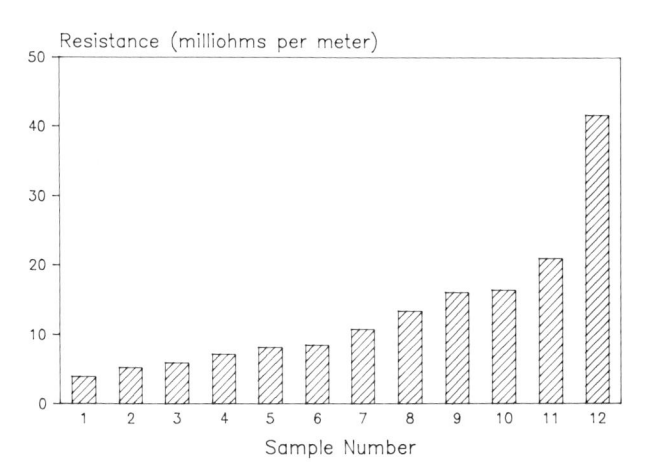

Fig. 1. Cable resistance in milliohms per meter. 1—HF10C; 2—jumper; 3—Krell; 4—Litz; 5—16LPC; 6—138-064; 7—9718; 8—CVT; 9—8LPC; 10—4PR; 11—191-036; 12—19123.

power levels, and reasonable lengths. The maximum current for 12 AWG wire with PVC insulation in an ambient temperature of 30°C, allowing for a 50°C temperature rise, is 36 A. This seems fine for audio applications, since 36 A into 8 Ω is greater than 7 kW rms (1.8 kW rms into 2 Ω).

Fig. 2 shows the cable capacitance. As expected, flat cables show the highest capacitance (samples 6 and 11), multiconductor cables less (samples 4, 5, 8, 9, and 10), and two-conductor cables the least (samples 1, 2, 3, 7, and 12).

Fig. 3 shows the cable inductance. Cables with only two separated conductors show the highest inductance, while most multiwire cables show the lowest inductance. An exception is Music Interface Technologies' CVT due to its construction.

3 CABLE IMPEDANCE VERSUS FREQUENCY

The impedance of a cable across the audio spectrum shows the influence of reactive and skin effects. Better cables will have a low impedance that remains constant with frequency which permits flatter voltage response.

A current of 1 A at a given frequency will cause a voltage difference equivalent to the magnitude of the cable's impedance in ohms at that frequency. For this test, a resistive load of 1.0 Ω (with approximately 0.06-μH inductance) was driven at a current of 1.0 A rms at 12 frequencies between 30 Hz and 20 kHz. All measurements were made with a Fluke 8050A digital voltmeter and waveforms monitored on a Tektronix 2215 oscilloscope. The amount of current was determined by driving the amplifier until the voltage across the load was 1.000 V rms at each frequency and for every cable, thus removing frequency response variations from signal source, attenuator, and amplifier. The voltage difference from the output of the amplifier to the load was then measured and recorded, and the impedance calculated.

The results of these measurements are shown in Figs. 4 and 5 as cable impedance versus frequency, where the value of impedance reflects the contribution of both conductors. Cables with the most constant impedance were the flat cables with higher capacitance (Fig. 4, 138-064; Fig. 5, 191-036). Other multiconductor cables such as Kimber 16LPC and AudioQuest Green Litz (Fig. 4, 16LPC and Litz) and the lighter gauge Kimber 8LPC and 4PR (Fig. 5, 8LPC and 4PR) display a small impedance rise. Of the two conductor cables tested,

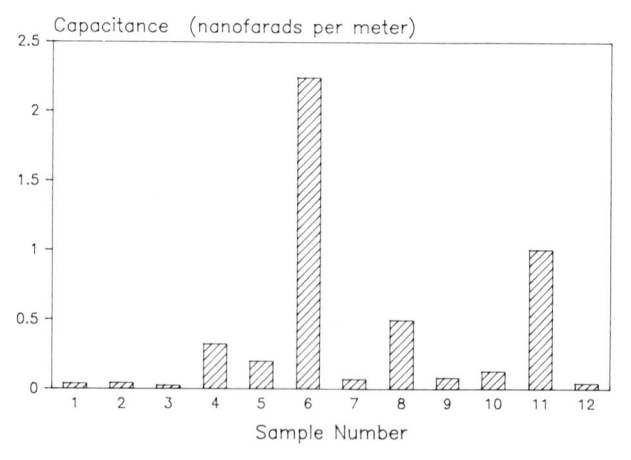

Fig. 2. Cable capacitance in nanofarads per meter. 1—HF10C; 2—jumper; 3—Krell; 4—Litz; 5—16LPC; 6—138-064; 7—9718; 8—CVT; 9—8LPC; 10—4PR; 11—191-036; 12—19123.

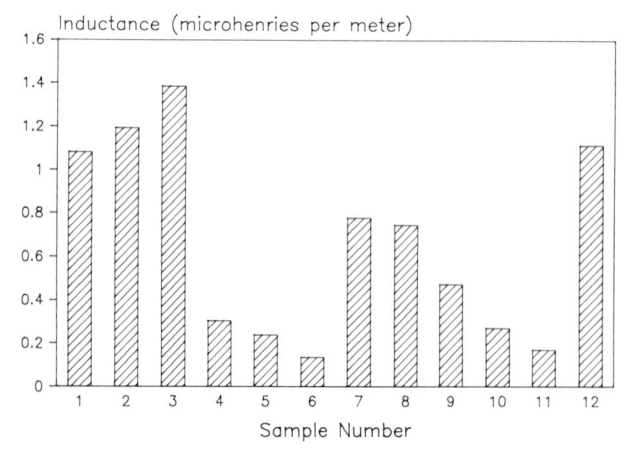

Fig. 3. Cable inductance in microhenrys per meter. 1—HF10C; 2—jumper; 3—Krell; 4—Litz; 5—16LPC; 6—138-064; 7—9718; 8—CVT; 9—8LPC; 10—4PR; 11—191-036; 12—19123.

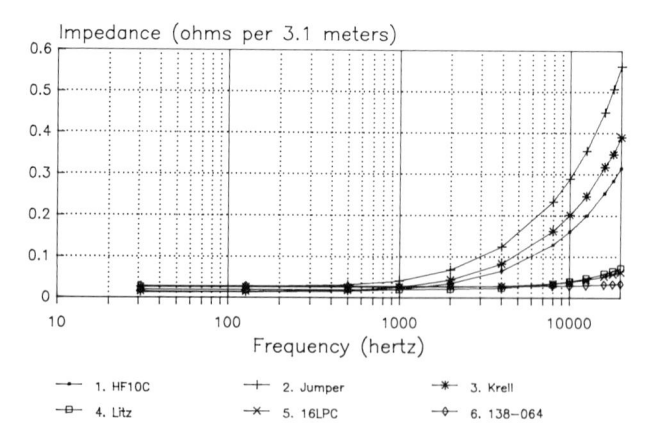

Fig. 4. Cable impedance versus frequency for cable samples 1–6.

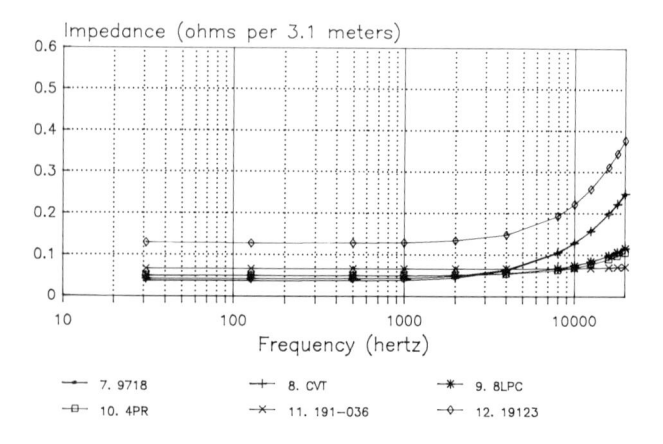

Fig. 5. Cable impedance versus frequency for cable samples 7–12.

12 AWG wires (Fig. 5, 9718 and CVT) performed the best, since both heavier and lighter gauges showed greater high-frequency impedance. The complex layer construction of the CVT cable has duplicated almost exactly the impedance characteristics of the 12 AWG Belden 9718 (Fig. 5, CVT and 9718; coefficient of correlation = 0.997).

The effect of inductive reactance in this sample of cables is far more significant than the skin effect. For example, 3.1 m of the largest diameter cable sampled, Levinson HF10C (sample 1), will show a 3.42 times increase in resistance at 20 kHz due to the skin effect, but the inductive reactance will be 9.8 times greater than resistance at that frequency. When driving 8 Ω at 20 kHz through 3.1 m, the skin effect alone would produce a drop of −0.044 dBV relative to 20 Hz, while the combined reactance and skin effects would produce a drop of −0.43 dBV.

Higher cable capacitance will tend to reduce the combined reactive component of the cable, thus lowering cable impedance at high frequencies and improving the high-frequency response. This effect is contrary to the popular belief that high frequencies will be attenuated more with higher cable capacitance [5], [8]. Such conclusions are drawn from a cable model consisting of series resistance and shunt capacitance, but no series inductance. Spectra-Strip 138-064 (sample 6) showed the highest capacitance (6.847 nF for 3.1 m), lowest inductance, and flattest cable impedance. Well designed amplifiers are not affected by this amount of capacitance, but some amplifiers may become unstable.

4 TEST LOUDSPEAKER AND AMPLIFIER CHARACTERISTICS

The impedance and phase characteristics of loudspeakers A and B used in these tests are shown in Figs. 6 and 7, measured at the same frequencies used in the cable impedance test. Please note that the lines connecting the data points in these graphs are intended to simplify reading the plot and do not reflect valid data between the sampled frequencies. Loudspeaker A is a three-way design with an acoustic suspension woofer, three dome midrange drivers, and three dome tweeter drivers. It exhibits mostly capacitive reactance (negative phase angle) at the frequencies sampled between 127 Hz and 12 kHz, with its lowest impedance of 4.8 Ω above 8 kHz. Loudspeaker B is a two-way system with a bass reflex enclosure and dome tweeter. It shows much more inductive reactance (positive phase angle) than loudspeaker A around 1 kHz, and a capacitive reactance peak around 8 kHz. Its lowest impedance is 5.8 Ω around 500 Hz.

Amplifier frequency response and damping factor are shown in Fig. 8. Amplifier A exhibits more significant frequency response variations and a large drop in damping factor above 1 kHz. Amplifier B has a flat frequency response and a high, almost linear damping factor. In Secs. 5 and 6, the effect of the amplifier is factored out, showing only the cable and loudspeaker interactions. In Sec. 7, amplifier effects are included with loudspeaker and cable effects for a total system response.

5 CABLE RESPONSE WITH LOUDSPEAKER LOADS

Obviously, a loudspeaker can only perform to the quality of the electrical input to its terminals, so the best cable will show the flattest frequency response

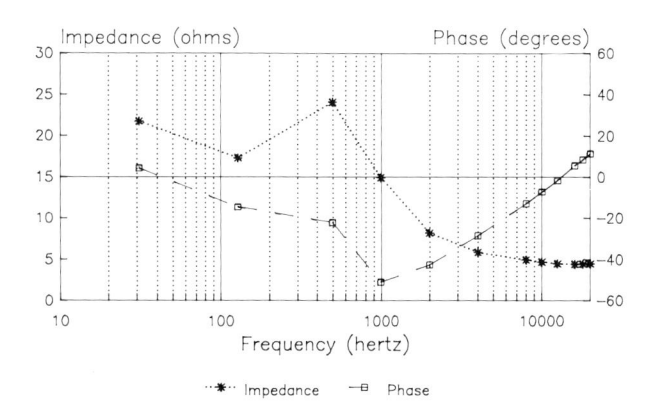

Fig. 6. Impedance and phase response of loudspeaker A. Note that frequencies are sampled and lines connecting data points do not reflect valid data.

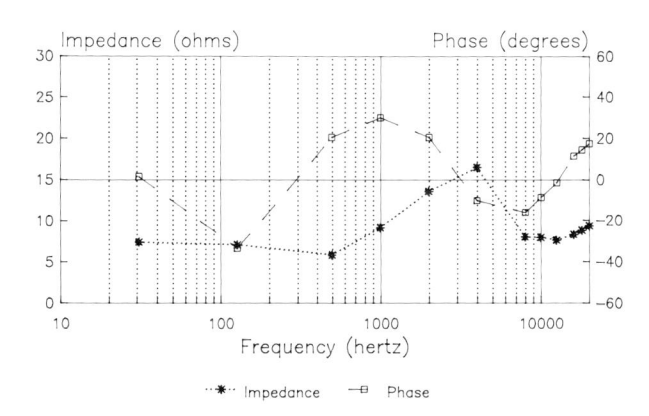

Fig. 7. Impedance and phase response of loudspeaker B. Note that frequencies are sampled and lines connecting data points do not reflect valid data.

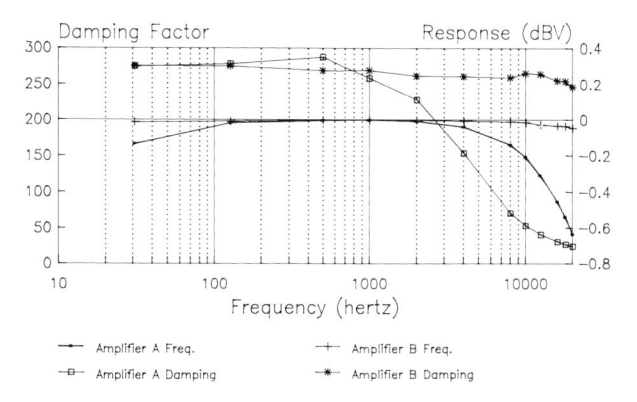

Fig. 8. Frequency response and damping factor for amplifiers A and B.

despite loudspeaker impedance or phase angle. The cable electrical response was measured using two commercial loudspeakers as a load.

A constant amplifier output of 1 V (0.00 dBV) was used at each frequency to remove any variations due to amplifier or signal source. The amplitude of the voltage at the loudspeaker terminals was measured in dBV and recorded.

The low-inductance multiconductor cables show the most linear response (Fig. 9, Litz, 16LPC, and 138-064; Fig. 10, 8LPC, 4PR, and 191-036). Also note the relatively flat response of the 12 AWG cable with both loudspeakers (Figs. 10 and 11, 9718) when compared to other two-wire cables (Figs. 9 and 11, HF10C and Krell). Another common effect is the high-frequency loss with the higher inductance two-conductor cables.

Fig. 9 also shows the interaction of a cable's inductive reactance with loudspeaker A's capacitive reactance where the level rises above 0 dBV in the 1-kHz to 10-kHz region. At this point the loudspeaker terminal voltage has exceeded the amplifier's output. The cause of this will become apparent with the loudspeaker cable model introduced in Sec. 6.

Four cables representing a variety of types were tested with loudspeaker B (Fig. 11). Loudspeaker B shows inductive reactance and low impedance between 300 Hz and 3 kHz and the response dips. When the reactance of loudspeaker B becomes capacitive around 8 kHz, it shows the same rise with the more inductive cables (HF10C and Krell).

6 LOUDSPEAKER CABLE MODEL

Expressions for transmission lines (such as characteristic impedance, impedance matching, reflections) do not fit audio applications, since the cable lengths involved are minute fractions of the shortest audio wavelength (about 16 km at 20 kHz in copper). This is discussed thoroughly in Greiner [1]–[3].

Therefore, cable and loudspeaker should be treated as lumped-circuit elements. The cable response model in this engineering report is simple and is based on the ratio of the vector sum of the loudspeaker's resistive and reactive components to the vector sum of both loudspeaker and cable resistive and reactive components together. The cable is modeled at each frequency as a resistance in series with an inductive reactance using the measured values of resistance and inductance. The skin effect was calculated and applied to the resistance where appropriate. The capacitive component of the cable is too small to have much influence at audible frequencies, and is thus omitted from the model. The loudspeaker is modeled at each frequency as a resistance in series with a reactance that can be either inductive or capacitive. The expression for the cable response at the loudspeaker terminals for a given frequency is

$$V_s(f) = V_a(f) \frac{\sqrt{R_s^2 + X_s^2}}{\sqrt{(R_w + R_s)^2 + (X_w \pm X_s)^2}}$$

where

$V_s(f)$ = voltage at loudspeaker terminals at frequency f
$V_a(f)$ = voltage at amplifier output at frequency f
R_w = cable resistance, including skin effect, at frequency f
X_w = cable inductive reactance at frequency f
R_s = loudspeaker resistance
$\pm X_s$ = loudspeaker reactance at frequency f, inductive (+) or capacitive (−).

The response in dBV was found by taking the log-

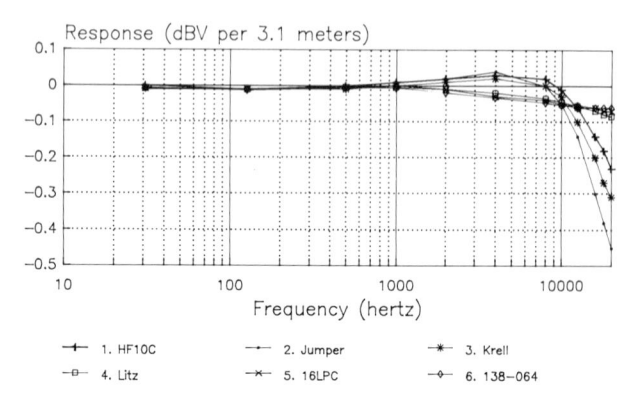

Fig. 9. Measured cable response with loudspeaker A for cable samples 1–6.

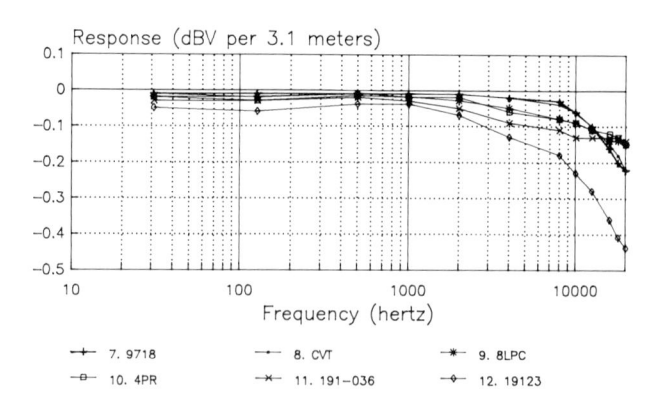

Fig. 10. Measured cable response with loudspeaker A for cable samples 7–12.

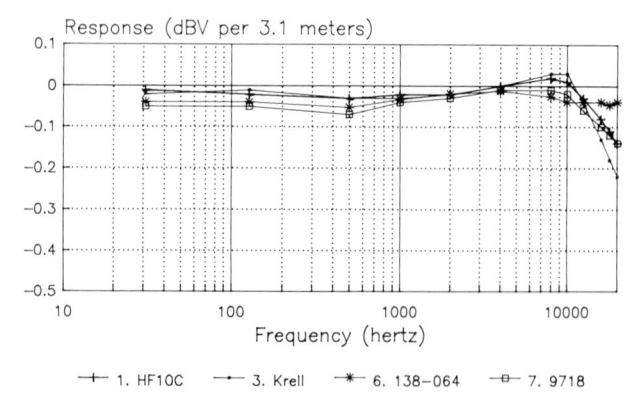

Fig. 11. Measured cable response with loudspeaker B for cable samples 1, 3, 6, and 7.

arithm of the ratio of the response at a test frequency and the 1-kHz response,

$$V_s(f)_{dBV} = 20 \log \frac{V_s(f)}{V_s(1\text{ kHz})}.$$

Three different styles of cables are modeled and compared to measured values in Fig. 12. The model gives a very good approximation to the measured responses (coefficient of correlation = 0.999, 0.948, and 0.997 for HF10C, 16LPC, and 19123, respectively). The results are for the full 3.1-m length of the cable since they are not directly scalable to other lengths.

The rise above 0 dBV in the measured responses occurs when the combined magnitude of the impedance of loudspeaker and cable (as seen by the amplifier) is lower than the loudspeaker's impedance alone. This results when the reactance of the loudspeaker is capacitive and subtracts from the cable's inductive reactance. The result is a lower total reactive component, which reduces the magnitude of the impedance seen by the amplifier. Since the amplifier output is held at a constant voltage for the cable impedance test, the current through the loop is higher than the loudspeaker's impedance alone would require. This higher current results in a voltage across the loudspeaker terminals that is higher than the amplifier output. Low-inductance cables will provide a more ideal response since cables whose inductive reactance is much less than the loudspeaker's capacitive reactance will reduce this "hump" effect and present little more than the loudspeaker's complex impedance to the amplifier as a load. When the effective impedance of cable and loudspeaker is lower, it should not prove difficult for a well-designed amplifier because the effect is small with short cables (approximately 0.6% for the worst case in these tests, sample 2, auto jumper cable). The lowest impedance seen by the amplifier and the greatest rise in loudspeaker voltage as a result of this effect occur at resonance, when $X_{cable} = -X_{speaker}$. The impedance will then be limited by the resistive components of both cable and loudspeaker. For example, loudspeaker A would require just over 12.4 m of Belden 9718 cable to provide enough inductance to achieve resonance at 10 kHz, where the resistance seen by the amplifier would be about 4.84 Ω.

7 AMPLIFIER EFFECTS

Now that the relationship between loudspeaker and cable is better understood, the effects of the amplifier will be considered. As seen with the cable model, added inductance will cause frequency response deviations due to interactions with the loudspeaker's reactive components. Therefore it would be desirable to minimize reactive effects from the amplifier as well. Most amplifiers include added inductance (typically 0.5–10 μH) paralleled with a resistance (typically 2.7–27 Ω) between the output of the amplifier (generally from the point that negative feedback is taken) and the amplifier's output terminals. This inductance is added to isolate phase shifts due to capacitive loads from causing instability in the feedback loop. Both amplifiers A and B include this network. Obviously, this inductance is in series with the cable inductance, and in some cases can exceed the cable inductance.

The damping factor of an amplifier can also shape the frequency response. The damping factor (and the output impedance of the amplifier) is controlled by the frequency-dependent loop gain of the amplifier, the degree of negative feedback, the impedance of the output devices, and any other components in series between the amplifier output and the output terminals. The amplifier output voltage will be lower where the damping factor is lower or where the load impedance is lower. An amplifier with low damping factor is less able to control back EMF and reactive effects of the loudspeaker.

The responses of all cables were tested with the same loudspeaker, but using two different amplifiers. Figs. 13 and 14 present the responses of loudspeaker A and amplifier A, while Figs. 15 and 16 present the responses of loudspeaker A with amplifier B. These graphs illustrate the combined responses of loudspeaker, cable, and amplifier. Immediately obvious is that the response of amplifier A overwhelms the individual cable effects (Figs. 13 and 14). The damping factor for amplifier A and the impedance of loudspeaker A both drop in the same frequency range, which exacerbates their inter-

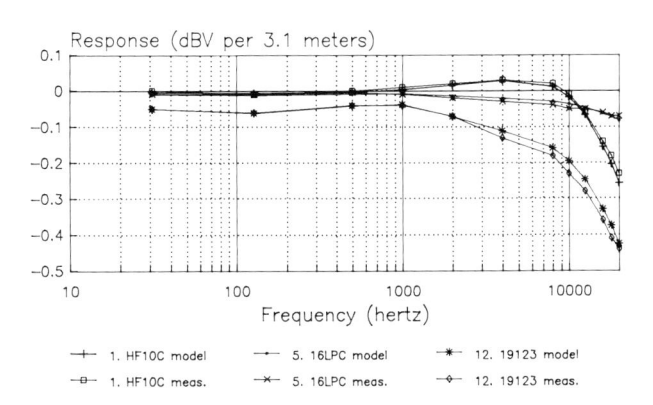

Fig. 12. Modeled and measured response with loudspeaker A for cable samples 1, 5, and 12.

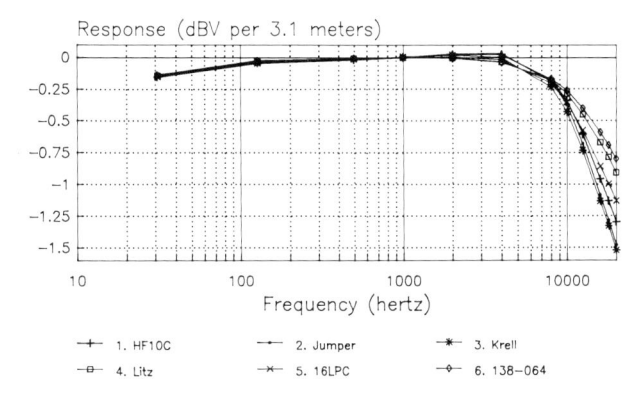

Fig. 13. Complete system response for amplifier A with loudspeaker A, cable samples 1–6.

action. The response with amplifier B (Figs. 15 and 16) closely resembles the response of the cable and loudspeaker alone (Figs. 9 and 10). The high damping factor of amplifier B maintains better control of reactive effects with the more inductive cables, producing a flatter response (Fig. 15).

The effect of the amplifier can be added to the cable response model by including the additional resistance and reactance of the amplifier's output:

$$V_s(f) = V_a(f)' \frac{\sqrt{R_s^2 + X_s^2}}{\sqrt{(R_a + R_w + R_s)^2 + (X_a + X_w \pm X_s)^2}}$$

where $V_a(f)'$ = amplifier voltage at frequency f.

Fig. 17 illustrates the results of this model, using amplifier B's voltage response with loudspeaker A's impedance and phase (converted to dBV relative to the 1-kHz response as before). The model fits well with the measured data (coefficient of correlation = 1.000, 0.997, and 0.999 for HF10C, 16LPC, and 19123, respectively). Because the model is very simple and amplifier dynamic responses are more complex, it does not fit as closely with all amplifiers, especially the ones that have a more complex output reactance (which may include capacitive effects). The model infers that the flattest response will occur by keeping the reactance of the amplifier and cable as low as possible.

8 CONCLUSIONS

If loudspeakers were only simple resistance, then large, low-resistance cables would not be a bad idea. However, loudspeaker systems exhibit a frequency-dependent complex impedance that can interact with the reactive components of amplifier and cable. The best response was obtained with low-inductance cables and an amplifier with low-inductance output and a high, frequency-independent damping factor.

These tests have shown that the best way to achieve adequately low resistance *and* inductance in a cable is by using many independently insulated wires per conductor rather than one large wire. Efforts to reduce the skin effect (such as Litz construction) will help, but due more to the reduction of inductance than the reduction of the skin effect. Inductive reactance is more significant in large cables than the skin effect. If an amplifier does not disagree, larger capacitance in a cable is not significant since this component is comparatively small and reduces amplifier and cable inductive reactance effects.

The best performance was measured with the multiconductor cables Spectra-Strip 138-064, Kimber 16LPC, and AudioQuest Litz. Smaller multiconductor

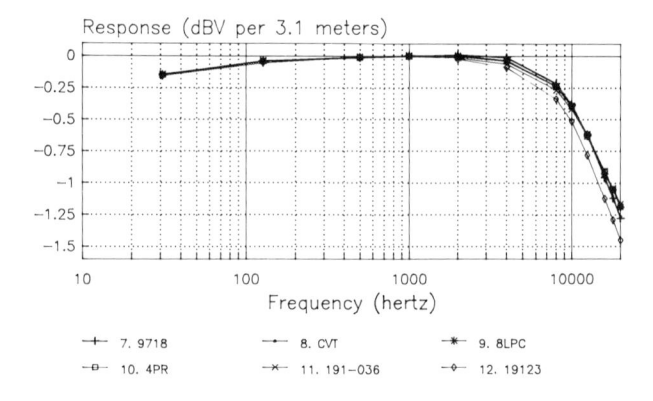

Fig. 14. Complete system response for amplifier A with loudspeaker A, cable samples 7–12.

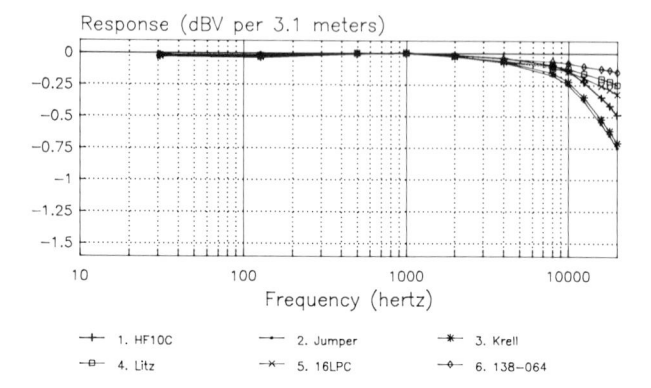

Fig. 15. Complete system response for amplifier B with loudspeaker A, cable samples 1–6.

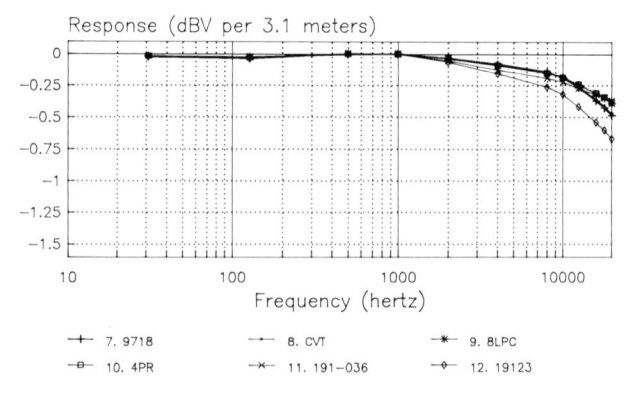

Fig. 16. Complete system response for amplifier B with loudspeaker A, cable samples 7–12.

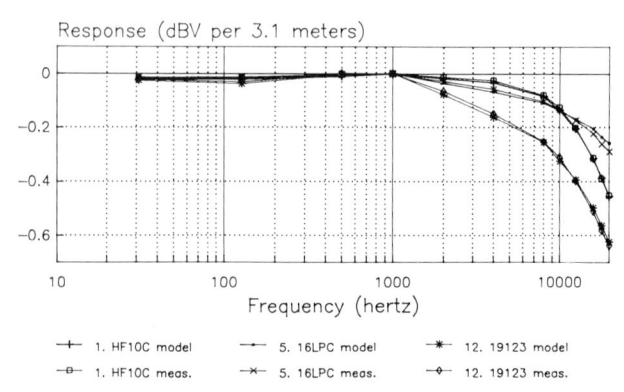

Fig. 17. Model of complete system response for amplifier B with loudspeaker A.

cables such as Kimber 8LPC, Kimber 4PR, and Spectra-Strip 191-036 also performed well.

Of the two-wire cables, 12 AWG provided the best performance with reactive loads, while both smaller and larger gauges (3–7 AWG and 18 AWG) showed greater high-frequency drop and interaction with capacitive reactance in a load. 12 AWG seems more than adequate, even for demanding systems, high power levels, and reasonable lengths.

The effects of 3.1-m cables are subtle, so many situations may not warrant the use of special cables. Low-inductance cables will provide the best performance when driving reactive loads, especially with amplifiers having low damping factor, and when flat response is critical, when long cable lengths are required, or when perfection is sought. Though not as linear as flat cables, 12 AWG wire works well and exceeds the high-frequency performance of other two-conductor cables tested. By the way, keep the auto jumper cables in the garage!

9 ACKNOWLEDGMENT

The author would like to thank Robert A. Pease for his invaluable suggestions on measurement techniques, Dave Sales for his amplifier and exotic cables, and Brian Converse, Dr. Robert Milstein, and Dianne Davis for their support and suggestions.

10 REFERENCES

[1] R. A. Greiner, "Amplifier–Loudspeaker Interfacing," *J. Audio Eng. Soc.*, vol. 28, pp. 310–315 (1980 May).

[2] R. A. Greiner, "Cables and the Amp/Speaker Interface," *Audio*, vol. 73, pp. 46–53 (1989 Aug.).

[3] R. A. Greiner, "Another Look at Speaker Cables," *BAS Speaker*, vol. 7, no. 3, pp. 1–4 appended (1978 Dec.); addenda, vol. 7, no. 6, pp. 6–7 (1979 Mar.).

[4] C. Ward, J. Thompson and M. Harling, "Speaker Cables Compared," *BAS Speaker*, vol. 8, no. 7, pp. 25–29 (1980 Apr.).

[5] R. Warren, "Getting Wired," *Stereo Rev.*, vol. 55, pp. 75–79 (1990 June).

[6] D. Olsher, "Cable Bound," *Stereophile*, vol. 11, pp. 107–118 (1988 July).

[7] B. Jones, "Speaker Cable Electrical Tests," ACSnet/UUCP: brendan@otc.otca.oz 1990; a series of discussions and rebuttals can be found referencing ⟨1857@otc.otca.oz⟩, newsgroup: rec.audio.

[8] D. Salz, "The White Paper on Audio Cables," Straight Wire Inc., Hollywood, FL (1988).

[9] B. Brisson, "How Phase Shift in Audio Cables Influences Musical Waveforms," Musical Interface Technologies, Auburn, CA (undated).

[10] "Cable Design, Theory Versus Empirical Reality," AudioQuest, San Clemente, CA (1990).

[11] "Sumiko Reports: OCOS—The Formula," Sumiko, Berkeley, CA (1989).

[12] J. H. Johnson, "Power Amplifiers and the Loudspeaker Load," *Audio*, vol. 61, pp. 32–40 (1977 Aug.).

[13] M. Otala and P. Huttunen, "Peak Current Requirement of Commercial Loudspeaker Systems," *J. Audio Eng. Soc.*, vol. 35, pp. 455–462 (1987 June).

THE AUTHOR

Fred E. Davis is currently a consulting electronics engineer, and previously has been director of engineering for machine vision and optical data storage companies. He has worked with biomedical electronics, audio, recording studio and video electronics, industrial controls, portable instrumentation and tools, computer hardware design and software, and consumer electronics. Besides a passion for music, his interests include photography, books (especially old and rare science books), and film.

Considerations in Grounding and Shielding Audio Devices*

STEPHEN R. MACATEE, AES Member

Rane Corporation, Mukilteo, WA 98275, USA

Many audio manufacturers, consciously or unconsciously, connect balanced shields to audio signal ground. This is the source of many audio interconnection hum and buzz problems. The options available to manufacturers who follow this improper practice are discussed. Both balanced and unbalanced schemes and their incompatibilities are covered. Many manufacturers may already follow proper interconnecting practices. Those who are not have many options, including doing nothing.

0 INTRODUCTION

Now that the Audio Engineering Society has adopted the "pin 2 is hot" standard, the question of what to do with pin 1 is being addressed. A recommended practices document is being created covering interconnection of professional audio equipment. How and where to connect pin 1 is too complex to be issued as a standard; thus only a recommended practice is being developed. The recommended practices may affect manufacturers who choose to follow them.

Many shield-wiring practices exist in the audio industry today. The majority of available literatue on the subject prescribes clear solutions to any wiring problem, yet problems are rampant due to inconsistent variations on the well-documented ideal. Two clear groups have developed on either side of a hard-to-straddle fence — the balanced world and the unbalanced world.

Over the past decade the declining cost of professional audio equipment has facilitated its use in more and more home studio environments. As home studios incorporate professional, balanced equipment into their systems, the unbalanced and balanced worlds collide. Home studios adding balanced equipment to their traditionally unbalanced gear also add connectivity problems. Professional users never consider unbalanced gear, yet still have connectivity problems.

The performance of any interconnection system is dependent on input/output (I/O) circuit topologies (specific balanced or unbalanced schemes), printed-circuit-board layout, cables, and connector-wiring practices. Only wiring practices, both in the cable and in the box, are covered here. The I/O circuit topologies are assumed ideal for this discussion to focus on other interconnection issues.

The AES recommendation will address a simple issue, the absurdity that one cannot buy several pieces of professional audio equipment from different manufacturers, buy off-the-shelf cables, hook it all up, and have it work hum- and buzz-free. Almost never is this the case. Transformer isolation and other interface solutions are the best solutions for balanced–unbalanced interconnections, though they are too costly for many systems. Even fully balanced systems can require isolation transformers to achieve acceptable performance. Some consider isolation transformers the *only* solution. These superior solutions are not covered here.

Another common solution to hum and buzz problems involves disconnecting one end of the shield, even though one cannot buy off-the-shelf cables with the shield disconnected at one end. The best end to disconnect is unimportant in this discussion. A one-end-only shield connection increases the possibility of radio-frequency (RF) interference since the shield may act as an antenna. The fact that many modern-day installers still follow the one-end-only rule with consistent success indicates that acceptable solutions to RF issues exist, though the increasing use of digital technology increases the possibility of future RF problems. Many reduce RF interference by providing an RF path through a small capacitor connected from the lifted end of the shield to the chassis.

The details of noise-free interconnections and proper

---

* Presented at the 97th Convention of the Audio Engineering Society, San Francisco, CA, 1994 November 10–13; revised 1995 April 3.

grounding and shielding are well covered in other literature. They are not revisited here. Readers are encouraged to review bibliography listings for further information. Most of these materials have been applicable in the audio industry for well over 50 years, though until now they have not been implemented or embraced by many.

0.1 Balanced versus Unbalanced Shields

For the ensuing discussion, the term shield is qualified with the description balanced or unbalanced. An unbalanced return conductor physically resembles a shield and provides shielding for electric fields, but magnetic fields are not shielded. Though this is also true for balanced shields, the twisted-pair construction of balanced cables provides much greater immunity to magnetic field interference. Unbalanced cable shields also carry signal in the form of return current, further alienating unbalanced shields from "true" shields. Shield is defined by Ott [1] as ". . . a metallic partition placed between two regions of space. It is used to control the propagation of electric and magnetic fields from one place to another." A balanced interconnection provides the superior interface of the two.

0.2 The "Pin 1 Problem"

Many audio manufacturers, consciously or unconsciously, connect balanced shields to audio signal ground—pin 1 for three-pin (XLR-type) connectors, the sleeve on 1/4-in (6.35-mm) jacks. Any currents induced into the shield modulate the ground where the shield is terminated. This also modulates the signal referenced to that ground. Normally great pains are taken by circuit designers to ensure "clean and quiet" audio signal grounds. It is surprising that the practice of draining noisy shield currents to audio signal ground is so widespread. Amazingly enough, acceptable performance in some systems is achievable, further providing confidence for the manufacturer to continue this improper practice—unfortunately for the unwitting user. The hum and buzz problems inherent in balanced systems with signal-grounded shields have given balanced equipment a bad reputation. This has created great confusion and apprehension among users, system designers as well as equipment designers.

Similar to the "pin 2 is hot" issue, manufacturers have created the need for users to solve this design inconsistency. Until manufacturers provide a proper form of interconnect uniformity, users will have to continue their struggle for hum-free systems, incorporating previously unthinkable practices.

0.3 The Absolute, Best Way to Do It

Clearly, the available literature prescribes balanced interconnections as the absolute, best way to interconnect audio equipment. The use of entirely balanced interconnection with both ends of the shield connected to chassis ground at the point of entry provides the best available performance.

The reasons for this are clear and have been well documented for over 50 years. Using this scheme, with high-quality I/O stages, guarantees hum-free results. This scheme differs from current practices in that most manufacturers connect balanced shields to signal ground, and most users alter their system wiring so that only one end of the shield is connected. Due to these varied manufacturer and user design structures, an all-encompassing recommendation with proper coverage of both balanced and unbalanced interconnections is essential.

Conceptually it is easiest to think of shields as an extension of the interconnected units' boxes (Fig. 1). Usually metallic boxes are used to surround audio electronics. This metal "shell" functions as a shield, keeping electromagnetic fields in and out of the enclosure. For safety reasons the enclosures in professional installations are required by law to connect to the system's earth ground (which in many systems is not the planet Earth—an airplane is a good example).

1 SPECULATIVE EVOLUTION OF BALANCED AND UNBALANCED SYSTEMS

One may ask: if the balanced solution is best, why is it that all equipment is not designed this way? Well, reality takes hold; unbalanced happens.

Back in the early days of telephone and ac power

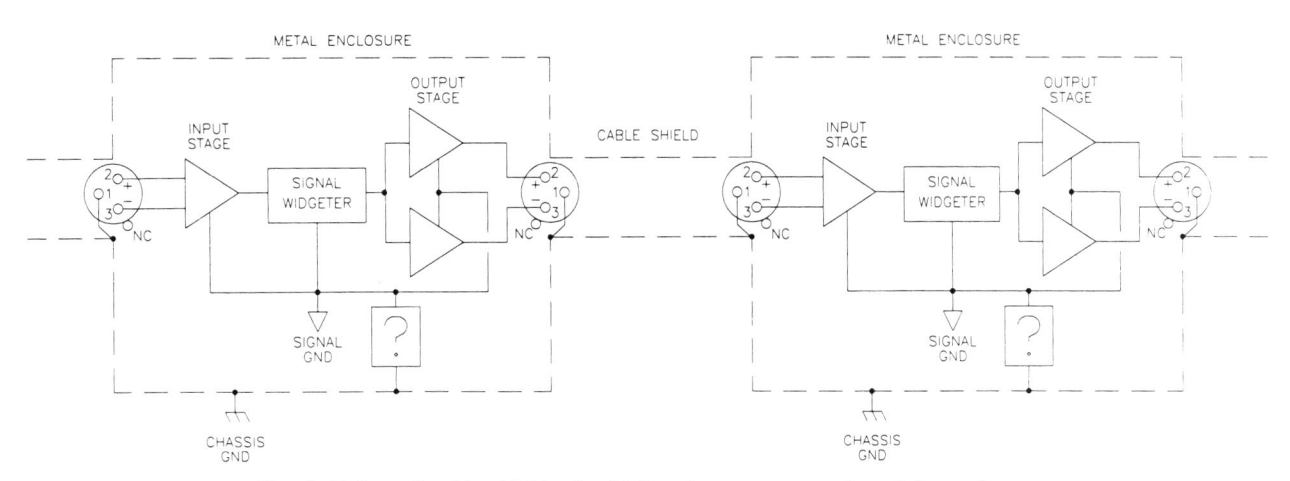

Fig. 1. Balanced cable shields should function as an extension of the enclosure.

distribution a specific class of engineers evolved. They learned that telephone and ac power lines, due to their inherently long runs, must be balanced to achieve acceptable performance. (To this day, many telephone systems are still balanced and unshielded.) In the 1950s hi-fi engineers developed systems that did not necessitate long runs, and they used unbalanced interconnections. The less expensive nature of unbalanced interconnections also contributed to their use in hi-fi. These two classes of engineers evolved with different mind-sets, one exclusively balanced, the other exclusively unbalanced. The differing design experiences of these engineers helped form the familiar balanced and unbalanced audio worlds of today.

Now add spice to the pot with the continued price decrease and praise devoted to balanced "professional" audio interconnections with the desire for better audio performance at home, and one sees the current trend of merging balanced and unbalanced systems arise. Home studio owners, previously on the unbalanced side of the fence, dream to jump but unfortunately straddle the fence, getting snagged on the fence's ground barbs when connecting their new balanced equipment (Fig. 2).

1.1 How Could This Happen?

To fulfill their users' desires to "go" balanced, hi-fi designers started upgrading equipment to balanced. From an unbalanced designer's mind-set, connecting the new balanced circuit's shield to ground is almost subconscious. This issue of which ground connects to the shield is alien or unknown. The old unbalanced "shield" (really the return signal conductor, not a true shield) is already "grounded." Without appropriate balanced interconnect research, this hi-fi mind-set may not think to *add* a chassis-grounded shield around the existing two-conductor cable. This redefines the "old" return conductor as a "new" negative signal carrier, not as a shield. It was perhaps the convenience of the situation and this mind-set that started improper signal grounding of balanced shields in the first place. Little treatment of this subject is given in educational institutions, and many systems happen to work satisfactorily even with improperly grounded shields.

Other designers, when upgrading to balanced interconnections, may have realized that by connecting the shield to signal ground, interfacing to unbalanced equipment is made simpler since signal ground (needed for unbalanced interconnections) will be available on the cable. [This, unfortunately, allows easy use of ¼-in (6.35-mm) mono connectors.] This still creates the same problem, signal-grounded balanced shields. Signal-grounded shields on balanced equipment create ground loops in the audio path and modulate the audio signal ground, wreaking havoc with most systems. This practice penalizes those who want to realize the superior performance of balanced interconnections and has given balanced interconnections a bad reputation.

A third possible reason for signal-grounded balanced shields arises if designers change phantom-powered microphone inputs to balanced line-level inputs and do not use caution. The phantom-power return currents travel through the shield, requiring shield connection to the signal ground. When changing this topology to line-level balanced inputs, the designer may not think to change the shield connection to the chassis ground. This issue is further complicated by manufacturers who incorporate ground lift switches in their products. Ground lift switches disconnect chassis and signal ground. Thus care should be taken to ensure that phantom-power return currents always have a return path to their power supply, regardless of the ground lift switch position.

Manufacturers who started in balanced fields, such as the telephone and broadcast industries, used chassis-grounded shields when maximum protection from electromagnetic interference (EMI), which includes RF, was necessary. Perhaps users from these balanced fields assumed that all balanced equipment had chassis-grounded shields. When improperly wired manufacturers' equipment was installed, they discovered hum and buzz problems. They solved them with isolation transformers, by disconnecting one end of the shield, or by simply not using that manufacturer's equipment. The feedback to inform manufacturers of their improper shielding practices never developed. Manufacturers may have suggested isolation transformers or cable rewiring solutions instead of addressing the cause of the problem—signal-grounded balanced shields. Again, some systems with signal-grounded shields work acceptable, causing further and future bewilderment.

1.2 The History Lesson

The lesson to be learned from this account involves keeping in mind these audio interconnection issues when specifying, designing, or upgrading other connectivity systems such as AES/EBU, SPDIF, and other electrical interfaces. Balanced and unbalanced systems are not designed to interface together directly. As the audio industry embraces more digital products, interconnection systems must be clearly designed and specified for use within the limits of their electrical interfaces. Multiple-conductor connectors, carrying either digital or analog

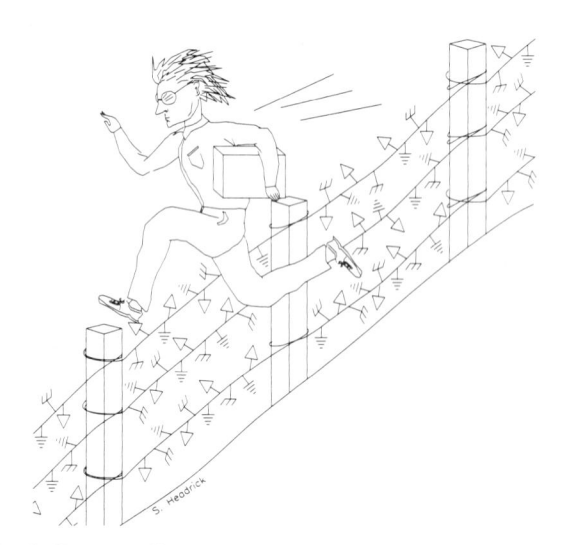

Fig. 2. Home studio owner trying to jump the balanced–unbalanced fence.

signals, present even more challenges. The distance between units is an important issue. Keeping interconnections balanced and chassis ground shielded provides the best possible immunity from EMI, regardless of cable lengths. Unbalanced interconnections may be less expensive to manufacture and sell, but perhaps are more expensive to install—hum- and buzz-free.

The Audio Engineering Society is to be applauded for assembling and disseminating this information to those who may be unfamiliar with it. Manufacturers and, more importantly, users will eventually be rewarded.

2 CHASSIS GROUND VERSUS SIGNAL GROUND

Let us examine the distinction between chassis and signal ground in audio devices. Chassis ground is generally considered any conductor that is connected to a unit's metal box, or chassis. The term chassis ground may have come about since units with three-conductor line cords connect the chassis to earth *ground* when plugged into a properly wired ac outlet. In units with a two-conductor line cord (consumer equipment) the chassis does not connect to earth ground, though it is normally connected to the signal ground in the box.

Signal ground is the internal conductor used as the 0-V reference potential for the internal electronics and is sometimes further split into digital and analog ground sections. Further signal ground splits are also possible, though it is important to remember that all "divisions" of signal ground connect together in one place. This is usually called a star grounding scheme.

It is easy to confuse chassis ground and signal ground since they are usually connected together, either directly or through one of several passive schemes. Some of these schemes are shown in Fig. 3. The key to keeping an audio device immune from external noise sources is knowing *where* and *how* to connect the signal ground to the chassis.

First let us examine why they must be tied together. We shall cover where and how in a moment. There are at least two reasons why one should connect signal ground and chassis ground together in a unit.

One reason is to decrease the effects of coupling electrostatic charge on the chassis and the internal circuitry. External noise sources can induce noise currents and electrostatic charge on a unit's chassis. Noise currents induced into the cable shields also flow through the chassis, since the shields terminate (or should terminate) on the chassis. Since there is also coupling between the chassis and the internal circuitry, noise on the chassis can couple into the internal audio. This noise coupling can be minimized by connecting the signal ground to the chassis, which will allow the entire grounding system to fluctuate with the noise, surprisingly providing a quiet system. Further coupling reduction is gained when the chassis is bonded solidly to a good earth ground—either through the line cord, through the rack rails, or with an independent technical or protective ground conductor. This provides a nonaudio return path for any externally induced noise.

The second reason for connecting signal ground to the chassis is the necessity to keep the signal grounds of two interconnected units at very nearly the same voltage potential. Doing so prevents the loss of system dynamic range where the incoming peak voltage levels exceed the power supply rails of the receiving unit.

Unbalanced units connect successive signal grounds together directly through each interconnecting cable, the sleeve of each RCA cable. This—and the fact that the chassis is generally used as a signal ground conductor—keeps the signal ground impedance of unbalanced systems very low. Many may agree that unbalanced systems are helped by the fact that the chassis are normally not earth grounded. This allows an entire unbalanced system to float with respect to earth ground. It eliminates the potential for multiple return paths for the audio grounding system, since there is no second path (ground loop) through the earth ground conductor. Low signal ground impedance between units is essential for acceptable operation of all non-transformer-isolated systems, balanced and unbalanced.

The design of balanced interconnections does not connect signal grounds directly together. The negative conductor provides the required signal return current. To avoid loss of dynamic range, balanced systems use a different method of keeping signal ground potentials small.

Since the cable shield already connects the two chassis together, simply connecting the signal ground to the chassis in each box keeps the signal ground potentials between units small. The key is how to connect them. Since the cables between units also provide the shortest (and therefore the lowest impedance) path between two units, using the cable shield to minimize the signal ground potentials between units is quite effective.

Now that we know why one must connect the signal ground to the chassis, let us discuss how to connect them. The schemes in Fig. 3 appear straightforward enough, but what is not shown is precisely where and how the conductors connect together. It all comes down to paying close attention to where currents flow. As discussed, the shield noise currents flow through the chassis and shunt to earth ground on units with three-conductor line cords. The key issue is that these noise currents do not flow through a path shared by any audio currents. It seems so simple, and is—especially to draw (see Fig. 3 again). The hard part is implementing the proper layout scheme.

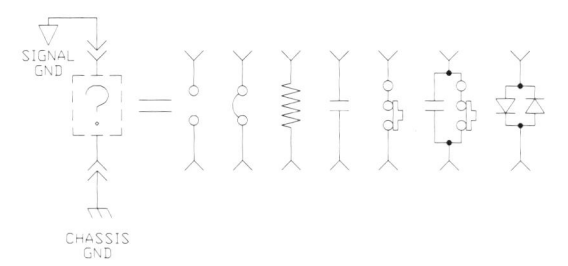

Fig. 3. Some passive schemes for connecting signal ground to chassis.

Connecting the signal ground to the chassis in each unit can only be done in *one* place in each unit. If done twice, one leaves the possibility open that the noise currents will flow through a path shared by audio.

There are two schools of thought on where to connect the signal ground to the chassis. They are both versions of the star ground scheme mentioned. The first connects a trace (or wire) directly from the audio power supply ground terminal to the chassis ground point (Fig. 4). It is important, in both schools, that no other signal currents be allowed to flow through this trace. Do not allow this trace to share any other return currents from other signal-grounded circuit points, such as the input or output circuit's ground. This keeps chassis noise currents from flowing through the same trace, which is a return path for an audio signal. Also keep in mind that this trace may contain noise currents and should be kept away from noise-sensitive circuitry. This is a star grounding scheme, which uses a point originating at the output of the power supply as the center of the star. There are two common locations in the power supply for the star's center—the output terminal of the power supply and the point between the ac filter capacitors.

Another school of thought on where to connect the signal ground to the chassis simply moves the center of the star ground to the input jack's ground. This scheme makes the most sense for unbalanced units and for balanced units equipped with ¼-in (6.35-mm) connectors where the use of mono plugs is possible.

3 MANUFACTURER ISSUES TO ADDRESS

Implementing their users' desires to "upgrade" to balanced, traditionally unbalanced manufacturers are faced with an important issue: How do you solve the balanced–unbalanced incompatibility problem? If you sell your product to a mixed balanced–unbalanced market, a suggested method of interconnection must be available. Isolation transformers and active interface boxes are the best solution and should be offered as the best interconnection alternative. However, persuading unbalanced customers to buy an expensive interface solution is much harder than the lower performance option of rewiring their cables. (The add-on transformer solution is analogous to a software company releasing a new software revision which renders your existing files incompatible unless an additional file conversion program is purchased.)

Through careful rewiring of the cables, acceptable interconnection solutions are achievable in some systems. (One of Rane's most popular technical notes, Rane Note 110 [2], now being rewritten, is one example of the "custom" wiring needed in some systems.) This same cable rewiring solution holds whether the equipment is wired with signal ground or with chassis ground on the balanced circuit's shields.

4 SOLUTIONS FOR MIXED BALANCED AND UNBALANCED SYSTEMS

It is obvious from the vast quantity of literature that for fully balanced operation, the shield should connect to the chassis ground at the point of entry. This is also true for unbalanced operation when a third shield conductor is available. Connect the shield to the chassis ground at the point of entry. However, this is only valid when two-conductor shielded cable is used.

4.1 Shielded Two-Conductor Connectivity

Fig. 5 shows recommended wiring for all combinations of balanced and unbalanced I/O interconnections when two-conductor shielded cable is used. It also includes the two most common manufacturer shield-grounding schemes—signal grounding the shield and chassis grounding the shield. Identifying these schemes for every unit in a system is essential to debug system hum and buzz. This is no simple tasks since chassis and signal grounds may be connected together. The goal is

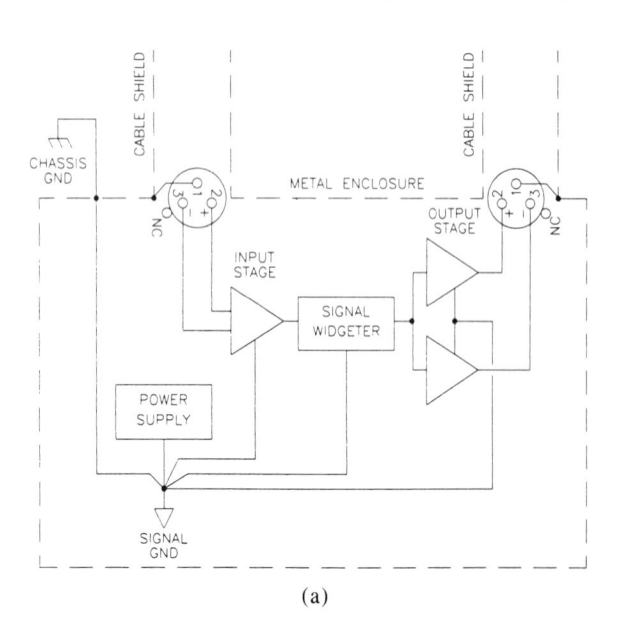

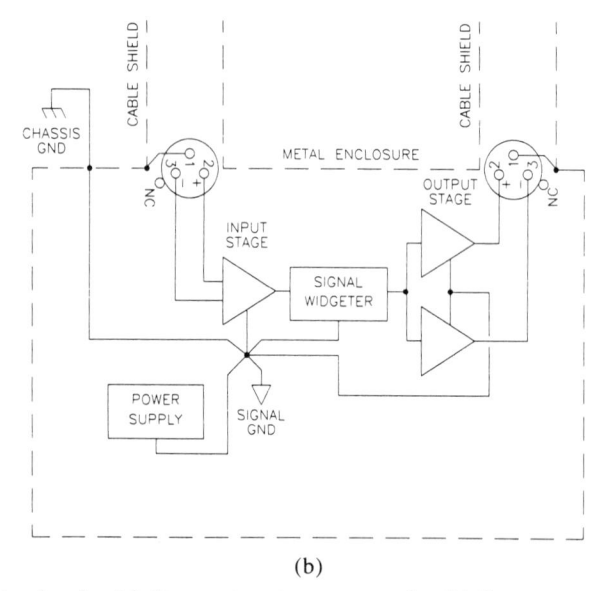

Fig. 4. Two star ground schemes for connecting signal ground to chassis. (a) Star center at power supply. (b) Star center at input ground.

to find out if the manufacturer connected them together in such a way that shield currents do not affect the audio signal. The dashed lines in Fig. 5 represents the units' chassis boundaries. Connections between dashed lines are functions of the cable. Connections outside these lines are the manufacturers' choosing, whether conscious or unconscious.

Fig. 5 is arranged such that the top and leftmost drawing [Fig. 5(a)] is the theoretical "best" way to connect equipment with optimal results. "Best" way means that everything is completely balanced with all shields (pin 1s) connected to the chassis ground at the point of entry. As one moves down or to the right, degradation in performance is expected. Whether a system operates acceptably or obeys these theoretical predictions is too system specific to predict accurately. However, one must start somewhere.

The quality and the configurations of the input and output circuits are omitted from Fig. 5 and the ensuing discussion to focus on cable wiring and the internal wiring of the units. The I/O circuitry is assumed ideal.

4.2 Fully Balanced

Fully balanced systems (left column in Fig. 5) provide the best performance when both ends of the shield connect to units with chassis-grounded shields [Fig. 5(a)]. When units with signal-grounded shields are encountered, disconnect the shields at the signal-grounded ends [Fig. 5(b) and (c)]. This keeps the induced shield currents out of the audio signal ground. If both units involved have signal-grounded shields, you have entered the twilight zone [Fig. 5(d)]. This is perhaps the most common scheme. Most disconnect one end of the shield. Specifically which end is disconnected creates strong political debates and is left for the individual user to decide. Never disconnect both ends of a shield.

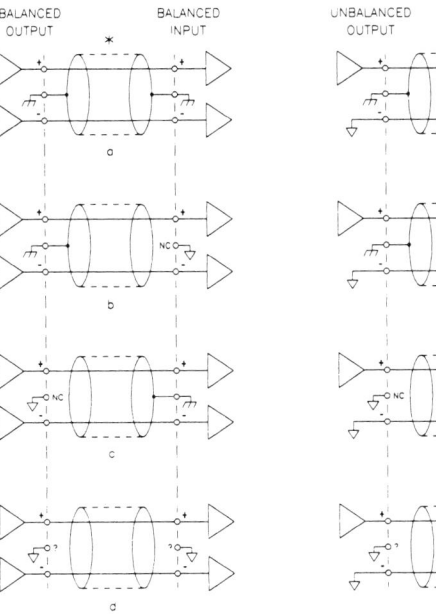

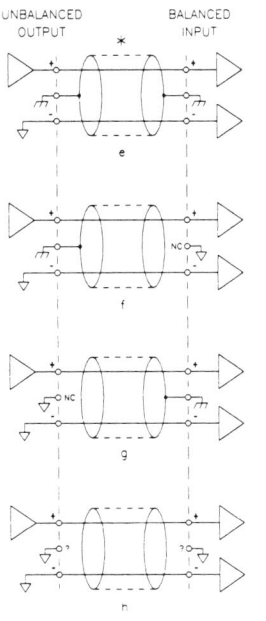

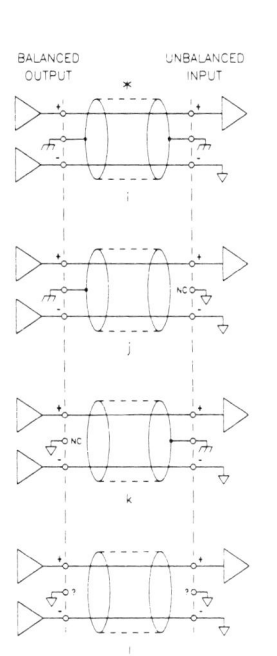

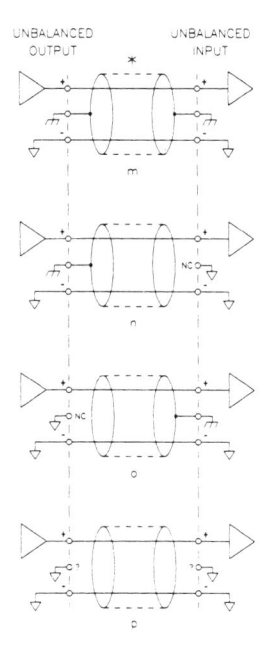

4.3 Unbalanced Output Driving Balanced Input

The second column in Fig. 5 shows unbalanced outputs driving balanced inputs. Again only shielded two-conductor cable is used. The best case here has both ends of the shield connected to units whose shield is chassis grounded [Fig. 5(e)]. Some may argue that the induced noise on the signal conductors may be injected into the "sending" unit through the unbalanced output stage. This is a function of the system and the output circuit and is quite likely. Disconnecting the shield at the unbalanced output may help reduce this problem.

Again, when units with signal-grounded shields are encountered, disconnect the shields at the signal-grounded ends [Fig. 5(f) and (g)]. This keeps the noisy shield currents out of the audio signal ground. If both units involved have signal-grounded shields, you have entered the twilight zone again [Fig. 5(h)]. Support your favorite one-end-only political party.

4.4 Balanced Output Driving Unbalanced Input

The third column in Fig. 5 is the most troublesome, balanced outputs driving unbalanced inputs. Since the input stage is not balanced, induced noise on the signal conductors is not rejected. If you must use an unbalanced input, use as short an input cable as possible. This reduces the induced noise. Fig. 5(i) shows both ends of the cable shield connected to units with chassis-grounded shields. If the units are far apart, the chance of the shield currents inducing noise on the signal conductors is greater. Keeping this cable very short reduces the shield current and therefore reduces the noise that is *not* rejected by the unbalanced input stage. Most systems may require disconnecting one end of the shield for the Fig. 5(i) case. Even a small current in the shield may prove too much for an unbalanced input stage. Again,

Fig. 5. Interconnectivity using shielded two-conductor cable only. Asterisks denotes usability with off-the-shelf cable.

support your favorite one-end-only position.

Disconnect the shields at units with signal-grounded shields [Fig. 5(j) and (k)]. If both ends have signal-grounded shields, run for your favorite one-end-only political party [Fig. 5(l)].

This scheme connects the balanced output's negative output to the signal ground rather than a high-impedance input. Many balanced output circuits will attempt to drive this signal ground, causing great distortion and potentially damaging the output stage. Other balanced output stages are termed "floating" balanced. (The Analog Devices SSM-2142 balanced line driver chip is one example.) Also called a cross-coupled output, these circuits mimic the performance of fully balanced transformer solutions and are designed so that the negative output can short-circuit to the signal ground. If you find or use this scheme, be sure that the balanced output stage can properly handle the signal ground on its negative output.

4.5 Full Unbalanced

Fully unbalanced systems rarely provide a three-conductor connector to enable proper use of a shield. In the unlikely event that you run across one, use the wiring in the fourth column. Again keeping cable lengths short will reduce noise problems, with or without a shield [Fig. 5(m)–(p)].

Most home audio systems are fully unbalanced. Millions of these systems work virtually hum- and buzz-free every day, due to their small nature, short cable runs, and two-conductor ac line cords. The headaches begin when one tries to add a balanced unit to such a system. In unbalanced home audio products neither of the line cord's conductors connects to the chassis, since plugging older nonpolarized ac plugs into an improperly wired outlet would place the hot wire on the unit's chassis. Lack of the third pin on the line cord prevents ground loops in home systems since a second path to ground, or between units, is unavailable. Professional audio equipment generally comes equipped with a three-wire line cord. The third wire (green wire) is required to connect to the chassis. This provides the second ground path (loop) from one unit to the next.

4.6 Connector Choice

The connector type was purposely left out of Fig. 5 and the preceding discussion since connector choice adds another layer of complexity to interconnection systems. The most troublesome culprit is the ¼-in (63.5-mm) connector. Mono ¼-in connectors are used on most musical instruments and in phone systems. Stereo ¼-in connectors are used for headphones, balanced interconnections, effect and insert send and return loops, relay switch closure points, and an extravagant collection of other miscellaneous connections. Murphy's law tells us that if you provide such a diverse selection of ¼-in interconnecting options, they will be hooked up improperly. The audio industry's problem is that many of these options are completely incompatible. A properly wired mono ¼-in connector has the signal ground on the sleeve, a properly wired balanced ¼-in connector has the chassis ground on the sleeve. Interconnecting this combination should not be achievable—much like trying to connect 120 V ac to an RCA jack (see Fig. 6). The ¼-in connector's low cost, high availability, and small size all contribute to its widespread and varied use. Undoubtedly the numerous interconnection uses of such a popular connector arose for these reasons.

Sadly, the possibility of including the connector type in a recommended practices document is slim. To duplicate connectors on many audio components contributes to higher costs and wastes millions of dollars worth of connectors that are never used. Some manufacturers are attempting to eliminate the ¼-in connector to avoid the confusion and problems when ¼-in jacks are used. This is a step in the right direction, though the high density allowed by these connectors requires less valuable rear-panel real estate. Most marketing departments prefer 30 connectors per inch, making the currently available three-pin (XLR) alternative markedly unpopular. What is needed is a three-pin connector solution that requires less space than the traditional XLR connector. A locking, stackable three-pin mini-DIN comes to mind.

Terminal block and Euroblock connector types are used when separate cable-end connectors are unnecessary or impractical. These connection solutions provide the user with the most wiring options when both signal and chassis ground terminals are available. It allows the user to decide which wiring practice to incorporate. This is the most desirable solution, though most studio equipment does not call for these connector types.

4.7 "Hidden" Balanced I/O Solution

An interesting solution for mono interconnection incorporates unshielded balanced stages, much like most telephone systems. Fig. 7 shows this configuration. This allows off-the-shelf mono cables to be used to connect unbalanced or unshielded balanced I/Os to a system. Though not as ideal as a shielded balanced interconnection, systems with mono connectors, such as home theater systems, benefit from this configuration. Keeping cable lengths short is essential and not difficult in a home environment.

One advantage of such a system—besides making it impossible, on fully balanced systems, to get the signal ground on an external cable—is that it provides an easy upgrade path to balanced signal connections. The manufacturer need only change the connector to a three-pin version. Also crucial for this solution is the need to have cross-coupled output stages, since the negative output

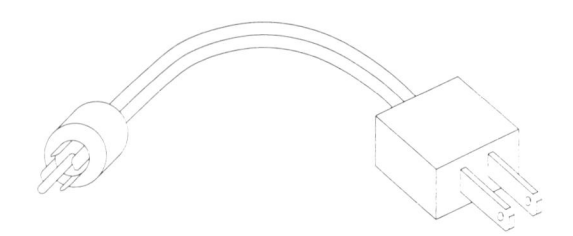

Fig. 6. Difficult-to-find connector type.

may connect to the signal ground.

A slight disadvantage lies with the common use of nontwisted-pair cables in off-the-shelf mono cables. Using twisted cable with this unshielded balanced scheme greatly improves the achievable performance.

4.8 The Muncy Solution

Neil Muncy, an electroacoustic consultant and veteran of years of successful system design, chairs the AES committee working on these interconnection issues. His long-standing solution to these issues provides real-world proof of the guaranteed performance achievable with fully balanced systems wired per the AES recommendation. Mr. Muncy implements what I call the Muncy solution and alters every piece of gear such that it has balanced inputs and outputs with both ends of the shield connected to the chassis ground at the point of entry. Years of this practice, and the early research and discipline to understand the basic physics required to implement it properly, have given Mr. Muncy the drive to tirelessly tour the country dispersing his findings. Mr. Muncy's seminars educate those who are ignorant of the "right" way to wire balanced equipment, and show the advantages gained when every piece of gear in the system is wired accordingly.

4.9 Current Manufacturer Solutions

Let us examine manufacturers' choices regarding signal grounding or chassis grounding balanced cable shields. The problems of signal grounding balanced shields have already been covered. Users choose to live with hum and buzz, alter off-the-shelf cables by disconnecting one end of the shield, or, even in fully balanced systems, use isolation transformers. All are senseless alternatives for inconsistent manufacturing methods. Their advantages and disadvantages are outlined in Tables 1 and 2.

For the manufacturer, several shield connectivity choices are available.

1) *Keep or change shield connections to the chassis ground*. Manufacturers who chassis-grounded balanced shields originally must still recommend isolation transformers, cable altering, and the technical support that go with these hum and buzz solutions. This is unfortunately necessary since not all balanced equipment has chassis-grounded shields. Ideally, if *all* balanced equipment were suddenly and miraculously chassis grounded on both ends, off-the-shelf cables could be used in every system, leaving only the I/O circuitry to dictate system performance.

2) *Change shield connections to the signal ground*. Though this would be a step backward, it is still a choice. Most equipment is connected this way, and most users have found their own costly add-on interconnection solutions.

3) *Offer the shield connection choice to the user*. Provide both options. With two independent screw terminals (one signal, one chassis), a switch or a jumper option will permit users to wire as they please. More on this later.

5 MANUFACTURER SOLUTIONS FOR EFFICIENTLY AND EFFECTIVELY CONNECTING BALANCED SHIELDS TO THE CHASSIS

5.1 Printed-Circuit-Board-Mounted Jacks

The printed-circuit-board-mounted jack provides manufacturers with the most cost-effective solution for transferring cable signals to a printed-circuit board. On the printed-circuit board most manufacturers connect the balanced shield conductor (to signal ground) with a board trace. For optimum balanced performance connect the shield to the chassis ground at the *point of entry*. This means that the shield conductor, to avoid spraying any induced RF energy into the box, never passes the chassis' outer plane. This is not a simple task. Currently no printed-circuit-board-mounted three-conductor connectors provide this optimum solution.

5.2 Terminal Strips

When both signal and chassis ground terminals are provided on terminal block or Euroblock connector types, the user decides which wiring practice to incorporate. This is a desirable solution, though a lot of equipment does not call for these connector types. Providing a Pem nut, screw, and toothed washer near the cable

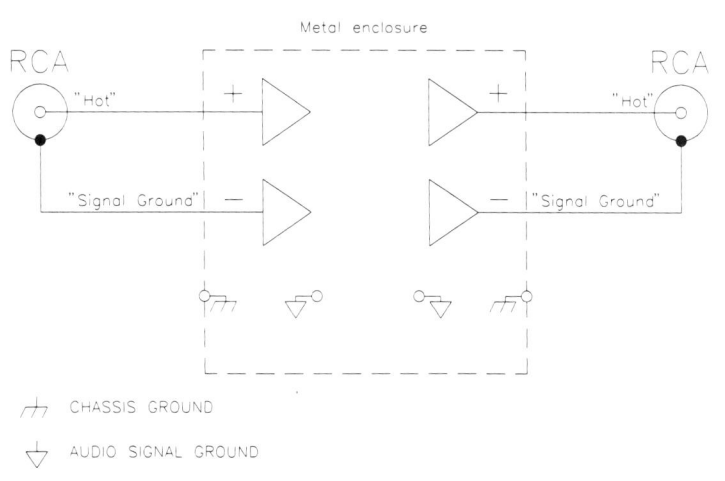

Fig. 7. "Hidden" balanced interconnection.

terminals, instead of an additional chassis-grounded screw terminal, prevents the shield conductor from entering the enclosure—supplying the ultimate interconnection solution. Users select their preferred wiring practice, and the shield cannot spray RF into the enclosure. Maintaining the shield around the signal conductors all the way to the I/O terminals is important. Keeping the Pem screw near the terminals is therefore essential.

5.3 Panel-Mount Jacks with Wires

Panel-mount jacks require the manufacturer to connect a wire from a terminal pin to the printed-circuit board or chassis. This is a good solution for chassis grounding a shield, though this allows the shield to enter the enclosure. Keep the wire short, the gauge large, and the path to the chassis away from sensitive circuits. "Wire" is a four-letter word to many manufacturers, and some consider wires too costly due to their labor-intensive nature. Achieving consistent results with hand-wired connections is difficult, making the printed-circuit-board-mounted jack solution more desirable.

5.4 L-Bracket or Standoff Solution

A circuit-board trace run to a nearby chassis-grounded point is another option. The use of an L bracket, standoff, or similar mechanical connection to the chassis provides mechanical stability, but also consumes valuable rear panel or printed-circuit-board real estate at the same time. Important here is avoiding long traces and keeping the trace away from sensitive areas since it acts as a noise source when shield currents are large or noisy.

5.5 Jumper Options

Not as "friendly" as the screw terminal solution, an internal jumper option provides user configuration of internal shield connection points. This allows the use of XLR or 1/4-in (6.35-mm) connectors, yet still gives the user control of shield wiring practices. Providing a separate, external swich for this function is not cost-effective. Two issues arise with this solution. The first is that there is no external visual indication showing the shield connection point. The second issue to address is which position to ship the jumpers in.

The first problem is nothing new. Most manufacturers do not specify where their shields are connected. The unit's manual or its schematic, if available, may indicate which ground connects to the shield. The schematic symbols used for grounds are not standardized, though there is an AES standards group addressing drafting symbols to solve the dangling-triangle mystery. Proper schematics indicate which symbols represent signal and chassis grounds. The second issue's answer is clear—chassis grounding the balanced shield is the "best" default option, though offering the choice supplies an elegant solution for parties on both sides of the fence. For fully balanced systems, defaulting the shield jumper to the chassis provides the best solution, but only when all interconnected units have chassis-grounded shields. Other units with signal-grounded shields short-circuit the shield currents to the signal ground when connected, causing potentially nasty modulation of the signal ground. This makes the other guy appear the culprit, but does nothing to solve the problem. Clearly users must be able to determine the manufacturer's shield wiring practices. In addition, to support both one-end-only shield connection parties, separate input and output jumpers must be provided (Fig. 8).

5.6 Neutrik Solution

Neutrik AG, Liechtenstein, offers snap-in printed-circuit-board-mounted jacks with metal brackets that pierce the inside of the chassis when external mounting screws are installed. This chassis-pierced bracket also has a separate pin available through the printed-circuit board. The sharp piercing tab provides the electrical connection between the chassis housing and the printed-circuit board. This solves the problems of the labor-intensive wiring and the need to connect to a chassis point, providing the best solution for manufacturers and users. [Neutrik's popular "combo" receptacles—combined female XLR and female 1/4-in (6.35-mm) connectors—provide this piercing tab feature.] Unfortunately, depending on the available height in a given unit, these jacks have trouble fitting into a single-rack space unit due to their slightly larger heights. Hopefully other jacks with this built-in feature will become available, providing manufacturers with a cost-effective solution to this grounding problem. The other problem with these connectors is that only the female connector has this piercing tab.

5.7 Other Suggestions

While doing research for this paper a few more important concepts and discussions were uncovered that warrant mentioning. Martin Glasband has written a series of articles on balanced ac systems [3], [4]. Glasband applies the same balanced concepts used in audio inter-

Table 1. Signal-grounded balanced shields.

Advantages	Disadvantages
Permits mono 1/4-in (6.35-mm) cable use if proper I/O stage is present.	Hum and buzz is present. Must alter cables to interface with many components. Use of isolation transformers and/or interface boxes is needed in some systems. Most manufacturers do it this way.

Table 2. Chassis-grounded balanced shields.

Advantages (with Ideal I/O Circuitry)	Disadvantages
Use of off-the-shelf cable is permitted. No hum and buzz occurs. No isolation transformers or add-on solutions are needed.	Mono 1/4-in (6.35-mm) cables cannot be used. Few manufacturers do it this way.

connection to ac power systems. Common 120-V ac single-phase circuits, like their unbalanced counterparts, have grounded (earthed in this case) "return" conductors. This creates differing impedances to ground on the two signal conductors—the 120-V ac "hot" lead and the neutral. Glasband suggests using a balanced ac system, where the power transformer's center tap is grounded and a balanced 120-V source is created from the positive and negative secondary windings (Fig. 9). The National Electrical Code does not mention properly grounded 120-V two-phase wiring systems [5], though several installations incorporating balanced alternating current of this type have been approved through careful education of the inspector (and the inspector's supervisor).

This balanced ac power scheme really strikes at the heart of another problem. Audio power supplies are not designed to operate from the unbalanced alternating current they are commonly fed. Power supply return currents simply build up on the neutral conductor, adding more noise to the ac ground system. Return currents in balanced ac systems appear out of phase at the transformer secondary and cancel. This balanced 120-V two-phase system is not much different from the common 240-V (dryer) system found in most homes. The transformer provides both 120 and 240 V, the 240 V from the out-of-phase positive and negative secondary windings. (This is similar to Fig. 9, except with the secondary voltages doubled.) Although these 120-V two-phase systems may be currently unorthodox and difficult to install legally, once some of the safety issues are solved, the benefits for audio systems may live up to the potential improvement achievable—unlike balanced audio interconnections. It is hoped that the bad reputation of balanced audio interconnections will not last much longer.

Many years ago RCA developed their own guidelines for rear-panel I/O practices. Some manufacturers and users practice their own methods of left-to-right interconnection customs. Alternating current and loudspeaker-level I/O circuitry on one side, microphone and lower level signals on the other side. This permits easier rack wiring and decreases crosstalk between cable runs in the rack and along cable paths. While the recommended practices document may not dictate product design at such a basic level, this type of thinking benefits everyone. With multimanufacturer standardized network-controlled products still near their infancy, now is the time to address these basic features. Users with "standardized" interconnection systems, designed by informed engineers with the user in mind, will spend less time debugging and installing systems. This allows more installations per day, generates better, quieter systems, and provides more business with smiling users *and* manufacturers.

6 FIBER IS THE FUTURE

Digital fiber-optic interconnections will solve all the foregoing problems of electrical interconnection systems, though one must face a new set of problems. Fiber is difficult to terminate and split, potentially fragile, sensitive to ambient light when receivers are not covered, and, for now, expensive. However, when one adds up the debugging costs of eliminating hum from electric systems, fiber may no longer seem as expensive.

7 CONCLUSION

Balanced and unbalanced interconnections are two very different beings. The incompatibility between these two configurations, whether using analog or digital sig-

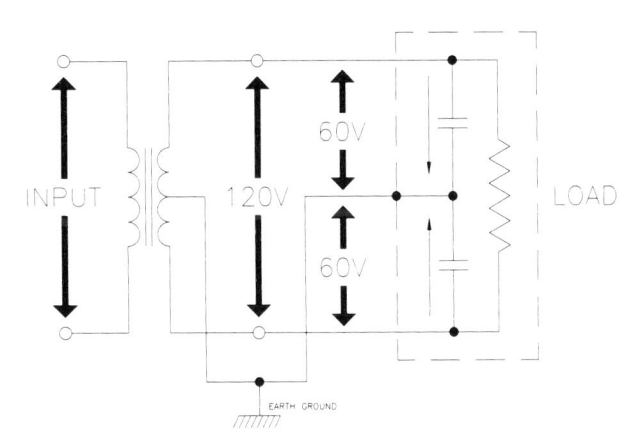

Fig. 9. 120-V two-phase (60-V) balanced power.

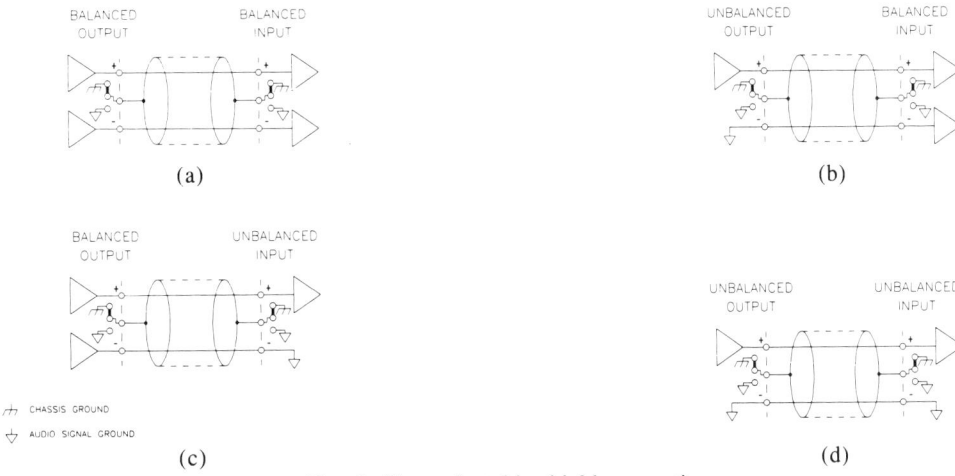

Fig. 8. User-selectable shield connections.

nals, must be considered when designing, specifying, installing, or upgrading equipment and systems. Literature on the subject of grounding and shielding audio devices dictates chassis grounding balanced shields. Most manufacturers, however, signal ground their balanced shields. Speculation about how and why this practice materialized was explored. The Audio Engineering Society is developing a recommended practices document which also condones chassis grounding balanced shields, among other things. It was shown that a manufacturer's choice of either signal grounding or chassis grounding balanced shields does not affect the cable rewiring and other technical support solutions normally recommended when interconnection of balanced and unbalanced equipment is needed. Therefore manufacturers need not hesitate in addressing their "pin 1 problems," and should provide users with the *real* benefits of balanced interconnections by providing chassis ground on balanced shields. Efficient and effective ways of doing this were also discussed.

Also covered was the importance of reducing signal ground voltages between interconnected units by carefully and properly connecting chassis ground to signal ground, in one place, in each unit. Vitally important is the manner in which one connects these two grounds together. The same care must be taken when connecting I/O cable shields to the chassis ground. One must avoid common impedance coupling in the shield-to-chassis trace to ensure optimum performance from balanced interconnections.

The goal of the Audio Engineering Society in recommending these balanced interconnection solutions is to reduce or eliminate the need for interconnection alternatives through education and information sharing. This is the mission statement of the Audio Engineering Society in the first place. Systems installed with chassis-grounded balanced shields on all units, with well-twisted interconnecting cables operate hum- and buzz-free, leaving only the I/O circuit topology specifications to dictate system performance.

The AES recommendation's purpose is not to create another "pin 2 is hot" war. In reality, users and installers have found acceptable solutions for the "pin 1 problem" of signal-grounded balanced shields and are unlikely, nor will they be able, to suddenly change over to not using alternatives. Manufacturers specify the I/O connector type on data sheets; similarly, we should specify shield connection practices in equipment specifications, on the chassis, or at least in the manual, thus providing users with the information required for proper system configuration.

8 ACKNOWLEDGMENT

The author wishes to thank Rick Jeffs of Rane Test Engineering and Neil Muncy of Muncy and Associates for keeping the author grounded to the real world during the hours of practical and sometimes frazzling discussions on grounding.

9 REFERENCES

[1] H. W. Ott, *Noise Reduction Techniques in Electronic Systems* (Wiley, New York, 1976).

[2] T. Pennington, *Rane Note 110* (Rane Corp., Mukilteo, WA, 1985).

[3] M. Glasband, "Lifting the Ground Enigma," *MIX* magazine, vol. 18., no. 11, pp. 136–146 (1994 Nov.).

[4] M. Glasband, "Audio Noise and AC Systems Revisited, Rethinking Studio Electrical Systems," *R•E•P*, vol. 23, no. 8, pp. 30–34 (1992 Aug.).

[5] M. Glasband, "Audio Noise and AC Systems," *R•E•P*, vol. 22, no. 6, pp. 42–45 (1991 June).

10 BIBLIOGRAPHY

Giddings, P., *Audio System Design and Installation* (Howard W. Sams, Indianapolis, IN, 1990).

Jung, W., and A. Garcia, "Op Amps in Line-Driver and Receiver Circuits, Part 2," *Analog Dialogue*, vol. 27, no. 1 (1993).

Metzler, B., *Audio Measurement Handbook* (Audio Precision, Portland, OR, 1993).

Morrison, R. *Grounding and Shielding Techniques in Instrumentation* (Wiley, New Tork, 1967).

Morrison, R., *Noise and Other Interfering Signals* (Wiley, New York, 1992).

Perkins, C., "Measurement Techniques for Debugging Electronic Systems and Their Interconnection," in *Proc. 11th Int. AES Conf.*, Portland, OR (1992 May).

THE AUTHOR

Stephen Macatee is a design engineer with Rane Corporation in Mukilteo, WA. He received a BSEE from Monmouth College, NJ, in 1986, and started working in Rane's manufacturing department in 1987, stuffing and soldering boards. He soon moved to Rane's computer-aided design department where he helped design and develop Rane's customized computer-aided design system, and spent several years designing printed-circuit boards. In this capacity, he was able to apply his engineering background and manufacturing experiences to optimize Rane's printed-circuit board designs for low-noise, high-performance operation and efficient manufacturing. Currently he focuses his efforts in the area of digital hardware and software design.

In his free time Mr. Macatee designs alternative MIDI controllers. His most recent project involved sensing a juggler's motions thus creating a bizarre new musical instrument from the juggler's movements. He also enjoys music, percussion, good food, working on his home and spending time with his wife, Jill.

Fundamentals of Grounding, Shielding, and Interconnection*

KENNETH R. FAUSE, AES Member

*Smith, Fause and Associates, Inc., El Segundo, CA 90245,
and Smith, Fause and McDonald, Inc., San Francisco, CA 94105, USA*

The ultimate performance of modern audio systems may be significantly constrained by signal contamination introduced by inappropriate grounding and interconnection practices. Fundamental principles of electromagnetism and linear circuits are reviewed. From this, a body of good engineering practice for grounding, shielding, and interconnection methods is developed. A logical, consistent grounding scheme is presented which strictly conforms to the requirements of the National Electrical Code, a model safety code widely adopted in jurisdictions in North America. Conventional grounded neutral conductor mains power distribution is used, which may be single phase, polyphase, or multiple separately derived polyphase systems without constraint. The fundamental principles may be applied to mains power distribution systems conforming to other internationally established safety standards.

0 INTRODUCTION

Let us begin by demolishing a myth. Techniques for shielding and grounding are not black magic. There are no incantations to chant while deciding where to connect pin 1. This report discusses principles of good engineering practice, not folklore or aspects of the occult arts.

None of the information in this report is new. The engineering practice described here dates from the early days of the telephone and cinema sound industries. A paper on this topic can be found in the first issue of this *Journal* [1]. Why then devote an entire day of the 97th Convention of the Audio Engineering Society to a workshop and invited papers on this topic, followed by an issue of the *Journal* presenting these reports?

With the increasing market penetration of audio digital storage media devices, many users have come to the alarming discovery that the actual maximum signal-to-rms-noise ratio realized in their audio signal system is significantly less than the available dynamic range of the incorporated digital storage media device. In practice, this system performance constraint is frequently due to signal contamination introduced by the use of inappropriate or ineffective grounding, shielding, and interconnection practices.

Pursuant to European Community (EC) directives, all apparatus offered for sale or placed in service in the EC beginning 1996 January 01 must bear a mark indicating conformance with applicable electromagnetic compatibility directives and from 1997 January 01 with the low-voltage (safety) directive as well [2], [3]. A significant proportion of audio apparatus presently in the market may fail to meet these requirements [4]–[6].

The failure of much audio apparatus to meet mandatory requirements or to function to reasonable user expectations within practical audio systems should be of urgent concern to the audio engineering community.

In order to discuss the topics of shielding and grounding, it is necessary to review basic principles of electromagnetism and linear circuits. Here rests most of the confusion about shielding and grounding—the principles are so fundamental that they are often neglected in practice. A disclaimer: this report will discuss general topics and provide suggested guidelines for use in a range of reasonably typical cases. Other reports treat in detail a variety of the topics discussed in the general case herein. For a complete treatment of the topic of shielding, grounding, and electromagnetic compatibility and for guidance on the resolution of specialized problems, the author recommends that the reader consult the references [7]–[9].

1 CAUTIONARY NOTE

The sale, construction, and operation of audio systems and their constituent apparatus elements may be subject

---

* Manuscript received 1994 November 30; revised 1995 March 5 and April 3.

to national, provincial, state, county, district, municipal, and other governing laws, ordinances, and regulations applicable to buildings, commerce, health and safety in the workplace, and the use of utility mains power. Insurance carriers and labor organizations may have additional requirements. In this report, such requirements are collectively referred to as requirements of the authorities having jurisdiction or, for simplicity of language, the code. Where any of the suggestions or guidelines of this report may conflict with the applicable requirements of the authorities having jurisdiction, the requirements of the latter must prevail.

Mains power presents a potentially lethal risk of shock and fire. Any construction of or alteration to mains power systems must be performed in strict conformance with the applicable requirements of the authorities having jurisdiction, by qualified personnel duly authorized to perform such work in the jurisdiction.

2 NOISE SUPPRESSION

Most professional audio activity takes place in modern, populated areas. Along with their many other attributes, populated areas are generally blanketed with electric, magnetic, and electromagnetic (radio) fields of varying intensity. Some lucky few will record birdcalls in the wild. The rest of us in the audio industry are unfortunately faced with the problem of transducing, amplifying, processing, recording, auditioning, and otherwise manipulating audio program signals while rejecting a broad range of electromagnetic interference (EMI) contributed by the surrounding environment or by elements of the audio system itself. For the purpose of this discussion, we shall refer to all such interference as noise, where we shall define noise to be *undesired signal*.

In any general case involving noise suppression, be it acoustic, mechanical, or, in this case, electrical and magnetic, we will be concerned with:
1) The source
2) The propagation path
3) The receiver.

In this case we will define the noise receiver to be either the complete audio system under consideration, or a given apparatus within that system. In theory one could achieve a desired noise reduction by manipulating any of the three elements mentioned. For the general noise suppression case it is usually preferred to perform the mitigation effort at the source. This avoids the complexity of treating multiple potential propagation paths and receivers. In audio system practice the choice is limited. Most practical sources of interference are external to the audio system and cannot be treated as variables. Most users prefer not to modify manufactured audio apparatus if at all possible.

1) Modification generally voids any safety and/or EMC agency listing the apparatus may carry.
2) Modification generally voids any manufacturer's warranty the apparatus may carry.
3) Modification generally renders the apparatus not interchangeable with a standard apparatus in the event of apparatus failure.

Other reports address recommendations for apparatus design considerations to reduce the susceptibility of individual audio apparatus to EMI.

From a practical audio system design standpoint, this leaves only the propagation path as a variable for the noise reduction process. It will be shown that a practical audio system noise reduction effort must consider the following:
1) Shielding and grounding
2) Mains power safety requirements
3) Signal interconnection methods
4) Apparatus design issues.

3 FUNDAMENTALS OF INTERFERENCE COUPLING

In order to devise an efficient engineering approach, we must recall some fundamental theory of electromagnetism and linear circuits.

1) *Ampere's law*. A varying electric field E induces a magnetic field H.

2) *Faraday's law of induction*. A varying magnetic field H induces an electric field E.

3) *By extension from 1) and 2)*. Whenever a charge moves in a conductor, that conductor will radiate both electric and magnetic fields.

4) *Kirchhoff's current law*. The algebraic sum of the currents at a junction will be zero.

5) *Kirchhoff's voltage law*. The algebraic sum of the voltages around a closed loop will be zero.

Virtually all practical cases of interference coupling arise from fundamental causes.

1) *Electric field or capacitive coupling*. An electric (charge) field of the source is coupled into the receiver circuits.

2) *Magnetic field or inductive coupling*. A magnetic field of the source is coupled into the receiver circuits.

3) *Electromagnetic field coupling*. An electromagnetic field is the vector product of an electric field and a magnetic field. The practical case will be concerned with a propagating electromagnetic field, known as an electromagnetic wave (Hertzian wave, radio wave) originating at a source and coupled into the receiver circuits.

4) *Common impedance coupling*. Currents from two or more circuits flow through a common impedance.

3.1 Field Shielding

A shield (screen) is generally defined as a conductive partition between two regions of space, intended to control the propagation of electric or magnetic fields from one region of space to another. Infinite planes of shield in space are of interest only as textbook examples. In practice the shield is generally formed into an enclosure for circuit conductors or circuit assemblies. For special cases, an entire room might be shielded.

A shield may be intended to contain the field (source shielding, Fig. 1) or to prevent an external field from acting on some circuit (receiver shielding, Fig. 2). In the case of receiver shielding, the essence of the idea is to provide a path for the field to travel other than through the receiver circuit or conductor.

3.2 Electric Field Shielding

Assume a closed hypothetical surface (a Gaussian surface) of conductive material. From Gauss's law it may be shown that if an excess charge is placed on the conductor, at equilibrium the charge will reside entirely on the surface. No charge will be enclosed [10, p. 692].

For effective electric field shielding:

1) The shield material must be an electrical conductor.
2) The number and size of openings in the shield must be minimized.
3) The portion of circuitry or conductor extending beyond the shield must be minimized.
4) The accumulated charge on the shield must be drained. For this to occur, a conductor must be provided to connect the charged body of the shield to a significantly larger body; the connected bodies will come to charge equilibrium. In most audio practice, the "large body" used is the earth, or a body connected to the earth.
5) A shield should not carry other than its own charge drain current. Any current in the shield will produce corresponding radiated electric and magnetic fields. These may couple into adjacent circuits as noise.

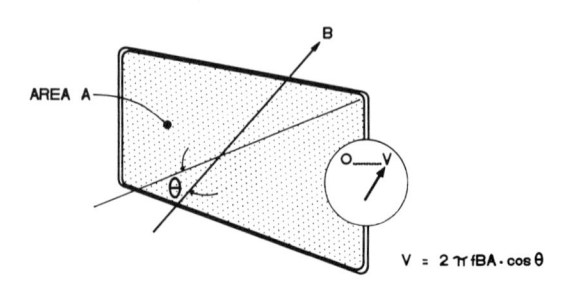

3.3 Magnetic Induction

Consider the situation shown in Fig. 3. Here define a simple geometry of a stationary single loop of conductor in a plane, enclosing an area A. Presume that we thread this loop with a sinusoidally time-varying magnetic flux of density B constant over the area A. According to Faraday's law of induction, an electromotive force is induced [10, p. 871]. For the simplified case indicated, Faraday's law may be stated (without dimensional constants) as

$$V = 2\pi f BA \cos \theta \tag{1}$$

where

V = rms value of induced electromotive force (voltage)
B = rms value of flux density varying sinusoidally at frequency f
A = area of closed loop
θ = angle between flux vector direction and normal to plane surface of area A.

Observe the key practical points:

1) Induced voltage is proportional to frequency f.
2) Induced voltage is proportional to loop area A.

Assume the object is to minimize V, the induced voltage. Everything on the right of Eq. (1) is a product. Thus we may minimize V by minimizing any or all of the terms on the right. In a practical case, the magnetic flux field is likely to be caused by stray ac mains radiation. Thus f is generally not a variable. Both B and θ may be changed by altering circuit conductor locations, moving transformers, and the like.

Observe a critical point. If the conductor loop is not closed, there is no enclosed area A. If $A = 0$, V must equal 0—no loop; no magnetically induced noise voltage. Of course there is a significant limitation. For a signal circuit to be useful, current must flow. From fundamental circuit laws, current can only flow in a closed loop. Any practical metallic signal circuit is therefore prone to magnetic inductive coupling of noise into the signal path. The primary mitigation method is to minimize the loop area A.

A final point about magnetic induced coupling. A grounded electric field shield enclosing a conductor does not significantly attenuate magnetic induction at audio frequencies. Only materials with high relative magnetic permeability will provide a useful degree of magnetic shielding in the audio-frequency region.

3.4 Electromagnetic Field Shielding

As previously indicated, an electromagnetic field is the vector product of an electric field E and a magnetic field H. For a propagating field, an electromagnetic wave, the ratio E/H is defined as the wave impedance Z_W. At a given receiver location, the characteristics of an electromagnetic field will be a function of the source,

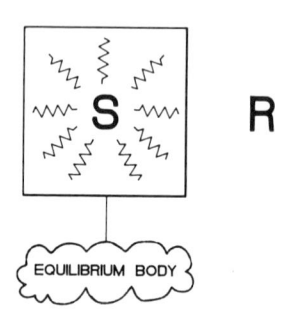

Fig. 1. Source shielding.

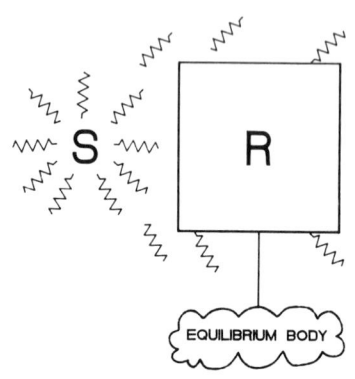

Fig. 2. Receiver shielding.

Fig. 3. Magnetic induction.

the distance between source and receiver, and the medium of propagation between source and receiver [7, p. 159].

When an electromagnetic wave propagating in free space impacts a sheet of conductive and magnetically permeable material, a portion of the incident wave energy is reflected, and a portion of the energy is absorbed within the material. If the shield is defined as thin relative to the incident frequency and wave impedance, additional energy may be absorbed due to multiple internal reflections. The remainder of the wave energy is transmitted. The ratio of incident to transmitted field strength is defined as shielding effectiveness. Shielding effectiveness varies with the properties of the shield material (conductivity and magnetic permeability), the incident frequency, the incident wave impedance, the geometry of the shield itself, and the geometry relative to the source [7, p. 165]. The conditions previously described for effective electric field shielding must be met. Further, it is essential to recall that any conductive body exposed to an incident electromagnetic wave also behaves as an antenna.

3.5 Common Impedance Coupling

Consider the situation shown in Fig. 4. Here define V_S as an ideal voltage source (voltage is constant and independent of current drawn) and the conductors as ideal (lossless). Assume two separate circuits having impedances Z_1 and Z_2 with a junction to a common impedance Z_C. Solve for V_1 and V_2 in terms of the other circuit parameters. From Kirchhoff's current law, the algebraic sum of the currents at each junction must be zero [11]. From Kirchhoff's voltage law, the algebraic sum of the voltages around a closed loop must be zero [11]. By inspection, these conditions are satisfied only when

$$I_C = I_1 + I_2 \tag{2}$$

$$V_1 = V_2 \tag{3}$$

$$V_S = V_C + V_1 . \tag{4}$$

Apply Ohm's law and substitute to determine V_1 or V_2,

$$V_1 = V_S - (I_1 + I_2) Z_C . \tag{5}$$

The supply voltage to each circuit is affected by the supply current drawn by the other through the common impedance. This situation may arise in power supply, signal, and ground circuits, and is likely to be undesirable in all cases.

4 GROUNDING

Ground is commonly defined as a zero signal reference point (ZSRP) for a complex of electronics. Notice from this definition that connections to ground and to earth are not necessarily the same thing. Of course reality is not so simple. A number of distinct functions all employ the term and concept of "ground." It is necessary to consider each.

1) *Shield ground.* As stated earlier, this drains the accumulated shield charge. For the drain to operate, it must be connected to some large body which will sink the charge by equilibrium. In audio practice this "large body" usually is the earth. Shield ground thus must eventually connect to earth ground.

2) *Frame, chassis, or rack ground.* It essentially performs the same function for an apparatus enclosure that shield ground does for a shield.

3) *Transmission ground.* When used, this provides a reference for signal voltages being transmitted or received. The most obvious case is a reference for unbalanced (asymmetric) circuits. Balanced (symmetric) circuits usually do not require a transmission ground, although in common audio practice they are often referenced, but not connected, to transmission ground. Transmission ground should not be the circuit shield unless the source and receiver apparatus are specifically designed for this interconnection. Several reports treat this topic in detail.

4) *Circuit power ground.* This is exactly as the name implies. It is the zero potential reference for the circuit-powering voltages.

5) *Safety ground.* Electrical codes in force, labor laws, and the insurance carriers are virtually certain to require that all conductive enclosures and surfaces be bonded to earth ground by an approved means. The reasoning is simple. An electrical fault, such as an insulation breakdown between the phase (hot) conductor of the ac mains supply and an enclosure will make that enclosure live. An individual touching both the enclosure and a grounded surface will then be connected across the ac power line. This is not a hypothetical scenario—several performers have been killed as a result of contact with "hot" microphones.

The solution is to bond the apparatus enclosure to ground. There are two results.

1) A low-impedance path to ground is established, much lower than the typical impedance through the human body. In the event of a hot-to-enclosure fault, this creates a voltage divider in the ratio of the bond path impedance to the body impedance. A relatively low body current will flow.

2) If a sufficiently low-impedance ground path is established, then a very high current will flow under fault conditions. The intent is to trip the mains circuit protection device (fuse or circuit breaker) rapidly and thus clear the fault.

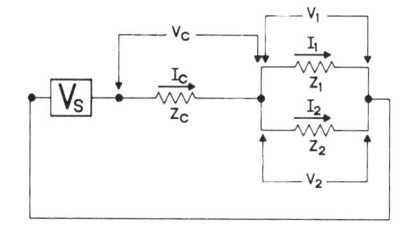

Fig. 4. Common impedance coupling.

Notice that the ampacity (current-carrying capacity) of the apparatus safety ground conductor is determined by the applicable code to withstand the maximum fault current that the related ac supply mains can deliver for a period long enough to clear the fault. Language from a model code widely adopted by jurisdictions in North America is as follows [12]:

> 250-51. Effective Grounding Path. The path to ground from circuits, equipment, and metal enclosures for conductors shall (1) be permanent and continuous; (2) have capacity to conduct safely any fault current likely to be imposed on it; and (3) have sufficiently low impedance to limit the voltage to ground and to facilitate the operation of the circuit protective devices.

It is essential to observe that due to reactance effects, any practical grounding scheme will behave differently at direct current (dc), at mains power frequencies, at low frequencies (below 150 kHz), and at high frequencies (above 150 kHz).

4.1 Multipoint (Mesh) Ground Scheme

Define an ideal ground plane as an extended plane of ideal conductive material connected to an equilibrium body by a lossless conductor. In multipoint grounding, circuits are connected to the ground plane by the shortest available path (refer to Fig. 5).

In practical apparatus, the ground plane may be the chassis. In the case of a facility, a continuous extended plane of conductor is generally impractical or uneconomic to achieve. A bonded grid or mesh of low-resistance conductors is often implemented to synthesize the ground plane. In this case it is preferred to make the grid spacing as small as practical to minimize the loop area, and to make the individual conductor resistance as low as possible. For the case of signal conductors, the shield is connected to the ground plane at every reasonable opportunity. Observe that this inherently creates loops. The multipoint ground scheme:

1) Is simple.
2) Is preferred at frequencies above 10 MHz, or where lengths of conductors exceed $1/20-1/10$ the wavelength of a disturbing electromagnetic field.
3) Is subject to magnetic induced noise due to loops.
4) Is subject to common impedance coupled noise due to finite resistance between any two points of the ground plane.

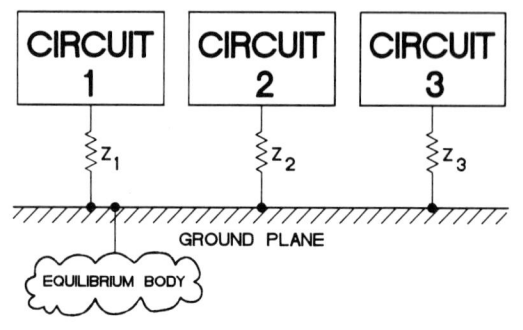

For interconnecting an audio system made up of two or more items of audio apparatus, the multipoint ground scheme may be an appropriate engineering choice when:

1) The scheme is required by national, organizational, or other norms.
2) The audio system is portable.
3) Signal interconnections are made by connectorized portable signal cables.
4) The system signal wiring must be reconfigured by nontechnical personnel.
5) Immunity from radio-frequency interference (RFI) is a strong objective.
6) The system signal interconnecting cables are not subject to strong magnetic fields.
7) The system signal interconnecting cables are selected to minimize shield current induced noise [13].
8) The individual items of apparatus are not prone to "pin 1 problems" as described by Muncy [13].

4.2 Single-Point (Star) Ground Scheme

Fundamental limitations of the multipoint ground scheme at low frequencies may be addressed by use of a single-point ground scheme (Fig. 6). The ground plane of the previous scheme is collapsed to a single defined point, which is connected to an equilibrium body by a lossless conductor. Each separate circuit is connected to the defined single ground point by a single path.

1) Use of a single ground path for each separate circuit avoids a closed loop, eliminating possible magnetic induction coupling.
2) Use of a single ground path for each separate circuit avoids common impedances, eliminating possible common impedance coupling.
3) At audio through radio frequencies, the concept of an extended equal-potential ground plane is a polite fiction. Finite resistance and reactance will exist. Two separate ground points are seldom at the same potential. A ground potential difference may, by common impedance coupling, couple into a signal circuit as a noise voltage.

For a practical audio system grounded by the single-point scheme, each piece of audio apparatus, and each type of ground associated with that apparatus, is connected to a single defined point by a single path. For the case of signal conductors, the shield is connected to the ground point at one end only. Where cable lengths

Fig. 5. Multipoint ground scheme.

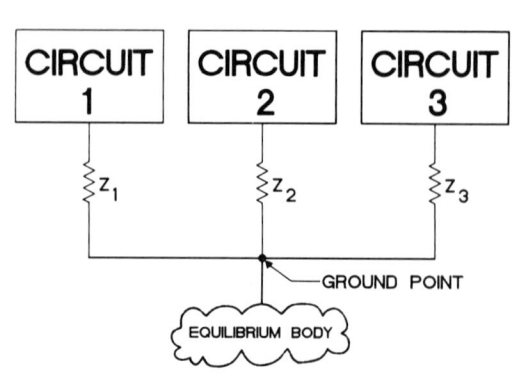

Fig. 6. Single-point ground scheme.

approach $1/20-1/10$ wavelength and where strong radio-frequency (RF) fields are present, it may be necessary to ground the shield at radio frequencies at more than one point to ensure that the entire shield remains at or near ground potential at all frequencies. A capacitor, or a capacitor in series with a low-value resistor, is used to establish the required RF ground. Disk ceramics of 10 nF or other capacitors with similar impedance characteristics have proven effective in practice.

Separate conductor bundles or groups of printed-wire traces should be formed for ground conductors of each functional type. Strict application of this recommendation in large systems becomes a matter of some rigor. Any deviations taken may result in reduced system performance. There are several very good reasons for this seemingly extreme approach.

1) Separate ground conductors for each ground type admittedly may present a clumsy mechanical layout problem. The object is to avoid common impedance noise coupling as was described earlier. For this reason, audio circuit power, logic power, relay power, and indicator-lighting power should *not* share the same ground lead. Switching transients are likely to wind up in the signal.

2) Separate ground-type conductor bundles or trace groups may also seem extreme. Once again, recall that any current in a conductor produces electric and magnetic fields. These in turn may induce noise in nearby circuits.

The single-point ground scheme:

1) Is potentially more complex to implement than multipoint grounding.

2) Is preferred at frequencies below 10 MHz.

3) Is not subject to magnetic induced noise due to loops.

4) Is not subject to common impedance coupled noise.

For interconnecting an audio system made up of two or more items of audio apparatus, the single-point ground scheme may be an appropriate engineering choice when:

1) The system is fixed in place.

2) Signal interconnections are made by fixed signal cables.

3) The system signal wiring need not be reconfigured; or if it is to be reconfigured, the reconfiguration may be effected by the use of jack fields, routing switchers, or similar methods.

4) Immunity from RFI is an objective.

5) The system signal interconnecting cables may be subject to strong magnetic fields.

6) The system signal interconnecting cables are not selected to minimize shield current induced noise [13].

7) The individual items of apparatus may be prone to "pin 1 problems" as described by Muncy [13].

4.3 Grounding and Safety

In determining the grounding scheme for any audio apparatus or system, the requirements of safety and electromagnetic compatibility must prevail.

A safe audio system which complies with an applicable code may be achieved when the audio apparatus and interconnection cable ground scheme is derived in a rigorous manner by way of the code-required equipment safety ground conductors. In part because of the resultant low ground impedance—recall that the safety ground conductor is sized by the code to handle fault current—audio system performance is often considerably superior to that regularly achieved with other, possibly unsafe, ground schemes.

Most audio apparatus is classed as cord-and-plug-connected apparatus, where the mains feed to the apparatus is via a flexible service cord terminating in a male mains attachment plug (cord cap). Depending on the design of the apparatus and applicable apparatus safety standards, the service cord may be two wire, containing supply conductors only, or three wire, containing supply conductors and an equipment safety grounding conductor. Alternatively, the apparatus may employ either an external mains transformer assembly or an external dc power supply assembly, either of which would connect to the powered apparatus via a multiconductor low-voltage flexible service cord. The significant proportion of professional audio apparatus also employs an enclosure design which permits the apparatus to be used either as portable apparatus or as rack-mounted apparatus conforming to ANSI/EIA 310-D, *1992, Racks, Panels and Associated Equipment*, and DIN 41 494.

It is necessary to mention the abhorrent, but sadly common, practice of removing the grounding connection from the mains attachment plug of a three-wire apparatus service cord, deleting the safety ground. A separate conductor—possibly the shield or drain wire of a signal cable—is then run from the apparatus to a "technical" ground. This is a flagrant code violation and is extremely dangerous. The ground impedance of such an arrangement is likely inadequate to sustain fault current, creating a major potential life hazard of fire and shock. It will be shown that such expedient schemes are completely unnecessary.

The following recommendations and related guidelines apply to the case of mains power circuits with a branch topology and where one of the supply conductors is grounded (identified conductor, neutral conductor). These recommendations and related guidelines *do not* apply to mains power circuits with a ring topology or where neither of the supply conductors is grounded. The specific recommendations herein apply to permanently installed systems using North American mains power practices in conformance with the National Electrical Code. This code is the model for most wiring practice in the United States and is adopted in its entirety by enabling legislation in many jurisdictions. The underlying principles may be applied to mains power systems conforming to other internationally adopted standards. Codes of record vary. It is essential to verify that a specific scheme is acceptable to the local authorities having jurisdiction.

The key to a coordinated mains power and audio technical ground scheme lies in the design of the mains power system. A practical mains power system is not an ideal voltage source; it exhibits a finite source impedance. Further, portions of a practical audio system may impose a significant time-varying current demand to the mains, such as a recorder in fast spooling mode or a large array of power amplifiers driven by a program signal.

Define a mains power distribution system where each branch circuit includes conductors for *hot* (ungrounded conductor, phase conductor, black or other color), *neutral* (identified conductor, white), and *equipment safety ground* (green or green/yellow), with the conductors continuous from the receptacle to the serving panel board, and with conductors of equal cross-sectional areas for the neutral and the equipment safety ground. The applicable code may require that the hot and neutral conductors be of equal ampacity. Notice that this scheme specifically excludes the use of a single neutral conductor acting as a common return path for two branch circuits in a single-phase grounded midpoint distribution system (120/240-V system in North America) or for two or three branch circuits in a polyphase grounded starpoint distribution system (120/208-V system in North America). It does *not* exclude the use of grounded midpoint or polyphase branch circuit panel boards.

1) The use of a separate neutral conductor for each individual branch circuit eliminates possible common-point impedance coupling at the branch circuit level.

2) The use of equal cross-sectional areas for the neutral and the equipment grounding conductor of each branch circuit results in reasonable equality of the impedance to ground via the neutral path and via the equipment grounding conductor path. This mitigates the impact of possible minor neutral-to-chassis leakage coupling.

4.4 Bonded Raceway Grounding Scheme

This approach is suitable for smaller audio systems subject to reasonable ambient electrical noise conditions. Conventional receptacles are used, bonded to the serving raceway. This is detailed in Article 250-74 of the National Electrical Code [12]:

> 250-74. Connecting Receptacle Grounding Terminal to Box. An equipment bonding jumper shall be used to connect the grounding terminal of a grounding-type receptacle to a grounded box.

Refer to Fig. 7 for typical receptacle wiring applicable to North America. By code, all conductive raceway (conduit, trunking) is bonded to ground. The resulting impedance is low at mains power frequencies and is adequate to withstand fault current. The impedance to ground may not remain low at audio and higher frequencies due to the multiple mechanical connections in a practical raceway system. Therefore it is strongly recommended that an additional grounding conductor (insulated, green or green/yellow) be provided to run continuous from the grounding terminal of each branch circuit receptacle to the serving panel board. This is defined as an additional bonding jumper. The individual bonding jumpers are connected to a terminal bar bonded to the panel board, which in turn, is bonded via the mains supply raceway. By code, the neutral (identified conductor) also must be bonded to earth ground at the service entrance point (refer to Fig. 8).

Straightforward application of this scheme requires that all mains power circuits be fed from the same neutral bus. A practical way to do this is to derive from a single panel board all mains circuits feeding the audio system. The more physically spread out the audio system topology, the more important is this requirement, as the ground impedance will increase, resulting in higher possible ground voltage differences between individual apparatus, thus a greater likelihood of induced noise.

Notice the primary limitation of this scheme. Stray currents may flow in the power raceway as a result of other, non-audio-system loads. Noise may couple into the audio system by common impedance coupling in the ground path.

4.5 Isolated Equipment Grounding Conductor Scheme

In the event that the raceway-derived ground at the panel board is expected to be excessively noisy or the audio system is large, a significant refinement may be made to the preceding scheme. This is detailed in Article 250-74, Exception No. 4, of the National Electrical Code [12]:

> Exception No. 4: Where required for the reduction of electrical noise (electromagnetic interference) on the grounding circuit, a receptacle in which the grounding terminal is purposely insulated from the receptacle mounting means shall be permitted. The receptacle grounding terminal shall be grounded by an insulated equipment grounding conductor run with the circuit conductors. This grounding conductor shall be permitted to pass through one or more panelboards without connection to the panelboard grounding terminal as permitted in Section 384-20, Exception, so as to terminate within the same building or structure directly at an equipment grounding conductor terminal of the applicable derived system of service.

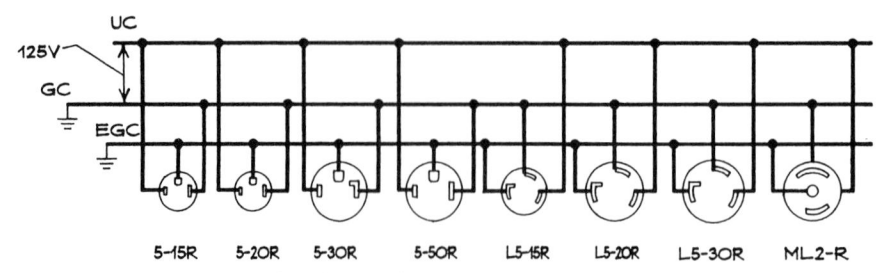

Fig. 7. Wiring diagrams for two-pole three-wire plugs and receptacles, National Electrical Manufacturers Association (NEMA) 125-V configurations, applicable to North America. UC—ungrounded conductor (hot, phase); GC—grounded conductor (neutral, system ground); EGC—equipment grounding conductor.

ENGINEERING REPORTS GROUNDING, SHIELDING, AND INTERCONNECTION

(FPN) [Fine Print Note]: Use of an isolated equipment grounding conductor does not relieve the requirement for grounding the raceway system and outlet box.

This approach, illustrated in Figs. 9 and 10, creates a "clean" or "technical" ground free from ground noises due to odd ground currents in the raceway system. The

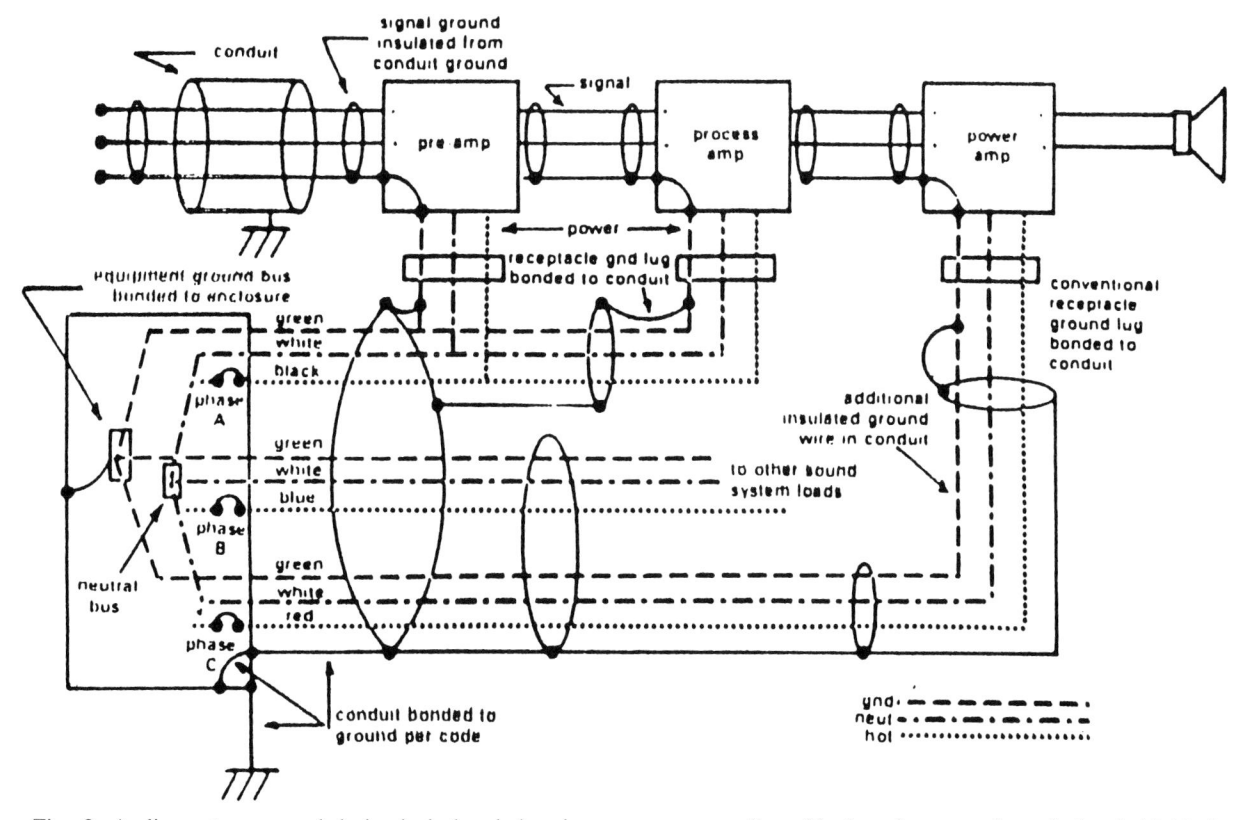

Fig. 8. Audio system ground derived via bonded mains raceway grounding. Single-point grounding of signal shields is indicated; multipoint grounding is optional.

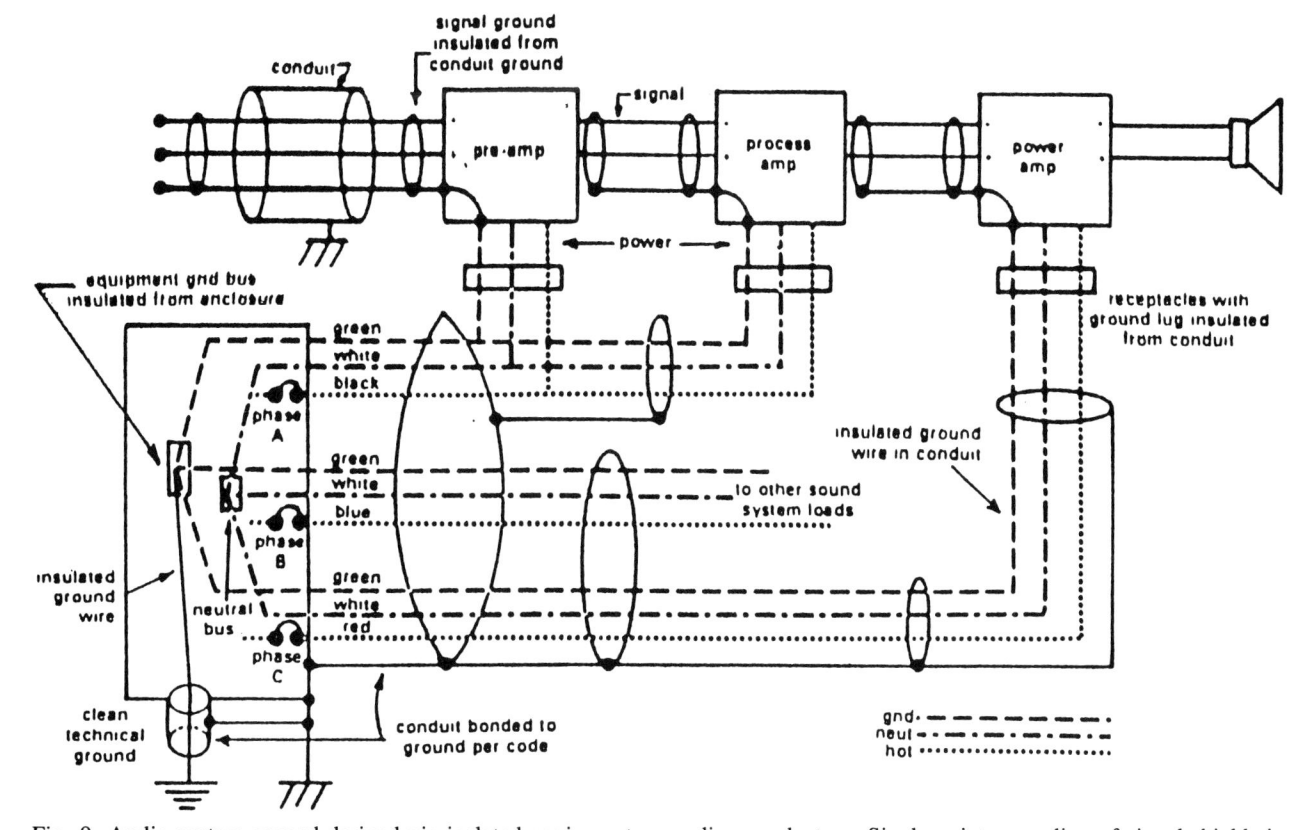

Fig. 9. Audio system ground derived via isolated equipment grounding conductors. Single-point grounding of signal shields is indicated; multipoint grounding is optional.

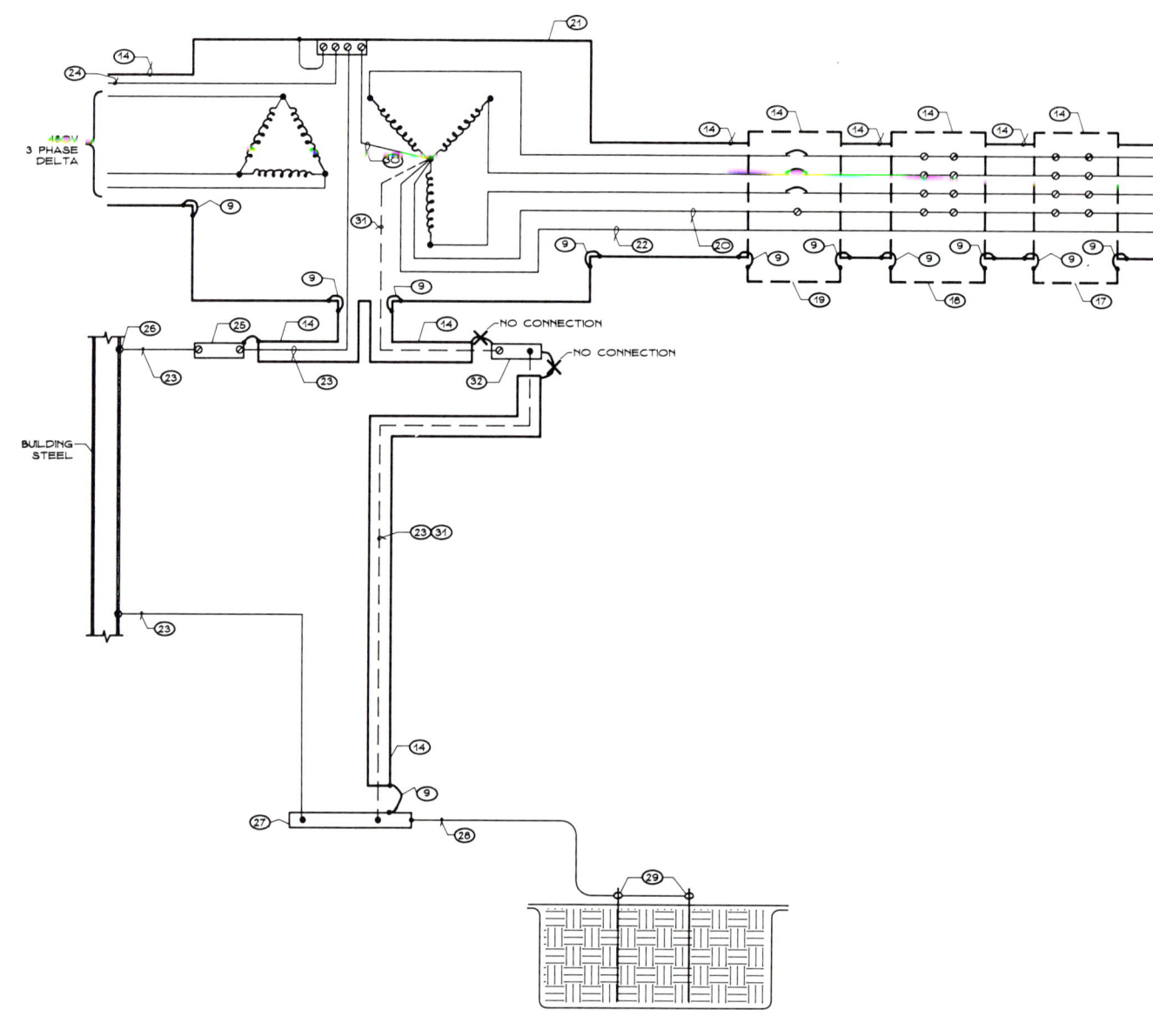

ABBREVIATIONS

UC   UNGROUNDED CONDUCTOR (HOT, PHASE).

GC   GROUNDED CONDUCTOR (NEUTRAL, SYSTEM GROUND).

EGC EQUIPMENT GROUNDING CONDUCTOR

IEGC INSULATED EQUIPMENT GROUNDING CONDUCTOR.

GENERIC TECHNICAL GROUND NOTES

1. "CORD AND PLUG CONNECTED EQUIPMENT", MECHANICALLY FASTENED TO EQUIPMENT RACK.
2. "CORD AND PLUG CONNECTED EQUIPMENT", PORTABLE.
3. EQUIPMENT WITH 3-WIRE SUPPLY CORD GROUNDED BY METHOD OF NEC 250-59 (B).
4. EQUIPMENT WITH 2-WIRE SUPPLY CORD GROUNDED BY METHOD OF NEC 250-59 (C).
5. ATTACHMENT PLUG (CORD CAP) WIRED PER NEMA/ANSI FOR APPLICABLE CONFIGURATION.
6. ISOLATED GROUND RECEPTACLE PER NEC 250-74, EXCEPTION NO. 4, WIRED PER NEMA/ANSI FOR APPLICABLE CONFIGURATION.
7. METAL OUTLET BOX(ES) AND RACEWAY OR METAL MULTI-OUTLET ASSEMBLY.
8. EQUIPMENT RACK: ELECTRICALLY CONTINUOUS FIXED METAL ENCLOSURE, GROUND PER NEC 250-57 (B).
9. BONDING PER NEC 250-75.
10. ISOLATED MOUNTING/FASTENING.

Fig. 10. Audio system apparatus ground derived via isolated equipment grounding conductors, for general case of a fixed system, powered by a separately derived polyphase power system to North American practices. Paragraph references to NFPA 70 [12] are provided to demonstrate strict conformance with applicable code. UC—ungrounded conductor (hot, phase); GC—grounded conductor (neutral, system ground); EGC—equipment grounding conductor; IEGC—insulated equipment grounding conductor.

ENGINEERING REPORTS   GROUNDING, SHIELDING, AND INTERCONNECTION

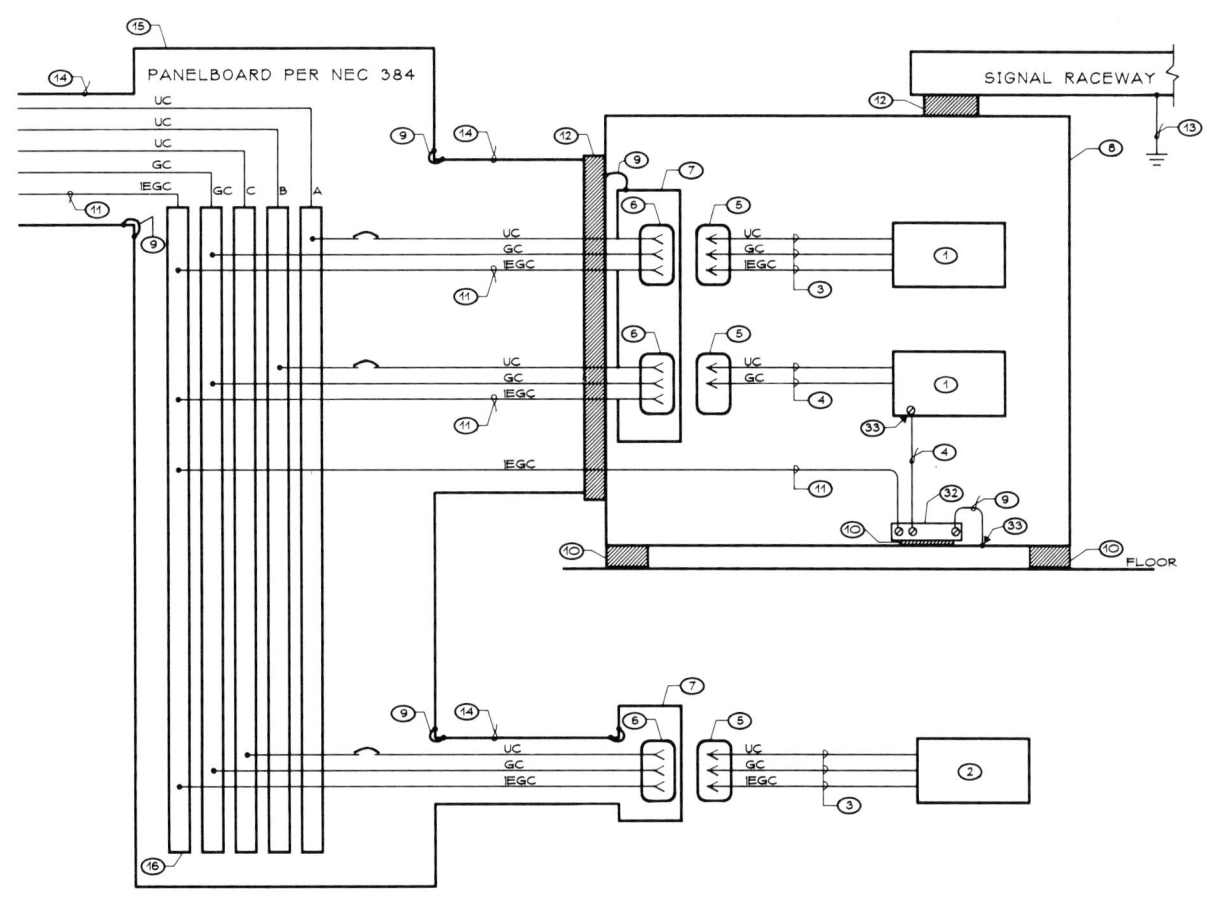

⑪ INSULATED EQUIPMENT GROUNDING CONDUCTOR PER NEC 250-75, EXCEPTION. COMPLY WITH NEC 250-51, NEC 250-57, NEC 250-92 (C), NEC 250-95 AND NEC 250-114.

⑫ NON-METALLIC SPACER OR FITTING PERMITTED PER NEC 250-75, EXCEPTION.

⑬ GROUND AND BOND PER NEC 640-4 (D) AND NEC 250-33. DO NOT CONNECT TO IEGC.

⑭ GROUND AND BOND PER NEC 250-32.

⑮ GROUND AND BOND PER NEC 384-20.

⑯ ISOLATED EQUIPMENT GROUNDING CONDUCTOR BUS PER NEC 384-20, EXCEPTION.

⑰ TRANSIENT OVERVOLTAGE SUPPRESSOR, WHERE OCCURS.

⑱ VOLTAGE REGULATOR WHERE OCCURS.

⑲ OVERCURRENT PROTECTION, WHERE OCCURS PER NEC ARTICLE 240.

⑳ CONDUCTOR TO BE GROUNDED PER NEC 250-25 (3).

㉑ TRANSFORMER ESTABLISHING SEPARATELY DERIVED ALTERNATING CURRENT SYSTEM. PER NEC 250-26.

㉒ BONDING JUMPER PER NEC 250-26 (A).

㉓ GROUNDING ELECTRODE CONDUCTOR PER NEC 250-26 (B).

㉔ PRIMARY GROUNDING CONDUCTOR, WHERE OCCURS.

㉕ GROUND BAR AT LOCAL ELECTRICAL EQUIPMENT ROOM.

㉖ GROUNDING ELECTRODE PER NEC 250-26 (C).

㉗ PROJECT MAIN GROUNDING BAR.

㉘ GROUNDING CONDUCTOR(S) PER NEC ARTICLE 250, PART J.

㉙ GROUNDING ELECTRODE SYSTEM PER NEC ARTICLE 250, PART H.

㉚ CONNECTION WHERE SEPARATELY DERIVED SYSTEM IS GROUNDED TO STANDARD PROJECT GROUNDING ELECTRODE CONDUCTOR (I.E. DIRTY GROUND).

㉛ ALTERNATE 'CLEAN GROUND' CONNECTION WHERE APPROVED BY AUTHORITIES HAVING JURISDICTION; BASED ON NEC 250-26 (C) (3). CONNECTIONS 30 AND 31 ARE MUTUALLY EXCLUSIVE.

㉜ 'CLEAN GROUND BAR'.

㉝ CONNECTION TO COMPLY WITH NEC 250-113, 250-117 AND 250-118.

clean-ground scheme has been found effective for extremely large sound systems demanding low-noise requirements, and for sites with otherwise difficult RFI.

Notice that under the National Electrical Code a *completely* separate ground system is permitted—including separate ground electrodes (rods driven in the earth). Separate ground electrodes are not permitted in many jurisdictions as, for example, in many countries in Europe. Even if permitted, experience indicates that a separate driven ground electrode field is neither required nor desirable in most cases. A service entrance ground field sized to code presents a sufficiently low impedance for most practical audio system purposes. Connection of the isolated equipment grounding conductor directly at the mains service entrance system ground point avoids a potential hazard when separate ground electrode fields are used. If earth currents flow due to lightning or electrical fault conditions, a significant voltage difference may exist between separate ground electrode fields. Raceway and apparatus grounded by separate systems will exhibit the same potential difference.

5 SIGNAL INTERCONNECTION

For the audio system to be functional, signal interconnection methods must be selected to conform with the apparatus grounding scheme, the topology of which now has been defined by mains power safety requirements. This reverses the usual expedient procedure where grounding is adjusted to meet interconnection requirements.

The signal interconnection design objective is to transfer the audio signal at the required level of system performance, while minimizing or eliminating noise voltage induced in the load (signal contamination) due to:

1) Differential voltage of the ground reference between signal source and load
2) Magnetic induction in the signal conductors
3) Other noise coupling mechanisms.

Ideal signal interconnection conditions would be as indicated in Fig. 11. A simplified version of a real case is presented in Fig. 12. Observe that for analysis both source and load may each be considered as two-port networks, and the signal interconnection means may be considered as a four-port network. The following are typical generic schemes.

1) *Unbalanced (asymmetric) circuit*. The impedances of the two signal conductors to ground or other circuit conductors are unequal. In the typical case, one conductor of the signal conductor pair is at or near ground potential.

2) *Balanced (symmetric, differential) circuit*. The impedances of the two signal conductors to ground or to other circuit conductors are equal. Under conditions of signal transmission, at all times the signals on the two conductors of the pair are equal in amplitude, opposite in polarity.

3) *Balanced (symmetric, differential) floating circuit*. The impedances of the two signal conductors to ground or other circuit conductors are equal, approaching infinity as a limit. Under conditions of signal transmission, at all times the signals on the two conductors of the pair are equal in amplitude, opposite in polarity.

4) *Balanced floating circuit, isolated*. The definition of a balanced floating circuit applies. The impedance between the source apparatus signal conductor terminals and the load apparatus signal conductor terminals approaches infinity as a limit.

In a practical audio system interconnection design, investigate the loads (signal inputs, receivers) for the following:

1) Input impedance versus frequency
2) Common-mode rejection ratio versus frequency
3) Common-mode rejection range
4) Dc offset
5) Isolation
6) "Pin 1 problems" as described by Muncy [13].

Investigate the sources (signal outputs) for the following:

1) Output impedance versus frequency
2) Balance versus frequency
3) Dc offset
4) Isolation
5) "Pin 1 problems" [13].

The audio load and source topology determines the applicable signal interconnection method for a given path. Detailed treatment of this topic is found in Muncy [13], Macatee [14], and Whitlock [15]. Assuming typical audio signal circuit transmission voltage levels:

1) Unbalanced interconnection is effective in relatively small systems or in subsystems of larger systems where the voltage differential in the ground references of the various interconnected devices may be held to a relatively small fraction of the signal transmission voltage.

Fig. 11. Ideal signal interconnection. V_S is an ideal voltage source, the conductors are lossless, R_L is a pure resistor, and there are no electric or magnetic fields present other than those due to charge motion in circuit.

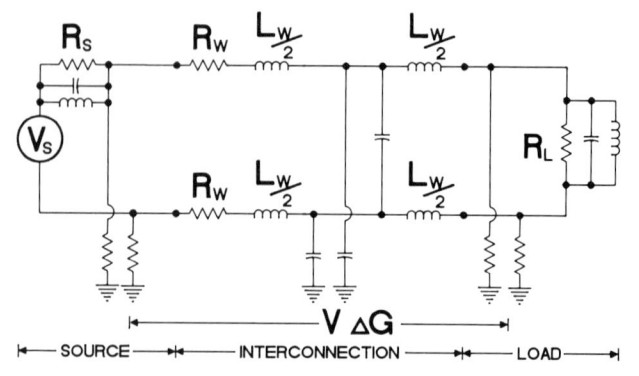

Fig. 12. Simplified real signal interconnection. Source, load, and interconnection each exhibit complex impedance. Electric or magnetic fields other than those due to charge motion in circuit may be present.

2) Balanced interconnection is effective in medium-scale systems where device ground reference potentials may vary by millivolts to tens of millivolts.

3) Balanced floating interconnection is effective in large-scale systems where device ground reference potentials may vary by millivolts to tenths of a volt.

4) Balanced, floating, isolated interconnection is effective in extreme-scale systems where device ground reference potentials may vary by volts to tens of volts.

5) The signal interconnection schemes may be intermixed in a single audio system with caution and with individual investigation of applicable source and load conditions [14], [15].

At this point it is (at last!) appropriate to consider the grounding treatment of shields on interconnection cables. Refer to Sections 4.1 and 4.2 for a general discussion of the engineering tradeoffs of multipoint (grounding the shield at both ends) versus single-point grounding (grounding the shield at one end only). For the single-point shield grounding case, refer to Whitlock [15] for a discussion of the engineering tradeoffs of establishing a single shield grounding point at the source versus at the load.

For systems with any reasonable quantity of interconnection wiring, organize the wiring mechanically to minimize potential signal contamination. In general, classify the wiring by signal type, nominal signal operating level, and other relevant parameters. In the absence of a specific investigation of electromagnetic compatibility, raceway and cable bundles should contain only wiring of a single classification. Cable bundles of different classifications should be spaced a prudent distance apart to minimize possible crosstalk. It is good practice to maximize the distance between cable bundles carrying low-level audio circuits (namely, microphone level) and mains power wiring.

6 SUGGESTED DESIGN PROCESS AND GUIDELINES—PRACTICAL AUDIO GROUNDING, SHIELDING, AND INTERCONNECTION PROCEDURES

6.1 Objectives and Conditions

Assess:
1) Performance objectives of audio system
2) Size of audio system
3) Conditions of system use
4) Potential disturbing electric and/or magnetic fields
5) Audio system mains power requirements.

Begin with the coordinated design of the mains power system and related safety grounding. Strictly comply with the applicable codes.

6.2 Independent Mains Power System (Technical Power System)

For audio systems requiring power from more than one mains branch circuit, strongly consider establishing an independent audio technical power system, inclusive of panel boards and branch circuit distribution. Do not permit this technical power system to support other loads, such as lighting, appliances, motors, or heating, ventilating, and air-conditioning (HVAC) equipment. Establish an exclusive low-impedance feeder from the service entrance equipment to the technical power panel boards.

6.2.1 Review the long-term voltage stability of the power mains with reference to the manufacturer's recommendations for the audio apparatus. Consider the application of voltage regulation equipment at either the panel board or the branch circuit level. If adopted, strongly consider the use of equipment with a low source impedance and a sinusoidal output waveform independent of load effects.

6.2.2 Review the mains for possible transients and conducted noise. Be aware of other loads to the mains which may induce transients or noise. Consider the application of transient overvoltage suppression equipment at either the panel board or the branch circuit level. If adopted, strongly consider the use of equipment with a low source impedance and a sinusoidal correction output waveform.

6.2.3 Depending on the audio system scale and performance objectives, select either the bonded raceway grounding scheme or the isolated equipment grounding conductor scheme for equipment grounding of the technical power system. All audio apparatus and related interconnections will be grounded exclusively by means of this equipment grounding path.

6.2.4 Stranded copper conductors are strongly preferred, regardless of the cross-sectional area.

6.2.5 For the fixed-installation case, a steel raceway (trunking) system is strongly preferred to minimize possible magnetic emission from the mains. Bond the raceway system to ground according to the most restrictive of either the applicable code or these recommendations. Where flexible metal conduit is used, provide a copper bonding jumper to ensure a low-impedance path.

6.2.6 Where steel raceway cannot be used, for the portable case or the case of the direct burial of raceway in corrosive soils, exercise care in the placement of power conductors relative to audio conductors.

6.2.7 Use mains power receptacles that will maintain contact pressure on apparatus attachment plugs over time. Strongly consider the use of "hospital-grade" wiring devices.

6.3 Isolated Ground

For the case of the isolated equipment grounding conductor scheme, the following is suggested.

6.3.1 The isolated equipment ground path is established by an insulated copper conductor isolated from the raceway, with a complete copper path to the project ground electrodes. No connection is permitted between the raceway (bonded) ground, the grounded building parts, and the isolated equipment ground system, except at the project ground electrode connection.

The following test of isolation of this ground path must be performed only by qualified personnel observing mains voltage safety precautions. Disconnect all apparatus from the mains power receptacles. Deenergize the mains power system and confirm the absence of any mains voltage. At the project ground electrode, cautiously disconnect the isolated equipment grounding conductor. Measure for possible voltage between the isolated equipment grounding conductor and the project ground electrode. In the absence of voltage, measure the impedance between the same points. The isolated equipment grounding conductor must measure "open" to project ground. If other than an extremely high impedance is observed, seek out and remove all inadvertent connections between the isolated equipment grounding system and other paths to ground.

If the project ground electrode connection point will be inaccessible after construction is complete, connect the isolated equipment grounding conductor to project ground at the project service entrance ground connection point.

6.3.2 At technical power panel boards, provide an isolated equipment ground bus isolated (insulated) from the panel board. Size this bus to allow the connection of conductors equal to at least 150% of the panel board maximum permitted branch circuit count. Size the panel board isolated equipment ground feeder conductor to the larger of either the code (for fault current) or as required to realize 0.10–0.15 Ω to the project ground field connection. Refer to Atkinson and Giddings [16] for detailed suggestions on specific cases.

6.3.3 For each branch circuit, provide conductors of the same cross-sectional area (gauge) for hot, neutral, and isolated equipment grounding (insulated green or green/yellow conductor). It is strongly preferred that no sharing of neutral or isolated equipment grounding conductors with any other branch circuit be permitted.

In the case of polyphase mains power with three branch circuits, one each on phases A, B, and C returning to the technical power panel board in the same raceway, provide nine conductors. With respect to nomal three-phase mains circuit operation, the additional conductors are considered to be neutral and ground impedance correction conductors operating at audio and higher frequencies. For the purpose of raceway fill calculations, count all conductors when establishing the summation of the cross-sectional areas. For the purpose of ampacity deration calculations, the two additional neutrals and the three isolated equipment ground conductors are generally not classified as current-carrying conductors.

6.3.4 Provide isolated ground type mains power receptacles with the isolation method being an integral part of the device. These are identified on the face with an isolated ground symbol (green triangle in North America), a unique color, or both. Where isolated ground type mains power receptacles are available in a range of colors, strongly consider the use of unique colored receptacles to more clearly define the technical power as distinct from normal appliance power.

6.3.5 At a central location to each ensemble of audio apparatus (equipment cluster, equipment rack, or group of equipment racks) provide an isolated equipment grounding conductor run with the associated circuit conductors in code size raceway. Size this conductor to the largest of:

6.3.5.1 The largest phase conductor of branch circuits feeding the ensemble.

6.3.5.2 As required by applicable code for bonding of a conductive equipment enclosure.

6.3.5.3 As required to realize 0.10 Ω to the isolated equipment ground terminal bar at the associated technical power panel board. See also [16].

6.4 Apparatus and Enclosures

Ground and bond audio apparatus, apparatus racks, and similar apparatus enclosures containing powered apparatus exclusively via the selected equipment grounding system conductors.

6.4.1 Where racks are fixed in place, isolate the rack mounting, anchorage, and raceway connections. Be aware that typical concrete may be a medium- to high-impedance conductor, depending on the properties of the local sand and aggregate.

6.4.2 At each equipment cluster or rack, provide a copper equipment grounding terminal bar. At conductive metal equipment racks, bond this terminal bar to the rack frame with a stranded copper conductor bonding jumper. For the case of the isolated equipment grounding conductor scheme, consider the use of a removable bonding jumper to simplify the ground isolation testing procedure.

6.4.2.1 For portable racks or equipment clusters fed by flexible mains cords, connect the terminal bar to the grounding contact of a three-wire mains attachment plug using a suitable insulated service cord. Connect this attachment plug to a technical power receptacle.

6.4.2.2 For fixed racks using the bonded raceway scheme, have a qualified person connect the terminal bar to the grounding means of a mains receptacle feeding the rack.

6.4.2.3 For a single equipment cluster or rack using the isolated equipment grounding conductor scheme, connect the terminal bar to the isolated equipment grounding conductor. Refer to Fig. 10.

6.4.2.4 For an ensemble of equipment racks using the isolated equipment grounding conductor scheme, provide a master terminal bar to land conductors from the individual rack equipment grounding terminal bars. Connect the master terminal bar to the isolated equipment grounding conductor. Refer to [16].

6.4.2.5 Where a facility grounding mesh has been selected and provided, connect the terminal bar to the mesh, preferably at or near an intersection.

6.4.3 Ground individual items of apparatus via the insulated equipment grounding conductor (green or green/yellow) of the apparatus three-wire mains power cord. If the apparatus uses a two-wire mains power cord, or an external low-voltage ac transformer or external dc power supply where an equipment grounding conductor is not carried through the low-voltage supply cord to the apparatus, provide a separate bond wire (green or green/yellow) to the local equipment grounding terminal bar. At the apparatus, provide a lug and suitable hardware for bonding. Size the separate bond wire to the largest of either of the following:

6.4.3.1 The largest phase conductor of the mains power cord feeding the apparatus.

6.4.3.2 As required by the applicable code for bonding of a conductive equipment enclosure.

6.4.3.3 As required to realize 0.05 Ω to the local equipment grounding terminal bar. See also [16].

6.4.4 Do not permit apparatus or enclosures to touch each other unless they are mechanically connected and electrically bonded. Be aware that rack panel mounting screws may not establish a low-impedance bond, especially if the apparatus panels are fabricated of hard anodized aluminum.

6.4.5 Observe that bonding both the apparatus and a metallic equipment rack (enclosure) is required by the applicable code but inherently creates a ground loop. This does not appear to be a problem in practice. The rack encloses the apparatus. If the rack is steel, the rack functions as a partial magnetic shield, minimizing the flux threading the possible loop.

6.5 Signal Raceway System

For fixed installations, some means is generally desirable to organize, support, and protect signal conductors running to locations distant from the main concentration of the audio equipment.

6.5.1 Consider the use of a fully enclosed metallic raceway system. This provides robust mechanical protection and serves as an overall electric field shield. Steel raceway may provide some amount of magnetic field shielding, but this is not very effective as the permeability of typical commercial steel is low at low field strengths.

6.5.2 For metallic raceway, bond the entire raceway system to project ground per the most restrictive of either the code or these provisions. Provide a bond jumper conductor at all flexible metal conduit, regardless of length.

6.5.3 Where metallic raceway is not feasible, and plastic or other nonmagnetic raceway is a requirement, increase the cross-sectional area of the raceway to permit the installation of audio cable types having low susceptibility to magnetic induction, such as star-quad constructions.

6.5.4 In the absence of a specific investigation of compatibility, space raceways for the audio signal system at least 75 mm from raceways for all other systems if both are steel, 250 mm if either or both are not steel.

6.6 Interconnection and Shield Treatment

6.6.1 Review the load and source parameters of the apparatus. Review the apparatus for "pin 1 problems" [13]. If a casual inspection is inconclusive, conduct the "hummer test" described by Windt [17], the more elaborate test described by Perkins [18], or paragraph B.2.2 of CENELEC test configuration 2 [5]. With this information established, select the appropriate interconnection schemes.

6.6.2 Review the conditions of audio system use. Select the multipoint or single-point ground scheme for shields and apply it rigorously throughout the system. Clearly document any expedient exceptions taken.

6.6.3 With the shield grounding scheme chosen, review candidate cable types with respect to shield current induced noise [13]. Review ambient conditions and cable routing for potential magnetic induction. Select appropriate cable types. All signal shields should be insulated from conductive raceway and enclosures.

6.6.4 If the single-point ground scheme is selected for shields, review the engineering tradeoffs of grounding at the load versus at the source, and make an informed choice. For audio systems with moderate performance objectives, such as permanently installed entertainment sound reinforcement systems in public buildings, the author has generally selected the compromise of grounding shields to the load apparatus. This optimizes the path for immunity from RFI rather than common-mode rejection ratio at the load apparatus. Whitlock [15] clearly indicates that this choice is not optimal for cases with high audio performance objectives.

6.6.5 If the preceding process is followed, the result is that all shielded cables will be grounded exclusively to the defined equipment grounding system via the audio apparatus mains cord or via connection to the local equipment grounding terminal bar. Refer to Figs. 13–15 for examples.

6.6.6 For microphone circuits with phantom (simplex) powering imposed, make the run electrically continuous from the microphone to the phantom powering source (the microphone preamplifier or a separate phantom power apparatus). Isolate such circuits from other grounding paths.

6.6.7 Where jack fields and similar cross-connection equipment are implemented, treat jacks in the same general manner as selected for the apparatus. For the rigorous case, for each jack frame to be grounded, provide individual ground conductors to the related rack equipment grounding terminal bar. Size this conductor for approximately $0.05-0.10$ Ω. For the typical case, bus each row of jack frames and provide a common ground conductor to the related rack equipment grounding terminal bar. Size the conductor for approximately 0.05 Ω. Refer to Fig. 16 for suggested wiring of analog audio jack fields for fixed installation and moderate audio performance objectives.

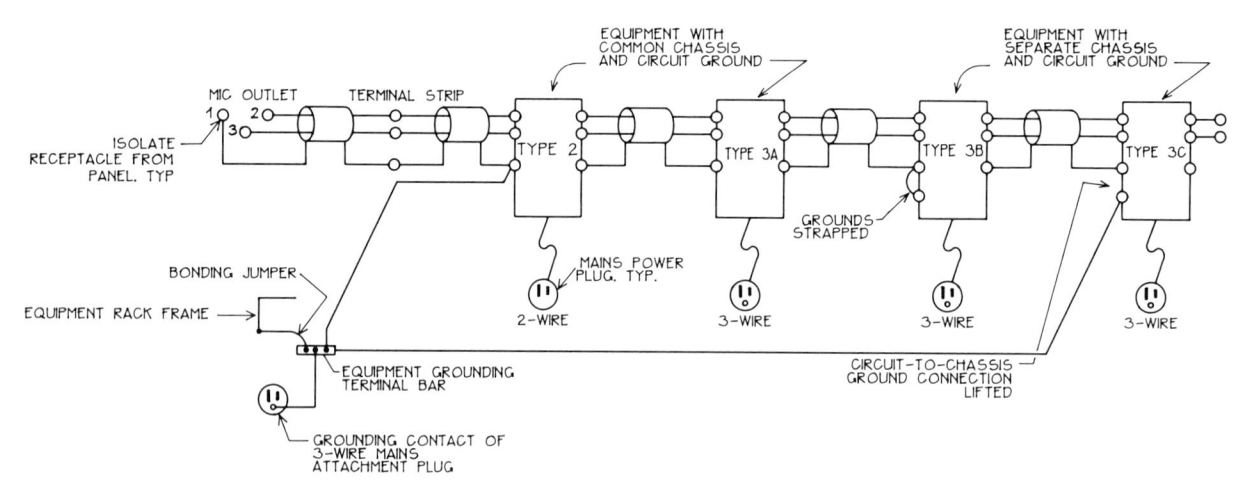

Fig. 13. Grounding of audio apparatus and cable shields, multipoint grounding scheme.

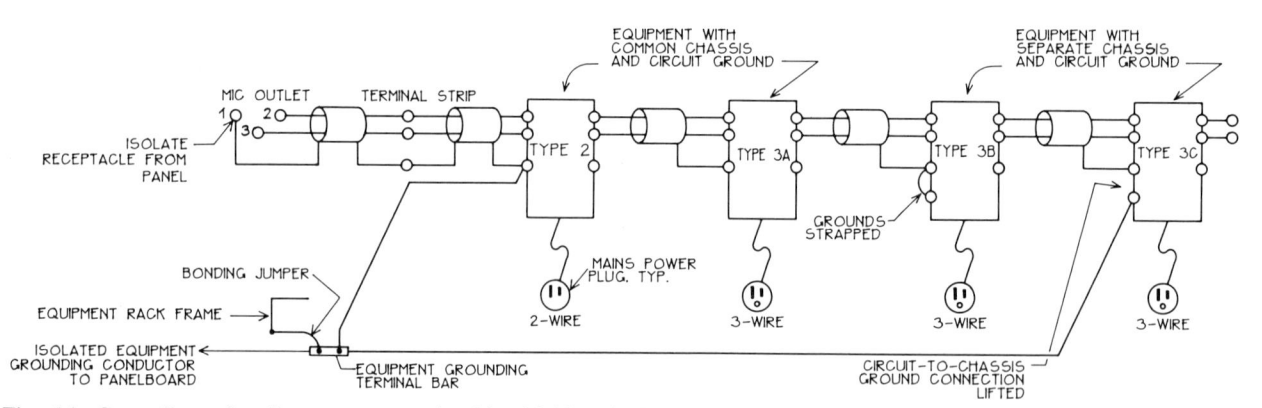

Fig. 14. Grounding of audio apparatus and cable shields, single-point grounding scheme, for moderate audio performance objectives.

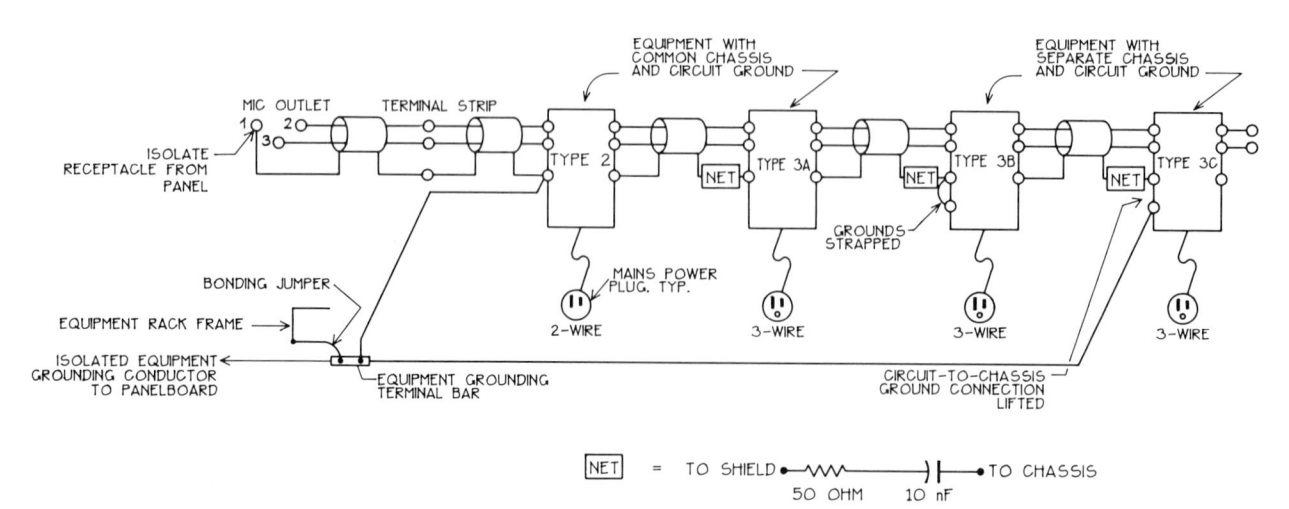

Fig. 15. Grounding of audio apparatus and cable shields, single-point grounding scheme, for high audio performance objectives.

6.6.8 For most applications, signal ground provisions should realize less than 0.15 Ω to the primary ground connection (the project ground field connection). Refer to [16] for detailed criteria.

7 FURTHER PRACTICAL MATTERS

7.1 *Bond to ground or isolate from ground.* Casual or intermittent grounds make troubleshooting a nightmare.

7.2 Test all receptacles providing audio system mains power for correct polarity of connection and ground continuity before connecting and energizing the audio system apparatus. Simple receptacle testers are available at wholesale electric-equipment suppliers. A surprising number of receptacles are found to be miswired in the field. Possible corrections to mains wiring should be performed only by a qualified person authorized to perform such work in the jurisdiction.

7.3 Portable signal cables for general use, such as typical cables using male and female circular audio connectors, should be assembled with electrical continuity for all active pins of the connectors. If a "pin 1 lifter" is required, a flagrantly labeled adapter device is strongly preferred.

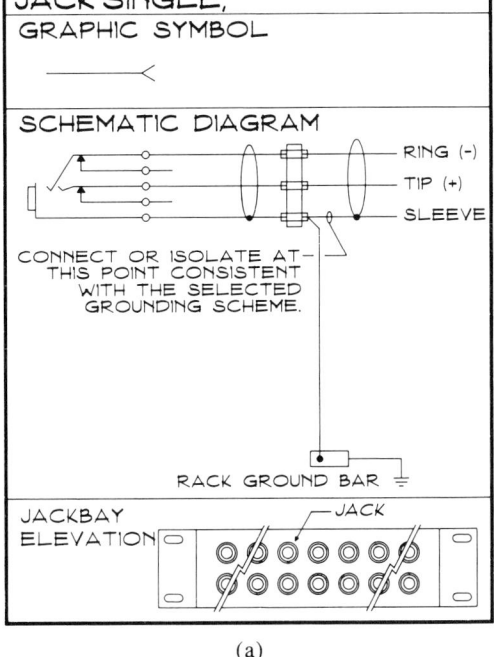

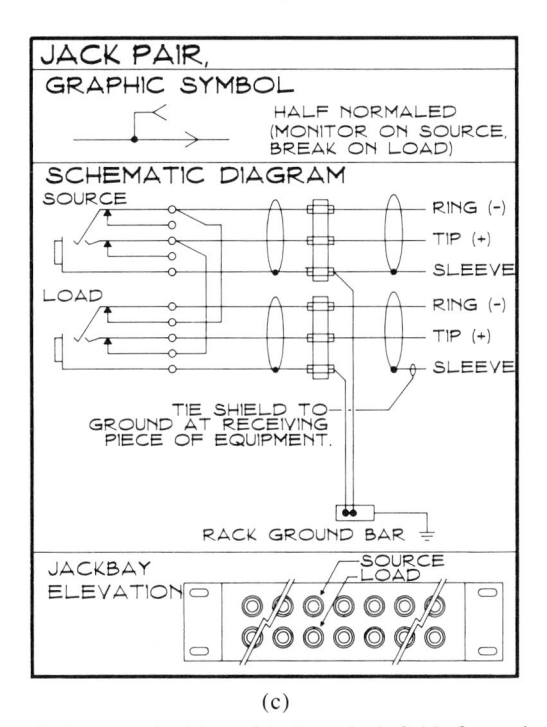

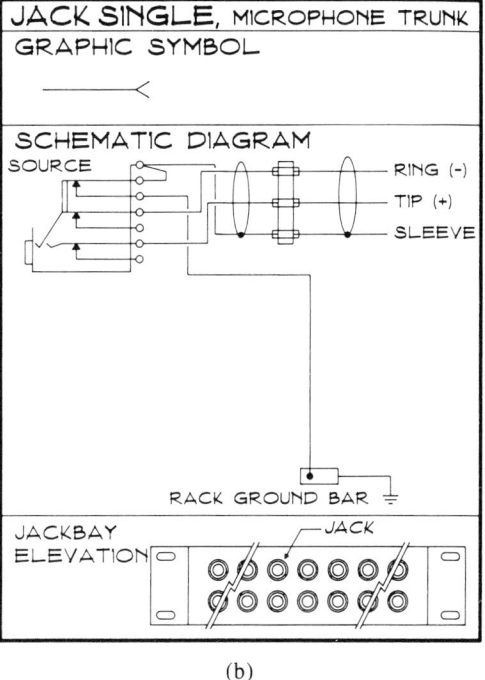

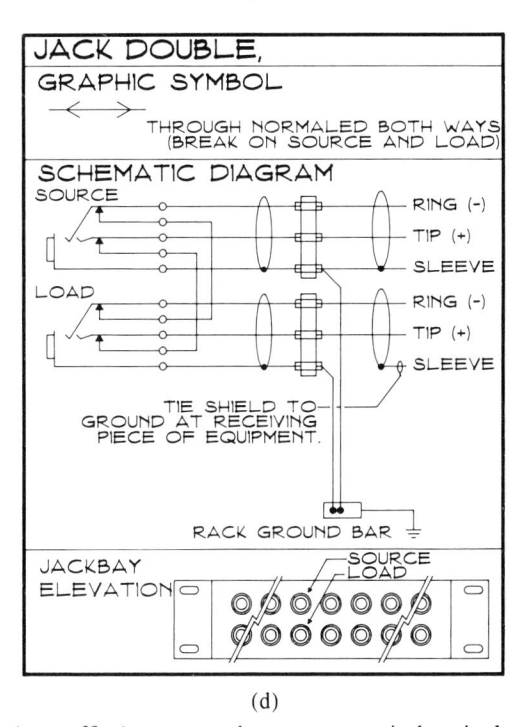

Fig. 16. Suggested wiring of jacks at jack fields for typical telephone (post office) pattern or bantam pattern jacks, single-point grounding scheme, for moderate audio performance objectives.

7.4 Isolate audio connector shells from conductive receptacle plates fastened to raceway system outlet boxes.

7.5 For portable audio cables, grounded connector shells of conductive material are strongly preferred to maintain the continuity of electric-field shielding.

Some users prefer to isolate the connector shell or float the shell with respect to ground. This is done with the intent of preventing possible intermittent ground loops due to the contact of connector shells with those of another cable or other grounded surfaces, such as outlet box covers, conduits, and the like. The tradeoff is a reduction in RF immunity. Further, there will be hum induction when anyone touches the shell.

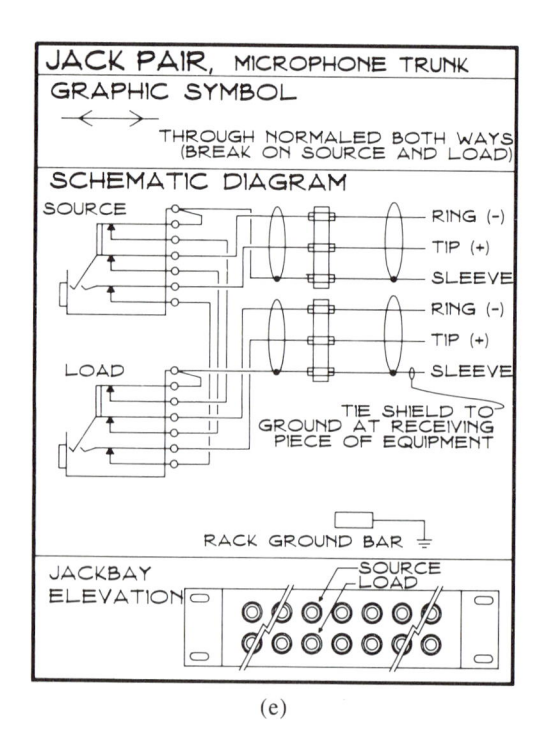

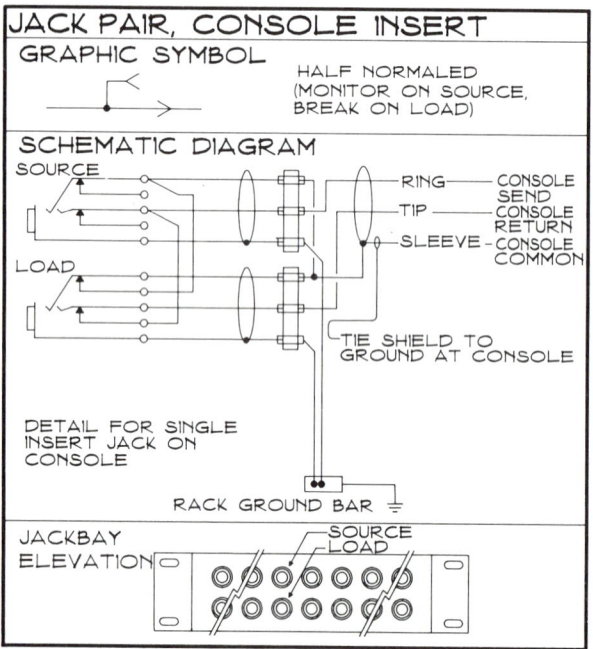

7.6 It is a paradox, but it is actually easier to establish a rational and safe ground scheme for a portable touring sound system than for a permanent installation. In the touring situation the users will generally carry their own power distribution system. With three-wire grounded portable cords and a portable distribution panel board, system ground paths are entirely predictable. It is only necessary to find an earth connection in the form of a driven rod, a cold-water pipe, building steel, or other means acceptable to the authorities.

In a pinch, a safe and functional portable power distribution system may be rented. In most jurisdictions construction power distribution systems approved for use in wet excavations will meet all the foregoing recommen-

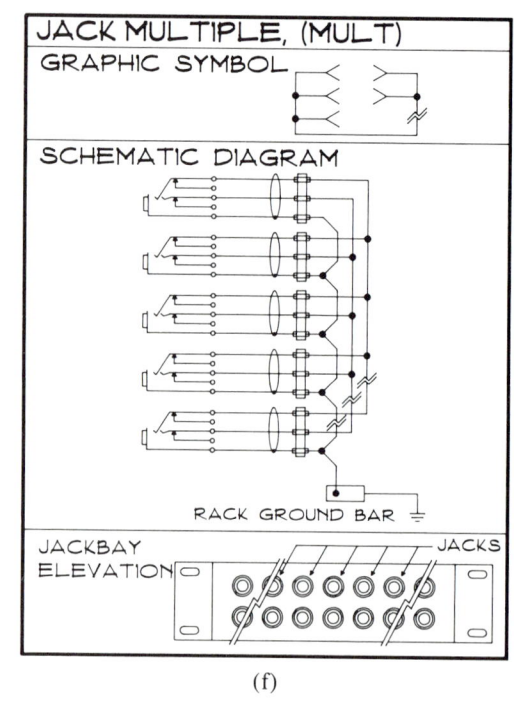

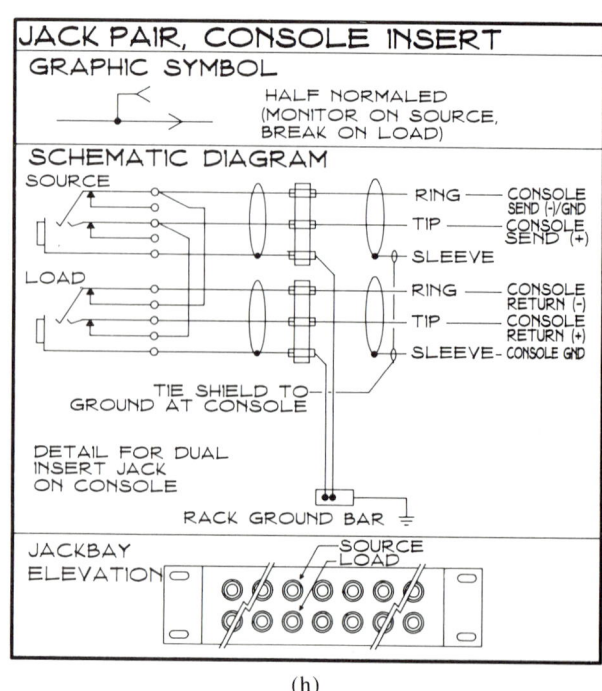

(e)

(f)

(g)

(h)

Fig. 16. continued

dations and are generally equipped with ground fault current interrupters (GFCIs). These are modified circuit breakers that will sense and trip out on a ground fault current condition, a comforting safety feature when you happen to be playing an outdoor gig in the rain.

7.7 Induced interference from production and architectural lighting control systems has plagued audio systems since the introduction of the phase-angle-firing silicon-controlled-rectifier dimmer. If a dimming system is of competent manufacture and is installed according to the applicable portion of the electrical code, and interference is observed in the audio system, *the problem lies in excessive susceptibility of the audio system*. The techniques described herein have proven totally effective in eliminating dimmer interference in live entertainment sound reinforcement systems operating in the immediate proximity to dimmer banks controlling theater lighting loads in the megawatt region.

7.8 If RFI is observed in an audio system, first investigate the system ground scheme, unless a given item of apparatus is obviously at fault.

The author's consulting practices employ a simple practical test for possible system susceptibility to RFI. A handheld 2-W business band transceiver is introduced inside an open equipment rack, the carrier is keyed, and modulation is applied. The audio system should exhibit little, if any, change in signal performance. Audio systems shielded and grounded according to the preceding discussion routinely pass this test. Clearly, the test is crude and lacks repeatability, but it provides a rapid real-world relative indication of system immunity.

7.9 When attempting corrections on an existing "scramble-grounded" system, it is often faster to select an appropriate coordinated ground scheme and to proceed methodically with implementation than to attempt corrective efforts on a "hunt and repair" basis. Especially when converting a system from scramble grounded to single-point grounded, have faith in fundamental physics and press on. The system noise floor may rise at intermediate points in the process before it falls sharply.

8 CONCLUSION

Fundamental principles have been reviewed. From these principles, schemes for achieving safe, technically effective audio grounding systems have been derived and described.

The proposed scheme for combining mains power and technical system grounds has been demonstrated to be a valid scheme for achieving a safe, functional, reliable audio ground scheme. Admittedly, it may be contrary to much present practice. It is not a panacea. Readers may discover situations where other approaches will better suit their needs.

Some practical experience in application should be mentioned. The author has been using the coordinated isolated equipment grounding scheme in work specified in his consulting practices since 1972. It was developed at that time to deal with the problems presented by five complex interconnected control rooms for a university research laboratory. He initially published his approach in 1978 [19]. Since 1972 the author has applied the scheme in projects for recording control rooms, discotheques, high-level entertainment sound systems, live theater sound effects systems, broadcasting facilities, large-scale video duplication plants, and large-scale video production and postproduction facilities, inclusive of several hundred completed systems. Mains power loads of the connected technical systems have ranged from less than 1 kW to in excess of 500 kW. Virtually all the related mains power systems have been polyphase.

The author makes no claim of precedence or originality of the schemes presented. Numerous audio system designers and system integrators have independently arrived at similar or identical schemes.

9 ACKNOWLEDGMENT

The author wishes to thank John W. S. Brooks, Peter Butt, Bill Isenberg, (the late) Deane Jensen, John Lanphere, Neil Muncy, Dick Rosmini, and John Windt for their helpful comments and suggestions during many discussions of the topic. Further, the author wishes to thank Gail S. Kai for preparation of the manuscript, and Christie T. Davis for preparation of the illustrations and proofreading of the manuscript. The author thanks the reviewer for helpful comments, which have benefited both the author and the reader. Any errors remain the sole responsibility of the author.

10 REFERENCES

[1] A. Davis, "Grounding, Shielding and Isolation," *J. Audio Eng. Soc.*, vol. 1, pp. 103–104 (1953 Jan.).

[2] A. Mornington-West, "Report by Allen Mornington-West, vice-chair of SC-10 and member of the SC-05-05 Working Group on grounding and EMC practice, on European standardization of electromagnetic compatibility testing," *J. Audio Eng. Soc. (AES Standards Committee News)*, vol. 42, pp. 703–704 (1994 Sept.).

[3] A. Mornington-West, "EMC and the Professional Audio Video and Lighting Industries," unpublished manuscript (1994 Nov.).

[4] Comité Européen de Normalisation Electrotechnique (CENELEC), "Electromagnetic Compatibility— Limits and Methods of Measurement of Radio Disturbance for Audio, Video, Audio-Visual and Entertainment Lighting Control Apparatus for Professional Use," Draft Standard prEN 55103-1:1994 (English transl.), pp. 1–16, unpublished (1994 Oct.).

[5] Comité Européen de Normalisation Electrotechnique (CENELEC), "Electromagnetic Compatibility— Immunity Requirements for Audio, Video, Audio-Visual and Entertainment Lighting Control Apparatus for Professional Use," Draft Standard prEN 55103-2:1994 (English transl.), pp. 1–29, unpublished (1994 Oct.).

[6] A. Mornington-West and K. Dibble, "Aspects of

EMC Performance of High Power Amplifiers," unpublished (1994 Nov.).

[7] H. W. Ott, *Noise Reduction Techniques in Electronic Systems*, 2nd ed. (Wiley, New York, 1988).

[8] R. Morrison, *Grounding and Shielding Techniques in Instrumentation*, 3rd ed. (Wiley, New York, 1986).

[9] R. Morrison and W. H. Lewis, *Grounding and Shielding in Facilities* (Wiley, New York, 1990).

[10] D. Halliday and R. Resnick, *Physics* pt. II (Wiley, New York, 1968).

[11] R. Scott and M. Essigman, *Linear Circuits*, pt. 1 (Addison-Wesley, Reading, MA, 1960), p. 19.

[12] NFPA 70, *National Electrical Code*, National Fire Protection Association (1993 ed.).

[13] N. A. Muncy, "Noise Susceptibility in Analog and Digital Signal Processing Systems," *J. Audio Eng. Soc.*, vol. 43, pp. 435–453 (1995 June).

[14] S. R. Macatee, "Considerations in Grounding and Shielding Audio Devices," *J. Audio Eng. Soc. (Engineering Reports)*, vol. 43, pp. 472–483 (1995 June).

[15] B. Whitlock, "Balanced Lines in Audio Systems: Fact, Fiction, and Transformers," *J. Audio Eng. Soc.*, vol. 43, pp. 454–464 (1995 June).

[16] C. Atkinson and P. Giddings, "Grounding Systems and Their Implementation," *J. Audio Eng. Soc. (Engineering Reports)*, vol. 43, pp. 465–471 (1995 June).

[17] J. Windt, "An Easily Implemented Procedure for Identifying Potential Electromagnetic Compatibility Problems in New Equipment and Existing Systems: The Hummer Test," *J. Audio Eng. Soc. (Engineering Reports)*, vol. 43, pp. 484–487 (1995 June).

[18] C. Perkins, "Automated Test and Measurement of Common Impedance Coupling in Audio System Shield Terminations," *J. Audio Eng. Soc. (Engineering Reports)*, vol. 43, pp. 488–497 (1995 June).

[19] K. Fause, "Audio Shielding, Grounding and Safety," *Record. Eng./Producer*, vol. 9, p. 54 (1978 June); *ibid (Letters)*, vol. 9, pp. 16 ff. (1978 Aug.).

THE AUTHOR

Kenneth R. Fause received the B.S. degree in engineering from Cornell University, Ithaca, NY, in 1971. After graduation he joined the Cornell staff as a research technician for the Social Psychology Laboratory, where he developed data acquisition and analysis systems for human communication behavior experiments. In 1970 he had served as a consultant with Bolt, Beranek and Newman, New York. He moved to Los Angeles, CA, in late 1973 to pursue graduate studies in theater, architecture, and engineering acoustics at UCLA, where he received an M.A. degree in threater arts in 1978. From 1975 to 1977 he was a chief engineer for Audio Concepts, Inc., Hollywood, CA. From July through December 1980 he served as president of Filmways Audio Services, Inc.

Mr. Fause served as principal consultant and owner of Fause & Associates prior to merging the firm with Howard G. Smith Acoustics in 1981 to form Smith, Fause & Associates, Inc., Los Angeles. In 1986, he joined with Howard Smith and Peter McDonald in forming Smith, Fause & McDonald, Inc., San Francisco. He is currently a principal consultant in presentation technology, as well as cofounder and joint owner, of both companies.

Mr. Fause is a member of the Audio Engineering Society and has served as committee member and chairman of the Los Angeles Section, facilities chairman of past conventions, alternate delegate to the Broadcast Television Stereo Committee, and presenter at numerous local section meetings and convention workshops.

B.
loudspeaker performance and the loudspeaker–room interface

Acoustical Measurements by Time Delay Spectrometry*

RICHARD C. HEYSER

California Institute of Technology, Jet Propulsion Laboratory, Pasadena, California

A new acoustical measurement technique has been developed that provides a solution for the conflicting requirements of anechoic spectral measurements in the presence of a reverberant environment. This technique, called time delay spectrometry, recognizes that a system-forcing function linearly relating frequency with time provides spatial discrimination of signals of variable path length when perceived by a frequency-tracking spectrum analyzer.

INTRODUCTION It is a credit to technical perseverance that the electronic subsystems which make up an audio installation have been brought to a high state of perfection. It is possible not only to predict the theoretical performance of a perfect electronic subsystem, but to measure the deviation of the performance of an existing subsystem from that perfect goal. The capability of measuring performance against an ideal model and of predicting the outcome for arbitrary signals is taken for granted. Yet the very acoustical signals which are both the source and product of our labors seldom have sufficient analysis to predict performance with comparable analytical validity. The measurement of even the simpler parameters in an actual acoustical system may be laborious at best. There exists, in fact, very little instrumentation for fundamental measurements which will allow prediction of performance under random stimuli. It is the intent of this paper to describe a new acoustical measurement technique which allows "on-location" measurement of many acoustical properties that normally require the use of anechoic facilities. A new acoustical model of a room is also introduced as a natural by-product of this technique. With this model substantial objects may be selectively analyzed for their effect on sound in the room.

The acoustic testing process to be described relies heavily on electronic circuit techniques which may not be familiar to acousticians. As a brief review of fundamental principles it will be recalled that in linear electronic circuit analysis the concept of superposition permits complete analytical description of circuit response under the influence of any driving function which is describable as a distribution of sinusoids. Each sinusoid in the distribution will possess a unique amplitude and time rate of change of angle or frequency, and will produce a network response which is in no way dependent on the existence of any other sinusoid. The response of a network at any particular frequency is therefore obtained quite simply by feeding in a sinusoid of the desired frequency and comparing the phase and amplitude of the output of a network with its input. The response of a network to all frequencies in a distribution will then be the linear superposition of the network response to each frequency. The response of a network to all possible sinusoids of constant amplitude and phase is called the frequency response of that network. The frequency response is in effect the spectrum of frequency distribution of network response to a normalized input. Cascading of linear networks will involve complex multiplication of the frequency response of each included network to obtain an overall response. An analytical solution to such a combination may then be readily obtained from this overall frequency response.

The equivalent frequency response of an acoustical system should, in principle, be the comparable value in analyzing resultant performance. The processes of sound generation, reflection, transmission, and absorption all have their counterparts in network theory. It is well known, however, that any attempt at utilizing a simple sinusoid driving function on a real-world acoustical system will lead to more confusion than insight. Any object with dimensions comparable to a spatial wavelength of the sinusoid signal will react to the signal and become an undesired partner in the experiment. Furthermore, since the velocity of sound in the various media prevents instantaneous communication, time enters into the measurement in the form of standing wave patterns which will be different for each applied frequency. For those acoustical subsystems for which a single frequency response might be meaningful, such as loudspeakers, microphones, or certain acoustical surfaces, special (and expensive) anechoic test areas are utilized in an attempt

---

* Presented October 16, 1967 at the 33rd Convention of the Audio Engineering Society, New York.

to remove acoustically any object that might interfere with the measurement. Unfortunately there are many acoustical situations for which such a measurement appears to be impossible or even meaningless. A single frequency response of an auditorium, for example, will be of no use in evaluating the "sound" of the auditorium as perceived by an observer. The large number of reflecting surfaces give a time-of-arrival pattern to any attempted steady-state measurement which prevents analytical prediction of response to time-varying sound sources. Clearly an auditorium has not a single frequency response, or spectral signature, but rather a linear superposition of a large number of spectral responses, each possessing a different time of arrival.

Here we have the essence of many real-world acoustical measurement problems. It is not a single frequency response that must be measured but a multiplicity of responses each of which possesses a different time delay. The room in which an acoustical measurement is made simply adds its own series of spectral responses, which may mask the desired measurement. Selection of the proper responses will yield a set completely defining the acoustical system under test so long as superposition is valid. Conversely, once the entire set of spectra are known along with the time delay for each spectrum it should be analytically possible to characterize the acoustical system for any applied stimulus. Traditional steady-state techniques of measurement are unable to separate the spectra present in a normal environment since a signal response due to a reflection off a surface differs in character from a more direct response only in the time of arrival following a deliberately injected transient. In the discussion to follow a method of time-delay spectrometry will be developed which allows separation and measurement of any particular spectral response of an acoustic system possessing a multiplicity of time-dependent spectral responses. A practical implementation of this technique will be outlined using presently available instruments. Analytical verification of the technique will be developed and a discussion included on acoustical measurements now made possible by this technique.

EVOLUTION OF MEASUREMENT TECHNIQUE

In evolving the concept of time-delay spectrometry it will be instructive first to consider a very simple measurement and then to progress by intuitive reasoning to the general case. Assume that it is desired to obtain the free-field response of a loudspeaker situated in a known reverberant environment. A calibrated microphone will be placed at a convenient distance from the speaker in the direction of the desired response. With the acoustical environment initially quiescent, let the speaker suddenly be energized by a sinusoid signal. As the speaker activates the air a pressure wavefront will propagate outward at a constant velocity. This pressure wavefront will of course not only travel toward the microphone but also in all other directions with more or less energy. Assume that the microphone is connected through a relatively narrow-bandwidth filter to an indicating device, and that this filter furthermore is tuned to the exact frequency sent to the loudspeaker. As the leading edge of the pressure wave passes the microphone, the only contributor to this wave could be the loudspeaker since all other paths from reflective surfaces to the microphone are longer than the direct path. The loudspeaker will take some time to build up to its "steady-state" excitation value and the microphone and tuned circuit will similarly have a time constant. If the system arrives at a steady-state value before the first reflected sound arrives at the microphone, this steady-state is in effect a free-field measurement at the frequency of the impressed sinewave. Note that because of the broad spectrum of a suddenly applied sinewave, a tuned filter circuit is necessary to prevent shock-excited speaker or microphone resonances from confusing the desired signal.

The measurement thus described is quite simple and has actually been used by some investigators.[1-3] As long as a fixed filter is used, the measurement must be terminated prior to receipt of the first reflected "false" signal and the system must be de-energized prior to a subsequent measurement. Suppose, however, that the fixed-frequency sinewave is applied to the speaker only long enough to give a steady-state reading prior to the first false signal, then suddenly shifted to a new frequency outside the filter bandwidth. Assume also that by appropriate switching logic, a filter tuned to this new frequency is inserted after the microphone at the precise time that the sound wave *perceived by the microphone* changes frequency. The microphone circuit will thus be tuned to this new frequency and the later reflected false signals of the first frequency will not be able to pass through the new filter. If one continues this process through the desired spectrum it is apparent that the indicator circuit will never "know" that the measurement was performed in a reverberant environment and a legitimate frequency response may thus be measured.

The practical economics of inserting fixed filters and waiting for the starting transient at each frequency to die down weigh heavily against such a system, so an alternative may be considered. Project a smooth glide tone to the speaker and utilize a continuous tracking filter after the microphone. If the tracking filter is tuned to the frequency of the emitted glide tone as perceived by the microphone and if the glide tone has moved in frequency by at least the bandwidth of the tracking filter before the first reflected signal is perceived, no buildup transient is encountered and the measurement will be anechoic even though performed in a reverberant environment. The nature of the glide tone may readily be ascertained by intuitive reasoning. If there is no relative motion between speaker and microphone it can be stated with absolute certainty that the time delay between speaker and microphone is a constant. There is, in other words, a unique and linear relationship between time and distance traveled by the pressure wave. Each reflecting surface will appear to be a new sound source with a time delay corresponding to path length. If we specify that the sweeping tone and the tracking filter combination be capable of maximizing response for all frequencies from any given apparent source then we have required an equivocation of frequency, room spacing, and time. The glide tone satisfying this requirement possesses a constant slope of frequency versus time. If all reverberant energy due to any given frequency has died to an acceptable level a fixed time following excitation, say T seconds, then the glide tone may be allowed to repeat its linear sweep in a sawtooth fashion with a period of no less than T seconds.

RICHARD C. HEYSER

While we began by postulating direct loudspeaker measurement it is apparent that we could by suitable choice of sweep rate, bandwidth, and time delay, "tune" in on at least first-generation reflections with the selective exclusion of others, even the direct loudspeaker response. Because the output of the tracking filter yields the spectral signature of the perceived signal with a frequency proportional to time and since selective spatial isolation of the desired signal is obtained by utilizing the fixed time delay between source and microphone, the rationale of the name *time-delay spectrometry* becomes apparent.

Some simple relationships may be directly derived from the basic geometry of a practical situation. Consider the representation of Fig. 1 in which a microphone is connected to a tracking filter "tuned" to perceive a source at a distance X on a direct path. The filter has a bandwidth B Hz; it is seen that the sweep tone will traverse some ΔX in space while within a band B of any given frequency. Define ΔX as the region in space, along the direction of propagation of the acoustic signal, within which the selected signal power will be no less than half the maximum selected value. This is the spatial analog of the half-power bandwidth B of the tracking filter, and will therefore be referred to as the space-equivalent bandwidth. This space-equivalent bandwidth ΔX is related to the tracking filter electrical bandwidth B, the velocity of sound c and the rate of change of frequency $\Delta F/\Delta t$, by

$$\Delta X = B[c/(\Delta F/\Delta t)] \approx c/B. \quad (1)$$

The last relation is based upon an optimized bandwidth which is the square root of the sweep rate. While not immediately obvious, this optimized bandwidth is common in sweeping analyzers of the variety recommended for this measurement.[4]

The signal perceived by the microphone is that emitted by the speaker some time in the past. To visualize the relationship consider Fig. 2 diagramming the behavior of a sweep tone repetitive in a time T. The signal emitted from the source or transmitter will be denoted by F_t while that received by the microphone is F_r. It is usually desired to sweep through zero frequency; this is shown in the diagram as a signal dropping in frequency uniformly with time until zero frequency and then rising again. We are thus going through zero beat. In this diagram the direct signal possessing the shortest time delay is shown dashed. It is clear that the relationship between distance, transmitted and received frequencies at any instant, and sweep rate is

$$X = (F_r - F_t)[c/(\Delta F/\Delta t)]. \quad (2)$$

This states the fact that to "tune" to the response of a signal X feet away from the microphone it is only necessary to offset the frequencies of glide tone and tracking filter by a fixed difference. Referring again to Fig. 1 we could check the response of the reflective surface A in one of two ways. First, we might offset the source and received frequencies to yield the primary sound signal as shown and then, without changing this frequency offset, physically transport the microphone to position A. It is assumed that a position may be found for which no contour of undesired reflected or direct sound comes into space tune at position A. The second way of adjusting for surface A would be to maintain the existing microphone position and change the offset between transmitted and received frequencies to account for the longer path of A. Both techniques allow for probing the effective acoustic surfaces in the room. The room is, of course, filled with sound, but since we have an instrument uniquely relating time, space, and frequency we have in effect "frozen" the space contours of reflected sound. We may probe in space merely by adjusting a frequency offset. Analytically we may state that we have effected a coordinate conversion which trades spatial offset for driving frequency offset, but which retains intact all standard acoustical properties including those due to the frequency of the system-driving function. This is the power of this technique, for so long as the acoustic properties remain substantially linear, a hopeless signal combination in normal spatial coordinates transforms to a frequency coordinate generally more tractable to analyze.

In considering the uniqueness of signals perceived by frequency offsetting it may be observed that a transmitted signal passing through zero frequency will reverse phase at the zero frequency point and continue as a real frequency sweep with reversed slope of frequency *vs* time.

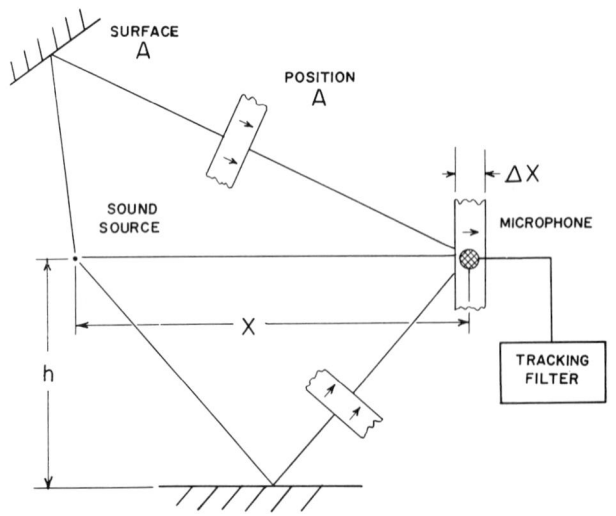

Fig. 1. Positional representation of direct and reflected acoustic pressure waves of constant frequency and fixed space-equivalent bandwidth as emitted by a swept frequency source and perceived by a microphone connected to a tracking filter tuned to maximize the response at a distance X.

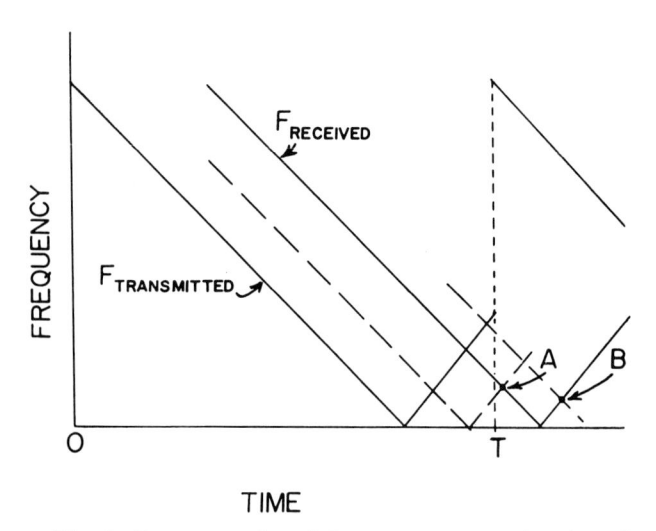

Fig. 2. Frequency plot of the sweep tone passing through zero frequency, showing room signals received as a function of time.

Thus, near zero there will be two transmissions of a given frequency during one sweep. The signals corresponding to the same frequency slope as that of the tracking filter will be accepted or rejected in total as the proper offset is entered for the appropriate time delay. The duplicate frequencies near zero possessing a different slope will cause a "ghost" impulse to appear when the instantaneous frequency perceived by the microphone corresponds to that of the tracking filter, regardless of the offset. Similarly, a tracking filter may be set to "look" for frequencies on both sides of zero. If the signal under analysis is a first-generation reflection, the main signal from the speaker will arrive sooner than this reflection and an impulse will occur at position A in Fig. 2. If, on the other hand, we are looking for a signal prior to the main signal, the impulse might appear at position B. Thus a repetitive sweep passing through zero frequency may contain image impulses due to signal paths other than that to which the tracking filter has been tuned. All sweeps with sufficiently long periods which do not pass through zero frequency will avoid any such ambiguities.

Another relationship worthy of consideration is the distance one may separate speaker and microphone without interference from a large object offset from the direct line of sight. Such considerations arise when measurements are made near floors, walls and the like. Surprisingly, an off-center reflecting object such as a floor limits the maximum distance of speaker-microphone separation. Consider Fig. 1 where the speaker and microphone are assumed to be a height h off the floor. The maximum distance will be that for which the path length from the image speaker is just equal to $X+(\Delta X/2)$. The maximum usable line-of-sight distance between speaker and microphone, X_{max}, is related to height h and space equivalent bandwidth ΔX by

$$X_{max} = (\Delta X/4)\{[h/(\Delta X/4)]^2 - 1\}. \qquad (3)$$

Note that if it is necessary to make lower frequency measurements or, what is the same thing, higher definition measurements, the distances scale up to values comparable to the wavelength of that frequency for which the period is the rise time of the tracking filter. This follows from observing that Eq. (1) may be rewritten as

$$\Delta X \cdot B \approx c \qquad (4)$$

PRACTICAL IMPLEMENTATION

While it might appear at first glance that the acoustical measurements outlined in the previous paragraph require specialized apparatus, equipment necessary to perform the described measurement is readily assembled from commercially available instruments. The tracking filter is a portion of an audio spectrum analyzer. This instrument is a narrow-band superheterodyne receiver tuned through the audio spectrum with a local oscillator swept linearly in time—the needed frequency characteristic. Depending upon the resolution required, the commercially available analyzers will sweep through a variety of given frequency dispersions in a repetitive sawtooth fashion. A sweep rate of one per second is typical for spectrum analyzers covering the full audio spectrum. The output of this audio tracking filter is rectified and applied to the vertical axis of a self-contained oscilloscope with the linear time axis swept horizontally. Because of the repetitive nature of the display, this instrument provides a visual presentation of signal energy *vs* frequency. The bandwidth of the spectrum analyzer is usually chosen to be the narrowest possible without losing information when swept past a complex spectrum.

The sweep tone for driving the loudspeaker may be obtained by down-converting the spectrum analyzer local oscillator to the audio band. If the local oscillator is heterodyned with another oscillator equal to the analyzer intermediate frequency, the difference frequency will always be at the precise frequency to which the analyzer is tuned. The proper offsetting frequency for spatial tuning of acoustical signals is obtained by detuning the fixed oscillator from the intermediate frequency. A down-converting synchronously sweeping generator is usually available as an accessory to the spectrum analyzer. To perform time-delay spectrometry it is only necessary to substitute a stable tunable oscillator for the fixed crystal oscillator. Where extreme accuracy is required it may be necessary to use a frequency counter to monitor the proper offset frequency. The remainder of the equipment consists of the usual power amplifier, microphone, and preamplifier.

Figure 3 diagrams a complete setup capable of performing any or all of the measurements outlined. While the cost of such assembled equipment is greater than

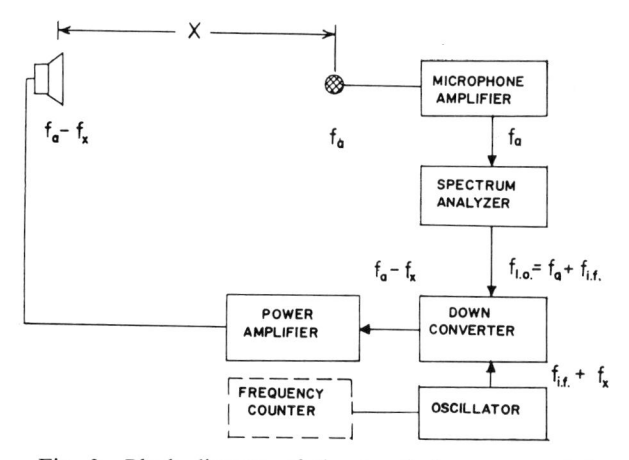

Fig. 3. Block diagram of the practical arrangement for the time-delay spectrometry measurements.

that of the typical instrumentation found in acoustical facilities, it is a small fraction of the investment for a self-contained anechoic facility capable of comparable measurements. The heart of the measurement is, of course, the spectrum analyzer. Several excellent commercial models are available; the analyzer used for these experiments is one which provides a continuously adjustable sweep width from 200 Hz to 20 KHz with a sweep center frequency separately adjustable from dc to 100 KHz. The bandwidth is tracked with the sweep width control to yield optimum resolution at the sweep rate of one per second. According to Eq. (1), this means that at maximum dispersion a 20 KHz spectrum may be obtained at 141 Hz resolution at 7.8 ft space-equivalent bandwidth. If measurements are desired valid to 32 Hz, for example, the sweep width would be set to 1000 Hz and the space-equivalent bandwith would be 34 ft. Where smaller space-equivalent bandwidths are desired and the

RICHARD C. HEYSER

decreased resolution can be tolerated, it is possible to increase the internal analyzer sweep rate by as much as a factor of five by minor circuit modification without compromising the validity of the measurement.

APPLICATIONS OF TIME DELAY SPECTROMETRY

Perhaps the best way of illustrating the use of time-delay spectrometry would be a detailed look at a typical measurement. Assume that an on-axis pressure response frequency characteristic is desired for a direct radiator loudspeaker. The entire spectrum from zero to 20 KHz is desired, with a frequency resolution of 140 Hz acceptable. The room selected for measurement has an 8 ft floor-to-ceiling height and a reverberation time of less than one second.

The space-equivalent bandwidth is approximately 7.8 ft (from Eq. 1), which means that no substantial object should be placed within 4 ft of the pickup microphone. An acceptable sweep rate is one 20 KHz sweep per second. From Eq. (3) it can be seen that if the speaker is placed halfway between floor and ceiling, the maximum distance the microphone should be placed from the speaker is six feet. This assumes, of course, a worst-case specular reflection from both floor and ceiling as well as a uniform polar response from the speaker.

The speaker under test is then placed 4 ft off the floor and more than 4 ft from any substantial object. The pickup microphone should be placed on-axis within 6 ft of the speaker and at least 4 ft from any reflective surface. The microphone is electrically connected to a suitable preamplifier which in turn feeds into the spectrum analyzer input. The down-converted local oscillator signal from the tracking oscillator is fed to an appropriate power amplifier driving the speaker. The offset oscillator which replaces the crystal in the tracking oscillator is connected to a frequency counter to complete the electrical setup.

With the deviation and deviation rate set to the desired test limits in the spectrum analyzer, a sweeping tone will be heard from the speaker. If the microphone-to-speaker distance is known, the offset oscillator should be set in accordance with Eq. (2) to achieve maximum deflection of the spectrum analyzer display. For example, a 6 ft separation will require a 114 Hz offset. It is at this point that the advantage of a tracking analyzer is evident: not only may the intensity of sound from the speaker be modest, but also it is not necessary to cease all sound and motion in the room while measurement is in progress. Any reasonable extraneous sound may be tolerated, and mobility need only be restricted to the extent that travel between loudspeaker and microphone is discouraged.

The spectrum analyzer display should remain stationary in vertical deflection. To ascertain that the proper frequency offset is used, the offset oscillator may be detuned on both the high and low side; the resultant display should show a reduction in amplitude. The peaked stationary pattern on the screen of the spectrum analyzer will be a plot of pressure response *vs* frequency with a smoothing bandwidth of 0.7% of the displayed dispersion.

Without altering the test setup several types of acoustical measurements may be made. Off-axis response of the speaker may be obtained by rotating the speaker by the required angle and immediately viewing the results on the analyzer. If it is desired to determine the sound transmission characteristic of a particular material, for example, one first obtains the on-axis speaker response and then interposes a 4 ft sample between speaker and microphone, noting the new response. If the spectrum analyzer is set for logarithmic deflection, the sound absorption in decibels is the difference in the two responses independent of speaker or microphone response.

The reflection coefficient may similarly be obtained by positioning the material behind the microphone by at least 4 ft and retuning the offset oscillator for the reflected signal. If the microphone is rotated 180° to remove its polar characteristics from the measurement, and if the proper space loss is entered, the reflection coefficient is obtained directly as a function of frequency, independent of microphone and loudspeaker characteristic. If one does not wish to compute the space loss, and space permits, the microphone may be physically transported to the position of maximum on-axis response with the reflective surface removed, and space loss obtained as response difference from the earlier position.

ROOM SPECTRUM SIGNATURES

The analysis up to this point has been directed at producing an electrical signal which when fed to a speaker will allow two very important measurements to be made on the sound picked up by a microphone. First, only the selected direct or reflected signal will be pulled out and displayed. Second, the spectral response of the selected signal will be directly presented. With this in mind, consideration will now be given to a characterization of sound in a room by time-delayed spectra.

Assume that an observer is positioned in a room with a localized source of sound such as a loudspeaker. This loudspeaker furthermore has a pressure frequency response characteristic, such as Curve A in Fig. 4. As the loudspeaker is energized by any arbitrary signal, the first sound heard by an observer will be characterized by Curve A. (This does not imply that all the frequencies

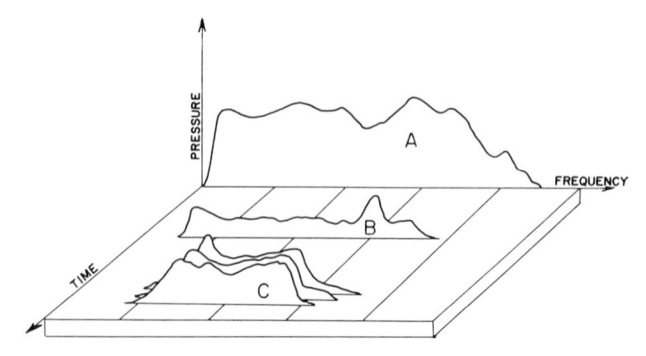

Fig. 4. Acoustic model of room spectrum signatures as perceived by an observer listening to a sound source with spectral distribution *A*.

are present, but only that those electrical signal components which are present are modified in accordance with Curve A.) Since the loudspeaker output varies as a function of time it will be assumed that Curve A is representative of the pressure output at a particular instant in time. We have in effect plucked out that signal which is characteristic of a given instant and will follow this pressure wave as it travels through the room and intercepts the observer.

A few milliseconds after the principal pressure wave A has passed, a second pressure wave B will be perceived by the observer. This is a reflected wave and is made up of the principal wave modified by the frequency response of the reflecting surface. As in electronic circuit theory, the observed resultant frequency response of B is the product of the frequency response of A and the frequency response of the reflecting surface. Beside the obvious time delay introduced by the longer path length, the surface creating the wave B may introduce a dispersive smearing of portions of the frequency range. This may be due to surface irregularities which act as local scattering centers or to acoustic impedance variations causing effective displacement of the position of optical and acoustic surfaces. If the reflecting surface has very little depth variation and is primarily specularly reflective, such as a hard wall or ceiling, the time-pressure profile might appear as wave B. If on the other hand the reflecting object has depth, such as a chair, the time-pressure profile may take on the character of the third pressure wave C. Here the effect is a distinct broadening of the time during which the observer perceives the signal. Perhaps a better name than reflection coefficient in the case of such an object might be scattering coefficient.

As time progresses the observer will perceive more scattering spectra with increasing density and generally lower amplitude until all sound due to the initial loudspeaker excitation has fallen below a predetermined threshold. The complete pressure-frequency-time profile will be a unique sound signature of the room as perceived by an observer listening to the loudspeaker. This is, after all, the way in which the sound reaches the observer. A more general characterization of the room itself would replace the loudspeaker and its unequal polar pressure response by an analytically uniform pressure transducer. Each position of observer and transducer will have its unique pressure-frequency-time profile. It quite frequently happens, however, that such a general characteristic is of little concern and what is desired is the observer-loudspeaker situation of Fig. 4.

The complete pressure-frequency-time profile represented by Fig. 4 is a useful acoustic model of a reverberant room in which an observer perceives a localized source of sound. As far as the observer is concerned there are many apparent sound sources. Each of these has the same "program content" as the primary source of sound but possesses its own spectral energy distribution and unique time delay, corresponding to its apparent location with respect to the observer. The effect of any given object in the acoustic environment may be determined in this model by noting the equivalent time delay from source to object to observer, and analyzing the scattering spectrum corresponding to this delay. The entire set of time-dependent scattering spectra should allow detailed analysis of this system for any time-dependent stimulus applied to localized source of sound.

This acoustic model of a room is seen to tie directly to time-delay spectrometry. The electronic sweep tone fed to the loudspeaker will in effect represent all possible frequencies in the chosen sweep range. The time axis of Fig. 4 could also be labeled offset oscillator frequency; the signal displayed on the spectrum analyzer will be seen to be the frequency spectrum of the selected time delay. The space-equivalent bandwidth, ΔX, will establish the ability to resolve independent scattering spectra since time-delay spectrometry forms an equivocation of time, offset frequency, and distance. It is immediately apparent that the selection of loudspeaker on-axis response used as an example for generation of time-delay spectrometry entails selection of Curve A and is only a special case of a more general acoustical measurement technique. It is possible by suitable choice of loudspeaker and microphone position to "pull out" important acoustic properties of a room without destroying the room or utilizing large-scale digital computer techniques.

Analytically, a surface at a distance corresponding to a time delay of t_k seconds may be characterized by a frequency spectrum multiplied by a linear phase coefficient. The cumulative distribution of these sources is represented by Fig. 4 and may be expressed as

$$R(\omega) = \sum_k S_k(\omega) e^{-i\omega t_k}, \quad (5)$$

where $R(\omega)$ is the cumulative distribution of all sources, both real and apparent, as perceived by an observer listening to a localized source of sound in a reverberant environment, and $S_k(\omega)$ is the spectral energy distribution of the source at a distance corresponding to a time delay of t_k seconds. The angular frequency ω, expressed in radians per second, will be used in this paper for analytical simplicity. Since the absolute magnitude of $S_k(\omega)$ is measured by time-delay spectrometry it is not necessary to introduce a space loss term. If the source has a spectral distribution $P(\omega)$, then the observer will perceive a signal which has a frequency spectrum that is the product of $P(\omega)$ and $R(\omega)$. The impulse response of the room, $r(t)$, is simply the Fourier Transform of $R(\omega)$:

$$r(t) = (1/2\pi) \int_{-\infty}^{\infty} R(\omega) e^{+i\omega t} d\omega \quad (6)$$

and the time-varying signal $o(t)$ perceived by an observer for a program source $p(t)$ is the convolution integral of program source and room response

$$o(t) = \int_{-\infty}^{\infty} p(\tau) r(t-\tau) d\tau \quad (7)$$

The concept expressed by Eqs. (6) and (7) is certainly not new to the field of acoustics. The difficulty in applying these relations lies in the massive amounts of data that must be processed before the effect of a particular object in a room may be evaluated. The acoustic model of Eq. (5) as represented in Fig. 4 leads to a simpler analysis since time-delay spectrometry isolates the $S_k(\omega)$ function of acoustic surfaces and presents the information in a meaningful form which requires no further reduction. While not a cure-all for acoustic analysis, time-delay spectrometry provides a good insight to actual acoustic situations.

EXPERIMENTAL RESULTS

Sufficient experimental evidence has been collected to assure validity of the technique. The examples below give some insight into the nature of the spectrum display.

RICHARD C. HEYSER

Loudspeaker Testing

A single-cone 8-in. loudspeaker housed in a ported box was chosen for test, in a room deliberately selected to be a poor environment for loudspeaker testing. The room measures 24 × 10 × 7 ft. The floor is hard vinyl on cement and the ceiling is plaster, while all other surfaces are hard wood. Miscellaneous objects throughout the room reduce the open area so that when the loudspeaker-to-microphone distance is at the calculated maximum of Eq. (3) the smallest separation from the microphone to any substantial object is one half the space-equivalent bandwidth. (The room spacing is chosen to be no larger than the calculated minimum dimension in order that the concept may be better demonstrated.) The spectrum measurement was made to include dc to 10 KHz at a dispersion rate of 20 KHz per second. Fig. 5a is a photograph of the on-axis pressure response in decibels as a function of linear frequency. Since the analyzer sweeps through zero-beat to bring dc at the left-hand edge, a mirror image response is seen for those frequencies to the left of dc.

In order to illustrate the nature of the response to be obtained for the steady-state application of sinewaves, the test setup was left intact and the spectrum analyzer driven at such a slow rate that the room acoustics entered into the measurement. Figure 5b is the result of this measurement. The sweep width, vertical sensitivity, and analyzer bandwidth are identical to those of Fig. 5a, but the sweep rate is such that it takes over three minutes to go from zero frequency to 10 KHz. The camera aperture was optimized to give a readable display, and consequently the large number of standing wave nulls and peaks did not register well. The wild fluctuation in reading as a function of frequency is quite expected and lends credence to the futility of steady-state loudspeaker measurement in the acoustic equivalent of a shower stall.

One certain way of verifying that a "free-field" measurement is made is to impose inverse square law.[5] Figure 5c is a time-delay spectrograph of an inverse square law measurement. The setup of Fig. 5a and 5b was spectrographed and appears as the lower trace in Fig. 5c. The loudspeaker was then moved toward the microphone to one half its former distance, and a second exposure was taken with the proper offset oscillator setting. Since the display is logarithmic, the two traces should be parallel and separated by 6 dB, and indeed this is seen to be the case. The minor discrepancies are due to the fact that the first measurement was made at 5 ft, which is the maximum theoretical separation for the room and dispersion rate. The closer measurement is made at 2½ ft where the loudspeaker does not exactly appear as a point source since the separation-to-diameter ratio is only four.

Isometric Display of Room Signature

The acoustic model of a room proposed above resulted in a three-dimensional plot of pressure, frequency, and time. Time-delay spectrometry is a means of probing and presenting this model. In order to visualize this phenomenon, the spectrum analyzer display has been modified to accept what may be called Isometric Scan. The offset oscillator is in this instance a linear voltage-controlled oscillator. Each sweep of a time delay spectrograph is presented normally as a plot of pressure vs frequency. Isometric scan advances the offset oscillator

a.

b.

c.

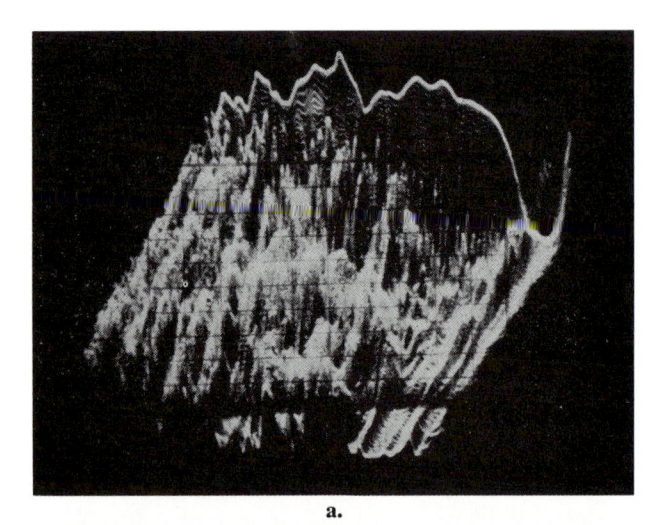

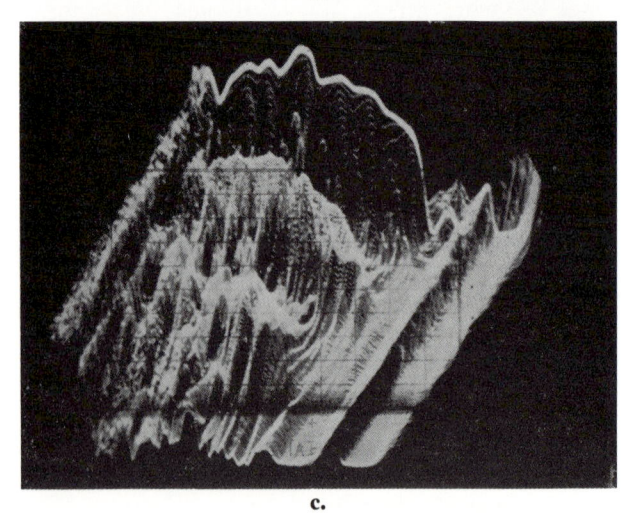

Fig. 5. Oscillographs of logarithmic pressure response measurements on a single-cone loudspeaker from dc to 10 KHz, measured in a reverberant environment by time-delay spectrometry. **a.** Using rapid sweep. **b.** Using slow sinewave sweep. **c.** Showing inverse square response as a function of distance.

following each sweep and creates the effect of the third dimension of offset frequency exactly as one illustrates a three-axis system on two-dimensional paper, that is, by a combined horizontal and vertical offset displacement of the sweep. The net effect is a presentation identical in form to Fig. 4.

ACOUSTICAL MEASUREMENTS BY TIME DELAY SPECTROMETRY

Figure 6a is an isometric scan of the room and setup of the preceding loudspeaker test. The time axis commences at zero when the principal on-axis signal is received and advances through 43 msec which is equivalent to twice the longest room dimension. All frequencies from zero through 10 KHz are displayed. The exceedingly large number of "alpine peaks" in evidence are only partially visible in the photograph since it is difficult to capture the visual impression an observer sees as this three-dimensional plot unfolds on the oscilloscope screen.

Figure 6b is an isometric scan of a multicellular horn from dc to 20 KHz and is a good representation of the principal pressure wave A of Fig. 4 with the finite space-equivalent bandwidth in this case equal to ½ ft. Figure 6c is an interesting example of multiple reflections. An acoustic suspension loudspeaker system was placed on a floor pointing upward at a hard ceiling, with an omnidirectional microphone halfway between speaker and ceiling. The sweep extends from dc to 20 KHz with a 20 KHz marker "fence" in evidence in the spectrograph. The space-equivalent bandwidth is ½ ft. Time starts at the receipt of the principal wave. The second peak corresponds to the ceiling reflection. The third peak is the reflection of the ceiling wave off the loudspeaker itself. Some of the energy is reflected by the loudspeaker grille and some by the speaker cones themselves yielding beautiful doppler data, and some energy penetrates through the loudspeaker to the floor.

ANALYTICAL VERIFICATION

The measurement technique of the previous paragraphs was developed in a highly intuitive manner. Although experimental measurements tend to verify the technique, it is still necessary to analyze two very important considerations: first, the nature of a repetitive glide tone, and second, the validity of identifying the spectrum analyzer display with the acoustic spectrum.

Fourier Spectrum of a Repetitive Linear Glide Tone

The system-forcing function developed for time-delay spectrometry has the characteristic shown in Fig. 7. Loosely speaking, it is a signal which has an instantaneous frequency linearly proportional to time for a total period of T seconds and then repeats the cycle indefinitely at this period. Clearly the signal must have a Fourier series spectrum of terms with periods which are integral submultiples of T. It cannot, in other words, be a continuum of frequencies. Yet the intuitive development presupposed a forcing function which not only did not have "holes" in the spectrum but also was of constant amplitude.

Readers familiar with frequency modulation will recognize that the signal of Fig. 7 is in reality a linear sawtooth frequency modulating a carrier halfway between the peak deviation frequencies. Let the maximum and minimum deviation frequencies be ω_2 and ω_1 respectively and the period of sweep T seconds. The function to be defined then has an instantaneous frequency ω_{inst} given by

$$\omega_{inst} = \left(\frac{\omega_2 - \omega_1}{T}\right)t + \left(\frac{\omega_2 + \omega_1}{2}\right) \triangleq \frac{D}{T}t + \omega_c \quad (8)$$

where the angular dispersion D and effective carrier frequency ω_c are introduced for simplicity. Since we are interested in time-dependent factors, the time phase dependence becomes

$$\phi(t) = \int_0^t \omega_{inst} dt = (D/2T)t^2 + \omega_c t. \quad (9)$$

The actual function of time which is to be expanded

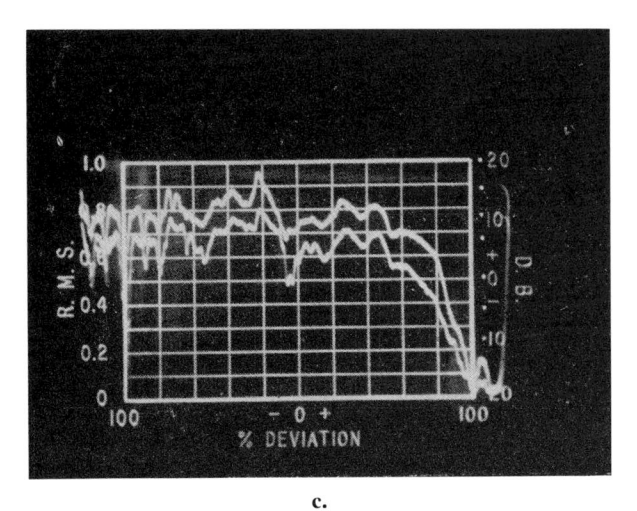

Fig. 6. Oscillographs of isometric displays of spectral signatures using time-delay spectrometry. **a.** Room signatures of the setup of Fig. 5. **b.** Principal response of a multicellular horn from dc to 20 KHz. **c.** Spectral signatures obtained with loudspeaker system pointed at ceiling, including direct response, ceiling reflection, and speaker reflection of ceiling wave.

RICHARD C. HEYSER

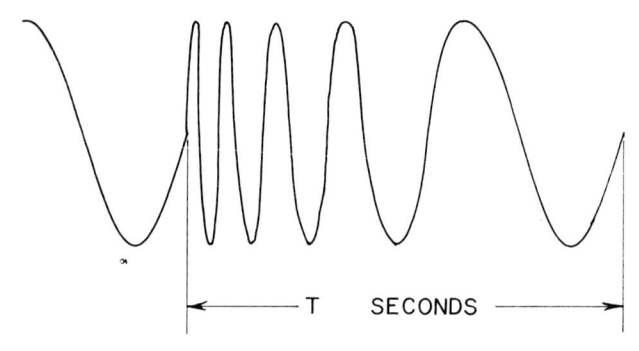

Fig. 7. Graphic representation of the repetitive sweep tone used as system-forcing function.

in a Fourier series is that signal possessing the phase $\phi(t)$, or

$$E(t) = \cos\phi(t) = \tfrac{1}{2}[e^{i\phi(t)} + e^{-i\phi(t)}]. \quad (10)$$

From an analytical standpoint it is simpler to use the exponential form. Since each complex exponential is separable into the product of a steady-state term and a term periodic in T, it is sufficient to expand this periodic term in a Fourier series. Furthermore, the positive frequency portion will be analyzed first and the negative frequency terms considered from this; thus the Fourier frequencies and coefficients become, in exponential form:

$$e^{i(D/2T)t^2} = \sum_{N=-\infty}^{\infty} C_N e^{iN\omega_0 t} \quad (11)$$

$$C_N = (1/T) \int_{-T/2}^{T/2} e^{i(D/2T)t^2} e^{-iN\omega_0 t} dt \quad (12)$$

where

$$\omega_0 = 2\pi/T.$$

By completing the square, multiplying and then dividing by a normalizing factor $(2/\pi)^{1/2}$, C_N becomes

$$C_N = \quad (13)$$

$$\frac{1}{T}\left(\frac{\pi T}{D}\right)^{1/2} e^{-i(T/2D)(N\omega_0)^2} \left(\frac{2}{\pi}\right)^{1/2} \int_{-T/2}^{T/2} e^{ix^2} dx$$

where

$$x = (D/2T)^{1/2}[t - (T/D)N\omega_0]$$

or, by noting Eq. (8),

$$C_N = \frac{1}{T}\left(\frac{\pi T}{D}\right)^{1/2} e^{-i(T/2D)(N\omega_0)^2} \quad (14)$$

$$\left[\left(\frac{2}{\pi}\right)^{1/2}\int_0^{\omega_2} e^{ix^2} dx - \left(\frac{2}{\pi}\right)^{1/2}\int_0^{\omega_1} e^{ix^2} dx\right]$$

$$x = (T/2D)^{1/2}[\omega_{inst} - (\omega_c + N\omega_0)].$$

Normally analysis would stop here because the definite integrals can be evaluated only as an infinite series. However, the complex expansion of the exponential integral

$$\left(\frac{2}{\pi}\right)^{1/2}\int_0^u e^{ix^2} dx = C(u) + iS(u) \quad (15)$$

is made up of the Fresnel cosine integral $C(u)$ and Fresnel sine integral $S(u)$, both of which have been tabulated.[6]

Noting the fact that the argument changes sign in the second integral, we can write the resultant Fourier coefficient for the Nth sideband term as

$$C_N = (1/T)(\pi T/D)^{1/2} e^{-i(T/2D)(N\omega_0)^2} \quad (16)$$
$$[C(\omega_1) + C(\omega_2) + iS(\omega_1) + iS(\omega_2)].$$

The C_n's are the coefficients of the Fourier components or, considered another way, are the sideband terms and consist of the product of the quantized spectrum of a continuous swept tone and a complex Fresnel integral modifying term involving the finite frequency terminations. Figure 8 illustrates how to calculate the coefficient for the Nth sideband; Fig. 9 is a plot of the locus of the complex value of the Fresnel integral term for a high deviation ratio, which is common for time-delay spectrometry. Note that the spectrum is completely symmetrical about the effective carrier and that while the dropoff is shown for one band edge, the other band edge is identical in form. It is seen that the magnitude is essentially constant throughout the spectrum, as expected, and that within the swept band the phase departure is limited to 15° and rapidly approaches 0° at center frequency.

The function

$$e^{ix^2} \quad (17)$$

goes through a first zero crossing for the real value, where $u = 1$ when

$$x^2 = \pi/2. \quad (18)$$

From Eq. (14), this occurs when

$$\Delta\omega = (2D/T)^{1/2}(\pi/2)^{1/2} = (\pi D/T)^{1/2}. \quad (19)$$

Since the relation between optimized bandwidth B and dispersion rate D/T for the analyzer is

$$B \cong (D/2\pi T)^{1/2} = [1/(2\pi)^{1/2}](D/T)^{1/2}, \quad (20)$$

then

$$\Delta\omega \cong B/\sqrt{2}. \quad (21)$$

In other words, even though Fig. 9 shows an overshoot at band edge of diminishing amplitude and increasing frequency as one approaches the effective carrier, these ripples are not distinguishable to a tracking analyzer since they are smoothed by the proper filter bandwidth and the analyzer cannot distinguish this distribution from a true

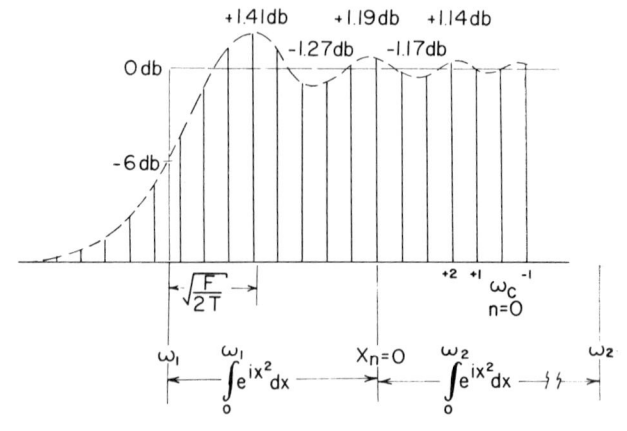

Fig. 8. Method of calculating sideband coefficients for linear sawtooth frequency-modulated carrier with peak angular frequencies ω_1 and ω_2, dispersion $D = 2\pi F = \omega_2 - \omega_1$, and sweep time T.

ACOUSTICAL MEASUREMENTS BY TIME DELAY SPECTROMETRY

constant amplitude. Thus the desire for an effective constant amplitude continuum is satisfied.

While the conventional rates of time-delay spectrometry dictate a high effective modulation index yielding the near band-edge distribution of Fig. 9, it may be of some interest to investigators to determine the distribution for low effective indexes. This may be done quite simply by noting that a complex plot of Eq. (15) yields a Cornu spiral. This is shown in Jahnke and Emde[6] and appears

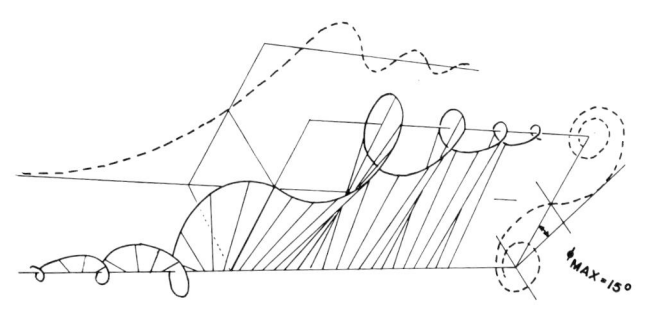

Fig. 9. Locus of the complex value, including magnitude and phase, of the Fresnel Integral term in the coefficient of the sweep tone Fourier series for high deviation ratios.

as the phase projection in Fig. 9. The desired sideband amplitude is the length of segment joining the appropriate points on the Cornu spiral. It should also be pointed out that for high indexes the spectral distribution will approach the amplitude probability distribution of the modulating waveform. For a perfect sawtooth this will be a square spectrum with the characteristic frequency terminating overshoot analogous to Gibb's phenomenon, while for a slightly exponential sawtooth the spectrum will appear to be tilted.

Several important points may now be established concerning this forcing function. First, the spectrum is quite well behaved for all modulation indexes. As one starts with an unmodulated carrier and begins modulation at progressively higher indexes, it will be noted that there is no carrier or sideband nulling as is the case with sinusoidal frequency modulation. Thus any deviation ratio is valid for time-delay spectrometry. Second, the spectral sidebands are down 6 dB at the deviation band-edge and are falling off at the rapid rate of about 10 dB per unit analyzer bandwidth. Thus the spectrum is quite confined and no unusual system bandwidth is required. Third, an expansion of the negative frequencies will show a comparable symmetric spectrum, which means that spectrum foldover effects such as those due to passing through zero beat do not in any way compromise the use of this type of modulation for time-delay spectrometry.

A final point in this spectral analysis: one tends to take for granted the validity of the analytical result without much consideration for the physical mechanism. Equation (16) shows that this glide tone can be generated by connecting a large number of signal generators with almost identical amplitude to a common summing junction. Furthermore, the amplitude of the resultant glide tone will be equal to that of any one of the separate generators. At first glance this would not only seem to violate conservation of energy but it would definitely be hard to convince an observer that when he heard a smooth glide tone with definite pitch as a function of time he was actually hearing all possible frequencies. Inspection of the phases as controlled by the partial Fresnel integrals reveals that at any given time all generators except those in the vicinity of the instantaneous glide tone frequency will cancel to zero, and that the distribution of those generators which are not cancelled is approximately Gaussian about the perceived tone. In addition, the "bandwidth" of this distribution is of the order of the analyzer bandwidth. The result of this is that an analyzer slightly detuned from the glide tone will not only be down in response but will be down by the same amount at all frequencies.

Spectrum Analyzer Response

In considering the nature of the display presented by a tracking spectrum analyzer used for time-delay spectrometry, the generalized circuit of Fig. 10 will be used. The portion within the dashed contour characterizes the general spectrum analyzer. The sweeping local oscillator signal is brought out and mixed in a balanced modulator with a fixed local oscillator which has a slight offset from the analyzer intermediate frequency. The filtered difference output of the mixing process is the audio glide tone previously discussed, which is used as a forcing function for the acoustical system under analysis. The acoustical response of this system as perceived by a microphone will consist of a multiplicity of time-delayed responses. The entire signal from the microphone is sent to the spectrum analyzer. The purpose of the analysis below is to demonstrate that any given delayed response may be selected from all inputs by appropriate selection of the glide tone generating offset oscillator, and to de-

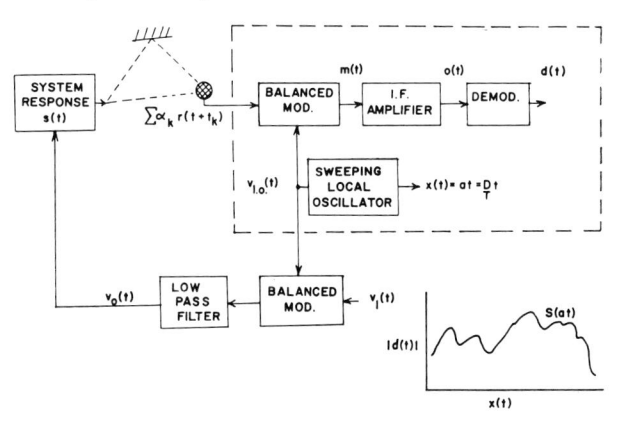

Fig. 10. Generalized block diagram of the acoustical system test using time-delay spectrometry.

velop conditions and limitations of analyzer display.

The analytical description of a sweeping tone used as a system-forcing function leads to a remarkable property: The sweep tone is the complex conjugate of its own Fourier Transform, with time in place of frequency. Just as a steady sinusoid applied to a network provides a narrow frequency window which pulls out the frequency response of the network, the sweep tone may be shown to provide a spectrum window pulling out the frequency spectrum of the network. As a consequence of this property this sweep function $w(t)$ will be designated as a window function,

$$w(t) \triangleq e^{i\frac{1}{2}at^2}, \quad (22)$$

where a represents the angular dispersion rate D/T.

RICHARD C. HEYSER

Proceeding to the analysis it will be observed that the analyzer local oscillator consists of a sweeping tone $w(t)$ heterodyned to the intermediate frequency ω_i. It is a real time function and is thus

$$v_{1.0}(t) = [w(t)e^{i\omega_i t} + w^*(t)e^{-i\omega_i t}]/2 \quad (23)$$

where the starred operation indicates complex conjugation. Similarly, the down-converting oscillator consists of the intermediate frequency ω_i offset by a fixed value ω_o and is

$$v_1(t) = [e^{i(\omega_i + \omega_o)t} + e^{-i(\omega_i + \omega_o)t}]/2. \quad (24)$$

After low-pass filtering, the system-driving function, neglecting constant gain terms, is

$$v_0(t) = w(t)e^{-i\omega_o t} + w^*(t)e^{+i\omega_o t} \quad (25)$$

Assuming the system response is linear, the system response $r(t)$ for this driving function is the convolution integral of the driving function $v_0(t)$ and the system time response $s(t)$:

$$r(t) = \int_{-\infty}^{\infty} s(\tau)v_0(t-\tau)d\tau \triangleq s(t) \otimes v_0(t). \quad (26)$$

The second symbolism will be utilized because of its simplified form.

Because of the finite time delay, t_k, between system output and microphone input for each probable path K, the microphone response $r(t)$ consists of the sum of all the signal path inputs,

$$r(t) = \sum_K a_K r(t + t_k) \quad (27)$$

where a_K is the strength of the Kth signal. Also, each separate path length signal is expressible as

$$r(t + t_k) = [s(t + t_k) \otimes w(t + t_k)e^{-i\omega_o(t + t_k)}] \quad (28)$$
$$+ [s(t + t_k) \otimes w^*(t + t_k)e^{i\omega_o(t + t_k)}].$$

The process of balanced modulation is a simple multiplication in the time domain. The modulator output $m(t)$ is then, neglecting constant multipliers,

$$m(t) = \sum_K a_K r(t + t_k) \cdot (w(t)e^{i\omega_i t} + w^*(t)e^{-i\omega_i t}). \quad (29)$$

The frequency spectrum of the modulator output $M(\omega)$, considering only positive frequencies, is

$$M(\omega) = \sum_K a_K \int e^{-i(\omega - \omega_i)t} w(t) r(t + t_k) dt. \quad (30)$$

(The convention followed for the purpose of this analysis is that lower-case functional notation designates time dependence while upper-case notation designates frequency dependence.)

From Eqs. (A5), (A6), and (A9) of the Appendix, Eq. (30) may be rewritten as

$$M(\omega) =$$
$$\sum_K a_k \int e^{-i(\omega - \omega_i + at_k - \omega_o)t} W(at_k - \omega_o)(a/2\pi i)$$
$$\{W(\omega_o)w(\xi)[s(\xi) \otimes w(\xi)]$$
$$+ W^*(\omega_o)w(\xi)[s(\xi) \otimes w^*(\xi)]\} dt \quad (31)$$

where $\xi = t + t_k - (\omega_o/a)$.

From Eqs. (A7) and (A8) in the Appendix, we may expand this into two integral summations

$$M(\omega) =$$
$$\sum_K a_K \int e^{-i(\omega - \omega_i + at_k - \omega_o)t} W(at_k - \omega_o)(a/2\pi i)^{3/2}$$
$$\{W(\omega_o)w(\xi)w(\xi)[S(a\xi) \otimes W(a\xi)]\} dt$$
$$+ \sum_K a_K \int e^{-i(\omega - \omega_i + at_k - \omega_o)t} W(at_k - \omega_o)(a/2\pi i)^{3/2}$$
$$\{W^*(\omega_o)[S(a\xi) \otimes W^*(a\xi)]\} dt. \quad (32)$$

The first integral is the balanced modulator "upper sideband" and is the transform of a scanned time spectrum multiplied by two window functions. The transform will therefore be in the vicinity of the intermediate frequency only for $\xi = O$ and will constantly retreat from ω_i for all other times at a rate twice that of the sweeping local oscillator. As a consequence, this integral need not be considered further.

The second integral is the "lower sideband" and will transform the scanned spectrum to lie directly on the frequency $(\omega_i + \omega_o - at_k)$. For both integrals the "bandwidth" of any substantial energy to be found in the transformed spectrum will be determined by the convolution of $W(at)$ and $W^*(at)$ respectively with the system frequency spectrum, with the frequency parameter replaced by time. Figure 11 symbolizes the spectral energy distribution given by Eq. (32).

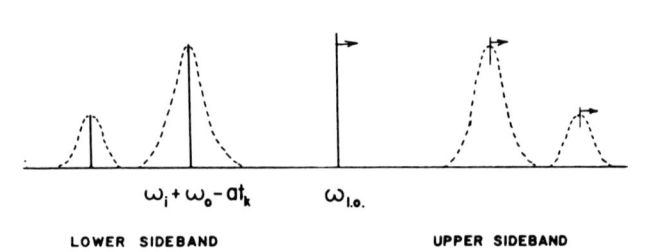

Fig. 11. Spectral energy distribution at the input to the intermediate frequency amplifier of the spectrum analyzer.

In the frequency domain the output of the intermediate frequency amplifier $O(\omega)$ will be the product of the intermediate frequency spectrum $I(\omega - \omega_i)$ and the balanced mixer spectrum:

$$O(\omega) = \sum_K a_k I(\omega - \omega_i) M(\omega - \omega_i + at_k - \omega_o). \quad (33)$$

If the delay times t_k are not so close together as to overlap the functions $M(\omega - \omega_i + at_k - \omega_o)$ in the vicinity of $(\omega - \omega_i)$, then any particular response may be selected by adjusting ω_o such that

$$\omega_o = at_k \quad (34)$$

for the desired path time delay t_k. Then Eq. (33) becomes

$$O(\omega) = G_K I(\omega - \omega_i) M(\omega - \omega_i). \quad (35)$$

The time spectrum of the output of the intermediate frequency amplifier is thus

$$o(t) = g_k [i(t) \otimes W^*(at) \otimes S(at)](a/2\pi i)^{3/2} \quad (36)$$

or, noting Eq. (A3) in the Appendix and regrouping in

accordance with the associative property of convolution,

$$o(t) = g_k(a/2\pi i)[i(t) \otimes w(t)] \otimes S(at). \quad (37)$$

The demodulator output $d(t)$ is the signal displayed on the vertical axis and is thus

$$d(t) = (constant) \cdot |o(t)|. \quad (38)$$

Inspection of Eqs. (37) and (38) discloses that the proper spectrum has indeed been displayed. It should be recognized that while the analysis was done on a continuous basis no loss of validity is experienced for repetitive sweeps.

In interpreting Eq. (37) it is apparent that the proper spectrum is modified by the window function and time response of the intermediate-frequency amplifier of the analyzer. The action of these latter terms is a smoothing upon the actual spectrum. This smoothing is similar in character to a simple low-pass filtering of the perfect spectrum.

To demonstrate that no spectrum bias is introduced in Eq. (37) by $w(t)$ and $i(t)$, consider the case of a perfectly flat spectrum where

$$S(at) = Constant \quad (39)$$

and where a finite time is assumed. In this case the convolution integrals collapse to simple integrals yielding

$$o(t) = Constant. \quad (40)$$

If, in other words, the system spectrum is independent of frequency, the spectrum analyzer display will show a straight line whose height above the baseline will be proportional to the gain or loss through the system.

Looking at the other end of the functional dependence, consider the case of a system spectrum response which is zero for all frequencies but one, and is infinite at that singular frequency. What, in other words, is the response to a network of infinite Q. Such a function is a Dirac delta

$$S(at) = \delta(t-t_o) \quad (41)$$

simple substitution into Eq. (37) yields

$$o(t) = K[i(t) \otimes w(t-t_o)]. \quad (42)$$

This is exactly the same response as experienced by a spectrum analyzer viewing a single sinewave spectrum.[4] This zero width spectrum is displayed as a smoothed function, closely approximating the intermediate frequency response for sweep rates such that

$$B^2 \gtrsim dF/dt. \quad (43)$$

As the sweep rate increases above the inequality of Eq. (43), the displayed function broadens and the "phase tail" of $w(t) \otimes i(t)$ evidences an increasing ring on the trailing edge. Thus for the normal spectrum analyzer the infinite Q response is smoothed to the order of the analyzer bandwidth.

One exceedingly important result follows from Eq. (37). This is that the spectrum is preserved not only in amplitude but also in phase. Thus the actual spectrum experiences both an amplitude and a phase smoothing. This gives us a measurement tool which could not have been predicted from the intuitive reasoning utilized. The implications are significant, since it is now possible to isolate the influence of an acoustic subsystem without removing that subsystem from its natural environment and analyze complex behavior, amplitude, and phase, in response to an applied stimulus.

CONCLUSION

What has been described is an acoustical testing procedure which allows selective spatial probing of a natural environment by commercially available equipment. The results are displayed immediately in the form of pressure response as a function of frequency, and require no further processing for interpretation. There are of course resolution limitations imposed by the dimensions of the testing area, but these are easily calculated and are small penalties to pay for the privilege of making formerly unobtainable on-location tests.

An acoustic model of reflecting objects has been introduced which identifies the object with an equivalent frequency response expressible as a frequency-dependent scattering coefficient. This frequency response modifies the frequency response of a sound source in a manner analogous to its electronic circuit counterpart, but possesses a time delay corresponding to its position relative to source and auditor. The effect of an assemblage of reflecting objects can be represented as a three-dimensional plot of pressure, frequency and time.

The utilization of time-delay spectrometry is limited primarily by the imagination and ingenuity of the experimenter. The work described in this paper attempted to yield detailed insight into the technique yet not prevent its application because specialized equipment was unavailable. It must be clear that there is still a substantial class of meaningful measurements which not only require extensive modification of the equipment but also demand further understanding and analysis, in particular the phase angle which is analytically available as well as the amplitude of response. Electronic circuit concepts were used to develop the technique as well as the acoustic model. Perhaps once we possess detailed complex spectra of various acoustic systems we may be able to utilize the existing wealth of circuit techniques to reduce the multi-dimensional real-world acoustic problems to a form more amenable to analysis. It may well be that time-delay spectrometry is a tool which will contribute to this goal.

APPENDIX

Summary of Important Relations

1. $f(t) = (1/2\pi)\int_{-\infty}^{\infty} F(\omega)e^{i\omega t}d\omega,$

   $F(\omega) = \int_{-\infty}^{\infty} f(t)e^{-i\omega t}dt$

2. $w(t) = e^{i\frac{1}{2}at^2} \quad W(\omega) = (2\pi i/a)^{\frac{1}{2}} e^{-i(\omega^2/2a)}$

3. $w(t) = (a/2\pi i)^{\frac{1}{2}} W^*(at)$

4. $\int_{-\infty}^{\infty} f(t)g(x-t)dt \triangleq f(x) \otimes g(x)$

5. $s(t+t_k) \otimes w(t+t_k) e^{-i\omega_o(t+t_k)} =$
   $(a/2\pi i)^{1/2} W(\omega_o)[s(t+t_k-\omega_o/a) \otimes w(t+t_k-\omega_o/a)]$

6. $s(t+t_k) \otimes w^*(t+t_k) e^{i\omega_o(t+t_k)} =$
   $(a/2\pi i)^{1/2} W^*(\omega_o)[s(t+t_k-\omega_o/a) \otimes w^*(t+t_k-\omega_o/a)]$

7. $w(t)[s(t) \otimes w(t)] =$
   $w(t)w(t)[S(at) \otimes W(at)](a/2\pi i)^{1/2}$

8. $w(t)[s(t) \otimes w^*(t)] = [S(at) \otimes W^*(at)](a/2\pi i)^{1/2}$

9. $w(t) =$
   $w(t+t_k-\omega_o/a) W(at_k-\omega_o) e^{-i(at_k-\omega_o)t} (a/2\pi i)^{1/2}$

REFERENCES

1. W. B. Snow, "Loudspeaker Testing in Rooms," *J. Audio Eng. Soc.* **9**, 54 (1961).

2. D. H. Bastin, "Microphone Calibrator," *Electronics* (Nov. 1948).

3. T. J. Schultz and B. G. Watters, "Propagation of Sound Across Audience Seating," *J. Acoust. Soc. Am.* **36**, 885 (1964).

4. H. W. Batten, R. A. Jorgensen, A. B. MacNee and W. W. Peterson, "The Response of a Panoramic Analyzer to C.W. and Pulse Signals," *Proc. I.R.E.*, 948 (1954).

5. G. E. Peterson, G. A. Hellwarth, and H. K. Dunn, "An Anechoic Chamber with Blanket Wedge Construction," *J. Audio Eng. Soc.* **15**, 67 (1967).

6. For example, E. Jahnke and F. Emde, *Tables of Functions* (Dover Publications, New York, 1945).

THE AUTHOR

Richard C. Heyser received his B.S.E.E. degree from the University of Arizona in 1953. Awarded the AIEE Charles LeGeyt Fortescue Fellowship for advanced studies he received his M.S.E.E. from the California Institute of Technology in 1954. The following two years were spent in post-graduate work at Cal Tech leading toward a doctorate. During the summer months of 1954 and 1955, Mr. Heyser was a research engineer specializing in transistor circuits with the Motorola Research Laboratory, Phoenix, Arizona. From 1956 until the present time he has been associated with the Cal Tech Jet Propulsion Laboratory in Pasadena, California. He is a senior member of the JPL Technical Staff presently engaged in the lunar television operation on Project Surveyor.

Mr. Heyser has presented several papers before the AES and is a member and fellow of the Audio Engineering Society, the Institute of Electrical and Electronic Engineers, as well as Tau Beta Pi, Pi Mu Epsilon, Phi Kappa Phi, Sigma Pi Sigma, and Sigma Xi. He is also a member of Theta Tau, National Professional Engineering Fraternity.

Determination of Loudspeaker Signal Arrival Times*
Part I

RICHARD C. HEYSER

Jet Propulsion Laboratory, California Institute of Technology, Pasadena, Calif.

Prediction has been made that the effect of imperfect loudspeaker frequency response is equivalent to an ensemble of otherwise perfect loudspeakers spread out behind the real position of the speaker creating a spatial smearing of the original sound source. Analysis and experimental evidence are presented of a coherent communication investigation made for verification of the phenomenon.

INTRODUCTION: It is certainly no exaggeration to say that a meaningful characterization of the sound field due to a real loudspeaker in an actual room ranks among the more difficult problems of electroacoustics. Somehow the arsenal of analytical tools and instrumentation never seems sufficient to win the battle of real-world performance evaluation; at least not to the degree of representing a universally accepted decisive victory. In an attempt to provide another tool for such measurement this author presented in a previous paper a method of analysis which departed from traditional steady state [1]. It was shown that an in-place measurement could be made of the frequency response of that sound which possessed a fixed time delay between loudspeaker excitation and acoustic perception. By this means one could isolate, within known physical limitations, the direct sound, early arrivals, and late arrivals and characterize the associated spectral behavior. In a subsequent paper [2] it was demonstrated how one could obtain not only the universally recognized amplitude spectrum of such sound but also the phase spectrum. It was shown that if one made a measurement on an actual loudspeaker he could legitimately ask "how well does this speaker's direct response recreate the original sound field recorded by the microphone?" By going to first principles a proof was given [3] that a loudspeaker and indeed any transfer medium characterized as absorptive and dispersive possessed what this author called time-delay distortion. The acoustic pressure wave did not effectively emerge from the transducer immediately upon excitation. Instead it emerged with a definite time delay that was not only a function of frequency but was a multiple-valued function of frequency. As far as the sonic effect perceived by a listener is concerned, this distortion is identical in form to what one would have, had the actual loudspeaker been replaced by an ensemble of otherwise perfect loudspeakers which occupied the space behind the position of the actual loudspeaker. Furthermore, each of the speakers in the ensemble had a position that varied in space in a frequency-dependent manner. The sonic image, if one could speak of such, is smeared in space behind the physical loudspeaker.

This present paper is a continuation of analysis and experimentation on this phenomenon of time-delay distortion. The particular emphasis will be on determining how many milliseconds it takes before a sound pressure wave in effect emerges from that position in space occupied by the loudspeaker.

---
* Presented April 30, 1971, at the 40th Convention of the Audio Engineering Society, Los Angeles.

APPROACH TO THE PROBLEM

The subject matter of this paper deals with a class of measurement and performance evaluation which constitutes a radical departure from the methods normally utilized in electroacoustics. Several of the concepts presented are original. It would be conventional to begin this paper by expressing the proper integral equations, and thus promptly discouraging many audio engineers from reading further. Much of the criticism raised against papers that are "too technical" is entirely just in the sense that common language statements are compressed into compact equations not familiar to most of us. The major audience sought for the results of this paper are those engineers who design and work with loudspeakers. However, because the principles to be discussed are equally valuable for advanced concepts of signal handling, it is necessary to give at least a minimal mathematical treatment. For this reason this paper is divided into three parts. The first part begins with a heuristic discussion of the concepts of time, frequency, and energy as they will be utilized in this paper without the usual ponderous mathematics. Then these concepts are developed into defining equations for loudspeaker measurement, and hardware is designed around these relations. The second part is a presentation of experimental data obtained on actual loudspeakers tested with the hardware. The third part is an Appendix and is an analytical development of system energy principles which form the basis for this paper and its measurement of a loudspeaker by means of a remote air path measurement. The hope is that some of the mystery may be stripped from the purely mathematical approach for the benefit of those less inclined toward equations, and possibly provide a few conceptual surprises for those accustomed only to rigorous mathematics.

TIME AND FREQUENCY

The sound which we are interested in characterizing is the result of a restoration to equilibrium conditions of the air about us following a disturbance of that equilibrium by an event. An event may be a discharge of a cannon, bowing of a violin, or an entire movement of a symphony. A fundamental contribution to analysis initiated by Fourier [4] was that one could describe an event in either of two ways. The coordinates of the two descriptions, called the domains of description, are dimensionally reciprocal in order that each may stand alone in the ability to describe an event. For the events of interest in this paper one description involves the time-dependent pressure and velocity characteristics of an air medium expressed in the coordinates of time, seconds. The other description of the same event is expressed in the coordinates of reciprocal time, hertz. Because these two functional descriptions relate to the same event, it is possible to transform one such description into the other. This is done mathematically by an integral transformation called a Fourier transform. It is unfortunate that the very elegance of the mathematics tends to obscure the fundamental assertion that any valid mathematical description of an event automatically implies a second equally valid description.

It has become conventional to choose the way in which we describe an event such that the mathematics is most readily manipulated. A regrettable consequence of this is that the ponderous mathematical structure buttressing a particular choice of description may convince some that there is no other valid mathematical choice available. In fact there may be many types of representation the validity of which is not diminished by an apparent lack of pedigree. A conventional mathematical structure is represented by the assumption that a time-domain characterization is a scalar quantity while the equivalent reciprocal time-domain representation is a vector. Furthermore, because all values of a coordinate in a given domain must be considered in order to transform descriptions to the other domain, it is mathematically convenient to talk of a particular description which concentrates completely at a given coordinate and is null elsewhere. This particular mathematical entity, which by nature is not a function, is given the name impulse. It is so defined that the Fourier transform equivalent has equal magnitude at all values of that transform coordinate. A very special property of the impulse and its transform equivalent is that, under conditions in which superposition of solutions applies, any arbitrary functional description may be mathematically analyzed as an ordered progression of impulses which assume the value of the function at the coordinate chosen. In dealing with systems which transfer energy from one form to another, such as loudspeakers or electrical networks, it is therefore mathematically straightforward to speak of the response of that system to a single applied impulse. We know that in so doing we have a description which may be mathematically manipulated to give us the behavior of that system to any arbitrary signal, whether square wave or a Caruso recording.

In speaking of events in the time domain, most of us have no reservations about the character of an impulse. One can visualize a situation wherein nothing happens until a certain moment when there is a sudden release of energy which is immediately followed by a return to null. The corresponding reciprocal time representation does not have such ready human identification, so a tacit acceptance is made that its characterization is uniform for all values of its parameter. An impulse in the reciprocal time domain, however, is quite recognizable in the time domain as a sine wave which has existed for all time and will continue to exist for all time to come. Because of the uniform periodicity of the time-domain representation for an impulse at a coordinate location in the reciprocal time domain, we have dubbed the coordinate of reciprocal time as frequency. What we mean by frequency, in other words, is that value of coordinate in the reciprocal time domain where an impulse has a sine wave equivalent in the time domain with a given periodicity in reciprocal seconds. So far all of this is a mathematical manipulation of the two major ways in which we may describe an event. Too often we tend to assume the universe must somehow solve the same equations we set up as explanation for the way we perceive the universe at work. Much ado, therefore, is made of the fact that many of the signals used by engineers do not have Fourier transforms, such as the sine wave, square wave, etc. The fact is that the piece of equipment had a date of manufacture and we can be certain that it will some day fail; but while it is available it can suffice perfectly well as a source of signal. The fact that a mathematically perfect sine wave does not exist in no way prevents us from speaking of the impulse response of a loudspeaker in the time domain, or what is the same thing, the frequency response in the reciprocal time domain. Both descriptions are spectra in that the event is functionally dependent upon a single-valued coordinate and is arrayed in terms of that coordinate. If we define, for any reason, a zero coordinate in one domain, we have defined the corresponding epoch in the other domain. Since each domain representation is a spectrum description we could state that this exists in two "sides." One side is that for which the coordinate is less in magnitude than the defined zero. The form of spectral description in the general case is not dependent on the coordinate chosen for the description. Thus we could, by analogy with communication practice which normally deals with frequency spectra in terms of sidebands, say that there are time-domain sidebands. The sideband phenomenon is the description of energy distribution around an epoch in one domain due to operations (e.g., modulation) performed in the other domain. This phenomenon was used by this author to solve for the form of time-delay distortion due to propagation through a dispersive absorptive medium [3].

Fourier transform relations are valid only for infinite limits of integration and work as well for predictive systems as they do for causal. There is no inherent indignation in these transforms for a world with backward running clocks. The clock direction must be found from some other condition such as energy transformation. As pointed out previously [3], this lack of time sense led some investigators to the erroneous conclusion that group delay, a single-valued property, was uniquely related to real-world clock delay for all possible systems and has led others to the equally erroneous conclusion that a uniform group delay always guarantees a distortionless system. When we consider working with causal systems, where our clocks always run forward at constant rate, we must impose a condition on the time-domain representation that is strongly analogous to what the communication engineer calls single sideband when he describes a frequency attribute. We must, in other words, say that the epoch of zero time occurs upon stimulation of the system and that no energy due to that stimulation may occur for negative (prior) time. The conditions imposed on the other domain representation, frequency domain, by this causal requirement are described as Hermitian [21]. That is, both lower and upper sidebands exist about zero frequency and the amplitude spectrum will be even symmetric about zero frequency while the phase spectrum is odd symmetric.

The mathematical simplicity of impulse (and its sinusoidal equivalent transform) calculations has led to a tremendously useful series of analytical tools. Among these are the eigenvalue solutions to the wave equation in the eleven coordinates which yield closed form [10]. However, these are mathematical expansions which, if relating to one domain wholly, may be related to the other domain only if all possible values of coordinate are assumed. Tremendous mathematical frustration has been experienced by those trying to independently manipulate expressions in the two domains without apparently realizing that each was a description of the same event. Having assumed one description, our ground rules of analysis prevent an arbitrary choice of the description in the other domain. Because the time- and frequency-domain representations are two ways of describing the same event, we should not expect that we could maintain indefinite accuracy in a time-domain representation if we obtain this from a restricted frequency-domain measurement, no matter how clever we were. If we restrict the amount of information available to us from one domain, we can reconstruct the other domain only to the extent allowed by the available information. This is another way of expressing the interdomain dependence, known as the uncertainty principle. Later we shall consider the process of weighting a given spectrum description so as to minimize some undesirable sideband clutter when reconstructing the same information in the complementing spectral description.

When dealing with very simple systems, no difficulty is encountered in using a frequency-only or time-only representation and interpreting joint domain effects. But the very nature of the completeness of a given domain representation leads to extreme difficulty when one asks such seemingly simple questions as, "what is the time delay of a given frequency component passing through a system with nonuniform response?" A prior paper demonstrated that there is a valid third description of an event [3]. This involves a joint time–frequency characterization which can be brought into closed form if one utilizes a special primitive descriptor involving first-order and second-order all-pass transmission systems. In some ways this third description, lying as it does between the two principal descriptions, may be more readily identified with human experience. Everyone familiar with the score of a musical piece would be acutely aware of a piccolo solo which came two measures late. A frequency-only or time-only description of this musical fiasco might be difficult to interpret, even though both contain the information. Other joint domain methods have been undertaken by other investigators [5].

Applying this third description to a loudspeaker provided the model yielding time-delay distortion. It was shown that the answer to the time of emergence of a given frequency component had the surprise that at any given frequency there were multiple arrivals as a function of time. The nature of the third description was such that one could envision each frequency arrival as due to its own special perfect loudspeaker which had a frequency-dependent time delay which was single valued with frequency. If the system processing the information (in this case a loudspeaker) has a simple ordered pole and zero expansion in the frequency domain, then the arrival times for any frequency are discrete. If the expansion has branch points, then the arrival times may be a bounded distribution. The general problem for which this provides a solution is the propagation of information through a dispersive absorptive medium. Even though this characterization of information-bearing medium best fits a loudspeaker in a room, as well as most real-world propagation problems, attempts at solutions have been sparse [6], [7].

The equipment available to us to make measurements on a loudspeaker, such as oscilloscopes and spectrum analyzers, work in either of the primary domains and so do not present the third-domain results directly. This does not mean that other information processing means, perhaps even human perception of sound, work wholly in the primary domains. Within the restrictions of the uncertainty principle, which is after all a mathematical limitation imposed by our own definitions, we shall take a given frequency range and find the time delay of all loudspeaker frequency components within that range. The nature of this type of distortion is illustrated schematically in Fig. 1. If a momentary burst of energy $E(t)$ were fed a perfect loudspeaker, a similar burst of energy $E'(t)$ would be observed by O some time later due to the finite velocity of propagation c. More generally an actual loudspeaker will be observed by O to have a time smeared energy distribution $\epsilon(t)$. As far as the observer is concerned, the actual loudspeaker will have a spatial smear $\epsilon(x)$.

ENERGY, IMPULSE AND DOUBLET

Anyone familiar with analysis equipment realizes that the display of Fig. 1 will take more than some simple assembly of components. In fact, it will take a closer scrutiny of the fundamental concepts of energy, frequency, and time. The frequency-domain representation of an event is a complex quantity embodying an amplitude and a phase description. The time-domain representation of the same event may *also* be expressed as a complex quantity. The scalar representation of time-domain performance of a transmission system based on impulse excitation, which is common coinage in communication engineering [9], is the real part of a more general vector. The imaginary part of that vector is the Hilbert transform [2], [4] of the real part and is associated with a special excitation signal called a doublet by this author. For a nonturbulent (vortex free) medium wherein a vector representation is sufficient, the impulse and doublet responses completely characterize performance under conditions of superposition. For a turbulent medium one must use an additional tensor excitation which in most cases is a quadrupole. For all loudspeaker tests we will perform, we need only concern ourselves with the impulse and doublet response.

Any causal interception of information from a remote source implies an energy density associated with the actions of that source. The energy density represents the amount of useful work which could be obtained by the receiver if he were sufficiently clever. For the cases of interest in this paper, the total energy density at the point of reception is composed of a kinetic and a potential energy density component. These energy density terms relate to the instantaneous state of departure from equilibrium of the medium due to the actions of the remote source.

If we wish to evaluate the amount of total work which could be performed on an observer, whether microphone diaphragm or eardrum, at any moment, such as given in Fig. 1, then we must evaluate the instantaneous total energy density. In order to specify how much energy density is available to us from a loudspeaker, and what time it arrives at our location if it is due to a predetermined portion of the frequency spectrum, we must choose our test signal very carefully and keep track of the ground rules of the equivalence of time and frequency descriptions. We may not, for example, simply insert a narrow pulse of electrical energy into a loudspeaker, hoping that it simulates an impulse, and view the intercepted microphone signal on an oscilloscope. What must be done is to determine first what frequency range is to be of interest; then generate a signal which contains only those frequencies. By the process of generation of the excitation signal for a finite frequency band, we have defined the time epoch for this signal. Interception of the loudspeaker acoustic signal should then be made at the point of desired measurement. This interception should include both kinetic and potential energy densities. The total energy density, obtained as a sum of kinetic and potential energy densities, should then be displayed as a function of the time of interception.

The foregoing simplistic description is exactly what we shall do for actual loudspeakers. Those whose experiments are conducted in terms of the time domain only will immediately recognize that such an experiment is commonly characterized as physically unrealizable in the sense that having once started a time-only process, one cannot arbitrarily stop the clock or run it backward. In order to circumvent this apparent difficulty we shall make use of the proper relation between frequency and time descriptions. Rather than use true physical time for a measured parameter, the time metric will be obtained as a Fourier transform from a frequency-domain measurement. Because we are thus allowed to redefine the time metric to suit our measurement, it is possible to alter the time base in any manner felt suitable. The price paid is a longer physical time for a given measurement.

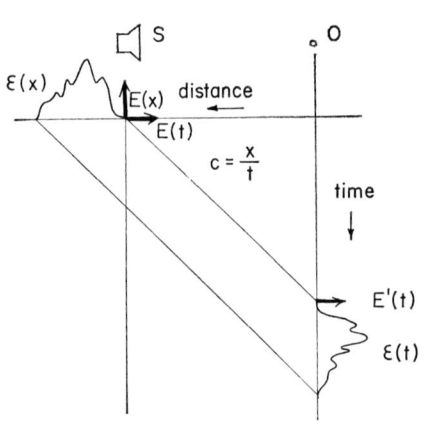

Fig. 1. Symbolic representation of time–distance world line for observer O perceiving energy from loudspeaker S.

A loudspeaker energy plot representing one millisecond may take many seconds of real clock time to process, depending upon the time resolution desired. The process utilized will coincide in large measure with some used in coherent communication practice, and the amount by which the derived time metric exceeds the real clock time will correspond to what is called filter processing gain. The basic signal process for our measurement will start with that of a time delay spectrometer (TDS) [1], [16]. This is due not only to the basic simplicity of instrumentation, but the "domain swapping" properties of a TDS which presents a complex frequency measurement as a complex time signal and vice versa.

ANALYSIS, IMPULSE AND DOUBLET

If we consider that a time-dependent disturbance $f(t)$ is observed, we could say that either this was the result of a particular excitation of a general parameter $f(x)$ such that

$$f(t) = \lim_{\lambda \to \infty} \int_{-\infty}^{\infty} f(x) \frac{\sin \lambda (t-x)}{\pi (t-x)} dx \qquad (1)$$

or it was the result of operation on another parameter $F(\omega)$ such that

$$f(t) = \int_{-\infty}^{\infty} F(\omega) e^{i\omega t} d\omega. \qquad (2)$$

Eq. (1) is known as Fourier's single-integral formula [4, p. 3] and is frequently expressed as

$$f(t) = \int_{-\infty}^{\infty} f(x) \delta(t-x) dx \qquad (3)$$

where $\delta(t-x)$ is understood to mean the limiting form shown in Eq. (1) and is designated as an impulse because of its singularity behavior [10].

The functions $f(x)$ and $F(\omega)$ of Eqs. (2) and (3), Fourier's two descriptions of the same event, are thus considered to be paired coefficients in the sense that each of them multiplied by its characteristic "driving function" and integrated over all possible ranges of that driving function yields the same functional dependence $f(t)$. This paired coefficient interpretation is that expressed by Campbell and Foster in their very significant work [8].

Eq. (3) may be looked upon as implying that there was a system, perhaps a loudspeaker, which had a particular characterization $f(x)$. When acted upon by the driving signal $\delta(t-x)$, the response $f(t)$ was the resultant output. Conversely, Eq. (2) implies that there was another equally valid characterization $F(\omega)$ which when acted upon by the driving signal $e^{i\omega t}$ produced the same response $f(t)$.

The functions $f(x)$ and $F(\omega)$ are of course Fourier transforms of each other. The reason for not beginning our discussion by simply writing down the transform relations as is conventional practice is that to do so tends to overlook the real foundations of the principle. To illustrate that these functions are not the only such relations one could use, consider the same system with a driving signal which elicits the response

$$g(t) = \lim_{\lambda \to \infty} \int_{-\infty}^{\infty} f(x) \left\{ \frac{\cos \lambda (t-x) - 1}{\pi (t-x)} \right\} dx \qquad (4)$$

which as before we shall assume to exist as the limiting form

$$g(t) = \int_{-\infty}^{\infty} f(x) d(t-x) dx. \qquad (5)$$

Also,

$$g(t) = \int_{-\infty}^{\infty} F(\omega) \{-i \, \text{sgn}(\omega)\} e^{i\omega t} d\omega. \qquad (6)$$

Obviously this is an expression of the Hilbert transform of Eqs. (2) and (3) [2], [4]. It is none the less a legitimate paired coefficient expansion of two ways of describing the same phenomenon $g(t)$. By observing the way in which $\delta(t-x)$ and $d(t-x)$ behave as the limit is approached in Eqs. (1) and (4), it is apparent that both tend to zero everywhere except in a narrow region around the value where $t=x$. Thus there is not one, but at least two driving functions which tend toward a singularity behavior in the limit. As we shall see, these two constitute the most important set of such singularity operators when discussing physical properties of systems such as loudspeakers. Because of the nature of singularity approached by each, we shall define them as impulse and doublet, respectively. The following definitions will be assumed.

The impulse operator is approached as the defined limit

$$\delta(t) = \lim_{a \to \infty} \frac{\sin at}{\pi t}. \qquad (7)$$

The impulse operator is not a function but is defined from Eq. (3) as an operation on the function $f(x)$ to produce the value $f(t)$. $\delta(t)$ is even symmetric. The application of an electrical replica of the impulse operator to any network will produce an output defined as the impulse response of that network. The Fourier transform $F(\omega)$ of this impulse response $f(t)$ is defined as the frequency response of the network and is identical at any frequency to the complex quotient of output to input for that network when excited by a unit amplitude sine-wave signal of the given frequency.

The doublet operator is approached as the defined limit

$$d(t) = \lim_{a \to \infty} \frac{\cos at - 1}{\pi t}. \qquad (8)$$

The doublet operator is not a function but is defined from Eq. (5) as an operation on the function $f(x)$ to produce the value $g(t)$. $d(t)$ is odd symmetric. The application of an electrical replica of the doublet operator to any network will produce an output defined as the doublet response of that network. The doublet response is the Hilbert transform of the impulse response. The Fourier transform of the doublet response is identical to that of the impulse response, with the exception that the doublet phase spectrum is advanced ninety degrees for negative frequencies and retarded ninety degrees for positive frequencies.

In addition to the above definitions, the Fourier transform of the impulse response will be defined as being of minimum phase type in that the accumulation of phase lag for increasing frequency is a minimum for the resultant amplitude spectrum. The Fourier transform of the doublet response will be defined as being of nonminimum phase.

It must be observed that the doublet operator defined here is not identical to that sometimes seen derived from the impulse as a simple derivative and therefore possessing a transform of nonuniform amplitude spectral density [9, p. 542]. The corresponding relation between the doublet operator $d(t)$ and the impulse operator $\delta(t)$ prior to the limiting process is

$$d(t) = -\frac{1}{\pi} \frac{d}{dt} \int_{-\infty}^{\infty} \delta(x) \ln \left| 1 - \frac{t}{x} \right| dx. \qquad (9)$$

The distinction is that the doublet operator defined here has the same power spectral density as the impulse operator. Furthermore as can be seen from Eq. (9), the doublet operator may be envisioned as the limit of a physical doublet, as defined in classical electrodynamics [10].

ANALYTIC SIGNAL

We have thus defined two system driving operators, the impulse and the doublet, which when applied to a system produce a scalar time response. Although the relation between the time-domain responses is that of Hilbert transformation, if one were to view them as an oscilloscope display, he may find it hard to believe they were attributable to the same system. However, the resultant frequency-domain representations are, except for the phase reference, identical in form. We will now develop a generalized response to show that it is not possible to derive a unique time behavior from incomplete knowledge of a restricted portion of the frequency response.

Symbolizing the operation of the Fourier integral transform by the double arrow ↔, we can rewrite Eqs. (2) and (6) as the paired coefficients

$$f(t) \leftrightarrow F(\omega) \qquad (10)$$

$$g(t) \leftrightarrow -i(\text{sgn } \omega)F(\omega). \qquad (11)$$

Multiplying Eq. (10) by a factor $\cos \lambda$ and Eq. (11) by $\sin \lambda$ and combining,

$$\cos \lambda \cdot f(t) + \sin \lambda \cdot g(t) \leftrightarrow F(\omega)[\cos \lambda - i(\text{sgn } \omega) \sin \lambda]. \qquad (12)$$

The frequency-domain representation is thus

$$\begin{array}{ll} F(\omega)e^{-i\lambda}, & 0 < \omega \\ F(\omega)e^{i\lambda}, & 0 > \omega. \end{array} \qquad (13)$$

In this form it is apparent that if we were to have an accurate measurement of the amplitude spectrum of the frequency response of a system, such as a loudspeaker, and did not have any information concerning its phase spectrum, we could not uniquely determine either the impulse response or doublet response of that loudspeaker. Such an amplitude-only spectrum would arise from a standard anechoic chamber measurement or from any of the power spectral density measurements using noncoherent random noise. This lack of uniqueness was pointed out in an earlier paper [2]. The best that one could do is to state that the resultant time-domain response is some linear combination of impulse and doublet response.

Because the time domain representation is a scalar, it is seen that Eq. 12 could be interpreted as a scalar operation on a generalized time-domain vector such that

$$\text{Re}[e^{-i\lambda}h(t)] \leftrightarrow F(\omega) \cdot e^{-i\lambda\{\text{sgn } \omega\}} \qquad (14)$$

where $\text{Re}(x)$ means real part of and where the vector $h(t)$ is defined as

$$h(t) = f(t) + ig(t). \qquad (15)$$

This vector is commonly called the analytic signal in communication theory, where it is normally associated with narrow-band processes [11], [12]. As can be seen from (14) and the Appendix, it is not restricted to narrow-band situations but can arise quite legitimately from considerations of the whole spectrum. This analytic signal is the general time-domain vector which contains the information relating to the magnitude and partitioning of kinetic and potential energy densities.

The impulse and doublet response of a physical system, which in our case is the loudspeaker in a room, is related to the stored and dissipated energy as perceived. This means that if one wishes to evaluate the time history of energy in a loudspeaker, it is better sought from the analytic signal of Eq. (15). It is not sufficient to simply use the conventional impulse response to attempt determination of energy arrivals for speakers. The magnitude of the analytic signal is an indication of the total energy in the signal, while the phase of the analytic signal is an indication of the exchange ratio of kinetic to potential energy. The exchange ratio of kinetic to potential energy determines the upper bound for the local speed of propagation of physical influences capable of producing causal results. We call this local speed the velocity of propagation through the medium. From the basic Lagrange relations for nonconservative systems [13] it may be seen that the dissipation rate of energy is related to the time rate of change of the magnitude of the analytic signal. If the system under analysis is such as to have a source of energy at a given time and is dissipative thereafter with no further sources, the magnitude of the analytic signal will be a maximum at that time corresponding to the moment of energy input and constantly diminishing thereafter. The result of this is that if we wish to know the time history of effective signal sources which contribute to a given portion of the frequency spectrum, it may be obtained by first isolating the frequency spectrum of interest, then evaluating the analytic signal obtained from a Fourier transform of this spectrum, finally noting those portions of the magnitude of the analytic signal which are stationary with time (Hamilton's principle) [10].

It becomes apparent that by this means we will be able to take a physical system such as a loudspeaker in a room and determine not only when the direct and reflected sound arrives at a given point, but the time spread of any given arrival. Because we will be measuring the arrival time pattern for a restricted frequency range, it is important to know what tradeoffs exist because of the spectrum limitations, and how the effects can be minimized.

SPECTRUM WEIGHTING

We will be characterizing the frequency-dependent time delay of a loudspeaker. The nature of the testing signal which we use should be such that minimum energy exists outside the frequency band of interest, while at the same time allowing for a maximum resolution in the time domain. This joint domain occupancy problem has been around for quite some time and analytical solutions exist [14]. For loudspeaker testing where we wish to know the time-domain response for a restricted frequency band, we can use any signal which has a frequency spectrum bounded to the testing band with minimum energy outside this band [19], [20]. An intuitive choice of a signal with a rectangular shaped frequency spectrum which had maximum occupancy of the testing band would not be a good one, because the time-domain characterization while sharply peaked would not fall off very rapidly on either side of the peak. The consequence of this is that a genuine later arrival may be lost in the coherent sideband clutter of a strong signal. A much better choice of band-limited spectrum would be one which places more energy in the midband frequency while reducing the energy at band edge. Such a spectrum is said to be weighted. The weighting function is the frequency-dependent multiplier of the spectral components. An entire uniform spectrum is spoken of as unweighted, and a uniform bounded spectrum is said to be weighted by a rectangular function.

The proper definition of spectrum weighting must take into account both phase and amplitude. If a rectangular amplitude, minimum phase weighting is utilized, the resultant time function will be given by Eq. (1) without the limit taken. The parameter λ will be inversely proportional to the bandwidth. If rectangular amplitude but nonminimum phase weighting is utilized, then the time function will be given by Eq. (4).

It should be obvious that one can weight either the time or frequency domain. Weighting in the time domain is frequently referred to as shaping of the pulse response. The purpose in either case is to bound the resultant distribution of energy. Two types of weighting will be utilized in this paper. The first is a Hamming weighting [14, p. 98] and takes the form shown in Fig. 2a. As used for spectral limitation, this is an amplitude-only weighting with no resultant phase shift. Although this type of weighting cannot be generated by linear circuits, it can be obtained in an on-line processor by nonlinear means. The second weighting is a product of two functions. One function is the minimum phase amplitude and phase spectrum of a tuned circuit. The other function is shown in Fig. 2b. This second weighting is that utilized in a TDS which will be used as a basic instrument for this paper.

Reference to the earlier paper and its analysis discloses that within the TDS intermediate frequency amplifier (which of course could be centered at zero frequency), the information relating to a specific signal arrival is contained in the form

$$o(t) = [i(t) \otimes w(t)] \otimes S(at) \qquad (16)$$

where $\otimes$ signifies convolution, $S(at)$ is the complex

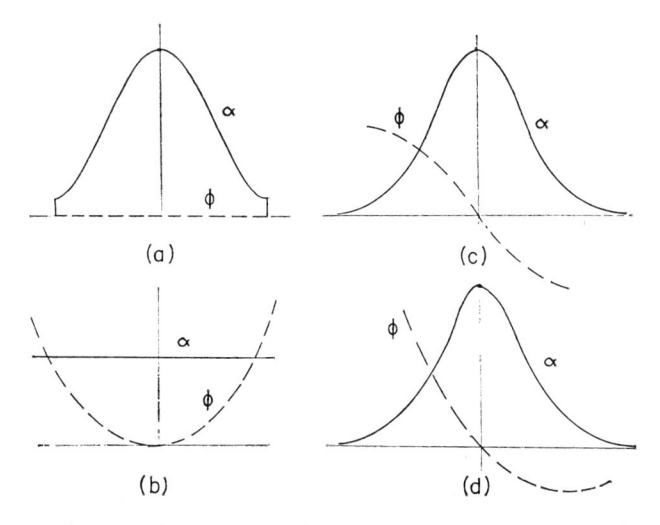

Fig. 2. Various system weighting functions including amplitude (solid line) and phase (dashed line) utilized to bound the resultant energy when taking a Fourier transform. **a.** Hamming weighting. **b.** Quadratic phase all-pass weighting. **c.** Passive simple resonance weighting. **d.** Simple TDS weighting formed as a product of **b** and **c**.

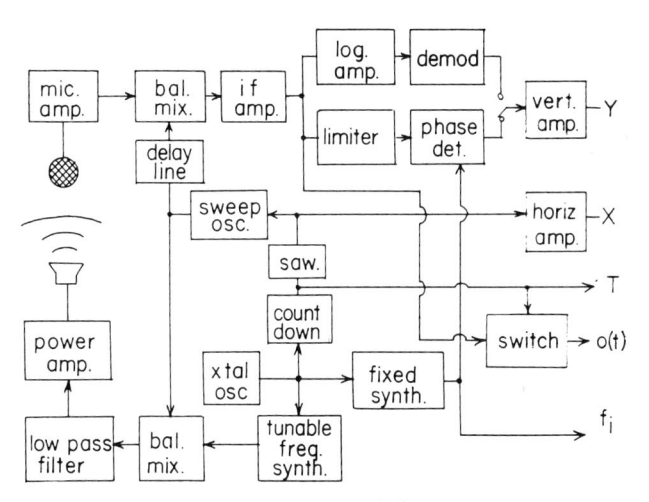

Fig. 3. Block diagram of simple TDS.

Fourier transform of the impulse response of the system under test, $i(t)$ is the impulse response of the intermediate frequency amplifier, and $w(t)$ is the window function defined as the impulse response of a quadratic phase circuit. It can be seen that the TDS provides a weighting of the time domain arrivals of that signal selected in order to give an optimum presentation of the frequency spectrum.

TRANSFORMATION TO TIME

We will now consider how to convert a measured frequency response to the analytic signal. We must start of course by having the loudspeaker frequency response available. Assume then that the system under test is evaluated by the TDS. The block diagram of this is reproduced in Fig. 3 from an earlier paper. As before, a tunable frequency synthesizer is used. Assume that the output of the intermediate frequency amplifier is taken as shown. This output signal will be of the form of Eq. (16), but of course translated to lie at the center of the intermediate frequency. Because the frequency deviation of the sweep oscillator is restricted to that portion of the frequency spectrum of interest (dc to 10 kHz, for example), the signal $o(t)$ is representative of this restricted range. The duration of sweep will be a fixed value of T seconds, so that $o(t)$ is a signal repetetive in the period T. Assume that we close the switch for $o(t)$ shown in Fig. 3 for a period of T seconds and open it prior to and following that time. The signal characterization out of this switch is

$$e^{i\omega_i t}[i(t) \otimes w(t) \otimes S(at)] \, \text{Rect}\,(t-T) \qquad (17)$$

where the rectangular weighting function is defined as,

$$\text{Rect}\,(t-T) = \begin{matrix} 1, & 0 < t < T \\ 0, & \text{elsewhere.} \end{matrix} \qquad (18)$$

Assume we now multiply the signal by the complex quantity

$$e^{-i\omega_i t} e^{i\Omega t} \qquad (19)$$

and the product is in turn multiplied by a weighting function $A(t)$. If we take the integral of the product of these functions, we have

$$\int_{-\infty}^{\infty} [\text{Rect}(t-T)] \cdot A(t) \cdot \{i(t) \otimes w(t) \otimes S(at)\} e^{i\Omega t} dt. \quad (20)$$

The infinite limits are possible because of the rectangular function which vanishes outside the finite time limits. The integral of Eq. (20) may now be recognized as a Fourier transform from the t domain to the Ω domain. This may be expressed in the Ω domain as

$$a(\Omega) \otimes [I(\Omega) \cdot e^{-i\Omega^2/2a} \cdot h(\Omega)] \quad (21)$$

where $a(\Omega)$ is the transform of the weighting in the t domain, $I(\Omega)$ is the frequency response of the intermediate frequency amplifier expressed in the Ω domain, the exponential form is the quadratic phase window function, and $h(\Omega)$ is the Ω domain form of the analytic function shown in Eq. (15). The reason that this is $h(\Omega)$ and not the impulse response $f(\Omega)$ is that we have assumed a TDS frequency sweep from one frequency to another, where both are on the same side of zero frequency.

What we have done by all this is instrument a technique to perform an inverse Fourier transform of a frequency response. The answer appears as a voltage which is a function of an offset frequency Ω. Even though we energized the loudspeaker with a sweeping frequency, we obtain a voltage which corresponds to what we would have had if we fed an infinitely narrow pulse through a perfect frequency-weighted filter to a loudspeaker. We now say that by changing an offset frequency Ω, the answer we see is what would have been observed had we used a pulse and evaluated the response at a particular moment in time. By adjusting the offset frequency we can observe the value that would be seen for successive moments in time.

SINGLE- AND DOUBLE-SIDED SPECTRA

We will be making measurements in the frequency domain and from this calculate the time-domain energy arrivals. If our measurement includes zero frequency, we have shown earlier how one could invoke the odd symmetry requirement to define "absolute" phase [2]. By so doing we have eliminated the parameter λ from Eq. (13). It is thus possible to calculate either the impulse or doublet time response in a unique manner. However, since the lower and upper sidebands are redundant in the frequency domain, care must be taken in using Eq. (20) that only one sideband, for example, positive frequencies, be used and the other sideband rejected. Failure to do this will result in an improper calculation not only of impulse and doublet response, but of the analytic signal as well. If one is aware of this single-sided versus double-sided spectrum pitfall, he may use it to advantage. For example, if a perfectly symmetric double-sided spectrum is used, the analytic signal calculation will yield the impulse response directly, as can be seen from Eq. (12).

Most loudspeaker measurements are made in a single-sided manner. For example, one may wish to know what time distribution arises from the midrange driver which works from 500 Hz to 10 kHz. In this case, because zero frequency is not available it may not be possible to define the parameter λ of Eq. (13), even if both amplitude and phase spectra are measured. Because of this an unequivocal determination of the impulse response (potential energy relation) or of the doublet response (kinetic energy relation) may be impossible. One may always determine the analytic signal magnitude (potential plus kinetic energy relation). The time position of effective energy sources can be determined by noting the moments when the effective signal energy is a maximum.

INSTRUMENTATION

The loudspeaker energy arrivals are obtained from the analytic signal. The signals of interest are first isolated from the remainder of the room reflections by means of a TDS. This measurement is the frequency domain description of the loudspeaker anechoic response, even though the loudspeaker is situated in a room. This loudspeaker description, although mathematically identical to a frequency-domain description, has been made available within the TDS in the time domain. In order to take a Fourier transform of this frequency-domain description to obtain the time-domain analytic signal, we must multiply by a complex sinusoid representing the time epoch, multiply this in turn by a complex weighting function, and then integrate over all possible frequencies (Eq. 20). This process would normally require substantial digital computational facilities for a frequency-domain measurement; however, the "domain swapping" properties of a TDS allow for straightforward continuous signal processing. Fig. 4 is a block diagram of the functions added to the TDS of Fig. 3 to effect this process. The signal from the intermediate frequency amplifier of the TDS is buffered and fed to two balanced mixers. By using an in-phase and quadrature multiplier of the same frequency as the intermediate frequency, the two outputs are obtained which are Hilbert transforms of each other and centered at zero frequency. Each of these is then isolated by low-pass filters and processed by identical switching multipliers controlled by a cosine function of the sweep time to effect a Hamming weighting. The net output of each weighting network is then passed through sampling integrators. At the start of a TDS sweep, the integrators are set to zero by

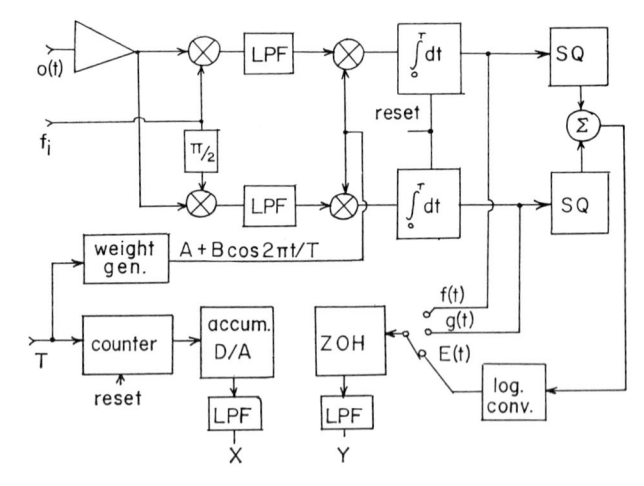

Fig. 4. Block diagram of Fourier transformation equipment attaching to a TDS capable of displaying time-domain plots. **a.** Impulse response $f(t)$. **b.** Doublet response $g(t)$. **c.** Total energy density $E(t)$.

the same clock pulse that phase locks the TDS offset frequency synthesizer so as to preserve phase continuity. Each integrator then functions unimpeded for the duration of the sweep. If the proper phase has been set into the offsetting synthesizer, the output of one integrator at the end of the sweep will correspond to the single value of the impulse response for the moment of epoch chosen, while the other will correspond to doublet response. If the proper phase has not been selected, then one integrator will correspond to a linear combination such as expressed in Eq. (12), while the other integrator will correspond to the quadrature term.

If one desires to plot the impulse or doublet response, the appropriate integrator output may be sent to a zero-order hold circuit which clocks in the calculated value and retains it during the subsequent sweep calculation. This boxcar voltage may then be recorded as the ordinate on a plotter with the abscissa proportional to the epoch. One trick of the trade which is used when horizontal and vertical signals to a plotter are stepped simultaneously, is to low-pass filter both channels with the same cutoff frequency. The plotter will now draw straight lines between interconnecting points.

If one is interested in the magnitude of the analytic signal (14), which from the Appendix, to be published December 1971, is related to energy history, then the most straightforward instrumentation is to square the output from each integrator and linearly add to get the sum of squares. A logarithmic amplifier following the sum of squares will enable a signal strength reading in dB without the need for a square root circuit. By doing this, a burden is placed on this logarithmic amplifier since a 40-dB signal strength variation produces an 80-dB input change to the logarithmic element. Fortunately, such an enormous range may be accommodated readily by conventional logarithmic elements. For graphic recordings of energy arrival, the output of the logarithmic amplifier may be fed through the same zero-order hold circuit as utilized for impulse and doublet response.

The configuration of Fig. 4 including quadrature multipliers, sampled integrators, and sum of squares circuitry is quite often encountered in coherent communication practice [17], [18]. This circuit is known to be an optimum detector in the mean error sense for coherent signals in a uniformly random noise environment. Its use in this paper is that of implementing an inverse Fourier transform for total energy for a single-sided spectrum. An interesting byproduct of its use is thus an assurance that no analytically superior instrumentation as yet exists for extracting the coherent loudspeaker signal from a random room noise environment.

CHOICE OF MICROPHONE

The information which our coherent analysis equipment utilizes is related to the energy density intercepted by the microphone. The total energy density in joules per cubic meter is composed of kinetic energy density E_T and potential energy density E_V, where [22, p. 356]

$$E_T = \frac{1}{2} \rho_0 v^2$$

$$E_V = \frac{1}{2} \rho_0 c^2 s^2. \quad (22)$$

The equilibrium density is ρ_0, v is the particle velocity, s is condensation or density deviation from equilibrium, and c is velocity of energy propagation.

At first glance it might be assumed that total energy may not be obtained from either a pressure responsive microphone which relates to E_V or a velocity microphone relating to E_T. The answer to this dilemma may be found in the Appendix. One can always determine one energy component, given the other. Hence a determination of acoustic pressure or velocity or an appropriate mixture of pressure and velocity is sufficient to characterize the energy density of the original signal.

This means that any microphone, whether pressure, velocity, or hybrid, may be used for this testing technique, provided that a calibration exists over the frequency range for a given parameter. This also means that any perceptor which is activated by total work done on it by the acoustic signal will not be particular, whether the energy bearing the information is kinetic or potential. There is some reason to believe that human sound preception falls into this category.

REFERENCES

[1] R. C. Heyser, "Acoustical Measurements by Time Delay Spectrometry," *J. Audio Eng. Soc.*, vol. 15, p. 370 (1967).

[2] R. C. Heyser, "Loudspeaker Phase Characteristics and Time Delay Distortion: Part 1," *J. Audio Eng. Soc.*, vol. 17, p. 30 (1969).

[3] R. C. Heyser, "Loudspeaker Phase Characteristics and Time Delay Distortion: Part 2," *J. Audio Eng. Soc.*, vol. 17, p. 130 (1969).

[4] E. C. Titchmarsh, *Introduction to the Theory of Fourier Integrals* (Oxford Press, London, 2nd ed., 1948).

[5] C. H. Page, "Instantaneous Power Spectra," *J. Appl. Phys.*, vol. 23, p. 103 (1952).

[6] L. Brillouin, *Wave Propagation and Group Velocity* (Academic Press, New York, 1960).

[7] H. G. Baerwald, "Über die Fortpflanzung von Signalen in dispergierenden Systemen," *Ann. Phys.*, vol. 5, p. 295 (1930).

[8] G. A. Campbell and R. M. Foster, *Fourier Integrals* (D. Van Nostrand, Princeton, N.J., 1961).

[9] E. A. Guillemin, *The Mathematics of Circuit Analysis* (M.I.T. Press, Cambridge, Mass., 1965).

[10] E. U. Condon and H. Odishaw, *Handbook of Physics* (McGraw-Hill, New York, 1967).

[11] R. S. Berkowitz, *Modern Radar* (John Wiley, New York, 1967).

[12] L. E. Franks, *Signal Theory* (Prentice-Hall, Englewood Cliffs, N.J., 1969).

[13] T. v. Karman and M. A. Biot, *Mathematical Methods in Engineering* (McGraw-Hill, New York, 1940).

[14] R. B. Blackman and J. W. Tukey, *Measurement of Power Spectra* (Dover, New York, 1959).

[15] A. W. Rihaczek, *Principles of High-Resolution Radar* (McGraw-Hill, New York, 1969).

[16] R. C. Heyser, "Time Delay Spectrometer," U.S. Patent 3,466,652, Sept. 9, 1969.

[17] A. J. Viterbi, *Principles of Coherent Communication* (McGraw-Hill, New York, 1966).

[18] H. L. VanTrees, *Detection, Estimation, and Modulation Theory* (John Wiley, New York, 1968).

[19] T. A. Saponas, R. C. Matson, and J. R. Ashley, "Plain and Fancy Test Signals for Music Reproduction Systems," *J. Audio Eng. Soc.*, vol. 19, pp. 294-305 (Apr. 1971).

[20] A. Schaumberger, "Impulse Measurement Techniques for Quality Determination in Hi-Fi Equipment, with Special Emphasis on Loudspeakers," *J. Audio Eng. Soc.*, vol. 19, pp. 101-107 (Feb. 1971).

[21] R. Bracewell, *The Fourier Transform and Its Applications* (McGraw-Hill, New York, 1965).

[22] D. H. Menzel, *Fundamental Formulas of Physics* (Prentice-Hall, Englewood Cliffs, N.J., 1965).

[23] E. A. Guillemin, *Theory of Linear Physical Systems* (John Wiley, New York, 1963).

[24] I. S. Gradshteyn and I. M. Ryzhik, *Table of Integrals, Series and Products* (Academic Press, New York, 1965), Eq. 3.782.2.

[25] B. B. Baker and E. T. Copson, *The Mathematical Theory of Huygens' Principle* (Oxford Press, London, 1953).

THE AUTHOR

Richard C. Heyser received his B.S.E.E. degree from the University of Arizona in 1953. Awarded the AIEE Charles LeGeyt Fortescue Fellowship for advanced studies he received his M.S.E.E. from the California Institute of Technology in 1954. The following two years were spent in post-graduate work at Cal Tech leading toward a doctorate. During the summer months of 1954 and 1955, Mr. Heyser was a research engineer specializing in transistor circuits with the Motorola Research Laboratory, Phoenix, Arizona. From 1956 until the present time he has been associated with the California Institute of Technology Jet Propulsion Laboratory in Pasadena, California where he is a senior member of the JPL Technical Staff.

Mr. Heyser has presented several papers before the AES and is a member and Fellow of the Audio Engineering Society, the Institute of Electrical and Electronic Engineers, as well as Tau Beta Pi, Pi Mu Epsilon, Phi Kappa Phi, Sigma Pi Sigma, and Sigma Xi. He is also a member of Theta Tau.

Determination of Loudspeaker Signal Arrival Times
Part II

RICHARD C. HEYSER

Jet Propulsion Laboratory, California Institute of Technology, Pasadena, Calif.

EXPERIMENT

The information to be determined is the time delay of total acoustic energy that would be received from a loudspeaker if fed from an impulse of electrical energy. Because we are interested in that energy due to a preselected portion of the frequency band, we may assume that the impulse is band limited by a special shaping filter prior to being sent to the loudspeaker. This filter would not be physically realizable if we actually used an impulse for our test; but since we are using a method of coherent communication technology, we will be able to circumvent that obstacle. Fig. 5 shows three responses for a midrange horn loaded loudspeaker. The frequency band is dc to 10 kHz, and each response is measured on the same time scale with zero milliseconds corresponding to the moment of speaker excitation. The driver unit was three feet from the microphone. Curve (a) is the measured impulse response and is what one would see for microphone pressure response, had the loudspeaker been driven by a voltage impulse. Curve (b) is the measured doublet response and is the Hilbert transform of (a). In both (a) and (b) the measured ordinate is linear voltage. Curve (c) is the total received energy on a logarithmic scale. Here the interplay of impulse, doublet, and total energy is evident.

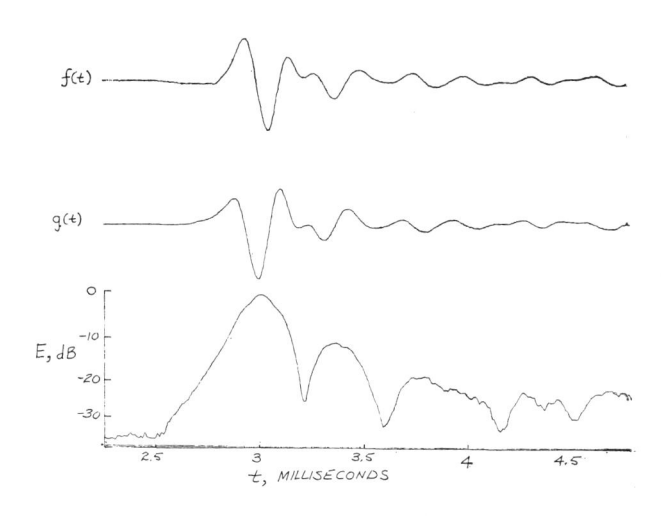

Fig. 5. Measured plots of impulse response $f(t)$, doublet response $g(t)$, and total energy density E for a midrange horn loudspeaker for spectral components from dc to 10 kHz.

---

* Presented April 30, 1971, at the 40th Convention of the Audio Engineering Society, Los Angeles. For Part I, please see pp. 734-743 of the October 1971 issue of this *Journal*.

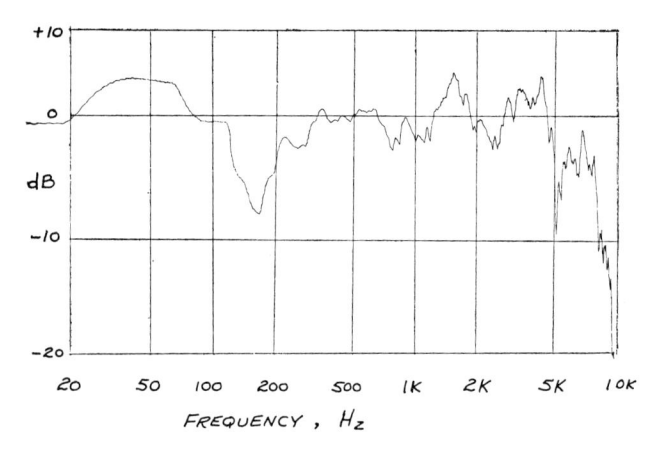

Fig. 6. TDS plot of frequency response (amplitude only) of an eight-inch open-back cabinet mounted loudspeaker with microphone to cone air path spacing of three feet.

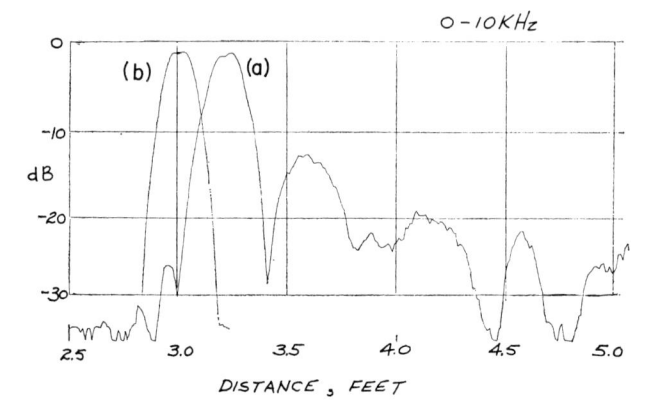

Fig. 7. Curve (a)—Energy-time arrival for loudspeaker of Fig. 6, taking all frequency components from dc to 10 kHz. Curve (b)—Superimposed measured curve to be expected if loudspeaker did not have time-delay distortion.

Fig. 6 is the TDS measured amplitude frequency response of a good quality eight-inch loudspeaker mounted in a small open-back cabinet. For simplicity the phase spectrum is not included. Fig. 7 is the time delay of energy for the speaker of Fig. 6. Superimposed on this record is a plot of what the time response would have been, had the actual loudspeaker position and acoustic position coincided and if there were no time-delay distortion. It is clear from this record that time-delay distortion truly exists. If one considers all response within 20 dB of peak, it is evident that this loudspeaker is smeared out by about one foot behind its apparent physical location. It should be observed that the response dropoff above 5 kHz coincides with a gross time delay of about three inches, as predicted by earlier analysis (Part I, [2]).

Fig. 8 is a plot of energy versus equivalent distance in feet for a high-efficiency midrange horn loaded driver. The band covered is 500 Hz to 1500 Hz and includes the region from low-frequency cutoff to midrange. The physical location of the driver phase plug is shown, and it may be seen that the acoustic and physical positions differ by nearly one foot. Fig. 9 is the same driver, but the band is 1000 Hz to 2000 Hz. The acoustic position is now closer to the phase plug and a hint of a double hump in delay is evident. Because the bandwidth is 1000 Hz, the spatial resolution available does not allow for more complete definition of acoustic position for this type of display.

Fig. 10 is a data run on an eight-inch wide range loudspeaker without baffle. A scaled pseudo cross section of the loudspeaker is shown for reference. The frequency range covered is dc to 20 kHz. Although the time-delay value may vary from one part of the spectrum to another, it is apparent that a wide frequency range percussive signal may suffer a spatial smear of the order of six inches.

Fig. 11 is a medium-quality six-inch loudspeaker mounted in an open-back cabinet. The position of main energy is, as predicted, quite close to that which would be assumed for a cutoff at about 5 kHz. The secondary hump of energy from 3 to 3.5 milliseconds is not due to acoustic energy spilling around the side of the enclosure, but is a time delay inherent in the loudspeaker itself.

Fig. 12 is the energy received from an unterminated midrange driver excited from dc to 10 kHz. The effect of untermination is seen as a superposition of an exponential time delay, due to a relatively high Q resonance, and internal reverberation with a 0.3-millisecond period.

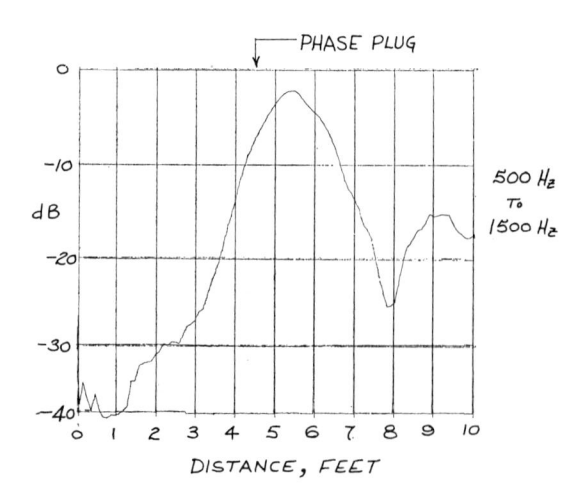

Fig. 8. Energy-time arrival for high-quality midrange horn loudspeaker for all components from 500 Hz to 1500 Hz. Measured position of phase plug is shown.

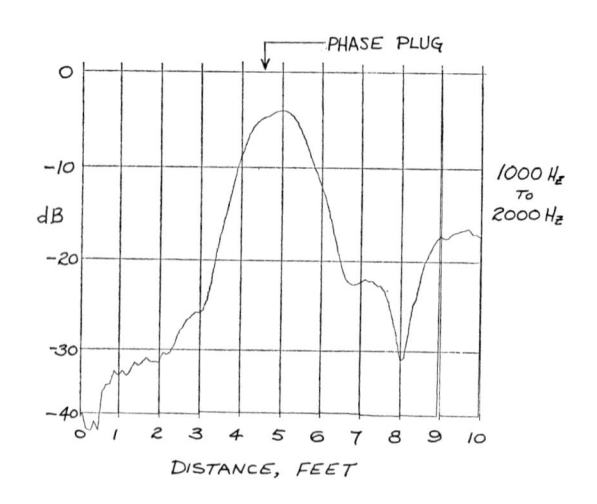

Fig. 9. Same loudspeaker as Fig. 8, but excitation is from 1000 Hz to 2000 Hz.

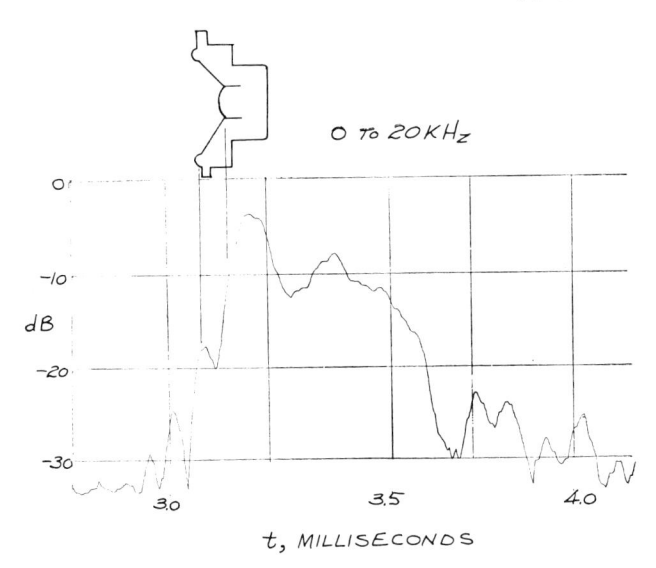

DETERMINATION OF LOUDSPEAKER SIGNAL ARRIVAL TIMES: PART II

Fig. 10. Energy-time arrivals for unbaffled high-quality eight-inch loudspeaker. Frequency band is dc to 20 kHz and phantom sketch of loudspeaker physical location is included for identification of amount of time-delay distortion relative to speaker dimensions.

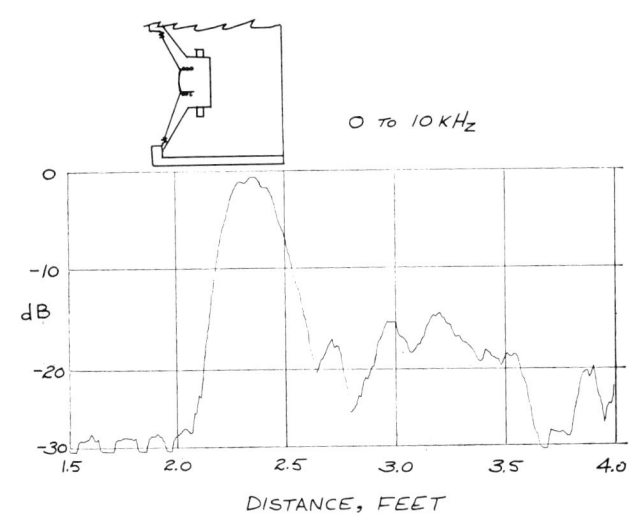

Fig. 11. Energy-time arrivals for open-back cabinet mounted medium-quality eight-inch loudspeaker. Frequency band is dc to 10 kHz and loudspeaker position shown to approximate scale.

Fig. 13 shows the effect of improper termination by a horn with too high a flare rate. The resonance is more efficiently damped, but the internal reverberation due to acoustic mismatch still exists. This is a low-quality driver unit.

Fig. 14 is the time delay distortion of the midrange driver discussed at some length elsewhere (Part I [2, Fig. 4]. The internal delayed voices are plainly in evidence.

Fig. 15 is a high-quality paper cone tweeter showing the time-delay distortion for the dc to 10-kHz frequency range. The multiplicity of reverberent energy peaks with about a 0.13-millisecond period is due to internal scattering within the tweeter. It is not at all clear from the frequency response taken alone that such an effect exists; however, by observing the time-delay characteristic it is possible to know what indicators to look for upon reexamination of the complete frequency response.

Fig. 16 is the time display of a multiple-panel high-quality electrostatic loudspeaker. This is a 1–5-kHz response taken along the geometric axis of symmetry, coinciding with the on-axis response. The physical position of the closest portion of radiating element occurs at a distance equivalent to 2 milliseconds air path delay. Fig. 17 is the same speaker 15 degrees off axis. Not only is the total energy down, but the contribution of adjacent panels is now evident.

Figs. 18 and 19 are dc to 25-kHz on-axis and 15-degree off-axis runs on a high-quality horn loaded compression tweeter. The positions of mouth, throat, and voice coil are shown in the on-axis record and several interesting effects are observable which do not show up in normal analysis. There appears to be a small acoustic contribution due to the horn mouth. This effect has been repeatedly seen by this author in such units. One possible explanation is that a compressional or shear body wave is actually introduced in the material of the horn (or cone in direct radiators) which travels at least as fast as the air compressional wave and causes an acoustic radiation from the bell of the horn itself. Also an

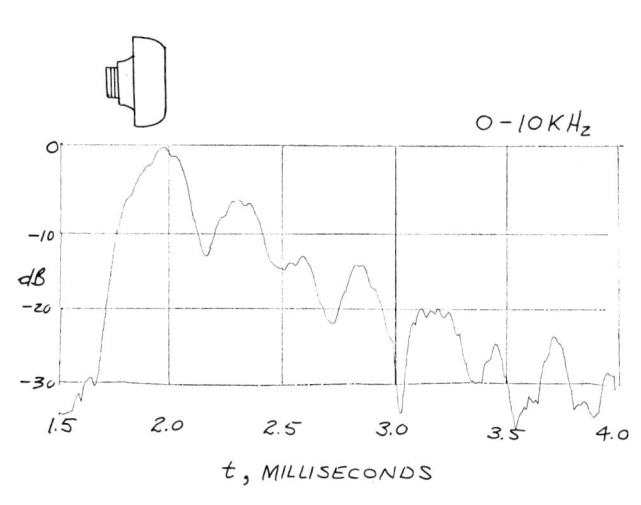

Fig. 12. Energy-time arrivals of unterminated low-quality midrange driver unit. Frequency range is dc to 10 kHz and physical position of driver shown.

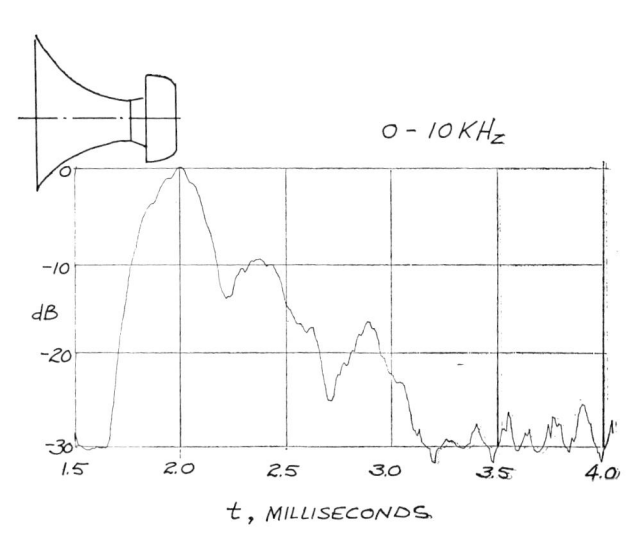

Fig. 13. Energy-time arrivals for improperly terminated driver unit of Fig. 12.

RICHARD C. HEYSER

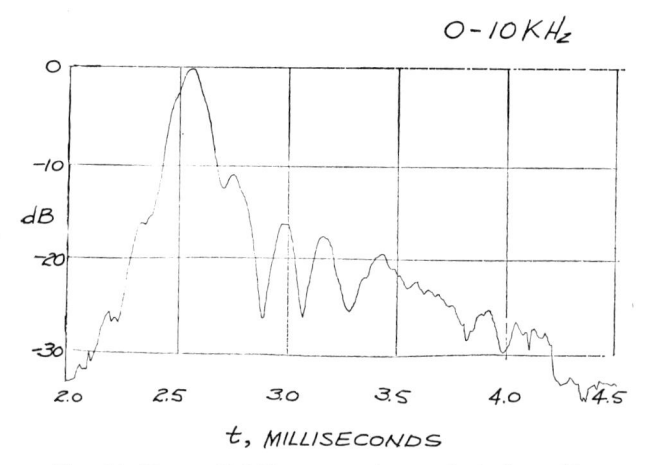

Fig. 14. Dc to 10 kHz energy-time arrivals for midrange horn loaded loudspeaker exhibiting distinct nonminimum phase frequency response.

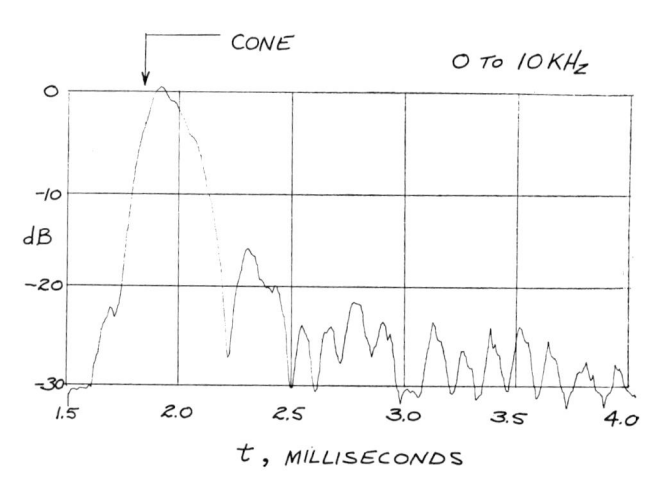

Fig. 15 Dc to 10 kHz energy-time arrivals of paper cone tweeter exhibiting distinct reverberation characteristic.

internal reverberation is observable following emergence of the main loudspeaker energy. This reverberation appears to be due to acoustic scattering off the sides of the internal structure of the horn itself. This may be inferred from the 0.12-millisecond period seen in Fig. 18, which coincides with the on-axis geometry, together with the replacement by a different behavior 15 degrees off axis as seen in Fig. 19. This suggests that closer attention might be paid to the details of mechanical layout of such horns whose acoustic properties may have been compromised for improved cosmetic appeal.

It has been noted by several authors that a network which introduces frequency-dependent phase shift, only without amplitude variation, quite often cannot be detected in an audio circuit, even when the phase shift is quite substantial. Because such networks create severe waveform distortion for transient signals while not apparently effecting the listening quality of such signals, it is assumed by inference that phase distortion must be inaudible for most systems. Fig. 20 is a measurement made through a nominal 2-millisecond electrical delay line with and without a series all-pass lattice. The network used is a passive four-terminal second-order lattice with a 1-kHz frequency of maximum phase rate. The frequency range is dc to 5 kHz, and an electronic delay is used to show the overall time delay on a scale comparable to that used for loudspeaker measurements. Although the lattice does indeed severely disturb the impulse response waveform, it is interesting to note that the total energy is not greatly effected when one considers a reasonable band of frequencies. Since this time-delay distortion, which agrees with calculated values, is due to an analytically perfect signal, it is not at all unlikely that a multimiked program heard over any loudspeaker possessing the degree of time-delay distortion measured in this paper would not appear to show this particular phase-only distortion. In view of the amount of time-delay distortion evident in most loudspeakers, it might be presumptuous to assume that this effect is totally inaudible in all systems.

SUMMARY AND CONCLUSION

A ground rule has been utilized in assessing the linear performance of a loudspeaker in a room. This rule is that the quality of performance may be associated with the accuracy with which the direct sound wave at the position of an observer duplicates the electrical signals presented to the loudspeaker terminals. Although it is realized that there are many criteria of performance,

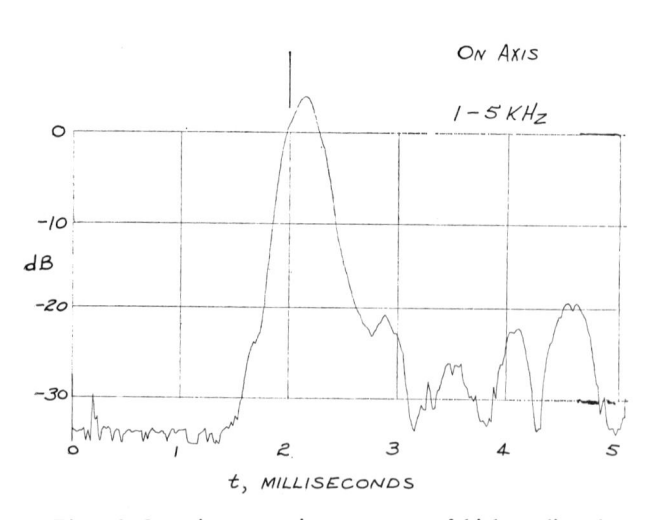

Fig. 16. On-axis energy-time response of high-quality electrostatic multi-panel midrange speaker with position of closest panel equivalent to air path delay of 2 milliseconds and 1–5-kHz excitation.

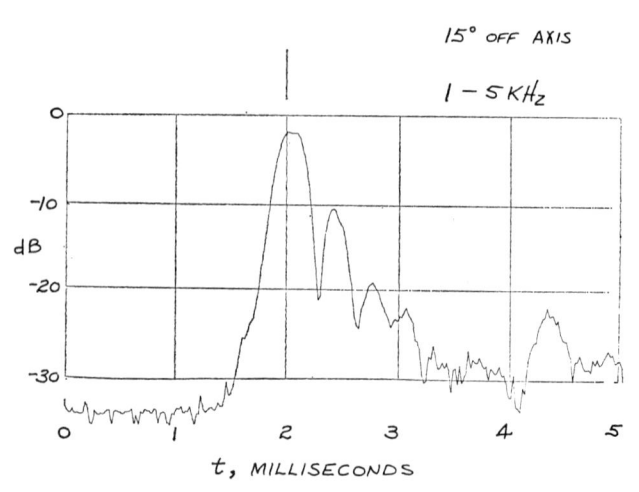

Fig. 17. 15-degree off-axis energy-time response of electrostatic loudspeaker of Fig. 16.

DETERMINATION OF LOUDSPEAKER SIGNAL ARRIVAL TIMES: PART II

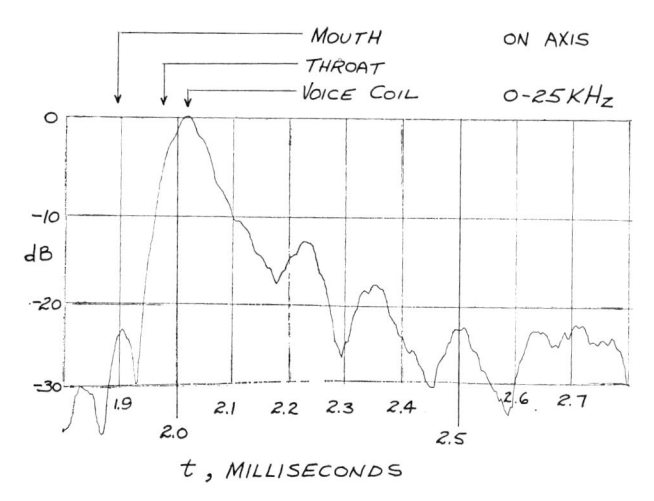

Fig. 18. On-axis energy-time response of a high-quality horn loaded compression tweeter. Frequency excitation is dc to 25 kHz and positions of mouth, throat, and voice coil shown.

this assumption of equivalence of acoustic effect resulting from an electrical cause has the advantage that it yields to objective analysis and test. The difference between the total sound due to a loudspeaker in a room and the same loudspeaker in an anechoic environment, for example, may be simplified to the following model. In an anechoic environment we have one loudspeaker at a fixed range, azimuth, and elevation with respect to an observer. In a room we have the original anechoic

Fig. 19. 15-degree off-axis energy-time response of speaker of Fig. 18.

loudspeaker, but in addition we have a multiplicity of equivalent loudspeakers assuming various positions of range, azimuth, and elevation. The additional loudspeakers, in this room model, all have the same program material as the anechoic loudspeaker, but of course suffer time delays in excess of the direct path delay of the anechoic loudspeaker. Also each room model loudspeaker has a frequency response unique to itself. The groundrule of loudspeaker quality may be applied to each equivalent source in turn and the composite effect analyzed for total quality of response in the room.

A purely mathematical analysis of any single loudspeaker in this room model disclosed that there is a direct tie between frequency response and time smear of signal received by an observer. The analysis showed that if we were to isolate any speaker to an anechoic environment, we could duplicate the acoustic response as closely as we desired for any given observer by replacing the original speaker and its frequency response aberrations with a number of perfect response loudspeakers. Each of these perfect response loudspeakers in this mathematical model occupies its own special frequency-dependent position in space behind the apparent physical position of the original imperfect loudspeaker. The result of this is that the acoustic image of a sound source is smeared in space behind the originating speaker. Perhaps another way of looking at this is that even in an otherwise anechoic environment an actual loudspeaker could be considered to be a perfect transducer imbedded in its own special "room" which creates an ensemble of equivalent sources. The type of distortion caused by this multiplicity of delayed equivalent sources has been called time-delay distortion.

A measurement of the amount of time-delay distortion in an actual loudspeaker in a room has now been made. The anechoic frequency response, both amplitude and phase, was first isolated by time-delay spectrometry for the specific portion of frequency spectrum of interest. The complex frequency response was then processed by real-time continuous circuitry to yield the complex time response.

Plots of the complex time vector components as a function of equivalent time of arrival for a variety of loudspeakers have been presented. The existence of time-delay distortion has been verified by this direct experimental evidence. It has been shown that the equivalent spatial smear for even the better class of loudspeaker may amount to many inches and that the equivalent acoustic source is always behind the apparent physical source location. It has not been possible to plot the individual joint time–frequency components predicted mathematically. This is because these components overlap in the time and frequency domains and a single-domain time presentation, even though band limited, cannot separate simultaneous arrival components. Sufficient experimental evidence has been presented to show that these components do exist to an extent necessary to create

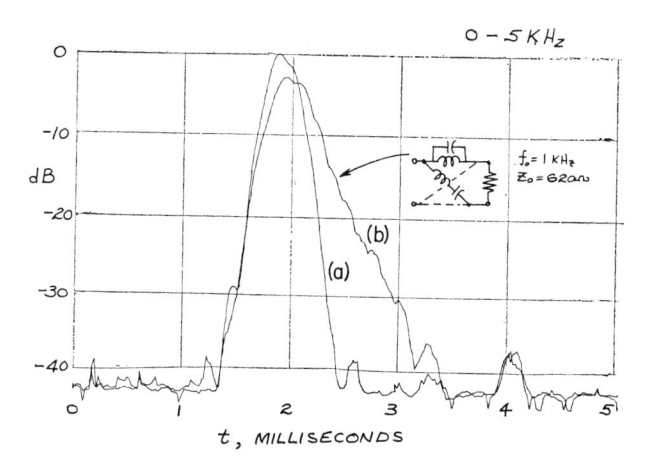

Fig. 20. Energy-time response. Curve (a)—Electrical delay line with 2-millisecond delay and excitation from dc to 5 kHz. Curve (b)—Delay line of (a) in series with second-order all-pass lattice which exhibits severe impulse response distortion due to rapid phase shift at 1 kHz.

the acoustic image smear detected by an observer.

Several energy principles have been originated and proved. While originally developed to determine techniques for investigating time-delay distortion, these principles reach far beyond simple loudspeaker testing. It has been shown that the unit impulse is but one component of a more generalized tensor. For nonturbulent systems the tensor becomes a simple two-component vector. This is the case for most acoustic and electronic situations of energy propagation. The conjugate term to the unit impulse is the unit doublet. In an acoustic field generated by a loudspeaker, one can associate the potential energy density with the impulse response of the loudspeaker. When one does this he may then associate the kinetic energy density with the loudspeaker doublet response. The total energy density may be associated with the vector sum of impulse and doublet response. Inasmuch as it is the total energy density which is available to perform work on an eardrum or microphone diaphragm, the majority of experimental data presented in this paper has been the time of arrival of this parameter.

It has also been shown that potential and kinetic energy densities are not mathematically independent if one is careful with his energy bookkeeping. What this means for acoustic radiation from a loudspeaker is that either the impulse or doublet response is sufficient to determine total performance if one has the proper tools at his disposal. But one should be cautious of gross simplification in the event that impulse or doublet response is utilized independently. As with any incomplete analysis, certain truths may not be self-evident.

An interesting area of speculation is opened up when one realizes that any reasonably well-behaved acoustic transducer placed in a sound field is capable of yielding information concerning the total energy density if associated with a suitable means of data processing. One cannot help but incautiously suggest that a closer look at the human hearing mechanism might be justified to determine whether total sound energy detection rather than potential energy (pressure) could shed a light on some as yet unexplained capabilities we seem to possess in the perception of sound.

Note: Mr. Heyser's biography appeared in the October 1971 issue.

Determination of Loudspeaker Signal Arrival Times*
Part III

RICHARD C. HEYSER

Jet Propulsion Laboratory, California Institute of Technology, Pasadena, Calif.

APPENDIX
Energy Relations as Hilbert Transforms

A fundamental approach to a complicated system may be made through that system's energy relations. Accordingly we present the following principles.

1) In a bounded system the internal energy density E is related to its potential and kinetic energy density components V and T by the vector relation

$$\sqrt{E} = \sqrt{V} + i\sqrt{T}$$

where the vector components are Hilbert transforms of each other.

2) In a bounded system a complete description of either the kinetic or potential energy density is sufficient to determine the total internal energy density.

3) By appropriate choice of coordinates within a bounded system, the available energy at a point of perception due to a signal source at a point of transmission may be partitioned as follows.

    a) The potential energy density is proportional to the square of the convolution integral of the signal with the system impulse response.

    b) The kinetic energy density is proportional to the square of the convolution integral of the signal with the system doublet response.

---

* Presented April 30, 1971, at the 40th Convention of the Audio Engineering Society, Los Angeles. For Parts I and II, please see pp. 734-743 and pp. 829-834 of the October and November issues of this *Journal*.

The first law of thermodynamics defines an exact differential function known as the internal energy (Part I, [10])

$$dE = dQ - dW, \quad \text{joules} \quad (23)$$

which equals the heat absorbed by the system less the work done by the system. By integration the energy may be obtained as a function of the state variables, and in particular for the class of electroacoustic situations of concern for this paper, it may be composed of kinetic energy and potential energy T and V,

$$E = T + V, \quad \text{joules.} \quad (24)$$

By taking the time rate of change of the components of (23) and expressing this in engineering terms, we have (Part I, [23, p. 124])

$$\frac{d(T+V)}{dt} = P - 2F, \quad \text{watts} \quad (25)$$

which asserts that the time rate of change of energy equals the power drawn from the system less the energy dissipated as heat within the system.

Properly speaking, the internal energy of a system is that property which is changed as a causal result of work done on or by that system. Energy, per se, is not generally measured. We may, however, describe and measure the energy density. Energy density is a measure of the instantaneous work which is available to be done by a system at a particular point in space and time if the total energy partitioned among the state variables could be annihilated. Energy density for state variables s is expressed as $E(s)$ and has the dimensions of joules per unit of s. The energy densities of joules per second and joules per cubic meter will be utilized in this paper. Energy density may be partitioned, for nonturbulent systems, into kinetic and potential densities. The methods by which we measure energy density, even for acoustic systems, may take the form of mechanical, electrical, or chemical means. The dynamical considerations which gave rise to Eqs. (23) and (24) naturally led to the terms kinetic and potential. When dealing with electrical or chemical characterizations, such terms are difficult to identify with the processes involved. This author has found it convenient to identify potential energy as the energy of coordinate configuration and kinetic energy as the energy of coordinate transformation.

Assume that the ratio of total kinetic to potential energy density at any moment is related to a parameter θ such that

$$\sqrt{T} / \sqrt{V} = \tan \theta. \quad (26)$$

From (24) and (26) it is possible to define the vector

$$\sqrt{E} = \sqrt{V} + i \sqrt{T}. \quad (27)$$

This is shown in Fig. A-1. We know from physical considerations that the internal energy of any bounded system is not only finite but traceable to a reasonable distribution of energy sources and sinks. If, for example, we measure the acoustic field radiated from a loudspeaker, we know that the value of that field at any point does not depend upon the way in which we defined our coordinate system. We can state, therefore, that $\sqrt{E}$ is analytic in the parameter t and is of class $L^2(-\infty, \infty)$ such that

$$\varepsilon = \int_{-\infty}^{\infty} |\sqrt{E}|^2 \, dt < \infty. \quad (28)$$

When conditions (27) and (28) are met it is known that the vector components of (27) are related by Hilbert transformation (Part I, [4, p. 122]). Furthermore,

$$\int_{-\infty}^{\infty} (\sqrt{T})^2 \, dt = \int_{-\infty}^{\infty} (\sqrt{V})^2 \, dt = \varepsilon/2 \quad (29)$$

which means that not only is it possible to express the kinetic and potential energy determining time components as Hilbert transforms, but when all time is considered, there is an equipartition of energy.

The relationship between kinetic and potential energy density is true for a bounded system, that is, one in which a boundary may be envisioned of such an extent as to totally enclose at any moment the total energy due to a particular signal of interest. A proper summation of the energy terms within that boundary for the signal of interest would then disclose a partitioning in accordance with principle 1). A measurement of the energy density at a microphone location due to a remote source will only yield a part of the total energy density of that source. The relation (27) will therefore not necessarily be observed by the microphone at any given moment. Thus, for example, the pressure and velocity components at a point in an expanding sound wave from a source will be related by what is called the acoustic impedance

$E = V + T$

$\sqrt{E} = \sqrt{V} + i\sqrt{T}$

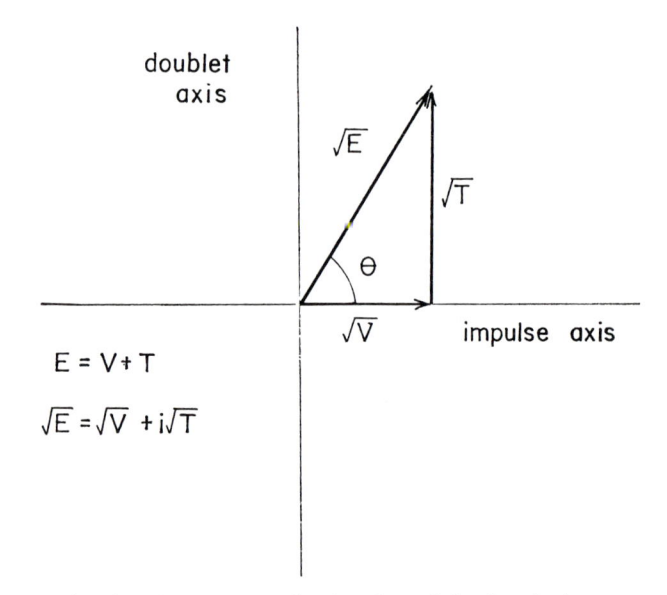

Fig. A-1. Root energy density plane defined such that one axis is system impulse response while quadrature (Hilbert) axis is doublet response.

of the medium and are not necessarily at that point related by Hilbert transformation. However, we know that the source of sound was, at the moment of energization, a bounded system and was therefore governed by the physics of (27). If the medium of propagation is such that a given energy component imparted by the source is preserved in form between source and microphone, we may take that microphone measurement and reconstruct the total energy-time profile of the source by analytical means. This observation is the basis for the measurements of this paper.

Any vector obtained from (Eq. 27) by a process of rotation of coordinates must possess the same properties. This surprisingly enough is fortunate, for although from dynamical arguments the components shown in (Eq. 27) are the most significant, it quite frequently happens that an experiment may be unable to isolate a purely kinetic component. This does not inhibit a system analysis based on total energy since we can obtain the total vector by adding our measured quantity to a quadrature Hilbert transform and be assured of a proper answer.

For verification of principle 3) we must consider that class of kinetic and potential energy related signals which could serve as stimulus to a system for resultant analysis. In particular we seek a signal form which when used as a system stimulus will suffice to define within a proportionality constant the system vector (27) by an integral process. This is done so as to parallel the analytical techniques which use a Green's function solution to an impulse (Part 1, [10]) and of course the powerful Dirac delta. Because we are dealing with quadrature terms we have not one but two possible energy stimuli. Consider the special representation of (27),

$$\sqrt{V(x)} = \frac{1}{\sqrt{2\pi a}} \frac{\sin ax}{x} \quad (30)$$

$$\sqrt{T(x)} = \frac{1}{\sqrt{2\pi a}} \frac{\cos ax - 1}{x}.$$

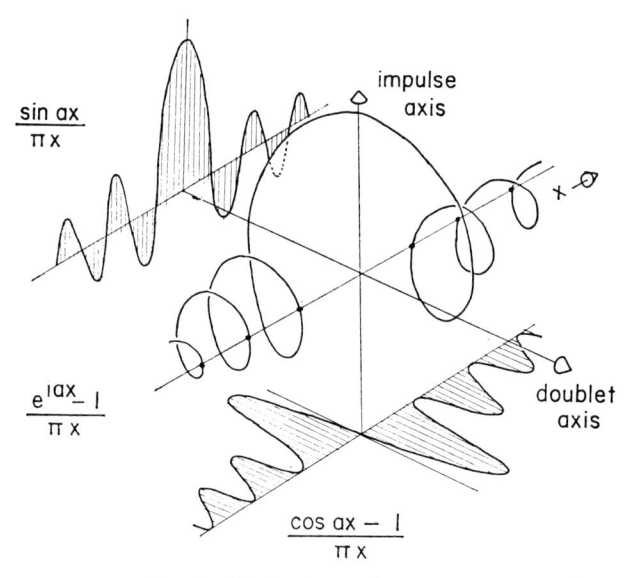

Fig. A-2. Sketch of defined complex energy vector prior to allowing the parameter a to become large without limit. Impulse (7) and doublet (8) are shown as orthogonal projections from this vector.

The energy density represented by this is obtained from (24) as

$$E(x) = \frac{\sin^2 ax + 1 - 2\cos ax + \cos^2 ax}{2\pi ax^2}$$
$$= \frac{1 - \cos ax}{\pi ax^2}. \quad (31)$$

The total energy represented by (31) as a becomes large without limit is (Part I, [24])

$$\varepsilon = \lim_{a \to \infty} \int_{-\infty}^{\infty} \frac{1 - \cos ax}{\pi ax^2} dx = 1. \quad (32)$$

Thus in the limit the quadrature terms of (30) produce a representation of unit total energy which exists only for $x = 0$ and is null elsewhere. To see this more clearly rewrite (27) with (30) components as

$$e(y) = \frac{1}{\sqrt{2\pi}} \frac{\sin \sqrt{a}y}{y} + i\frac{1}{\sqrt{2\pi}} \frac{\cos \sqrt{a}y - 1}{y} \quad (33)$$

where $y = \sqrt{a}\,x$. The vector (33) is as shown in Fig. A-2 with its quadrature components as projections. If we took the limiting form of (33) as $\sqrt{a}$ became large without limit, this would approach the impulsive vector

$$\epsilon(y) = \delta(y) + id(y) \quad (34)$$

where by definition

$$\delta(y) = \text{unit impulse} = \lim_{\lambda \to \infty} \frac{\sin \lambda y}{\pi y}$$
$$d(y) = \text{unit doublet} = \lim_{\lambda \to \infty} \frac{\cos \lambda y - 1}{\pi y} \quad (35)$$

This impulsive vector is symbolized in Fig. A-3 as its quadrature projections. It may be readily seen that $\delta(y)$ is identical to the impulse commonly referred to as the Dirac delta (Part I, [10, p. I-168]). To this author's knowledge this particular unit doublet has not received previous recognition.

In order to justify the designation of the energy-related vector $\epsilon(y)$ as impulsive, consider the magnitude squared form shown in (31). It is known that (Part I, [4, p. 35])

$$\lim_{\lambda \to \infty} \int_{-\infty}^{\infty} f(y) \frac{1 - \cos \lambda(x-y)}{\pi \lambda (x-y)^2} dy = f(x). \quad (36)$$

In the limit, utilizing (33),

$$f(x) = \int_{-\infty}^{\infty} f(y)\{\epsilon(x-y) \cdot \epsilon^*(x-y)\} dy \quad (37)$$

where the asterisk denotes complex conjugation. Thus the magnitude squared of the vector (34) is an impulse in the Dirac delta sense, although the generating vector is composed of an impulse and a doublet.

We know from classical analysis that the response at a receiving point due to injection of a Dirac delta at a transmitting point is a general system describing function. If the system is such as to allow superposition of solutions, then we can state that the total energy density at the receiving point due to an arbitrary forcing function $x(t)$ at the transmitting point is obtained from

$$\sqrt{E(t)} = \int_{-\infty}^{\infty} x(\tau) h(t - \tau) d\tau \quad (38)$$

where the describing function $h(t)$ is the normalized system response to the Dirac delta of total energy (34). Likewise the potential energy component $V(t)$, also obtainable from a Dirac delta, has a similar form with its own describing function. It must therefore follow that there is a kinetic energy describing function obtained as the response to the unit doublet as assumed from the generating form of (30). By this argument $\epsilon(y)$ could be regarded as a unit energy impulsive vector composed of equal portions of potential energy producing impulse and kinetic energy producing doublet. The assumption that the impulse is related to potential energy is drawn by analogy of form from classical mechanics in the assumption that the difference in state following an off-setting impulse of position is positional displacement, while the difference of state following the doublet is velocity. Relating to circuit theory, suppose a single resonance circuit is excited by a unit impulse of voltage.

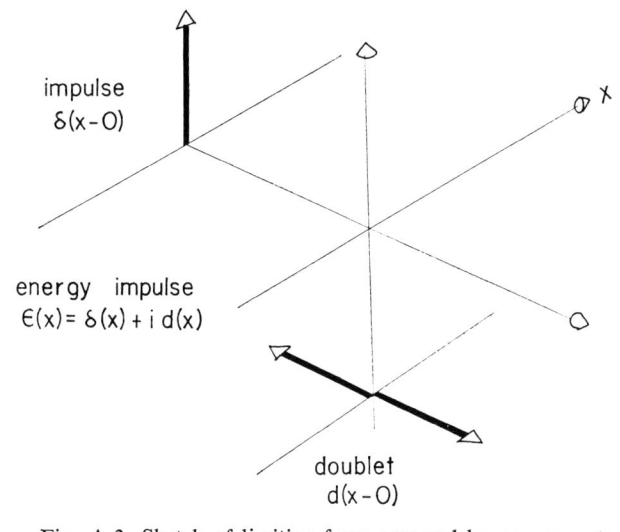

Fig. A-3. Sketch of limiting form assumed by components of Fig. A-2 when a is allowed to go to the limit. Note that the defining envelope of both impulse and doublet even as the limit is approached is proportional to the reciprocal of coordinate x.

At the instant following application of the impulse the capacitor has a stored charge (potential energy $\frac{1}{2}CV^2$) while the inductor has no current (kinetic energy $\frac{1}{2}LI^2$). Thereafter, the circuit exchanges energy under the relations (24) and (26). Should a unit doublet of voltage be applied, one would have as initial conditions a current in the inductor with no net charge in the capacitor. If one did not choose to identify the impulse with potential energy solely, he could multiply (34) by the unit vector of Eq. (14) to obtain

$$e^{i\lambda}\{\delta(t) + id(t)\} \qquad (39)$$

so as to redistribute the initially applied energy in the proper manner. Regardless of how one does this, it should be evident that a general description of system energy density must involve both the impulse and doublet response, not just the impulse response.

It might logically be asked why the need for a doublet response has not been previously felt with sufficient force to generate prior analysis. The answer is found in principle 2). An analysis based on either the impulse or doublet can be used to derive a complete system analysis by appropriate manipulation. The physical reason why one cannot use solely the impulse response or doublet response is that a measurement made on one system parameter, such as voltage, velocity, or pressure, can only express the momentary state of energy measured by that parameter. One scalar parameter of the type available from linear system operation does not represent the total system energy. A complete mathematics of analysis could be generated based completely on the doublet driving function and obtain the same results as a mathematics based on the impulse. This is because in order to get a complete answer, the complementing response must be calculated for either approach. Among the examples which spring to mind for the need of impulse and doublet analysis jointly is Kirchoff's formulation of Huygens' principle for acoustics (Part I, [25, p. 43]) and the impulse and doublet source solutions for electric and magnetic waves (Part I, [10]).

The response of a system $h(t)$ to the unit energy operation (34) is, from Eqs. (3), (5), and (15),

$$h(t) = f(t) + ig(t) = \int_{-\infty}^{\infty} f(x)\epsilon(t-x)\,dx. \qquad (40)$$

The system response $h(t)$ is the analytic signal composed of the impulse response $f(t)$ and the doublet response $g(t)$. From (25) the time position of energy sinks and sources is found from the local minima and maxima of

$$\frac{d}{dt}|h(t)|. \qquad (41)$$

While it is readily proved that the analytic signal $h(t)$ has a single-sided spectrum, this fact is of little value to our present consideration of energy. We assume that the parameter under analysis is a scalar or may be derived from a scalar potential. We assert that the sources of energy, which relate to the effective sources of sound, may be determined by considering both the kinetic and potential energies. These may be obtained separately as scalar components of the vector analytic signal. A local maximum in the magnitude of the analytic signal is due to a local source of energy and not an energy exchange.

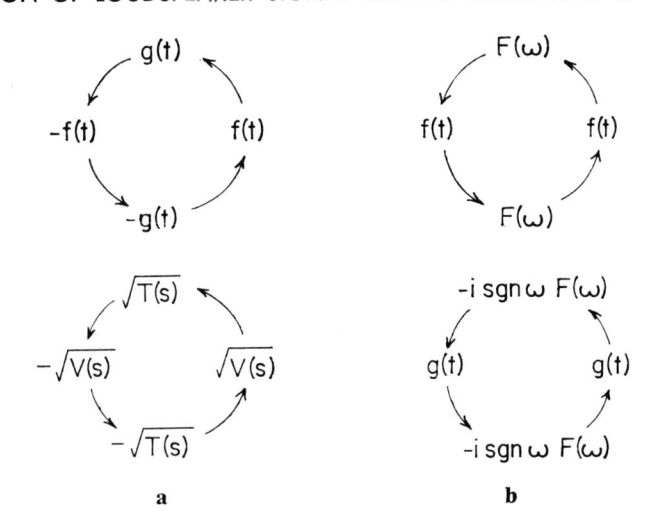

The total energy of (24) is a scalar obtained by squaring the defined vector components of (27). The Hilbert transform relations exist between the vector components of (27) and subsequently of (40). Although two successive applications of Hilbert transformation produce the negative value of the original function (skew reciprocity), the energy being obtained as a square is uneffected. The Fourier transform relating two descriptions of the same event is reciprocal in order that no preference be displayed in converting from one domain to the other (Fig. A-4). The Hilbert transform, being skew reciprocal, does show a preference. This is also separately derived from the Cauchy–Riemann relations for the analytic function (27). It should be observed that the geometric relation between analytic functions derived in [2, Appendix A] of Part I must hold between the impulse and doublet response. Hence it is possible to generate a reasonably accurate sketch of the form of a doublet response from an accurate impulse response measurement. From these one could infer the form of total energy of a given system.

There is a strong generic tie between the imaginary unit $i = \sqrt{-1}$ and the generalized Hilbert transform in that two iterations of the operation produce a change of sign while four iterations completely restore the original function. One must surely be struck by the analogy of the energy related vectors (27) and (40) to the quadrature operation of the imaginary unit which is known to be related to the system-describing operations of differentiation and integration.

Fig. A-4. Symbolic representation of functional changes brought about by successive applications of **a.** Hilbert transformation to conjugate functions of same dimensional parameter; **b.** Fourier transformation to functions of reciprocal dimensional parameter.

Note: Mr. Heyser's biography appeared in the October 1971 issue.

Editor's note: *The important historic papers by Richard C. Heyser, reproduced on the previous pages, were not included in the first Sound Reinforcement anthology because they appeared to be of interest primarily to those working in room acoustics instrumentation, not sound reinforcement systems. In retrospect, I regret that decision because, since then, they have proven to be the papers that first presented an instrumentation technique of remarkable value. It formed the basis of much sound system intelligibility testing and evaluation and new techniques for measuring sound system frequency response, digital delay requirements, loudspeaker alignment, and loudspeaker directivity. I am happy to correct a past error and include the papers in this volume for these reasons.*

David Lloyd Klepper

Acoustical Impulse Response of Interior Spaces *

RICHARD G. CANN

Grozier Technical Systems, Inc., Brookline, MA 01246, USA

AND

RICHARD H. LYON

Massachusetts Institute of Technology, Department of Mechanical Engineering, Cambridge, MA 02139, USA

The impulse response of interior spaces is measured simply with special instrumentation. Lack of reverberance in certain frequency bands, highly structured echo patterns indicative of flutter, or other undesirable characteristics may be revealed. The same equipment can be used to study scale models of the space and the effects of changes in the geometries of the materials within the space.

0 INTRODUCTION

Interior spaces are often designed for multiple uses. A shopping mall may be used as a connecting link between stores and a trade show area. An auditorium may be designed for political meetings and a concert hall. But whatever the projected use, the space must be evaluated to ensure that the acoustical criteria of the space are met in each regard.

The echo structure of the space (echogram) is a useful tool for revealing potential acoustical problems and gives a clear insight into how potential solutions might perform. Echograms may be obtained in a space by placing an impulsive source such as a small cannon, or even a bursting balloon, at typical source locations and picking up the echoes with a microphone.

Echograms are also very useful in evaluating scale models of a planned space. The impulsive source is a spark located at typical locations. A miniature microphone receives the sound at typical listener locations. But whether model of full-scale echograms are to be obtained, the signal-processing instrumentation is identical for each; an amplifier, a filter, a waveburst processor and an oscilloscope form the instrumentation chain.

This echogram system discussed here is currently in use by industry, government agencies, and consultants both in the United States and in several European countries.

---
* Manuscript submitted 1978 September; revised 1979 May 8.

1 ECHO STRUCTURE OF A RECTANGULAR ROOM

Before delving into the instrumentation details, it is helpful to understand how echo structures are formed. A rectangular model room serves as an example. If an impulsive noise is produced in a space, it propagates at a speed of approximately 0.3 m/ms, reflecting from any obstruction in its path. In a very large space where the first reflection arrives at the listener 100 ms after the direct sound, an echo is clearly heard. However, in smaller spaces or where multiple echoes exist, echogram instrumentation such as that described here is required to resolve direct sound from echoes. One trace of the display shown in Fig. 1 shows the echo from the floor of a model space. Both the sound traveling along the direct route and the echo are clearly distinguished.

The second trace on the display (energy) is important too. Most sounds for which the spaces are designed are not impulsive, but quasi steady. Therefore the impulsive noise used here must be converted to an equivalent steady noise level. The computation of energy does just that.

Another useful feature of the instrumentation is its ability to measure the energy of a pulse singled out from a train of impulses. This assists in the detailed examination of echoes and the contribution of each to the total energy level.

The usefulness of the modeling technique is illustrated here in a series of figures showing some echograms of a rectangular model with reflective surfaces in various stages of construction. Fig. 1 shows the response of just a floor.

Without walls and ceiling, the sound travels by only two paths; one direct and one reflected from the floor. The direct sound has a level of 18.5 dB and the reflected sound adds 2.5 dB to give a total of 21 dB.

Fig. 2 shows the addition of one wall: the number of arriving pulses are doubled. Fig. 3 shows the effect of adding a second wall at right angles: doubling the number of impulses again to eight. By contrast, moving the second wall to be parallel to the first clearly shows the well-known phenomenon of "flutter": an infinite string of images appears in Fig. 4. On completion of the room (Fig. 5) the typical reverberant decay is clearly seen. Multiple images from three pairs of parallel surfaces have filled in most of the gaps to obtain a reverberation decay curve. Note that the energy level increased 16 dB from an open space level of 21 dB to 37 dB by adding hard surfaces to complete the enclosure. The energy level has, in fact, reached 2 dB of its final level at only 10% of the reverberation time.

Using similar echograms the acoustics of an auditorium can be evaluated. Early arriving echoes can be studied for "direct loudness" and "flutter." Later echoes can be examined for "overall loudness," "diffusion," "focusing," and "frequency coloration" effects. A combination of the two sets of echoes can be used as shown below to estimate the articulation index (AI).

In a reverberant space the audibility of speech is increased by echoes or reflections that occur within 50 ms of the first arrival. Beyond that time reverberation causes later speech sounds to be progressively masked. The difference in decibels between the late (masking) energy level and the 50-ms energy level measures the degree of masking. Both

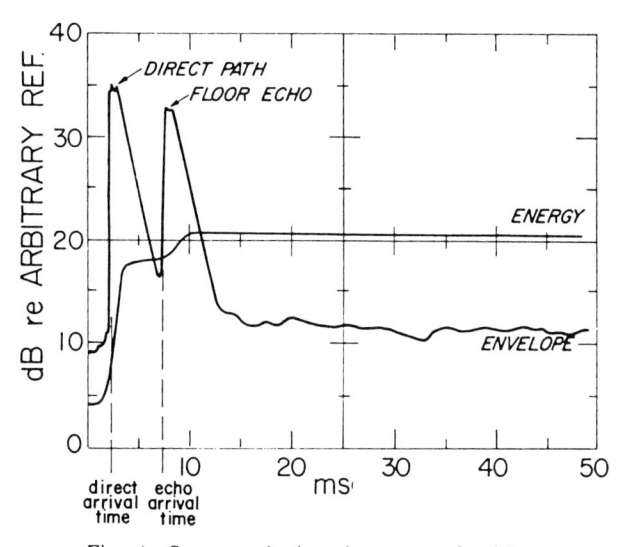

Fig. 1. Source and microphone over a hard floor.

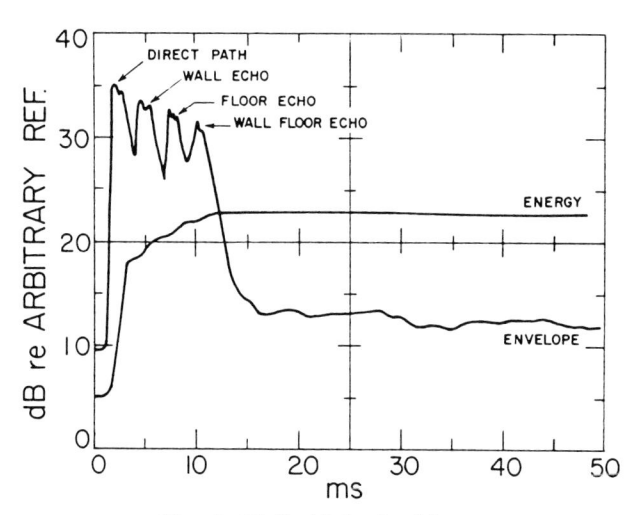

Fig. 2. Wall added to hard floor.

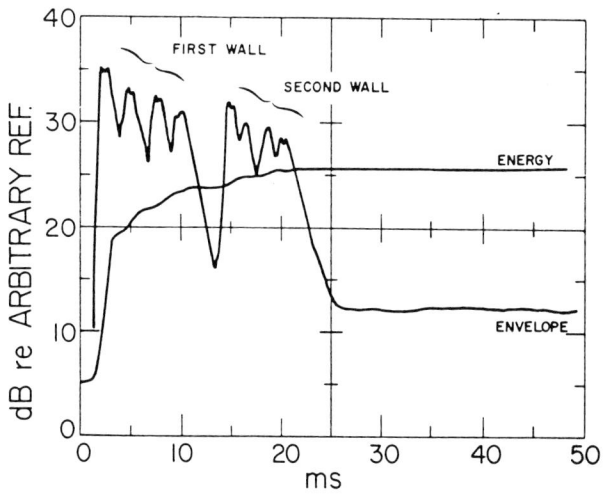

Fig. 3. Floor and two walls at right angles.

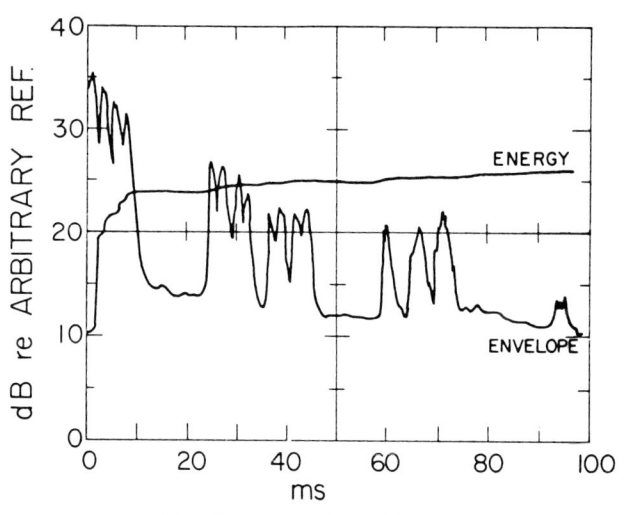

Fig. 4. Floor and parallel walls.

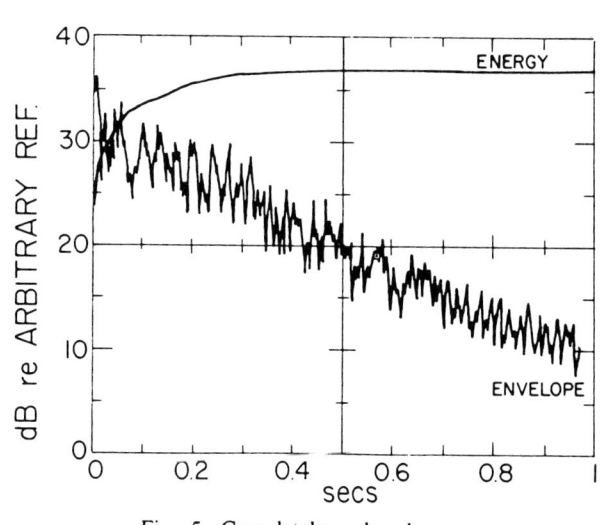

Fig. 5. Completely enclosed space.

the early and late energy levels are readily measured by the system described in the following section.

The articulating index is obtained by dividing the degree of masking by the range of speech signal levels which has been found by researchers to be 30 dB [1]. For example, in a reverberant space where the reverberant masking energy level is measured to be 6 dB below the 50-ms energy level the articulation index is AI = 6/30 = 0.2.

This simple technique is appropriate for individual octave bands. The articulation index over the full bandwidth is most readily obtained using the "dot area" method as described in [2]. Speech intelligibility can be derived from the articulation index through curves shown in [1] and [2].

2 INSTRUMENTATION SYSTEM

The modeling system is shown in Fig. 6. The miniature microphone and the oscilloscope are not shown. On the right is the model noise source, the spark generator, and to the left are the amplifier and the waveburst processor. A block diagram is shown in Fig. 7.

For scale modeling, impulsive noise is generated by triggering a spark at electrodes located at the tip of a wand. A remote push button controls the high-voltage discharge. Pressing the button charges a capacitor and releasing generates a small energy, but very high voltage which causes the capacitor to discharge across the electrode tips. The result is an acoustic impulse about 100 μs long, which contains a broad frequency band of noise from 2 kHz to 200 kHz. If a model of an interior space is built at a scale of 10:1, the full-scale frequency range represented becomes 200 Hz to 20 kHz, for a 20:1 model 100 Hz to 10 kHz. The spark generator is designed to produce acoustic pulses, highly repeatable in amplitude and waveform.

For full-scale testing a lower frequency impulsive source must be used. A small cannon, intended for use in starting sailing races, can be used in environments that have high ambient noise. Usually a quieter source is sufficient. A starter's pistol covers the midfrequencies well, but lacks low-frequency energy; a bursting balloon is often satisfactory and has the benefit of being less startling to neighbors.

2.1 Microphones

For auditorium measurements a good quality sound level meter works well. It must have an ac output connection to the analysis equipment. For scale models a miniature microphone is required, one that is easily adaptable for use in small spaces. The BBN 1/8-inch (3-mm) microphone is a good selection. Remember to consider the full size of the

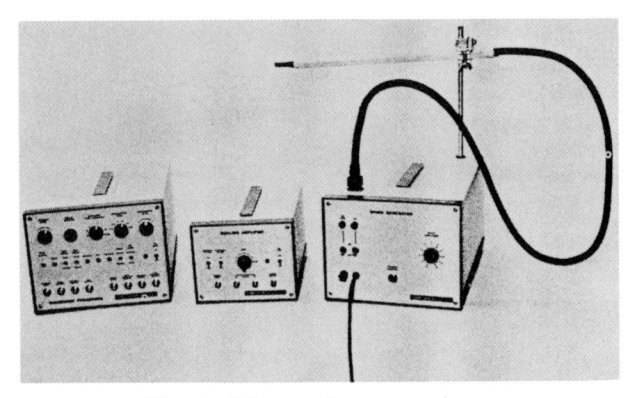

Fig. 6. Echogram instrumentation.

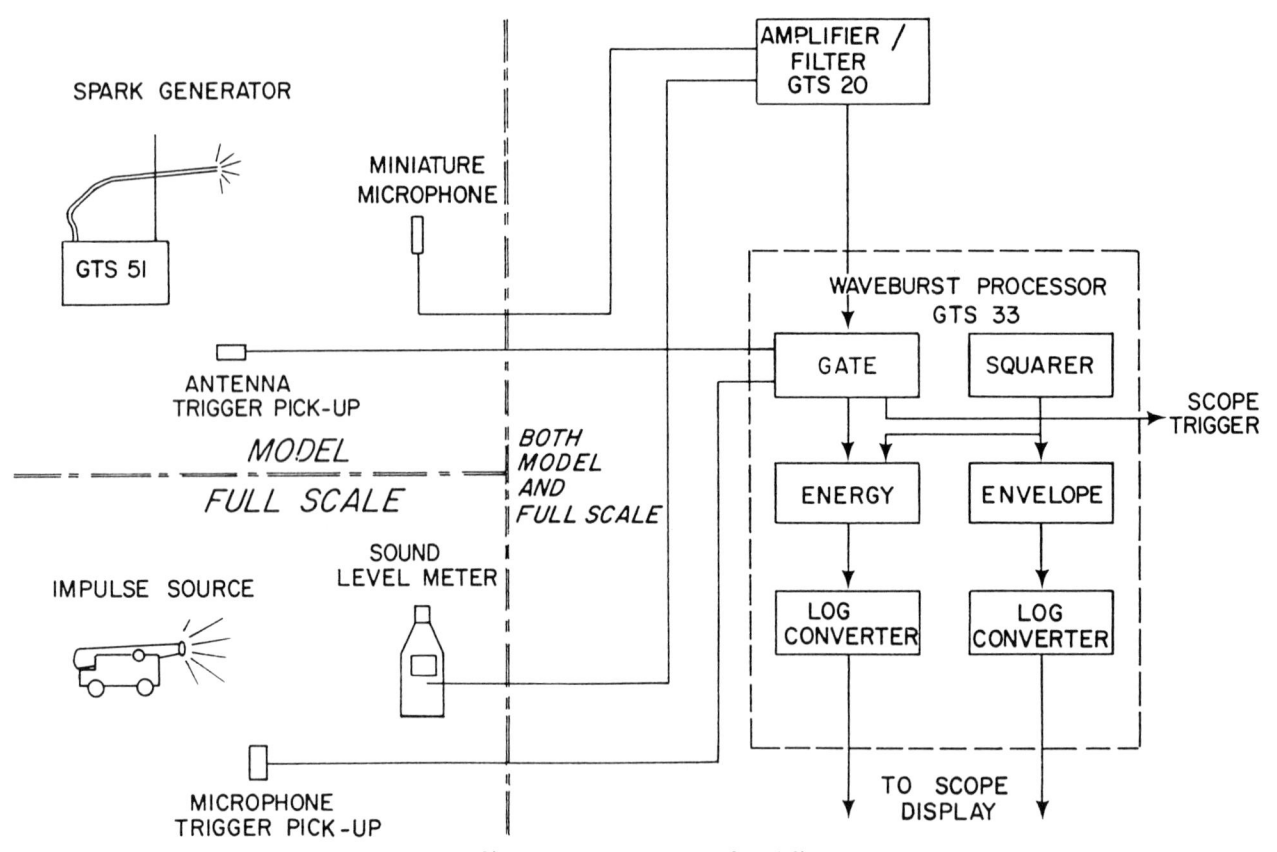

Fig. 7. Simplified block diagram of modeling system.

microphone used. In a 20:1 model the 1/8-inch (3-mm) microphone becomes 2½ inches (63.5 mm) diameter full scale. At high frequencies such a microphone may interfere with the sound field to be measured.

2.2 Echogram Analyzer

From this point on in the instrumentation chain (Fig. 7) the same equipment is used for both full-scale and model measurements. A filter first selects the frequency band of interest, amplifies it, and feeds it to the waveburst processor. There the signal is gated until an external trigger permits it through to the squarer. The trigger pulse is generated at an antenna pickup in the scale model and by another microphone in full scale. After squaring, the signal proceeds on two parallel paths. One produces the short-term average and displays its logarithm. This is the trace that is used for identifying arrival times or showing reverberant decay. The other signal path goes on to an integrator which is controlled in conjunction with the gate. The trigger causes the integrator to be released from its grounded state after a time preset on the front panel. At the instant of release a delayed trigger output is available for triggering the oscilloscope. Thus the display can be set to begin just prior to the first impulse arrival for a full echogram presentation. If any particular echo needs closer examination, a delay can be dialed in to permit the energy computation and display to begin just prior to its arrival.

Another feature of the circuit is that before the impulse arrival the ambient noise is being continuously monitored. This background of noise is automatically canceled in the energy computation. Thus when the system is operated with a bandpass filter, the varying amounts of background noise are automatically canceled when switching from band to band.

2.3 Auxiliary Equipment

The system may also be operated in conjunction with either a tape recorder or a digital event recorder. The latter instrument repetitively plays back a signal recorded in digital memory. This permits the use of a nonstorage oscilloscope. However, for typical full-scale applications only 50 ms of signal can be recorded for each 1000 words of memory. Thus the investigation of a full echogram is not easy without substantial memory.

3 CONCLUSION

The echogram is another useful tool helping to identify potential problems in the acoustic design of interior spaces. Acoustical scale models have been used in the design stages. A complete bibliography containing more than 100 references on modeling is available from the authors.

4 REFERENCES

[1] "Methods for the Calculation of Articulation Index, ANSI Std. S3.5-1969 (R1973).

[2] W. J. Cavanaugh, W. R. Farrell, P. W. Hirtle, and B. G. Watters, "Speech Privacy in Buildings," *J. Acoust. Soc. Am.*, vol. 34, pp. 475–492 (1962 Apr.).

[3] L. L. Beranek, Ed., *Noise Reduction* (McGraw-Hill, New York, 1960).

[4] L. L. Beranek, *Music Acoustics and Architecture* (Wiley, New York, 1962).

[5] M. Barron, "Early Lateral Reflection and Cross-Section Ratio in Concert Halls," *Proc. 8th Int. Cong. in Acoustics* (London, 1974), p. 602.

[6] S. Strom, "Acoustical Design of a Multi-Purpose Hall in 'Iben House'," *Proc. 8th Int. Cong. in Acoustics* (London, 1974), p. 605.

[7] P. I. Newman and P. K. Mackenzie, "Studies of Impulse Response on Model Auditoria," *Proc. 8th Int. Cong. in Acoustics* (London, 1974), p. 606.

[8] "Acoustical Scale Modeling Bibliography," Grozier Technical Systems Inc., Brookline, MA 02146.

THE AUTHORS

R. G. Cann

Richard G. Cann is presently with the Boston-based company of Grozier Technical Systems, Inc. He specializes in the application of acoustical scale modeling instrumentation to the solution of both indoor and outdoor sound propagation problems. The technique is becoming increasingly popular, particularly in Europe.

Prior to his present position, Mr. Cann was a senior engineer with the acoustical consulting company of Cambridge Collaborative. He has presented —together with

R. H. Lyon

Dr. Lyon—several acoustical modeling workshops for the Center for Advanced Engineering Studies at Massachusetts Institute of Technology. He is the author of many publications on modeling.

•

Richard H. Lyon received an A.B. degree from Evansville College (now University of Evansville) in 1952 and a Ph.D. from the Massachusetts Institute of Technology in 1955. He is a professor of mechanical engineering at M.I.T., working in the areas of sound and vibration. In addition, he is a consultant to government and industry, a founder and principal in Cambridge Collaborative, Inc., a consulting firm, and in Grozier Technical Systems, Inc., an acoustical instrumentation company.

For 10 years Dr. Lyon worked for Bolt, Beranek and Newman, Inc. where he was active as a department head, corporate vice president and director of the physical sciences division. Earlier in his career, he was an assistant professor of electrical engineering at the University of Minnesota where he taught courses in acoustics and communication theory. During that time, he was a National Science Foundation postdoctoral fellow at the University of Manchester in England. There he did research in statistical analysis of interacting vibrating systems. Among the many professional societies in which he holds membership, Dr. Lyon is on the editorial board and executive council of the Acoustical Society of America and a member of ASME. A prolific writer, he is the author of numerous publications in the field.

COMMENTS ON "ACOUSTICAL IMPULSE RESPONSE OF INTERIOR SPACES"

The above-mentioned paper[1] by Cann and Lyon caused me both surprise and concern. While the method described by them has legitimate use as a teaching tool in modeling techniques, it is this writer's opinion that several misapplications of the technique are presented in this paper.

My surprise stems from the fact that the work in this paper is a superficial repeat of Dick Heyser's work with regard to energy vs. time plots.[2] Not only has the basic concept of energy vs. time measurements already been exhaustively covered by Heyser, but Cann and Lyon suggest a sub-optimal instrumentation methodology compared to Heyser's patented method.

Specifically, the Heyser technique, by insuring that total signal energy is always present in the center of the tracking bandwidth, obtains a 20,000 to 1 energy advantage over impulse sources attempting to maintain the same "time window." The processing of impulse signals through a loudspeaker, microphone and FFT analyzer is invalid whenever the loudspeaker happens to have nonlinear response. The fast Fourier Transform is only valid for linear signals. The illustration in the Cann & Lyon article of the use of a 10-gauge yachting cannon is again, to be charitable, a highly questionable practice.

This writer would regret it if their article encouraged investigators of full-sized acoustic spaces to choose an inefficient system.

DON DAVIS
Synergetic Audio Concepts
San Juan Capistrano, CA 92693

AUTHORS' REPLY

Echograms have been successfully and productively used for years by many researchers, as our references clearly indicate. However, simple equipment is now available for individuals to make their own energy-time plots without concern for patent infringement or need for highly sophisticated processing equipment.

Mr. Davis correctly points out that impulsive signals may be nonlinear. However, researchers using sparks, pistol shots, and other such sources have demonstrated that results are not compromised on this account, if reasonable care is taken in selection of source strength and microphone locations.

The impulsive source is best selected according to the background noise level. For example, for a noisy industrial environment, a small ceremonial cannon is useful, but for a quieter space, a bursting balloon is quite sufficient.

Full details of source selection and a complete detailing of the do-it-yourself echogram technique may be found in "Acoustical Scale Modeling: a Practical Course for Architects and Engineers" published by Grozier Technical Systems, 157 Salisbury Road, Brookline, MA.

RICHARD H. LYON
Massachusetts Institute of Technology
Cambridge, MA 02139

and

RICHARD G. CANN
Grozier Technical Systems
Brookline, MA 02146

---

[1] R. G. Cann and R. H. Lyon, *J. Audio Eng. Soc.*, vol. 27, pp. 960-964 (1979 Dec.).

[2] D. Heyser, "Determination of Loudspeaker Signal Arrival Times," Parts I, II and III, *J. Audio Eng. Soc.*, vol. 19, nos. 9, 10, 11 (1979 Oct., Nov., Dec.).

Central Cluster Design Technique for Large Multipurpose Auditoria*

E. T. PATRONIS, JR., AND CATHARINA DONDERS

Georgia Institute of Technology, School of Physics, Atlanta, GA 30332, USA

A step-by-step central cluster design procedure is given which is applicable to large multipurpose spaces. The procedure takes into account the factors of speech intelligibility, uniformity of coverage, required sound pressure level, and the necessity for electrically reconfiguring the cluster for different events.

0 INTRODUCTION

The growing number of large multipurpose auditoria, both presently in existence and in the planning stage, are presenting acoustical and sound system designers with new and challenging problems. These problems are associated chiefly with the large distances involved, the complexity of the arrays needed to cover all listening spaces, the reverberant character of the environment, and the necessity for reconfiguring the sound system for different types of events or performances.

Little work has appeared in the literature directed toward these particular problems when treated as a whole, though much published work has been directed at these problems individually. Hopkins and Stryker [1] have studied the power requirements of loudspeakers when employed in reverberant spaces. Kaye and Klepper [2] have pointed out the necessity of having tight but reasonable specifications as well as the desirability of having designs which can be fulfilled by more than one vendor. Peutz [3] has studied speech transmission in rooms and has given formulas for calculating the articulation loss of consonants as a function of the acoustical parameters of the space. Klein [4] has employed the work of Peutz in evolving a design procedure for sound reinforcement systems which invokes

---
* Presented at the 69th Convention of the Audio Engineering Society, Los Angeles, 1981 May 12–15.

speech intelligibility requirements. Chaudiere [5] has pointed out the agreement between Klein and the criterion as given by Davis [6] for determining the maximum source-to-listener distance for which speech will remain sufficiently intelligible. Davis and Davis [7] have collected and presented the formulas useful in sound system design and have evolved design procedures for several types of systems. It is the intent of the present work to draw upon the above sources as well as others to formulate a central cluster design procedure which is applicable to large multipurpose auditoria. Such a procedure must take into account such factors as room acoustical parameters, cluster location, uniformity of coverage, speech intelligibility, sound pressure capability, and the necessity of having variable coverage patterns as dictated by the nature of the performance to be reinforced. Throughout the present work reference will be made to examples based on a 12 000 seat coliseum which is located at Clemson University. This coliseum is used for basketball, college commencements, concerts, convocations, and other special events.

1 ROOM ACOUSTICAL PARAMETERS

If the facility for which the sound system is being designed is an existing one, accurate room acoustical parameters can be obtained by direct measurement. The parameters required are the total surface area S, the total volume V, the reverberation time T, measured in octave bands, and the ambient noise level. It will be pointed out later in the present work that it is particularly important to determine T in the octave bands centered on 500 Hz, 1 kHz, and 2 kHz. After the reverberation times have been measured, values of the average absorption coefficient $\bar{\alpha}$ for either an occupied or an unoccupied space may be calculated using the formulas given by Beranek [8] or Davis and Davis [7]. Once the values of $\bar{\alpha}$ are at hand, the room constant R can be calculated from

$$R = \frac{\bar{\alpha} S}{1 - \bar{\alpha}} \tag{1}$$

again as a function of the frequency band and the number of occupants of the space.

If the facility exists only on paper, $\bar{\alpha}$ and also T must be calculated according to the procedures outlined by either Davis and Davis [7] or Beranek [8].

Table 1 lists the values associated with the Clemson facility. As this is an existing facility, the reverberation times, from which the average absorption coefficients were calculated, were determined by direct measurement with the facility unoccupied.

2 CLUSTER LOCATION

If the facility features live performances, it is essential for naturalness of source identification that the cluster be located immediately above the area where the performance occurs. Higher elevations will in general allow increased gain before feedback. Elevations in excess of 18 m above the performers should be avoided, however, because of the confusion presented to the performers by the delayed sound arriving from the cluster. The stage location for the Clemson facility is indicated in the plan view of Fig. 1 and an interior photograph is given in Fig. 2. The bottom of the cluster is located 16 m above the stage. If this facility were to be used for basketball alone, a much simpler and more economical cluster could be centered above the playing area.

3 CLUSTER CONFIGURATION

In arriving at the cluster configuration it is first suggested that a scale model be constructed of the facility. Such a model is invaluable in the role of analog computer from which one can readily determine sight lines, distances, and angles as various loudspeaker arrays are explored for their viability. Malmlund and Wetherill [9] have used such a model in conjunction with a special light projector to determine loudspeaker coverage patterns. Second, attention should be centered on the loudspeaker parameters as determined by using a one-octave band of pink noise centered on 1 kHz. This band is chosen for the following reasons. The constant-directivity horns presently being manufactured require higher crossover frequencies than was formerly the case. (This band has −3 dB points at 707 Hz and 1414 Hz. The new recommended crossover frequencies are

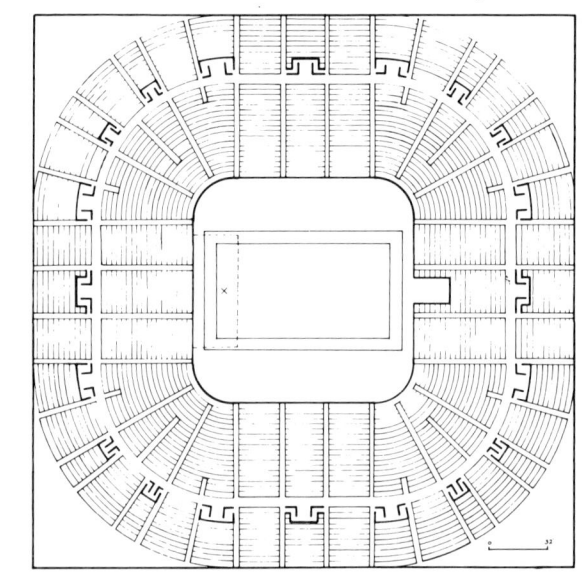

Fig. 1. Plan view of example. Dashed area—stage location; ×—cluster location.

Fig. 2. Interior view of example coliseum.

Table 1

Number of Occupants	500 Hz $\bar{\alpha}$	R	1 kHz $\bar{\alpha}$	R	2 kHz $\bar{\alpha}$	R
0	0.26	$6.21 \times 10^3 \text{m}^2$	0.31	$7.93 \times 10^3 \text{m}^2$	0.34	$9.10 \times 10^3 \text{m}^2$
2000	0.30	$7.56 \times 10^3 \text{m}^2$	0.35	$9.48 \times 10^3 \text{m}^2$	0.38	$1.08 \times 10^4 \text{m}^2$
4000	0.33	$8.70 \times 10^3 \text{m}^2$	0.38	$1.08 \times 10^4 \text{m}^2$	0.41	$1.23 \times 10^4 \text{m}^2$
6000	0.37	$1.04 \times 10^4 \text{m}^2$	0.42	$1.28 \times 10^4 \text{m}^2$	0.45	$1.44 \times 10^4 \text{m}^2$
8000	0.41	$1.23 \times 10^4 \text{m}^2$	0.46	$1.51 \times 10^4 \text{m}^2$	0.49	$1.70 \times 10^4 \text{m}^2$
10 000	0.44	$1.38 \times 10^4 \text{m}^2$	0.49	$1.70 \times 10^4 \text{m}^2$	0.52	$1.91 \times 10^4 \text{m}^2$
12 000	0.48	$1.63 \times 10^4 \text{m}^2$	0.53	$1.99 \times 10^4 \text{m}^2$	0.56	$2.25 \times 10^4 \text{m}^2$

$S = 1.77 \times 10^4 \text{m}^2$; $V = 9.12 \times 10^4 \text{m}^3$. Ambient noise level 52 dBA.

800 Hz or higher.) Both the low-frequency loudspeakers and the high-frequency horns are participating in this band so that the entire cluster is producing acoustic power. Finally the articulation loss formulas of Peutz [3] involve reverberation times which are the average of the values for the 1-kHz and 2-kHz octave bands.

The loudspeaker parameters required are the axial directivity factor Q, the coverage angles between the half pressure points, and the electroacoustic conversion efficiency. Most manufacturers do not supply data on the conversion efficiency, but rather furnish sensitivity ratings. If S_e is the direct sound pressure on axis expressed in decibels at some distance r for a loudspeaker being energized with 1 electrical watt of octave-band limited pink noise, centered at the frequency of interest, and if the appropriate axial Q is also known, then the conversion efficiency η is given by

$$\eta = \frac{16\pi r^2 \times 10^{(S_e/10 - 10)}}{1 \text{ W} \times \rho_0 CQ} \quad (2)$$

where ρ_0 is the static air density and C is the velocity of sound.

Consideration is now given to the cluster configuration necessary for the performance that requires the largest coverage area. In the Clemson facility this occurs for a concert situation in which seating is allowed on the basketball playing area as well as in all of the fixed seating positions. A useful though not unique procedure for configuring the cluster is the following. Beginning with the horn which must have the longest throw on axis and for which the highest Q is required, one proceeds to build up a cluster configuration of both high- and low-frequency elements to cover the required seating spaces using the general guide that the ratio of the square of the throw distance on axis to the axial directivity for the various elements is as small and is as similar as possible. In addition the patterns for the adjacent elements should cross at the half pressure points after having originated as closely as possible from a common acoustic center.

After a possible cluster arrangement has been arrived at, it is necessary to examine its performance in the reverberant space, taking into account the factors of uniformity of coverage, speech intelligibility, and sound pressure capability. The following analysis sheds light on all these factors.

4 ANALYSIS

Consider that the longest throw horn is energized with 1 electrical watt of octave band limited pink noise centered on the frequency of interest. Let P_1 be the total acoustical power output from this horn. The acoustic intensity both direct and reverberant on the axis of this horn at its throw point is given by

$$I_1 = P_1 \left(\frac{Q_1}{4\pi r_1^2} + \frac{4}{R} \right). \quad (3)$$

Now examine each loudspeaker in the cluster considered, one at a time, and adjust the electrical power into each so that each loudspeaker on axis at its throw point produces individually the same total intensity as horn 1. The acoustical power output from the ith loudspeaker needed to produce this intensity will then be

$$P_i = P_1 \left(\frac{Q_1}{4\pi r_1^2} + \frac{4}{R} \right) \left(\frac{Q_i}{4\pi r_i^2} + \frac{4}{R} \right)^{-1}. \quad (4)$$

The total acoustical power radiated by the entire cluster with all elements operating will be

$$P = \sum_{i=1}^{n} P_i$$

$$= P_1 + \sum_{i=2}^{n} P_1 \left(\frac{Q_1}{4\pi r_1^2} + \frac{4}{R} \right) \left(\frac{Q_i}{4\pi r_i^2} + \frac{4}{R} \right)^{-1} \quad (5)$$

and the total reverberant intensity will be

$$I_R = \frac{4}{R} \sum_{i=1}^{n} P_i$$

$$= \frac{4}{R} P_1 \left[1 + \sum_{i=2}^{n} \left(\frac{Q_1}{4\pi r_1^2} + \frac{4}{R} \right) \left(\frac{Q_i}{4\pi r_i^2} + \frac{4}{R} \right)^{-1} \right]. \quad (6)$$

The direct intensity from an individual loudspeaker, say the jth loudspeaker, is

$$I_j = P_j \left(\frac{Q_j}{4\pi r_j^2} \right) \quad (7)$$

the average direct intensity over all loudspeakers is

$$I_D = \frac{1}{n} \sum_{j=1}^{n} P_j \left(\frac{Q_j}{4\pi r_j^2} \right) \quad (8)$$

and the ratio of the total reverberant intensity to the average direct intensity can be shown to be

$$\frac{I_R}{I_D} = \frac{\dfrac{16\pi n}{R} \sum_{i=1}^{n} \dfrac{r_i^2}{Q_i} \dfrac{1}{1 + 16\pi r_i^2/RQ_i}}{\sum_{j=1}^{n} \dfrac{1}{1 + 16\pi r_j^2/RQ_j}}. \quad (9)$$

Alternatively one could proceed by matching the direct intensities on axis for the individual loudspeakers at their respective throw points. In this event,

$$I_1 = \frac{P_1 Q_1}{4\pi r_1^2} = \frac{P_i Q_i}{4\pi r_i^2} \quad (10)$$

and

$$P_i = \frac{P_1 Q_1}{Q_i} \frac{r_i^2}{r_1^2}. \quad (11)$$

The total acoustic power radiated by the entire cluster with all elements radiating is now

$$P = \sum_{i=1}^{n} P_i = P_1 \sum_{i=1}^{n} \frac{Q_1}{Q_i} \frac{r_i^2}{r_1^2}. \quad (12)$$

and the total reverberant intensity is

$$I_R = \frac{4}{R} P_1 \sum_{i=1}^{n} \frac{Q_1}{Q_i} \frac{r_i^2}{r_1^2} .\quad (13)$$

The ratio of the total reverberant intensity to the average direct intensity is now

$$\frac{I_R}{I_D} = \frac{16\pi}{R} \sum_{i=1}^{n} \frac{r_i^2}{Q_i} . \quad (14)$$

It is apparent that in order to minimize the reverberant-to-direct intensity ratio for a given number of elements one needs to make r_i^2/Q_i as small as possible. In any event upon examining Eqs. (9) and (14), one concludes that they yield identical results under two different circumstances. If R is sufficiently large such that

$$\frac{16\pi r_i^2}{RQ_i} \ll 1$$

for all i, the two expressions are identical. They are also identical, independent of the size of R, when

$$\frac{r_1^2}{Q_1} = \frac{r_2^2}{Q_2} = \cdots = \frac{r_n^2}{Q_n} .$$

Under any other circumstances Eq. (9) predicts a lower ratio than does Eq. (14), and the difference between the predictions increases as the room constant R decreases. The conclusion appears to be that if one is faced with a loudspeaker assortment with unequal r_i^2/Q_i ratios in a reverberant environment, it is better to adjust the power supplied to the individual loudspeakers so as to match their individual total intensities (direct plus reverberant) on axis. This will lead to a lower reverberant-to-average-direct intensity ratio and hence higher intelligibility. The design goal, of course, should be to require both Eqs. (9) and (14) to yield identical small results, but this can seldom be accomplished in practice because of the limited available Q values in commercial loudspeakers.

At this point it is possible to determine if a proposed cluster is workable. One simply applies the method leading to Eq. (9) or that leading to Eq. (14) to an empty hall. If the total reverberant-to-direct intensity ratio is 4 or less, then the articulation loss of consonants will be less than 15%, even for difficult halls possessing speech range reverberation times up to 4 s. This limit is obtained by a conservative interpretation of the work of Peutz [3] and Klein [4]. The method of Eq. (9) or (14) is then applied to a completely filled facility. If the total reverberant-to-direct intensity ratio is greater than 1, then the intensity variations throughout the hall will be 3 dB or less under worst-case conditions. Values outside of these limits will suggest cluster redesign, acoustical parameter change, cluster augmentation with delayed supporting loudspeakers, or a combination of these possibilities.

If the cluster design is found to be viable in the 1-kHz band, a fine-tuning procedure which will ease equalization problems is suggested. One proceeds as before, except that now the parameters associated with the 2-kHz octave band are used for the high-frequency elements, while those associated with the 500-Hz octave band are employed for the low-frequency elements. The power applied to the individual low-frequency elements is now adjusted so as to produce a match with the total intensity (direct plus reverberant) of the reference high-frequency horn. This procedure brings about an improved spectral balance between the low-frequency elements and the high-frequency elements in the reverberant field.

5 ELECTRICAL RECONFIGURATION FOR DIFFERENT EVENTS

A cluster which performs adequately according to the stated criteria when covering all seating spaces will probably perform adequately with a reduced seating space. The only modification required is to silence those loudspeaker elements, both high and low, which are directed toward unoccupied spaces. The direct intensity on axis of the operational loudspeakers will not be affected. The ratio of the total reverberant to direct intensity as calculated for an empty hall will be lowered as a result of the reduced acoustical power input to the facility. This is of no concern as it indicates higher intelligibility than before in an unoccupied space. The ratio which needs to be examined is that of total reverberant to direct intensity with full occupancy of the reduced seating space. The criterion previously stated requires this ratio to be in excess of 1. Two partially offsetting effects are occurring in the reduced seating space. The reduced acoustical input tends to reduce the reverberant intensity, while the reduced occupancy tends to lower the room constant from the value which it had at maximum occupancy in the complete space. As a consequence, with the possible exception of some very unusual room geometry, this ratio will vary only a small amount from the value which it had in the total coverage situation.

6 RESULTS

When the suggested cluster configuration design procedure is applied to the Clemson facility in the 1-kHz band, assuming commercially available loudspeakers, one possible cluster is found to consist of five low-frequency elements and ten high-frequency elements. The five low-frequency elements have nominal coverage angles of 90° × 40° and a Q of 13. There are three high-frequency horns with nominal coverage angles of 40° × 20° and a Q of 80, five high-frequency elements with nominal coverage angles of 60° × 40° with a Q of 32, and two high-frequency elements with nominal coverage angles of 90° × 40° and a Q of 13. The electroacoustic conversion efficiencies vary between 3.5 and 20%. The r_i^2/Q_i ratios range between 22 and 93 m². In the worst case (completely empty hall) the method of Eq. (9) yields a ratio of total reverberant to average direct intensity of 3.53, while the method of Eq. (14)

yields 3.72. For a completely filled facility the ratios are 1.41 and 1.49, respectively. The room constant for the Clemson facility is quite reasonable for a space of this size. Many similar facilities have room constants which are considerably smaller. If $\bar{\alpha}$ empty had been 0.2 instead of 0.31, the method of Eq. (9) would give 6.2, while that of Eq. (14) would yield 6.7 for this same cluster.

The sound pressure capability is readily calculated. In Section 4 it was assumed that 1 electrical watt was applied to the longest throw horn. Assuming that the driver on this horn has a 50-W electrical capacity, all acoustical powers will be scaled upward by a factor of 50 when operating at full power. The total acoustical power is now multiplied by $4/R$ in order to determine the total reverberant intensity. The intensity IL level in decibels is then given by

$$\text{IL} = 10\,\text{dB}\,\log \frac{50 \sum_i P_i \times 4/R}{10^{-12}\,\text{W/m}^2}. \quad (15)$$

For the Clemson facility with full cluster operation the values obtained are 105 dB for an unoccupied space and 101 dB with 12 000 occupants.

In the basketball configuration one low-frequency and one high-frequency element are silenced, and the maximum seating capacity becomes 10 000. The reverberant-to-direct intensity ratios are 3.25 empty and 1.52 filled. The maximum intensity levels are 105 and 101 dB, respectively.

In the commencement configuration three high-frequency and two low-frequency elements are silenced, and the maximum seating capacity becomes 6000. The reverberant-to-direct intensity ratios are 2.60 empty and 1.61 with 6000 occupants. The maximum intensity levels in the reverberant field are 104 dB when empty and 102 dB with 6000 occupants.

7 CONCLUSION

A central cluster design and analysis technique has been presented along with criteria for determining the cluster's performance in large multipurpose reverberant spaces. The technique has been applied to a large multipurpose facility and has been found to produce useful results.

8 REFERENCES

[1] H. F. Hopkins and N. R. Stryker, "A Proposed Loudness-Efficiency Rating for Loudspeakers and the Determination of System Power Requirements for Enclosures," *Proc. IRE* (1948 Mar.).

[2] D. H. Kaye and D. L. Klepper, "Sound System Specifications," *J. Audio Eng. Soc.*, vol. 10, p. 167 (1962 Apr.).

[3] V. M. A. Peutz, "Articulation Loss of Consonants as a Criterion for Speech Transmission in a Room," *J. Audio Eng. Soc.*, vol. 19, pp. 915–919 (1971 Dec.).

[4] W. Klein, "Articulation Loss of Consonants as a Basis for the Design and Judgment of Sound Reinforcement Systems," *J. Audio Eng. Soc.*, vol. 19, pp. 920–922 (1971 Dec.).

[5] H. Chaudiere, "Critical Distance and Critical Radius in the Design of Sound Reinforcement Systems," *J. Audio Eng. Soc.* (*Letters to the Editor*), vol. 20, p. 401 (1972 June).

[6] D. Davis, "Analyzing Loudspeaker Location for Sound Reinforcement Systems," *J. Audio Eng. Soc.*, vol. 17, p. 685 (1969 Dec.).

[7] D. Davis and C. Davis, *Sound System Engineering* (Howard W. Sams, Indianapolis, IN 1975).

[8] L. L. Beranek, *Acoustics* (McGraw-Hill, New York, 1954).

[9] W. A. Malmlund and E. A. Wetherill, "An Optical Aid for Designing Loudspeaker Clusters," *J. Audio Eng. Soc.*, vol. 13, p. 57 (1965 Jan.).

THE AUTHORS

E. Patronis, Jr.

C. Donders

Eugene T. Patronis, Jr., was born in Quincy, FL, in 1932. He received a B.S. degree from the Georgia Institute of Technology in 1953, joined the Brookhaven National Laboratory as research associate in 1957, and, in 1958, returned to the Georgia Institute as assistant professor. In 1961 he received a Ph.D., and in 1968 he became a professor on the faculty of that institute.

Dr. Patronis has published many scientific papers and is a member of the Society of Motion Picture and Television Engineers, the American Physical Society, Sigma Xi, the American Association for the Advancement of Science, and the Audio Engineering Society.

•

Catharina Donders was born in 1960 in Paramaribo, Surinam. She was educated in Holland and the U.S. and earned a B.S. degree from The Georgia Institute of Technology in 1981. In addition to being a physicist, Ms. Donders is also an accomplished musician and dancer. She is presently employed by Bruel and Kjaer in Marlborough, MA.

Loudspeaker Coverage by Architectural Mapping*

TED UZZLE

Altec Lansing Corporation, Anaheim, CA 92803, USA

A new technique uses rectangular-to-spherical coordinate transforms to generate an angular map, displaying auditorium seating in latitude and longitude angles as seen by the loudspeaker. Loudspeaker coverage and inverse-square losses can be instantly determined by this method, and intelligibility can be calculated easily for an entire area of seating. Loudspeaker radiation angular contours are presented for a number of high-frequency horns, as well as software for automatic computation of map point locations.

0 INTRODUCTION

All loudspeaking horns radiate sound in all directions, into full space. The greatest pressure amplitude in this radiation will be in two bundles subtending small angles, and subtending a small angle between them. Exactly midway between these two bundles is the direction, the vector, we associate with the axis of the horn. As we move toward or away from the horn, we will measure changes in pressure amplitude which we think we understand very well. If we maintain our distance and move to another direction of radiation off the horizontal and vertical axes, we will measure changes in pressure amplitude which not only do we not understand completely, but which have probably never been measured before.

How are we to measure the geometry of loudspeaker radiation accurately? Once measured, how are we to display it? Once displayed, how are we to use it effectively?

1 MAPPING LOUDSPEAKER RADIATION

Loudspeaker radiation is ordinarily measured by means of a polar graph generated by rotating the loudspeaker[1] in an anechoic chamber and recording the amplitude output of a microphone of known properties. The result is usually a vertical and horizontal polar pattern. We can think of this as a topographical cross section of the Earth, sliced at the Greenwich meridian and along the equator, hardly comprehensive and not necessarily characteristic of the entire planet. We thus blaze two trails and, with the medieval cartographer, are reduced to putting dragons and curlicues into the large unknown quadrants.

To make matters worse, we proceed to reduce these limited data to two numbers, a "vertical coverage angle" and a "horizontal coverage angle." Within these coverage angles we know that the direct sound varies at least 6 dB at a given distance; we suspect that it varies much more in the corners of the coverage area. Malmlund and Wetherill [2] have devised a very elegant method for using rated coverage in the design of loudspeaker arrays. Any limitation of their technique is purely extrinsic and is imposed by the limited information available to them from loudspeaker manufacturers.

The most obvious way to display horn radiation is with a globe, a three-dimensional model. A number of such models are known to exist in the laboratories of the manufacturers. The most obvious problem with the model is bulk and impracticality.

Evidently we are to map a sphere of radiation onto a flat paper. How much of the sphere do we have to show? This turns out to be a subtle problem. What parts of a loudspeaker's radiation are useful? What parts are inconsequential? What parts are—or may be—harmful?

---

* Presented at the 69th Convention of the Audio Engineering Society, Los Angeles, 1981 May 12–15; revised 1982 April 8.

[1] Some important work has been done recently to improve the geometry of polar charting; see in particular [1].

If we are designing solely for intelligibility of speech, the only areas of interest to us are those actually at the ears of the listeners. We do not need to separate each ear. The very high absorption of seating areas [3] assures us that any sound radiated between ears will not harm intelligibility. We can thus enlarge our definition of useful radiation to include an entire bank of seating. The sound going directly to walls and ceiling will eventually reach the listeners, and be of great importance if musical quality is our criterion. To assist intelligibility, however, such reflected sound must meet severe restrictions of path lengths and (thus) transit delays. In general, we are best off defining as useful to intelligibility only the sound coming down directly into the seating area.

We have analyzed the geometry of the 54 performance halls reported in Beranek [4]. We eliminated those few obviously unsuitable for speech reinforcement by central loudspeaker array (usually because of low ceiling or very deep balconies), and located a central array within the rest, either where we knew loudspeakers would be hung or in the most likely such locations. We then determined the inverse-square difference between the distance from the loudspeaker to the nearest seat and the distance from the loudspeaker to the farthest seat. Very few halls have the farthest seat only 6 dB farther away than the nearest; rather, the typical range is 8–10 dB.[2]

The universal quotation of the "−6-dB angles" as the "coverage" of a loudspeaking horn implies that the radiation within that solid angle bounded by the −6-dB pressure contour is the useful radiation and that the remainder is useless. We can say with some architectural evidence that this is incorrect.

It would seem apparent that the kind of map we need to display horn radiation should show latitude and longitude angles as a grid, on which attenuation contours are drawn. That is, along the horizontal axis we map the angles at which the sound pressure is reduced 3, 6, and 9 dB below the axis; along the vertical axis we map the angles at which the sound pressure is reduced 3, 6, and 9 dB; and then we fill in the corners. If we go 20 degrees horizontally *and then* 20 degrees vertically, what sound pressure attenuation do we measure?

Such a map was drawn, apparently for the first time, by Seeley [5],[3] who described it as sound power response (not pressure) nested in 1-dB increments from −1 to −6 dB. Six months later McCarthy [6] showed a similar depiction of six nested contours, this time of pressure response over solid angle. Neither author described his measurement technique explicitly.

In this paper we are publishing a set of nested angular contours for the high-frequency horns manufactured by Altec Lansing for sound reinforcement and the motion picture industry. The reader may have reasonable confidence that they are candidly measured and presented. As the first such to be published they have had to endure no marketing pressure to look better than any competitor's contours. Figs. 1–11 present contours for Mantaray™ constant-directivity horns, sectoral horns, and multicell horns.

These contours are drawn from data measured in Altec's Anaheim, CA, anechoic chamber. Each horn was placed on the Bruel & Kjaer turntable, and a horizontal polar pattern was taken from the measurement distance d. d was 6 ft (1.8 m) for every horn except the MR42, for which d was 10 ft (3 m). Then the horn's radiation was sliced at the latitude θ above the horizontal axis by raising the microphone $d \sin \theta$ and advancing it a horizontal component of $d(1 - \cos \theta)$ toward the horn.

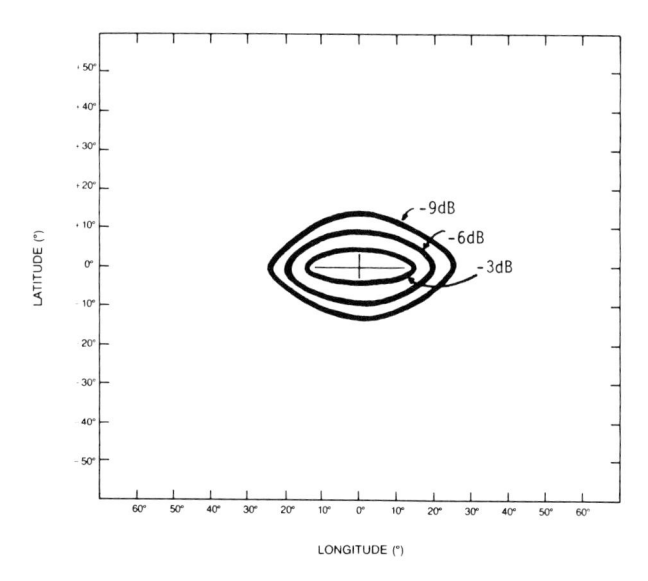

Fig. 1. Nested −3-, −6-, and −9-dB angular contours for the Altec Lansing MR42 Mantaray horn. Crosshairs locate the axis.

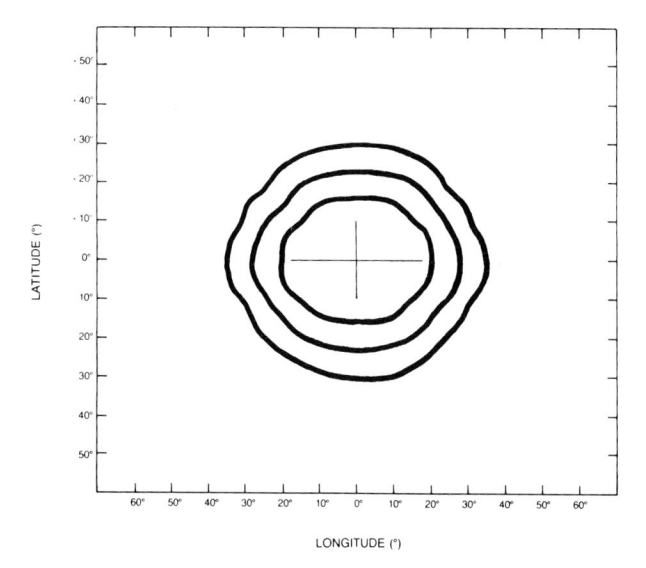

Fig. 2. Nested −3-, −6-, and −9-dB angular contours for the Altec Lansing MR64 Mantaray horn.

---

[2] Traditionally speaking, decibels are composed of power ratios only. Changes in pressure, voltage, or distance are "impure" uses of the decibel and require a series of implicit assumptions which users must understand for their arguments to have validity.

[3] Mel Sprinkle assisted in the development of the technique presented here. The author is indebted to Don and Carolyn Davis of San Juan Capistrano, CA, for direction to this reference.

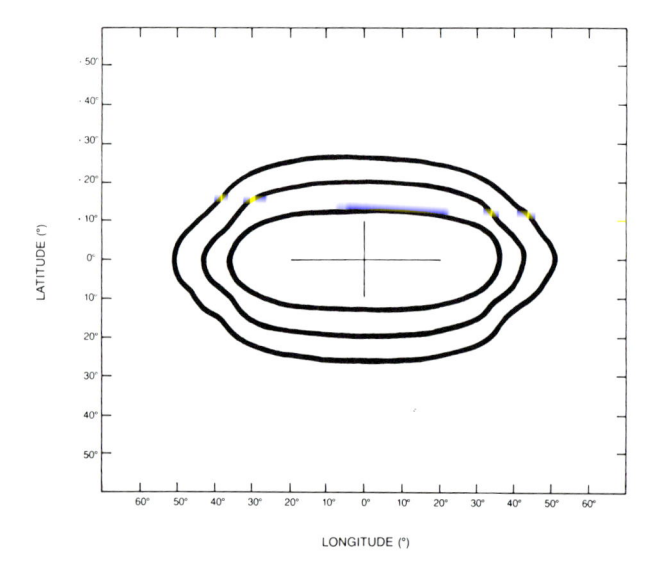

Fig. 3. Nested −3-, −6-, and −9-dB angular contours for the Altec Lansing MR94 Mantaray horn.

The horn was rotated around the horizontal apparent apex. The measurement microphone was raised and advanced with d measured to the vertical apparent apex [1]. θ was selected at 5° increments (see Fig. 12). The −3-, −6-, and −9-dB points were picked off and charted, and connected by a smooth curve. As a check, the horn was usually rotated 90° about its axis, and a vertical polar plot was measured.

The test signal used was a band of pink noise between 1 and 2 kHz. This band of the audio spectrum is most useful for the intelligibility of running speech,[4] which we have selected as our coverage criterion. Dispersion performance in this band will accurately char-

---

[4] We believe this to be justified by the work of French and Steinberg [7] and by ANSI S3.5-1969(R 1978) [8], work conducted by a group chaired by Karl Kryter.

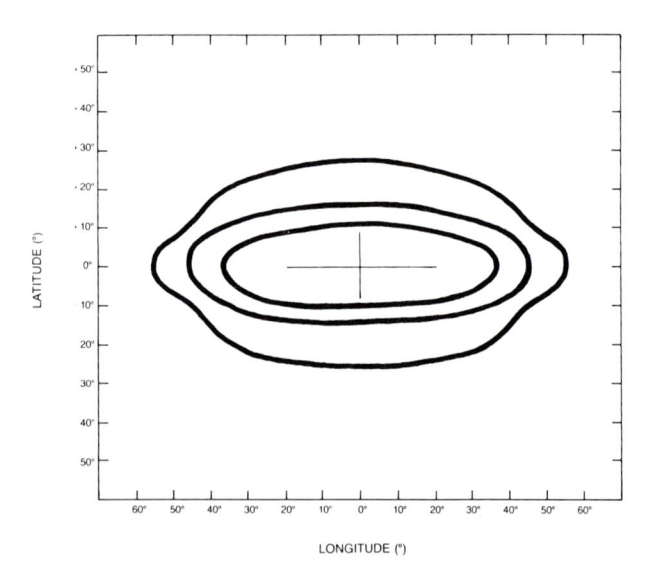

Fig. 4. Nested −3-, −6-, and −9-dB angular contours for the Altec Lansing MR94-8 Mantaray horn.

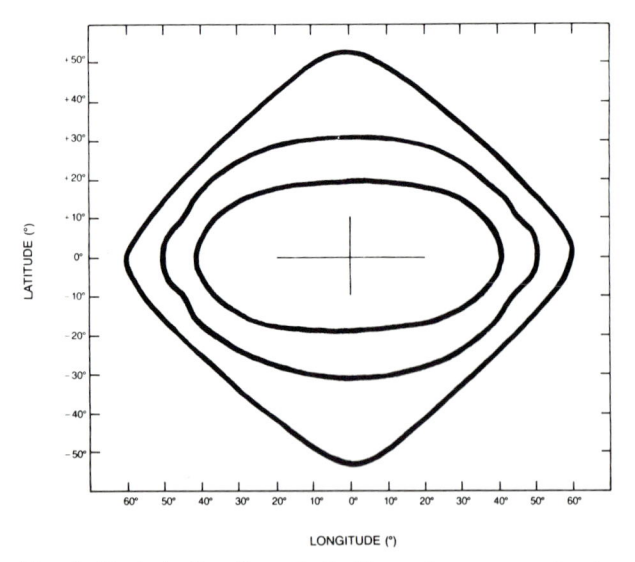

Fig. 5. Nested −3-, −6-, and −9-dB angular contours for the Altec Lansing 311-90 sectoral horn.

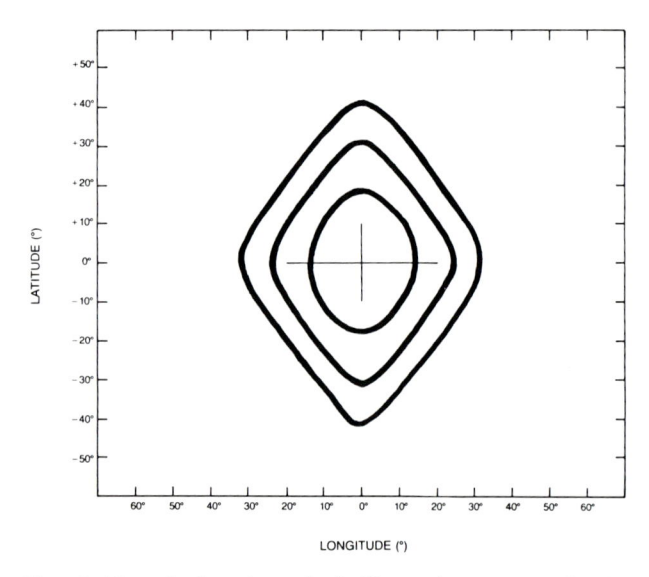

Fig. 6. Nested −3-, −6-, and −9-dB angular contours for the Altec Lansing 311-60 sectoral horn. At the test frequencies used, this horn has a greater vertical than horizontal angle. (For more information on angle versus coverage for this horn, see Altec Lansing Tech. Letter 221.)

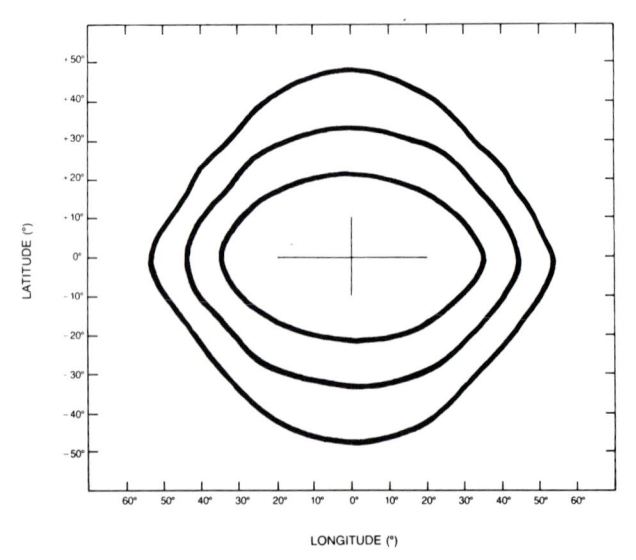

Fig. 7. Nested −3-, −6-, and −9-dB angular contours for the Altec Lansing 511B sectoral horn.

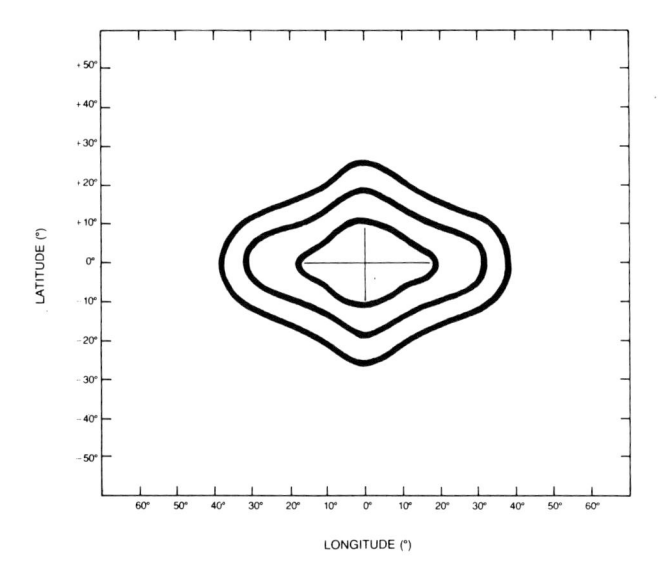

Fig. 8. Nested −3-, −6-, and −9-dB angular contours for the Altec Lansing 203B multicellular horn.

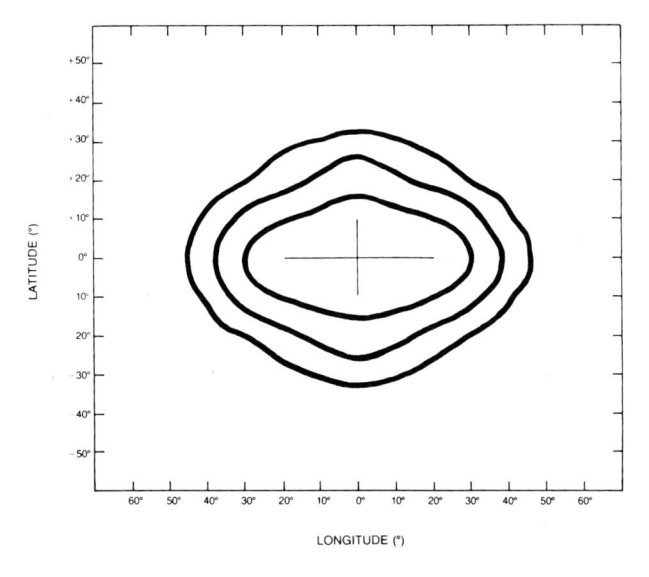

Fig. 9. Nested −3-, −6-, and −9-dB angular contours for the Altec Lansing 805B multicellular horn.

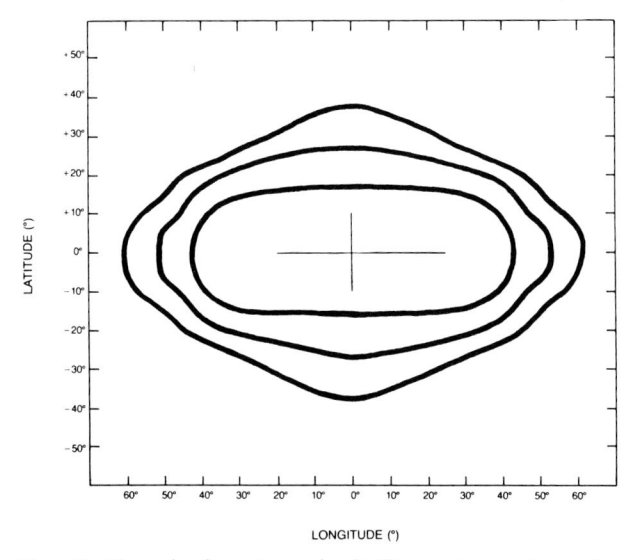

Fig. 10. Nested −3-, −6-, and −9-dB angular contours for the Altec Lansing 1005B multicellular horn.

acterize the dispersion of Mantaray™ horns throughout their useful range. Multicellular horns will conform to these contours over most of their useful bandwidth, while sectoral horns will undergo drastic changes in coverage contour over frequency.[5]

2 MAPPING THE ROOM

Once we have these solid angle maps of the principal lobe of sound radiation, what do we do with them? These radiation maps are of instant utility to those designing loudspeaker clusters with the optical device of Malmlund and Wetherill. Instead of the hypothetical rectangular mask used at present, the light projector can be masked with a small transparency, showing −3-, −6-, and −9-dB solid angles with perhaps different colors. We will not dwell on the mechanics of applying these maps to the optical device, expecting that sophisticated users of it will immediately see these employments, and probably other, more subtle, utilities as well.

What of those without a light projector? Or without the patience to build a cardboard model of the room? If we are mapping so many odd-shaped steradians of loudspeaker radiation onto a flat piece of paper, can we not also map onto that same piece of paper those odd-shaped steradians subtended by banks of seating? More directly, can we map the room on graph paper and then slide over it transparencies of horn radiation?

Several things are apparent. The map of the room must retain angular integrity throughout. A 5° × 5° pixel[6] of loudspeaker radiation must match a 5° × 5° group of seats anywhere in the map where the loudspeaker transparency may be slid. As we will presently see, this requires a nonconformable map in which loxodromes are curves near the poles, and roll angles are

---

[5] Compare these contours with the coverage angle versus frequency graphs in [9].

[6] This word is used analogously to electronic video, where it is a contraction of *pic*ture *el*ements and means the smallest resolvable area of a video image [10].

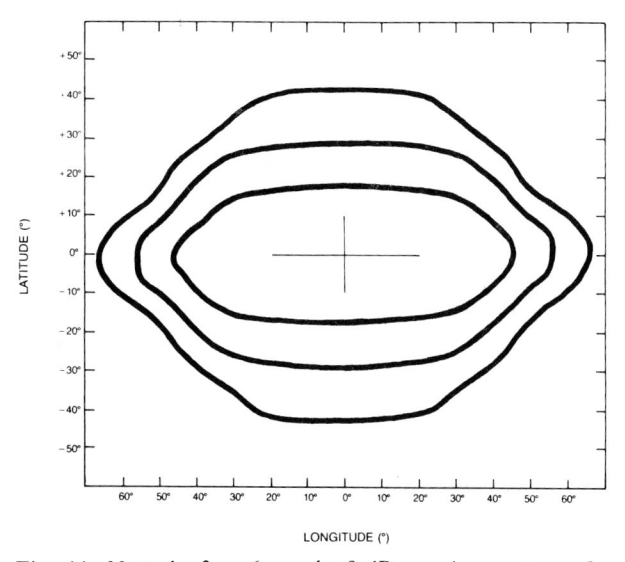

Fig. 11. Nested −3-, −6-, and −9-dB angular contours for the Altec Lansing 1505B multicellular horn.

compromised. We will show why we believe these compromises acceptable in real-world situations.

The map will have to show distance ratios in decibels, from which we will obtain inverse-square losses to compare with horn attenuation over angle. We will also obviously need automatic computation to assist us in plotting data points for the map.

It is a truism that when the radiation from a loudspeaker strikes an oblique plane, the shape of the beam is distorted when viewed on the plane. One loudspeaker manufacturer has published [11] an interesting set of contours of loudspeaker radiation cut by a plane at various angles. Our approach with this technique is exactly the opposite: we will distort the room to the mapping system used for the loudspeaker. The architectural map we thus produce will not be a view of the room with painterly perspective. This is a point of great importance which has been insufficiently appreciated in the prior work. Any attempt to map three dimensions onto two will inevitably introduce distortions [12] (For a somewhat less technical discussion see also [13].)

The naive sometimes assume that the great map of 1569 by Gerard Mercator used rectilinear coordinates because such lines of latitude and longitude are not very sophisticated. In fact, the world maps of Ptolemy in antiquity (at least as they have been handed down to us through medieval copies) used seemingly sophisticated polyconic and sinusoidal projections. The actual geography shown was quite wrong, of course. Mercator very deliberately introduced distortions of size and shape into his maps with the celebrated "waxing latitudes" in order that the rhumb lines, or loxodromes, should be straight lines.

Why so? What were people doing in the sixteenth century? They were building wooden ships and sailing all over the globe. Mercator's map projection shows true heading. Select any point on the compass, sail off in that direction, and your route will be a straight line on a Mercator map. Draw a straight line from your present position to your desired landfall, and read your desired compass heading with a protractor.[7] In 1569 mariners acknowledged and accepted the distortions because the map had practical utility [14, pp. 102–111].

For exactly the same reason, we adopt the cylindrical projection of the Portuguese cartographer Pedro de Lemos [14, p. 114]. In a cylindrical projection the sphere, as shown in Fig. 13, is mapped onto a cylinder so that each parallel is equidistant and the north and south poles are the circles at the top and bottom of the cylinder, as shown in Fig. 14. Cut the cylinder along any convenient meridian and unroll it into a plane, as is being done in Fig. 15. All lines of latitude and longitude are perfectly rectilinear and equally spaced. While shapes and distances may be severely distorted, angular subtense between any two points along latitude and longitude lines is always preserved. This is exactly the map projection we require.

---

[7] Do not confuse true compass heading with great circle route, which is the shortest route, requires a constantly changing heading, and which is a curve on a Mercator projection.

Fig. 13. Sphere erected around the loudspeaker location.

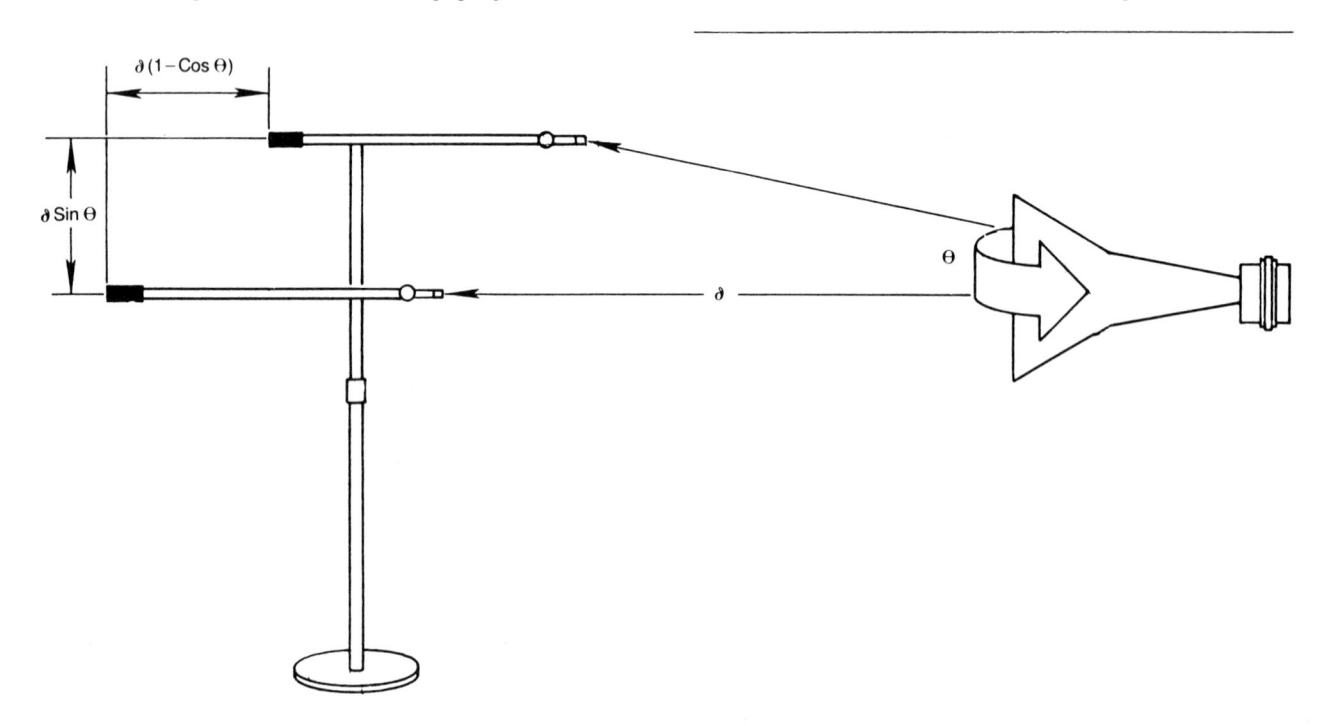

Fig. 12. Microphone positioning to measure data used to draw contours.

3 MODELING THE ROOM

We will not be mapping a sphere; we will in fact be mapping architecture, and the present author has previously published a useful mathematical model that may be applied easily and quickly to most real-world architecture [15]. Some architecture is of course *sui generis*, and there are no shortcuts around laborious measurement and computation.

We postulate a bank of seating that is trapezoidal in plan, of width W_F in the front and of width W_R in the rear. It is sloping, and a horizontal ceiling is of height H_F in the front and H_R in the rear. The seating length, first to last row, is L. These five dimensions give us our seating bank. We need two more to give us our loudspeaker location. Let F designate the horizontal component of the distance from the front row to the loudspeaker and let h designate the height of the loudspeaker above the floor level of the first row. In other words, drop a plumb line from the loudspeaker to the floor level at the front row. F is the distance from the front row to the plumb bob, and h is the length of the plumb line. It follows that the distance from the loudspeaker to the center seat of the front row is $\sqrt{F^2 + h^2}$. It is useful to imagine a hypothetical front wall of the auditorium, enclosing the loudspeaker (see Fig. 16).

We can locate any seat in the auditorium in this way. The seat is the distance d from the hypothetical front wall and the distance m from the nearer side of the seating.

The latitude angle is given by

$$-\arctan \frac{h - \frac{(d - F)(H_F - H_R)}{L}}{\sqrt{d^2 + \left(\frac{(W_R - W_F)(d - F)}{2L} + \frac{W_F}{2} - m\right)^2}}$$

and the longitude angle is given by

$$\arctan \frac{\frac{(W_R - W_F)(d - F)}{2L} + \frac{W_F}{2} - m}{\sqrt{d^2 + \left(h - \frac{(d - F)(H_F - H_R)}{L}\right)^2}}.$$

The distance from the loudspeaker to that seat is given by

$$\sqrt{d^2 + \left(h - \frac{(d - F)(H_F - H_R)}{L}\right)^2 + \left(\frac{(W_R - W_F)(d - F)}{2L} + \frac{W_F}{2} - m\right)^2}.$$

To delineate the bank of seating on our map we can compute a set of points starting with the center seat in the rear row, going around the perimeter of the seating, and arriving at the center seat of the front row. Using the above equations we successively use the values for d and m given in Table 1.

If we adopt the distance between the loudspeaker and the center seat in the last row as a reference distance, we can determine that the other calculated points are so many decibels nearer or farther away.

4 SAMPLE PROBLEM

To illustrate some of the techniques available with architectural mapping, let us take as a sample problem the Bayreuth Festspielhaus, designed by Otto Brückwald for the performance of Richard Wagner's operas. Fig. 16 shows the ordinate dimensions of the bank of seating in this auditorium. We will select a loudspeaker location within the proscenium, just above the fire curtain. We use the equations given above to generate a set of data points, which are mapped in Fig. 17. Next to a

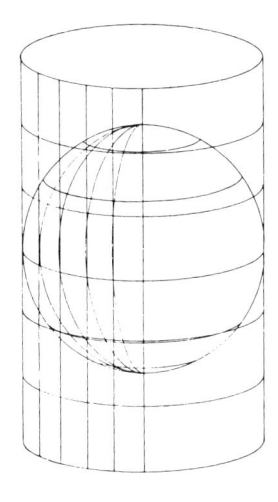

Fig. 14. Each point on the sphere is mapped onto a cylinder.

number of data points, decibel levels relative to the center rear seat are marked. These are representative of inverse-square distances only, and negative numbers here indicate points closer to the loudspeaker.

In Fig. 18 we see the seating bank of the Festspielhaus covered very neatly and economically with a single MR64 Mantaray™ horn. By the traditional concept of horn coverage this works very well, but note that direct sound will vary by 12 or 13 dB across the seating. Since the reverberant sound level is the same throughout, there will be large variations in intelligibility and listening quality.

This is how we arrive at direct sound distribution. The axis of the horn is pointed about 22° down, at a seat about one quarter the way back. This seat is 4 dB closer (distance decibels) than the last row, which lies along the −3-dB contour of the horn. Thus the direct sound decreases 7 dB from the horn's axis to the rear row, by a combination of polar losses and inverse-square losses. The rear corner seats are 4 or 5 dB farther away than the seat at the horn's axis, but they lie just within the −9-dB contour of the horn. We subtract inverse-square loss from polar loss and discover the difference in direct sound levels. Look at the end seats of the front row. They are 2 dB closer than seats at the axis of the horn, but lie along the horn's −9-dB contour. Therefore direct sound there is 7 dB below that at the axis of the horn.

Fig. 19 shows significantly improved evenness of di-

Table 1.

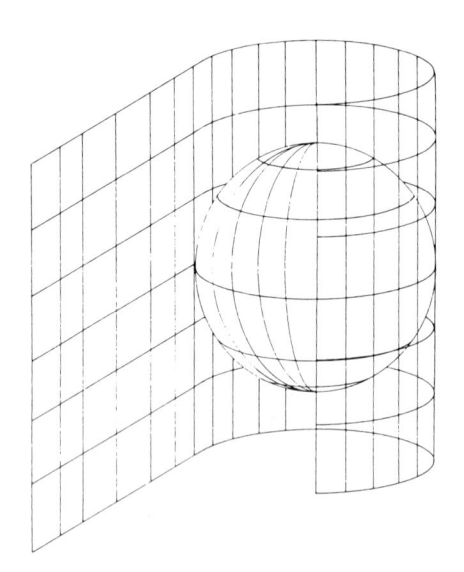

Point Number	d	m
1	$F + L$	$\dfrac{W_R}{2}$
2	$F + L$	$\dfrac{W_R}{4}$
3	$F + L$	$\emptyset$
4	$F + \dfrac{3L}{4}$	$\emptyset$
5	$F + \dfrac{L}{2}$	$\emptyset$
6	$F + \dfrac{L}{4}$	$\emptyset$
7	F	$\emptyset$
8	F	$\dfrac{W_F}{4}$
9	F	$\dfrac{W_F}{2}$

Fig. 15. The cylinder is cut and unrolled into a plane.

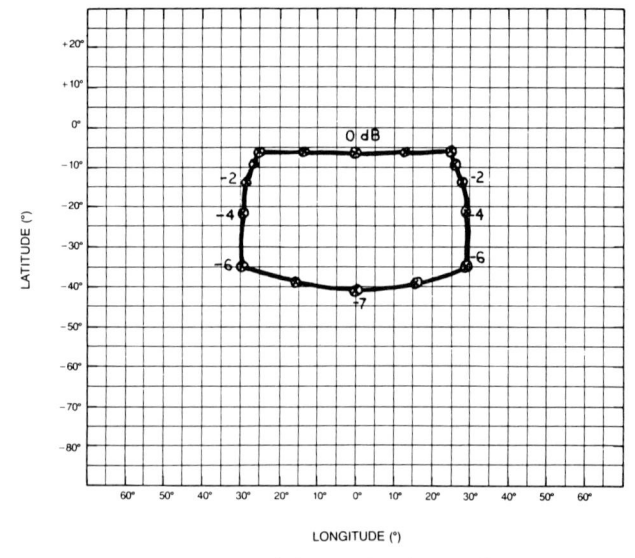

Fig. 17. Angular map of the seating bank of the Bayreuth Festspielhaus, as viewed from the loudspeaker location.

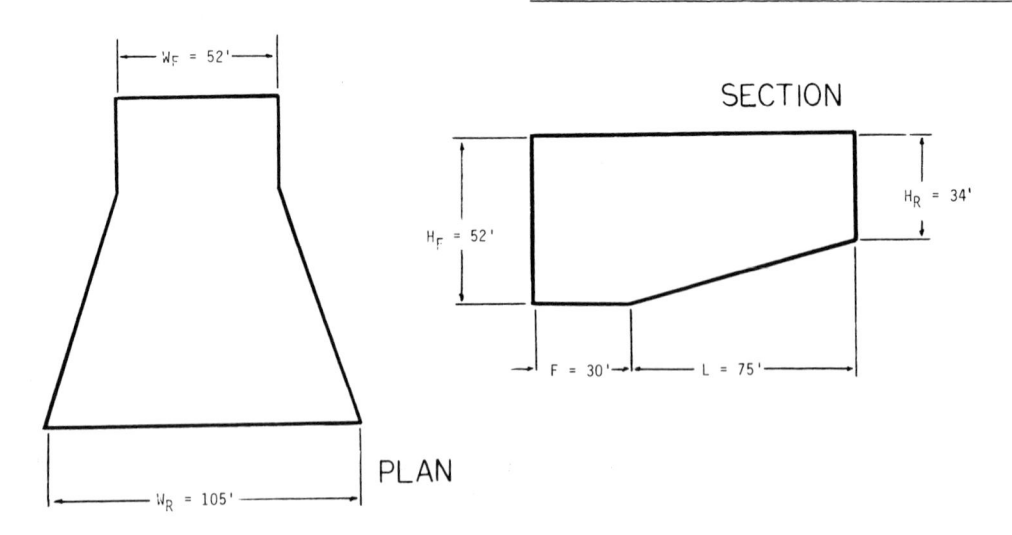

Fig. 16. Plan and section of the main seating bank of the Bayreuth Festspielhaus.

rect sound coverage side to side with the use of a wider pattern horn, the MR94 Mantaray™ horn. This is achieved at the price of much more sound radiation on the side walls. In Fig. 20 a 1505B multicellular horn is aimed at the last row, center seat. As we can see, uniformity of sound throughout the seating is vastly improved, and in fact is well within a 2-dB window everywhere. You can see, however, how much of the horn's radiation is going on the rear wall of the Festspielhaus. In order to obtain much better uniformity of direct sound, we may have introduced other problems that may be worse than degraded direct-to-reverberant ratios in the corner seats. We will want to ask some hard, specific questions: How absorptive is that rear wall? How specular will be reflections from it? Is it concave, and, if so, does it focus? What are the path-length differences between direct sound and an echo from the rear wall as it propagates back down the seating?

These are not intended as outstanding examples of sound system design, but only as illustrations of the wealth of new information and flexibility available with this graphical technique. The large number of permutations and combinations of horns suggest that the diligent sound system designer may achieve almost any compromise desired among the various coverage criteria: uniformity of direct sound, avoidance of reflective walls, avoidance of open microphones in the audience area, and so on.

5 CONCLUSION

This technique deals with the geometry of loudspeaker radiation into architecture, and that is all. It leaves untouched the issues that only statistical acoustics can resolve: total level and intelligibility. The prediction of feedback in reinforcement systems is a combination of geometric and statistical acoustics.

The sound system designer who uses statistical techniques need no longer design for only the "magic seat." Once the performance at one seat (at the axis of the horn) is calculated, we can now know the directivity factor Q along any axis and predict direct sound level, reverberant sound level, total sound level, critical distance, and articulation loss of consonants (an intelligibility figure of merit for direct versus reverberant sound) for any and every seat in the house, with either one horn or an array of them [16].

Although this technique is easy to manipulate with calculator or computer software, and the horn contours on transparencies, its novelty and the distortions introduced by the mapping are sometimes intimidating to those taking it up for the first time. Despite this, it is better for the loudspeaker user to employ this method than imitate the sea captain in *The Hunting of the Snark* [17] (see Fig. 21):

He had bought a large map representing the sea,
   Without the least vestige of land:
And the crew were much pleased when they found it to be
   A map they could all understand.

"What's the good of Mercator's North Poles and Equators,
   Tropics, Zones, and Meridian Lines?"
So the Bellman would cry: and the crew would reply,
   "They are merely conventional signs!"

"Other maps are such shapes, with their islands and capes!
   But we've got our brave Captain to thank,"
(So the crew would protest) "that he's bought us the best—
   A perfect and absolute blank!"

6 ACKNOWLEDGMENT

The author wishes to acknowledge his indebtedness to the cited paper by Mr. Thomas G. McCarthy of North Star Sound, Inc., Minneapolis, MN, which was the starting point for the method presented here.

Translations of the program in the Appendix into BASIC, FORTRAN IV, and for HP-67/97 have been generously worked by Mr. Peter More, Dr. Rex Sinclair, and Mr. Larry Lutz of Altec Lansing. In addition, Mr. Thomas Bouliane has written a valuable program which automatically locates a series of evenly spaced points

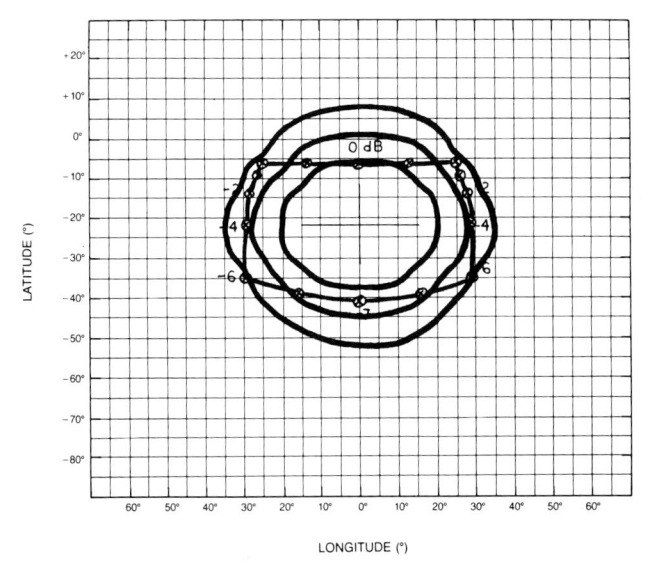

Fig. 18. Coverage of the Bayreuth Festspielhaus seating by a 60° × 40° horn.

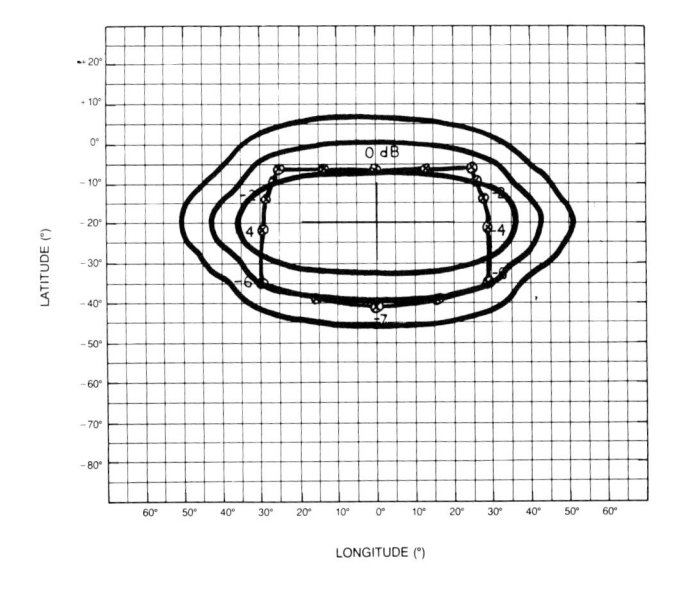

Fig. 19. Coverage of the Bayreuth Festspielhaus seating by a 90° × 40° horn.

within the seating area [18]. In addition, Mr. Bouliane pointed out to the author an improved equation for the calculation of the latitude angle. Dr. Sinclair also conducted the contour measurements, with the assistance of the author. Mr. Robert Trabue Davis and Mr. Lee Savoit, of Altec, supplied many basic ideas and suggestions and the nifty drawings.

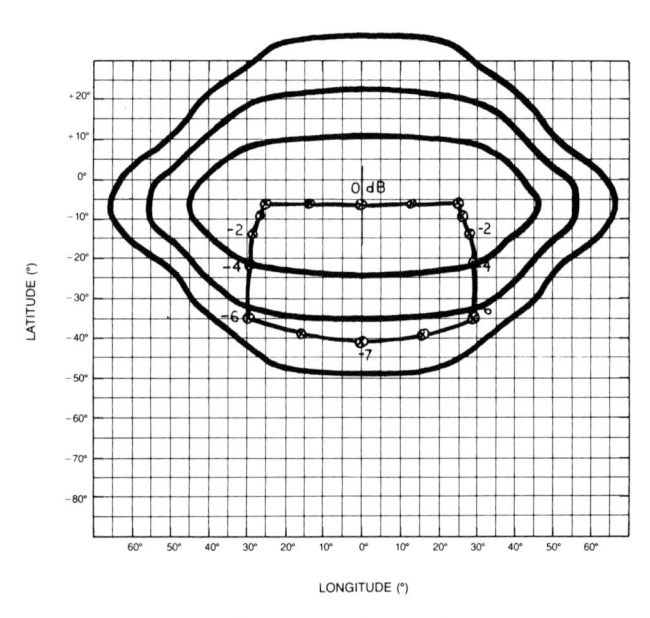

Fig. 20. Coverage of the Bayreuth Festspielhaus seating by a large-pattern multicellular horn.

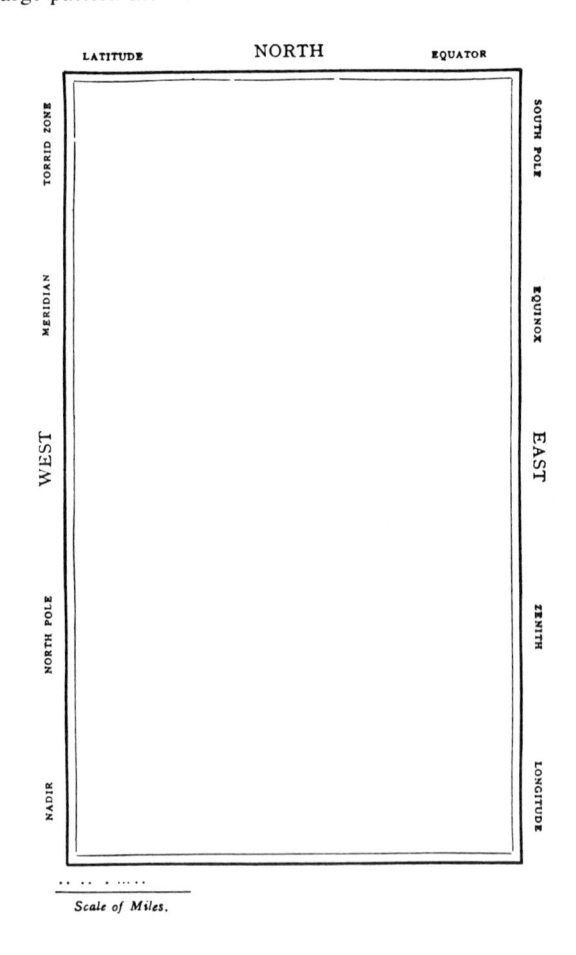

Fig. 21. Ocean chart.

7 REFERENCES

[1] M. Ureda, "Apparent Apex Theory: Far-Field Polar Characteristics at Close Proximity," presented at the 61st Convention of the Audio Engineering Society, *J. Audio Eng. Soc. (Abstracts)*, vol. 26, p. 988 (1978 Dec.), preprint no. 1403.

[2] W. Malmlund and E. Wetherill, "An Optical Aid for Designing Loudspeaker Clusters," in D. L. Klepper, Ed., *Sound Reinforcement, an Anthology* (Audio Engineering Society, New York, 1978), p. D-26, and *J. Audio Eng. Soc.*, vol. 1, 1965 January, p. 57.

[3] T. Schultz and B. Watters, "Propagation of Sound across Audience Seating," *J. Acoust. Soc. Am.*, vol. 36, p. 885 (1964).

[4] L. Beranek, *Music, Acoustics and Architecture* (Wiley, New York, 1962), pp. 83–392.

[5] E. Seeley, "Innovations in a Stadium Sound System Design," presented at the 60th Convention of the Audio Engineering Society, Los Angeles, 1978 May 2–5.

[6] T. McCarthy, "Loudspeaker Arrays—A Graphic Method of Designing," presented at the 61st Convention of the Audio Engineering Society, *J. Audio Eng. Soc. (Abstracts)*, vol. 26, p. 992 (1978 Dec.), preprint no. 1398.

[7] N. R. French and J. Steinberg, "Factors Governing the Intelligibility of Speech Sounds," *J. Acoust. Soc. Am.*, vol. 90, p. 90 (1949).

[8] ANSI S3.5-1969(R 1978), "Methods for the Calculation of the Articulation Index."

[9] M. Engebretson, "Directivity of Altec Loudspeakers," Altec Corp., Anaheim, CA., Altec Tech. Letter 221 (1974).

[10] L. Diamant, *The Broadcast Communications Dictionary* (Hastings House, New York, 1974), p. 88.

[11] "Engineers' and Architects' Design Guide," Bose Corp., Framingham, MA, 1980, pp. 15–19.

[12] E. N. Gilbert, "Distortion in Maps," *SIAM Rev.*, vol. 16, p. 47 (1974).

[13] W. Chamberlin, *The Round Earth on Flat Paper*, National Geographic Society, Washington, DC, 1947.

[14] G. R. Crone, *Maps and Their Makers* (Hutchinson, London, 1968).

[15] T. Uzzle, "Room Geometry for Acoustics," *Syn-Aud-Con Tech Topic*, vol. 6, no. 1 (1978).

[16] R. Sinclair and T. Uzzle, "Off-Axis Performance of Multiple Loudspeakers," presented at the 70th Convention of the Audio Engineering Society, *J. Audio Eng. Soc. (Abstracts)*, vol. 29, p. 926 (1981 Dec.), preprint no. 1825.

[17] C. Dodgson, *The Collected Verse of Lewis Carroll* (Macmillan, New York, 1933), pp. 278–279.

[18] Thomas Bouliane, Audio Contractors, Inc., Buffalo, NY, personal correspondence.

APPENDIX

SOFTWARE

USER INSTRUCTIONS "AP"

A.1 Introduction

"AP" generates a table of mapping points which, when plotted on Altec Lansing array perspective graph paper, allows immediate comparison with the angular radiation contours of Altec Lansing large-format horns.

A.2 Method

The equations are published elsewhere.

A.3 User Instructions for HP-41C and HP-41CV

This program requires that the HP-82143A peripheral printer be attached and switched to the NORM position. In the case of the 41C, one memory module must be installed. Load the cards sides 1 through 5. SIZE between 012 and 050; the initial allocation of 017 will do nicely. Execute "AP"; the printer will produce

XEQ "AP"

ALTEC

LANSING

ARRAY PERSPECTIVE

L = ?

This prompt will also appear in the display. You will notice in the display that flag 1 has been set. Enter the seating length, front row to last row, and touch R/S. The display and printout will prompt

WF = ?

Enter seating width at the front and touch R/S.

WR = ?

Enter seating width at the rear and touch R/S.

HF = ?

Enter ceiling height at the front and touch R/S.

HR = ?

Enter ceiling height at the rear and touch R/S.

ARRAY H = ?

Enter the array height above the front row and touch R/S. This program does not make an ear height correction for seated or standing listeners, so users must do this themselves. This way the program will work with either English or metric (or any other) units.

F = ?

If you are using this program to calculate room volume and surface area, enter the horizontal distance from the front row to the front wall, and touch R/S. If you are using this program for array design, enter the horizontal distance from the front row to a point on the floor directly beneath the array position you wish to consider, and touch R/S.

WANT V + S ?

Answer the question in digital language: 0 R/S for no, 1 R/S for yes. If you answer yes, the printer will list

V = aaa,aaa.a
S = bb,bbb.b

in the cubic and square units you used for the room dimensions. If you answer "WANT V + S?" with a no, these will be skipped and the calculator will come directly to

WANT AUTO?

Again, 0 R/S for no and 1 R/S for yes. If you answer yes, the printer will produce a tape thus:

VERT ∡ = cc.c
HOR ∡ = dd.d
RANGE = eee.e
Ld = ∅. dB

VERT ∡ = cc.c
HOR ∡ = ff.f
RANGE = ggg.g
Ld = h. dB

These will be followed by seven more blocks of data. Within each block, VERT ∡ and HOR ∡ are used as data points to graph the seating on Altec Lansing array perspective analysis paper. RANGE is the distance from the loudspeaker array to the given point, in whatever units the user employs, and Ld is the change in range in decibels, according to the inverse-square law, with the range to the first point as the standard.

At the end of the table, the calculator will ask

START OVER?

Answer yes by touching 1 R/S if you wish to change the position of the array (up or down, backward or forward) to try to get a more favorable match of room geometry to horn radiation. Answer no by touching 0 R/S, whereupon the calculator will ask

WANT MANUAL?

This routine is to plot irregularly shaped rooms, or to locate points within the seating area (open microphones or whatever). If you answered no to the question "WANT AUTO?," the program skipped directly to this point. The calculator will prompt

FRONT = ?

Assume that the front wall is built around the array. Now measure perpendicularly from this hypothetical wall to the given point of interest. Enter the distance and touch R/S.

SIDE = ?

Now measure the distance in from the nearest side edge of the seating, enter the distance, and touch R/S.

VERT ∡ = ii.i
HOR ∡ = jj.j
RANGE = kkk.k
Ld = ll. dB

FRONT = ?

The program will continue prompting for front and side distances until you finish.

A4 Sample Problem

Say that we have an audience seating bank 75 ft from front to last row, 52 ft wide at the front, and 105 ft wide at the rear, with the horizontal ceiling 52 ft above the front row and 34 ft above the rear row. We wish to place the loudspeaker array 30 ft in front of the front row and 35 ft above the front row. This position is in the proscenium. What are the volume and surface area of this auditorium? What are the map data points? Where in the map should there be located a microphone position 50 ft from the curtain and 18 ft from the side edge of the seating? This sample problem is the Festspielhaus in Bayreuth. Table 2 shows the printed results of the computation.

A5 Program Information

"AP" requires 12 data registers and 80 registers for program memory (270 program steps, 555 bytes). The memory assignments are as follows:

00—Range reference
01—L
02—F
03—W_f
04—W_r
05—H_f
06—H_r
07—h
08—Used
09—Used
10—Used
11—Used

Flag 1 is used to identify the reference range and flag 12 is set and cleared for double-wide printing of the word ALTEC. If you terminate execution before the first block of data has been printed, flag 1 will remain set, and this will be indicated with the annunciator. You must clear it manually thus:

SHIFT, CF, 0, 1

The calculator is set to FIX 0 and FIX 1 at various times during execution. After the program has run, it is left at FIX 1. The program list is shown in Table 3.

Table 2. Calculator tape for the Bayreuth Festspielhaus.

```
           XEQ "AP"                                          VERT ∠ = -34.5
                                  VERT ∠ = -6.2             HOR  ∠ = 29.4
    ALTEC                         HOR  ∠ = 0.0              RANGE  = 52.9
      LANSING                     RANGE  = 110.7            Ld = -6. dB
                                  Ld = 0. dB
                                                            VERT ∠ = -38.8
ARRAY PERSPECTIVE                 VERT ∠ = -6.1             HOR  ∠ = 15.7
                                  HOR  ∠ = 13.3             RANGE  = 47.9
L = ?                             RANGE  = 113.7            Ld = -7. dB
          75.0    RUN             Ld = 0. dB
WF = ?                                                      VERT ∠ = -40.6
          52.0    RUN             VERT ∠ = -5.6             HOR  ∠ = 0.0
WR = ?                            HOR  ∠ = 25.4             RANGE  = 46.1
         105.0    RUN             RANGE  = 122.5            Ld = -7. dB
HF = ?                            Ld = 0. dB
          52.0    RUN                                       START OVER?
HR = ?                            VERT ∠ = -9.2                      0.0    RUN
          34.0    RUN             HOR  ∠ = 26.3             WANT MANUAL?
ARRAY H = ?                       RANGE  = 103.5                     1.0    RUN
          30.0    RUN             Ld = 0. dB
F = ?                                                       FRONT = ?
          35.0    RUN             VERT ∠ = -14.3                    50.0    RUN
                                  HOR  ∠ = 27.5             SIDE = ?
WANT V + S?                       RANGE  = 85.1                     18.0    RUN
           1.0    RUN             Ld = -2. dB               VERT ∠ = -27.0
V = 377,615.0                                               HOR  ∠ = 13.2
S = 32,256.0                      VERT ∠ = -22.1            RANGE  = 58.1
                                  HOR  ∠ = 28.7             Ld = -5. dB
WANT AUTO?                        RANGE  = 67.9
           1.0    RUN             Ld = -4. dB               FRONT = ?
```

Table 3. Program list for HP-41.

```
              PRP "AP"

"CP1A"

  02◆LBL "AP"
FIX 1  ADV  SF 12
" ALTEC"  PRA  CF 12
"    LANSING"  PRA  ADV
ADV  "ARRAY PERSPECTI"
"-VE"  PRA  ADV

  17◆LBL 00
SF 01  "L = ?"  PROMPT
STO 01  "WF = ?"  PROMPT
STO 03  "WR = ?"  PROMPT
STO 04  "HF = ?"  PROMPT
STO 05  "HR = ?"  PROMPT
STO 06  "ARRAY H = ?"
PROMPT  STO 07  "F = ?"
PROMPT  STO 02  ADV
"WANT V + S?"  PROMPT
X=0?  XEQ 01  RCL 01
RCL 05  RCL 06  +
STO 10  *  2  /  RCL 05
RCL 02  *  +  RCL 03  *
RCL 04  RCL 03  -
RCL 01  *  RCL 05  4  *
RCL 06  -  *  6  /  +
"V = "  ARCL X  AVIEW
RCL 03  RCL 04  +
STO 11  RCL 10  +
RCL 01  *  RCL 10
RCL 11  *  2  /  +
RCL 03  RCL 05  +
RCL 02  *  2  *  +
"S = "  ARCL X  AVIEW
ADV

 102◆LBL 01
"WANT AUTO?"  PROMPT
X=0?  GTO 04  ADV
RCL 01  RCL 02  +
STO 08  RCL 04  2  /
STO 09  XEQ 06  RCL 04
4  /  STO 09  XEQ 06
CLX  STO 09  XEQ 06
RCL 02  RCL 01  .75  *
+  STO 08  XEQ 06
RCL 02  RCL 01  2  /  +
STO 08  XEQ 06  RCL 02
RCL 01  4  /  +  STO 08
XEQ 06  RCL 02  STO 08
XEQ 06  RCL 03  4  /
STO 09  XEQ 06  RCL 03
2  /  STO 09  XEQ 06

 159◆LBL 02
"START OVER?"  PROMPT
X=0?  GTO 03  GTO 00

 165◆LBL 03
"WANT MANUAL?"  PROMPT
X=0?  GTO 05

 170◆LBL 04
ADV  "FRONT = ?"  PROMPT
STO 08  "SIDE = ?"
PROMPT  STO 09  XEQ 06
GTO 04

 180◆LBL 05
"END"  PRX  STOP

 184◆LBL 06
RCL 08  RCL 02  -
STO 11  RCL 05  RCL 06
-  *  RCL 01  /  RCL 07
X<>Y  -  STO 10  RCL 11
RCL 04  RCL 03  -  *
RCL 01  /  2  /  RCL 03
2  /  +  RCL 09  -
STO 11  X↑2  RCL 08  X↑2
+  SQRT  RCL 10  X<>Y  /
ATAN  CHS  "VERT ∠ = "
ARCL X  AVIEW  RCL 11
RCL 10  X↑2  RCL 08  X↑2
+  SQRT  /  ATAN
"HOR ∠ = "  ARCL X
AVIEW  RCL 08  X↑2
RCL 10  X↑2  RCL 11  X↑2
+  +  SQRT  STO 10
"RANGE = "  ARCL X
AVIEW  FS?C 01  STO 00
RCL 10  RCL 00  /  LOG
20  *  FIX 0  INT
"Ld = "  ARCL X  "⊢ dB"
AVIEW  ADV  FIX 1  END
```

THE AUTHOR

Ted Uzzle received an A.B. degree from Harvard College in 1971. As an undergraduate he studied the history of architecture. He conducted a consulting practice from 1973 to 1980 in theater design and architectural acoustics, specializing in motion-picture facilities. In June 1980 he joined Altec Corporation in Anaheim, California, as manager, market development. Mr. Uzzle is a member of the Acoustical Society of America, the AES, the British Kinematograph Society, the SMPTE, and the U.S. Institute for Theatre Technology.

Editor's note: This is the earliest technical paper describing an attempt to model seating area boundaries within a calculator or computer, as viewed from the location of a central loudspeaker cluster, to permit near-optimum loudspeaker selection and aiming. It has errors, because accuracy is compromised in any global-to-flat mapping technique, even the one employed, and it was bypassed with the development of the far more accurate Prohs-Harris PHD, the Altec-Mark IV Acousticad, JBL Professional CADP, and Renkus-Heinz Ease programs. These four programs keep the three-dimensional model data within the computer with accurate flat projections displayed and printed. Ted Uzzle's paper is included for historical reasons because the designers of the later techniques have said they were influenced and inspired by his work.

David Lloyd Klepper

The Design of Distributed Sound Systems from Uniformity of Coverage and Other Sound-Field Considerations*

REX SINCLAIR

Altec Lansing Corporation, Anaheim, CA 92803, USA

The sound distribution on a plane normal to the axis of a single loudspeaker can be approximated by a simple equation with known constants. The sound distribution on the listening plane of distributed systems can then be found. Maximum and minimum direct sound pressure levels, relative to the axial values of a single loudspeaker, and their differences can also be expressed as simple equations with known constants along with the density in the number of loudspeakers per −6-dB coverage circle area. A method is given for designing distributed sound systems to satisfy a given maximum or minimum sound pressure level relative to the axial level of a single loudspeaker or a given uniformity of coverage. A previous restriction of only six available combinations of loudspeaker patterns and degrees of overlap is removed by replacing such combinations by loudspeaker density, even though it is not necessary to know the density to use the method, and there need be no direct reference to it or the degree of overlap. Two transparent overlays are used with appropriately scaled room drawings giving a short design time and design flexibility. Values of all constants needed for the simple calculations and the planar −6-dB coverage angles and radii are tabulated for a selection of loudspeakers. Worked examples are included.

0 INTRODUCTION

For certain applications, particularly the design of distributed systems, it is desirable to know the distribution of sound on a plane normal to the axis of a loudspeaker. A further advantage is gained if this distribution can be expressed as simple equations with known constants for each loudspeaker of interest. Of particular interest are the angles and radii of coverage on the irradiated plane to −6 dB.

It is useful to know the uniformity of direct sound coverage at ear level for different combinations of loudspeaker type, pattern, and degree of overlap. Conversely, it is also useful to know which pattern and overlap to use for a given loudspeaker in order to achieve a predetermined or acceptable uniformity of coverage. As the choices of loudspeaker and pattern depend not only on uniformity of coverage but also on other design factors, such as maximum sound pressure level (SPL) and cost, it is also useful to know typical uniformities of coverage prior to loudspeaker selection. Uniformity of coverage can be related to total loudspeaker cost for a given loudspeaker by showing the dependence of the maximum SPL variation on the density in the number of loudspeakers per coverage circle area.

Until recently, distributed systems in rooms with parallel ceilings and floors have been designed using square or hexagonal loudspeaker patterns with traditional amounts of overlap [1], [2], referred to as edge to edge, minimum, and edge to center (or center to center). The pattern and overlap were selected from the six combinations above on the basis of an increasing uniformity of coverage with increasing loudspeaker density. As only two regular patterns (square and hexagonal) are available in a plane, these are retained here. The restriction of only six combinations of pattern and overlap can be removed by replacing the combination choice by

---

* Presented at the 70th Convention of the Audio Engineering Society, New York, 1981 October 30–November 2; revised 1982 July 27.

loudspeaker density σ determined by preselected acceptable values of the maximum direct SPL L_{max} or the minimum direct SPL L_{min} relative to the axial SPL of a single loudspeaker or of $L_{max} - L_{min}$ (ΔL). With this method, the levels can be checked at each of two frequency bands for systems designed at one of them.

1 SOUND DISTRIBUTION ON A PLANE NORMAL TO A LOUDSPEAKER AXIS

1.1. The Equations

In the configuration shown in Fig. 1 the SPL L_P at P relative to the on-axis value can be taken from a measured polar and can be corrected to the level L_T on the plane OR at T by using the inverse square law to yield

$$L_T = L_P + 20 \log \cos \theta \text{ [dB]} . \quad (1)$$

This procedure can be carried out at different values of the off-axis angle θ. Examples are shown in Fig. 2 as ×. For the various loudspeakers treated in this way, attempts at best-fit equations were made for different types of equations. The overall best equation type was found to be

$$L_T = -d\theta^g \text{ [dB]} . \quad (2)$$

An example is shown as the solid line in Fig. 2. Values of d and g are tabulated in Appendix I for a number of loudspeakers for the 2-kHz and 4-kHz third-octave bands. The values obtained from Eq. (2) show good agreement to about −12 dB, and sometimes better.

Letting OS = h and OT = r (the radius of the circle at L_T),

$$\frac{r}{h} = \tan \theta \quad (3)$$

or

$$\theta = \arctan \left(\frac{r}{h} \right) \text{ [deg]}. \quad (4)$$

From Eqs. (2) and (4) L_T can be found for a given ratio $r:h$, that is,

$$L_T = -d \left[\arctan \left(\frac{r}{h} \right) \right]^g \text{ [dB]} . \quad (5)$$

Frequently it is desired to find θ and r/h for a given value of L_T. From Eq. (2),

$$\theta = \left(\frac{-L_T}{d} \right)^{1/g} \text{ [deg]} \quad (6)$$

and from Eqs. (3) and (6),

$$\frac{r}{h} = \tan \left[\left(\frac{-L_T}{d} \right)^{1/g} \right] \text{ [deg]} . \quad (7)$$

A more accurate determination of θ (and hence r/h) for a given L_T can be made by interpolation between points given by Eq. (1). The most commonly needed value of L_T is −6 dB. Values of θ at −6 dB [$\theta(-6)$] and of r/h at −6 dB [$r(-6)/h$] found by this method are tabulated in Appendix II.

1.2. Examples

1) Find L_T (relative to the on-axis SPL) for 45° off axis using a 5-in (127-mm) loudspeaker at 2 kHz. From Eq. (2), using values of a and b from Table 3,

$$L_T(45°) = 0.372 \times 10^{-3} \times 45^{2.432} = 3.89 \text{ dB}.$$

2) Find L_T (relative to the on-axis SPL) for $r/h = 0.4$ using a 5-in (127-mm) loudspeaker at 4 kHz. From Eq. (5),

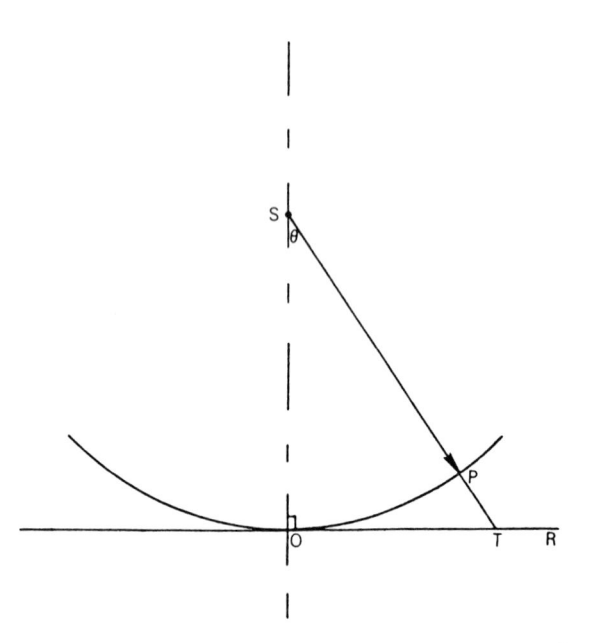

$$L_T(0.4) = -0.0205 \times 10^{-3} (\arctan 0.4)^{1.620}$$
$$-3.02 \text{ dB}.$$

Fig. 1. Loudspeaker S radiating onto a plane OR normal to the loudspeaker axis OS.

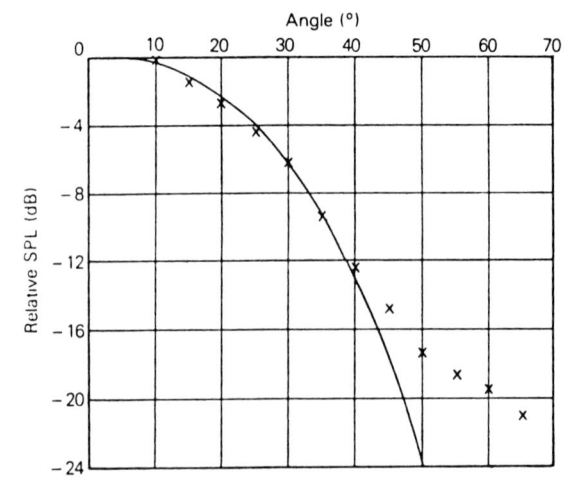

Fig. 2. Example of SPL values from polar curves corrected to radiation on a normal plane (indicated by ×) and best fit power curve (solid line). SPL = $d\theta^g$ dB. $d = 0.933 \times 10^{-3}$; $g = 2.5915$.

3) Find the angle θ off axis for -3.01 dB using an 8-inch (203-mm) loudspeaker at 2 kHz. From Eq. (6),

$$\theta(-3.01) = \left(\frac{3.01}{0.5 \times 10^{-3}}\right)^{1/2.544} = 30.60$$

$$= 30°\ 36'\ .$$

4) Find r/h for -4.5 dB using an 8-in (203-mm) coaxial loudspeaker at 4 kHz. From Eq. (7),

$$\frac{r(4.5)}{h} = \tan\left[\left(\frac{4.5}{0.0119}\right)^{1/1.772}\right] = 0.54\ .$$

5) Find the angle θ and r/h for -6 dB for a 12-in (305-mm) coaxial loudspeaker at 2 kHz. From Table 3,

$$\theta(-6) = 42.2°$$

$$\frac{r(-6)}{h} = 0.91\ .$$

2 UNIFORMITY OF COVERAGE IN DISTRIBUTED SOUND SYSTEMS

2.1 Theory

For a regular array of n ceiling loudspeakers shown, for clarity, diagrammatically as a one-dimensional model in Fig. 3, the SPL at a general point P in the ear level plane is assumed to be caused by the addition of the intensities at P from all n loudspeakers, that is,

$$\text{SPL}(P) = 10 \log\left(\sum_{i=1}^{n} 10^{L_i/10}\right)\ .\tag{8}$$

L_i for a given loudspeaker type can be calculated from the distance r between P and the projection of the center of the ith loudspeaker onto the ear level plane from

$$L_i = -d\left[\arctan\left(\frac{r}{h}\right)\right]^g\ [\text{dB}]\tag{9}$$

as explained in Section 1, where values of d and g are tabulated for a selection of loudspeakers. The parameter h is the loudspeaker height above ear level. To render such a system more tractable, it is further assumed that for all distant loudspeakers for which r/h is greater than some limiting value r'/h, the total intensity from these distant loudspeakers will only show little variation with the position of P and hence can be regarded as a constant low-level contribution to the direct sound.

For loudspeakers at distances much greater than $r(-6)$, the contributions are not only at a low level, but are delayed, and it can be argued that they more closely resemble early arrivals of the reverberant field. The value $r(-6)$ is the radius on the ear level plane for which the SPL from a single loudspeaker is 6 dB lower than the on-axis value. The calculations were made on approximately square arrays of almost one thousand loudspeakers.

2.2 Results

Values of the SPL at P were calculated from Eqs. (1) and (2) for different positions of P within one unit cell of the loudspeaker pattern for five different loudspeakers in two frequency bands. This was carried out for the two traditional patterns (square and hexagonal) at three different degrees of overlap, namely, edge to edge, minimum, and edge to center (sometimes referred to as center to center). The sound levels for a typical loudspeaker relative to the value on axis of a single loudspeaker are given for one unit cell in Figs. 4–9. These values are the mean of the five loudspeaker types and two different one-third-octave bands (2 kHz and 4 kHz). For specific loudspeakers in these frequency bands, individual unit cell sound patterns have been determined and are given elsewhere [3].

Values of the maximum SPL $L_{\max}$ and the minimum SPL $L_{\min}$ relative to the axial SPL of a single loudspeaker for the typical loudspeaker are given in Table 1 for for different combinations of pattern and overlap. Table 1 also lists the difference between $L_{\max}$ and $L_{\min}$ ΔL along with the areal density σ in the number of loudspeakers per coverage circle area. Similar values for individual loudspeakers are given in Appendix III.

The values in Table 1 can be approximated by an empirical equation of the form

$$L = a + b \ln \sigma\ [\text{dB}]\ .\tag{10}$$

Fig. 10 shows the dependence of $L_{\max}$ and $L_{\min}$ from Table 1 on density. Also shown are curves of the form given by Eqs. (3) and (4). For $L_{\max}$,

$$L_{\max} = -0.085 + 3.22 \ln \sigma\ [\text{dB}]\tag{11}$$

For $L_{\min}$ the equation found was

$$L_{\min} = -2.53 + 5.27 \ln \sigma\ [\text{dB}]\ .\tag{12}$$

The dependence of the values of ΔL from Table 1 on density is shown in Fig. 11. The approximate empirical equation also shown in Fig. 11 was found to be

$$\Delta L = 2.44 - 2.05 \ln \sigma\ [\text{dB}]\ .\tag{13}$$

The value of these equations is that they permit determining $L_{\max}$, $L_{\min}$, and ΔL for loudspeaker densities other than those associated with the traditional loud-

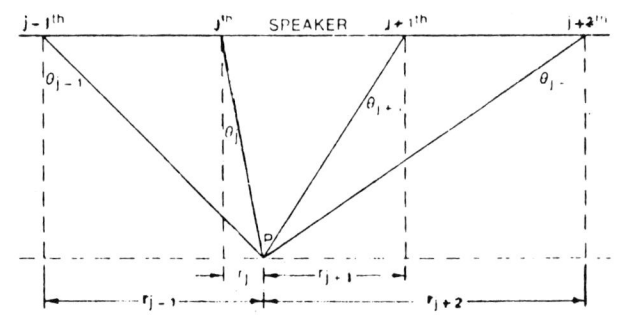

Fig. 3. One-dimensional representation of a regular ceiling array.

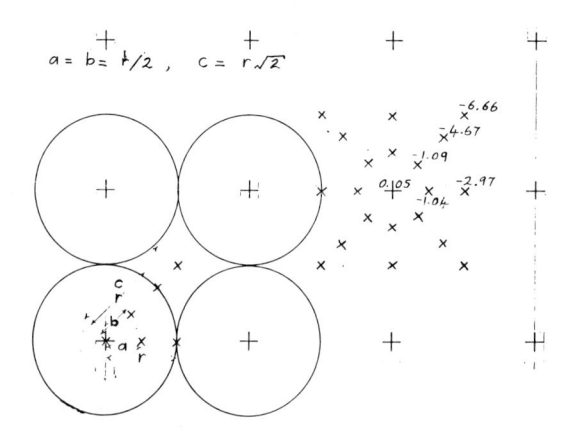

Fig. 4. Typical relative SPL values for edge-to-edge square configuration.

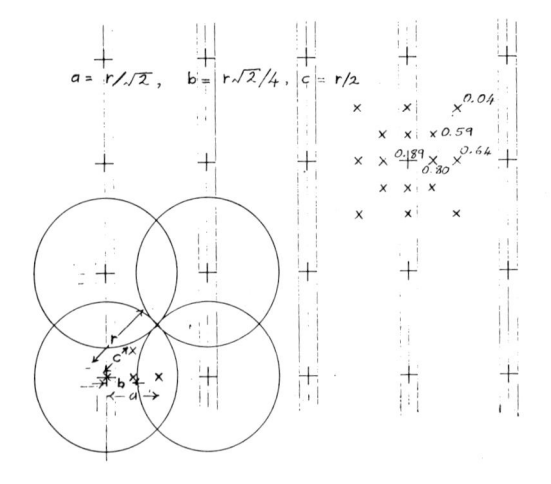

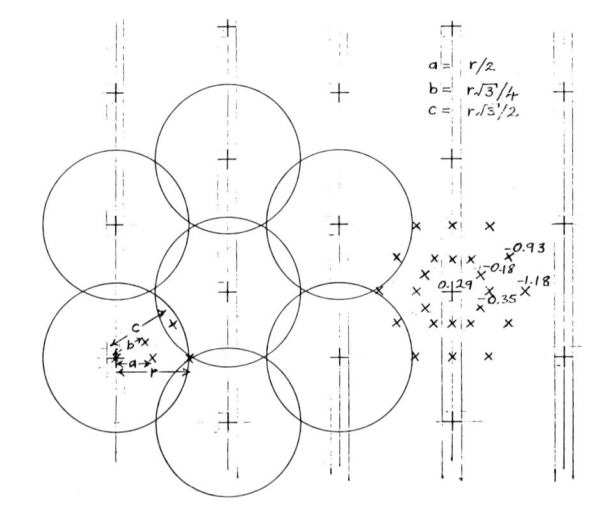

Fig. 5. Typical relative SPL values for minimum-overlap square configuration.

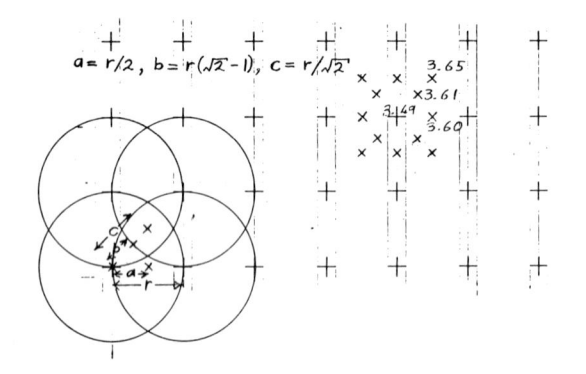

Fig. 6. Typical relative SPL values for edge-to-center square configuration.

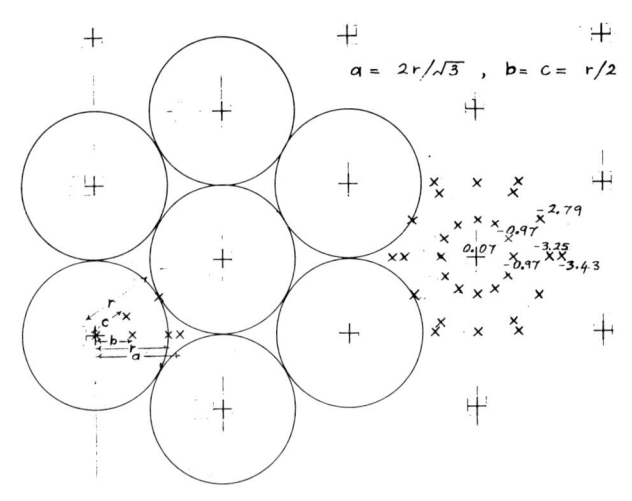

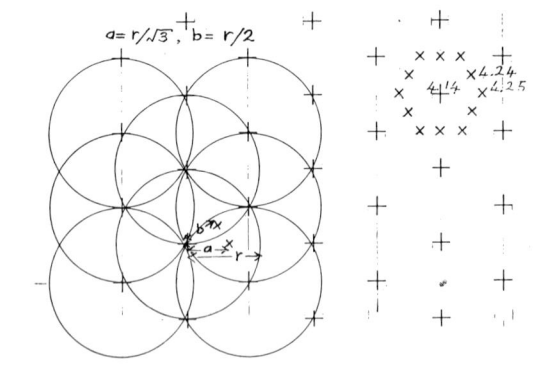

Fig. 7. Typical relative SPL values for edge-to-edge hexagonal configuration.

Fig. 8. Typical relative SPL values for minimum-overlap hexagonal configuration.

Fig. 9. Typical relative SPL values for edge-to-center hexagonal configuration.

Table 1. Values of loudspeaker density (number per coverage circle area), L_{max}, L_{min}, and ΔL for different combinations of pattern and overlap using a typical loudspeaker.

Pattern	Density	L_{max}[dB]	L_{min}[dB]	ΔL[dB]
Edge to edge, square	0.785	0.05	−6.66	6.71
Edge to edge, hexagonal	0.907	0.07	−3.43	3.50
Minimum overlap, square	1.571	0.89	0.04	0.85
Minimum overlap, hexagonal	1.209	0.29	−1.18	1.47
Edge to center, square	3.142	3.65	3.49	0.16
Edge to center, hexagonal	3.628	4.25	4.14	0.11

speaker pattern and overlap combinations. For example, configurations might be useful which lie in the present large gap between densities of $\pi/2$ and π. This gap is obvious from Fig. 10. Methods of calculating the density are indicated in Appendix IV.

So that relationships of the kind given in Eq. (10) can be used for specific loudspeaker types, values of a and b for calculating L_{max}, L_{min}, and ΔL are tabulated in Appendix V for several loudspeakers for the 2-kHz and 4-kHz one-third-octave bands.

3 THE DESIGN OF DISTRIBUTED SOUND SYSTEMS

3.1 Theory

Eq. (10) can be written in the form

$$\sigma = e^{(L-a)/b} . \tag{14}$$

For a square pattern with loudspeaker spacing x

$$\sigma = \frac{\pi r^2}{x^2} \tag{15}$$

as shown in Appendix IV, Eq. (27), where r is the radius of the -6-dB coverage circle. Hence

$$x = r\sqrt{\frac{\pi}{\sigma}} \tag{16}$$

Substituting for σ from Eq. (14),

$$x = r\sqrt{\pi\, e^{-(L-a)/b}} \tag{17}$$

which can be written as

$$x/h = c\, e^{-L/2b} \tag{18}$$

where

$$c = \left(\frac{r}{h}\right)\sqrt{\pi\, e^{a/b}} \tag{19}$$

h being the ear-to-ceiling height.

Using values of a and b given in Appendix V and of r/h given in Appendix II, values of c can be found. In fact, for a typical loudspeaker (the mean of several) expressions for x/h for given values of L_{max}, L_{min}, and ΔL become

$$\frac{x}{h} = 1.43\, e^{L_{max}/6.44} \tag{20}$$

$$\frac{x}{h} = 1.14\, e^{-L_{min}/10.54} \tag{21}$$

$$\frac{x}{h} = 0.98\, e^{\Delta L/4.10} . \tag{22}$$

Values of c for a selection of loudspeakers are also tabulated in Appendix V for the 2-kHz and 4-kHz one-third-octave bands, enabling the evaluation of x/h in Eq. (18) for given L_{max}, L_{min}, or ΔL.

The basis of the overlays is a unit cell of 1 in² (645 mm²). For the first, which is for the square pattern, the separation between adjacent loudspeakers must be 1 in (25 mm). For the second, which is for the hexagonal pattern, the separation y between adjacent loudspeakers is given by

$$y^2\, \frac{\sqrt{3}}{2} = 1\ \text{in}^2 . \tag{23}$$

Hence

$$y^2 = \frac{2}{\sqrt{3}} . \tag{24}$$

or

$$y = \left(\frac{4}{3}\right)^{0.25} \tag{25}$$

$$= 1.075\ \text{in}\ (27.3\ \text{mm}).$$

3.2 Design Method

A loudspeaker type is first selected and the design criterion (L_{max}, L_{min}, or ΔL) chosen. Values of c and b are then found from the tables. These values are substituted into Eq. (18). The resulting value of x/h is multiplied by h, yielding x in the dimensions of h (typically

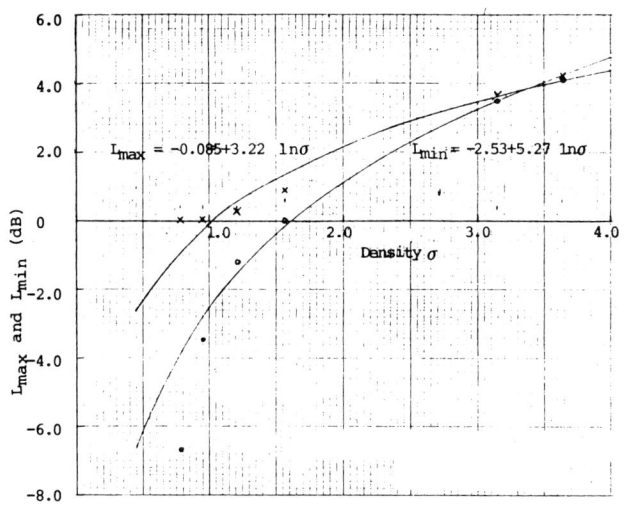

Fig. 10. Dependence of L_{max} and L_{min} on density.

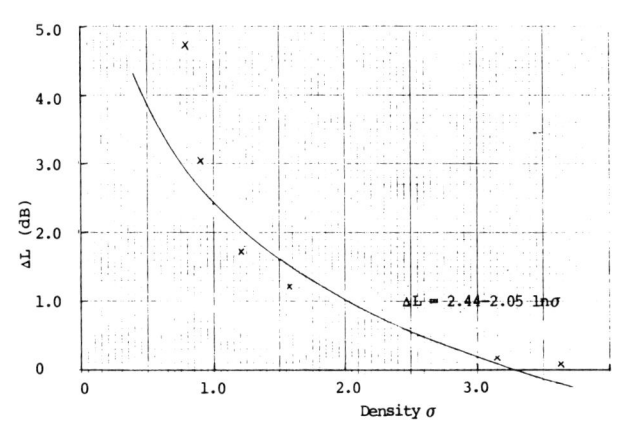

Fig. 11. Dependence of ΔL on density.

feet or meters). A scale drawing including major obstructions of the subject room is then made with 1 in representing length x (feet or meters). Each of two transparent overlays is in turn laid on the drawing and moved until a configuration is found which is considered suitable. The overlays are shown in Figs. 12 and 13.

If desired, parameters other than those selected can be checked using rearrangements of Eq. (18), that is,

$$L = -2b \ln\left[\frac{1}{c}\left(\frac{x}{h}\right)\right] . \qquad (26)$$

If these values are satisfactory, the design is complete.

As an option, the density can be found from Eq. (14). It can be seen that it is not necessary to know the density and that there is no direct reference to the extent of the overlap of the pattern.

3.3 Examples

1) Design a distributed system using 8-in (203-mm) coaxial loudspeaker for a room measuring 100 ft by 60 ft (30 m by 18 m) with a 14-ft (4.2-m) ceiling. The occupants are normally seated and the sound is to be uniform within approximately 1.5 dB at ear level (4 ft) for the 4-kHz octave band. From Eq. (18) and values of b and c from Table 11,

$$x/(14 - 4) = 0.63\, e^{-1.5/[2(-1.06)]}$$

$$x = 6.3\, e^{0.71} = 12.78 \text{ ft (3.9 m)} .$$

The scale drawing of the room is now made to the scale of 1 in for every 12.78 ft, that is, the room is drawn 7.82 in by 4.69 in. Each of the two overlays can now be placed in turn over the drawing, moved around, and rotated until acceptable placements are found. Four of many possibilities are shown in Figs. 14–17. The spacings of the square patterns are, of course, 12.78 ft. The hexagonal spacing is given as

$$12.78y = 12.78 \times 1.075 = 13.74 \text{ ft (4.2 m)} .$$

The final choice of configuration will depend on the number of loudspeakers (cost) and the edge coverage considered acceptable.

If desired, L_{max} or L_{min} can be calculated from Eq. (26). For example, to find L_{max} from Eq. (26) and Table 9 values,

$$L_{max} = -2 \times 2.63 \ln\left[\frac{1}{2.33}\left(\frac{12.78}{10}\right)\right]$$

$$= -5.26 \ln\left(\frac{1.278}{2.33}\right) = 3.16 \text{ dB} .$$

This means that the maximum SPL under the pattern is 3.16 dB higher than the axial value for a single loudspeaker at the same distance. As ΔL was originally selected as 1.5 dB, then

$$L_{min} = 3.16 - 1.5 = 1.66 \text{ dB}.$$

Another option is to calculate density. From Eq. (14), using the specified value of ΔL with values from Table 11,

$$\sigma = e^{(1.5 - 1.36)/(-1.06)} = e^{0.14/1.06} = 0.88 .$$

2) Design a distributed system using 5-in (127-mm) loudspeakers for a room measuring 60 ft by 40 ft (18 m by 12 m) with a ceiling height of 12 ft (3.6 m); minimum SPL 6.0 dB above single loudspeaker axial value at 4 kHz. Find density and ΔL at 2 kHz and 4 kHz. From

Fig. 12. Square pattern overlay with unit cell area of 1 in² (645 mm²).

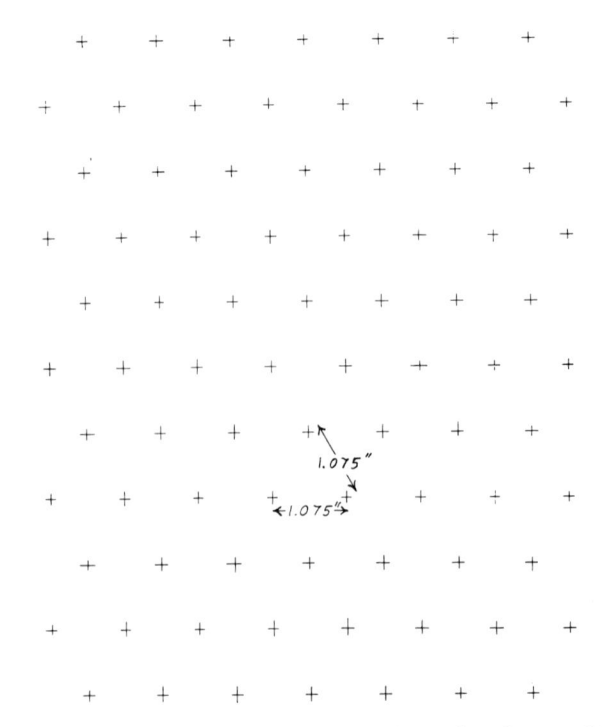

Fig. 13. Hexagonal pattern overlay with unit cell area of 1 in² (645 mm²).

Eq. (18) and Table 10,

$$x/(12 - 4) = 2.68\, e^{6/5.70}$$

$$x = 7.68 \text{ ft (2.3 m)}.$$

The scale drawing should be made to the scale of 1 in representing 7.68 ft, so that the scaled room measures 7.81 in by 5.55 in, which is shown in Figs. 18 and 19 with one possible example each of the square and hexagonal patterns. The square loudspeaker spacing is 7.68 ft and the hexagonal spacing is given as

$$7.68y = 7.68 \times 1.075 = 8.25 \text{ ft (2.5 m)}.$$

From Eq. (14) and Table 10,

$$\sigma = e^{(6-4.48)/2.85} = 1.70$$

ΔL at 2 kHz can be found from Eq. (26) and Table 11,

$$\Delta L(2 \text{ kHz}) = -2(-0.16) \ln \left[\frac{1}{0.85} \left(\frac{7.68}{8} \right) \right]$$

$$= 0.04 \text{ dB}.$$

Similarly, from Eq. (26) and Table 11,

$$\Delta L(4 \text{ kHz}) = -2(-0.61) \ln \left[\frac{1}{0.61} \left(\frac{7.68}{8} \right) \right]$$

$$= 0.55 \text{ dB}.$$

4 DISCUSSION

There are a few points in the above treatment which need some clarification. They are presented here at the end rather than as they occur to minimize disruption of the main arguments presented. The first relates to the range of agreement between Eq. (1) and the approximation in Eq. (2) as shown in Fig. 2. At the lower SPL

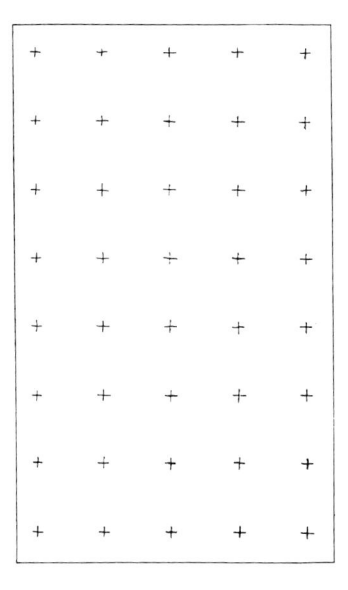

Fig. 14. Square configuration of loudspeakers with unit cell sides parallel to walls for Example 1.

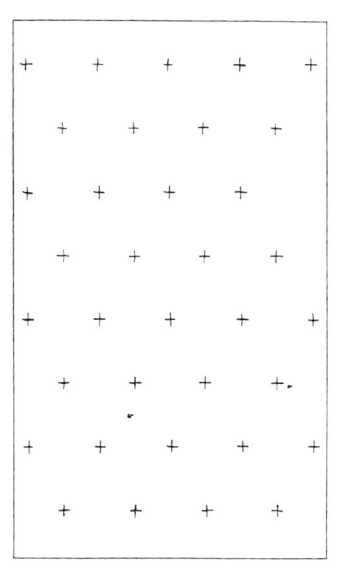

Fig. 16. Hexagonal configuration of loudspeakers with two unit cell sides parallel to longest walls for Example 1.

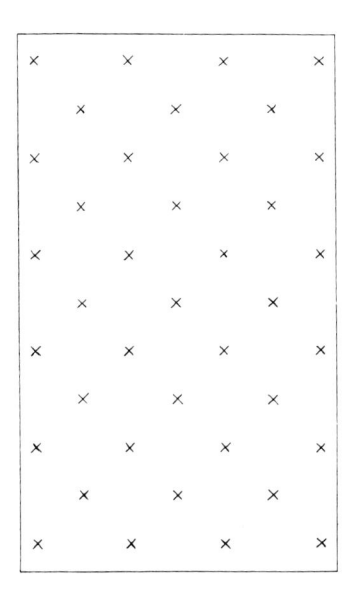

Fig. 15. Square configuration of loudspeakers with unit cell diagonals parallel to walls for Example 1.

Fig. 17. Hexagonal configuration of loudspeakers with two unit cell sides parallel to shortest walls for Example 1.

values the measurements corrected by Eq. (1) and depicted by × are approaching the anechoic chamber noise levels.

The second point requiring clarification is the premise that intensities rather than instantaneous excess pressures should be added for a listening point in the field of an array. Pressure adding would yield frequency-dependent interference patterns (e.g., [4], [5]). All interference phenomena depend on coherent sources. The most important function of sound systems is generally speech reproduction, and the most critical speech characteristics for comprehension are consonants. It follows that transient reproduction is of paramount importance. As the range of coherence is small for transients, intensities are assumed here to add. The situation is further complicated by the listeners not hearing at a single point but at two.

In some cases the empirical curves for L_{max} and L_{min} will intersect, as shown in Fig. 10, yielding negative values of ΔL which cannot be so. This situation can be improved by using best fit equations of different forms for L_{max} and L_{min} (and hence for ΔL). However, this spoils the simplicity of the method and complicates computational methods, particularly where computer or calculator programs are used. When using the method given here, no problems should be encountered as long as it is remembered that $\Delta L \geq 0$ dB and $L_{max} \geq 0$ dB.

5 ACKNOWLEDGMENT

The author would like to take this opportunity to acknowledge the work of Larry R. Lutz [6] in developing the original method of using transparent overlays and scale drawings for the design of distributed systems.

6 REFERENCES

[1] D. Davis and C. Davis, *Sound System Engineering* (Howard W. Sams and Co., Indianapolis, IN, 1975), p. 101.

[2] C. Enerson, "Distributed System Pattern Analysis," *Syn-Aud-Con Tech Topics*, vol. 5, no. 1, pp. 1–8.

[3] R. Sinclair, "Uniformity of Coverage in Distributed Sound Systems," Altec Lansing, Anaheim, CA, Tech. Lett. 258, 1981.

[4] R. Sinclair, "Stacked and Splayed Acoustical Sources, Pt. I," presented at the 61st Convention of the Audio Engineering Society, *J. Audio Eng. Soc. (Abstracts)*, vol. 26, p. 994 (1978 Dec.), preprint no. 1389.

[5] R. Sinclair, "Stacked and Splayed Acoustical Sources, Pt. II," presented at the 63rd Convention of the Audio Engineering Society, *J. Audio Eng. Soc. (Abstracts)*, vol. 27, p. 608 (1979 July/Aug.), preprint no. 1515.

[6] "Design Layout of Distributed Sound Systems," Altec Lansing, Anaheim, CA, Training Manual TMII, 1981.

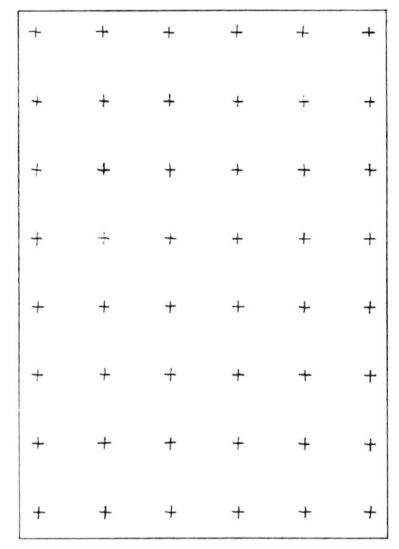

Fig. 18. Example of a square loudspeaker configuration for Example 2.

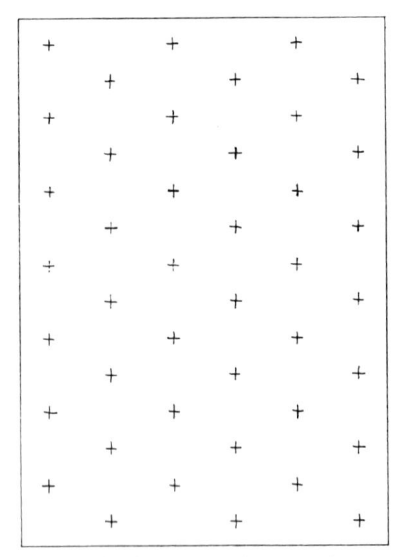

Fig. 19. Example of a hexagonal loudspeaker configuration for Example 2.

APPENDIX I

Table 2. Values for d and g for several loudspeakers, 2-kHz and 4-kHz octave bands.

Loudspeaker	2 kHz d	2 kHz g	4 kHz d	4 kHz g
8 in (203 mm)	9.064×10^{-3}	1.786	0.0219	1.879
5 in (127 mm)	0.372×10^{-3}	2.432	0.0205	1.620
8 in (203 mm), coaxial	0.5×10^{-3}	2.544	0.0119	1.772
12 in (305 mm), coaxial	0.679×10^{-3}	2.388	0.0641	1.383
16 in (406 mm), coaxial	0.032×10^{-6}	4.657	0.6776×10^{-3}	2.491

APPENDIX II

Table 3. Values for the off-axis angle and r/h for -6 dB on the normal plane for several loudspeakers, 2-kHz and 4-kHz octave bands.

Loudspeaker	Frequency 2 kHz $\theta(-6)$ [deg]	$r(-6)/h$	4 kHz $\theta(-6)$ [deg]	$r(-6)/h$
8 in (203 mm)	38.5	0.80	19.3	0.35
5 in (127 mm)	53.4	1.35	34.5	0.69
8 in (203 mm), coaxial	39.8	0.83	34.3	0.68
12 in (305 mm), coaxial	42.2	0.91	30.7	0.59
16 in (406 mm), coaxial	58.2	1.61	38.2	0.79

APPENDIX III

Table 4. L_{max}, L_{min}, and ΔL for a typical 8-in (203-mm) loudspeaker.

Pattern	2 kHz L_{max}	L_{min}	ΔL	4 kHz L_{max}	L_{min}	ΔL
Edge to edge, square	5.93	5.07	0.86	0.28	−4.65	4.93
Edge to edge, hexagonal	6.12	5.43	0.69	0.39	−2.71	3.10
Minimum overlap, square	7.01	6.60	0.41	1.49	0.30	1.19
Minimum overlap, hexagonal	6.55	6.05	0.50	0.82	−0.71	1.53
Edge to center, square	8.57	8.45	0.12	3.94	3.79	0.15
Edge to center, hexagonal	8.96	8.88	0.08	4.53	4.46	0.07

Table 5. L_{max}, L_{min}, and ΔL for a typical 5-in (127-mm) loudspeaker.

Pattern	2 kHz L_{max}	L_{min}	ΔL	4 kHz L_{max}	L_{min}	ΔL
Edge to edge, square	10.71	10.44	0.26	4.93	3.82	1.11
Edge to edge, hexagonal	10.84	10.61	0.23	5.14	4.26	0.88
Minimum overlap, square	11.32	11.16	0.16	6.15	5.64	0.51
Minimum overlap, hexagonal	11.08	10.90	0.18	5.63	4.99	0.64
Edge to center, square	12.20	12.15	0.05	7.90	7.74	0.26
Edge to center, hexagonal	12.43	12.41	0.02	8.32	8.22	0.10

Table 6. L_{max}, L_{min}, and ΔL for an 8-in (203-mm) coaxial loudspeaker.

Pattern	2 kHz L_{max}	L_{min}	ΔL	4 kHz L_{max}	L_{min}	ΔL
Edge to edge, square	0.78	−3.37	4.15	3.03	1.14	1.89
Edge to edge, hexagonal	0.94	−1.81	2.75	3.25	1.82	1.43
Minimum overlap, square	2.09	1.04	1.05	4.42	3.68	0.74
Minimum overlap, hexagonal	1.42	0.09	1.51	3.80	2.85	0.95
Edge to center, square	4.58	4.53	0.05	6.49	6.31	0.18
Edge to center, hexagonal	5.20	5.11	0.09	6.98	6.88	0.10

Table 7. L_{max}, L_{min}, and ΔL for a 12-in (305-mm) coaxial loudspeaker.

Pattern	2 kHz L_{max}	L_{min}	ΔL	4 kHz L_{max}	L_{min}	ΔL
Edge to edge, square	4.04	2.58	1.46	3.92	2.48	1.44
Edge to edge, hexagonal	4.24	3.13	1.11	4.16	3.02	1.14
Minimum overlap, square	5.24	4.64	0.57	5.34	4.69	0.65
Minimum overlap, hexagonal	4.71	3.97	0.74	4.73	3.91	0.82
Edge to center, square	7.09	7.02	0.07	7.31	7.12	0.19
Edge to center, hexagonal	7.24	7.54	0.	7.78	7.61	0.17

Table 8. L_{max}, L_{min}, and ΔL for a 16-in (406-mm) coaxial loudspeaker.

	Frequency					
	2 kHz			4 kHz		
Pattern	L_{max}	L_{min}	ΔL	L_{max}	L_{min}	ΔL
Edge to edge, square	1.36	−2.11	3.47	0.62	−3.82	4.44
Edge to edge, hexagonal	1.53	−0.83	2.36	0.77	−2.12	2.89
Minimum overlap, square	2.65	1.75	0.90	1.90	0.82	1.08
Minimum overlap, hexagonal	2.00	0.72	1.28	1.24	0.33	1.57
Edge to center, square	5.21	4.96	0.25	4.42	4.36	0.06
Edge to center, hexagonal	5.81	5.54	0.27	5.03	4.95	0.08

APPENDIX IV

To calculate the mean loudspeaker density of any regular array, a unit cell with a loudspeaker axis at the center must first be identified. This cell will always be a square or a regular hexagon (for the same spacing along rows in different directions). The cell sides are the perpendicular bisectors of line segments joining adjacent loudspeaker positions. The area S of the cell can be found in terms of r^2. A diagonal or a perpendicular to one side can usually be related to r. This area, when divided by the area of the coverage circle, yields the area (in coverage circle area units) per loudspeaker. The required density is the reciprocal of this, that is,

$$\sigma = \frac{\pi r^2}{S(r^2)} . \qquad (27)$$

Examples

1) Edge-to-edge hexagon. From Fig. 20 the unit cell is a regular hexagon with perpendicular of length r from the center to each side. The unit cell area is given by

$$S = 2r^2 \sqrt{3} . \qquad (28)$$

Substituting into eq. (27) yields

$$\sigma = \frac{\pi \sqrt{3}}{6} . \qquad (29)$$

2) Edge-to-center square. From Fig. 21, the unit cell is a square of side length r:

$$S = r^2 . \qquad (30)$$

Therefore

$$\sigma = \pi . \qquad (31)$$

3) A hexagonal pattern with centers at one third of a diameter spacing. From Fig. 22 the unit cell of this nonstandard configuratiion is a hexagon with a diagonal length of $2r/3$:

$$S = \frac{2r^2 \sqrt{3}}{3} . \qquad (32)$$

Therefore

$$\sigma = \frac{3 \pi r^2}{2 r^2 \sqrt{3}} = \frac{3 \pi \sqrt{3}}{2} = 8.162 . \qquad (33)$$

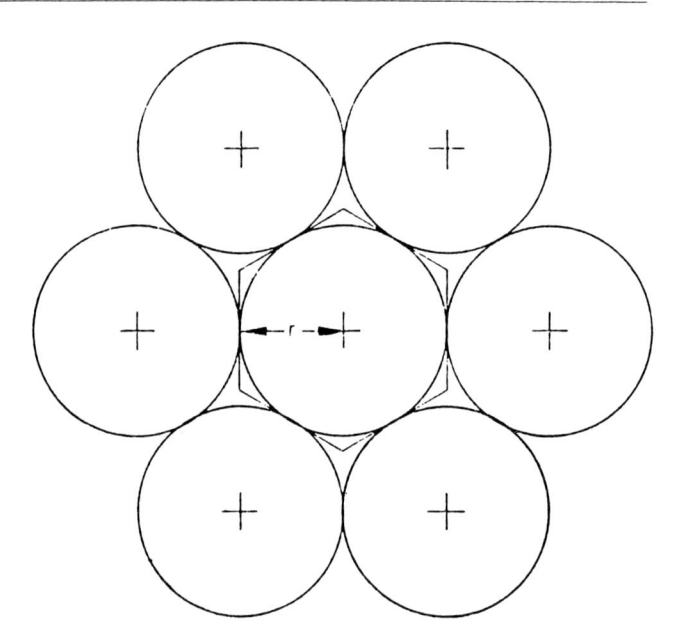

Fig. 20. Unit cell for edge-to-edge hexagonal pattern.

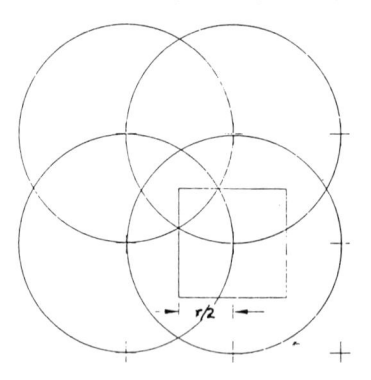

Fig. 21. Unit cell for edge-to-center square pattern.

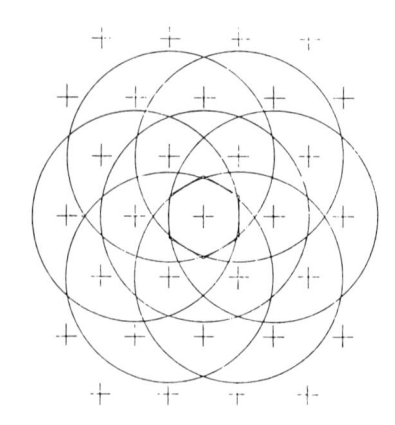

Fig. 22. Unit cell for hexagonal pattern with one third of a diameter spacing.

APPENDIX V

Values of a, b, and c are given here to permit finding L_{max}, L_{min}, or ΔL from such equations as Eq. (10) and x/h from Eq. (18) for typical loudspeakers in the 2-kHz and 4-kHz one-third-octave bands.

As shown in Appendix IV, these values can also be used to calculate density and hence spacing for a required L_{max}, L_{min}, or ΔL. This time the calculations are for specific loudspeakers.

Table 9. Values of a, b, and c for evaluating L_{max} and x/h.

Loudspeaker	2 kHz a	b	c	4 kHz a	b	c
Typical 8 in (203 mm)	6.28	2.00	6.82	0.61	2.69	0.69
Typical 5 in (127 mm)	10.92	1.12	2.21	5.32	2.24	4.01
8 in (203 mm), coaxial	1.15	2.97	1.79	3.66	2.63	2.33
12 in (305 mm), coaxial	4.42	2.32	4.18	4.37	2.56	2.46
16 in (406 mm), coaxial	1.73	3.00	3.81	0.98	2.97	1.65

Table 10. Values of a, b, and c for evaluating L_{min} and x/h.

Loudspeaker	2 kHz a	b	c	4 kHz a	b	c
Typical 8 in (203 mm)	5.62	2.48	4.40	−2.45	5.54	0.50
Typical 5 in (127 mm)	10.70	1.28	110.0	4.48	2.85	2.68
8 in (203 mm), coaxial	−1.51	5.29	1.28	2.10	3.69	1.60
12 in (305 mm), coaxial	3.36	3.20	2.73	3.28	3.34	1.70
16 in (406 mm), coaxial	−0.52	4.80	2.70	−1.84	5.44	1.18

Table 11. Values of a, b, and c for evaluating ΔL and x/h.

Loudspeaker	2 kHz a	b	c	4 kHz a	b	c
Typical 8 in (203 mm)	0.66	−0.48	0.71	3.06	−2.85	0.36
Typical 5 in (127 mm)	0.22	−0.16	0.85	0.84	−0.61	0.61
8 in (203 mm), coaxial	2.66	−2.32	0.83	1.36	−1.06	0.63
12 in (305 mm), coaxial	1.06	−0.88	0.88	1.09	−0.78	0.52
16 in (406 mm), coaxial	2.25	−1.80	1.53	2.82	−2.47	0.79

THE AUTHOR

Rex Sinclair studied physics at Birkbeck College, London, U.K., while employed by Kodak Ltd., graduating in 1961. From 1962 to 1967 he continued graduate studies in physical acoustics at Chelsea College of Science and Technology and at Imperial College, London. He taught in the Mechanical Engineering Department at the University of Houston from 1968 until 1973. Then followed positions with Wyle Laboratories and L. M. Cox Manufacturing Co. He has consulted on noise and vibration problems, and has been senior engineer, acoustics, with Altec Lansing Corporation, Anaheim, California, since 1976.

CORRECTIONS

We wish to correct the following errors which appeared in "The Design of Distributed Sound Systems from Uniformity of Coverage and Other Sound-Field Considerations," by Rex Sinclair (*J. Audio Eng. Soc.*, vol. 30, 1982 Dec.). On p. 872, col. 2, line 20, "a and b from Table 3" should read d and g from Table 2. On p. 873, col. 2, lines 14 and 15, "Eqs. (1) and (2)" should be Eqs. (8) and (9). On p. 873, col. 2, line 42, "Eqs. (3) and (4)" should be Eq. (10). On p. 876, col. 2, line 21 the correct expression is:

$$\sigma = e^{(1.5-1.36)/(-1.06)} = e^{-0.14/1.06} = 0.88 \ .$$

An Accurate and Easily Implemented Method of Modeling Loudspeaker Array Coverage*

JOHN R. PROHS AND DAVID E. HARRIS

Ambassador College, Pasadena, CA 91129, USA

A procedure for modeling a loudspeaker array for any room configuration is presented. The method easily enables the sound system designer to manipulate and observe the interaction of individual components in relation to the entire space. Sound-level requirements as well as every angular and rotational orientation are immediately apparent.

0 INTRODUCTION

Many developments have been made in measurement techniques that describe the acoustical environment. By utilizing these developments, such factors as distortion, acoustical gain, and intelligibility can be measured and calculated relatively easily [1]. But one of the most critical design factors of all, the orientation of the loudspeaker within the array, has lagged woefully behind until the advent of the concept of two-dimensional acoustical mapping [2], [3]. These mapping techniques are a great step forward in overcoming the problem but still have limitations and inaccuracies.

The two-dimensional angular mapping techniques came about as an outgrowth of increased information on sound coverage available from loudspeaker manufacturers coupled with the advent of powerful programmable calculators and microcomputers [4], [5]. The aim is to display the room as viewed from the loudspeaker cluster and to provide greater accuracy in component positioning and in prediction of sound dispersion. The typical procedure is to measure the room, compute the data, and make the necessary spherical to rectangular coordinate transformations, and to map the room on a polar plot or graph paper. Commercially available transparent patterns, or the designer's own, are shifted over the room plot until the best coverage ascertainable from the method is achieved.

However, an axiom of cartography is that the only true map is a globe (a sphere). Transformations from a sphere to a flat two-dimensional surface attempt to minimize distortions as much as possible, but there is no such thing as a distortion-free flat map [6]. If one attempts to flatten a spherelike object (such as a child's rubber ball with surface designs), a clear idea of what happens in flat mapping can readily be seen. In any type of two-dimensional mapping one or more of the following errors will occur: the scale of the map will be inaccurate except along only one or two parallels or meridians; angular relationships are not retained; relative sizes or shapes are distorted. With the two-dimensional mapping of loudspeaker arrays, the inevitable distortions cause the generated loudspeaker overlay to be accurate at only the area for which it is generated (and therefore inaccurate at all other positions), or to require awkward and complex manipulations over the discontinuous two-dimensional map in order to see its true coverage pattern. Fig. 1 illustrates the distortion present in a cylindrical-type projection. Notice that when a loudspeaker coverage pattern overlay is moved from the equator toward one of the poles, it must be distorted to show the actual coverage.

The three-dimensional mapping technique described in this article was developed to provide the following advantages:

1) To enable the sound system designer to "map" the *entire* listening space in a distortion-free manner from the perspective of the loudspeaker cluster.

---

* Presented at the 72nd Convention of the Audio Engineering Society, Anaheim, CA, 1982 October 23–27; revised 1983 September 13.

2) To make possible the immediate visualization of the composite loudspeakers' coverage in the listening space and to provide instant visualization of the results of the design at various stages of its conception.

3) To make apparent *accurate* sound-level requirements of the interrelating loudspeaker contours to the system's acoustical environment.

4) To make instantly apparent the angular and rotational orientations needed for the installation of each loudspeaker without compromising the accuracy of the design.

5) To provide a fast, practical, accurate, and portable method for field designing.

6) To provide methods for final documentation of the loudspeaker cluster design.

7) To provide a means of visualization of the system to which a nontechnical person can easily relate.

8) To further provide a way to project accurate loudspeaker coverage patterns onto a scale model (for designers or instructors desiring to do so) and thereby allowing the simultaneous loudspeaker interrelationships to be seen as well as the sound intensity and angular relationships on the scale model.

9) To provide a way of seeing the effects of first-order reflections from the loudspeakers.

1 THE CONCEPT

Manufacturers have for years published data about their loudspeakers giving the angle of coverage in the vertical and horizontal directions. All recognized that the sound wave that comes out of a loudspeaker is a spherical wavefront or a segment thereof and some have given complete horizontal and vertical polar plots.

Since this wavefront is accurately depicted only on a sphere, some manufacturers have reputedly used spheres for such representations. However, the technique described here not only relates the loudspeaker to the sphere but it also relates the room to the sphere. Since both are on the sphere, the sound radiations emanating from the loudspeaker can be envisioned just as they will be in the actual room, or acoustical environment. This becomes an extremely useful tool for cluster designing.

Everything is viewed from the loudspeaker cluster's vantage. To grasp this concept, envision yourself standing at the center of a transparent sphere etched with a grid of meridians and parallels. Standing at the center of the sphere, you are actually standing at the center of the loudspeaker cluster. As you look through the sphere, each location in the room can be marked as a unique point on the surface of the sphere [6]. As you view the room, imagine holding a pen in your hand and tracing the outline of all the listening areas onto the sphere's interior. You would then have a complete picture drawn on the sphere showing how the area appeared to you from this viewpoint (Fig. 2). This is in essence how the seating area would appear from the loudspeaker cluster's perspective.

A way to relate the loudspeaker to the room map is now needed. To understand this, envision one of the loudspeakers with its acoustical apex at the center of the sphere (where you are standing) radiating sound intensities which can be measured. The pattern of sound intensity can be drawn with contour lines very much like elevation is depicted with contour lines on a topographical map. Now trace this pattern onto the sphere's surface.

At this point the sphere has on its surface both the room map and the contour pattern of the loudspeaker. However, since the angular position of this contour changes as the loudspeaker's position is varied, a way is needed to move the contour over the outline of the room. In order to do this, the contour pattern is transferred to a thin clear spherical overlay that conforms to the surface of the sphere. (The attenuation contours on the overlay are normally represented as concentric -3-dB, -6-dB, and -9-dB contour lines.) The overlay can then be moved to any location to allow the designer to choose optimum positioning.

Now all that is needed is to add the element of distance to the angular information already on the sphere. Distance for sound can be thought of in several ways: time, attenuation (reduction in direct sound), or range (as expressed in feet or meters). All are different ways of expressing the same thing. It takes sound a specific time to travel a given distance. One could say something is a certain number of milliseconds away. Likewise, a spherical wavefront attenuates direct sound levels a certain amount in decibels from a reference; a point can be said to be a certain number of decibels away. Keeping the units in decibels makes it very convenient to relate the room to the loudspeaker cluster. With this information traced on the sphere, our view from the loudspeaker cluster's perspective is complete.

A reasonably small plastic sphere with an outside

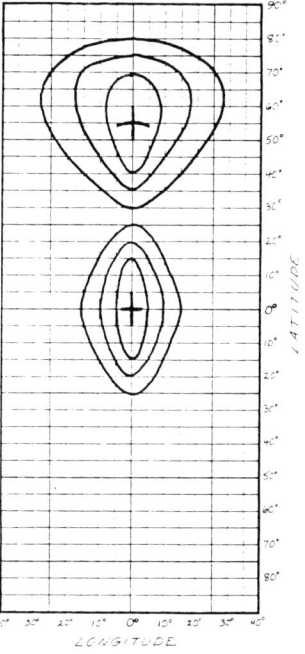

Fig. 1. Cylindrical projections of a sphere make necessary the distortion of a loudspeaker coverage overlay as it is moved toward the pole of the projection.

circumference of about 5¾ in (146 mm) was chosen to model listening area(s). The sphere is comprised of two separable hemispheres which easily fit into a standard attaché case. Each is marked with meridians and parallels every 5°.

2 PROCEDURE

The method for plotting on the sphere is described next.

Step 1: Obtain the data needed for plotting the room boundaries onto the sphere. Usually only the areas where people will be listening (hereafter referred to as seating banks) need to be plotted. Initially measurements are made of the room with conventional rectangular methods. These measurements can be taken from architectural blueprints or made in the room itself. Such distances as the distance from the loudspeaker cluster to the front row of a seating bank, the width of the seating bank, the depth of the seating bank, the height of the seating bank, and so on, are entered into a programmable calculator or small computer.

Step 2: Perform the calculations which solve for the desired data. (This information can be obtained with a nonprogrammable calculator. However, it is very time consuming.) These calculations supply a horizontal angle, a vertical angle, and the range to the points of interest. Also supplied is the attenuation to these points (as the log of a ratio of the range to a reference distance).

These angles are given from the perspective of the loudspeaker cluster and correlate to the calibrations on the plastic sphere, the same calibrations which are on a globe of the world (Fig. 3). The angle straight ahead of the loudspeaker cluster (level with the earth) is 0° where the prime meridian intersects the equator. The angle from the prime meridian to any point on the equator is its longitude, or horizontal angle, and contains the horizontal angles 0–180° east and west of the prime meridian. The western hemisphere is expressed as positive and the eastern one as negative.

The arc from the equator to either of the poles is divided into 90°. The smallest angle from the equator to any point is its latitude or vertical angle. The northern hemisphere is expressed as positive degrees and the southern hemisphere as negative degrees.

The formulas solving for the horizontal or vertical angle and range to any given point in a seating bank are derived from standard spherical trigonometry. Fig. 4 shows the variables that need to be solved.

Following is the definition of the terms as depicted in Fig. 4, which are used in the formulas. See Fig. 3 for the definition of the 0° horizontal and the 0° vertical angle planes.

R = range, the distance from the loudspeaker cluster to any given point in a seating bank

dBR = dB reference, the reference distance from the loudspeaker cluster (or an individual loudspeaker) with which the range is compared to determine the attenuation in decibels to a given point in the seating bank (For example, if you choose 4 ft as your dB reference, the attenuation at 4 ft will be zero and at 8 ft it will be −6 dB.)

AL = azimuth line, a line originating from a point on the vertical axis through the loudspeaker cluster and perpendicular to the front edge of a seating bank

The following are measurements taken from the top or plan view utilizing the 0° vertical angle plane as the plane of projection:

A = azimuth, the horizontal angle between AL and the 0° horizontal angle plane

F = front, the distance from the loudspeaker cluster to the front edge of the seating bank parallel to AL

D = depth, the depth of the seating bank as measured parallel to AL

DF = distance to front, the distance of any given point in the seating bank from the loudspeaker cluster parallel to AL

DS = distance to side, the perpendicular distance from AL to a given point in a seating bank

H∡ = horizontal angle, the angle the loudspeaker is swung to the side from the 0° horizontal angle plane so that it points (in conjunction with the V∡) at a desired location in a seating bank

χ = chi, the distance from the loudspeaker cluster to the intersection of H_0 in the 0° vertical angle plane

The following are measurements taken from the elevation view (the plane of projection being perpendicular to the 0° vertical angle plane):

H_0 = height, the distance from the 0° vertical angle plane to a given point in a seating bank

FC = front to cluster, the distance from the front edge of the seating bank to the 0° vertical angle plane

RC = rear to cluster, the distance from the rear edge of the seating bank to the 0° vertical angle plane

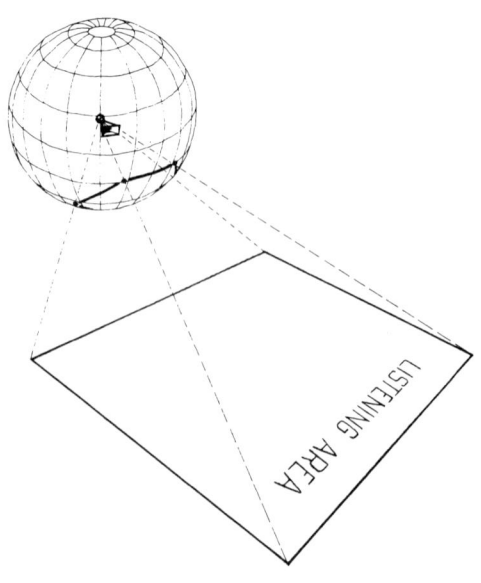

Fig. 2. Listening area mapped on a sphere from the perspective of a loudspeaker cluster. The acoustical center of the cluster is located at the center of the sphere.

Regarding FC and RC, edges of the seating bank above the 0° vertical angle plane are positive; edges below are negative.

- EH = ear height, the height above the floor to the average ear level of the listeners [Usually 5¼ ft (1.60 m) for people standing, 4 ft (1.2 m) for people seated in chairs, and 3½ ft (1 m) for people seated in bleachers.]
- θ = theta, the slope of a seating bank floor as measured parallel to AL
- Φ = phi, the apparent slope of the floor of a seating bank measured as it appears in the horizontal angle plane of a given horizontal angle (Angle Φ can therefore be less than angle θ.)
- V∡ = vertical angle, the angle the loudspeaker is tipped from the 0° vertical angle plane so that it points (in conjunction with H∡) at a desired location in a seating bank

With the terms defined, let us examine the mathematical steps necessary to locate any point in a seating bank in the terms of vertical and horizontal angles, range, and attenuation.

The first task is to solve for the floor slope:

$$\theta = \arctan\left(\frac{RC - FC}{D}\right). \quad (1)$$

Next the height of the loudspeaker cluster from the point is found:

$$H_0 = \tan\theta(DF - F) + FC. \quad (2)$$

Chi is then calculated:

$$\chi = [(DF)^2 + (DS)^2]^{0.5}. \quad (3)$$

Now find the range:

$$R = [(\chi)^2 + (H_0)^2]^{0.5}. \quad (4)$$

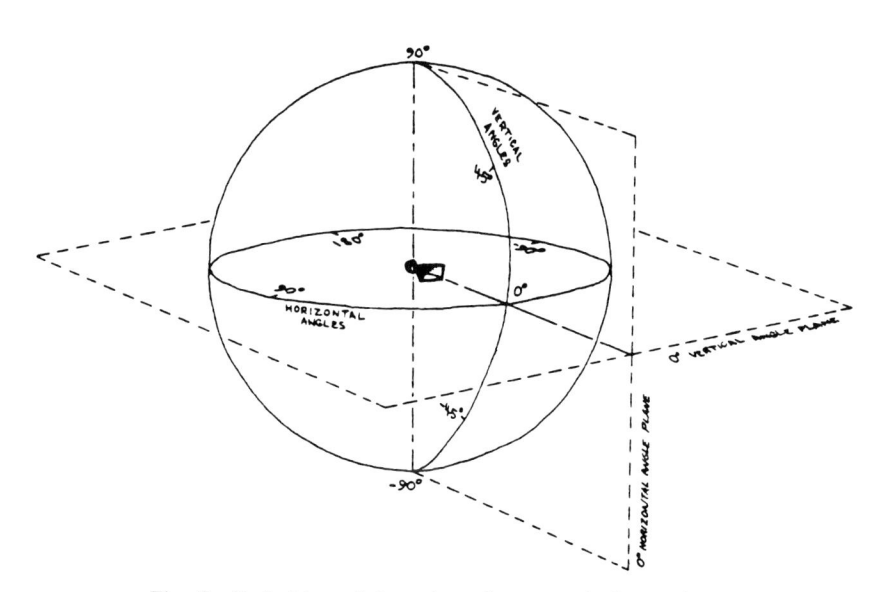

Fig. 3. Definition of the sphere for acoustical mapping.

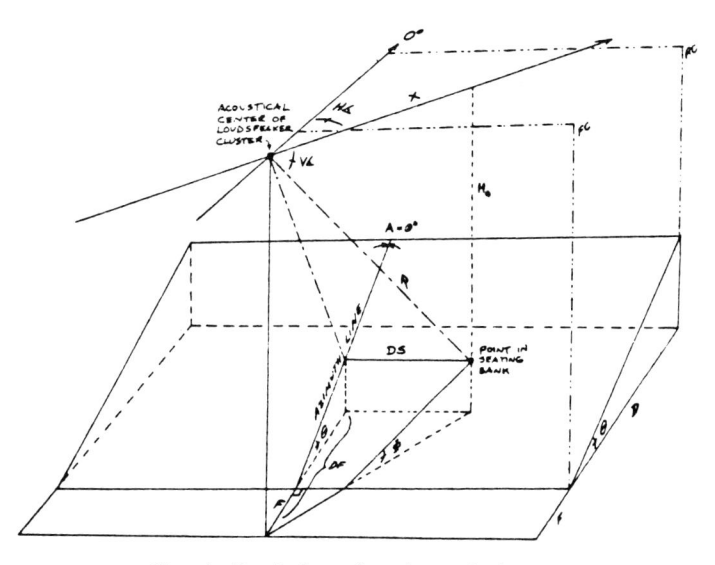

Fig. 4. Depiction of mathematical terms.

The attenuation is then easily solved:

$$dB = 20 \log\left(\frac{dBR}{R}\right). \quad (5)$$

Finally the vertical and horizontal angles are found:

$$V\measuredangle = \arctan\left(\frac{H_0}{\chi}\right) \quad (6)$$

$$H\measuredangle = \arctan\left(\frac{DS}{DF}\right). \quad (7)$$

In most cases 12 points give adequate definition of an area to plot it clearly. These points need to include the corner of each boundary. More or fewer points can be calculated depending on the resolution desired. Each point is solved for the horizontal and vertical angles, the range, and the attenuation. (Fig. 5 shows the 12 points for which the authors solve.)

The contours of sound attenuation are plotted across the entire seating bank, not just on the boundary points, so they may be more easily visualized. To accomplish this the amount of attenuation and the horizontal angle are specified. (Note that the attenuation parameters and horizontal angle information needed will be apparent from the previous calculations.) This information allows one to see the "topography" of the seating area(s) as expressed in attenuation of sound energy.

It is best to start with the greatest attentuation within the seating bank that is divisible by 3 and plot its contour at the horizontal angle interval (H∡ NTRVL), which the designer chooses. After this contour is plotted, successively closer contours to the loudspeaker cluster are plotted at 3-dB increments until the least attenuation divisible by 3 that falls within the boundary of the seating bank is reached. If the least possible attenuation dB_L does not fall on the boundary,

$$dB_L = 20 \log\{dBR/([(FC)^2 + (F)^2]^{0.5}(\sin[\arctan\left(\frac{F}{-FC}\right) + 90° - \theta]))\}. \quad (8)$$

Since the attenuation and the horizontal angle are given, the designer must find the vertical angle. This is accomplished by first transposing Eq. (5) to find the range:

$$R = \frac{dBR}{10^{dB/20}}. \quad (9)$$

Next the apparent floor slope at the given horizontal angle is found:

$$\Phi = \arctan[\cos(A - H\measuredangle) \tan \theta]. \quad (10)$$

Now the vertical angle can be resolved:

$$V\measuredangle = \Phi - \arcsin\left\{\frac{[(F \tan \theta) - FC]\cos \Phi}{R}\right\}. \quad (11)$$

With the horizontal and vertical angles known, any point on an attenuation (or dB) contour can be located on the sphere. Since the attenuation will be known, the dB contour for the point will also be known.

In some rare cases a dB contour may contain two vertical angles (one above and the other below). The following equation identifies this second vertical angle:

$$V\measuredangle_2 = 2\Phi - V\measuredangle_1 - 180°. \quad (12)$$

In summary, it was first shown how the information necessary to plot any point in a seating bank can be found. This included the horizontal and vertical angles of that point as well as its range and attenuation. From this the designer can plot points comprising the boundary of a seating bank and thereby outline the seating bank on the spherical model. Usually 12 boundary points, as shown in Fig. 5, are sufficient to outline a seating bank. Next the topography within the seating bank was plotted with dB contours. The dB contours are spaced 3 dB apart so that they easily relate to loudspeaker coverage patterns. Computer programs are available which automate the above step for multiple seating banks, greatly enhancing and speeding the system design.

Step 3: Transfer the calculated angular and attenuation information regarding the seating banks onto the plastic model sphere. Mapping is more easily done first on the outside of the sphere with a pen using water-soluble ink. Then the sphere is taken apart along its seam at the equator and the data are transferred to the inside of each hemisphere by tracing what was already mapped on the outside. The two hemispheres are put back together to form a complete sphere with seating banks plotted safely on the inside where they cannot be smudged. The outside is cleaned by wiping it off with a damp cloth.

To plot a point on the sphere find the vertical angle of the point along one of the meridians of longitude. (Remember that above the equator is positive and below is negative.) Then staying at that vertical angle, swing across the sphere along a parallel of latitude until you reach the desired horizontal angle. (Remember that the western hemisphere is positive and the eastern, negative.)

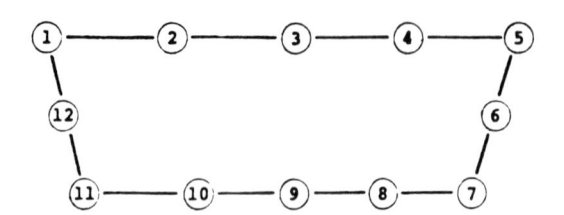

Fig. 5. An easy method to plot points around a seating bank boundary is to plot 12 points starting in the left rear corner and proceeding clockwise around the boundary.

Fig. 6 shows where a point with the coordinates $-10°$ V∡, $15°$ H∡, is located. Do this for all of the boundary points. Connect the points with lines or curves to outline the entire seating bank boundary. Next using a different color ink, plot the points for each dB contour. Connect all the points for each contour with an appropriate line or curve, and the seating bank is complete. A sample seating bank might appear like the one in Fig. 7. If desired, points can be plotted identifying any object within a seating bank such as a stage or a scoreboard.

Now the entire seating area can be viewed and all the features seen from the loudspeaker's perspective. The attenuation of sound due to inverse-square losses (as a result of distance from the sound source) is immediately apparent.

Step 4: Select and position the loudspeaker coverage overlays on the sphere until the desired sound coverage of the seating areas is obtained. Usually the system designer will want to start with the loudspeaker that has the longest distance on axis to project sound (and therefore requiring the highest Q or directivity ratio) [7]. What loudspeaker overlay will best match the dB contours in this distant location is first determined. The type of loudspeaker needed is evident from observing the spacing of the dB attenuation lines on the room map. Once the final position of the overlay is resolved, it is then fastened to the sphere with two small pieces of clear double-stick tape.

Next determine the maximum wattage requirement for the loudspeaker. The direct sound level that the loudspeaker produces on axis at a specified distance with a specified power level is used for calculating this requirement. Take, for example, a loudspeaker which is specified to deliver 103 dB at 4 ft (1.2 m) with 1 W.

This loudspeaker is pointed toward a room attenuation contour of -18 dB [referenced to 4 ft (1.2 m)]. The design specification requires that the direct sound level be 100 dB. Since the attenuation is 18 dB, the loudspeaker would have to produce 118 dB at 4 ft (1.2 m) to have the resultant 100 dB in the listening area. The voltage driving the loudspeaker must be 15 dB greater than that which produced 1 W in order to increase the level the loudspeaker produces from its 1-W reference of 103 dB to the 118 dB needed. This increase in voltage represents an increase from 1 W to 31.62 W in power.

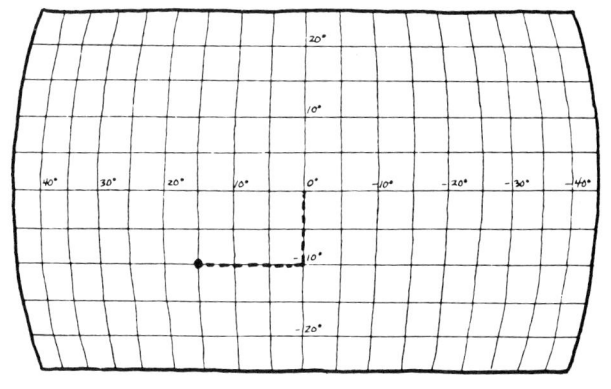

Fig. 6. A point with spherical coordinates $-10°$V∡, $15°$H∡

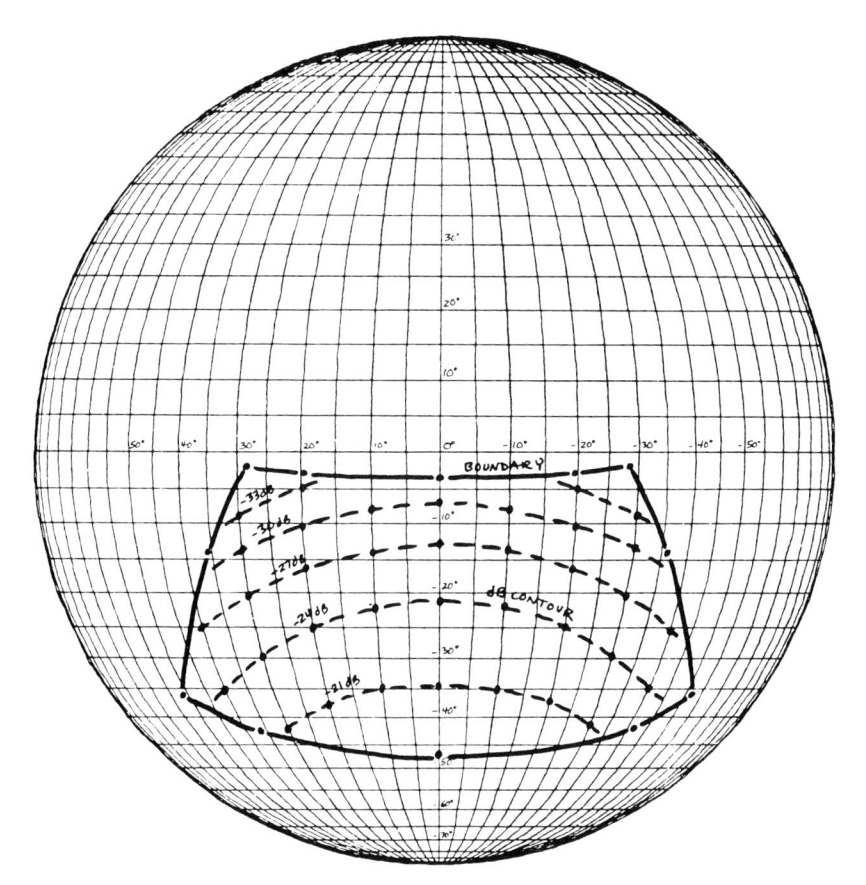

Fig. 7. Typical seating bank plotted on a sphere. Its boundary is outlined, and attenuation is represented by the dB contours.

Once the level is determined for the first loudspeaker, it can then be used as a reference to which all others may be compared.

Next the contour pattern for an adjacent loudspeaker is positioned and taped to the sphere. It is positioned so that, in the overlap zone, the addition of sound pressure is within a predetermined variation (usually 3 dB). Adding the overlapping sound levels in decibels can be done with a calculator, a chart, or a graph. This loudspeaker then has its wattage requirement determined in the same manner as the first loudspeaker. Different loudspeakers produce different levels with the same wattage at the same reference distance so each type of loudspeaker must be considered individually. This process is repeated for each loudspeaker until all listening areas are properly covered. As the loudspeakers are added one by one, via their overlays, the level loss due to the attenuation of sound, interaction between loudspeakers due to positioning, and power levels required can all be determined.

The exact position of each loudspeaker in the final cluster configuration can easily be read directly from the calibrated sphere. The vertical and horizontal angles read directly off the sphere are the same ones to which each real loudspeaker in the room or building would be oriented to achieve the same uniformity and pattern of coverage that have been designed on the spherical model. The loudspeaker coverage overlay may be rotated to better fit the acoustical environment. The rotation angle can be directly read off the sphere as measured from the meridian of longitude passing through the overlay's center. Fig. 8 illustrates how a loudspeaker is angled using the spherical coordinate data.

3 ADDITIONAL APPLICATIONS

For the average sound system designer the preceding steps would be very adequate for designing a superior loudspeaker cluster. However, the following applications show the versatility of the design techniques when applied to more intensive acoustical investigation or when desired as a teaching or selling aid.

The spherical model can be used as a projector with a scale architectural model or floor plan. If used with a scale model even the effects of acoustical reflections can be seen. (Malmlund and Wetherill are credited with the first use of a light projector to depict loudspeaker coverage [8].)

By using the projection capability on a scale model with mirrored surfaces the effects of acoustic reflections become apparent. Projection onto a model also serves as a method for exact verification of all plotting.

In order to use the sphere as a projector, a lamp assembly is clipped into the sphere and the sphere is attached to a stand which can support it over the scale model. (Note that the sphere at this point already has the room mapped on it and the loudspeaker coverage overlays in their intended positions.) The sphere's center, where the filament of the bulb has been adjusted to be, is then put into the exact scale position where the loudspeaker cluster will be positioned (Fig. 9). (If the loudspeaker cluster is to be 40 ft (12 m) high in the room, and the model uses a scale of ¼ in equal to 1 ft, the center of the sphere would be positioned 10 in (0.25 m) above the model.)

When the lamp is turned on, all of the seating bank boundaries will trace *exactly* around the model's seating bank boundaries. All loudspeaker patterns will be seen, and attenuation contours will be clearly seen. By using colored loudspeaker coverage overlays, each loudspeaker's contribution can be seen separately. Mirrored surfaces can be positioned at any location on the model, and the effect of reflections can be determined. Tinted mirrors may also be used to help trace reflections. (Note that if the sphere were large enough to accommodate a very bright bulb, strobe, or arc lamp, the projection device could be hung in the actual room at the proposed loudspeaker location, and all coverage patterns could be seen as they are mapped onto the building itself.)

Any overlay can be moved while watching the exact effect its new position has on the coverage and reflections, thus allowing the best engineering decision to be made about the position of the loudspeaker.

The projector also serves as a dramatic method of demonstrating the loudspeaker cluster coverage to a layman.

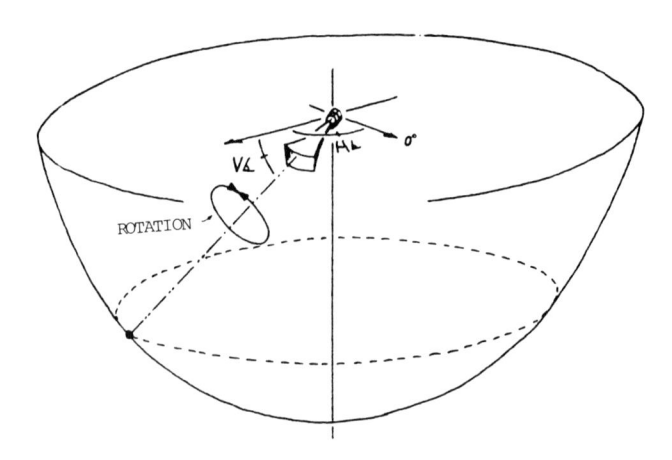

Fig. 8. Orientation of a loudspeaker using spherical angle information.

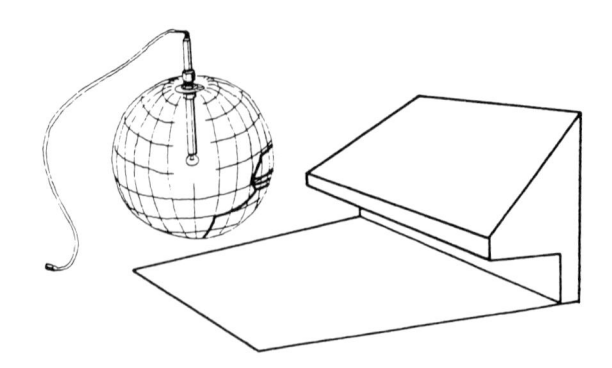

Fig. 9. Sphere with its attached and centered light assembly is positioned at a scale height and position over a scale model in order to project the loudspeaker coverage patterns onto the model.

4 DOCUMENTATION

There are two fundamental ways the final loudspeaker cluster configuration can be documented.

First and most obvious, the sphere itself contains complete documentation. It shows the acoustical environment, the actual angle of each loudspeaker, and the interrelationships between the coverage patterns of the loudspeakers. Labeling the wattage or relative decibel level for each loudspeaker on its respective overlay is recommended. In addition, the height and location of the loudspeaker system in the room or building, as well as the name and location of the building, should be labeled on the sphere itself. The overlays can be securely glued directly to the sphere. Then the sphere can be permanently glued together if desired. This sphere with its base becomes a complete documentation model. With the information on the sphere, a sound system installer has complete data for determining how to position each element of the loudspeaker cluster.

A second method of documentation enables a "flat" copy to be mailed or filed. It utilizes the projection capabilities of the sphere, a screen (consisting of rear projection material and filters), and a camera (any camera may be used, but an instant type such as a Polaroid is recommended for immediate results).

The sphere with light assembly attached is placed on its support base and turned on. The angle between the screen and the lamp assembly is adjusted for the best symmetry on the screen. Then a picture is taken of the sphere through the screen at the four cardinal points and the south pole (Fig. 10). The purpose of the screen is to focus the image of the sphere projected onto the screen for best possible reproduction. Filters within the screen cause the image photographed to be of relatively equal light intensity and provide color correction.

Although the shapes on the sphere are distorted on the screen, so are the shapes of the overlays, preserving all interrelationships. The design is now completely documented with five color photographs. These photographs can be mounted side by side on a flat sheet of paper, or they can be put together in a cube to give a better sense of direction. (Note that by moving the source of light within the sphere, by using different screen sizes, by varying the angles of projection, or by projecting via a mirror, various forms of projection can be made.)

5 SAMPLE PROBLEM

To illustrate the design techniques described in this paper, let us use the Blythe Arena in Squaw Valley as a sample problem. The Blythe Arena was constructed for the 1960 Winter Olympic Games. Fig. 11 gives a simplified perspective of the arena. The arena as it is to be used is layed out asymmetrically. Behind and to the right of the stage stretch two bleachers. Directly before the stage is the main floor which extends farther to the left than to the right. Located directly ahead of the stage is a large elevated grandstand. The bleachers, all of the main floor, and the grandstand area will be used for seating.

The task is to design a speech reinforcement system for a convention capable of delivering a modest 105-dB sound-pressure level ± 3 dB throughout the arena. Since this system design primarily concerns the speech intelligibility region (centered at approximately 2 kHz), we will not examine the bass horn components of the cluster. (In fact bass horns were used, but they were not as critical to the design. The crossover frequency used was 500 Hz.)

It is determined that a good location for a central loudspeaker cluster is 40 ft (12 m) above the stage. The stage itself is 4 ft (1.2 m) above the main floor. Measurements were made from blueprints, and data like the following sample were printed for each seating bank using the available software:

Boundary

Number	H∢	V∢	R	dB
1	−53	−13	175	−33
2	−38	−17	135	−31
3	−12	−21	112	−29
4	20	−20	116	−29
5	43	−16	145	−31
6	61	−20	115	−29
7	89	−23	103	−28
8	87	−47	55	−23
9	−85	−62	45	−21
10	−89	−27	89	−27
11	−89	−16	143	−31
12	−69	−15	152	−32

dB Contour

dB	H∢	V∢
−30	−90	−18
	−50	−18
	−10	−18
	30	−18
	70	−18
−27	−90	−27
	−50	−27
	−10	−27
	30	−27
	70	−27
−24	−90	−39
	−50	−39
	−10	−39
	30	−39
	70	−39
−21	−90	−62
	−50	−62
	−10	−62
	30	−62
	70	−62

This information represents the main floor of the arena. With similar data from each of the other three seating banks the arena was mapped onto the sphere as illustrated in Fig. 12. Fig. 12(a) views the sphere directly at 0°. Fig. 12(b)–(d) shows the arena as mapped on the sphere when looking at it from the other three cardinal points (90, 180, and −90°).

Now that the architectural measurements have been taken, the computing completed, and the arena mapped onto the sphere, we must choose the best loudspeakers and their optimum orientation within the cluster. Because Blythe Arena is constructed mostly of cement, steel, glass, and contains wood seats or bleachers, it is a very reverberant environment. As a result, constant-directivity or pattern-control type horns were chosen in order to minimize excitement of the reverberant field. These horns consisted of the standard 40° × 20°, 60° × 40°, and 90° × 40° patterns. Fig. 13 shows where the coverage pattern overlays were positioned on the sphere. Notice how the attenuation contour lines of each loudspeaker overlay were fit to match as closely as possible the dB contours in the building itself. The 40° × 20° (long-throw) horns were positioned first since the long distance they project allows them the least possible room for error. Next the areas starting from the long-throw patterns were covered, working back toward the front (closer to the cluster).

When overlapping the coverage patterns of the loudspeaker horns, the relative on-axis sound-pressure levels were first compared. A 40° × 20° horn pointing at a −33-dB contour line has been chosen as the 0-dB reference for comparison. A 60° × 40° horn pointing at a −30-dB contour will require 3 dB less direct sound-pressure level on axis than the 40° × 20° horn. This means that if the 60° × 40° horn pattern is overlapped with that of the 40° × 20° horn for even coverage, you need to remember that the −3-dB contour of the 60° × 40° horn is really 3 dB less, or −6 dB, when comparing it to the sound-pressure level of the 40° × 20° horn. Those comparisons are made between all the loudspeaker coverage patterns so they can be overlapped for even coverage (in this case ±3 dB).

Finally the power requirements for each horn are calculated. Again directly examine the 40° × 20° horn. The long-throw horn with its driver produces 128 dB at 4 ft (1.2 m) with 1 W. It is pointing at the −33-dB contour so the attenuated direct sound-pressure level that would actually arrive to that distance in the building would be 95 dB, 10 dB short of the design goal. Therefore the voltage is increased 10 dB to increase the power to the driver to almost 10 W, achieving 105 dB at the −33-dB contour. Each horn has its wattage requirement solved similarly. The power requirements and the driver model numbers should be written on each overlay. The information now contained on the sphere completely documents the design. Simply read the position and rotation (if any), power level, horn type, and driver type of each loudspeaker from the sphere, and the loudspeaker cluster can be accurately installed.

6 CONCLUSION

A sound system designer can now map the entire listening area (including related architecture) onto a sphere without the distortion inherent in a two-dimensional transform of the same. Because overlays depicting the actual coverage of a loudspeaker can now be placed on this spherical map and moved to any position or rotation with complete accuracy, it is possible to visualize immediately the loudspeaker's coverage anywhere. The interaction of all loudspeaker components of a cluster can be readily seen. The attenuation contours of each loudspeaker can be compared to the inverse-square losses in the architectural surroundings to angle the loudspeaker into its optimum position and determine how much power is needed to deliver a desired direct sound-pressure level. With the software available, a system can be designed very rapidly either in the field or in the office.

All of the sound design equipment needed fits easily into a standard attaché case, creating a highly portable package for the sound designer (Fig. 14). The sphere itself can be used as documentation which a non-technical person can easily understand, or photographic flat documentation can be made.

Instructors or designers can use the projection capabilities of the spherical model with blueprints or even scale architectural models. When projecting onto an architectural model, first-order reflections can be seen by using mirrored surfaces.

The ultimate benefits of this technique expand even beyond a more accurate sound system to a more cost-effective sound system and to a minimal-component sound system.

The technique can be optionally formatted by totally automating the plotting procedure (steps 2 and 3) and by utilizing the three-dimensional color graphics packages available for the more powerful computers. The three-dimensional spherical model is then depicted on the video screen. The room and loudspeaker coverage overlays are modeled on the image of the sphere on the screen. The sphere can then be rotated or tipped to any viewing angle allowing total visualization. Loudspeaker coverage overlays are automatically moved,

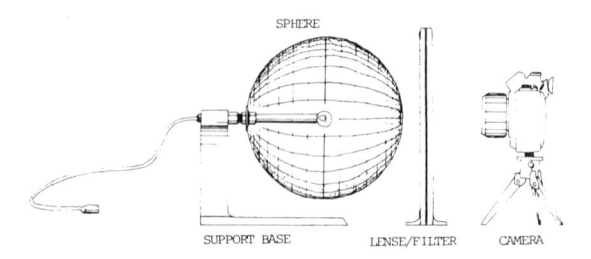

Fig. 10. Photodocumentation setup with camera focused on the south pole of the sphere. In addition, photographs would be taken of the four cardinal points along the equator.

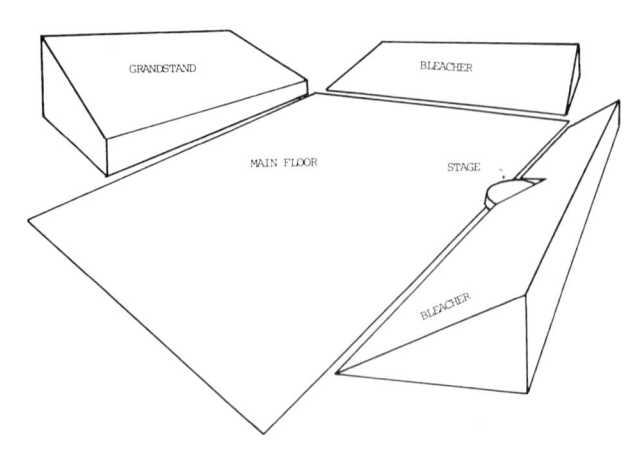

Fig. 11. Perspective of Blythe Arena in Squaw Valley, CA.

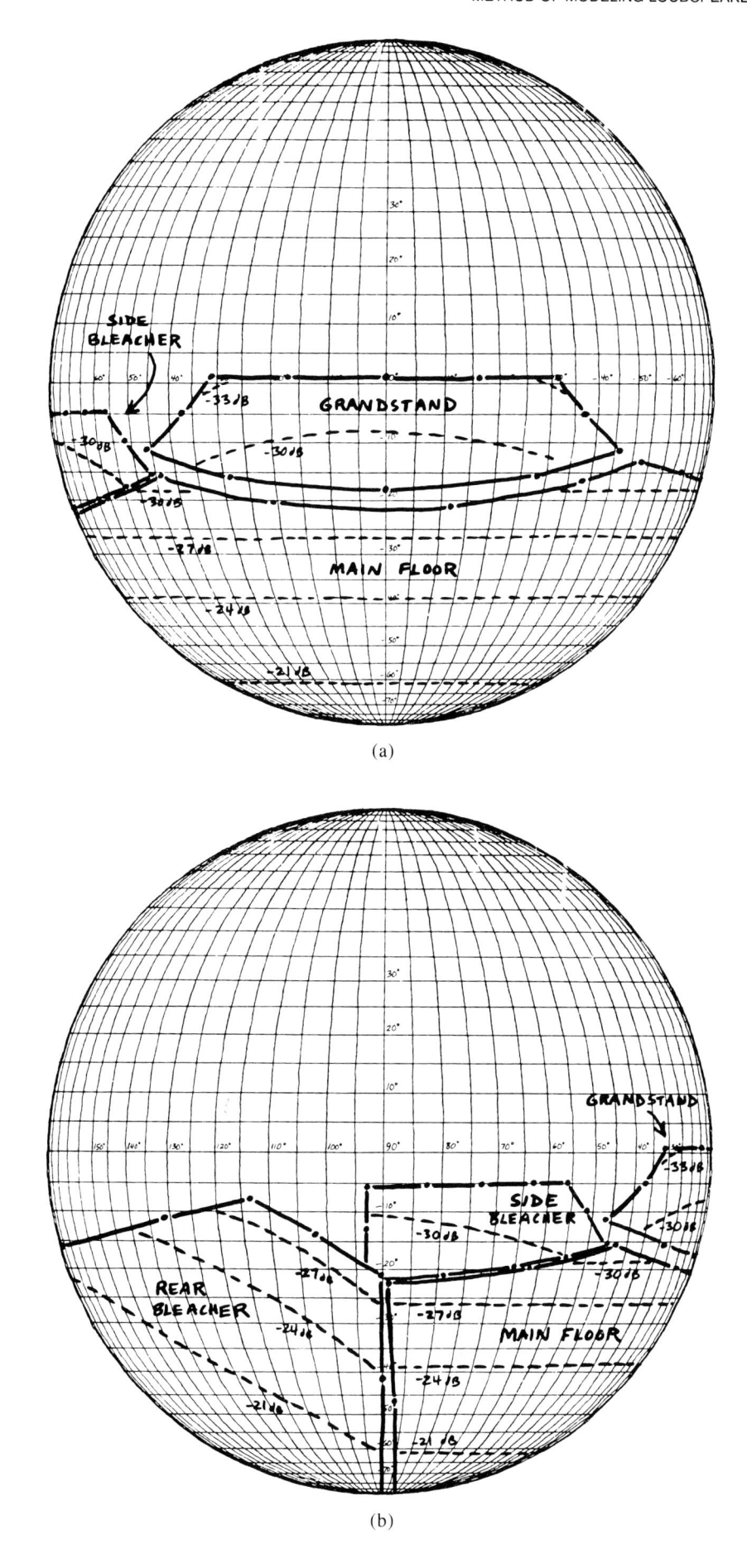

Fig. 12. Blythe Arena mapped on sphere as it would appear when looking directly on the sphere. (a) At 0° H∡. (b) At 90° H∡.

(c)

(d)

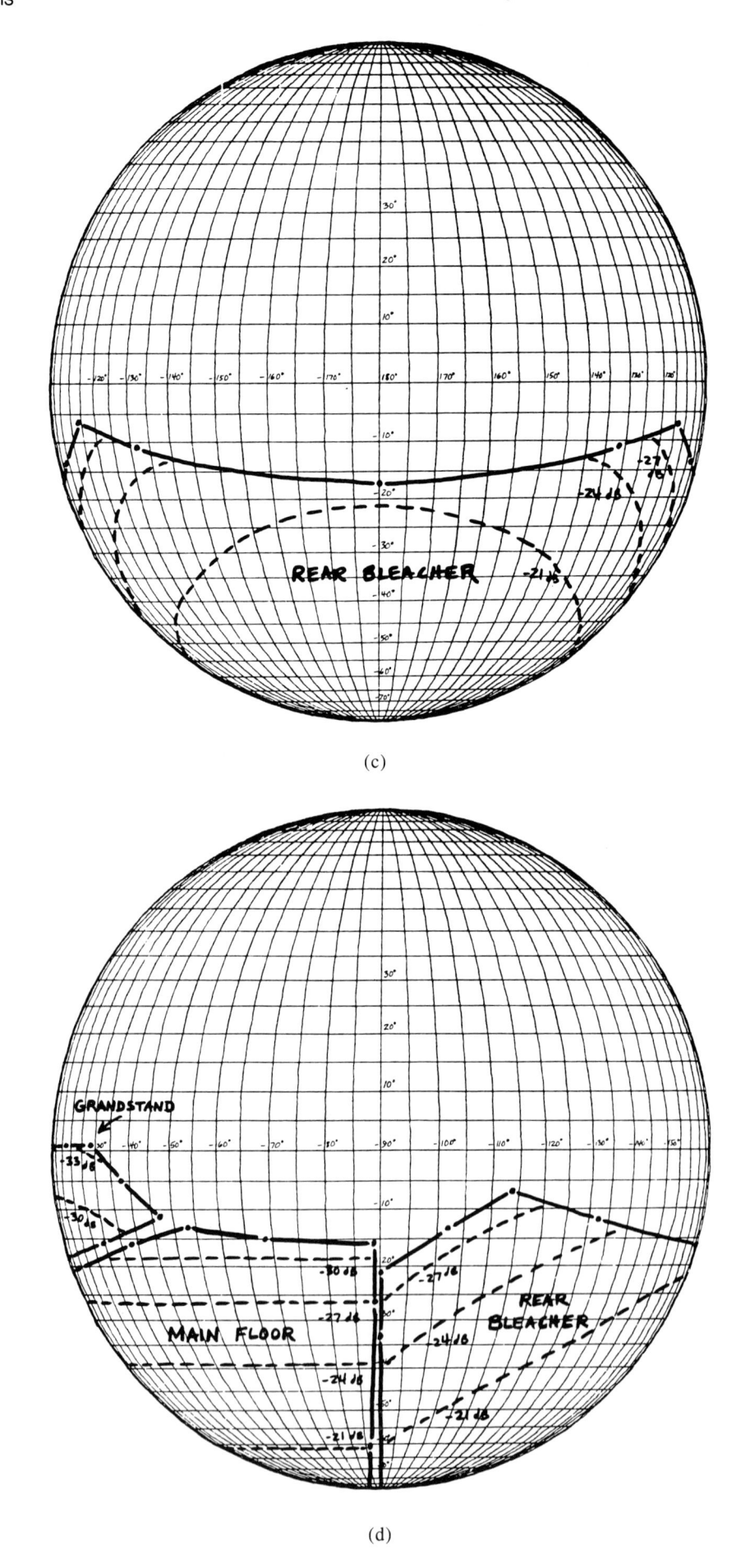

Fig. 12. Blythe Arena mapped on sphere as it would appear when looking directly on the sphere. (c) At 180° H∢. (d) At −90° H∢.

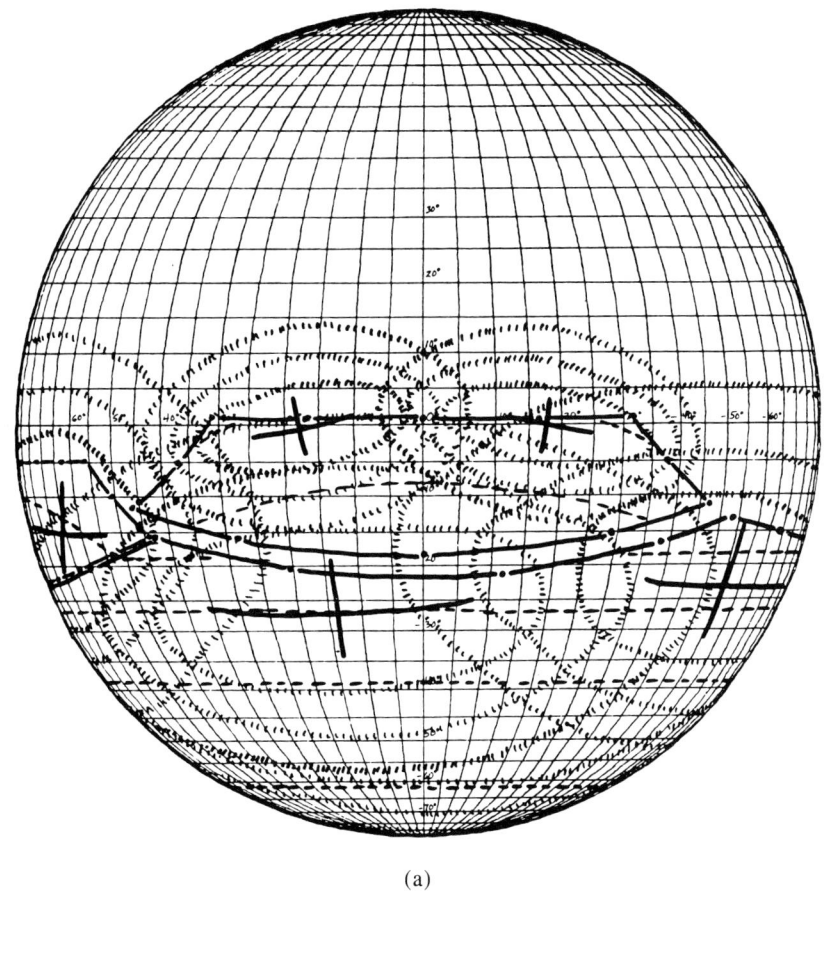

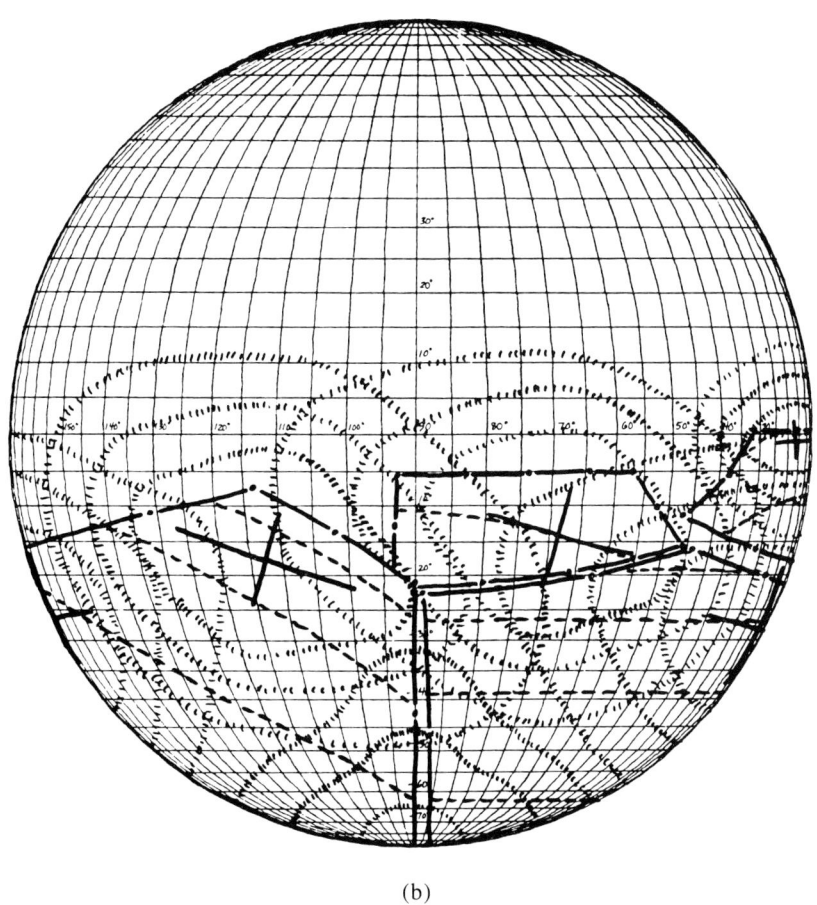

Fig. 13. Blythe Arena with loudspeaker coverage pattern overlays in position. (a) At 0° H∡. (b) At 90° H∡.

(c)

(d)

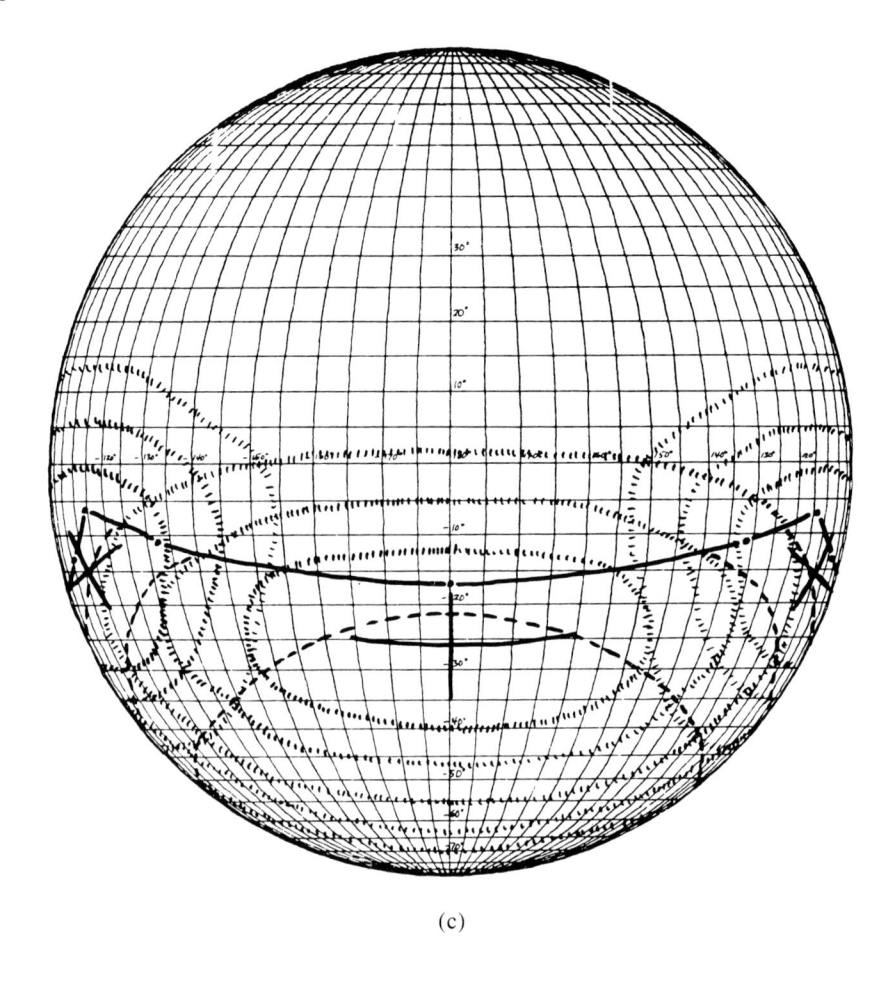

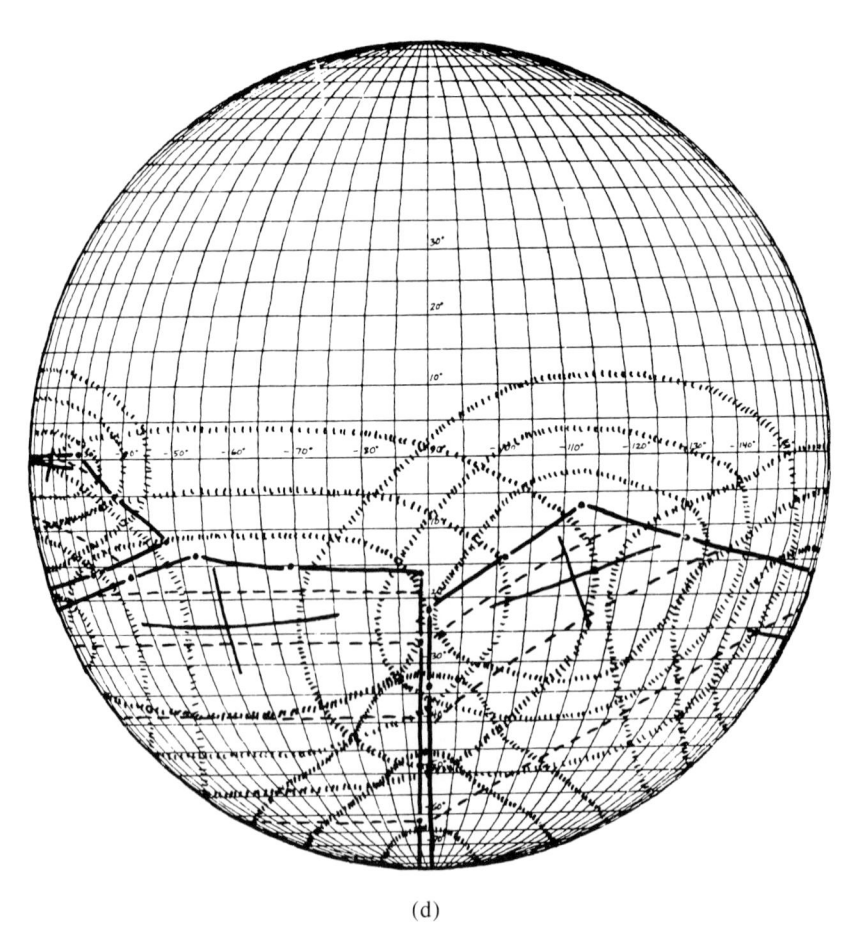

Fig. 13. Blythe Arena with loudspeaker coverage pattern overlays in position. (c) At 180° H∡. (d) At −90° H∡.

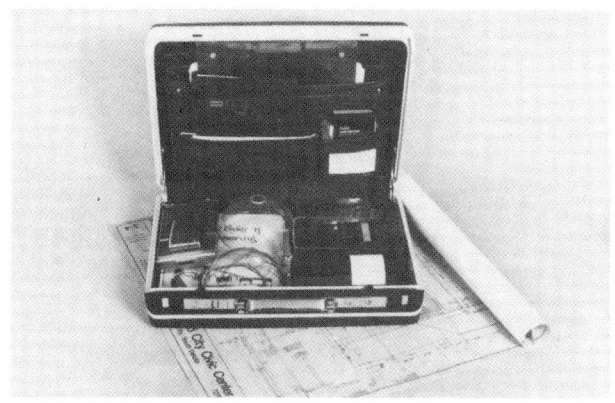

Fig. 14. Highly portable design package easily fits into a standard attaché case.

and the computer distorts the overlays commensurate with the spherical image. (As of this writing, software and spherical materials are available through Community Light & Sound, Inc., Chester, PA 19103, USA.)

7 REFERENCES

[1] D. Davis and C. Davis, *Sound System Engineering* (Howard W. Sams, Indianapolis, IN, 1975).

[2] E. Seeley, "Innovations in a Stadium Sound System Design," presented at the 57th Convention of the Audio Engineering Society, *J. Audio Eng. Soc. (Abstracts)*, vol. 25, p. 519 (1977 July/Aug.).

[3] T. G. McCarthy, "Loudspeaker Arrays—A Graphic Method of Designing," presented at the 61st Convention of the Audio Engineering Society, *J. Audio Eng. Soc. (Abstracts)*, vol. 26, p. 992 (1978 Dec.), preprint 1398.

[4] T. Uzzle, "Loudspeaker Coverage by Architectural Mapping," *J. Audio Eng. Soc.*, vol. 30, pp. 412–424 (1982 June).

[5] F. M. Becker, "A Polar-Plot Method of Loudspeaker Array Design," *J. Audio Eng. Soc.*, vol. 30, pp. 425–430 (1982 June).

[6] *Great International Atlas* (Prentice-Hall, Englewood Cliffs, NJ, 1981).

[7] E. T. Patronis, Jr., and C. Donders, "Central Cluster Design Technique for Large Multipurpose Auditoria," *J. Audio Eng. Soc.*, vol. 30, pp. 407–411 (1982 June).

[8] W. A. Malmlund and E. A. Wetherill, "An Optical Aid for Designing Loudspeaker Clusters," in D. L. Klepper, Ed., *Sound Reinforcement, an Anthology*, (Audio Engineering Society, New York, 1978), and *J. Audio Eng. Soc.*, vol. 13, p. 57 (1965 Jan.).

THE AUTHORS

J. Prohs

D. Harris

John R. Prohs was born in Gering, Nebraska, in 1942. He studied electrical engineering on a Regents scholarship at the University of Nebraska. In 1968 he received a B.A. degree from Ambassador College and immediately joined the staff of that institution in a technical capacity.

Mr. Prohs has developed a voice-actuated communications system in 1970 and a prototype time-gated spectrum analyzer in 1977. A paper was presented on the analyzer at the AES convention in California that year. The concept has since been incorporated into commercially available microprocessor-based audio test equipment.

In addition to extensive experience in working with radio, television, and communications audio systems, Mr. Prohs has designed and installed major sound systems at a number of resort areas across the US, including convention buildings at Wisconsin Dells, Mt. Pocono, Lake of the Ozarks, Squaw Valley, and Cape Cod.

He supervised the design and installation of the technical systems at Ambassador Auditorium, and is currently in charge of the audio engineers and all other technical aspects. He acts as consultant to other departments, and is US technical coordinator for worldwide Church of God convention sites.

Mr. Prohs has been a guest lecturer at Syn-Aud-Con Workshops on Loudspeaker Array Design. He is a member of the Audio Engineering Society, and he has held several offices in the Los Angeles Section including section chairman in 1980.

●

David E. Harris was born in 1957 in Lakeland, Florida. He studied nuclear physics at Central Florida Community College and received an A.S. degree in radiological health technology in 1977. He also obtained a B.A. degree with honors from Ambassador College in 1982.

Since 1979 Mr. Harris has been working in the Technical Support Department of Ambassador College as an audio engineer. There he has received diversified training in many areas including recording, circuit analysis, programming, graphics, and sound system design. He has participated in numerous Syn-Aud-Con seminars and workshops on subjects such as sound system design, loudspeaker array design, and T.E.F. instrumentation.

Mr. Harris is a member of the Audio Engineering Society and the International Oceanographic Foundation.

Computer Simulation of Loudspeaker Directivity*

DAVID G. MEYER

Purdue University, School of Electrical Engineering, West Lafayette, IN 47907, USA

Little work has been done in the study of loudspeaker directivity dynamics, particularly with arrays of drivers suitable for sound reinforcement applications. We have developed a mathematical model which allows the three-dimensional directivity characteristics of an arbitrarily configured arrangement of "real" drivers (that is, loudspeakers with defined amplitude, phase, and horizontal/vertical polar directivity characteristics as a function of frequency) to be simulated using a set of interactive computer programs. The mathematical basis of the author's model is derived and the set of computer programs in the simulation package is described. Results of several case studies illustrate its utility.

0 INTRODUCTION

While much analytical work has been done toward finding closed-form solutions to describe the directivity characteristics of various acoustic array configurations [1]–[4], the results obtained have limited utility. The underlying assumption in the derivation of all these closed-form solutions is that each array element acts as a point source. A point source, or "simple" source, is a source whose radius a is small compared with one-sixth wavelength (that is, $2\pi a/\lambda \ll 1$). Sources which meet this requirement are said to be omnidirectional. This implies the underlying assumption of a frequency-independent spherical directivity characteristic for each array element.

Virtually all "real" loudspeakers, however, have a frequency-dependent polar directivity characteristic: as the frequency of excitation increases, the beam width, or major lobe, of the polar directivity characteristic typically decreases. A typical 8-in (200-mm) cone loudspeaker, for example, will have a beam width of less then 25° at 8000 Hz. This is further complicated by an inherent frequency-dependent amplitude characteristic (commonly referred to as frequency response) as well as an inherent frequency-dependent phase characteristic (often referred to as time response) [5], [6].

In order to predict the behavior of loudspeaker systems constructed using real drivers, we have extended the work of Wolff and Malter [1] to develop a mathematical model which includes these frequency-dependent amplitude, phase, and polar directivity characteristics. This mathematical model has been incorporated into a computer simulation system which allows interactive creation of loudspeaker array data files, interactive analysis of horizontal/vertical directivity characteristics (with screen-size plots on a standard CRT terminal), interactive perturbation of parameters (such as loudspeaker-element type, mounting location, element spacing, mounting skew angles, signal filter parameters, and delay characteristics), and the production of various forms of hard-copy graphical output (including three-dimensional directivity balloons, both logarithmic and linear horizontal/vertical directivity "slice" plots, and a perspective drawing of the loudspeaker configuration being analyzed).

In this paper we first show the derivation of the mathematical model. We then briefly describe the interactive computer simulation system and present the results of several case studies.

1 MATHEMATICAL BASIS OF SIMULATION MODEL

In deriving the mathematical basis of the simulation model, we start with a discussion of simple source combinations in two dimensions, and define some of the terminology associated with directivity patterns. We then present a detailed example illustrating the operation of a dipole (two simple sources in phase), and extend this result to n in-line simple sources. The effects

---

* Presented at the 72nd Convention of the Audio Engineering Society, Anaheim, CA, 1982 October 23–27, as "Development of a Model for Loudspeaker Dispersion Simulation"; revised 1983 August 9.

of signal delay and phase modification are discussed next, followed by further generalizations of the formulation which allow an arbitrary three-dimensional arrangement of transducers, each having defined directivity, amplitude, and phase characteristics. We also present the pistonic approximation used to simulate the behavior of commonly available cone loudspeakers.

1.1 Combinations of Sources

We can write for the pressure observed at an arbitrary point (x_0, y_0, z_0) due to a harmonic point source of sound i at location (x_1, y_1, z_1)

$$P_i = \frac{B_i}{r_i} \cos\left(\omega t - \frac{2\pi r_i}{\lambda} - \Phi_i\right) \tag{1}$$

where
- ω = frequency of excitation, in radians per second
- λ = wavelength of reproduced signal
- r_i = distance from point source i to observation point
- B_i = proportional to intensity of radiation of point source i
- Φ_i = phase angle between motion at i and a standard of phase.

The resultant pressure for n point sources vibrating at the same frequency and arbitrarily positioned in space is given by

$$P = \sum_{i=1}^{n} P_i = \sum_{i=1}^{n} \frac{B_i}{r_i} \cos\left(\omega t - \frac{2\pi r_i}{\lambda} - \Phi_i\right). \tag{2}$$

In order to add the vectors analytically, it is simplest to express each one as the sum of its projections on two mutually perpendicular axes:

$$\overline{P}_i = \frac{B_i}{r_i} \cos\left(\omega t - \frac{2\pi r_i}{\lambda} - \Phi_i\right)$$

$$+ j \frac{B_i}{r_i} \sin\left(\omega t - \frac{2\pi r_i}{\lambda} - \Phi_i\right) \tag{3}$$

$$= \frac{B_i}{r_i} e^{j(\omega t - 2\pi r_i/\lambda - \Phi_i)}. \tag{4}$$

The vector sum is then

$$\overline{P} = \sum_{i=1}^{n} \overline{P}_i = \sum_{i=1}^{n} \frac{B_i}{r_i} e^{j(\omega t - 2\pi r_i/\lambda - \Phi_i)}. \tag{5}$$

An illustration for the case $n = 4$ is provided in Fig. 1.

For the moment we limit our interest to the directional characteristics of radiators, where the difference in distance from various parts of the radiator to the listener is small compared to the distance from the radiator to the listener. In other words, we assume that our observation point is in the *far field* of the radiator. (We will later relax this constraint.)

Referring again to Fig. 1, note that if r_0 (the distance from the radiator origin, here the origin of the coordinate system to the observation point) is large compared to the overall dimensions of the radiator, then $r_0 \simeq r_i$ for all i. This implies that the pressure contribution from each source in the radiator will be subject to the same $1/r_0$ attenuation. The directional effect of a radiator, that is, the variation in pressure intensity as a function of observation point, is due to the phase relationship of the contributions from each source in the radiator. This phase relationship is primarily a result of the differences in path length from each source in the radiator to a given observation point. Given a far-field observation point, then, the expression in Eq. (5) can be put into the form

$$\overline{P} = \frac{1}{r_0} \sum_{i=1}^{n} B_i \, e^{j(\omega t - 2\pi r_i/\lambda - \Phi_i)}. \tag{6}$$

Our main interest is in the magnitude of this summation, which is given by

$$|\overline{P}| = \frac{1}{r_0} \left| \sum_{i=1}^{n} B_i \, e^{j(\omega t - 2\pi r_i/\lambda - \Phi_i)} \right| \tag{7}$$

$$= \frac{1}{r_0} \left[\left(\sum_{i=1}^{n} B_i \cos\left(\omega t - \frac{2\pi r_i}{\lambda} - \Phi_i\right) \right)^2 \right.$$

$$\left. + \left(\sum_{i=1}^{n} B_i \sin\left(\omega t - \frac{2\pi r_i}{\lambda} - \Phi_i\right) \right)^2 \right]^{1/2}. \tag{8}$$

The maximum value of $|\overline{P}|$, that is, the maximum intensity of which the radiator is capable, will be produced if the vectors representing the radiation from the individual point sources are all in phase. The vector from the radiator origin to the point in space at which this condition occurs is referred to as the principal axis of the radiator. For simple radiators this vector is usually the central (or Z) axis of the radiator (hence the terminology "on axis"). We denote the maximum value

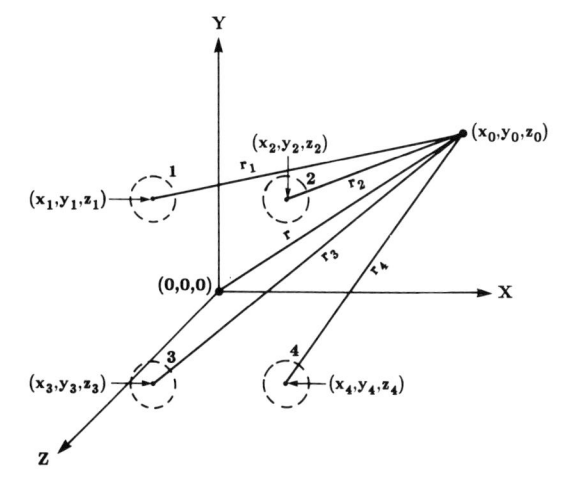

Fig. 1. Four-element radiator observed from (x_0, y_0, z_0).

of $|P|$ obtained along the principal axis of the radiator by M,

$$M = |P|_{max} = \frac{1}{r_0} \sum_{i=1}^{n} B_i . \quad (9)$$

1.2 Directivity Functions

In order to determine the directional characteristics of a radiator, it is necessary to find the *ratio* of the pressure intensity $|P|$ observed at each point of interest to the maximum pressure intensity M of which the radiator is capable. A set of pressure intensity ratios corresponding to a set of observation points is referred to as a directivity function. The set of observation points normally of interest are those at fixed radius from the radiator origin (where the central axis of the radiator corresponds to the Z axis of the coordinate system), lying in either the XZ plane or the YZ plane. A spherical coordinate system is most convenient for describing such sets of observation points; the one we use is illustrated in Fig. 2. Referring to this figure, note that an observation point (x_0, y_0, z_0) can be expressed in terms of spherical coordinates using the following transformation:

$$r_0 = [x_0^2 + y_0^2 + z_0^2]^{1/2} \quad (10a)$$

$$\theta_0 = \tan^{-1}\left[\frac{z_0}{(x_0^2 + y_0^2)^{1/2}}\right] \quad (10b)$$

$$\psi_0 = \tan^{-1}\left[\frac{x_0}{y_0}\right] \quad (10c)$$

where r_0 is the radius of observation (that is, the distance from the radiator origin to the observation point), ψ_0 is the clockwise angle of rotation in the XY plane about the Z axis, and θ_0 is the angle of observation with respect to the Z axis. (In the $\psi_0 = 0$ plane, θ_0 is the clockwise angle of rotation about the X axis.) Similarly, note that an observation point (r_0, θ_0, ψ_0) can be expressed in terms of Cartesian coordinates using the inverse transformation:

$$x_0 = r_0 \sin \theta_0 \sin \psi_0 \quad (11a)$$

$$y_0 = r_0 \sin \theta_0 \cos \psi_0 \quad (11b)$$

$$z_0 = r_0 \cos \theta_0 . \quad (11c)$$

Using the latter transformation in conjunction with the results of Eqs. (7) and (9) allows us to express a general far-field directivity function as

$$R(r_0, \theta_0, \psi_0) = \frac{|P|}{M} = \frac{1}{\sum_{i=1}^{n} B_i}$$

$$\left| \sum_{i=1}^{n} B_i e^{j(\omega t - 2\pi r_i/\lambda - \Phi_i)} \right| \quad (12)$$

where the path length r_i from each source to the observation point is

$$r_i = [(x_0 - x_i)^2 + (y_0 - y_i)^2 + (z_0 - z_i)^2]^{1/2} \quad (13)$$

with the coordinates of the observation point obtained using Eqs. (11).

The set of points at fixed radius from the central axis of a radiator lying in the YZ plane (that is, those points for which $\psi_0 = 0$) is referred to as the vertical directivity function, while the set lying in the XZ plane (that is, those points for which $\psi_0 = \pi/2$) is referred to as the horizontal directivity function. We refer to these particular polar directivity patterns as horizontal and vertical directivity "slice" plots, respectively.

Of primary interest in analyzing the characteristics of various radiator configurations is the extent of the main lobe of the directivity pattern. Also referred to as the beam width of a directivity pattern, the extent of the main lobe is defined as the angular distance between the two points on either side of the principal axis where the sound pressure level is down 6 dB[1] from its maximum possible level M [2].

We start our discussion of directivity patterns by applying the theory developed in the previous section to the case of two simple sources in phase.

1.3 Two Simple Sources in Phase

The geometric situation for this case is shown in Fig. 3. Note that observation points are restricted to the YZ plane. Also, it is assumed that the distance r from the two point sources to the point A at which the pressure is being measured is large compared with the separation b between the two sources.

The spherical sound wave arriving at point A from source 1 will have traveled a distance $r - (b/2) \sin \theta$, and the corresponding sound pressure will be

$$P_1 = \frac{B_1}{r} e^{j\{\omega t - (2\pi/\lambda)[r - (b/2)\sin\theta]\}} . \quad (14)$$

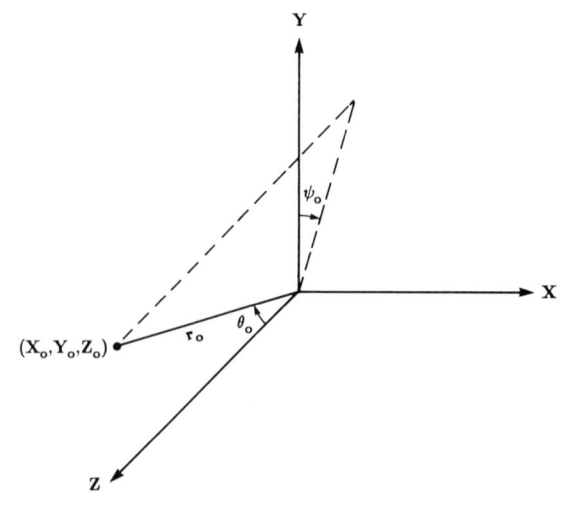

Fig. 2. Spherical coordinate system and positive angle convention.

---
[1] Some use 10 dB.

The wave from source 2 will have traveled a distance $r + (b/2) \sin \theta$, so that

$$P_2 = \frac{B_2}{r} e^{j\{\omega t - (2\pi/\lambda)[r - (b/2)\sin\theta]\}} . \quad (15)$$

The sum of $P_1 + P_2$, assuming $r >> b$ (that is, we have a far-field observation point) and that the sources are of equal strength, that is, $B_1 = B_2 = B$, gives

$$P_{1+2} = \frac{B}{r} e^{j(\omega t - 2\pi r/\lambda)} [e^{j(\pi b/\lambda)\sin\theta} + e^{-j(\pi b/\lambda)\sin\theta}] . \quad (16)$$

Multiplication of the numerator and the denominator of Eq. (16) by

$$e^{j(\pi b/\lambda)\sin\theta} - e^{-j(\pi b/\lambda)\sin\theta} \quad (17)$$

yields

$$P_{1+2} = \frac{B}{r} e^{j(\omega t - 2\pi r/\lambda)} \frac{[e^{j(2\pi b/\lambda)\sin\theta} - e^{-j(2\pi b/\lambda)\sin\theta}]}{[e^{j(\pi b/\lambda)\sin\theta} - e^{-j(\pi b/\lambda)\sin\theta}]} . \quad (18)$$

Using the identity

$$\frac{e^{jx} - e^{-jx}}{2j} = \sin x \quad (19)$$

and replacing the exponentials by sines, the expression becomes

$$P_{1+2} = \frac{B}{r} e^{j(\omega t - 2\pi r/\lambda)} \frac{\sin[(2\pi b/\lambda)\sin\theta]}{\sin[(\pi b/\lambda)\sin\theta]} . \quad (20)$$

Here the directivity pattern is calculated by evaluating the directivity function given in Eq. (12) at the set of observation points for which ψ is fixed at 0 and θ is an arbitrary angle θ_0. Since we are assuming a far-field observation point, this function is independent of the radius of observation. We therefore denote this particular directivity function by $R(\theta_0, 0)$. The numerator of this expression is the sum of the individual contributions from the two sources observed from angle $\theta = \theta_0$, while the denominator is the maximum pressure intensity of which this particular radiator is capable, here occurring at $\theta = 0$. So for this case

$$R(\theta_0, 0) = \frac{|P_{1+2}|_{\theta=\theta_0}}{|P_{1+2}|_{\theta=0}} . \quad (21)$$

Working on the denominator,

$$P_{1+2}\Big|_{\theta=0} = \frac{B}{r} e^{j(\omega t - 2\pi r/\lambda)} \frac{\sin[(2\pi b/\lambda)\sin\theta]}{\sin[(\pi b/\lambda)\sin\theta]}\Big|_{\theta=0} . \quad (22)$$

Here we must find

$$\lim_{\theta \to 0} \frac{\sin[(2\pi b/\lambda)\sin\theta]}{\sin[(\pi b/\lambda)\sin\theta]} . \quad (23)$$

Applying l'Hôpital's rule, we obtain

$$\frac{\cos[(2\pi b/\lambda)\sin\theta](2\pi b/\lambda)\cos\theta}{\cos[(\pi b/\lambda)\sin\theta](\pi b/\lambda)\cos\theta}\Big|_{\theta=0} = 2 . \quad (24)$$

Therefore,

$$P_{1+2}\Big|_{\theta=0} = \frac{2B}{r} e^{j(\omega t - 2\pi r/\lambda)} \quad (25)$$

and

$$|P_{1+2}|_{\theta=0} = \frac{2B}{r} . \quad (26)$$

Note that this is the same as the denominator for Eq. (12) evaluated with $n = 2$.

Working on the numerator,

$$P_{1+2}\Big|_{\theta=\theta_0} = \frac{B}{r} e^{j(\omega t - 2\pi r/\lambda)} \frac{\sin[(2\pi b/\lambda)\sin\theta_0]}{\sin[(\pi b/\lambda)\sin\theta_0]} . \quad (27)$$

Therefore,

$$|P_{1+2}|_{\theta=\theta_0} = \frac{B}{r} \frac{\sin[(2\pi b/\lambda)\sin\theta_0]}{\sin[(\pi b/\lambda)\sin\theta_0]} . \quad (28)$$

Finally,

$$R(\theta_0, 0) = \frac{|P_{1+2}|_{\theta=\theta_0}}{|P_{1+2}|_{\theta=0}} = \frac{\sin[(2\pi b/\lambda)\sin\theta_0]}{2\sin[(\pi b/\lambda)\sin\theta_0]} . \quad (29)$$

Referring to Fig. 3, we see that if b is very small compared to a wavelength, the two sources essentially coalesce and the pressure at distance r at any angle θ_0 is double that for one source acting alone. If we assume

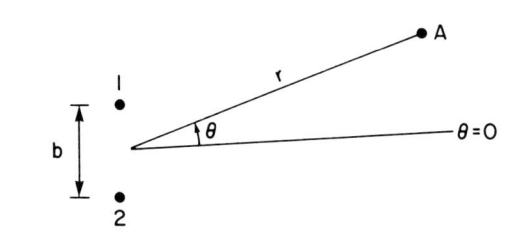

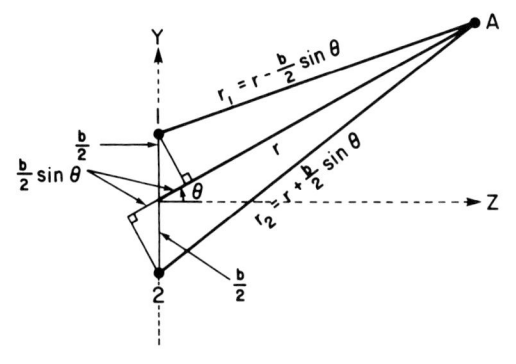

Fig. 3. Two simple sources in phase.

(as we will throughout the remainder of this paper) that these two point sources are mounted in an infinite baffle, the directivity pattern obtained will be hemispherical and therefore have $Q = 2$. (The corresponding directivity index D_I is defined as 10 log Q; here $D_I = 3$ dB [7].) As b gets larger, however, the pressures arriving from the two sources will differ in phase. Consequently the directivity pattern will vary as a function of θ. In particular, the beam width will narrow as b is increased.

An important observation can be made from this simple type of radiator that applies to all types of acoustic radiation. As the size of the radiator increases (in this case as b increases), the principal lobe (that is, the lobe centered on the $\theta = 0$ axis) will, for any given frequency, become sharper (that is, the Q of the radiator will increase). The number of side lobes will also increase. It is possible to suppress side lobes, however, by increasing the number of elements. By mathematical derivations similar to those used earlier in this section it can be shown that for a series of n in-line simple sources spaced a distance b apart along the Y axis (again with observation points limited to the YZ plane),

$$R(\theta_0, 0) = \frac{\sin[(n\pi b/\lambda)\sin\theta_0]}{n \sin[(\pi b/\lambda)\sin\theta_0]} . \tag{30}$$

This is the classic equation for describing the directivity characteristics of a line source array [1]–[4].

1.4 Signal Delay and Phase Modification

In the previous section we determined the directivity pattern of a dipole (two simple sources located on the Y axis separated by distance b) based on the far-field sum of the pressure contribution from each source. As determined in Eq. (14), the contribution of the source located on the Y axis is a distance $h = b/2$ from the origin is

$$P_i = \frac{B}{r} e^{j[(\omega t - 2\pi r/\lambda)(r - h \sin\theta)]} . \tag{31}$$

In this section we examine the effect of moving a source behind the XY plane. An illustration of such a configuration is provided in Fig. 4. Referring to this figure, note that here, since $r \gg h$ and $r \gg q$,

$$r'' \simeq r' + q \cos\theta' \simeq r - h \sin\theta + q \cos\theta' \tag{32}$$

and since the observation point A is far away from the radiator, $\theta' \simeq \theta$. Therefore,

$$r'' \simeq r - h \sin\theta + q \cos\theta . \tag{33}$$

The contribution of this source is then

$$P_i \simeq \frac{B}{r} e^{j[(\omega t - 2\pi/\lambda)(r - h \sin\theta + q \cos\theta)]} . \tag{34}$$

Note, however, that the $2\pi(q \cos\theta - h \sin\theta)/\lambda$ term corresponds to Φ, used in Eq. (4) to denote the relative phase angle of the pressure contribution from a given source. For a given source-mounting height h on the Y axis and a given far-field angle of observation θ (confined to the YZ plane), the effect of q can be determined by rewriting the expression as

$$\frac{\Phi\lambda}{2\pi} = q \cos\theta - h \sin\theta$$

$$= qK_1 + K_2 \tag{35}$$

where K_1 and K_2 are constants. The relation between the change in source distance behind the XY plane Δq and the change in the relative phase angle of the pressure contribution from the source $\Delta\Phi$ can then be expressed as

$$\frac{\lambda}{2\pi} \Delta\Phi = \Delta q \tag{36}$$

or

$$\Delta\Phi = \frac{2\pi}{\lambda} \Delta q = \frac{\omega}{c} \Delta q \tag{37}$$

where c is the speed of sound at ambient conditions. Noting that pressure contributions emanating from a source located a distance q behind the XY plane will be delayed $d = q/c$ seconds relative to pressure contributions emanating from a source located directly on the XY plane, we can write the following relationship:

$$\left(\frac{\omega}{c}\right)\Delta q = \omega\Delta d \tag{38}$$

or

$$\Delta d = \frac{1}{c} \Delta q . \tag{39}$$

For observation points confined to the YZ plane, then, moving the source a distance q behind the XY plane corresponds to electrically delaying the signal applied to the source by $d = q/c$ seconds. Therefore a frequency-dependent electrical delay of the signal applied to each source, denoted by $d_i(f)$, may be incorporated into the expression for the path length given in Eq. (13):

$$r_i' = [(x_0 - x_i)^2 + (y_0 - y_i)^2$$
$$+ (z_0 - z_i - cd_i(f))^2]^{1/2} . \tag{40}$$

Time response, which provides a measure of the acoustic center of a loudspeaker, is therefore analogous to frequency-dependent signal delay [6].

1.5 The "Not-So-Far" Field

To this point we have only examined cases involving far-field observation points confined to the YZ plane. We now generalize our observation point to an arbitrary position in space. Before we proceed, some consideration should be given to the far-field stipulation we have been using thus far. A distinction must be made between the far field of an individual driver and the far field of a conglomerate of drivers which we have been referring to as a radiator. Usually an observation distance on the order of 10 times the largest dimension of an individual driver will ensure measurement of its far field [3]. This stipulation is necessary due to interference effects which occur close to the diaphragm of most drivers. Once in the far field of a driver, attenuation proportional to the inverse of distance may be assumed.

An important point sometimes overlooked in directivity measurement of multielement radiators, however, is that while the measuring microphone may be at sufficient distance to be in the far field of each individual driver, it might not be at sufficient distance to be in the far field of the entire radiator. For example, consider a line source of length L constructed using 10 individual drivers. Assume for the moment that r_0, the radius of observation, is $2L$. While this is sufficient distance to be in the far field of each individual driver, one of our underlying assumptions is violated: unlike the cases previously discussed, r_i is not approximately r_0 for all i and for all angles of observation. An r_0 on the order of $10L$ would be needed to reduce the pathlength-induced attenuation differential among the individual drivers to a negligible quantity.

For lack of any other name, we designate radii of observation which are in the far field of each individual driver but not necessarily in the far field of the composite radiator as the not-so-far field. Note that there are many practically occurring examples where this designation applies. One is the large stacks used for concert sound reinforcement. For an observer sitting near the stack, there will be significant variation in r_i for each driver. Worse yet, however, are split-source systems. If operated monaurally, split-source systems act as a rather large dipole. Here a large percentage of the listening area is subject to significant variation in r_i for each individual driver.

To maintain accuracy for radii of observation in the not-so-far field and thus allow as much generality as possible, we use a form of the vector representation for pressure contributions given earlier in Eq. (4):

$$\bar{P}_i = \frac{B_i}{r_i} e^{j(\omega t - 2\pi r_i'/\lambda)} . \qquad (41)$$

Note that r_i' from Eq. (40) is used in the exponent to specify the electrical phase relationship among the signals applied to the individual sources (as well as to delineate differences in pathlength) while r_i from Eq. (13) is used in the denominator to specify the attenuation of the pressure contribution from each source.

Substituting $\omega = 2\pi f$ and $\lambda = c/f$, this expression becomes

$$\begin{aligned}\bar{P}_i &= \frac{B_i}{r_i} e^{j[2\pi ft - 2\pi f/c)r_i']} \\ &= \frac{B_i}{r_i} e^{j2\pi f(t - r_i'/c)} \\ &= \frac{B_i}{r_i} e^{j2\pi ft} e^{-j2\pi f(r_i'/c)} . \end{aligned} \qquad (42)$$

Further generalizations of this formulation are made in the next section.

1.6 Incorporation of Directivity, Amplitude, and Phase Characteristics

Now instead of assuming a simple source possessing omnidirectional directivity, we allow each transducer to have defined directivity characteristics which vary as a function of frequency and angle of observation. Let $V_i(\theta_i, f)$ denote the directivity of the vertical polar pressure intensity of source i as a function of θ_i, the angle at which the source is observed with respect to its central axis in the plane $\psi_i = 0$, and let f be the frequency of excitation. Similarly let $H_i(\theta_i, f)$ denote the directivity of the horizontal polar pressure intensity as a function of θ_i, the angle at which the source is observed with respect to its central axis in the plane $\psi_i = \pi/2$. These directivity functions represent the normalized pressure intensity of the transducer observed at unit distance.

As illustrated in Fig. 5, the pressure intensity contribution of a given transducer observed from an arbitrary position in space will be a function of θ_i as well as the angle of rotation ψ_i about its central axis from which it is observed. From the perspective of the observation point (r_0, θ_0, ψ_0) the angles θ_i and ψ_i at which each individual driver is viewed may be calculated as follows:

$$\theta_i = \tan^{-1}\left[\frac{z_i - z_0}{[(x_i - x_0)^2 + (y_i - y_0)^2]^{1/2}}\right] \qquad (43a)$$

$$\psi_i = \tan^{-1}\left[\frac{x_i - x_0}{y_i - y_0}\right] . \qquad (43b)$$

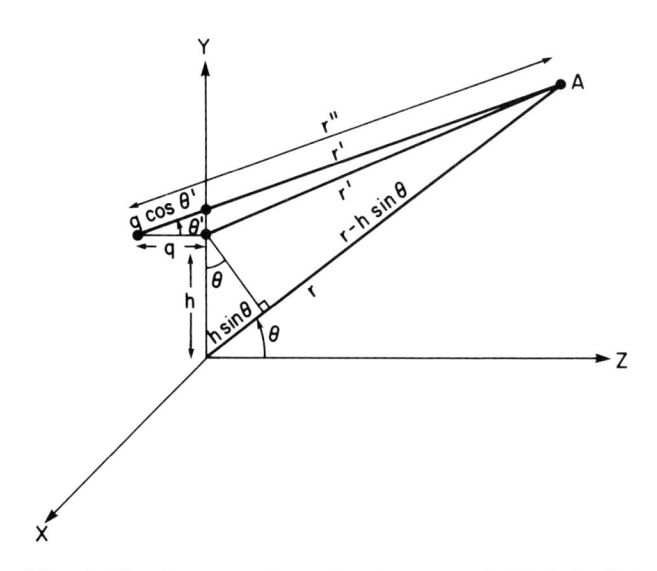

Fig. 4. Simple source located a distance q behind the XY plane.

For a three-dimensional representation of loudspeaker directivity, however, more information is needed than merely the vertical and horizontal directivity slices. In addition we need to know the planar shape of the wavefront emanating perpendicular to the central axis of each driver. As illustrated in Fig. 6, this shape may be viewed as a function of the angle of rotation ψ_i about the central axis of a given driver. We call this shape the wavefront aperture.

Also illustrated in Fig. 6 are the two basic types of wavefront apertures: circular/elliptical and square/rectangular. The relationship between the vertical and horizontal polar directivity functions at a given angle of rotation ψ_i about the central axis of each driver dictates the eccentricity of its wavefront aperture. If the horizontal and vertical directivity functions are identical, the wavefront aperture of a radiator will be circular (or square); otherwise it will be elliptical (or rectangular). The angle of the observation point θ_i with respect to the central axis of the radiator prescribes the degrees off axis with which its horizontal and vertical directivity functions are observed.

We denote the wavefront directivity function for a given transducer i by $W_i^k(V(\theta_i, f), H(\theta_i, f), \psi_i)$, where k indicates the wavefront directivity type: circular and elliptical wavefront apertures will be designated type 0, while square and rectangular wavefront apertures will be designated type 1.

The wavefront directivity function for a type 0 radiator can be expressed as

$$W_i^0 = [[\sin \psi_i \, V_i(\theta_i, f)]^2 + [\cos \psi_i \, H_i(\theta_i, f)]^2]^{1/2} \; . \quad (44)$$

Note that if $H_i(\theta_i, f) = V_i(\theta_i, f)$ for all θ_i and f, then W_i^0 is independent of ψ_i and therefore represents a circular wavefront aperture. For the general case in which the horizontal and vertical directivity functions are different, W_i^0 represents an elliptical wavefront aperture.

The wavefront directivity function for a type 1 radiator may be expressed as:

$$W_i^1 = \begin{cases} [[\sin \psi_i \, V_i(\theta_i, f)^2 + [H_i(\theta_i, f)]^2]^{1/2}, \\ \qquad\qquad \text{for } 0 \leq \psi_i \leq \gamma_i \\ [[V_i(\theta_i, f) + [\cos \psi_i H_i(\theta_i, f)]^2]^{1/2}, \\ \qquad\qquad \text{for } \gamma_i < \psi_i \leq \dfrac{\pi}{2} \end{cases}$$

(45a)

where γ_i is the corner angle of the rectangular aperture,

$$\gamma_i = \tan^{-1}\left[\frac{H_i(\theta_i, f)}{V_i(\theta_i, f)}\right] . \quad (45b)$$

An illustration of a square wavefront aperture, where $\gamma_i = \pi/4$, is provided in Fig. 6.

If a transducer is mounted such that the axis of its major lobe is not normal to the XY plane, we can use $\Delta\theta_i$ to denote the amount of tilting skew with respect to the Z axis and $\Delta\psi_i$ to denote the rotational skew about the Z axis. An alternate scheme, which is often more convenient, involves specifying the amount of skew with respect to the XZ plane along with the amount of skew with respect to the YZ plane. We designate these skew angles as $\Delta\alpha_i$ and $\Delta\beta_i$, respectively. In Fig. 7 the relationship between the spherical coordinate systems using (θ, ψ) and (α, β) is illustrated. The following transformation may be used to convert skew angles in the (α, β) system to the (θ, ψ) system:

$$\Delta\theta = \tan^{-1}\left[\frac{\cos(\Delta\alpha)\cos(\Delta\beta)}{[\cos^2(\Delta\alpha)\sin^2(\Delta\beta) + \sin^2(\Delta\alpha)]^{1/2}}\right] \quad (46a)$$

$$\Delta\psi = \tan^{-1}\left[\frac{\cos(\Delta\alpha)\sin(\Delta\beta)}{\sin(\Delta\alpha)}\right] . \quad (46b)$$

Regardless of how they are initially specified, the skew-modified angles from which a given driver will be observed may be represented as

$$\theta_i' = \theta_i - \Delta\theta_i \quad (47a)$$

$$\psi_i' = \psi_i - \Delta\psi_i . \quad (47b)$$

The wavefront directivity function may then be expressed as $W_i^k(V_i(\theta_i', f), H_i(\theta_i', f), \psi_i')$. This function will be used to replace the constant B_i used previously. Substituting this expression for B_i in Eq. (42) we obtain

$$\overline{P}_i = \frac{W_i^k(V_i(\theta_i', f), H_i(\theta_i', f), \psi_i')}{r_i}$$
$$\times \; e^{j2\pi ft} \, e^{-j2\pi f(r_i'/c)} . \quad (48)$$

Incorporation of a frequency-dependent normalized amplitude $a_i(f)$, characteristic for the signal applied

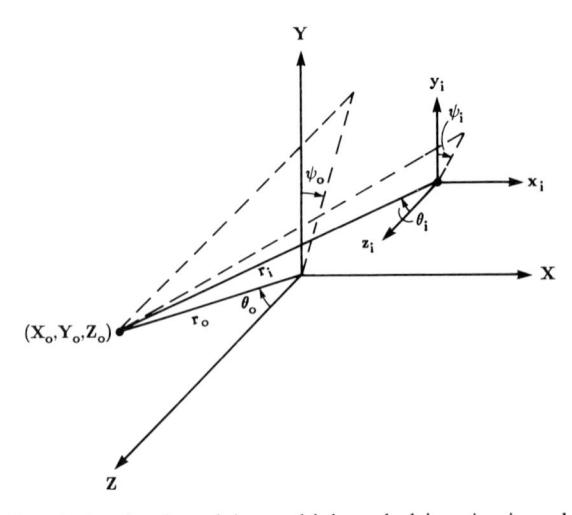

Fig. 5. Angles θ_i and ψ_i at which each driver is viewed.

to each source along with a composite broadband gain factor g_i modifies Eq. (48) as follows:

$$\bar{P}_i = \frac{g_i a_i(f) W_i^k(V_i(\theta_i', f), H_i(\theta_i', f), \psi_i')}{r_i}$$

$$\times \; e^{j2\pi ft} e^{-j2\pi f(r_i'/c)}. \qquad (49)$$

Allowing each transducer to have a defined frequency-dependent amplitude response $A_i(f)$ and a phase response $D_i(f)$, the expression for r_i from Eq. (40) becomes

$$r_i'' = [(x_0 - x_i)^2 + (y_0 - y_i)^2 + (z_0 - z_i c[d_i(f) + D_i(f)])^2]^{1/2} \qquad (50)$$

and the corresponding expression for $\bar{P}_i$ from Eq. (49) becomes

$$\bar{P}_i = \frac{g_i a_i(f) A_i(f) W_i^k(V_i(\theta_i', f), H_i(\theta_i', f), \psi_i')}{r_i}$$

$$\times \; e^{j2\pi ft} e^{-j2\pi f(r_i''/c)}. \qquad (51)$$

The vector sum of the individual pressure contributions from n transducers is

$$\bar{P} = \sum_{i=1}^{n} \bar{P}_i \qquad (52)$$

$$= \sum_{i=1}^{n} \frac{g_i a_i(f) A_i(f) W_i^k(V_i(\theta_i', f), H_i(\theta_i', f), \psi_i')}{r_i}$$

$$\times \; e^{j2\pi ft} e^{-j2\pi f(r_i''/c)} \; .$$

The magnitude of this vector sum, required for determining the directivity function, then becomes

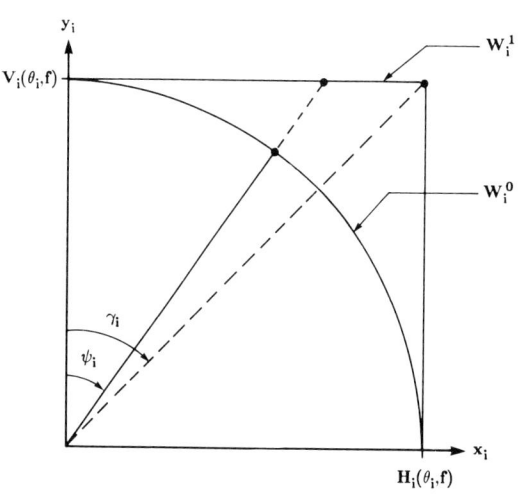

Fig. 6. Wavefront aperture types.

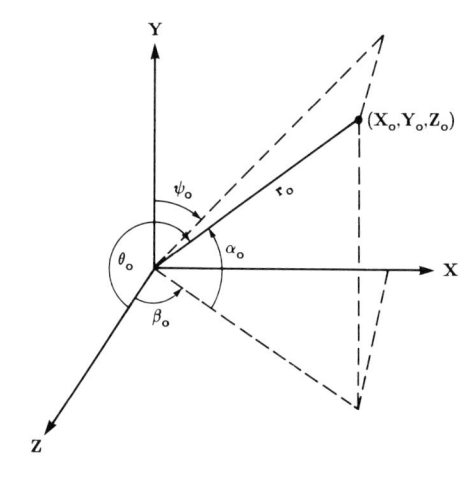

Fig. 7. Relationship between (θ, ψ) and (α, β) spherical coordinates.

$$|\bar{P}| = \left| \sum_{i=1}^{n} \frac{g_i a_i(f) A_i(f) W_i^k(V_i(\theta_i', f), H_i(\theta_i', f), \psi_i')}{r_i} \; e^{j2\pi ft} e^{-j2\pi f(r_i''/c)} \right|$$

$$= |e^{j2\pi ft}| \cdot \left| \sum_{i=1}^{n} \frac{g_i a_i(f) A_i(f) W_i^k(V_i(\theta_i', f), H_i(\theta_i', f), \psi_i')}{r_i} \; e^{-j2\pi f(r_i''/c)} \right|$$

$$= \left| \sum_{i=1}^{n} \frac{g_i a_i(f) A_i(f) W_i^k(V_i(\theta_i', f), H_i(\theta_i', f), \psi_i')}{r_i} \; e^{-j2\pi f(r_i''/c)} \right|$$

$$= \left| \left[\sum_{i=1}^{n} \frac{g_i a_i(f) A_i(f) W_i^k(V_i(\theta_i', f), H_i(\theta_i', f), \psi_i')}{r_i} \cos\left(\frac{2\pi f r_i''}{c}\right) \right] \right.$$

$$+ j \left[\sum_{i=1}^{n} \frac{g_i a_i(f) A_i(f) W_i^k(V_i(\theta_i', f), H_i(\theta_i', f), \psi_i')}{r_i} \sin\left(\frac{2\pi f r_i''}{c}\right) \right] \bigg|$$

$$= \left\{ \left[\sum_{i=1}^{n} \frac{g_i a_i(f) A_i(f) W_i^k(V_i(\theta_i', f), H_i(\theta_i', f), \psi_i')}{r_i} \cos\left(\frac{2\pi f r_i''}{c}\right) \right]^2 \right.$$

$$+ \left[\sum_{i=1}^{n} \frac{g_i a_i(f) A_i(f) W_i^k(V_i(\theta_i', f), H_i(\theta_i', f), \psi_i')}{r_i} \sin\left(\frac{2\pi f r_i''}{c}\right) \right]^2 \right\}^{1/2}. \qquad (53)$$

Also required for determining the directivity function is the maximum intensity of which the radiator is capable, designated earlier as M. Since we are now dealing with the vector sum of pressure contributions from arbitrarily configured drivers, each possessing frequency-dependent amplitude and phase characteristics, we can no longer use the simplistic approach of Eq. (9) to calculate M, which assumed that the maximum radiator intensity occurred directly on axis. Rather M must be determined based on the maximum value of $|\bar{P}|$ obtained from Eq. (53) for all θ_0 and ψ_0,

$$M = |\bar{P}|_{max} \quad \text{for all } \theta_0, \psi_0 . \quad (54)$$

Combining the results of Eqs. (53) and (54) yields the directivity function

$$R(r_0, \theta_0, \psi_0, f) = \frac{1}{M}\left\{\left[\sum_{i=1}^{n} \frac{g_i a_i(f) A_i(f) W_i^k(V_i(\theta_i', f), H_i(\theta_i', f), \psi_i')}{r_i} \cos\left(\frac{2\pi f r_i''}{c}\right)\right]^2 + \left[\sum_{i=1}^{n} \frac{g_i a_i(f) A_i(f) W_i^k(V_i(\theta_i', f), H_i(\theta_i', f), \psi_i')}{r_i} \sin\left(\frac{2\pi f r_i''}{c}\right)\right]^2\right\}^{1/2} \quad (55)$$

where

(x_0, y_0, z_0)	=	coordinates of the observation point, obtained from Eqs. (11)
(x_i, y_i, z_i)	=	coordinates of source i
r_i	=	pathlength from source to observation point, defined in Eq. (13)
r_i''	=	phase-modified pathlength, defined in Eq. (50)
W_i^k	=	wavefront directivity function, defined in Eqs. (44) and (45)
M	=	maximum pressure intensity of which a given radiator configuration is capable, defined in Eq. (54)
$d_i(f)$	=	delay characteristic of signal applied to source i
$a_i(f)$	=	normalized signal filter characteristic applied to source i
g_i	=	gain of composite signal applied to source i
$A_i(f)$	=	amplitude response of source i
$D_i(f)$	=	phase characteristic of source i
$V_i(\theta, f)$	=	normalized vertical polar response of source i
$H_i(\theta, f)$	=	normalized horizontal polar response of source i
θ_0	=	angle of observation with respect to Z axis
ψ_0	=	angle of rotation about Z axis of observation point
θ_i	=	angle with respect to central axis of source i at which it is observed
ψ_i	=	angle of rotation about central axis of source i at which it is observed
$\Delta\theta_i$	=	mounting skew of rotation about central axis of source i
$\Delta\alpha_i$	=	mounting skew of source i with respect to XZ plane (alternate representation)
$\Delta\beta_i$	=	mounting skew of source i with respect to YZ plane (alternate representation)
f	=	frequency of applied signal
c	=	speed of sound at ambient conditions (typically 341 m/s).

Eq. (55) is used by the simulation package to calculate directivity slices and balloons for arbitrary three-dimensional arrangements of transducers.

1.7 Pistonic Approximation for Cone Loudspeakers

A common loudspeaker type is the circular cone variety. Here the directivity characteristics can be approximated for use in the stimulation package by assuming that the transducer acts as a circular piston.

The defining equation for the directional characteristics of a circular piston source mounted in an infinite baffle with all parts of the surface vibrating with the same strength and phase are

$$H(\theta, f) = V(\theta, f) = \frac{2J_1((2\pi a f/c)\sin\theta)}{(2\pi a/\lambda)\sin\theta} \quad (56)$$

where

$H(\theta, f), V(\theta, f)$	=	horizontal and vertical directivity functions, respectively
J_1	=	Bessel function of first order
a	=	radius of circular piston
θ	=	angle between axis of circle and line joining center of circle
f	=	frequency of excitation
c	=	speed of sound at ambient conditions [3].

Letting $x = (2\pi a f/c)\sin\theta$, we can find $J_1(x)$ by using a Taylor series approximation,

$$J_1(x) = \frac{x}{2} - \frac{(x/2)^3}{1!2!} + \frac{(x/2)^5}{2!3!} - \frac{(x/2)^7}{3!4!} + \cdots . \quad (57)$$

The expression for $J_1(x)/x$ needed for the circular piston directivity pattern approximation is therefore

$$\frac{J_1(x)}{x} = \frac{1}{2} - \frac{x^2}{2^3 1!2!} + \frac{x^4}{2^5 2!3!} - \frac{x^6}{2^7 3!4!} + \cdots \quad (58)$$

2 COMPUTER SIMULATION SYSTEM

Having derived our mathematical model, we now describe the set of computer programs written which allow creation of loudspeaker array structures, simulation of directivity characteristics, and modification of control parameters.

To simulate the directivity pattern of an array of loudspeakers, the following data need to be available for each element i: the central coordinates (x_i, y_i, z_i), the mounting skew $\Delta\theta_i$ and $\Delta\psi_i$ (or, alternately, the mounting skew $\Delta\alpha_i$ and $\Delta\beta_i$), the amplitude characteristic of the applied signal $a_i(f)$, the delay characteristic of the applied signal $d_i(f)$, the gain of the composite signal g_i, the amplitude response of the loudspeaker $A_i(f)$, the phase response of the loudspeaker $D_i(f)$, the polar directivity characteristics of the loudspeaker, $V_i(\theta, f)$ and $H_i(\psi, f)$, and the wavefront directivity function W_i^k. This constitutes a considerable amount of data for each element. In order to simplify data entry, we created a data base containing amplitude, phase, and directivity characteristics of commonly used loudspeaker types. The utility program MAKESPK used to generate this data base allows interactive entry of "raw" (laboratory measured) data for given user-specified types as well as a pistonic approximation for circular cone loudspeakers. Loudspeaker types currently entered in the data base are listed in Table 1.

In addition to entering the data for a given array structure to be simulated, a facility for storing, retrieving, and editing these data is required. The interactive terminal editor and plotter program TERPLT that performs this task constitutes the heart of our simulation system. As well as allowing interactive entry, modification, storage, and retrieval of array configuration data files, this program also prints screen-size horizontal and vertical directivity slice plots. A typical plot (which can be saved for later output on a line printer) is shown in Fig. 8. The commands currently available for the terminal plotter and editor program are listed in Table 2. A list of simulation parameters with their units as well as minimum and maximum values allowed is provided in Table 3.

Since regularly structured homogeneous and nonhomogeneous arrays were commonly dealt with in our stimulation study, the need arose for some frontend programs which automatically generated $n \times m$ arrays of loudspeakers. Two different frontend processors have been written, one for creating homogeneous arrays (FRONTENDH) and the other for creating nonhomogeneous arrays (FRONTENDNH). To facilitate investigation of the control mechanisms suggested by the mathematical model, a series of spatial functions have been incorporated into these programs. The particular spatial function desired for each control parameter is selected using an index number (0 is the default for each case, indicating no spatial modification of the given parameter). Thus additional functions can be de-

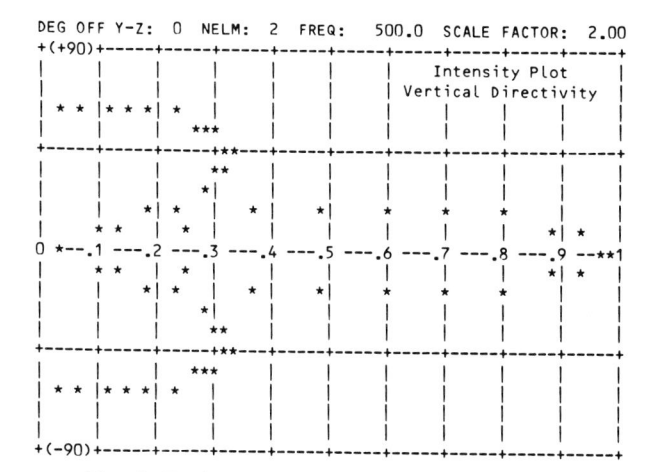

Fig. 8. Typical screen-size directivity slice plot.

Table 1. Loudspeaker types available in data base.*

Loudspeaker Type	Diameter [in (mm)]	Octave-Band Frequency Response (dB)									
		31Hz	63Hz	125Hz	250Hz	500Hz	1kHz	2kHz	4kHz	8kHz	16kHz
01CA	1 (25)	−30	−30	−30	−30	−24	−18	−12	−6	0	0
02CA	2 (50)	−30	−24	−18	−12	−6	0	0	0	0	−6
04CA	4 (100)	−24	−18	−12	−6	0	0	0	0	−6	−12
08CA	8 (203)	−12	−6	0	0	0	0	−6	−12	−18	−24
15CA	15 (380)	0	0	0	0	−6	−12	−18	−24	−30	−30
02CB	2 (50)	−21	−18	−15	−12	−9	−6	−3	0	0	0
04CB	4 (100)	−18	−12	−9	−6	−3	0	−3	−6	−9	−12
08CB	8 (203)	0	0	0	0	−3	−6	−9	−12	−15	−18
0.5CF	0.5 (12)	0	0	0	0	0	0	0	0	0	0
01CF	1 (25)	0	0	0	0	0	0	0	0	0	0
02CF	2 (50)	0	0	0	0	0	0	0	0	0	0
03CF	3 (75)	0	0	0	0	0	0	0	0	0	0
04CF	4 (100)	0	0	0	0	0	0	0	0	0	0
05CF	5 (125)	0	0	0	0	0	0	0	0	0	0
06CF	6 (152)	0	0	0	0	0	0	0	0	0	0
08CF	8 (203)	0	0	0	0	0	0	0	0	0	0
10CF	10 (254)	0	0	0	0	0	0	0	0	0	0
12CF	12 (305)	0	0	0	0	0	0	0	0	0	0
15CF	15 (380)	0	0	0	0	0	0	0	0	0	0

* All types listed have constant phase response and circular wavefront apertures.

fined by the user and added to the list of those available. The frontend processors currently allow spatial functions for depth (structural curvature), element tilt (mounting skew), filter (spectral content), composite gain, and signal delay.

Also of interest in our study was the capability of analyzing the interaction among more than one array of loudspeakers. This necessitated the facility for defining subarrays of loudspeakers having central coordinates (x_j', y_j', z_j'), where j corresponds to the subarray index. Array configuration data file editing commands are available on the terminal editor and plotter program TERPLT for manipulationg subarray loudspeaker elements as a unit.

A final programming consideration is the variety of hard-copy displays necessary for efficient analysis of the results obtained. In addition to high-resolution logarithmic and linear directivity slice plots (generated by SLICE), the need arose for three-dimensional directivity balloons (generated by BALLOON). Also readily apparent was the need for a perspective drawing (generated by PERSPEC) of the loudspeaker array to provide a physical interpretation of the masses of tabular data. These are all batch programs which utilize the Graphics Compatibility System (GCS) plotting package available on many large mainframe computers. Both Versatek and Calcomp plotting are possible with these programs. Examples of all these types of displays appear in the case studies that follow.

A summary of all the programs which comprise our simulation package along with a list of features and capabilities is provided in Table 4.

3 CASE STUDIES

We have examined a number of case studies over the past year in order to verify the accuracy of the simulation model. We have looked at various simple cases having known solutions (single loudspeakers using a pistonic approximation model, line source arrays with point source elements, and rectangular arrays) as well as cases for which there is no known analytical solution (nonhomogeneous arrays, arbitrary driver configurations, and so on). Further we have successfully simulated various line source configurations for which directivity plots have been published [8]–[10]. Included here are several case studies to illustrate the effectiveness and utility of the simulation system.

Before proceeding with the case studies, an explanation of the notation used in the various types of figures is in order. First, in the perspective drawings the number at the center of each loudspeaker element (represented

Table 2. Commands available for loudspeaker array editing.

Mnemonic	Function
ae	Add an element to last subarray
as	Add a new element to existing data
ax	Add a new element after deleting current data
ca	Change angle off axis of slice plot
cb	Change octave band of analysis
cc	Change absolute coordinates and mounting skew of element
cd	Change delay time of element signal
cf	Change filter characteristics of element signal
cg	Change gain of composite element signal
cl	Change location of subarray
cn	Change name of title to be printed with plot
co	Change radius of observation
cr	Change relative coordinates of element
cs	Change size (width and height) of subarray
ct	Change element type
cz	Change mounting skew of all subarray elements
dc	Display TERPLT commands
dd	Display data currently loaded in memory
dt	Display types of loudspeakers available
fr	Read an array configuration data file
fw	Write an array configuration data file
ph	Plot horizontal directivity slice
pv	Plot vertical directivity slice

Table 3. Units and ranges of simulation parameters.

Parameter	Unit	Minimum Value	Maximum Value
$A_i(f)$	dB	−50	0
$a_i(f)$	dB	−30	30
g_i	None	0	1
$D_i(f)$	Milliseconds	0	10
$d_i(f)$	Milliseconds	0	100
$V_i(\theta, f)$	None	0	1
$H_i(\theta, f)$	None	0	1
θ	Degrees	−90	90
$\Delta\theta_i$	Degrees	0	90
ψ	Degrees	−90	90
$\Delta\psi_i$	Degrees	0	90
(x_i, y_i, z_i)	Meters	0	10
f	Hz	31.25*	16,000*

* This corresponds to the 10 ISO octaves within this range.

Table 4. Simulation package programs.

Name	Mode	Description
BESSEL	Batch	Bessel function table generator
MAKESPK	Interactive	Loudspeaker-type data-base generator
FRONTENDH	Interactive	Homogeneous array-frontend processor
FRONTENDNH	Interactive	Nonhomogeneous array-frontend processor
TERPLT	Interactive	Loudspeaker-array-configuration file generator, editor, and terminal plotter
SLICE	Batch	Horizontal and vertical linear/logarithmic directivity slice plotter
SLICEFAM	Batch	Directivity slice family plotter
BALLOON	Batch	Three-dimensional directivity balloon plotter
BALLOONX	Batch	Extended-resolution three-dimensional directivity balloon plotter
PERSPEC	Batch	Loudspeaker-array-perspective drawing plotter

by a circle) indicates the element number, the rectangle drawn around a collection of elements represents a set of loudspeakers defined as a subarray, and the number which appears at the upper right-hand corner of the subarray indicates the subarray number. The size of the rectangle drawn around a set of loudspeaker elements is based on the width and height specifications for the subarray of which they are members.

Directivity slice plots are used to indicate the ratio of the pressure intensity calculated at an observation point (r_0, θ_0, ψ_0) [or, alternately, (x_0, y_0, z_0) in Cartesian coordinates] confined to the vertical ($\psi_0 = 0$) or horizontal ($\psi_0 = \pi/2$) plane, to the maximum intensity of which a given radiator configuration is capable. This maximum possible pressure intensity is indicated by the scale factor SF of the plot. Also indicated on these intensity plots are the number of elements NELM, the frequency of analysis FREQ, and the angle of view α, with respect to the XZ plane for horizontal directivity slices or the angle of view β with respect to the YZ plane for vertical directivity slices (refer again to Fig. 7). The Q of the radiator and the corresponding directivity index D_I are calculated from these data as defined in [7].

The logarithm of the normalized pressure intensity is also plotted for both the vertical and the horizontal planes. This allows us to calculate the 6-dB beam width, which defines the main lobe of the radiator.

Finally the directivity balloons which appear represent the pressure intensity distribution in three dimensions. The only information included with these plots is the scale factor (as defined above) and the frequency of analysis; the Q, D_I, and 6-dB beam width may be obtained from the corresponding directivity slice plots.

3.1 Directivity Balloon Family for a 4-in (100-mm) Loudspeaker

Prerequisite to the study of arrays of loudspeakers is an analysis of the performance of single loudspeakers. Of primary utility in our simulation study are circular cone loudspeakers because of the ease in modeling their directivity characteristics. As indicated earlier in Eq. (56), the circular piston approximation of cone loudspeaker behavior allows directivity simulation using a series of Bessel functions. The loudspeaker-type database generator MAKESPK, however, allows optional entry of raw data corresponding to virtually any transducer available.

One of the salient features of the response characteristic shared by all circular cone loudspeakers is a narrowing of the main lobe as the frequency of excitation increases. Furthermore the larger the diameter of the loudspeaker, the greater the effect of this so-called beaming.

A perspective drawing of a single 4-in (100-mm) cone-type radiator, generated by the PERSPEC program, is illustrated in Fig. 9. A family of directivity balloons for this radiator, at 500, 1000, 2000, and 4000 Hz, is illustrated in Figs. 10–13.

3.2 Line Source Beam Tilting

A desirable capability normally not found in conventional sound reinforcement loudspeaker clusters is a mechanism for dynamically steering the main lobe of the radiated energy. In many applications this loudspeaker cluster is placed high above the stage or source location, and has to be physically tilted in order to cover the audience area properly. The typical large

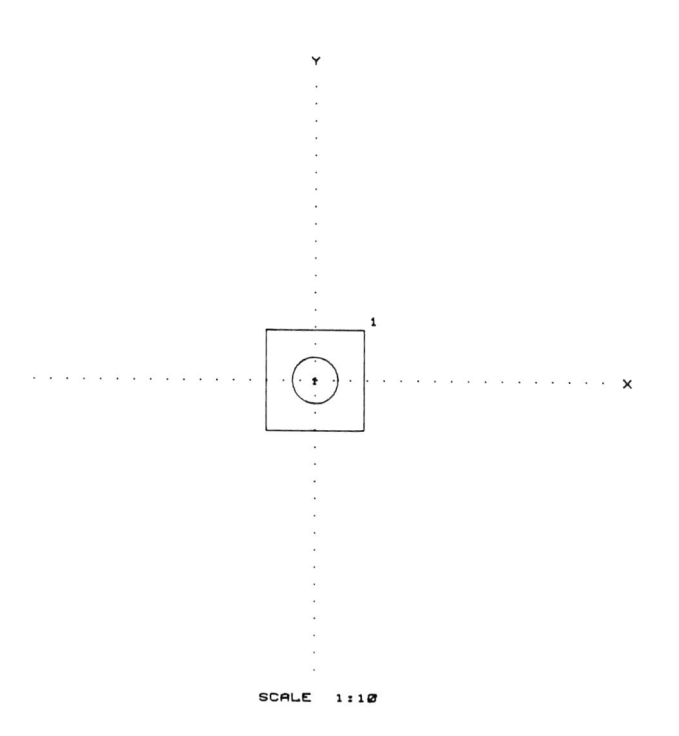

Fig. 9. Perspective of a single 4-in (100-mm) loudspeaker.

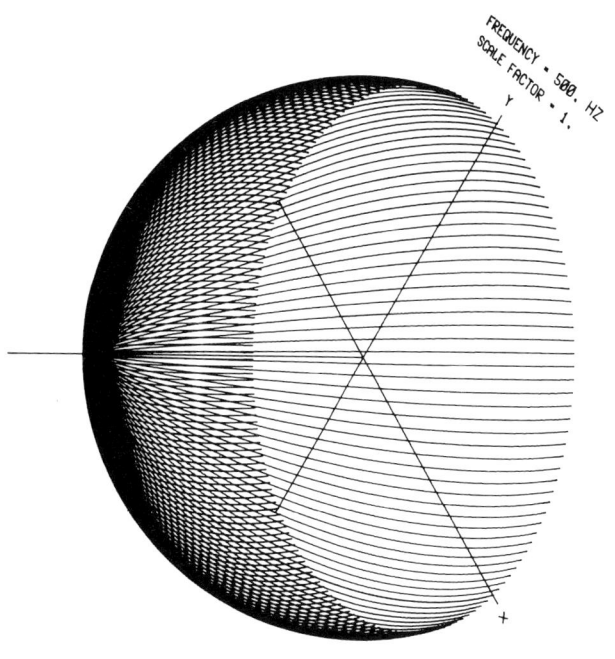

Fig. 10. Directivity balloon for a single 4-in (100-mm) loudspeaker; 500 Hz.

cluster usually consists of a number of separate cabinets and horns; tweaking the tilt of this entire structure is therefore a nontrivial operation. Often involved are human feats of daring and severe tests of ambidextrous abilities.

Further, recall that one of the design objectives of sound reinforcement systems is to minimize the amount of energy directed at the ceiling, the walls, and any unfilled seating. This errant sound energy contributes to the reverberant sound field and consequently degrades intelligibility as well as overall system performance. Ideally we would like to be able to specify the coverage angles necessary for a given pattern of auditorium fill and steer the loudspeaker main lobe so that it is centered over this audience area. We will discuss the possibility of directivity angle programmability in a later paper; here we address the problem of beam tilting and steering.

A number of relatively simple reinforcement applications utilize vertical line sources. One problem is that, especially in portable setups, the quantity and depth of the seating are often variable (the typical ballroom or gymnasium setup). Here the angle of tilt of the line source must be physically changed in order to accommodate varying seating depths. This is obviously not very convenient to do on a regular basis, particularly if the line source is permanently mounted high above the stage area. A perspective drawing of a physically tilted line source is illustrated in Fig. 14. Here the angle of tilt is 45°; note that dashed lines are used to indicate the tilt of each element. The directivity balloon for this case is illustrated in Fig. 15.

In an attempt to find a technique for electronically modifying the wavefront so that the radiated beam can be tilted, we recall that delaying the signal applied to a loudspeaker element is analogous to translating it in the negative Z direction. Thus it stands to reason that by applying increasing amounts of delay to successive elements of the line source, we should be able to tilt the beam at virtually any desired angle. (The theoretical basis for this is presented in Section 1 as well as in [4].) A perspective drawing of the electrically tilted line source is shown in Fig. 16. Here a successive delay of approximately 0.3 ms per element is applied in order to achieve a 45° tilt. (This corresponds to delays of 0 ms for element 1, of 0.3 ms for element 2, . . . , of 3.0 ms for element 10.) Note that while none of the elements are physically tilted, the corresponding directivity balloon (Fig. 17) is virtually identical to the case illustrated previously. As one might imagine,

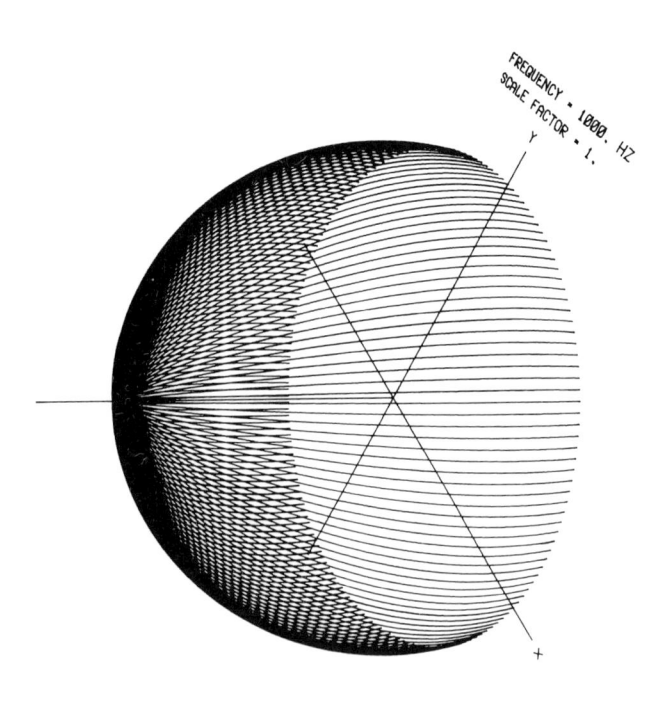

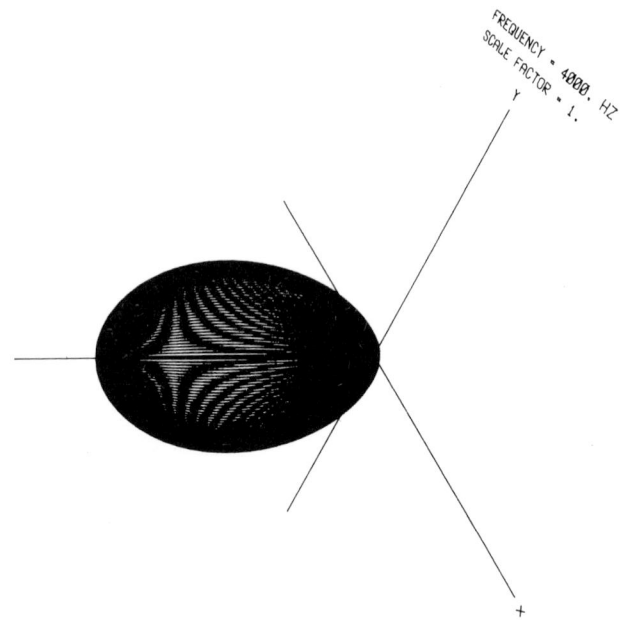

Fig. 11. Directivity balloon for a single 4-in (100-mm) loudspeaker. 1 kHz.

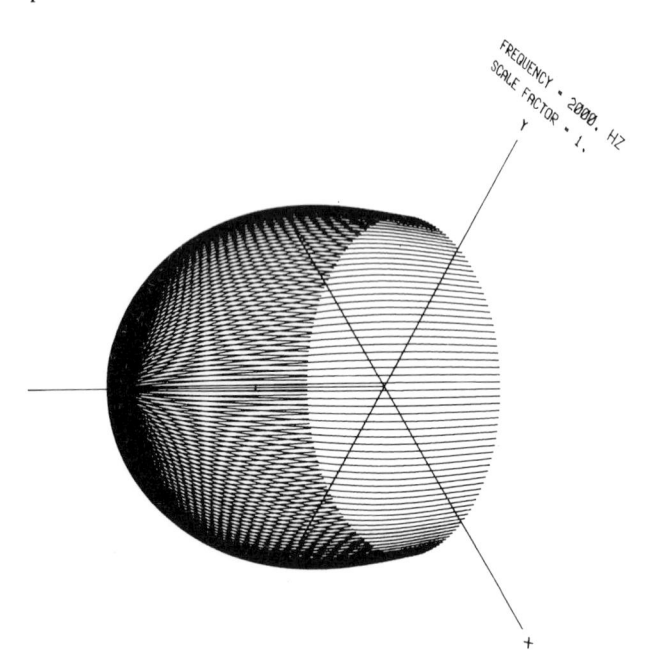

Fig. 12. Directivity balloon for a single 4-in (100-mm) loudspeaker: 2 kHz.

Fig. 13. Directivity balloon for a single 4-in (100-mm) loudspeaker: 4 kHz.

electrical delay can also be used for other forms of wavefront modification.

3.3 Pairs of Line Sources Spread Varying Distances

A common loudspeaker arrangement used in many portable sound reinforcement systems (such as for rock concerts) is some form of split source configuration. This type of loudspeaker placement is usually dictated by the arrangement of musical instruments or performers on the stage. Here the belief is that each of the loudspeaker systems will cover more or less half the audience area. Little consideration is given, however, to the possible interaction of the signals produced by these two sources.

Traditionally split source arrangements for professional reinforcement applications have been argued against on the basis of possible echo problems, perpetrated by the disparity in time delay of the two signals arriving at a given point in the listening area [11], [12]. If the two sources are far enough apart such that the signal which arrives from one is 50 ms behind the other, the delayed signal is perceived by the human ear as an echo. At ambient conditions (that is, with $c = 341$ m/

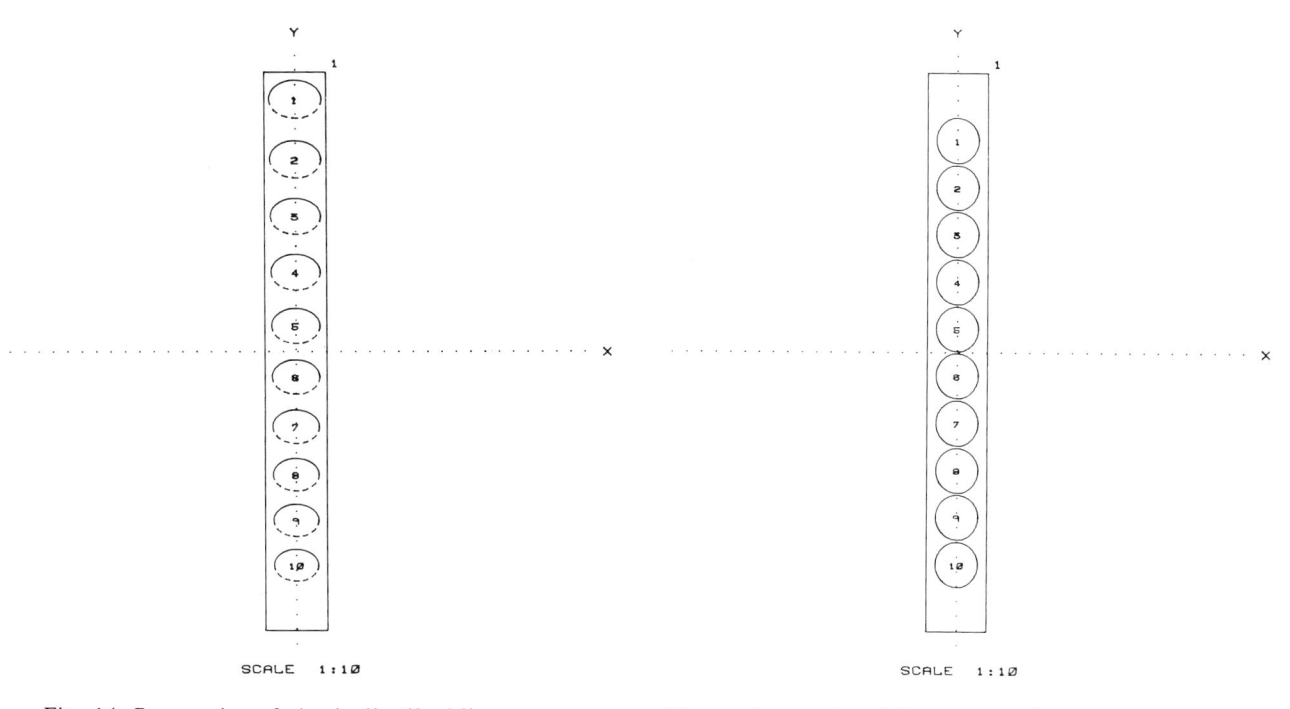

Fig. 14. Perspective of physically tilted line source.

Fig. 16. Perspective of line source with successive time delay.

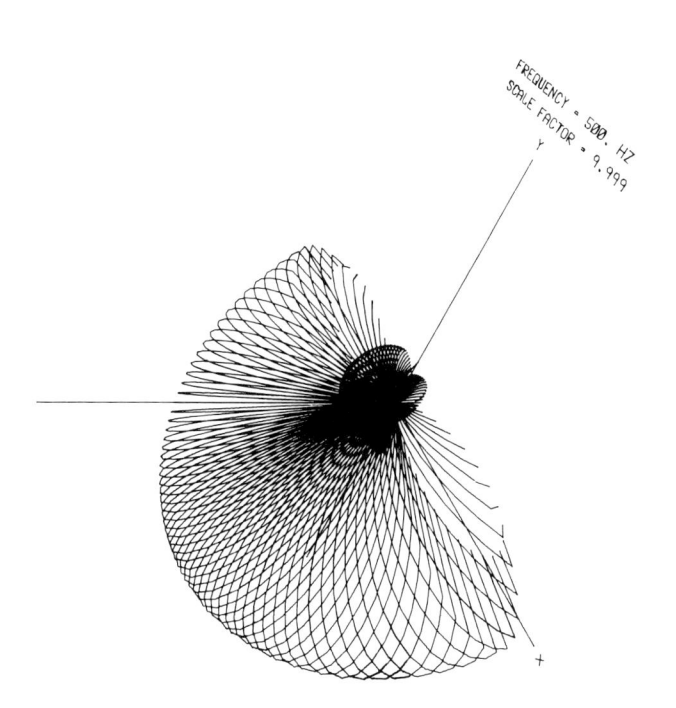

Fig. 15. Directivity balloon for physically tilted line source.

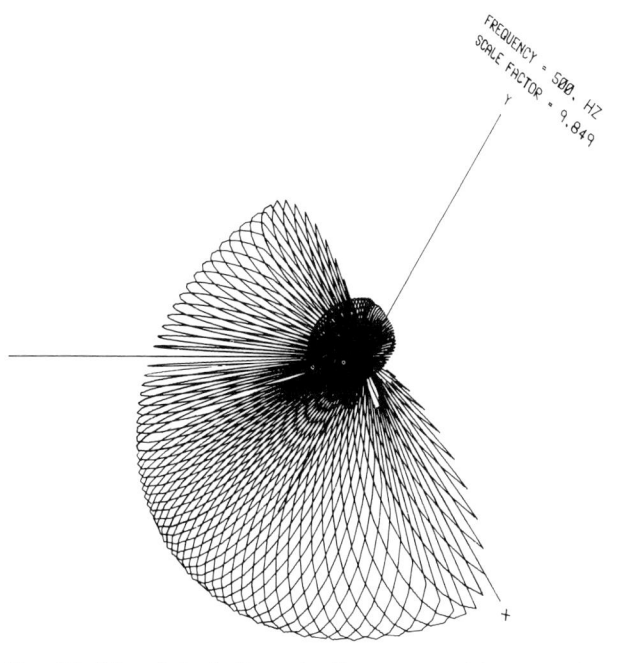

Fig. 17. Directivity balloon for line source with successive time delay.

s), this translates into a maximum spacing between the two sources of about 17 m.

In addition to the echo problem, however, is the problem of the far-field interaction of the signals generated by the two sources. Assuming that both sources are fed the same in-phase signal (that is, the reinforcement system is operated monaurally), the two sources in theory act as a rather large dipole. As discussed in Section 1, the larger the extent of this dipole, the greater in number and amplitude will be the side lobes generated.

As an example of how the simulation system can predict the directivity pattern that will be generated (note that there are few anechoic chambers large enough to accommodate such configurations), we will use a pair of line-source arrays spread varying distances apart: 0.5, 2, and 8 m. All cases analyzed here are for 500 Hz.

A single-line-source perspective and its corresponding directivity slice plots are illustrated in Figs. 18 and 19, respectively. The perspectives and the slices of the dual-line-source cases are provided in Figs. 20–25 for the 0.5-, 2-, and 8-m cases, respectively.

In the 0.5-m case (Figs. 20 and 21) note that the contributions of the two sources coalesce, and the horizontal directivity is sharpened while the vertical di-

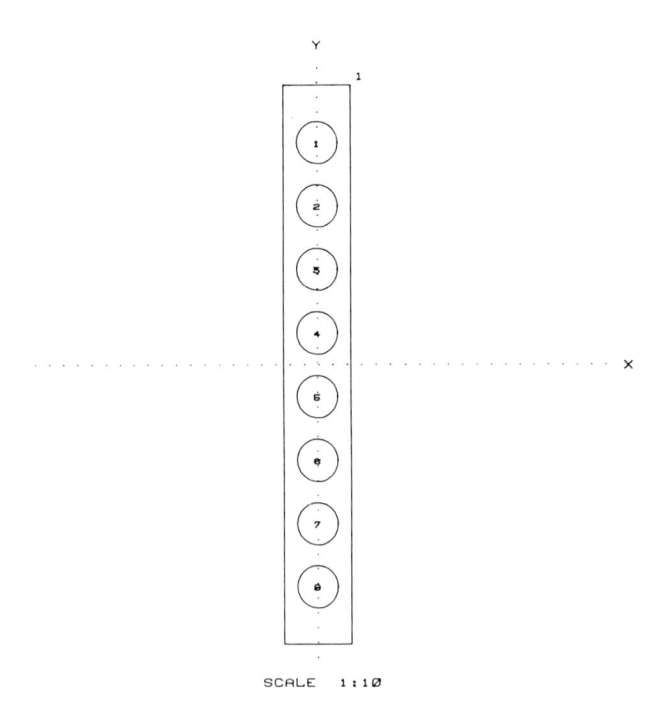

Fig. 18. Perspective of single line source.

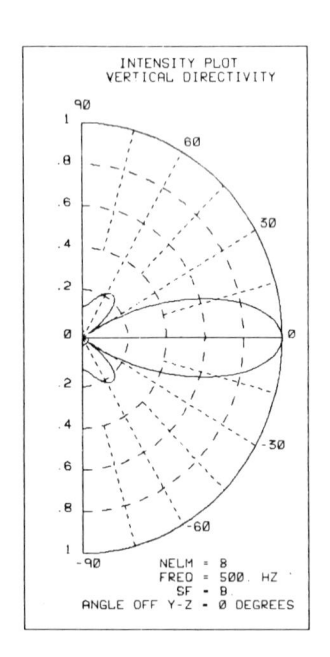

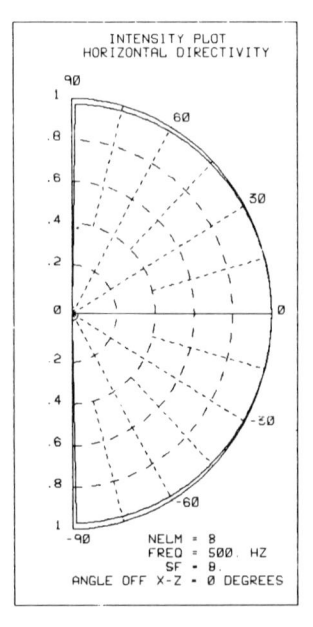

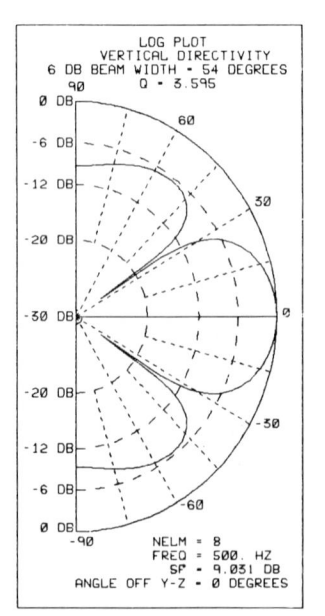

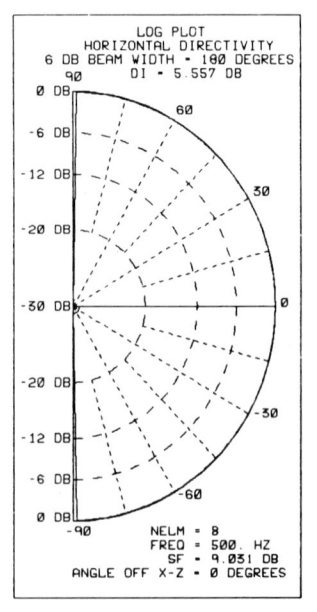

Fig. 19. Directivity slice plots for single line source.

rectivity remains unchanged, as expected.

Since the spacing is greater in the 2-m case (Fig. 22), it follows that there should be more "fingers" as well as sharper fingers in the horizontal pattern, while the vertical pattern should remain the same as in the previous case. This is verified by the directivity slice plots in Fig. 23.

Finally, as might be expected, examination of the 8-m case (Figs. 24 and 25) reveals a further sharpening of the horizontal fingers as well as an increase in the number of these fingers. Here note that the horizontal directivity takes on a noiselike appearance.

An important point to remember, in summary, is that the interference patterns discussed all vary as a function of frequency. This results in a virtually unpredictable pattern of nodes and antinodes throughout the listening area for typical source material. If equalization of the sound reinforcement system is attempted (as is often the case), it is difficult to imagine exactly what segment of the listening area is actually being "equalized." It is of little wonder, then, that this type of reinforcement system sounds so bad to the average off-axis listener.

3.4 Typical Multiway Loudspeaker System

In the final case study we illustrate how the simulation system can be used to avert possible design errors in multiway loudspeaker systems. The case in point il-

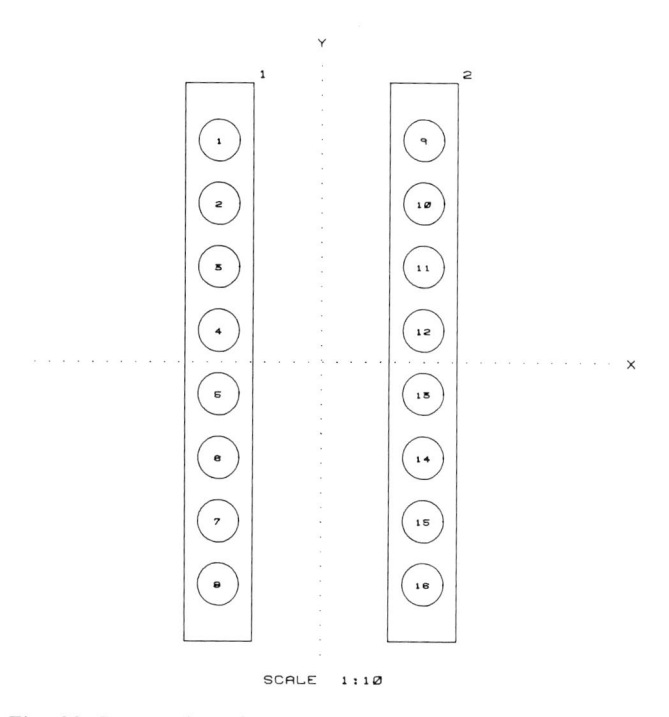

Fig. 20. Perspective of two line sources separated 0.5 m.

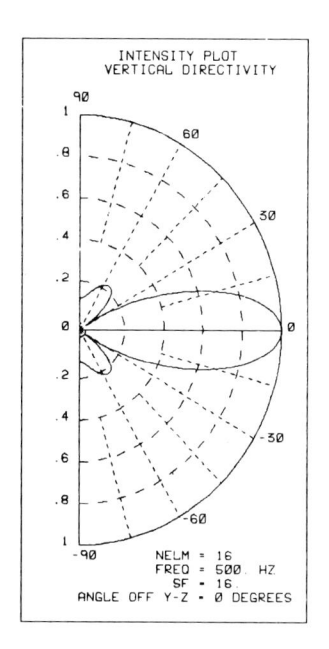

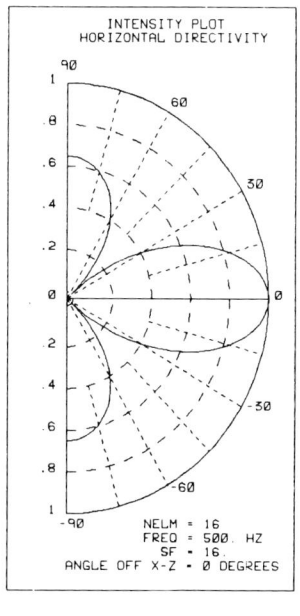

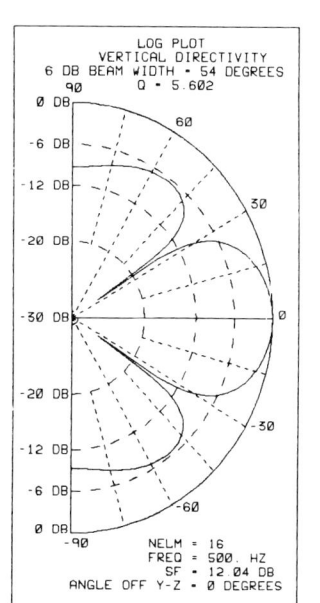

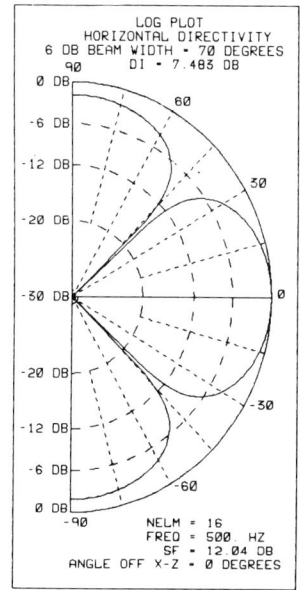

Fig. 21. Directivity slice plots for two line sources separated 0.5 m.

lustrated here is reminiscent of a pair of loudspeakers the author *used* to own. These multiway systems have a pair of 1-in (25-mm) tweeters ("if one is good, then two must be better"), a 5-in (125-mm) midrange, and a 15-in (380-mm) woofer. A perspective of this configuration is illustrated in Fig. 26. Here the crossover points are specified via filter parameters: the −3-dB point for the woofer is 500 Hz, the −3-dB bandwidth for the midrange is 500–4000 Hz, and the −3-dB point for the tweeters is 4000 Hz.

Just from looking at this particular driver arrangement one might guess that the two high-frequency drivers will act as a dipole at higher frequencies. Note that to ensure good coupling among drivers at their upper frequency limit, the spacing between driver centers should be less than or equal to $2a + a/2$, where a is the radius of the driver [10]. Examination of the perspective drawing reveals that this condition is violated, and problems can therefore be expected. As anticipated, the directivity slice family, which appears in Figs. 27–31, verifies this phenomenon. Note the lumpy horizontal pattern which occurs at both 4000 and 8000 Hz.

4 CONCLUSIONS

We derived a mathematical model which allows analysis of three-dimensional loudspeaker system di-

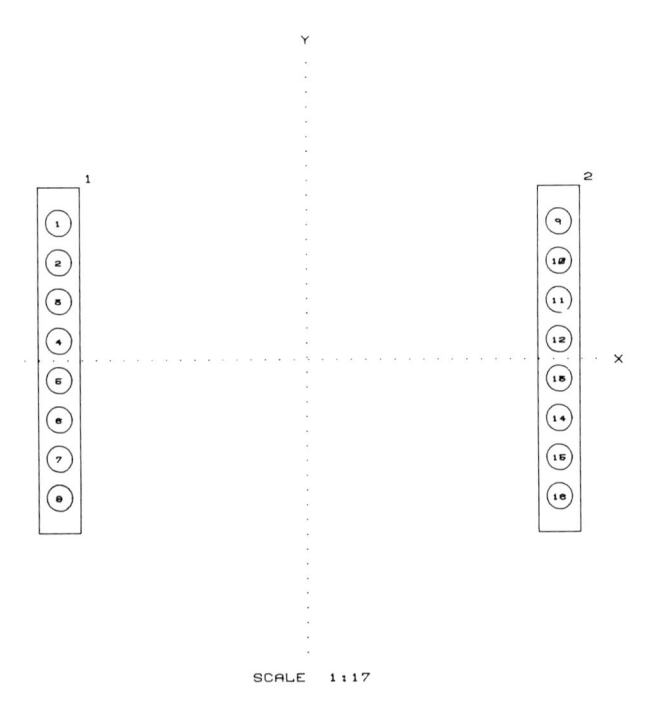

Fig. 22. Perspective of two line sources separated 2 m.

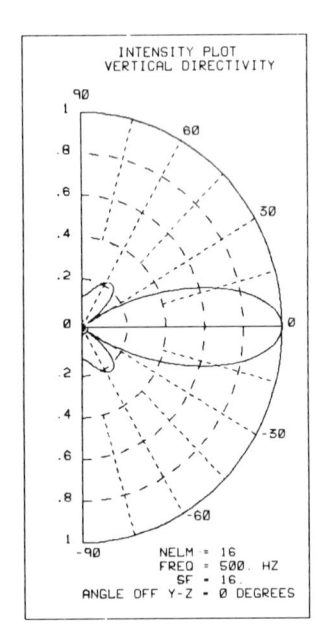

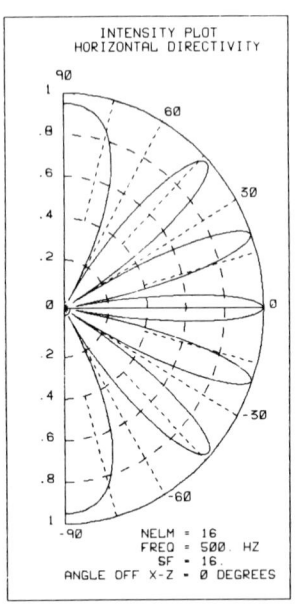

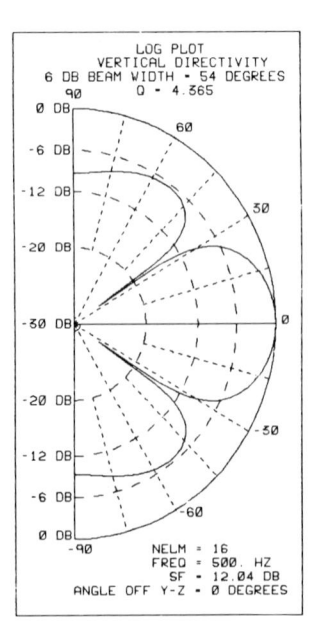

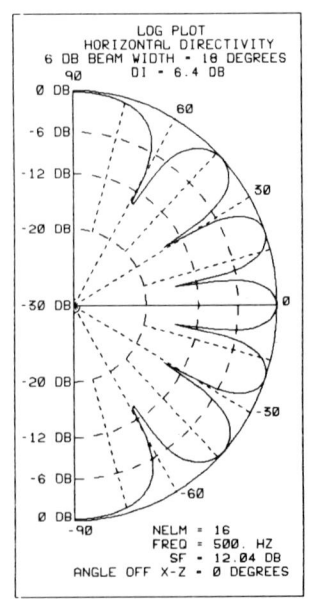

Fig. 23. Directivity slice plots for two line sources separated 2 m.

rectivity dynamics as a function of frequency, described a set of interactive computer programs which utilize this mathematical model to simulate both two- and three-dimensional polar directivity characteristics of arbitrarily configured loudspeaker drivers, and presented a series of case studies to illustrate the effectiveness as well as the utility of our simulation system. The simulation system developed permits the analysis of the directivity characteristics of a loudspeaker system design without actually physically building or measuring it. It is therefore a viable alternative to other methods which have been devised for modeling loudspeaker array coverage [13], [14]. As a synthesis tool, the simulation system can be used to create an arbitrary loudspeaker configuration, analyze the performance of the proposed design, and determine the effects of parameter perturbations as well as analyze the sensitivity to component variations. These capabilities are of particular utility in the design of loudspeaker systems with optimal directional patterns [15]. Finally for sound reinforcement systems, the simulator can be used to design arrays with defined directivity characteristics for each frequency of interest as well as predict the performance of various spatial arrangements of drivers.

5 ACKNOWLEDGMENT

The author would like to thank Professor Frederic Mowle who supervised the thesis work associated with

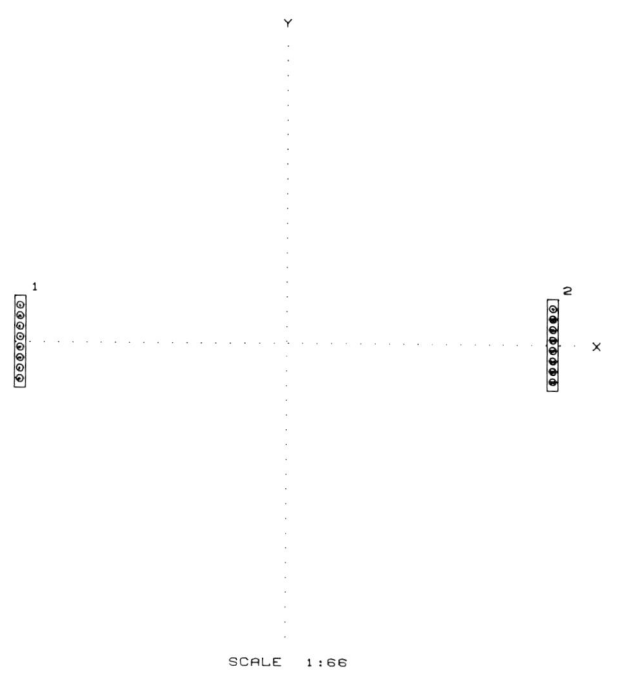

Fig. 24. Perspective of two line sources separated 8 m.

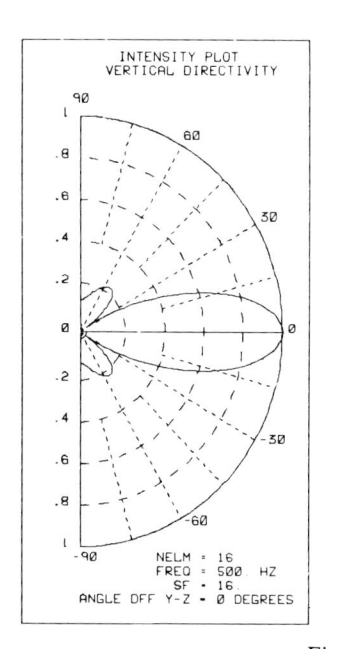

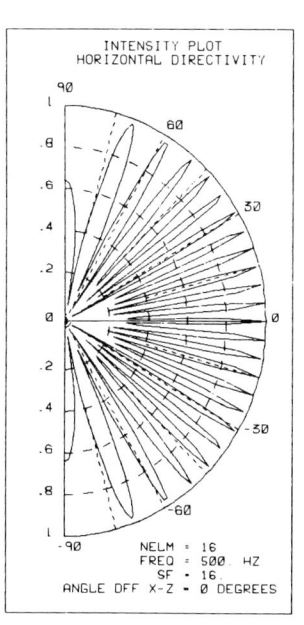

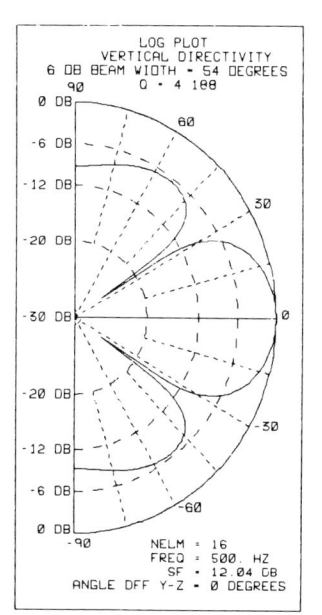

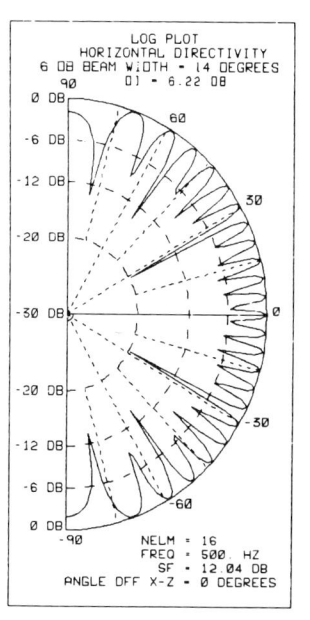

Fig. 25. Directivity slice plots for two line sources separated 8 m.

this project and provided many helpful suggestions toward development of the simulation system. The author would also like to thank Seth Hutchinson who provided many helpful comments and suggestions toward completion of this paper.

6 REFERENCES

[1] I. Wolff and L. Malter, "Directional Radiation of Sound," *J. Acoust. Soc. Am.*, vol. 2, pp. 201–241 (1930 Feb.).

[2] L. Beranek, *Acoustics* (McGraw-Hill, New York, 1954).

[3] L. Kinsler, A Frey, A. Coppens, and A. Sanders, *Fundamentals of Acoustics*, 3rd ed. (Wiley, New York, 1980).

[4] H. Olson, *Acoustical Engineering* (Van Nostrand, Princeton, NJ, 1967).

[5] R. C. Heyser, "Loudspeaker Phase Characteristics and Time Delay Distortion, Parts I and II," *J. Audio Eng. Soc.*, vol. 17, pp. 30–41 (1969 Jan.); pp. 130–137 (1969 Apr.).

[6] R. C. Heyser, "Determination of Loudspeaker Signal Arrival Times, Parts I–III," *J. Audio Eng. Soc.*, vol. 19, pp. 734–743 (1971 Oct.); pp. 829–834 (1971

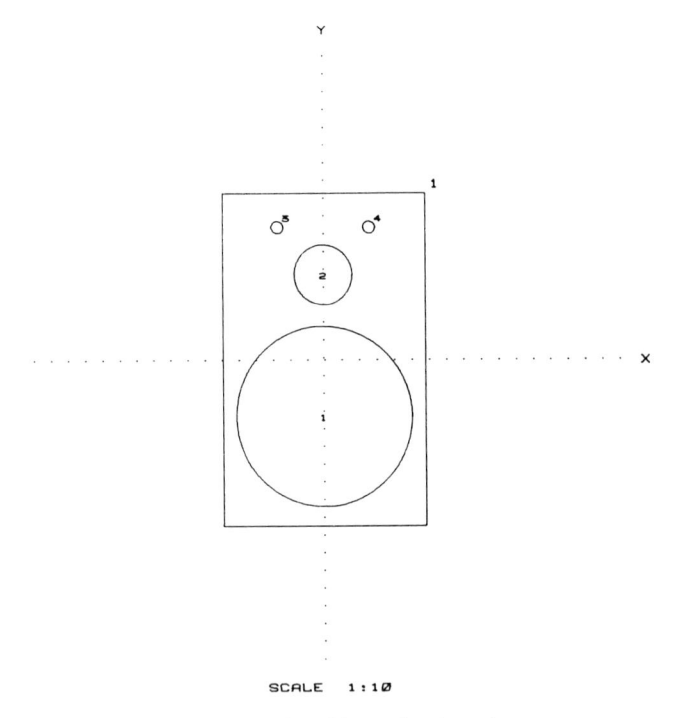

Fig. 26. Perspective of multiway loudspeaker system.

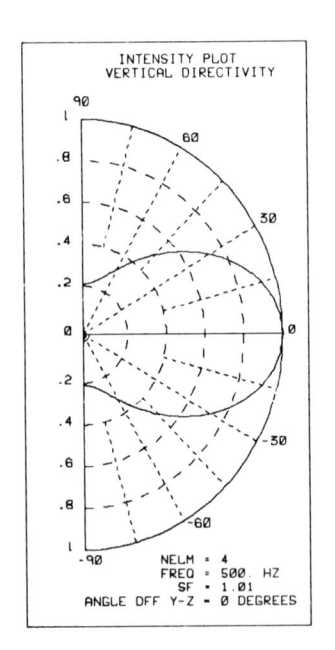

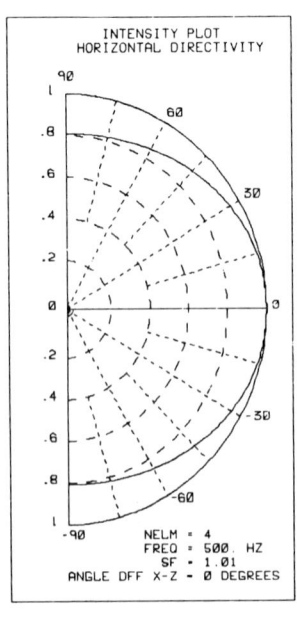

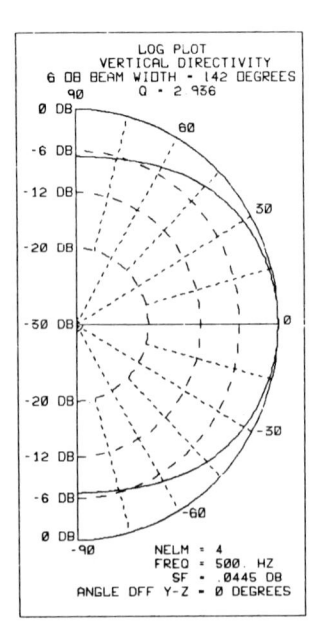

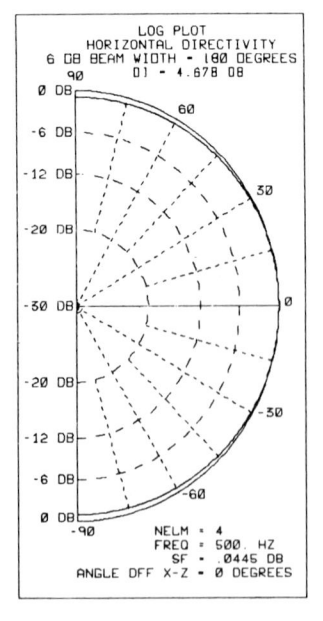

Fig. 27. Directivity slice plots for multiway loudspeaker system: 500 Hz.

Nov.); pp. 902–905 (1971 Dec.).

[7] D. Davis and C. Davis, *Sound System Engineering* (Howard W. Sams, Indianapolis, IN, 1976).

[8] L. Schaudinischky, A. Schwartz, and S. Mashiah, "Sound Columns—Practical Design and Applications," *J. Audio Eng. Soc.*, vol. 19, pp. 36–40 (1971 Jan.).

[9] D. L. Klepper and D. W. Steele, "Constant Directional Characteristics from a Line Source Array," *J. Audio Eng. Soc.*, vol. 11, pp. 198–202 (1963 July).

[10] A. P. Smith, "A Three-Way Columnar Loudspeaker for Reinforcement of the Performing Arts," *J. Audio Eng. Soc.*, vol. 19, pp. 213–219 (1971 Mar.).

[11] D. L. Klepper, "Sound Systems in Reverberant Rooms for Worship," *J. Audio Eng. Soc.*, vol. 18, pp. 391–401 (1970 Aug.).

[12] D. L. Klepper, "Requirements for Theater Sound and Communication Systems," *J. Audio Eng. Soc.*, vol. 20, pp. 642–649 (1972 Oct.).

[13] W. Malmlund and E. Wetherill, "An Optical Aid for Designing Loudspeaker Clusters," *J. Audio Eng. Soc.*, vol. 13, pp. 57–61 (1965 Jan.).

[14] J. R. Prohs and D. E. Harris, "An Accurate and Easily Implemented Method of Modeling Loudspeaker Array Coverage," presented at the 72nd Convention of the Audio Engineering Society, *J. Audio Eng. Soc. (Abstracts)*, vol. 30, p. 942 (1982 Dec.), preprint 1910.

[15] J. M. Kates, "Optimum Loudspeaker Directional Patterns," *J. Audio Eng. Soc.*, vol. 28, pp. 787–794 (1980 Nov.).

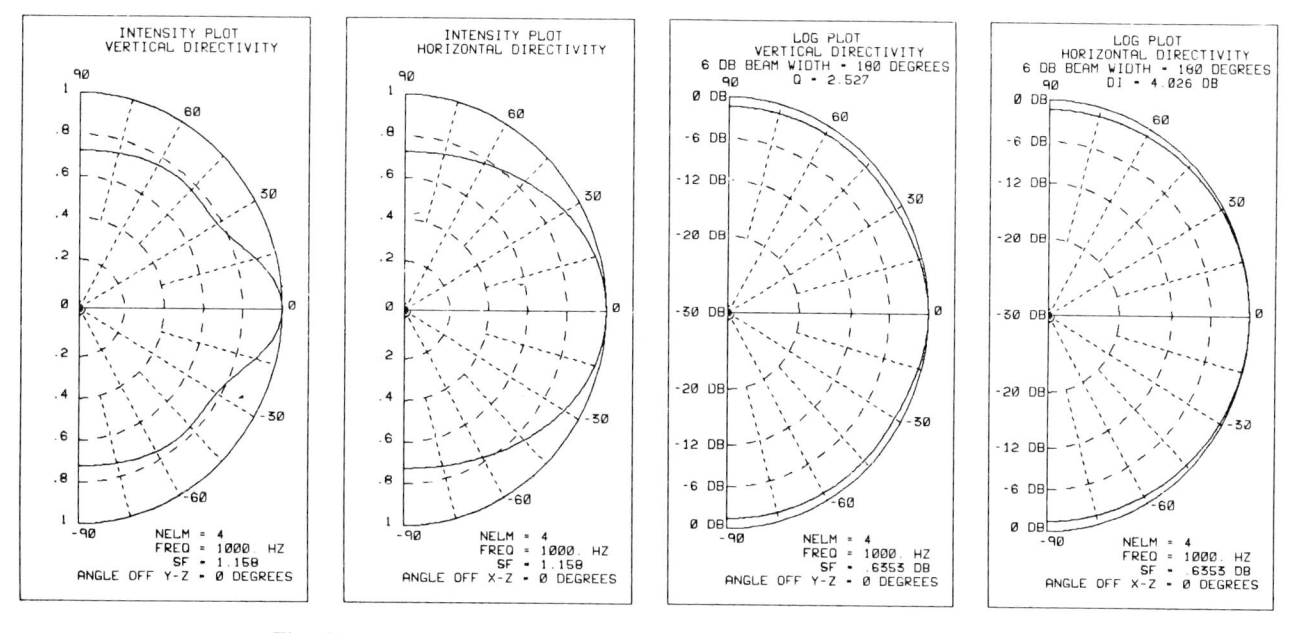

Fig. 28. Directivity slice plots for multiway loudspeaker system: 1 kHz.

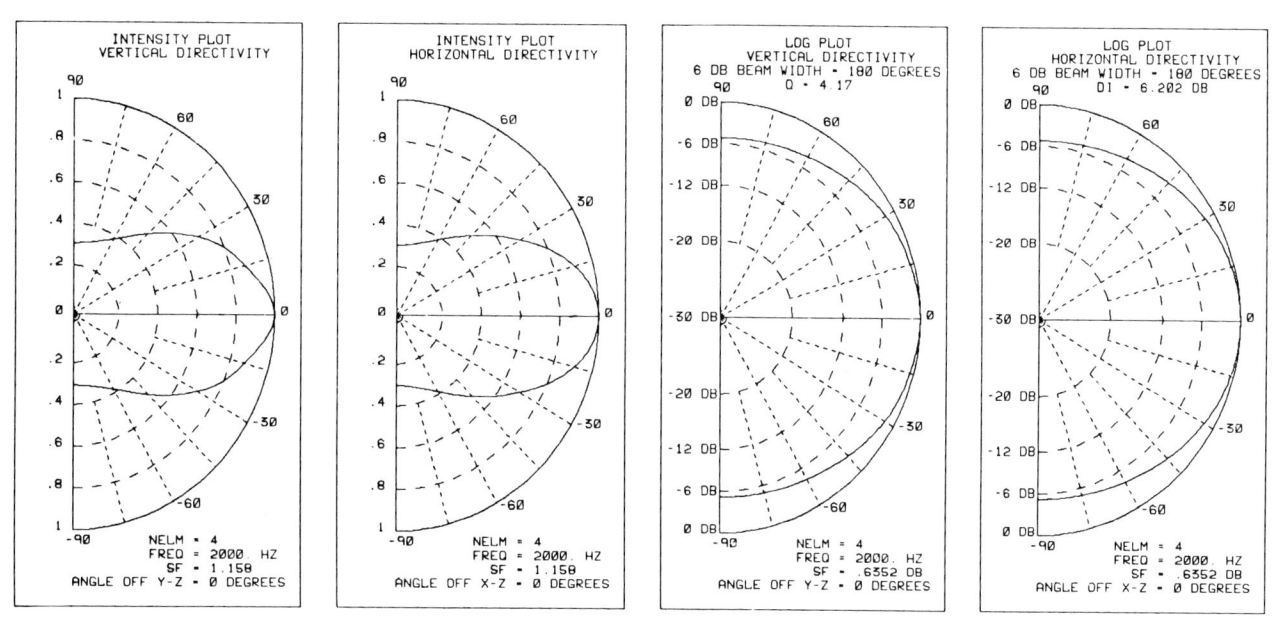

Fig. 29. Directivity slice plots for multiway loudspeaker system: 2 kHz.

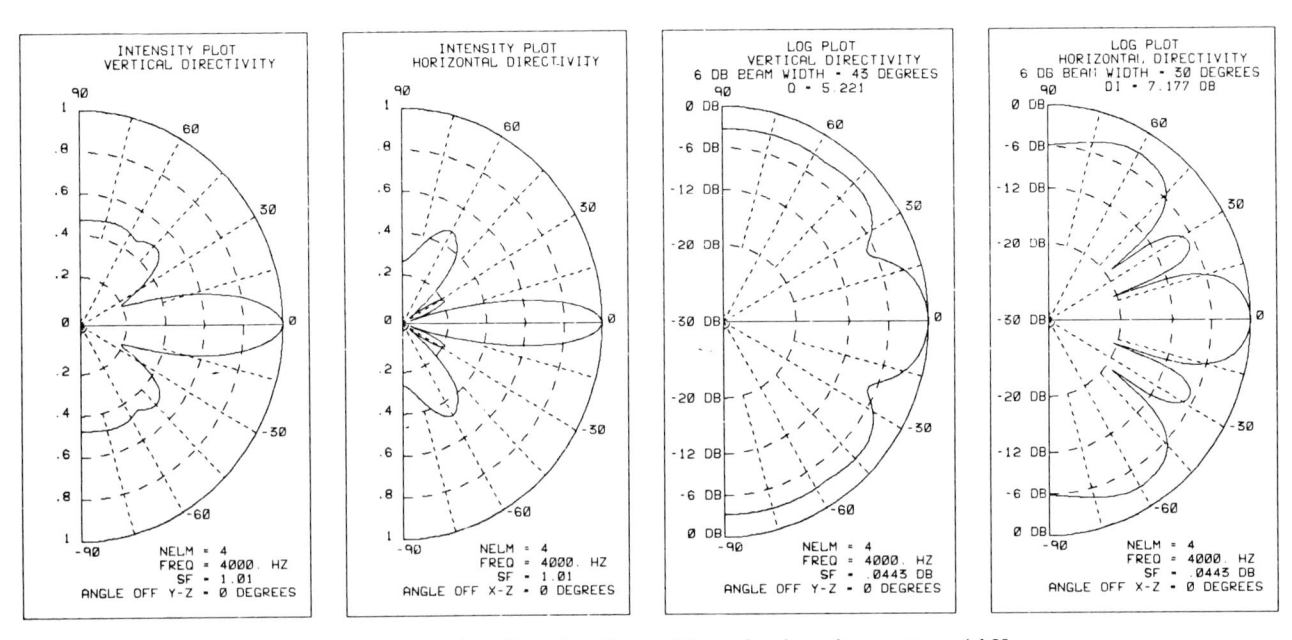

Fig. 30. Directivity slice plots for multiway loudspeaker system: 4 kHz.

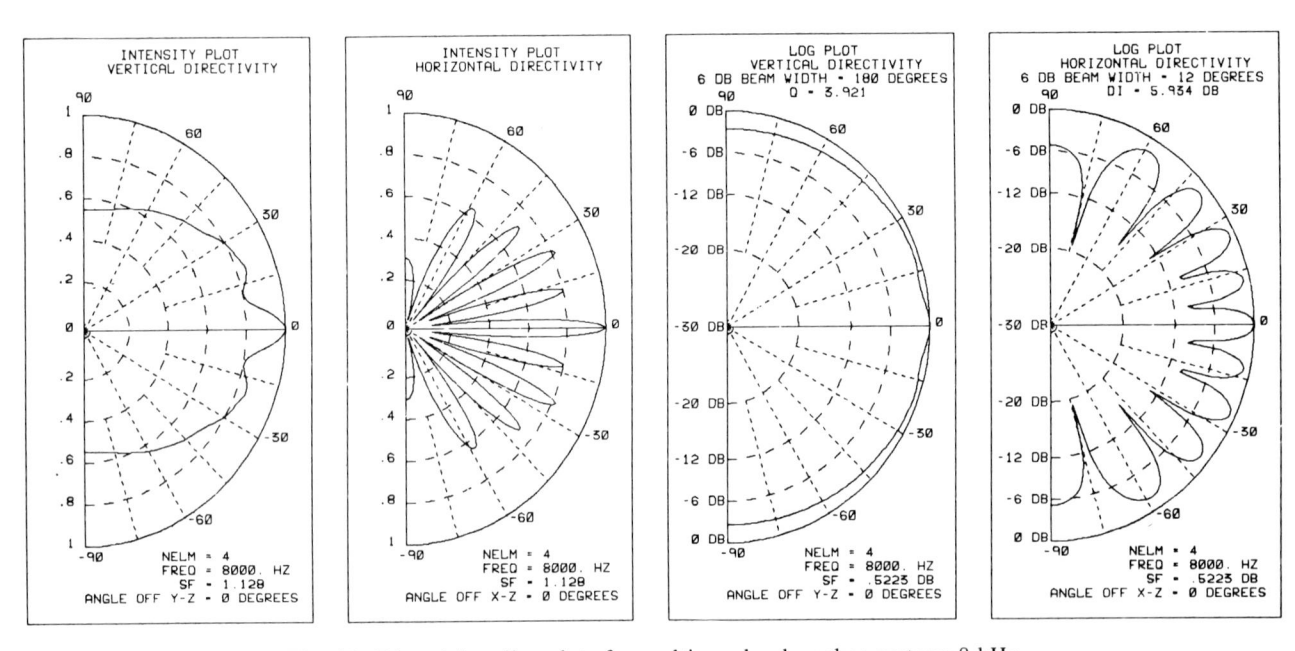

Fig. 31. Directivity slice plots for multiway loudspeaker system: 8 kHz.

THE AUTHOR

David Meyer was born in Indianapolis, Indiana, in 1951. He received a B.S.E.E. degree in 1973, an M.S.E. degree in electrical/biomedical engineering in 1975, an M.S. degree in computer science in 1979, and a Ph.D. in electrical engineering in 1981, all from Purdue University, West Lafayette, Indiana. During his stay at Purdue he served as both a research assistant and a teaching assistant, authoring several technical reports as well as developing instructional materials and media aids for several different courses. He is currently an assistant professor of electrical engineering at Purdue in the computer engineering area, where his primary responsibilities include teaching courses on digital logic design, digital hardware engineering, and microprocessors.

He is also actively pursuing research in digital signal processing and parallel computer architectures along with loudspeaker system modeling and directivity simulation. He is a member of the Audio Engineering Society, the IEEE, the ASEE, and the ACM.

ENGINEERING REPORTS

Digital Control of Loudspeaker Array Directivity*

DAVID G. MEYER

Purdue University, School of Electrical Engineering, West Lafayette, IN 47907, USA

In an earlier paper a mathematical model was described which allows computer simulation of the directivity characteristics of sound reinforcement loudspeaker arrays constructed using "real" drivers, that is, drivers possessing defined amplitude, group delay, and horizontal/vertical polar directivity characteristics. In the present paper a loudspeaker array implementation is presented which allows digital control of the parameters suggested by the mathematical model. Simulation study results illustrate the effects of the various control parameters to which this implementation allows access. Plans for a prototype digitally controlled loudspeaker array are outlined.

0 INTRODUCTION

A mathematical model and a computer simulation system for analyzing the directivity characteristics of arbitrarily configured loudspeakers were described previously [1, 2]. One of the salient features of this simulation system is the incorporation of amplitude, group delay, and horizontal/vertical polar directivity characteristics typical of "real" drivers. The directivity function used in the computer simulation package is as follows:

$$R(r_o, \theta_o, \psi_o, f)$$

$$= \frac{1}{M} \left\{ \left[\sum_{i=1}^{n} \frac{g_i a_i(f) A_i(f) W_i^k(V_i(\theta_i', f), H_i(\theta_i', f), \psi_i')}{r_i} \right. \right.$$

$$\left. \times \cos\left(\frac{2\pi f r_i''}{c}\right) \right]^2$$

$$+ \left[\sum_{i=1}^{n} \frac{g_i a_i(f) A_i(f) W_i^k(V_i(\theta_i', f), H_i(\theta_i', f), \psi_i')}{r_i} \right.$$

$$\left. \left. \times \sin\left(\frac{2\pi f r_i''}{c}\right) \right]^2 \right\}^{1/2} \quad (1)$$

where

---

* Presented at the 74th Convention of the Audio Engineering Society, New York, 1983 October 8–12.

(x_o, y_o, z_o) = coordinates of observation point, obtained using

$= x_o = r_o \sin\theta_o \sin\psi_o$ (2a)

$= y_o = r_o \sin\theta_o \cos\psi_o$ (2b)

$= z_o = r_o \cos\theta_o$ (2c)

(x_i, y_i, z_i) = coordinates of source i

r_i = pathlength from source to observation point, defined by

$r_i = [(x_o - x_i)^2 + (y_o - y_i)^2 + (z_o - z_i)^2]^{1/2}$ (3)

r_i'' = phase-modified pathlength, defined by

$r_i'' = [(x_o - x_i)^2 + (y_o - y_i)^2 + (z_o - z_i - c[d_i(f) + D_i(f)])^2]^{1/2}$ (4)

W_i^k = wavefront directivity function, defined by

$W_i^o = [[\sin\psi_i V_i(\theta_i, f)]^2 + [\cos\psi_i H_i(\theta_i(\theta_i, f)]^2]^{1/2}$ (5a)

for circular and elliptical wavefront

apertures and

$$W_i^1 = \begin{cases} [[\sin \psi_i V_i(\theta_i, f)]^2 + [H_i(\theta_i, f)]^2]^{1/2}, & 0 \leq \psi_i \leq \gamma_i \\ [[V_i(\theta_i, f)] + [\cos \psi_i H_i(\theta_i, f)]^2]^{1/2}, & \gamma_i < \psi_i \leq \pi/2 \end{cases} \quad (5b)$$

for square and rectangular wavefront apertures

M = maximum pressure intensity of which a given radiator configuration is capable, defined by

$$M = |\bar{P}|_{\max} \quad \text{for all } \theta_o, \psi_o \quad (6)$$

$d_i(f)$ = time delay characteristic of signal applied to source i
$a_i(f)$ = normalized signal filter characteristic applied to source i
g_i = gain of composite signal applied to source i
$A_i(f)$ = amplitude response of source i
$D_i(f)$ = group delay characteristic of source i
$V_i(\theta, f)$ = normalized vertical polar response of source i
$H_i(\theta, f)$ = normalized horizontal polar response of source i
θ_o = angle of observation with respect to z axis
ψ_o = angle of rotation about z axis of observation point
θ_i = angle with respect to central axis of source i at which it is observed
ψ_i = angle of rotation about central axis of source i at which it is observed
$\Delta\theta_i$ = mounting skew with respect to central axis of source i
$\Delta\psi_i$ = mounting skew of rotation about central axis of source i
$\Delta\alpha_i$ = mounting skew of source i with respect to x–z plane (alternate representation)
$\Delta\beta_i$ = mounting skew of source i with respect to y–z plane (alternate representation)
f = frequency of applied signal
c = speed of sound at ambient conditions (typically 341 m/s).

In this paper we first present some simulation study results to help illustrate the effect of the control parameters suggested by the mathematical model. We then present one possible implementation which allows digital control of the amplitude characteristic, group delay characteristic, and composite gain of the signal applied to each array element.

1 ILLUSTRATION OF CONTROL MECHANISMS

While a fair amount of study has been done toward analytically determining the directivity of arrays consisting of point sources [3]–[6], the closed-form solutions obtained do not incorporate the characteristics typical of "real" loudspeakers: amplitude response, group delay response, and variation of polar directivity as a function of frequency. The mathematical model and the associated computer simulation system we have developed incorporate these characteristics, however.

Before examining a possible implementation of a loudspeaker control mechanism, it is instructive to analyze the effects of the various control parameters suggested by the mathematical model. As an illustrative example for which there is a known analytical solution, we will examine a rectangular array of drivers operated at a frequency low enough such that each of the individual sources approximates the characteristics of a point source. In particular, we will examine a 5 × 9 (that is, 45-element) array of 4-in (0.1-m) drivers operated at 500 Hz. A perspective of this configuration is illustrated in Fig. 1, while a directivity balloon and "slice" plots appear in Figs. 2 and 3, respectively. Here each element is operated at equal intensity, with constant (or flat) group delay. Note that the results obtained closely agree with analytical solutions.

Several authors have dealt in detail with one of the major problems with such an array, namely, its beaming effect at high frequencies, that is, the very narrow 6-dB beam width obtained with increasing frequency [7]–[9]. One technique which can be used to compensate for this effect is that of tapering, that is, spatially mod-

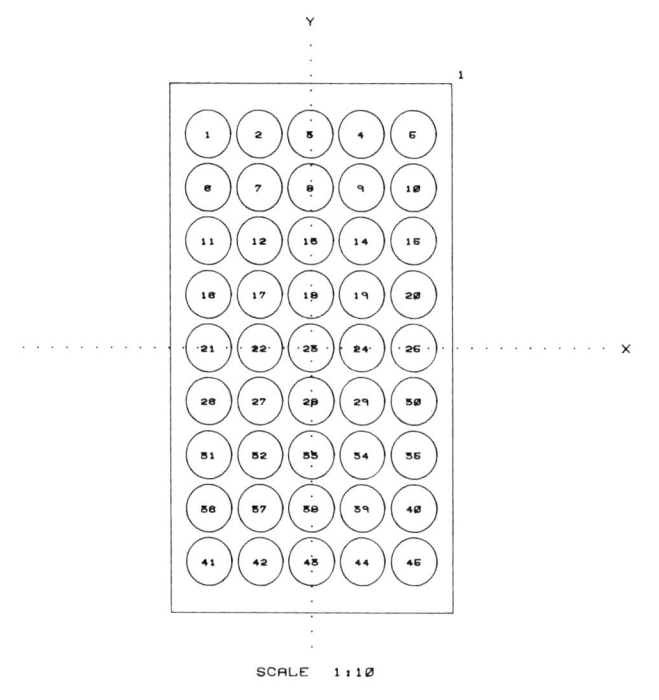

Fig. 1. Perspective of 5 × 9 array.

ifying the signal applied to each element and thereby changing the effective length of the radiator. The control mechanism of interest is the amplitude characteristic a_i of the signal applied to each array element. The effect of tapering, illustrated by our computer simulation system, appears in Figs. 4–7. In Figs. 4 and 5 the effect of broad-band (or "composite gain") tapering is illustrated. The linear tapering utilized here is a spatial function of the coordinates of each loudspeaker:

$$g_i = 2\left(1 - \frac{2|x_i|}{\text{width}}\right)\left(1 - \frac{2|y_i|}{\text{height}}\right). \quad (7)$$

Note the substantial difference in 6-dB beam width compared with that of the reference configuration.

Tapering can also be a function of frequency, in which case it is referred to as frequency tapering. Figs. 6 and 7 illustrate an example of frequency tapering. Here the amplitude characteristic a_i at frequency band B_n for each element is given by

$$a_i(B_n) = -\frac{2(n + 1)(100x_i^2 + 100y_i^2)^{1/2}}{10},$$

$$0 \leq n \leq \text{bands} \quad (8)$$

where a_i is in decibels, referenced to 0 dB. The intent is to vary the amount of tapering as a function of fre-

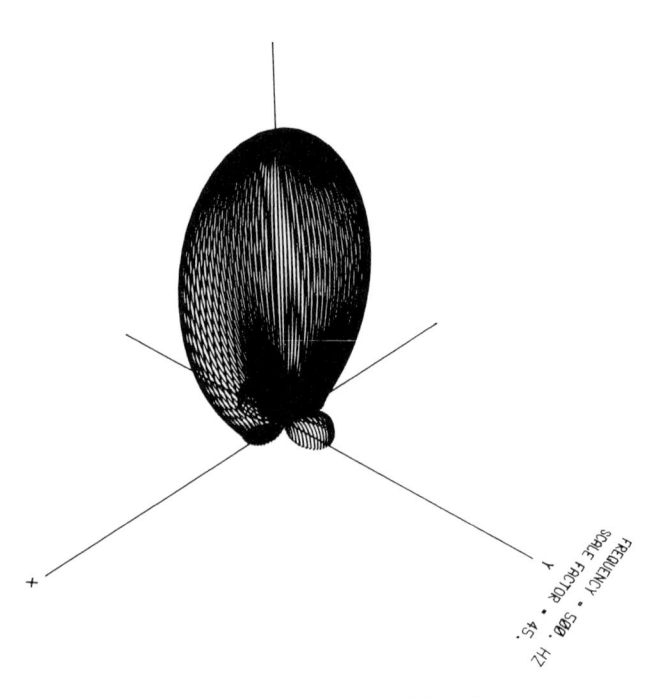

Fig. 2. Directivity balloon of 5 × 9 array.

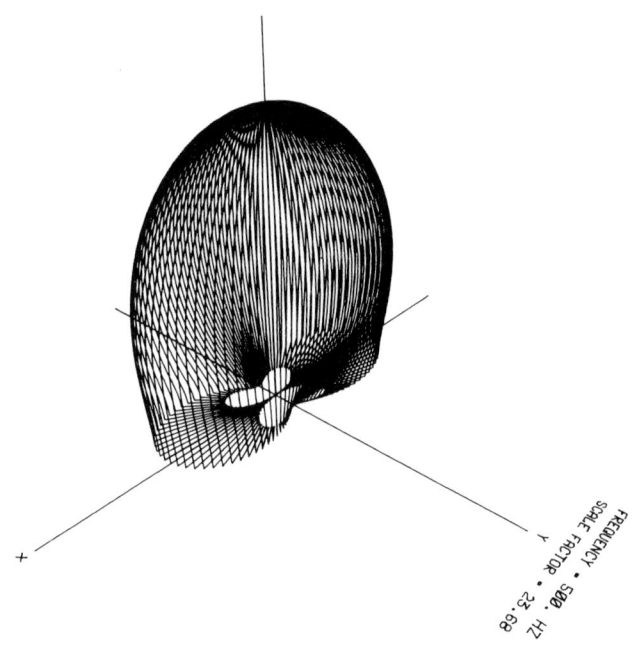

Fig. 4. Directivity balloon of 5 × 9 array with linear tapering.

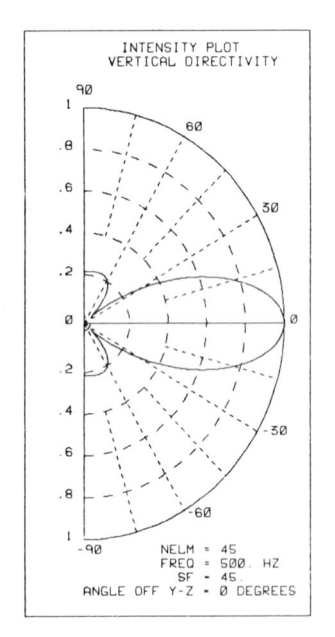

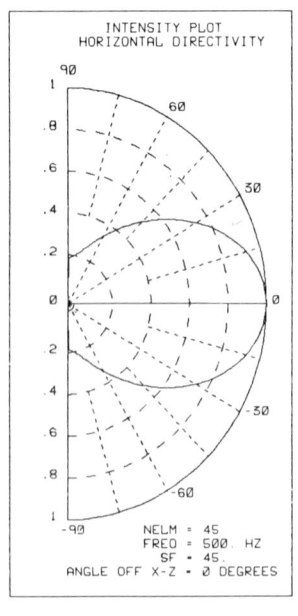

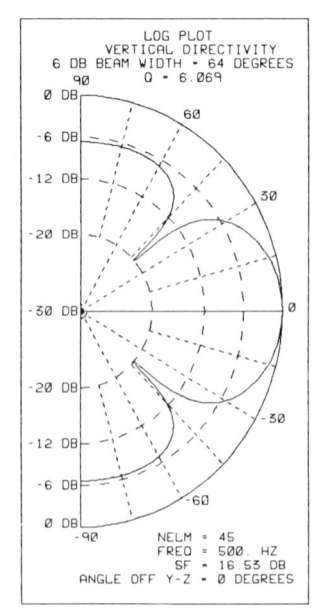

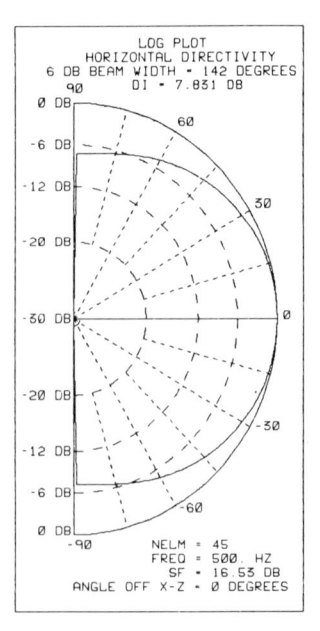

Fig. 3. Directivity slice plots of 5 × 9 array.

quency and thereby help alleviate the beaming effect, that is, help make directivity independent of frequency.

Another control mechanism suggested by the mathematical model is related to the group delay characteristics of the signal applied to each array element. Several possibilities are afforded via group delay control. One might be to obtain a flat (or constant) group delay characteristic for the array as a whole. Another might be to compensate for the naturally occurring element-by-element variation in group delay response. Yet another might be to emulate structural curvature or physical tilting by spatially applying successive amounts of group delay to each array element. This latter case is illustrated in Figs. 8 and 9. Here the rather arbitrary spatial group delay function utilized is

$$d_i(B_n) = 0.05(n + 1)(100x_i^2 + 100y_i^2)^{1/2},$$
$$0 \leq n \leq \text{bands} \qquad (9)$$

where d_i is in milliseconds. In effect, this emulates the effect of mounting the individual drivers on a convex surface; hence the increased beam width compared with the reference configuration. Many of the possibilities afforded by this particular control mechanism are as of yet unexplored.

Some rather interesting results are obtained if all the above-mentioned control mechanisms are combined.

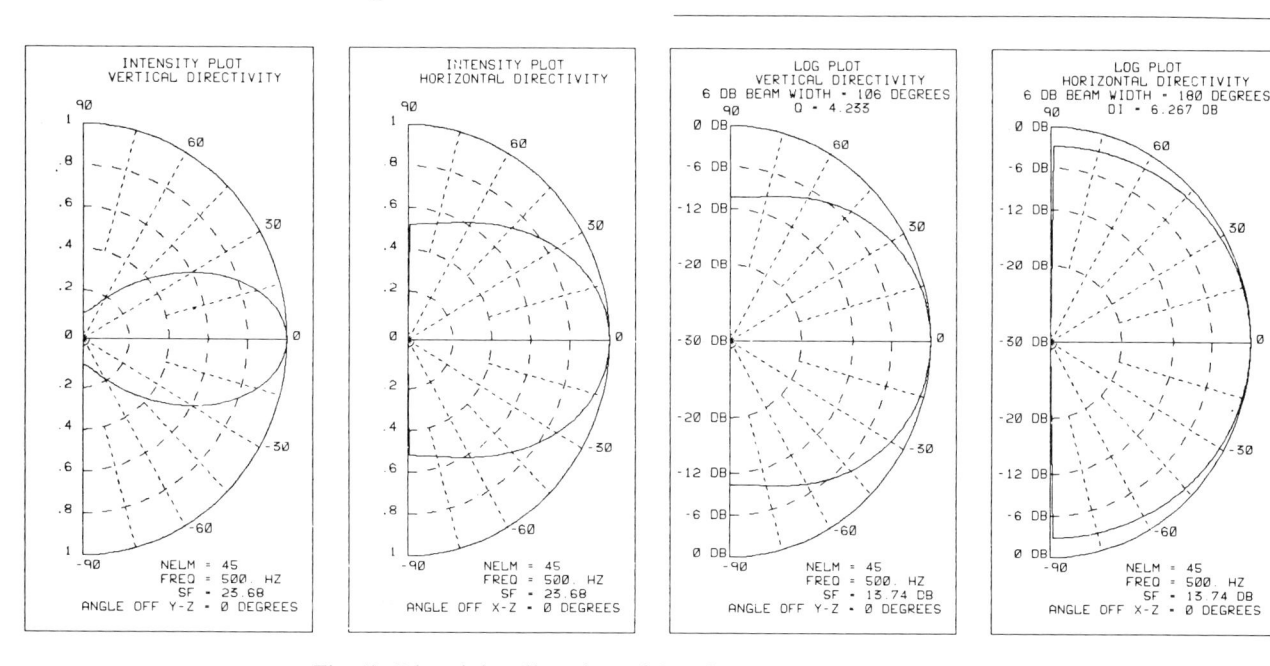

Fig. 5. Directivity slice plots of 5 × 9 array with linear tapering.

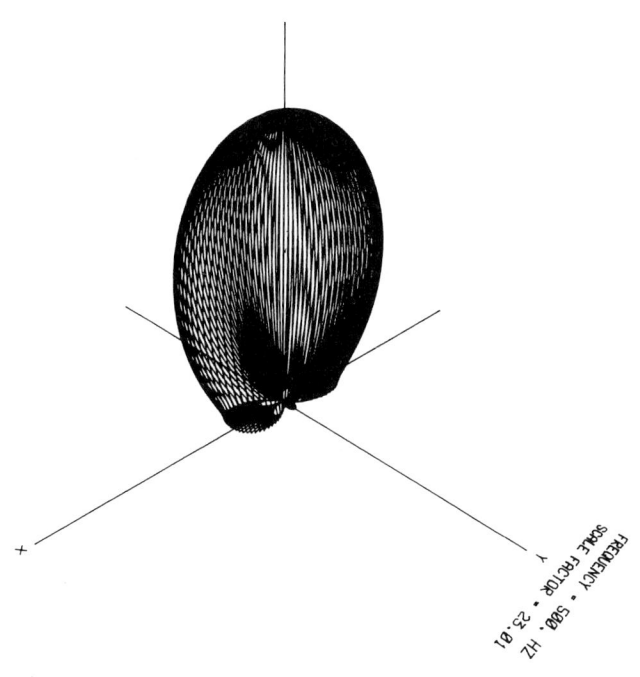

Fig. 6. Directivity balloon of 5 × 9 array with frequency tapering.

The directivity balloon and slices plots for this case are illustrated in Figs. 10 and 11, respectively. Superposition of the "broadening" effects appears to apply.

2 IMPLEMENTATION OF CONTROL MECHANISMS

As discussed previously, the mathematical model suggests three primary control mechanisms: varying the amplitude of the elemental signal as a function of frequency, varying the group delay of the elemental signal as a function of frequency, and varying the gain of the composite elemental signal. In the previous section we have briefly illustrated the effects of these control parameters. In this section we would like to discuss a possible implementation which allows access to these control parameters.

One possible implementation of these control mechanisms is illustrated in Fig. 12. Here in order to obtain amplitude and group delay control as a function of frequency, the audio input signal is divided into n bands, $B_0, \ldots, B_{n-1}$, by passing the signal through a bank of parametric bandpass filters. (Note that the center frequencies of these bands may be arbitrarily chosen; typically these would be standard ISO octaves or one-third octaves. Also note that if fixed octave or one-third-octave bands are used, the bandpass filters need not be parametric.) Each of these n frequency bands is routed through a tapped delay line, the outputs of which are bused to the loudspeaker array. (Note that these delay lines may be of the CCD or digital variety.) Thus there are n bands of audio available at each loudspeaker element, each with selectable group delay.

The function of the signal path circuitry at each element, then, is to select a specified group delay for each frequency band, provide a specified amount of gain for each band, sum the contributions from each band into a composite signal, and finally amplify this signal to an appropriate level for the loudspeaker element. As illustrated in Fig. 12, the group delay selection for each band is accomplished using an analog multiplexer connected to each delay bus. The contribution of each of the n frequency bands to the overall signal applied to a given array element (that is, the spectral content of the signal) is dictated by individual voltage-controlled amplifiers. These gain-controlled bands are combined into a composite signal using a conventional summer circuit. A final voltage-controlled amplifier is used to adjust the gain of the composite signal before it is applied to the power amplifier module and loudspeaker element.

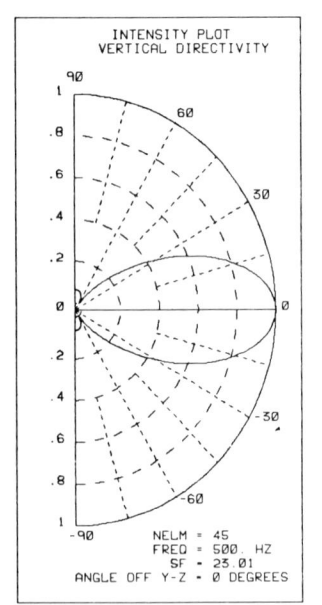

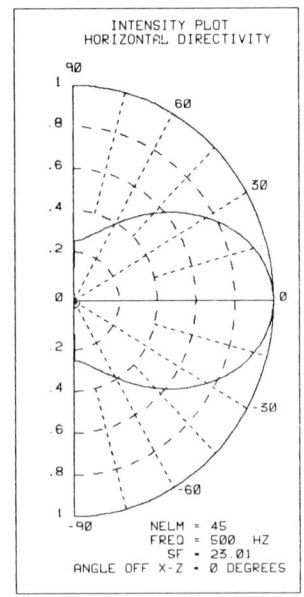

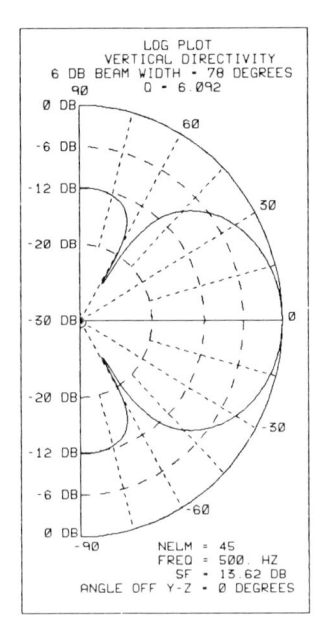

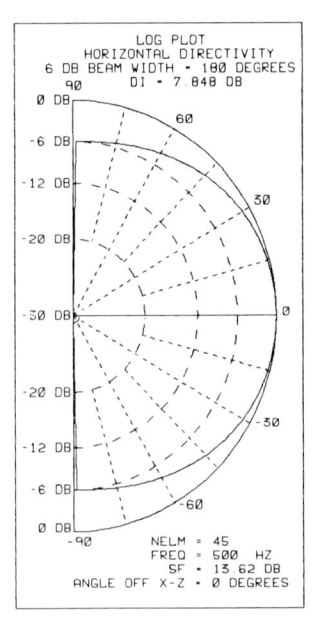

Fig. 7. Directivity slice plots of 5 × 9 array with frequency tapering.

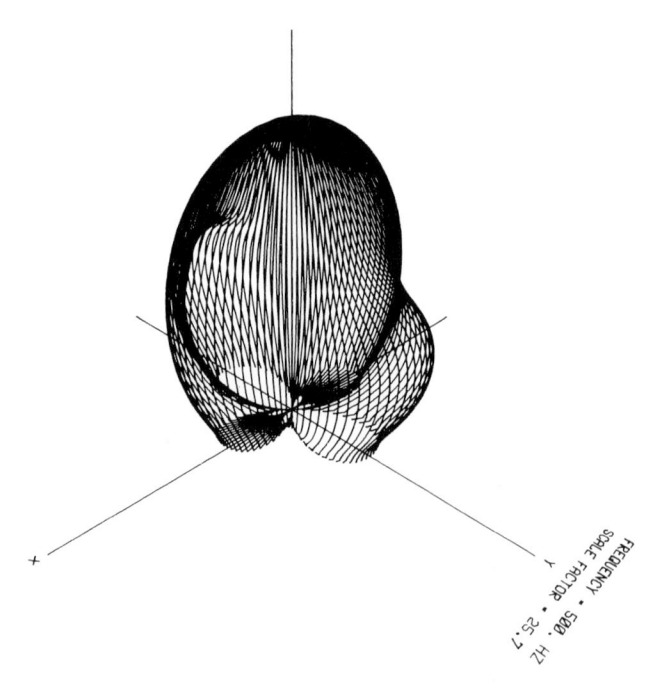

Fig. 8. Directivity balloon of 5 × 9 array with group delay tapering.

Next we must consider how the control signals for each element i which specify group delay $\{d_i(B_0) \ldots d_i(B_{n-1})\}$, spectral content $\{a_i(B_0) \ldots a_i(B_{n-1})\}$, and composite gain $\{g_i\}$ are passed to the signal path circuitry. This can be accomplished using the element control subsystem illustrated in Fig. 13. Here it is assumed that tables of gain and group delay values corresponding to a given array configuration are stored in a shared memory. These array element control parameters are loaded by the supervisor processor from a mass storage device or generated locally by the supervisor processor through execution of a control algorithm (Fig. 14).

However generated, these values are accessed from the shared memory in a periodic fashion by the element control processor. This "slave" processor is assumed to have a limited amount of scratchpad RAM, while program execution occurs out of a control ROM. Also accessed are control status packets. These are transferred to and from the supervisor processor via fixed locations in the shared memory. Using three separate buses, the element control processor routes gain and group delay parameters to the loudspeaker elements through an array element bus interface. At each element, control information is extracted from the buses using an element select decoder (to determine if that element is being addressed), a device select decoder (to indicate which parameter is being sent), and data latches for each control parameter. Note that $\{d_i(B_0) \ldots d_i(B_{n-1})\}$ are digital control signals (connected to the analog multiplexer select lines) while $\{a_i(B_0) \ldots a_i(B_{n-1})\}$ and $\{g_i\}$ are analog control signals (connected to the voltage-controlled amplifiers); hence the need for digital-to-analog converters on these latter control signals. Thus for an implementation with n frequency bands, $2n + 1$ control signals are derived at each array element.

The path from tables of gain and group delay control parameters stored in memory established, we must finally consider the generation of these parameters. Ideally we would like to specify a given arrangement of drivers (either homogeneous or nonhomogeneous), each having known amplitude, group delay, and horizontal/vertical directivity characteristics [denoted by $A_i(f)$, $D_i(f)$, $H_i(\theta, f)$, and $V_i(\psi, f)$, respectively], and be able to generate a specified directivity pattern which can be maintained over a relatively wide band of frequencies. We might also want to perform beam tilting and steering operations or other special effects.

One solution is to use a medium-size mainframe computer to simulate typical "useful" array configurations and calculate corresponding tables of control parameters for each of these configurations. These val-

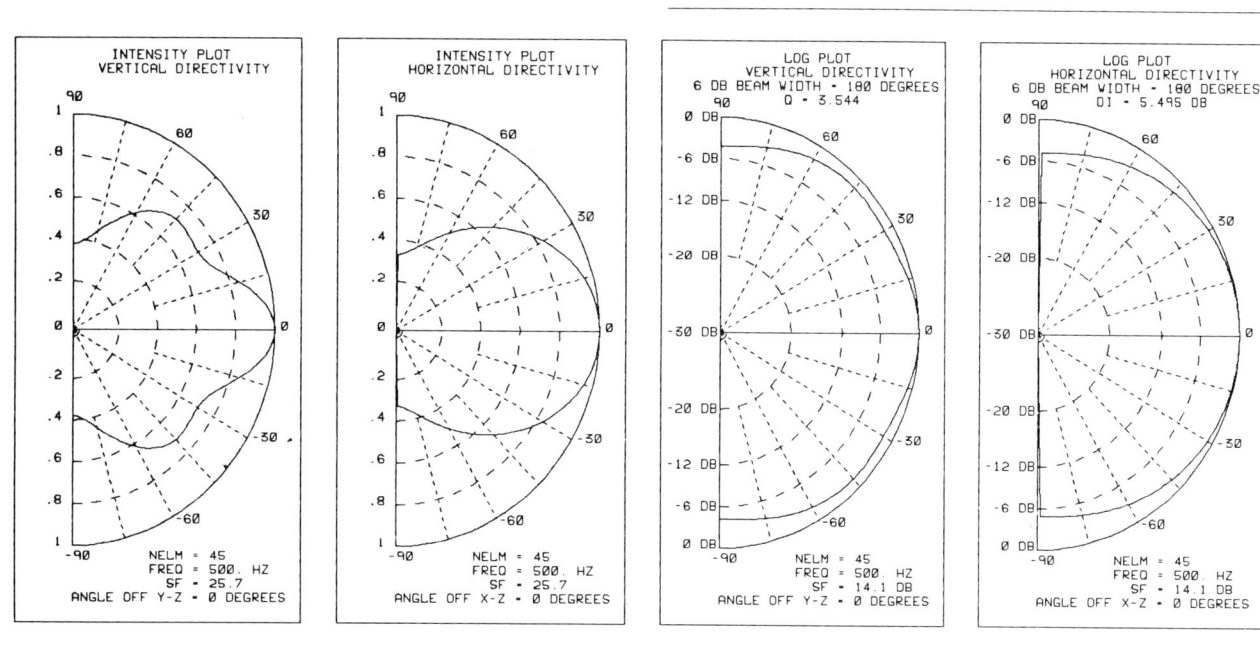

Fig. 9. Directivity slice plots of 5 × 9 array with group delay tapering.

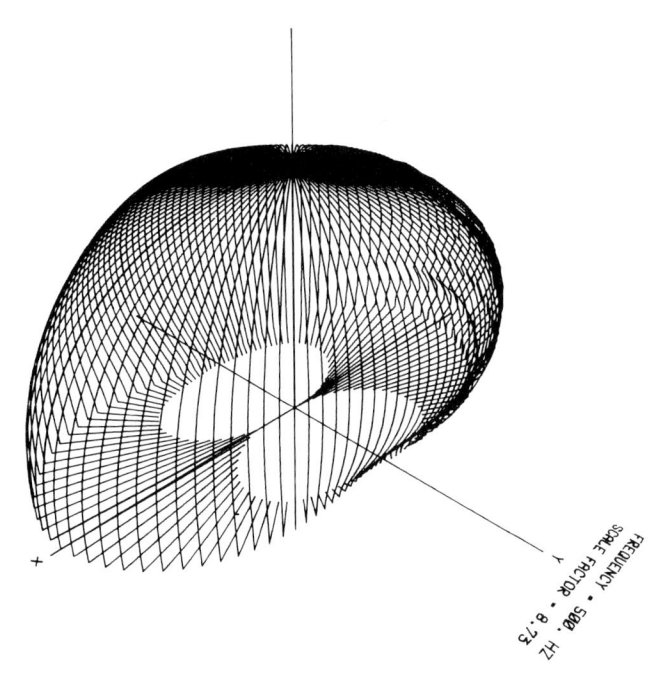

Fig. 10. Directivity balloon of 5 × 9 array with all controls.

ues could be stored on a mass storage device (say, floppy disk) and subsequently loaded via the supervisor processor, as illustrated in Fig. 14. These "canned" array configurations could then be interactively pulled from the mass storage device and sent to the element control subsystem through the shared memory.

Another solution is to generate the element control parameters locally. With the trend toward increasing processing power of microprocessor systems, this is a viable solution despite the amount of "number crunching" required. This solution would greatly facilitate in situ adjustments and interactive fine-tuning of loudspeaker array performance. Here a series of calibrated microphones placed at strategic locations throughout the listening area would allow the supervisor to monitor the directivity tracking performance of the loudspeaker array (and possibly correct for error) as well as automatically equalize the loudspeaker system frequency response. Analysis of system performance could be displayed on the supervisor console or some other type of performance monitoring device.

3 CONCLUSIONS

In this paper we have presented a loudspeaker array implementation which allows digital control of spectral content and group delay at each element. As illustrated by simulation studies, the suggested implementation allows dynamic reconfiguration of the directivity and orientation of the loudspeaker array main lobe. This opens the possibility for developing "intelligent" sound reinforcement arrays which have programmable, constant directivity as well as the capability for special effects such as beam tilting or a "swirling" sound field. The capability of spectral content and group delay control for each loudspeaker in an array also makes compensation for naturally occurring element-by-element variations possible. Further, the suggested implementation allows both amplitude and group delay equalization to be performed on an element-by-element basis.

4 REFERENCES

[1] D. G. Meyer, "Development of a Model for Loudspeaker Dispersion Simulation," presented at the 72nd Convention of the Audio Engineering Society, *J. Audio Eng. Soc. (Abstracts)*, vol. 30, p. 946 (1982 Dec.), preprint 1912.

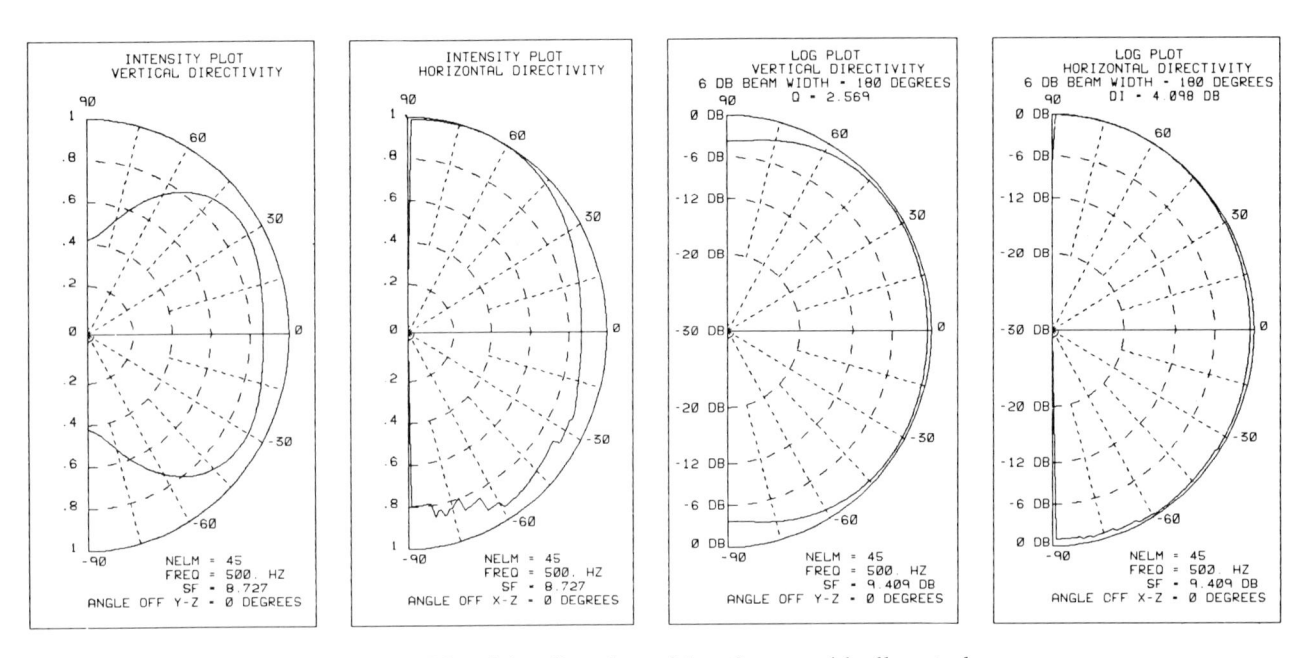

Fig. 11. Directivity slice plots of 5 × 9 array with all control.

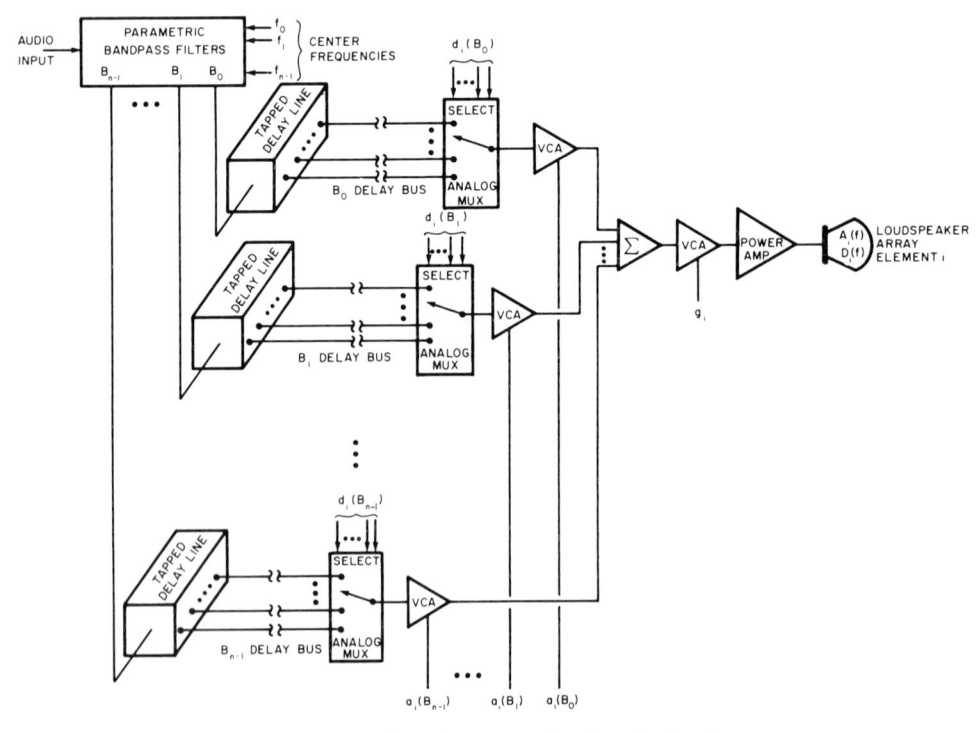

Fig. 12. Loudspeaker array signal path circuitry.

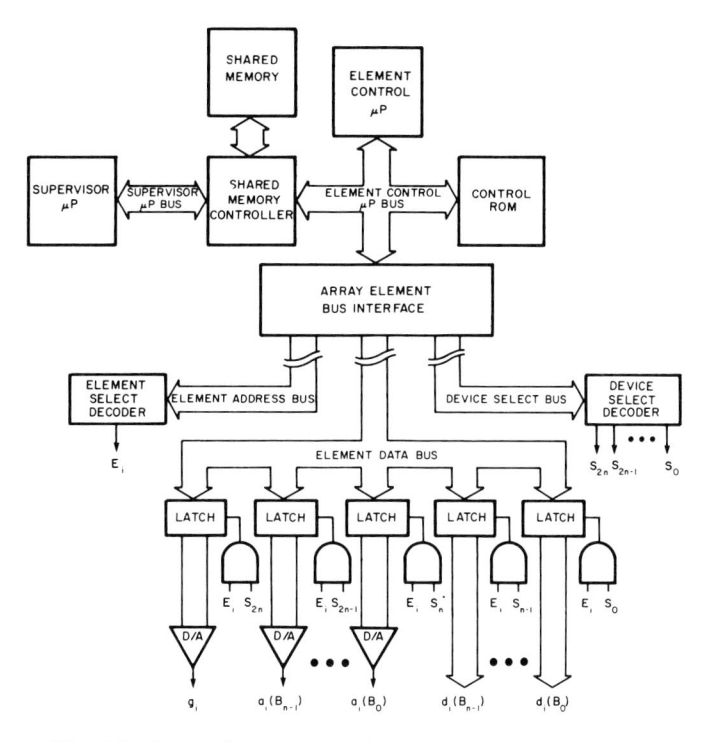

Fig. 13. Array element control processor and bus interface.

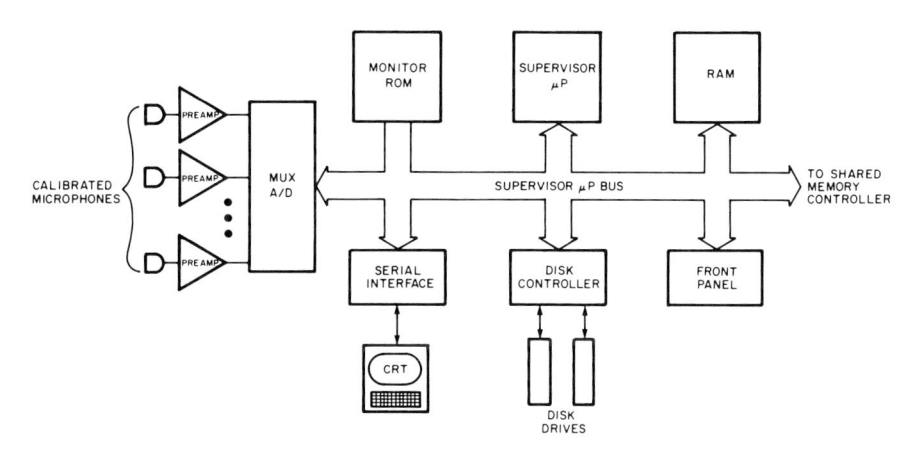

Fig. 14. Loudspeaker array supervisor processor.

[2] D. Meyer, "Computer Simulation of Loudspeaker Directivity," *J. Audio Eng. Soc.*, vol. 32, pp. 294–315 (1984 May).

[3] I. Wolff and L. Malter, "Directional Radiation of Sound," *J. Acoust. Soc. Am.*, vol. 2, pp. 201–241 (1930 Feb.).

[4] L. Beranek, *Acoustics* (McGraw-Hill, New York, 1954).

[5] L. Kinsler, A. Frey, A. Coppens, and A. Sanders, *Fundamentals of Acoustics*, 3rd ed. (Wiley, New York, 1980).

[6] H. Olson, *Acoustical Engineering* (Van Nostrand, Princeton, NJ, 1967).

[7] L. H. Schaudinischky, A. Schwartz, and S. T. Mashiah, "Sound Columns—Practical Designs and Applications," *J. Audio Eng. Soc.*, vol. 19, pp. 36–40 (1971 Jan.).

[8] D. L. Klepper and D. W. Steele, "Constant Directional Characteristics from a Line Source Array," *J. Audio Eng. Soc.*, vol. 11, pp. 198–202 (1963 July).

[9] A. P. Smith, "A Three-Way Columnar Loudspeaker for Reinforcement of the Performing Arts," *J. Audio Eng. Soc.*, vol. 19, pp. 213–219 (1971 Mar.).

David Meyer's biography was published in the 1984 May issue.

A Graphic Method for Choosing and Aiming Loudspeakers for Reinforcement*

PETER W. TAPPAN

R. Lawrence Kirkegaard & Associates, Inc., Downers Grove, IL 60515, USA

A simple transparent template laid over an architectural plan drawing shows required angular coverage, loudspeaker aim angles, and variations of sound pressure level due to distance. Then, after a tentative choice of loudspeakers and aiming, the resulting relative sound pressure levels at various seats may be plotted quickly if the dispersion characteristics of the loudspeakers are known.

0 INTRODUCTION

Over the last 20 years various optical, graphical, and mathematical aids have been devised to assist the reinforcement system designer in laying out loudspeaker systems for desired audience coverage. Each method has its advantages and drawbacks, as does the method to be explained.

In 1965 Malmlund and Wetherill [1] described a simple optical device that projects an outline of the nominal coverage pattern of a selected loudspeaker onto a model of a listening plane. After manually adjusting the aiming of the projector for optimum audience coverage, and verifying that the coverage pattern of the selected loudspeaker is appropriate, the aiming angles are read from calibrated dials on the projector mechanism. The main disadvantages of this method are the necessity for a darkened room, the need for a model of the listening planes (except when an architectural plan drawing serves that purpose because the floor is flat or nearly flat), and the fact that the results can be misleading because the projected pattern does not show the inverse-square-law falloff of sound pressure with distance. Moreover, as originally constructed, only the −6-dB angular coverage contour was projected, but additional contours could easily be added.

It is interesting to note that the technique could theoretically be expanded to include distance effects, such that an accurate display of the sound pressure at any location is obtained. To accomplish this, it would be necessary to use a projection slide or template that is completely transparent in the center, becoming less so toward the edges, in exact accordance with the angular dispersion characteristics of the loudspeaker. Then the light intensity projected onto the model would be exactly proportional to the sound intensity projected onto the real listening plane, and could be measured with a photocell.

Much more recently, Uzzle [2] and then Prohs and Harris [3] have described methods wherein the listening plane is remapped onto a cylindrical or spherical surface, and then transparent templates or overlays representing the angular coverage contours of particular loudspeakers are slid around over the remapped listening plane for a "best fit." These schemes might be regarded as the light projection method in reverse, that is, the listening plane is "projected" onto the loudspeaker contour template instead of vice versa. The principal drawbacks are the tedium of the remapping process, and the resulting distorted view of the venue, which can sometimes lead to errors in judgment.

Eargle and Kalmanson [4] devised a computerized method which is accurate and certainly the most versatile of all, displaying a wide range of data including listening-plane sound pressure levels, direct-to-reverberant ratios, intelligibility ratings, and even loudspeaker drawings. Moreover, it can show combined levels of multiple loudspeakers, with or without phase effects taken into account. Even this method has its disadvantages, however, the principal ones being the tedium of

---

* Presented at the 78th Convention of the Audio Engineering Society, Anaheim, California, 1985 May 3–6.

entering coordinates for the listening plane and loudspeaker location and the fact that the method provides no help in the initial selection of loudspeakers and tentative aiming.

The technique to be described is more limited in its practical applicability than some of the previous methods, and in some instances less accurate, but it is much faster and easier to use. Most of the tedium occurs during the construction of templates which, once made, can be used again and again. The transparent templates are applied directly over architectural plan drawings, so that no remapping or model building is required.

Incidentally, a paper describing a different graphic technique was presented at the 76th Convention of the Audio Engineering Society by Thurmond [5]. It has its merits, but in some ways yields less information than the method to be described.

1 THE BASIC TEMPLATE

Fig. 1 is an example of the basic template. It represents a plan view of a portion of an infinite flat-floor listening plane, with a loudspeaker system suspended over the dot at the middle of the top of the drawing. For a given architectural scale, such as $1/8'' = 1'0''$, a different template is required for each different loudspeaker height. However, the only difference is one of magnification. In other words, a template for a 40-ft (12-m) loudspeaker height at $1/8'' = 1'0''$ is identical to one for a 20-ft (6-m) height at the same scale except that it is enlarged by a factor of 2. Moreover, the number of templates required can be kept within reasonable bounds if the loudspeaker heights are quantized in 5-ft (1.6-m) increments or so. Besides, a template for a loudspeaker height of 20 ft (6 m) at $1/8'' = 1'0''$ also works for a height of 40 ft (12 m) at $1/16'' = 1'0''$, etc. About 90% of our work is done with only four templates.

The vertical line down the middle of the template represents the azimuth aim angle of the loudspeaker or loudspeaker system. This line is intersected by horizontal lines, one for every 5° increment in down angle (over most of the range). In other words, if a line were drawn from the loudspeaker at an angle 30° below horizontal, down toward the azimuth aim line in the plane of the template, it would intersect that line where the horizontal line marked 30° crosses the azimuth aim line.

The gently curved, angled lines represent the intersection of various horizontal coverage angles with the plane of the template. Imagine a point source of light located above the template at the loudspeaker position. Imagine further that the light is centered and on axis between two parallel vertical black disks, one to the left and one to the right of the light. If the spacing between the disks is such as to permit light to escape in a 50°-wide beam, for example, the light would illuminate the portion of the template between the 25°

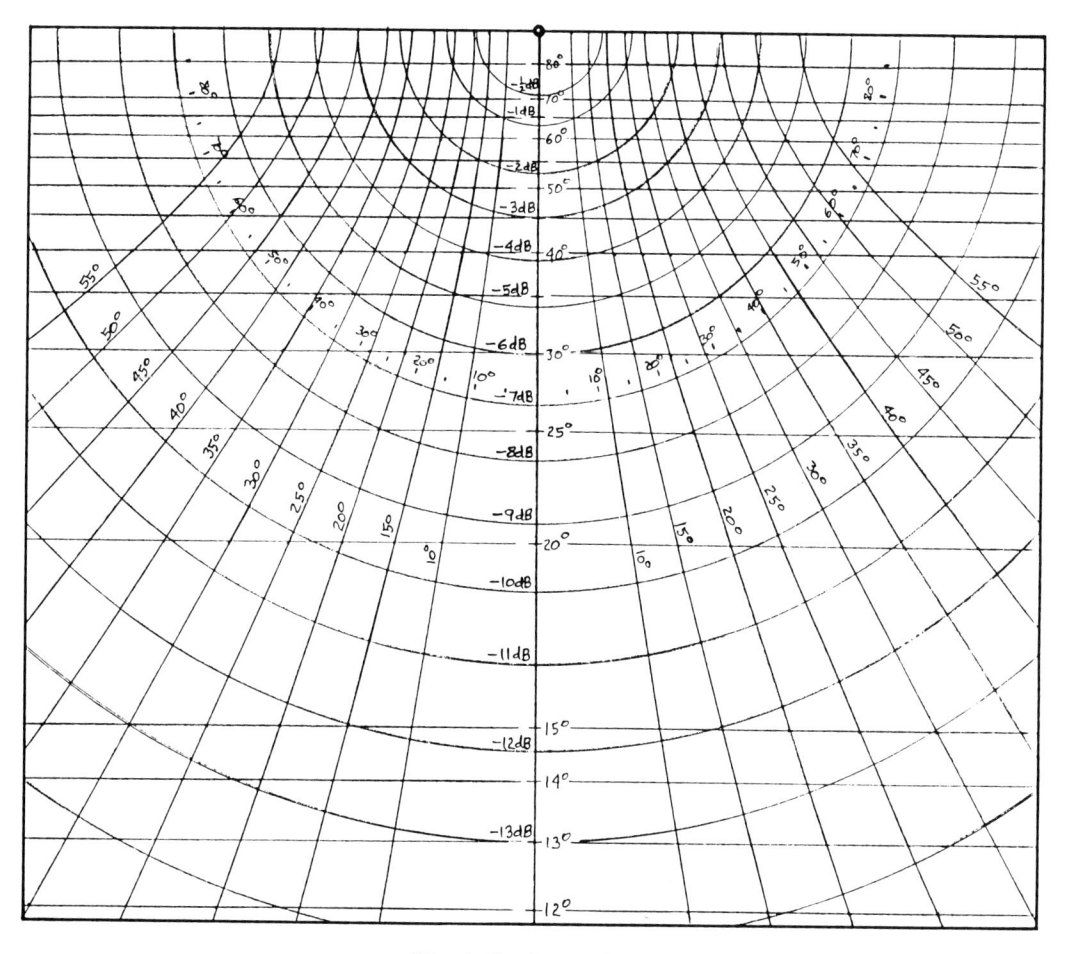

Fig. 1. Basic template.

left and 25° right lines, and the remainder of the template would be in shadow. In other words, the 25° left and 25° right lines represent the limits of coverage of a loudspeaker with a horizontal coverage of 50° and a vertical coverage of 360°.

The protractor angle marks concentric with the dot simply permit the azimuth aim angle of the loudspeaker with respect to the centerline of the listening plane to be read without using a separate protractor, in instances where than angle is not 0°.

Finally, the concentric circular arcs indicate inverse-square-law attenuation, that is, they show the sound pressure level in the plane of the template with respect to that at the dot, due to an omnidirectional source at the loudspeaker position.

2 USING THE BASIC TEMPLATE

Suppose we wish to design a horn loudspeaker cluster for the church plan shown in Fig. 2. The first step is to select a basic template for the proper loudspeaker system height (within a few feet) at the scale of the plan drawing, or to construct a new template if we do not already have a suitable one. Then we lay the template over the plan drawing with the dot over the loudspeaker location, as shown in Fig. 3. In this example the listening plane is symmetrical and the loudspeaker system is to be located on the church centerline, so the template is oriented so that its axis or azimuth aim line coincides with the room centerline. Thus we shall not have any use for the protractor markings in this instance.

The template immediately shows us the required horizontal coverage at every down angle. The front row of pews is at a down angle of 60° and requires 90°

Fig. 2. Plan of church pews, with loudspeaker cluster above dot.

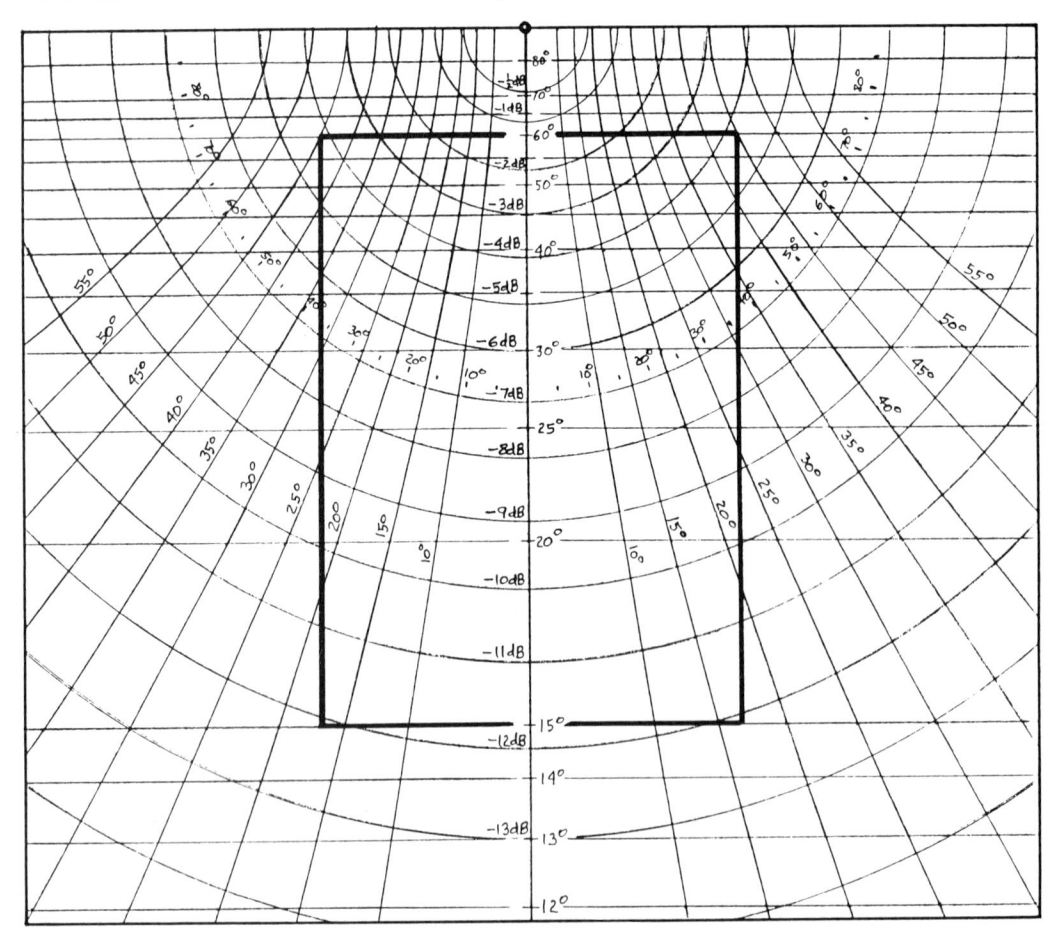

Fig. 3. Basic template superimposed on church pews.

horizontal coverage (because the ends of the row are at 45° left and 45° right). The last row is at a down angle of 15°, and requires 34° horizontal coverage. The template also tells us that if our loudspeaker system were omnidirectional, the range of direct sound levels at the pews would be approximately 10½ dB (from −12 dB at the rear corners to −1½ dB at front center).

Obtaining this same information by usual methods requires a scale, a protractor, a calculator, and much more time.

Let us suppose that we have available three horn models, with −6-dB beamwidths of 40° × 26°, 65° × 48°, and 94° × 48°. Can we cover the pews with adequate uniformity using only one 94 × 48° horn? The template shows that we cannot, for even if we aimed the horn at the center of the front row, the front corners would be down about 8 dB compared to the front center (approximately 5 dB due to horn beamwidth, plus approximately 4½ − 1½ = 3 dB due to the inverse square law). Moreover, the rear would then be even farther down. If we aimed the horn farther back to improve the coverage of the rear, the front corners would be down even more.

Thus we conclude that we probably must use three horns, one for the rear and two side by side for the front. (Another possibility worth investigating at this stage is using two of the 65° × 48° horns side by side, aimed somewhere near the rear corners. Further analysis shows, however, that this would not be as good as three horns in this example.) Carrying the three-horn concept further, we surmise that the front horns should be 65° × 48° units, since two of them side by side and angled apart should be capable of covering the 90° span of the front row quite uniformly, whereas the other models would cover either too little or too much. We can also conclude that the rear horn should likewise be a 65° × 48° unit, for although the rear row spans only 34°, the angular span increases rapidly as we move forward, and a 40° × 26° horn would run out of horizontal coverage close to the rear.

3 THE SECOND PHASE

Before we can aim the front horns, we must aim the rear one and see what it does. Maximum uniformity of coverage, from front to rear, is usually achieved by aiming the horn at the rear row, so that inverse-square-law attenuation is partly counteracted by on-axis beaming. In doing this, we should keep in mind that half of the horn's power output then sails over the heads of the audience, increasing reverberation and perhaps generating a bad echo by reflection from the rear wall. Often it is better to sacrifice some uniformity of coverage in order to minimize echo and/or reverberation. Let us assume, however, that in this example the room acoustics allows us to aim the horn at the rear, so we shall try a down angle of 15°.

There are at least three good ways of finding out the results. One is to enter the listening plane and the loudspeaker in the computer program developed by Eargle and Kalmanson. The template analysis has provided us with a good starting point for using that program.

A second method is to employ that same program to generate a display of sound levels not on the specific listening plane of the example, but on a generic plane that is larger. In other words, a printout is obtained of sound levels created by that horn at that elevation and down angle, at that drawing scale, on a plane area somewhat larger than the pew area in the example. From the printout of numerical decibel values, a transparent template is made, preferably converting the numbers to isobar contours. The template can then be laid over the example listening plane drawing to show the results. While this is more work than just entering the example listening plane in the computer, the advantage is that a template has been generated that can be used not only for this example but also for any future project where this horn is used at this elevation and down angle. Moreover, the template can be rotated about the loudspeaker location to other azimuth angles, and can be enlarged or shrunk to fit other scales or other loudspeaker elevations.

The third alternate method is like the second except that no computer program is needed, and results are obtained faster but with less accuracy. Using the basic template and knowledge of the −3-, −6-, and −9-dB beamwidths of the loudspeaker, a template is made that shows sound levels projected on the listening plane. This takes only a few minutes' work and, as with the second method, the resulting template can be used on other projects.

To construct a template for our 65° × 48° horn at a 15° down angle, we lay a sheet of translucent paper or plastic over the basic template and trace the loudspeaker position, the azimuth aim line, and the horizontal line corresponding to a 15° down angle. Let us suppose that the manufacturer tells us that the −3-dB beamwidth of the horn is 55° × 37°. In other words, the level is down 3 dB, 27½° off axis horizontally and 18½° off axis vertically. At the 15° down line, we draw short line segments parallel to the nearest horizontal coverage lines at positions corresponding to 27½° left and right (Fig. 4). This represents the −3-dB horizontal beamwidth. We also draw a short horizontal line segment through the azimuth aim line at a position corresponding to a down angle of 15° + 18½° = 33½°. This represents the lower limit of the −3-dB vertical beamwidth. (If the loudspeaker were aimed farther down, we could also mark the upper limit of the −3-dB vertical beamwidth, but in this instance it is off the paper and of no interest.) Now we draw a smooth curve through these three line segments, forming an approximately elliptical arc. The result is an approximation of where the −3-dB angular coverage contour of the loudspeaker meets the listening plane.

We now repeat this for the −6- and −9-dB beamwidths of the loudspeaker. The results are shown in Fig. 5.

The final step in the preparation of the template is to fill in relative sound pressure level values, based on the angular coverage contours we have just drawn and on inverse-square-law attenuation values from the basic template. To do this, we begin at point A (Fig. 5) which is the aim point of the loudspeaker, and write down two numbers. One is zero because we shall use the sound pressure level at this point as our reference for other locations. The other, in parentheses, is the inverse-square-law value for that point read from the basic template, −11.7 dB in this instance. This will be useful later in deciding how much power to feed this loudspeaker relative to other loudspeakers, as we shall see.

Now we proceed to the left along a circular arc centered at the loudspeaker location (so that the inverse-square-law attenuation remains the same) until we reach the −3-dB angular coverage contour (point B in the figure). Since this point has the same inverse-square-law attenuation as point A, but lies on the −3-dB angular coverage contour, the sound pressure level here is 3 dB lower than at point A. Thus we write −3 at this point. Similarly, points C and D, which lie on the −6- and −9-dB angular coverage contours, are labeled −6 and −9, respectively. These values are duplicated at the corresponding points on the right side of the template.

Now we move up from point B along the −3-dB angular coverage contour until we reach a point where the inverse-square-law attenuation is 2 dB less than at A or B (point E). In other words, we cross two of the circular arcs which are at 1-dB increments. Since we are still on the −3-dB angular coverage contour, the sound pressure level here is 2 dB greater than at point B, so we label this point −1. In this manner, we can calculate relative sound pressure level values at all points on the −3-, −6-, and −9-dB angular coverage contours.

Values for points inside the −3-dB angular coverage contour can be approximated by estimating the attenuation due to angle and reading off the inverse-square-law attenuation. For example, the sound pressure level at point F is 4 dB greater than at A due to the inverse square law, but perhaps 1 dB less due to the angle, so we label it +3.

After as many sound pressure level values as desired have been filled in, the completed loudspeaker coverage template is removed from the basic template and placed over the architectural plan drawing, as in Fig. 6. We see that, as expected, the rear two-thirds of the pews are covered reasonably uniformly. To do any better, we would have to use a horn with less vertical coverage. Since we have assumed that the only available model with less vertical coverage has too narrow a horizontal coverage, we would have to use two of them angled somewhat apart to provide the needed lateral uniformity. For purposes of the example, let us assume we are satisifed with the single 65° × 48° horn for the rear.

The next step is to guess at a suitable down angle for the front horns, and to construct a loudspeaker cov-

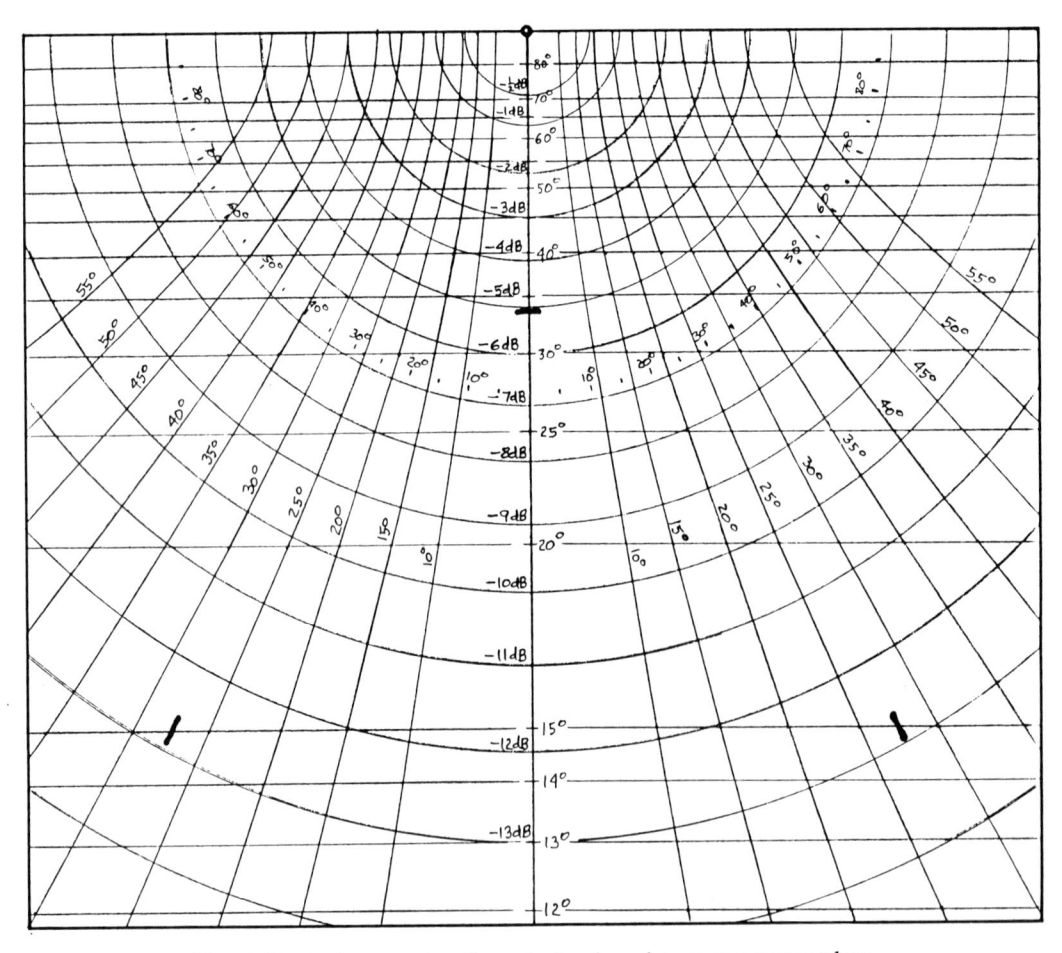

Fig. 4. Beginning construction of a loudspeaker coverage template.

erage template for that angle if we do not already have one on file. We can use the basic template again to help us with that guess. Placing it on the architectural plan, with the rear horn template in place as in Fig. 6, we rotate it to one side about the loudspeaker location so that the azimuth aim line runs through the middle of the front pews, on that side, which are not within the nominal coverage of the rear horn. In other words, we aim it very roughly as one of the front horns might be aimed. We then select a reasonable-looking down angle from the calibrations along the azimuth aim line of the template. In this example, 40° seems appropriate.

After a loudspeaker coverage template has been made or found for a 40° down angle, it is placed on the drawing and rotated to optimize the aiming. The result is shown in Fig. 7. The aiming is adjusted so as to provide a smooth overlap on the pew centerline between the two front horns. To accomplish this, the aiming should be such that the sound pressure level due to one horn is 3–7 dB lower on the centerline than it is some distance away from the centerline. We can now measure the azimuth aim angle with a protractor.

If the results indicated that the down angle might not be optimum, templates for other down angles could be tried.

Finally, we examine the matter of relative power levels. In this example, all horns are the same, so no sensitivity difference needs to be taken into account. Obviously, the two front horns should be operated at the same level, but we must determine the proper level for these relative to that of the rear horn. Looking at the numbers in parentheses on the two templates, we see that the zero reference level for the rear horn is −11.7 dB, while that for the front horn is −3.8 dB. Thus we would need to drive the rear horn 7.9 dB harder than either front horn to make the zero references the same. In many instances, the number obtained this way is an appropriate one to use. In this example, however, because of the very shallow down angle of the rear horn and the flat floor, we are getting relative sound pressure levels as high as +3½ dB near the front of the horn's coverage. In addition, there is appreciable overlap between this horn and the front horns, so the combined sound pressure levels in this region would be even higher. Thus it would be wise here to decrease the rear horn's level by perhaps 2 to 4 dB. Hence we conclude that the rear horn should be driven 4–6 dB harder than either front horn.

After completing this graphic analysis, the careful or cautious designer may wish to check the results and/or refine the design by means of the Eargle–Kalmanson computer program.

4 BEAMWIDTH VARIATIONS WITH FREQUENCY

No loudspeaker has the same beamwidth at all frequencies. Generally we have based our loudspeaker coverage templates on an eyeball-weighted average of

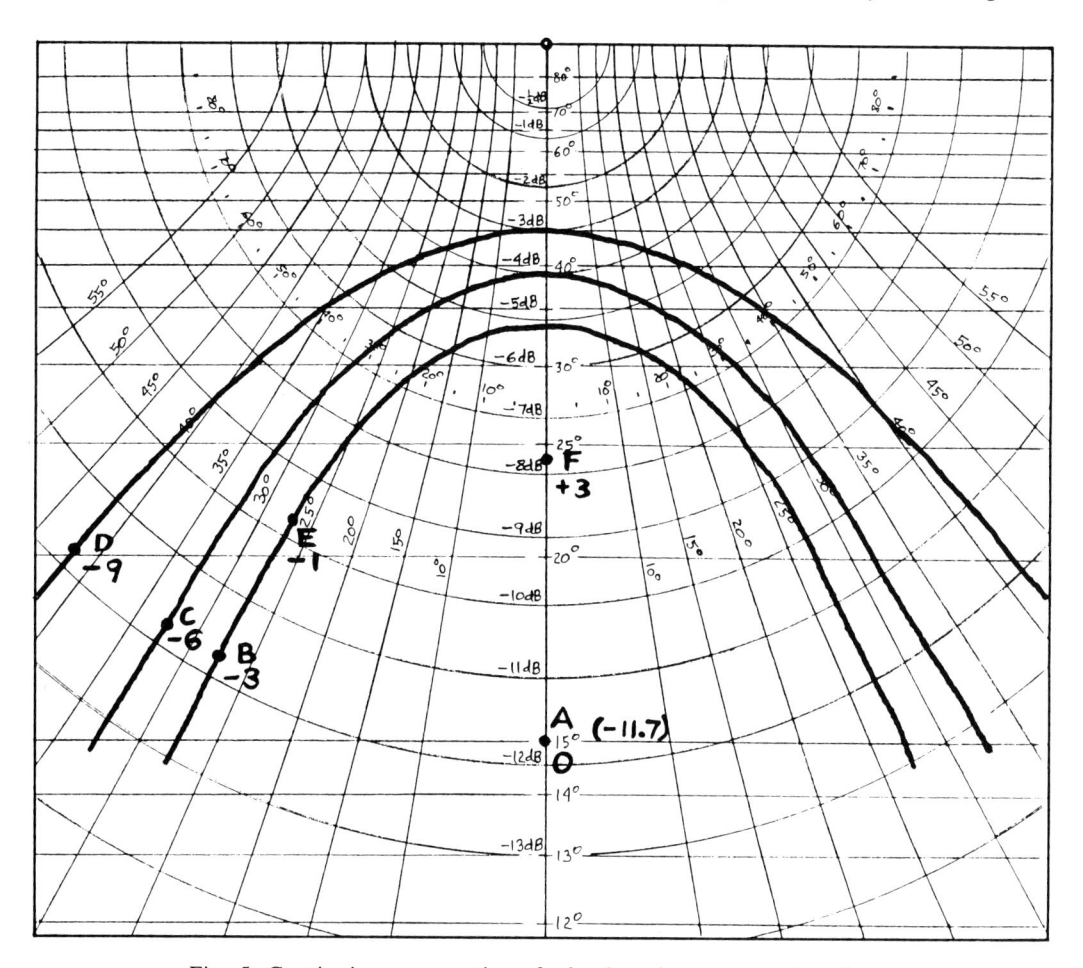

Fig. 5. Continuing construction of a loudspeaker coverage template.

beamwidths over a frequency band of 1000–8000 Hz, giving most weight to the 2000-Hz values. Fortunately most modern constant-directivity horns have reasonably uniform beamwidths over that range.

5 SKEWED LOUDSPEAKERS

It is just as easy to construct a coverage template for a loudspeaker skewed (rotated about its axis) 90° as it is for one in the normal orientation. At this writing, we have not attempted to devise a graphic technique for constructing templates for loudspeakers at other skew angles. The Eargle–Kalmanson computer program could, however, be used for that purpose.

6 SLOPED LISTENING PLANE

A listening plane that is sloped upward toward the rear, as in a typical auditorium or theater, adds a minor complication. A basic template or loudspeaker coverage template that is correct for one value of loudspeaker height above the ears is not correct for another.

There are at least two ways of dealing with this problem. One is to do the analysis in steps, using a different template for every 5-ft (1.6-m) or so increment in ear height. A better method, usually, is to tilt the building, so to speak, to make the listening plane level.

Consider the auditorium longitudinal section shown in Fig. 8. The actual locus of the listeners' ears is represented by the heavy three-segment line, but may be satisfactorily approximated by a single-segment line L with a slope S. The loudspeaker is at O, at a height h above the front row of ears. It projects a ray R of sound at a down angle ϕ, which intersects the listening plane L at point P, a horizontal distance p from the loudspeaker.

If we were to use one of the templates as already described for a level listening plane situation, with a loudspeaker height h, results would be correct at the front of the listening plane, but the template would show ray R intersecting the listening plane at point Q at a horizontal distance q from the loudspeaker, instead of at point P. As is evident from the figure, the error can be considerable. Specifically, the fractional error is $(q - p)/p = \tan S/\tan \phi$. At a down angle of 15° and a floor slope of 10°, this amounts to a 65% error.

To overcome this, we "tilt the building" so that the listening plane L is level. The loudspeaker is then at a height h' above the listening plane, and over a point C' which is closer to the front row by a distance d. We can measure h' and d from the architectural section drawing. Using templates for the height h', with their loudspeaker location dots over C', we then obtain correct results, except for one factor. The architect, unaware that we were going to tilt the building, has made the plan drawing (on which we shall use our templates) based on the old vertical direction. To make this drawing correct for our leveled listening plane, we would have to stretch the drawing longitudinally by the ratio $1/\cos S$. Since the paper is not likely to cooperate with

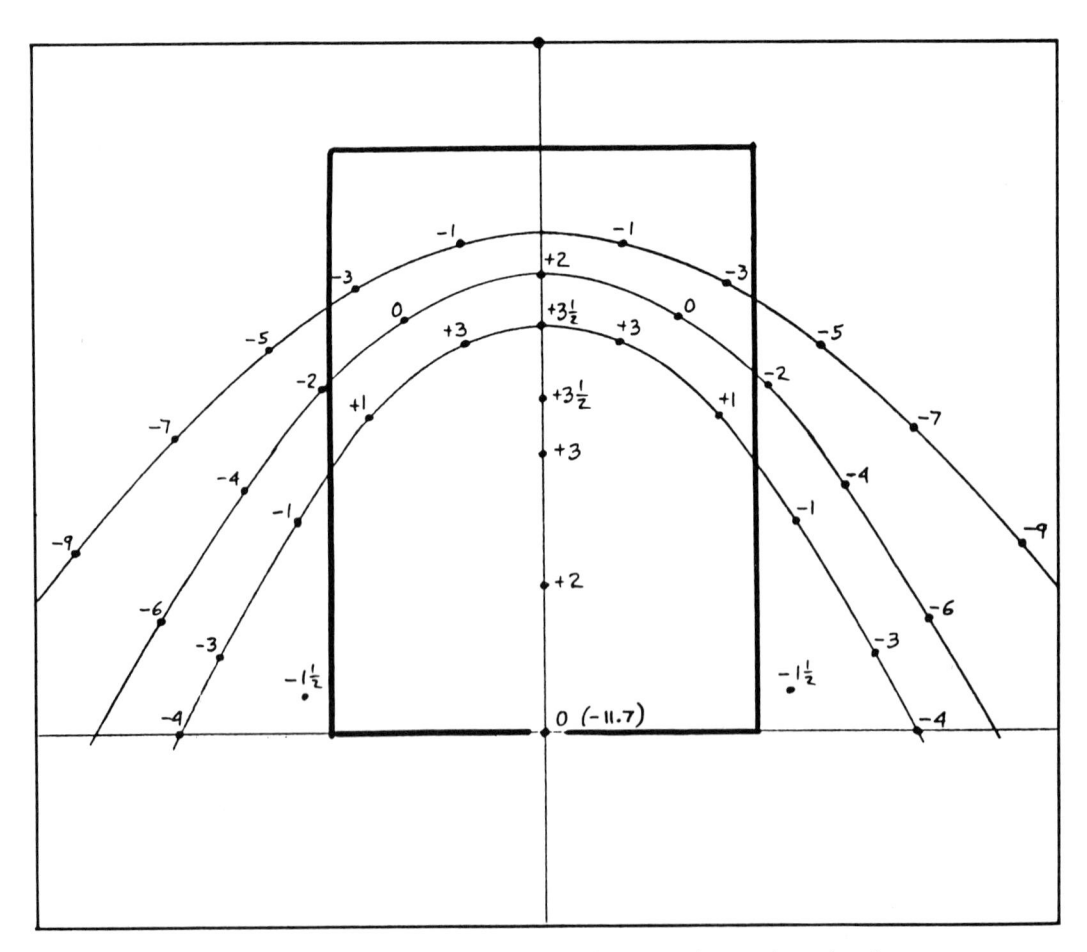

Fig. 6. Finished loudspeaker coverage template superimposed on church pews.

this effort, however, we can choose to ignore the error if it is small enough, as it will be in most practical instances. Specifically, for a 10° floor slope, the error is less than 2%, and rises to 5% only for slopes of 20° or greater.

After we have determined loudspeaker down angles by this technique, we must remember to subtract S from them for loudspeakers with azimuth aim angles nearly parallel to the direction in which the building is tilted, if we are going to untilt it when we are finished. Down angles of loudspeakers aimed at the sides, that is, parallel to the axis of tilt, do not change, although strictly speaking they skew by the amount of the floor slope, normally a negligible effect. The down angle correction for intermediate values of the azimuth angle A is equal to arc tan (tan S cos A).

The foregoing has assumed that the listening plane, though tilted, is essentially a flat plane. When the plane is also substantially curved horizontally around the loudspeaker location, as in a bowl configuration, a different direction of building tilt is needed for each loudspeaker pointing in a different azimuth direction. Obviously this can get complicated, and it may be simpler to do the analysis in incremental steps of elevation.

7 CONCLUSION

Some of the explanations herein seemed long-winded in writing, but once the concepts are understood, the technique is relatively quick and easy to use. It was conceived five or six years ago, and we have been employing it in most of our cluster designs with good results. As a file of loudspeaker coverage templates accumulates, it becomes less and less necessary to make new ones.

We have no objection to anyone making basic templates for his or her own personal use (see instructions in the appendix), but we wish to reserve all commercial reproduction rights to the basic template. If enough interest is expressed, we may make a set of basic templates available at a nominal charge.

8 REFERENCES

[1] W. A. Malmlund and E. A. Wetherill, "An Optical

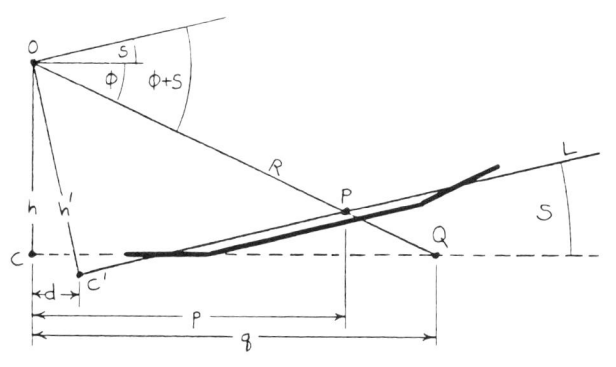

Fig. 8. Longitudinal section of a sloped listening plane.

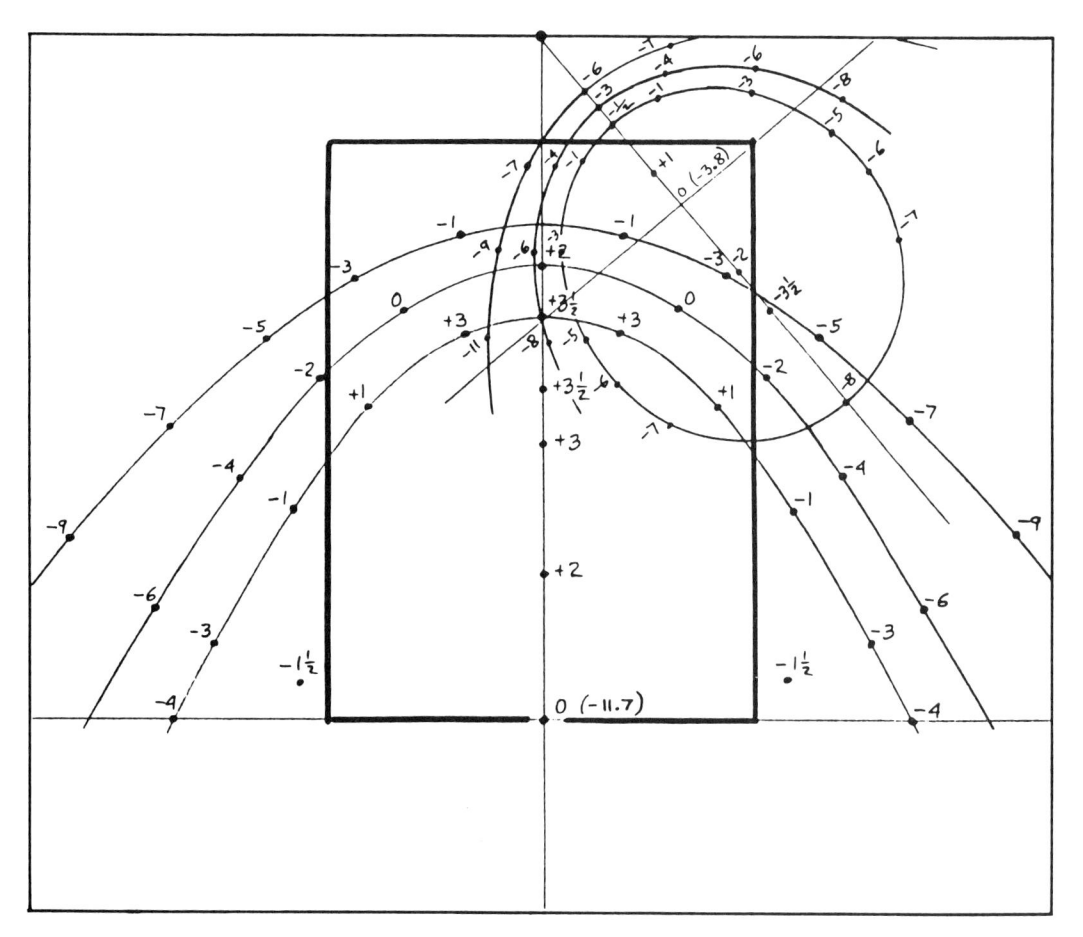

Fig. 7. Church pews with loudspeaker coverage templates for rear horn and one front horn.

Aid for Designing Loudspeaker Clusters," *J. Audio Eng. Soc.*, vol. 13, p. 57 (1965 Jan.).

[2] T. Uzzle, "Loudspeaker Coverage by Architectural Mapping," *J. Audio Eng. Soc.*, vol. 30, pp. 412–424 (1982 June).

[3] J. R. Prohs and D. E. Harris, "An Accurate and Easily Implemented Method of Modeling Loudspeaker Array Coverage," *J. Audio Eng. Soc.*, vol. 32, pp. 204–217 (1984 Apr.).

[4] J. Eargle and M. Kalmanson, "Design by Computer," *Sound & Video Contractor*, vol. 2, no. 7 (1984 July 15).

[5] G. R. Thurmond, "Horn Layout Simplified," presented at the 76th Convention of the Audio Engineering Society, *J. Audio Eng. Soc. (Abstracts)*, vol. 32, pp. 1014, 1015 (1984 Dec.), preprint 2167.

APPENDIX
CONSTRUCTION OF THE BASIC TEMPLATE

Fig. 9 depicts a loudspeaker system at point O at a height h above point C on a listening plane L. The system is aimed horizontally parallel to a line y in the listening plane. A sound ray from the loudspeaker at a down angle ϕ intersects y at point P, at a distance or "throw" of t from the loudspeaker. (The up angle of the loudspeaker as viewed from P is the same as the down angle as viewed from the loudspeaker, and is so marked in the figure.) Points Q_l and Q_r, at distances x to the left and right of point P, subtend angles of θ left and right of the line t.

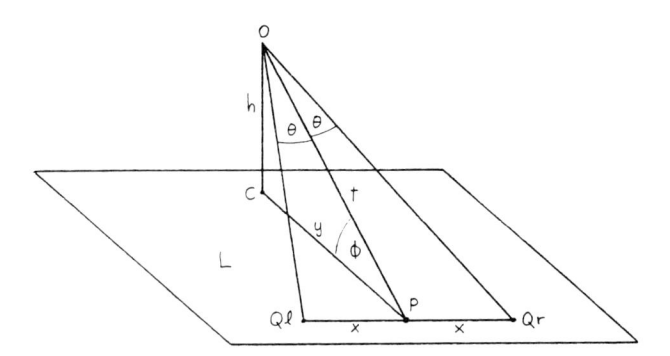

Fig. 9. Geometry for constructing the basic template.

The distance y from point C to point P, for any down angle, is given by $y = h \cot \phi$. This relationship is used to plot the down angle lines.

The throw to point P is given by $t = h/\sin \phi$. The sound level at point P relative to that at point C, due to inverse-square-law attenuation, is given by dB = $20 \log (h/t) = 20 \log \sin \phi$. Conversely, the down angle for any relative sound level is given by ϕ = arc sin antilog (dB/20). This relationship is used to plot the concentric circles.

The lateral displacement from any point P on the azimuth aim line, as a function of the angle θ subtended by that displacement at the loudspeaker, is given by $x = t \tan \theta = h \tan \theta / \sin \phi$. This relationship is used to plot the horizontal coverage lines.

THE AUTHOR

Peter W. Tappan was born in 1928 in New York City. He received B.S. and M.S. degrees in physics in 1952 and 1958 from the Illinois Institute of Technology. He was employed by Motorola, Inc. in 1951 and later that year joined the physics department of Armour Research Foundation (now I.I.T. Research Institute), performing research in magnetic recording and electroacoustical devices.

Mr. Tappan moved to Warwick Manufacturing Corporation in 1956, where he was responsible for research in consumer audio equipment. In 1962, he joined Industrial Research Products, Inc., where he designed miniature microphones. From 1963 to 1976, he was an acoustical consultant with Bolt Beranek and Newman, Inc. For the past ten years, he has been vice president of R. Lawrence Kirkegaard & Associates, Inc., where he has designed sound reinforcement systems for many performing arts facilities, churches, sports buildings, courtrooms, etc., as well as consulting in noise control and acoustics.

Mr. Tappan is a fellow of the Audio Engineering Society and has served as central vice president and admissions chairman. He is also a member of the Acoustical Society of America, past president of the Chicago Acoustical and Audio Group, past secretary of the Executive Committee of the Midwest Acoustics Conference, and past editor of the *IEEE Transactions on Audio*. He has written papers for several technical journals and lectured for numerous organizations including the University of Wisconsin.

LETTERS TO THE EDITOR

COMMENTS ON "A GRAPHIC METHOD FOR CHOOSING AND AIMING LOUDSPEAKERS FOR REINFORCEMENT"

Upon publication of my above paper,[1] I received a letter from T. G. McCarthy pointing out that in my citing of previous work in this field and in my references, I failed to mention his paper, "Loudspeaker Arrays: A Graphic Method of Designing," presented at the 61st Convention of the Audio Engineering Society, *J. Audio Eng. Soc. (Abstracts)*, vol. 26, p. 992 (1978 Dec.), preprint 1398.

While I was aware that Mr. McCarthy had done work along this line, through carelessness I overlooked the fact that he had published some of it. My paper definitely should have mentioned his presentation and preprint. In fact, the later paper by Uzzle, which I did mention, credits McCarthy's paper as its starting point, and it appears that McCarthy is the real pioneer in remapping rooms to fit loudspeaker isobar contours. (My work takes the opposite approach.)

I regret this oversight and hope that publication of this letter will, in some measure, correct it.

PETER W. TAPPAN
R. Lawrence Kirkegaard & Associates, Inc.
Downers Grove, IL 60515, USA

---

[1] P. W. Tappan, *J. Audio Eng. Soc.*, vol. 34, pp. 269–277 (1986 Apr.).

PAPERS

Horn Layout Simplified*

BOB THURMOND

G. R. Thurmond and Associates, Austin, TX 78756, USA

The difficulty of predicting the off-axis behavior of high-frequency horns often leads to grossly inaccurate system designs. A straightforward graphic technique, developed to solve this problem, yields a wealth of information on the performance of high-frequency horns in real applications. Accurate information is obtained very easily and quickly without computer or calculator. Actual design examples and their measured results are presented.

0 INTRODUCTION

It is very common practice for a single high-frequency horn to supply sound to hundreds of listeners at a time. Since very few of these listeners can be located on or near the horn axis, the horn's off-axis characteristics become more important than those on axis. Tests reveal that a horn's response and output levels both change with the angle off axis (Fig. 1). Furthermore, the rate of this change also increases in a nonlinear fashion. All this means that a horn's output is highly nonuniform within its stated coverage area and does not cut off sharply at the boundaries. These characteristics vary with the horn design, but are present in all designs.

Horns with a true constant-directivity design exhibit much less output variation with angle than earlier designs, but still have problems at frequency extremes and near the edges of their coverage pattern.

A further practical complication is the fact that the various listeners are almost always located at various distances from the horn. Because of this, even if the horn had perfectly uniform characteristics, the listeners would not all receive the same level of sound.

Altogether these problems make it difficult to select the best horn type, location, and aiming to provide the most uniform coverage to all listeners. When several horns must be used together, as is often the case, the problems become even more difficult. While some installations permit fairly extensive experimenting and adjusting, most do not. The design must be very nearly correct before construction even begins.

1 DEVELOPMENT

Until recently designers were typically limited to simple angle-of-coverage designs, which were notoriously inaccurate, or light-model techniques [1], which can be highly accurate and informative, but which are time consuming and therefore expensive to implement. Recently other techniques [2]–[4] have been developed which provide relatively good results for a reasonable effort. All these techniques, however, require either extensive calculations, specialized apparatus, or a personal computer. Furthermore, each has difficulty in providing some aspect of the design information.

Around 1970 the author developed another approach, which has proven to be simple, accurate, and very useful. The approach was to determine the real-world equal-level dispersion curve for a given horn. Such a curve has a linear radial scale when drawn on polar graph paper rather than the logarithmic scale normally used. This is correct because it properly represents the drop-off of 6 dB per double distance of direct sound. The resulting curve has the same shape as an equal-level contour in free space. This can be verified experimentally, for example, by setting up a horn in a free field, driving it with an appropriate signal, and plotting the location, at all angles in front of the horn, of all points having the same measured level as a given point on axis.

It is ironic that many older texts presented horn dispersion characteristics as polar plots with this useful linear scale. Virtually all contemporary measurements are made with a logarithmic radial scale. Fig. 2 shows a comparison of these two scales. Fortunately the polar plots now published by several manufacturers may be converted rather easily to linear scale, which will make them more useful. If desired, the correct contours may be plotted automatically by means of the apparatus shown in Fig. 3.

The next step is to plot a family of curves for a particular type of horn, spaced at 3-dB intervals, as in Fig. 4. These isolevel contours are drawn on transparent film and may be extended as needed to cover any size

---

* Presented at the 76th Convention of the Audio Engineering Society, New York, 1984 October 8–11; revised 1987 July 27.

drawing. It has been found that the contours for horns of similar design are also very similar, so generic plots can be used.

What makes these isolevel plots so useful is that when they are overlaid on the drawing of a room, they show the direct sound pressure levels (SPL) that would be obtained from a particular horn location and aiming in that room. The overlay is then moved to explore various horn locations and aimings to see the coverage effects of each. This is continued until the best fit of an isolevel contour with the seating (or, more exactly, the ear level) contour is obtained. If a good fit cannot reasonably be obtained, then good coverage will not be obtained, and a different horn or a different loudspeaker arrangement should be investigated.

2 EXAMPLES

Some examples will clarify this procedure and illustrate its benefits. Fig. 5 shows a section through a large theater with a balcony. It was desired to locate a horn cluster above the proscenium. It is usually best to start a design in this way and to consider the vertical aspect first, beginning with the most distant seats.

Fig. 6 shows the section with the isolevel contours for the vertical dispersion of a 40° by 90° horn overlaid. The origin is placed at the proposed horn location and the aiming is as indicated by the axis line. The isolevel contours are then compared with the ear level line, and the aiming is adjusted until the best fit occurs, that is, until the isolevel contours are parallel to the ear level line for the greatest part of its length. If they cannot be made parallel, which is usually the case, then the coverage will not be even in this plane. However, the amount of unevenness may be acceptable. To determine the amount, it is only necessary to inspect the curves.

As usual, the best fit occurs when the horn is aimed at the most distant seat. With that aiming here we see that one contour passes just beyond the lower end of the balcony ear level line, while the next one passes just beyond the upper end of this line. Since these contours are at 3-dB intervals, the sound level at the top of the balcony is indicated to be about 3 dB lower than that at the bottom. This may be acceptable; cer-

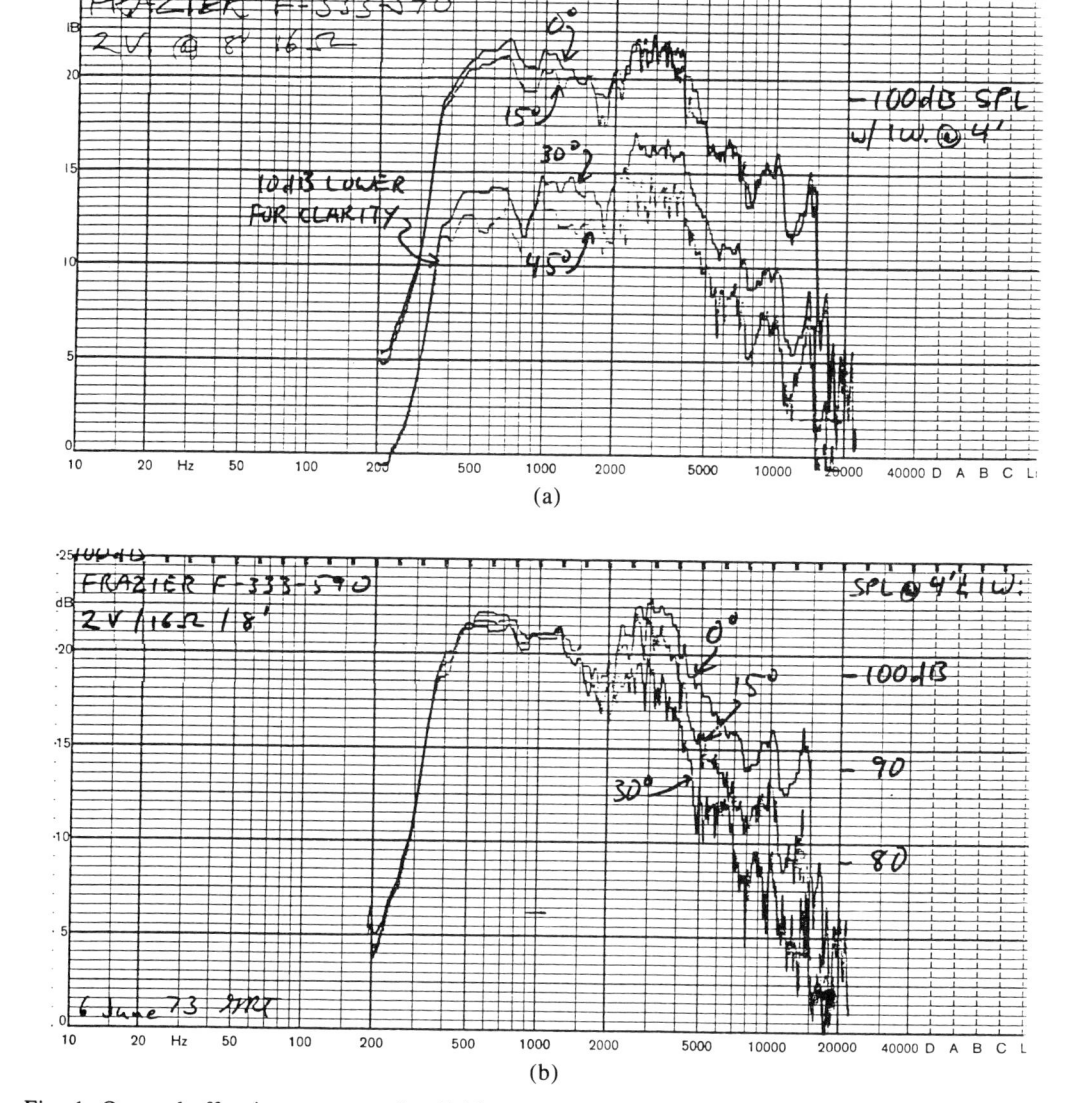

Fig. 1. On- and off-axis responses of radial horn. (a) Horizontal dispersion. (b) Vertical dispersion.

tainly, worse coverage than this is commonly found. However, there are other considerations.

In this case the top of the balcony is just over 40 m from the horn location, which will cause its level to be about 32 dB lower than it is at 1 m. Since a good 90° horn and driver combination will produce about 122 dB SPL maximum at 1 m, this reduces to about 90 dB maximum SPL at the top of the balcony. (Note that this is direct sound; reverberant levels will be substantially higher, depending on room acoustics. However, this reverberant sound contributes little to speech intelligibility, so it may be wise, and it certainly is safe, to consider direct sound levels only.) Also, this is peak level; maximum average program levels will be closer to 80 dB SPL. This may be adequate for a speech system, but probably not so for music.

Let us assume that the system will be used for music and that somewhat higher levels are desired. Before we consider moving the horn closer to the balcony (such as in a secondary location fed through a delay unit), let us see whether a different horn would work better.

A 40° by 60° horn will provide only about 2 dB higher level, and as we note from the isolevel contours, has nearly the same vertical dispersion as the 90° horn. Thus it would not provide better coverage and only slightly higher level.

A 20° by 40° horn is a different matter, however. Such a long-throw horn will produce at least 6 dB higher level on axis than a 90° horn, a significant improvement. Furthermore, if we examine the coverage as shown by the appropriate isolevel contours (Fig. 7), we see that the contour is nearly parallel to the ear level line over all its length, indicating a maximum level variation of less than 2 dB.

But what about the horizontal coverage? Fig. 8 shows the horizontal isolevel contours of two 20° by 40° horns overlaid on the plan of the balcony. The horns are angled so that their coverage patterns cross over at −3 dB, a conservative approach. If we examine the coverage

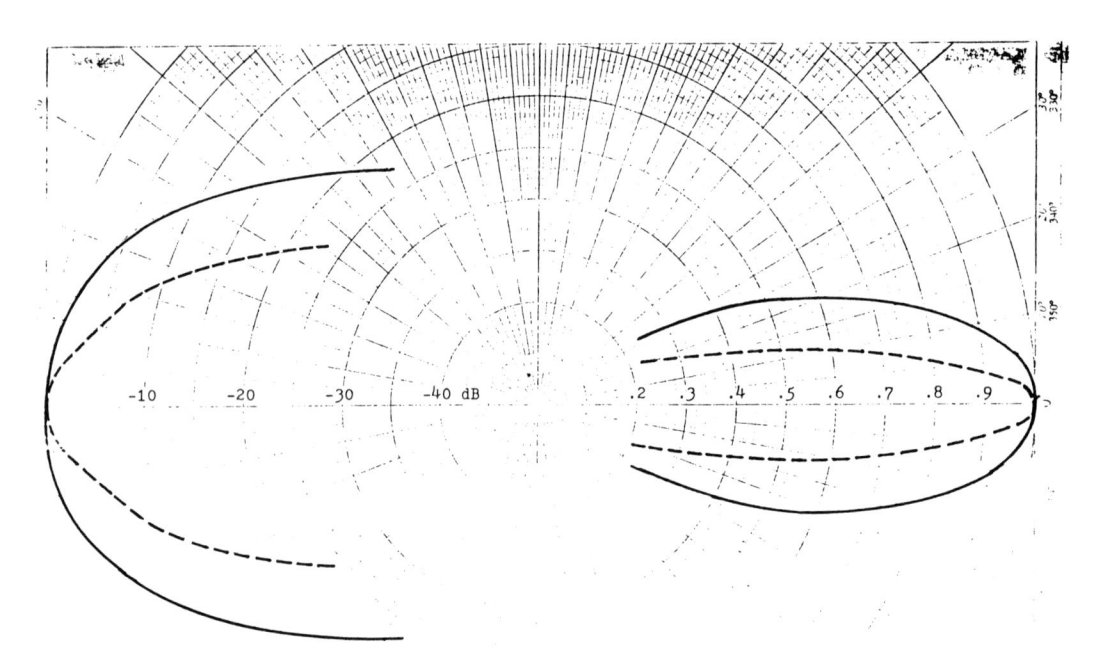

Fig. 2. Comparison of logarithmic (left) and linear (right) radial scales on dispersion polar plots of a 20° by 40° horn. —— horizontal; − − − vertical.

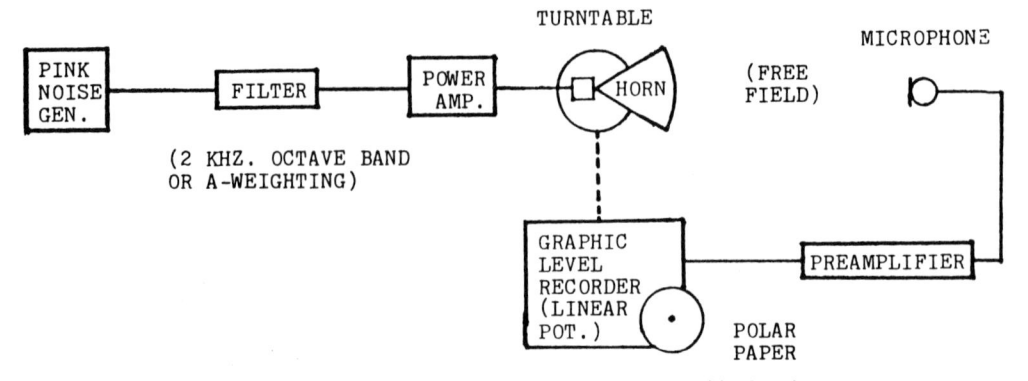

Fig. 3. Block diagram of instrumentation for recording of isolevel contours.

across the front row of the balcony (the worst case), we see that it will be very even between the two horn axis lines (because of the addition of levels at overlap), dropping off to −4 dB at the end of the row. Perhaps the horns should be spread slightly; only a few seats at the ends of the first two or three rows experience a drop-off of more than 2 dB.

Remember that Fig. 7 showed us that the level in the first row is perhaps 2 dB lower than the maximum, which occurs about two-thirds of the way to the top. If we add to this the 2-dB drop-off to the side (ignoring the front corner seats), we have an overall seat-to-seat variation of 4 dB, a very good performance. In fact, it may turn out better than this because of beneficial room reflections.

Experience has shown that two or three long-throw horns, such as these, usually provide the best coverage in a balcony. Often a similar configuration is best to reach under a balcony or to the rear of a long seating area. Experience and measurements [5] have shown that a close side-by-side placement of these horns results in the smoothest overall coverage.

Now let us consider the coverage of the floor seating, again starting with the rear. Fig. 9 shows a 20° by 40° horn aimed just under the lip of the balcony. This aiming will provide maximum level in the rear seats. Several things are revealed by the isolevel contours in this case. First the front-to-rear coverage will not be uniform because the isolevel contours do not parallel the ear level line. Second, the point of maximum level will occur somewhat back under the balcony. This point can be located exactly, if desired, by noting the points on the isolevel contours where they are most nearly parallel (or tangent) to the ear level line. A line is then drawn through these points and the horn location point. Where this line intersects the ear level line is the point of maximum level.

The levels at other locations, relative to this maximum level, can then be determined quickly by inspection.

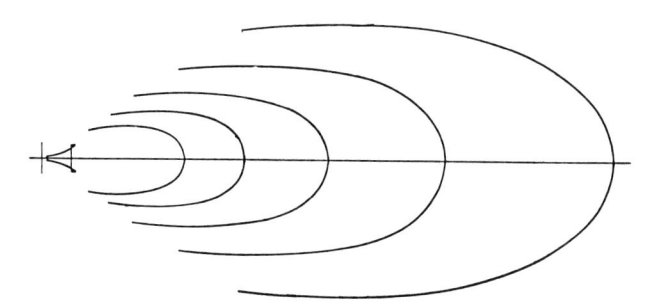

Fig. 4. Family of isolevel contours.

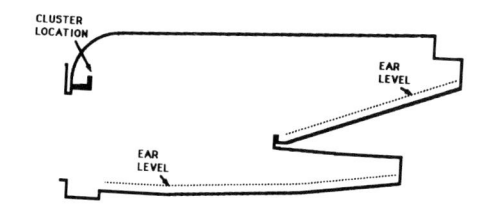

Fig. 5. Section view of large auditorium.

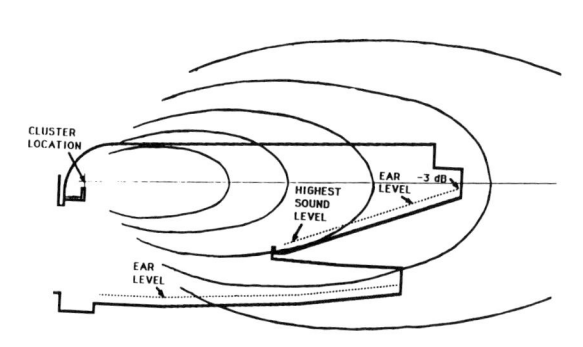

Fig. 6. Vertical isolevel contours of 90° radial horn aimed at top of balcony.

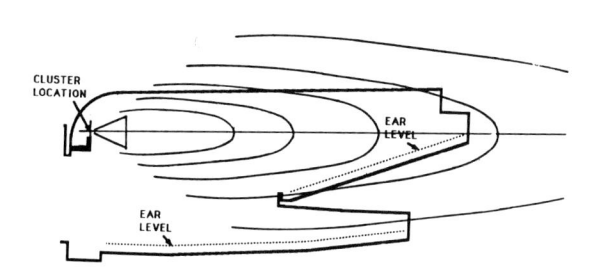

Fig. 7. Vertical isolevel contours of 40° by 20° horn aimed at top of balcony.

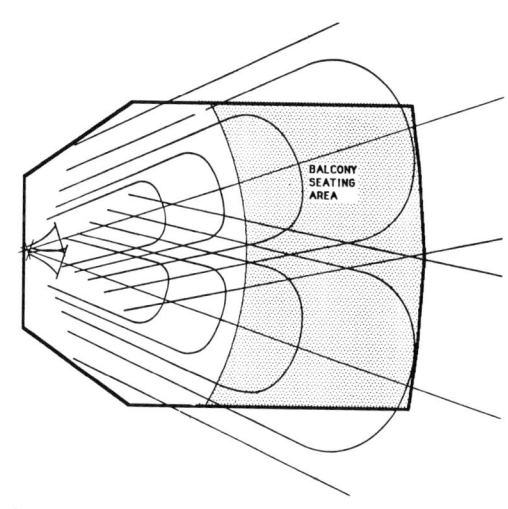

Fig. 8. Horizontal isolevel contours of two 40° by 20° horns aimed into balcony.

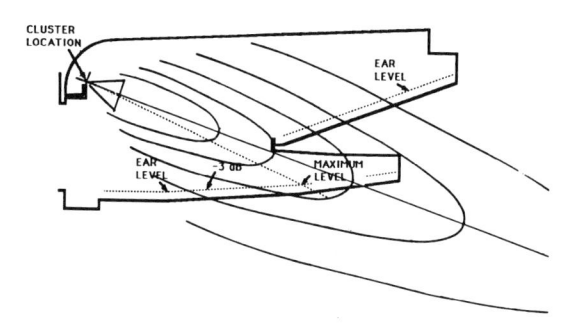

Fig. 9. Vertical isolevel contours of 40° by 20° horn aimed at rear floor.

For example, the level at the point where the horn is aimed will be less than 1 dB below maximum. Further back, the balcony overhang will cut off the coverage sharply, so that supplemental coverage, such as by balcony soffit loudspeakers fed through a delay unit, will be necessary. The area to be covered is clearly indicated.

Furthermore, the −3-dB point toward the front of the auditorium is easy to locate. Let us take this point as the maximum forward extent of the horn coverage and consider horizontal coverage within this area.

To do so accurately, we must take into account the fact that the actual distance from the horn to the −3-dB point is considerably greater than just the horizontal distance between these two points. Thus if the origin (horn location) of an isolevel contour is placed at the point corresponding to the cluster location on the floor plan, an incorrect distance to the −3-dB point, and incorrect horizontal dispersion at that distance, will be indicated. The correct procedure is to place the origin at the same distance from the −3-dB point on the plan as the horn actually will be from that point. This location is a virtual source point, which will yield the correct horizontal dispersion information. In other words, we assume that an actual horn, located at the virtual source point in the room and aimed horizontally, will produce the same transverse dispersion at the −3-dB distance as the same horn in the cluster location aimed as previously determined. This may not be strictly true, as we shall see, but it comes acceptably close.

Fig. 10 shows this configuration with two long-throw horns as used for the balcony. Note that the −3-dB point is extended horizontally as a line. Remember also that the isolevel information is valid only along this line, because the virtual source point changes for each front-to-back point selected for study. Normally it is sufficient to examine only the worst case, which is usually the one closest to the cluster. The method is so quick, however, that it is quite feasible to examine other points as well.

In this case the two horns fall badly short of providing the desired side-to-side coverage. We could use three horns, or we could see if we can get acceptable results with two horns of a different type.

Fig. 11 shows the results of using a 40° by 60° horn instead of the long throw. Aimed, again, just under the balcony, it produces a maximum level point somewhat forward of the balcony face, with the level being −2 dB at the aiming point. The forward −3-dB point is a few rows forward of that from the long-throw horns. This front-to-rear coverage is at least as good as that produced by the long-throw horns.

Fig. 12 shows the horizontal coverage to be expected with two horns of this type. We see that the level along the −3-dB line drops smoothly by 2 dB from the center to the sides, making the level at the sides 5 dB lower than the maximum. This is again a worst case; the center-to-side drop-off will be less toward the rear of the room. Furthermore, it may turn out to less than this in practice because of beneficial reflections off the sidewalls.

It should be emphasized that this center-to-side drop-off is entirely caused by the side seats being farther from the cluster than the center; the horn coverage pattern is quite uniform over this area. Thus we see that this isolevel overlay technique inherently includes

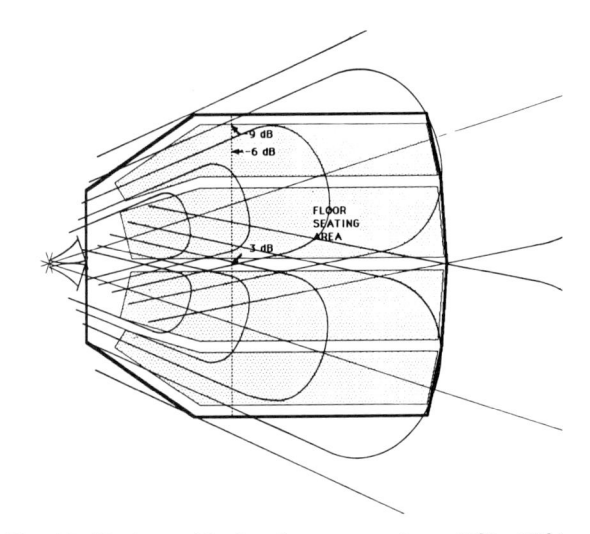

Fig. 10. Horizontal isolevel contours of two 40° by 20° horns aimed at rear floor.

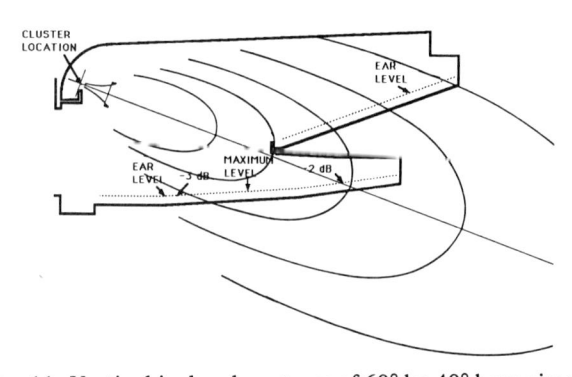

Fig. 11. Vertical isolevel contours of 60° by 40° horn aimed at rear floor.

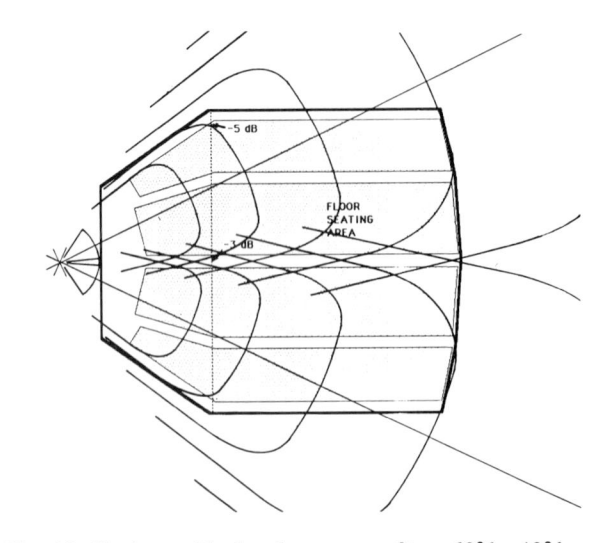

Fig. 12. Horizontal isolevel contours of two 60° by 40° horns aimed at rear floor.

the effects of distance as well as horn dispersion in the results. This represents a strong advantage over some other techniques.

Now we have only the front floor seats to cover. Fig. 13 shows the vertical isolevel contours for the 40° by 60° horn previously selected, plus those for a 40° by 90° horn aimed into the front floor area. The two coverages overlap at their mutual −3-dB points and, thereby, provide coverage uniform to ±1 dB from the front row to the cutoff point under the balcony.

Fig. 14 shows the transverse coverage of the 40° by 90° horn at the front row of seats. Here the drop-off to the sides is only 4 dB. Thus the side seats here will receive sound levels only about 5 dB lower than the maximum.

Another noteworthy implication of the isolevel contours shown in Fig. 13 is that the front section of the balcony will receive as much sound level from the 40° by 60° horns as the floor seats will. This will completely override the slight drop-off of the long-throw horns toward the front of the balcony and will also fill in the weak areas in the front corners of the balcony. This side effect is beneficial in this case, but it may not always be so. In any case, it is immediately obvious from the isolevel contours, while it may be much more obscure with other techniques.

Fig. 13 also shows that the level at the edge of the stage is about 7 dB below the maximum in the seats. This confirms that the direct sound level on stage will not be so high as to cause undue feedback.

Fig. 15 shows the actual layout of the horns in this cluster, which is almost exactly what we have just designed. Ideally they should be positioned much more closely together, but the architecture of the room made this impossible, as is shown in Fig. 16. Nevertheless, the measured coverage, after the drive levels were adjusted for equal sound levels in the respective coverage areas, was highly uniform, measuring ±2 dB in the 2-kHz octave band in all seats. The excellent results are almost exactly as predicted by the isolevel contours.

Coverage through the overlap zones between horns is never smooth, and it was worse than usual between the front and rear floor horns because of the gross misalignment of the drivers. For perfectly smooth coverage, the distances from any listener in the overlap zone to each driver must match to less than about one-quarter wavelength. This is possible at lower frequencies, but not at high frequencies because of the configurations involved. The best approach is to minimize the problem by having as few overlaps as possible, and by placing the drivers very close together and carefully aligned.

Just for fun, let us consider some other horn configurations in this auditorium, especially on-stage locations. Fig. 17 shows the effects of placing a 40° by 90° horn on stage (actually about 3 m above the stage, as if it were in a loudspeaker stack) and aiming it straight out. It will reach the rear seats, of course, but the level there will be 9 dB lower than in front. Note also that the front of the balcony gets 3 dB more level than the rear of the floor. This is not a very good performance, but is widely accepted without question.

Fig. 18 shows the same situation but with a 20° by 40° horn. Here the level is very even from the front to halfway back, and is down only 5 dB at the back. This is much better coverage, but who would think that a long-throw horn is needed in a situation such as this?

Fig. 19 depicts a 40° by 90° horn aimed into the balcony. Not only does the level drop 5 dB from the front of the balcony to the rear, but the level at the front of the floor is 5 dB higher than even at the front

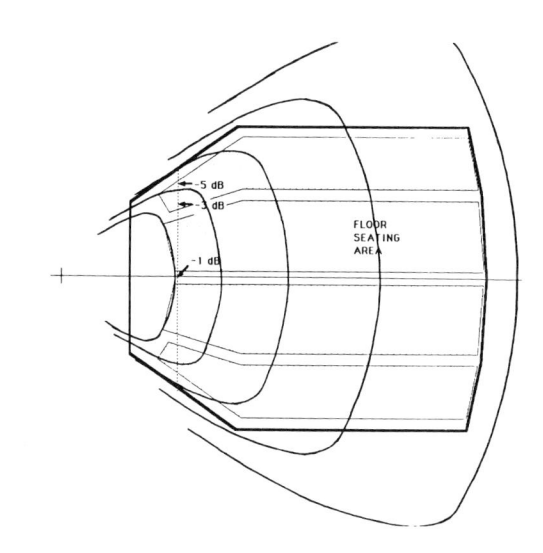

Fig. 14. Horizontal isolevel contours of 90° by 40° horn aimed at front floor.

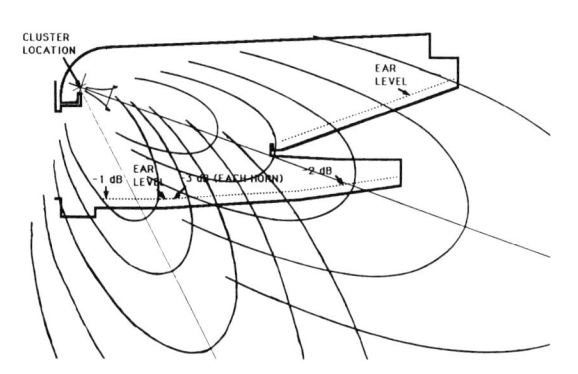

Fig. 13. Vertical isolevel contours of 60° by 40° horn aimed at rear floor and 90° by 40° horn aimed at front floor.

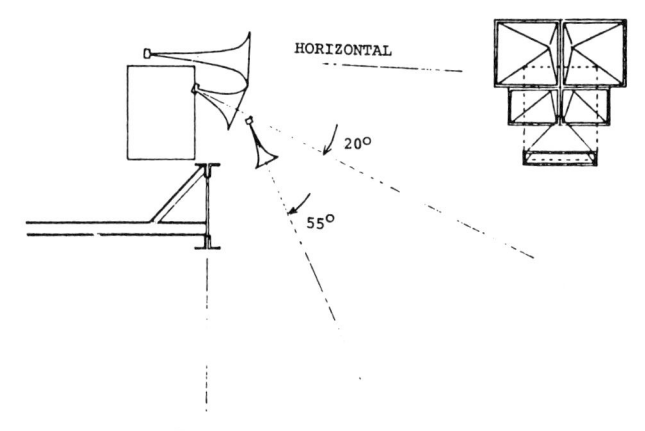

Fig. 15. Final horn layout.

of the balcony. Fig. 20 shows a long-throw horn in the same orientation. Here the drop-off from the front to the rear of the balcony is reduced to 4 dB, only a slight improvement. However, the level at the front of the floor is now the same as at the front of the balcony, a 5-dB improvement. Again, this bit of information might never surface with other design techniques.

These examples clearly show the importance of correct horn cluster location. Trying various locations to see which works best is very clumsy with most design techniques, but the isolevel contour overlays make it quite easy, with the results being immediately obvious.

Another design example, for a room that needed 360° coverage from a central cluster, was given in a previous

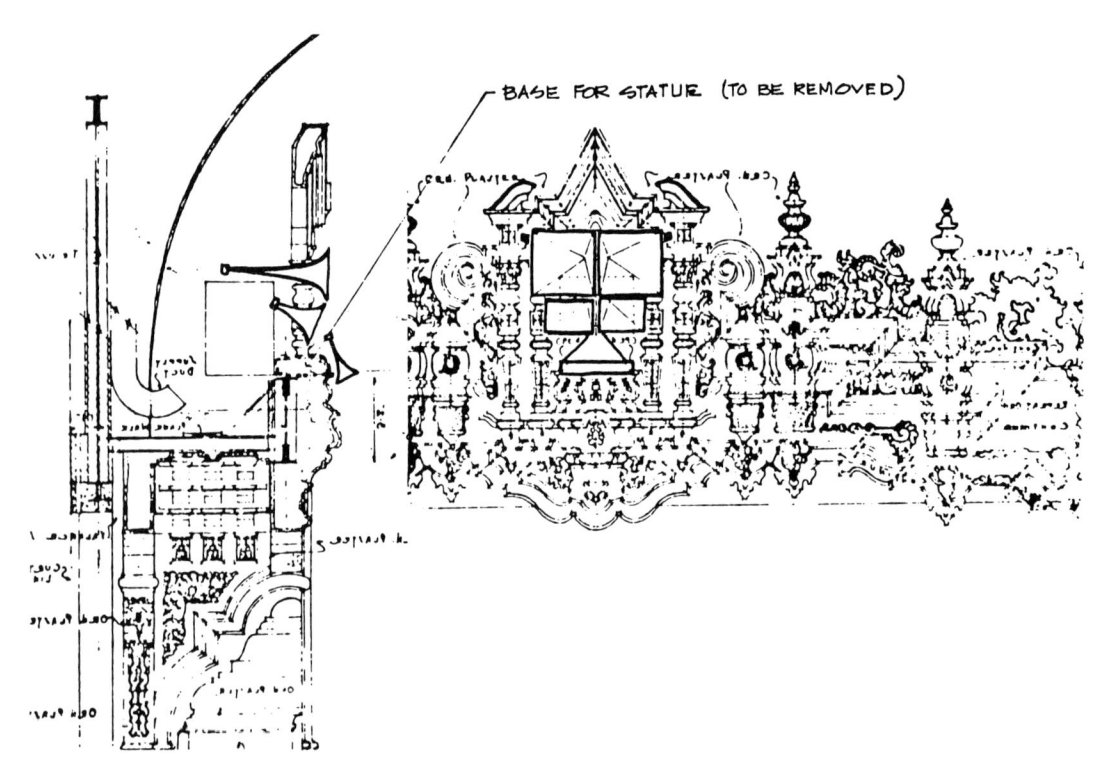

Fig. 16. Final horn layout showing surrounding architecture.

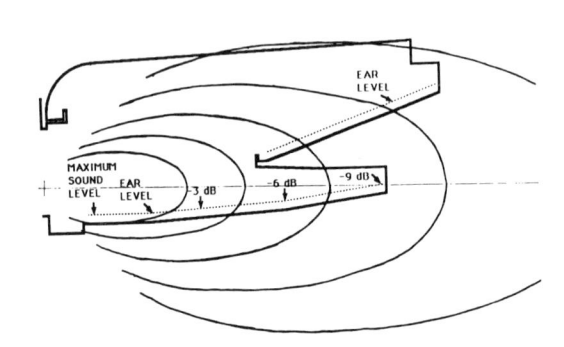

Fig. 17. Vertical isolevel contours of 90° by 40° horn located on stage and aimed straight out.

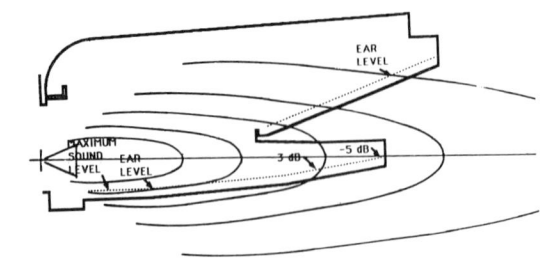

Fig. 18. Vertical isolevel contours of 40° by 20° horn located on stage and aimed straight out.

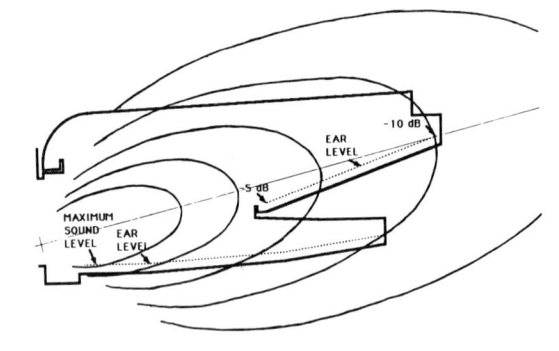

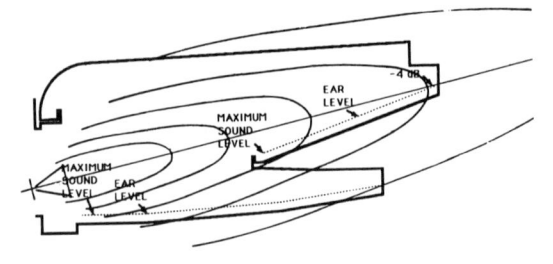

Fig. 19. Vertical isolevel contours of 90° by 40° horn located on stage and aimed at balcony.

Fig. 20. Vertical isolevel contours of 40° by 20° horn located on stage and aimed at balcony.

paper [6]. In that paper, in addition to detailed design procedures, extensive data on the actual performance of the completed system are also given.

3 LIMITATIONS

It is clear that the accuracy of the isolevel contours along a horn's horizontal and vertical axes is simply a matter of measurement technique. The levels between these axes, however, are inferred by interpolation. Undoubtedly this will result in some degree of error, but how much? A few manufacturers now provide data on dispersion over the entire frontal area of their horns, in addition to that on the two axes. In several such cases the coverage predicted by the isolevel contours, which translates into elliptical "footprints" on a plane normal to the horn axis, was compared with published data, which show more irregular "footprints." Most horns' measured characteristics fall within 1 dB or so of the appropriate isolevel predictions, but some have obvious aberrations, which may deviate locally by 2–3 dB.

There are a few cases where the isolevel contour technique does not work well, such as when a horn is aimed diagonally into a sloped seating area. To predict the sound levels accurately in such a case, the virtual horn location may need to be displaced both horizontally and vertically from the real location, and the horn rotated about its main axis. This is possible to do, but it is cumbersome and confusing.

No situation has yet been encountered which precluded the use of the isolevel contour technique. Some cases, such as those just noted, require additional effort to achieve the desired results, but most are quite straightforward. In fact, isolevel contours have been used successfully in hundreds of sound system designs, performed by contractors as well as consultants for the past 10 years.

4 CONCLUSIONS

Isolevel dispersion contours provide extensive, accurate information on actual sound coverage from horns quickly and easily. They can be used for single or multiple horns, and encourage experimentation to determine the best possible configuration. They are intuitive and informative because they closely represent real-world horn performance.

5 REFERENCES

[1] W. A. Malmlund and E. A. Wetherill, "An Optical Aid for Designing Loudspeaker Clusters," *J. Audio Eng. Soc.*, vol. 13, p. 57 (1965 Jan.).

[2] T. Uzzle, "Loudspeaker Coverage by Architectural Mapping," *J. Audio Eng. Soc.*, vol. 30, pp. 412–424 (1982 June).

[3] J. R. Prohs and D. E. Harris, "An Accurate and Easily Implemented Method of Modeling Loudspeaker Array Coverage," *J. Audio Eng. Soc.*, vol. 32, pp. 204–217 (1984 Apr.).

[4] J. Eargle and M. Kalmanson, "Design by Computer," *Sound & Video Contrac.*, vol. 2 (1984 July 15).

[5] B. Thurmond, "Measured Performance of Loudspeaker Combinations," presented at the 58th Convention of the Audio Engineering Society, *J. Audio Eng. Soc. (Abstracts)*, vol. 25, p. 1074 (1977 Dec.), preprint 1284.

[6] B. Thurmond, "A Legislative Sound System," presented at the 63rd Convention of the Audio Engineering Society, *J. Audio Eng. Soc. (Abstracts)*, vol. 27, p. 604 (1979 July/Aug.), preprint 1483.

THE AUTHOR

Bob Thurmond is a 1962 graduate of The University of Texas at Austin. Before joining Tracor, Inc. of Austin in 1965, he had experience in a recording studio, a language laboratory, and the U.S. Army Corps of Engineers. An acoustical consultant since then, he has also been with McCandless Consultants, Inc. (which he helped found in 1969), and G. R. Thurmond and Associates (which he formed in 1971). His experience includes sound reinforcment system design and equalization, room acoustics, instrumentation, and circuit design, and he has written more than 20 technical articles on these topics.

In recent years Mr. Thurmond has devoted considerable time to research and development in the fields of test instruments and equalization hardware, and to testing of professional sound equipment. He has been a member of the Audio Engineering Society since 1965 and is a member of the AES Standards Working Group for Sound Reinforcement Products. He is also the Chairman of the Austin Section.

Cluster Suitability Predictions Simplified*

BOB THURMOND

G. R. Thurmond and Associates, Austin, TX 78756, USA

A central-cluster-type loudspeaker is preferred for sound reinforcement whenever it is suitable. Of the several factors affecting its suitability, the one most difficult to evaluate is room reverberation. Now a simplified technique provides an accurate prediction of the speech intelligibility potential in a given room. If there is too much reverberation, this technique will predict how much additional room absorption is needed.

0 INTRODUCTION

Experience has shown that a central cluster loudspeaker will usually outperform any other type for several reasons. Such a single, small sound source provides directional information for our ears and minimizes frequency- and location-dependent irregularities due to cancellation effects. Under some circumstances, however, a central cluster cannot be made to work adequately. Such circumstances include the following.

1) Acceptably even coverage of the audience is not possible with available loudspeaker devices.

2) The distance between cluster and listener is too great to obtain adequate sound levels with available devices.

3) An acceptable cluster location is not available.

4) The appearance of a loudspeaker cluster is unacceptable.

5) Source and cluster locations differ excessively.

6) Cost is too great.

7) Room reverberation is excessive.

All these factors are easy to consider and evaluate, except the last one. In fact, no way has been found to predict with any real accuracy whether a given cluster would work acceptably in a given reverberant room. An attempt was made to fill in this gap.

1 BACKGROUND

Other than frequency response, evenness of coverage, and signal-to-noise ratio, probably the most significant evaluator of a sound system is speech intelligibility. In a highly reverberant room, the first three factors can be excellent and the last completely unacceptable. Fortunately intelligibility in a reverberant situation is closely related to the ratio of direct to reverberant sound energy, which is rather easy to measure and even calculate under certain circumstances. The formula for calculating this ratio is

$$\frac{\text{SPL}_{\text{dir}}}{\text{SPL}_{\text{rev}}} = 20 \log \frac{D_C}{D_R} \qquad (1)$$

where D_R in the source (cluster) to listener distance and D_C the critical distance. D_C is given (in feet) by [1]

$$D_C = 0.031 \sqrt{\frac{R_\theta V}{T}} \qquad (2)$$

---

* Presented at the 81st Convention of the Audio Engineering Society, Los Angeles, 1986 November 12–16; revised 1987 December 2 and 1988 February 24.

where V is the room volume in cubic feet and T is reverberation time. The volume can be calculated easily from appropriate measurements, and reverberation can be measured or calculated by conventional means, or estimated by a simplified technique [2]. This leaves only R_θ, the directional characteristic of the source, as the problem.

Nowadays R_θ values for individual horns are routinely published by major manufacturers, but no method has been found for calculating the R_θ of a cluster of horns, except for a few very simple cases. Therefore no means has existed for calculating the intelligibility of speech over most sound systems. Clearly, such a means is needed.

Earlier researchers have devised methods for predicting speech intelligibility in a reverberant room, but these do not include a sound system. Lochner and Burger [3], for example, require that certain measurements be taken in the actual room or a scale model of it, using a specific acoustic source. Peutz [4] describes a purely analytical method, which is based on a nondirectional source. It is widely assumed that this or another technique could be applied to a reinforced situation by simply figuring in the directivity of the particular source. This may be true, but it has never been done except in simple cases because there was no way to predict the directivity of a complex source.

2 A METHOD

The R_θ of any source is defined as the ratio of the output in a reference direction to the average output in all directions. If the source were a cluster of ideal horns, that is, horns whose output is uniform over a specified solid angle and zero in all others, then we could calculate the cluster R_θ by knowing only the R_θ and the output level of each horn [5]. By definition,

$$R_\theta = \frac{\mathrm{SPL}_\theta^2}{\overline{\mathrm{SPL}^2}} \qquad (3)$$

and since the sound power level L_W is proportional to the sound pressure level squared,

$$R_\theta = \frac{L_{W\theta}}{\overline{L_W}} \qquad (4)$$

we can find $\overline{L_W}$ by

$$\overline{L_W} = \frac{L_{W1}a_1 + L_{W2}a_2 + \cdots + L_{Wn}a_n}{a_{\mathrm{TOTAL}}} \qquad (5)$$

for n sources (horns), each covering a solid angle a with a sound power level L_W, all measured at the same distance. If we define a_{TOTAL} (the surface of a sphere with its center at the center of the cluster) to be 1, then each individual a represents a fraction of the total surface area, which is given by $1/R_\theta$ for that horn. Therefore,

$$\overline{L_W} = \frac{L_{W1}}{R_{\theta 1}} + \frac{L_{W2}}{R_{\theta 2}} + \cdots + \frac{L_{Wn}}{R_{\theta n}}. \qquad (6)$$

A convenient trick becomes possible if we assume that every horn in the cluster produces the same SPL and, therefore, L_W at the point where its axis line intersects the audience plane. In most good cluster designs this is very nearly true. Let us call D the distance between this point and the cluster for each horn. Then, because of the inverse square rule,

$$\frac{L_{W2}}{L_{W1}} = \frac{D_2^2}{D_1^2}. \qquad (7)$$

Since we almost certainly know D for each horn, we can now find all the L_W ratios between the various horns. Let us now designate the horn with the highest output to be horn 1. Let its axis also be the reference direction θ for the cluster. This axis will pass very close to the most distant seat, which is where speech intelligibility will be the worst. Thus the distance from the cluster to this seat will be D_1, and also $D_{\max}$.

Let us now assign a value of 1 to L_{W1}. Then all the other L_W will have values between 0 and 1, which we can find easily. We now insert the values into Eq. (5) to find $\overline{L_W}$. Since we designated $L_{W\theta} = L_{W1} = 1$, we now have the cluster R_θ. Next we calculate D_C and find the ratio $D_{\max}/D_C$. From this we could then calculate $\mathrm{SPL}_{\mathrm{dir}}/\mathrm{SPL}_{\mathrm{rev}}$, but it is not really necessary to do so, as we shall see.

It should be noted that D_C varies with direction around a directional source since the direct sound level varies with direction and the reverberant level is nearly constant everywhere. Since a well-designed horn cluster will produce an $\mathrm{SPL}_{\mathrm{dir}}$ that is nearly constant throughout the seating area, the $\mathrm{SPL}_{\mathrm{dir}}/\mathrm{SPL}_{\mathrm{rev}}$, ratio and, therefore, the speech intelligibility will be nearly the same at all seats, no matter what their distance from the cluster.

3 PRACTICAL CONSIDERATIONS

Since many of the terms in our equations have values that are frequency dependent, we must select a frequency range for specific calculations. Since we are concerned with speech intelligibility, our frequency range will certainly be within the primary speech range of 300 Hz to 3 kHz. The consonant sounds, which are most important and most easily lost, are concentrated in the upper part of this range. On the other hand, reverberation time is often significantly shorter at higher frequencies, especially in highly reverberant spaces, making the middle frequencies more difficult. A definitive investigation would include calculations at several frequencies, but if we want to make a "worst-case" calculation, we should probably select the octave band centered at 1 kHz.

Our derivations included two major assumptions, one explicit and one tacit. The explicit assumption was that the horns were perfect—which, of course, they are not. Real horns have an output that is somewhat irregular over their defined coverage angle, and some output outside this angle. The former imperfection will cause the direct sound level to vary somewhat from one location to another. This effect can be minimized by careful selection and layout of the horns [6]. Reflections and obstructions will have similar effects, and are often impossible to avoid. In other words, there will always be some seat-to-seat variations in coverage and intelligibility, a fact that experience has already taught us. These calculations should yield a reasonable approximation, however.

The latter imperfection is not very great at all on the best horns currently available [7]. Others are not so good, however, so we should consider what this means. Essentially, a significant amount of energy will be radiated outside their rated coverage angle, adding to the reverberant energy and lowering the actual cluster R_θ. However, such poorer horns generally have a lower R_θ rating to begin with, so our calculations should not be that far off. Furthermore, such excess radiation can also increase SPL_{dir} in the reference direction, which would act to raise the actual cluster R_θ. In most cases this imperfection can probably be ignored.

The tacit assumption is that all the energy from the cluster will become diffused throughout the room, turning into reverberant energy. This will be approximately true if the area at which the cluster is aimed is no more absorptive than the average of all the room surfaces. In practice this means hard seats and little or no audience. This is exactly the case in many rooms where reverberation is a serious problem, so our results should be accurate. However, this will not always be the case, so we should consider accordingly.

The sound energy which strikes an absorptive area first will be reduced by a fraction equal to the absorptive coefficient of that area. A typical value for an occupied audience area is 0.8, which means that if all the sound from the cluster were directed into such an audience area, the reverberant level would be reduced by 80%, or 7 dB. This will never be achieved in reality, however, because not all of the sound energy is directed at the seating area.

In fact, the horns with the greatest output will usually be aimed at the top or last row of seating, so that half their output misses the seating area. If the surfaces above the seating have low absorption, the energy striking them becomes reverberant, and the total reverberant level is reduced by only 40%. Other horns may be aimed more into the seating area, but they are usually operated at lower levels, thus contributing less to the reverberant field. Typically an audience reduces the reverberant level by 3 or 4 dB, compared to an empty seating area of low absorption, which represents an increase in D_C of about 1.5. This value is certainly appropriate for a first estimate. It should be multiplied by the new value of D_C which was found recalculating using the new T value obtained as a result of the added absorption of the audience.

4 EXAMPLES

To test this method, several examples were selected where the necessary characteristics were known and where long reverberation time caused speech intelligibility problems. The first was an auditorium with about 2000 seats, a volume of 200 000 ft^3 (5664 m^3), and almost no absorption. The measured empty reverberation time was 3.0 s at 1 kHz. The cluster consisted of two 60° radial horns aimed at the rear row of balcony seats, a distance of 120 ft (36 m), and two 90° radial horns aimed at the floor at a distance of 100 ft (30 m). The R_θ of these particular horns were 16 and 8, respectively, at 1 kHz. From Eq. (6),

$$\overline{L_W} = \frac{1}{16} + \frac{1}{16} + \frac{0.833^2}{8} + \frac{0.833^2}{8} = 0.2985$$

and, from Eq. (4),

$$R_\theta = \frac{1}{0.2985} = 3.35 \ .$$

From Eq. (2),

$$D_C = 0.031 \sqrt{\frac{3.35 \times 400\,000}{3.0}} \approx 21 \text{ ft (6.3 m)} \ .$$

This gives a D_{max}/D_C ratio of 5.8, much greater than 4, which has long been recommended [8] as the maximum acceptable value for this ratio. In fact, speech was quite unintelligible over this system. With a full audience the reverberation time dropped to 1.3 s, which brought the uncorrected D_C up to about 33 ft (10 m), and the corrected value to 49 ft (15 m). This gives a new D_{max}/D_C ratio of 2.4, which should have been quite acceptable. In fact, it was just barely so, which gives us pause.

A second example is also instructive. This one is a large arena, almost square in plan, with a central cluster. The cluster was rather poorly designed, with considerable overlap between several horns and quite uneven coverage, so our proposed R_θ calculation method might not be accurate enough. However, the overall coverage was roughly a hemisphere, so a reasonable estimate of the cluster R_θ would be 2. The maximum D_R was 160 ft (48 m), V was 5 000 000 ft^3 (141 600 m^3), and T at 1 kHz was 4.4 s, a definite problem.

Calculations reveal a D_C of almost 47 ft (14 m) and a D_{max}/D_C ratio of 3.4. Conventional wisdom indicates that this should have been acceptable, but it was not. In fact, speech was almost unintelligible over this system. A full audience of 10 000 people, a rare occurrence, brought the 1-kHz T down to 2.6 s and the corrected D_{max}/D_C to about 1.8. This was considered acceptable, but the near-empty conditions were not. Absorption was applied to the ceiling, which brought the 1-kHz T down to 1.8 s and the empty D_{max}/D_C ratio to 2.2. This proved to be quite satisfactory.

Several other examples were checked, and a pattern began to emerge. One arena with poor intelligibility was found to have a D_{max}/D_C ratio of 3.2. Two others, where the intelligibility was fair, had D_{max}/D_C ratios of 2.4 and 2.7. These suggest a possible rating scale.

D_{max}/D_C	Intelligibility
2.0	Good
2.5	Acceptable
3.0	Marginal
3.5	Unacceptable

A high-quality system, well equalized, might produce an intelligibility somewhat higher than these ratings would indicate, but not much so. Conversely, a poorly executed system could easily have lower intelligibility.

There are probably other valid methods for calculating speech intelligibility over a sound system, but all must take the directivity of the source into account. In the case of a complex horn cluster, the method just described makes such calculations possible for the first time.

5 IMPLICATIONS

What if these calculations indicate that a particular situation is unsatisfactory? What can be changed to improve it? There are only four parameters—D_{max}, R_θ, V, and T—which affect the situation at all. D_{max} is a function of the room configuration, which probably cannot be changed, and of the cluster location, which may be variable. The impulse is to reduce D_{max} by moving the cluster location closer to the center of the seating area. However, this will require a broader coverage, which means a lower cluster R_θ, which tends to offset the advantage of the lower D_{max}. In fact when all things are properly considered, the cluster location will have almost no effect on its intelligibility.

The cluster R_θ is largely a function of the cluster location and the audience area configuration as seen from that location. All cluster layouts which fulfill the design goals of even coverage within the audience area and minimal coverage outside of it will be nearly identical overall. Thus the cluster R_θ can be changed significantly only by changing the cluster location, as we have already discussed.

It is very unlikely that the room volume can be changed. This leaves the reverberation time as the only parameter that might be changed significantly. Knowing that this is our only option, we can start with a desired D_{max}/D_C ratio of 2.0 or 2.5 and work backward through the calculations to obtain the necessary reverberation time, and even the necessary additional absorption to produce it. If this cannot be achieved, then a central loudspeaker cluster simply will not work satisfactorily, and another approach must be found.

6 CONCLUSIONS

It is possible to predict whether a central loudspeaker cluster will provide satisfactory speech intelligibility in a given reverberant space. This is done by designing a cluster to provide proper audience coverage, calculating (or estimating) the R_θ of that cluster, then calculating and evaluating the resulting D_{max}/D_C ratio. If this ratio is too high, then speech intelligibility will be unsatisfactory over the system. The only effective correction is to reduce the room reverberation.

7 REFERENCES

[1] H. G. Smith, "Acoustic Design Considerations for Speech Intelligibility," *J. Audio Eng. Soc.*, vol. 29, pp. 408–415 (1981 June).

[2] G. R. Thurmond, "Count Your AEP's," *db Mag.* (1975 Dec.).

[3] J. P. A. Lochner and J. F. Burger, "The Influence of Reflections on Auditorium Acoustics," *J. Sound Vib.*, vol. 1, p. 426 (1964).

[4] V. M. A. Peutz, "Articulation Loss of Consonants as a Criterion for Speech Transmission in a Room," *J. Audio Eng. Soc.*, vol. 19, pp. 915–919 (1971 Dec.).

[5] C. T. Molloy, "Calculation of the Directivity Index for Various Types of Radiators," *J. Acoust. Soc. Am.*, vol. 20, p. 387 (1948).

[6] G. R. Thurmond, "Horn Layout Simplified," *J. Audio Eng. Soc.*, vol. 35, pp. 976–983 (1987 Dec.).

[7] G. L. Augspurger, *Sound System Design Reference Manual*, JBL Inc., 1982.

[8] D. Davis, "Calculating and Using Critical Distance," Altec Lansing Tech. Lett. 196 (1979).

THE AUTHOR

Bob Thurmond is a 1962 graduate of The University of Texas at Austin. Before joining Tracor, Inc. of Austin in 1965, he had experience in a recording studio, a language laboratory, and the U.S. Army Corps of Engineers. An acoustical consultant since then, he has also been with McCandless Consultants, Inc. (which he helped found in 1969), and G. R. Thurmond and Associates (which he formed in 1971). His experience includes sound reinforcement system design and equalization, room acoustics, instrumentation, and circuit design, and he has written more than 20 technical articles on these topics.

In recent years Mr. Thurmond has devoted considerable time to research and development in the fields of test instruments and equalization hardware, and to testing of professional sound equipment. He has been a member of the Audio Engineering Society since 1965 and is a member of the AES Standards Working Group for Sound Reinforcement Products. He is also the Chairman of the Austin Section.

INTRODUCTION

The Technical Committee on Acoustics and Sound Reinforcement was formed to promote, report on, and focus AES activity in the areas of reinforced sound and acoustics. The production, transmission, and perception of sound are emphasized. In addition, the participation of academia, manufacturers, and practitioners in committee-organized events is encouraged.

One function of the Technical Committee on Acoustics and Sound Reinforcement is to organize special paper sessions on topics identified as particularly important or timely. A session chairman is chosen or volunteers from the committee. The special session chairman, with the help of the committee chairman, compiles a list of prospective authors who are invited to submit abstracts for papers for the special session. Thus, in contrast to the open paper sessions, these special sessions are intended to focus the latest research in a specific area through special invitation.

In 1988, David Klepper organized the special session on Speech Intelligibility presented at the AES 85th Convention in Los Angeles. In 1989, at the 87th Convention in New York, the special session was on Large-Array Systems. For 1990 at the 89th Convention in Los Angeles, Michael Klasco is planning a special session on the modeling of acoustic spaces, including the latest advances in both architectural scale modeling and computer modeling.

The Technical Committee on Acoustics and Sound Reinforcement decided at its 1988 committee meeting to organize and sponsor a special session on Large-Array Systems. Twelve authors were invited to participate and six affirmative responses were received. The session contained papers from three manufacturing companies, two universities, and one consultant. This special issue of the *Journal* contains all six papers from the special session, plus one paper from the 86th Convention held in Hamburg in 1989. The papers are published here in the same order as they were presented, followed by the paper from Hamburg. The committee hopes that, by focusing special sessions on the various areas under the purview of the committee, society members may avoid—as George Augspurger succinctly said in one of his famous reviews—imaginary solutions to nonexistent problems.

The 1989 Large-Array Systems session was particularly successful not only because attendance was outstanding, but because each author took the time to prepare an official preprint. Due to the interest in the presentations, the editors of the *Journal* agreed to publish all of the papers originating in the special session. Each manuscript has been reviewed and revised accordingly, as with all *Journal* papers.

Large arrays have been studied since the early 1930s, and by the 1950s many texts and compilations contained descriptions of colinear arrays which included the well-known description of the directional characteristics of a line array of n equal point sources, separated by equal distances d, the ubiquitous

$$R_\theta = \frac{\sin(n\pi d/\lambda \sin\theta)}{n\sin(\pi d/\lambda \sin\theta)}$$

where, at large fixed distances from the source, R is the ratio of the pressure at an angle θ to the pressure for an angle $\theta = 0$, the direction $\theta = 0$ being at right angles (normal) to the line, and where λ is the wavelength.

However, the behavior of actual arrays was not well understood. This is now changing as many manufacturers, academics, and consultants have begun extensive measurement and simulation programs. In addition, many of the techniques that have been used successfully in sonar and radar can now be implemented in audio frequency arrays as advances in signal processing equipment are accompanied by reductions in the cost of the signal processing hardware.

The Technical Committee on Acoustics and Sound Reinforcement would like to thank the authors for their work and the editors of the *Journal of the Audio Engineering Society* for the presentation of these papers in this special issue.

NEIL A. SHAW
Paul S. Veneklasen and Associates
Chairman, 87th Convention Special Session on Large-Array Systems

---

AES Technical Committee on Acoustics and Sound Reinforcement

Chairman: KENNETH D. JACOB

WOLFGANG AHNERT
SØREN BECH
MENDEL KLEINER

DAVID L. KLEPPER
JOEL LEWITZ
NEIL A. SHAW

Measurement and Estimation of Large Loudspeaker Array Performance*

MARK R. GANDER AND JOHN M. EARGLE

JBL Incorporated, Northridge, CA 91329, USA

While the individual elements used in large loudspeaker arrays may be well documented, the performance characteristics of the complete arrays themselves have not generally been measured in detail. Measurements on limited arrays, in conjunction with array theory and advanced modeling techniques, correlate well with observations made of large arrays and thus form the basis for accurate estimation of the performance of arbitrary arrays in large spaces.

0 INTRODUCTION

Concert-sound reinforcement systems are generally composed of multiple full-range loudspeaker systems, often supplemented below 50–100 Hz by large subwoofer arrays. The reasons for this are flexibility in system logistics and layout and, above all, ease in assembly and teardown.

Traditional sound reinforcement has stressed the necessity of tight clustering of like elements and the alignment of acoustical centers in overlap zones. Such considerations are often of secondary importance however, when the primary requirements are for flat power bandwidth at very high levels for all patrons.

High-level music reinforcement has developed empirically, and there is little in the way of general documentation that will enable the performance of a large system to be accurately estimated beforehand. This paper is concerned with the degree to which measurements on relatively small arrays can be extrapolated to provide guidelines for the specification of larger arrays.

---

* Presented at the 87th Convention of the Audio Engineering Society, New York, 1989 October 18–21; revised 1990 January 22.

1 MEASUREMENT METHODS

Fig. 1 shows a typical use of the class of components measured in these tests. Ground plane conditions were used [1], and plan and section views of the testing environment are shown in Fig. 2. The ground surface was smoothly graded packed gravel. Gently sloping edges of the plane provide a "soft" boundary, minimizing reflections back toward the measurement microphone.

The decision was made early in the testing procedure to limit measurements to on- and off-axis angles on the ground plane itself, inasmuch as such measurements would correspond to the actual listening plane in large outdoor venues, taking scale factors into account. The measurement distance was 10 m for off-axis measurements on all arrays of three or more elements. Care was taken to minimize frequency response aberrations due to thermal gradients over the large measurement distance.

1.1 Polar Data

While small arrays can easily be mounted on a polar turntable, arrays of the size targeted for study here would be extremely difficult to be so mounted. As an example of the kind of detail that polar measurements provide, we present the data shown in Figs. 3–8, focusing only on high-frequency horn interaction. The devices measured here were small-format horns with nominal horizontal coverage angles of 60° (Flat Front Bi-Radial[1] model 2385A). They were mounted in integral enclosures (JBL Concert Series model 4862), with side relief angles of 15°, creating a splay angle between horns of 30°. Spacing between horns was approximately 0.7 m (28 in), and the horns were offset from the center of the enclosure by approximately 0.1 m (4 in).

Fig. 3 shows the polar response of two horns mounted as described. The input signal is an 8-kHz sine wave. In many ways, a pair of loudspeakers represents the worst interference case, since the pair can easily produce peaks of 6 dB and nulls of 20 dB or more. The data are, if anything, too detailed.

Fig. 4 shows the same loudspeakers, this time with a one-third-octave band of pink noise centered at 8 kHz. Here it is clear that too much data have been lost. When the applied signal is a one-octave noise band centered at 8 kHz, the loss of detail is even greater, as shown in Fig. 5.

Figs. 6–8 show sine wave, one-third-octave, and octave wide pink noise polars for a side-by-side array of three units of the same type measured in Figs. 3–

---

[1] Bi-Radial is a registered trademark of JBL Incorporated.

Fig. 1. Typical application of full-range loudspeakers as array elements for high-level music reinforcement.

5. The driving signals are again at 8 kHz. Since the splay angles between enclosures are 30°, we note that there are effectively two sets of lobes, each occurring when the microphone is in line with adjacent pairs of loudspeakers.

Selection of the measurement signal is critical. Octave-band pink noise does not provide sufficient resolution, and one-third-octave noise bands, while informative in a general sense, tend to mask the degree of signal interference and cancellation which is inherent in all large arrays. We therefore decided to make all measurements on the limited arrays using swept sine wave signals.

Polar graphs present great detail as a function of angular variation, and they are essential in describing the fine directional response of individual horns and transducers. If we were to use polar graphs to describe the performance of a large array, we would need many of them, and the task would be a difficult one.

1.2 Data Measured at Discrete Angles

By comparison, off-axis frequency response curves present great detail as a function of frequency, and the angular interval can be chosen for the need at hand. It is possible to combine up to five or six curves on a single graph if color coding is used. The directional performance of a large array, over the normal listening angle, can often be fairly well described in a single graph, as shown in Fig. 9. Here we have measured a pair of loudspeaker systems using the same horns as were measured in Figs. 3–8, with the addition of two 300-mm (12-in) low-frequency transducers (JBL Concert Series model 4852). Frontal dimensions of the trapezoidally shaped enclosures are 1.0 m (39 in) by 0.6 m (24 in). Angular increments were taken every 7.5° from 0 to 30°, covering an included angle of 60°. Studying the data shown in Fig. 9, we note that there are three main areas in frequency response. We can see that in the region below about 200 Hz there is little response difference between the various off-axis curves.

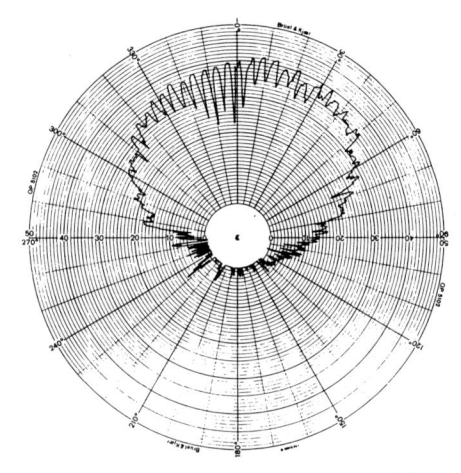

Fig. 3. Polar response; two loudspeakers (8-kHz sine wave input).

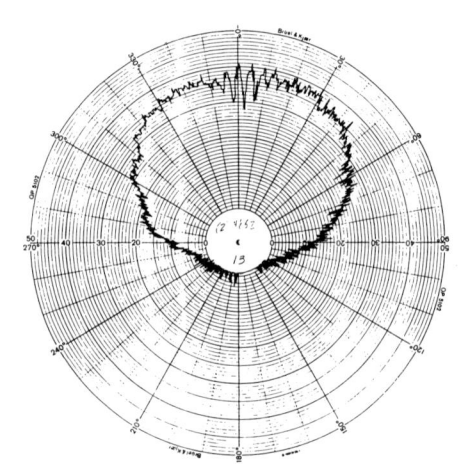

Fig. 4. Polar response; two loudspeakers (8-kHz one-third-octave-band noise input).

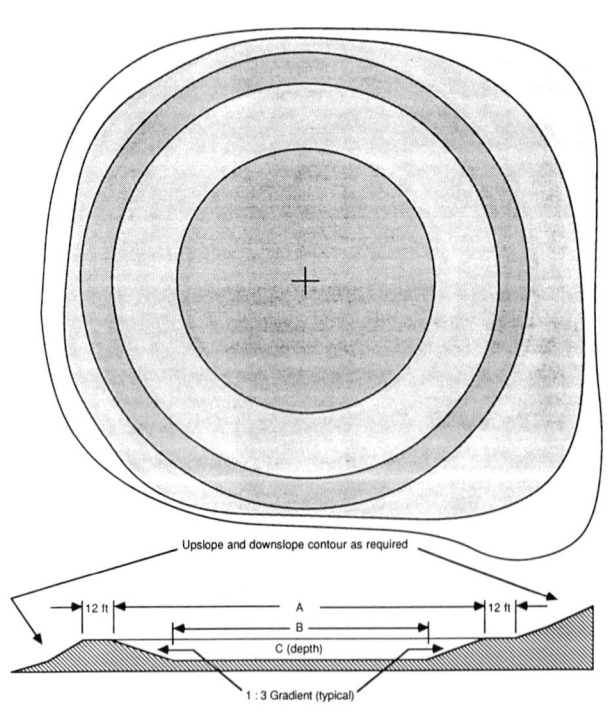

Fig. 2. Plan and section views of ground plane measuring environment. $A = 148$ ft (45 m); $B = 100$ ft (30.5 m); $C = 8$ ft (2.4 m).

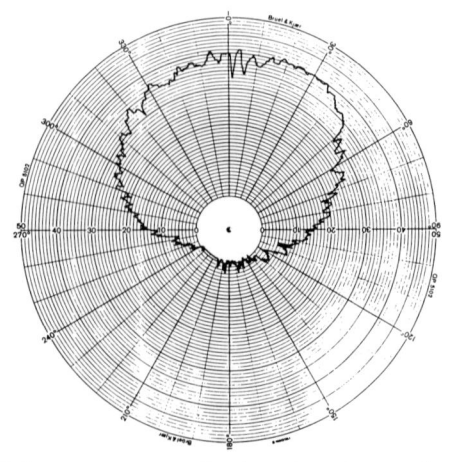

Fig. 5. Polar response; two loudspeakers (octave-band noise at 8 kHz).

In the region from 200 Hz to about 8 kHz the response clearly shows combing effects in the off-axis curves, while above 8 kHz the response variations are less and the combing intervals small.

Fig. 10 shows a similar set of curves from an array of nine (three by three) of the same loudspeakers. Note that the frequency above which lobing becomes apparent has been scaled downward due to the increased size of the array. Further, the frequency above which the combing interval has become small has moved down to about 5 kHz. Note further that the overall high-frequency response has decreased relative to the mid- and low-frequency response as a consequence of a mutual coupling and directivity increase at low frequencies, and interference effects at middle and high frequencies.

2 DETAILED MEASUREMENTS ON FULL-RANGE ELEMENTS

2.1 Flat Arrays

In this section we present off-axis measurements on groups of two, four (two by two), and nine (three by three) loudspeaker systems in various array configurations. Both flat and curved profiles were measured, since both details are used in array design. We assess the various measurements in terms of midrange lobing, drawing relative conclusions about the effectiveness of the various design options.

2.1.1 Two-Loudspeaker Arrays

Fig. 11(a) shows an array with both high-frequency elements located at the top. The off-axis measurements are shown in Fig. 11(b). The data here are the same as those presented in Fig. 9. For clarity in black and white presentation, the data are presented with no more than

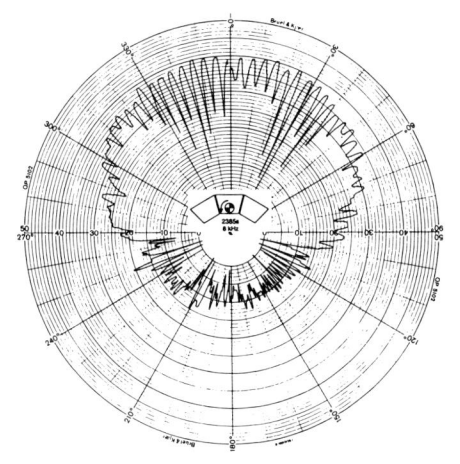

Fig. 6. Polar response; three loudspeakers (8-kHz sine wave input).

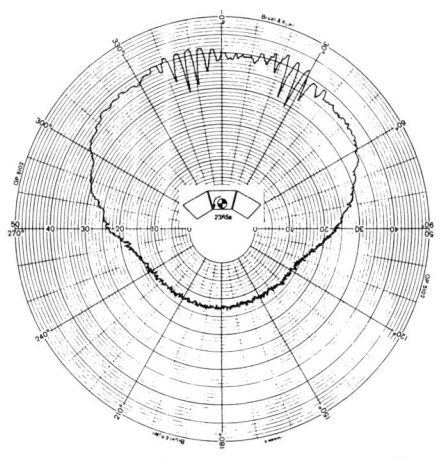

Fig. 7. Polar response; three loudspeakers (8-kHz one-third-octave-band noise input).

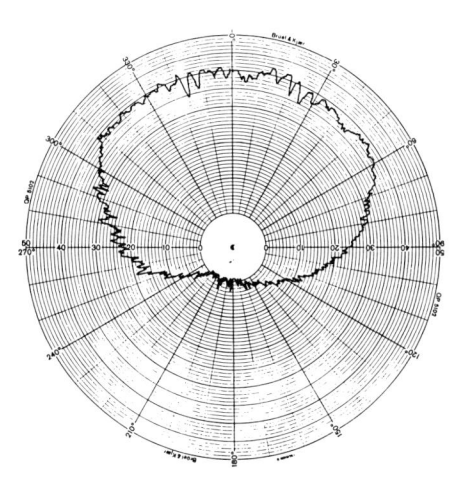

Fig. 8. Polar response; three loudspeakers (octave-band noise at 8 kHz).

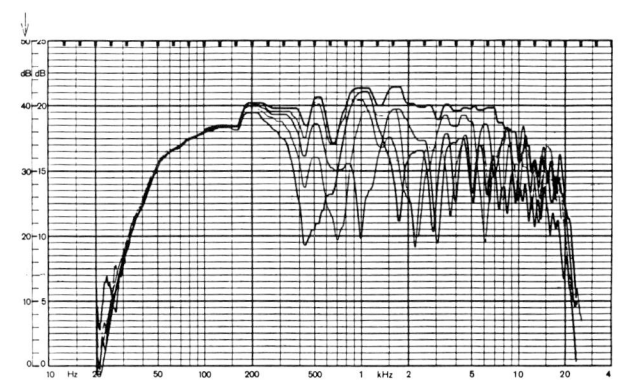

Fig. 9. Off-axis curves for two loudspeakers (0, 7.5, 15, 22.5, and 30°).

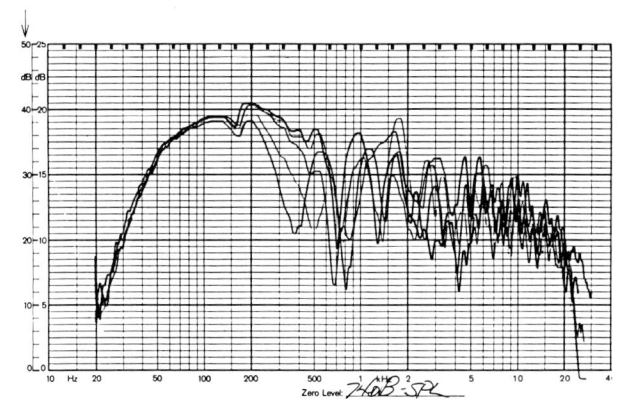

Fig. 10. Off-axis curves for nine loudspeakers (0, 7.5, 15, 22.5, and 30°).

three curves per graph. Note the pronounced lobing in the range from 400 Hz to 6 kHz.

When one of the loudspeakers is inverted, as shown in Fig. 12(a), the lobing pattern is minimized, as shown in Fig. 12(b). Here the lobing is only significant in the range from 700 Hz to about 3 kHz.

2.1.2 Four-Loudspeaker Arrays

Fig. 13(a) shows a four-loudspeaker array with all high-frequency elements in the same orientation. Fig. 13(b) shows the off-axis response curves. Note that lobing is significant in the range from 400 Hz to 4 kHz.

When two of the loudspeakers are inverted, as shown in Fig. 14(a), the off-axis response is as shown in Fig. 14(b). Note here that the lobing has improved slightly.

2.1.3 Nine-Loudspeaker Arrays

When the array size is increased to nine, as shown in Fig. 15(a), the degree of lobing is as shown to Fig. 15(b). Inverting the center loudspeakers, as shown in Fig. 16(a), improves the lobing slightly [Fig. 16(b)].

In general, plane arrays benefit from alternating, or staggering, loudspeakers in a "checkerboard" array, but the effect seems to diminish somewhat as the arrays become larger.

The balance between high- and low-frequency response in large arrays appears to be a function of array size. In the arrays measured here, the difference between 200-Hz response and the response in the region of 10 kHz averaged about 10 dB. With larger arrays we would expect to see a larger difference.

2.2 Curved Arrays

Curved arrays are used for wider dispersion than that provided by plane arrays. Accordingly, the off-axis curves were run at multiples of 12.5° increments up to 50°, indicating coverage over a total included angle of 100°. Fig. 17(a) shows details of a curved array with the elements staggered. The response is shown in Fig. 17(b).

When the elements in the array are not staggered,

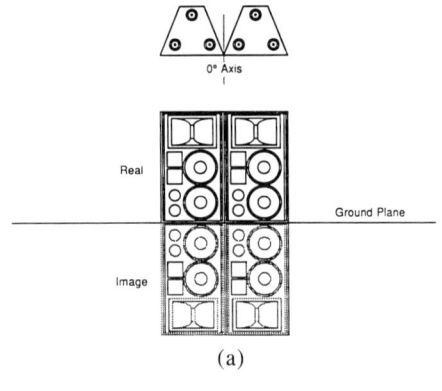

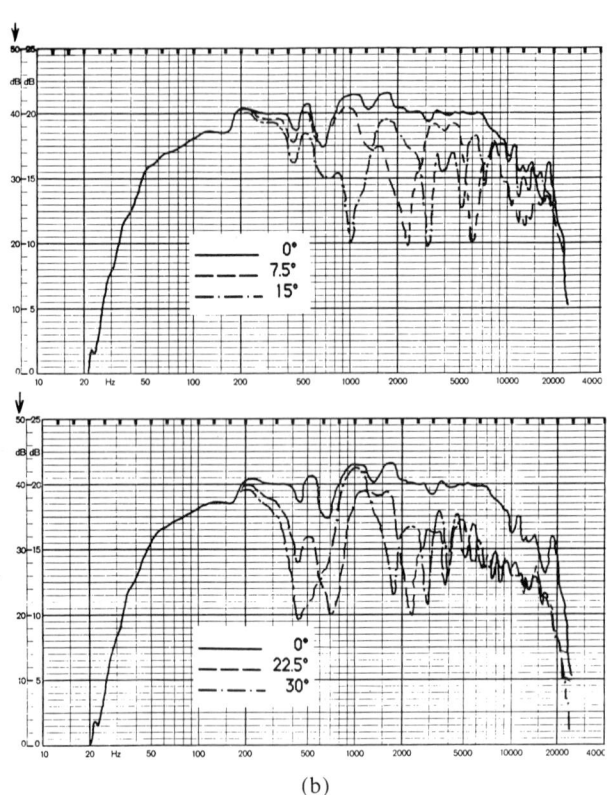

Fig. 11. Two-loudspeaker array. (a) Elevation view. (b) Off-axis response.

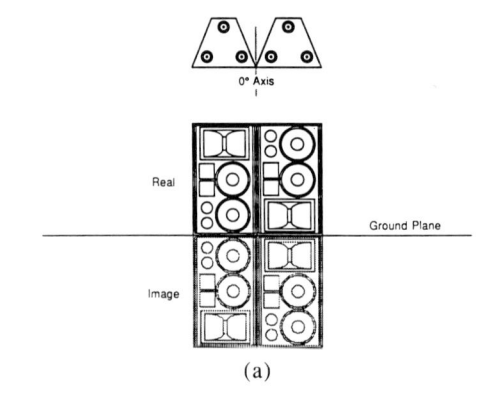

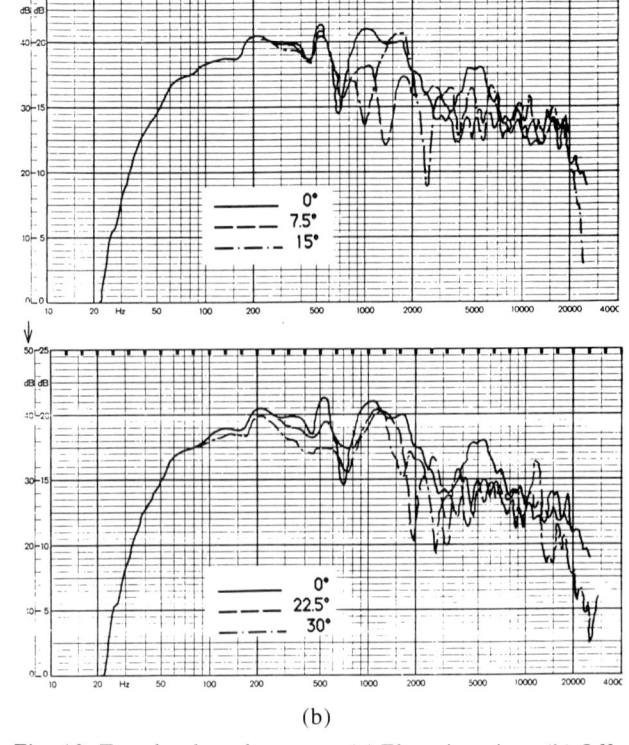

Fig. 12. Two-loudspeaker array. (a) Elevation view. (b) Off-axis response.

as shown in Fig. 18(a), the response is as shown in Fig. 18(b). Note that there is little difference between the two sets of measurements. The reason for this appears to be that the wide angular splay of 45° between adjacent systems results in less interference between them to begin with, hence the limited effect of staggering as opposed to nonstaggering.

Fig. 19(a) shows a curved arrangement in which the systems are separated by small standoffs of about 100 mm (4 in) in length. This detail is presented here since the technique is often used as a method to ensure safety in rigging. The response, shown in Fig. 19(b), produces slightly more lobing at high frequencies than the array shown in Fig. 18 due to the increased spacing between elements. The tradeoff here is some degradation in response versus choice of rigging method for ease, redundancy, and safety.

Fig. 20 shows another orientation of the curved array in which the 0° reference angle has been rotated by 90°. The off-axis measurements are made at increasing counterclockwise angles, as seen in plan view. Since the array is the same as the one measured in Fig. 18, the data shown here provide a view of the array at even greater angles off the axis of symmetry.

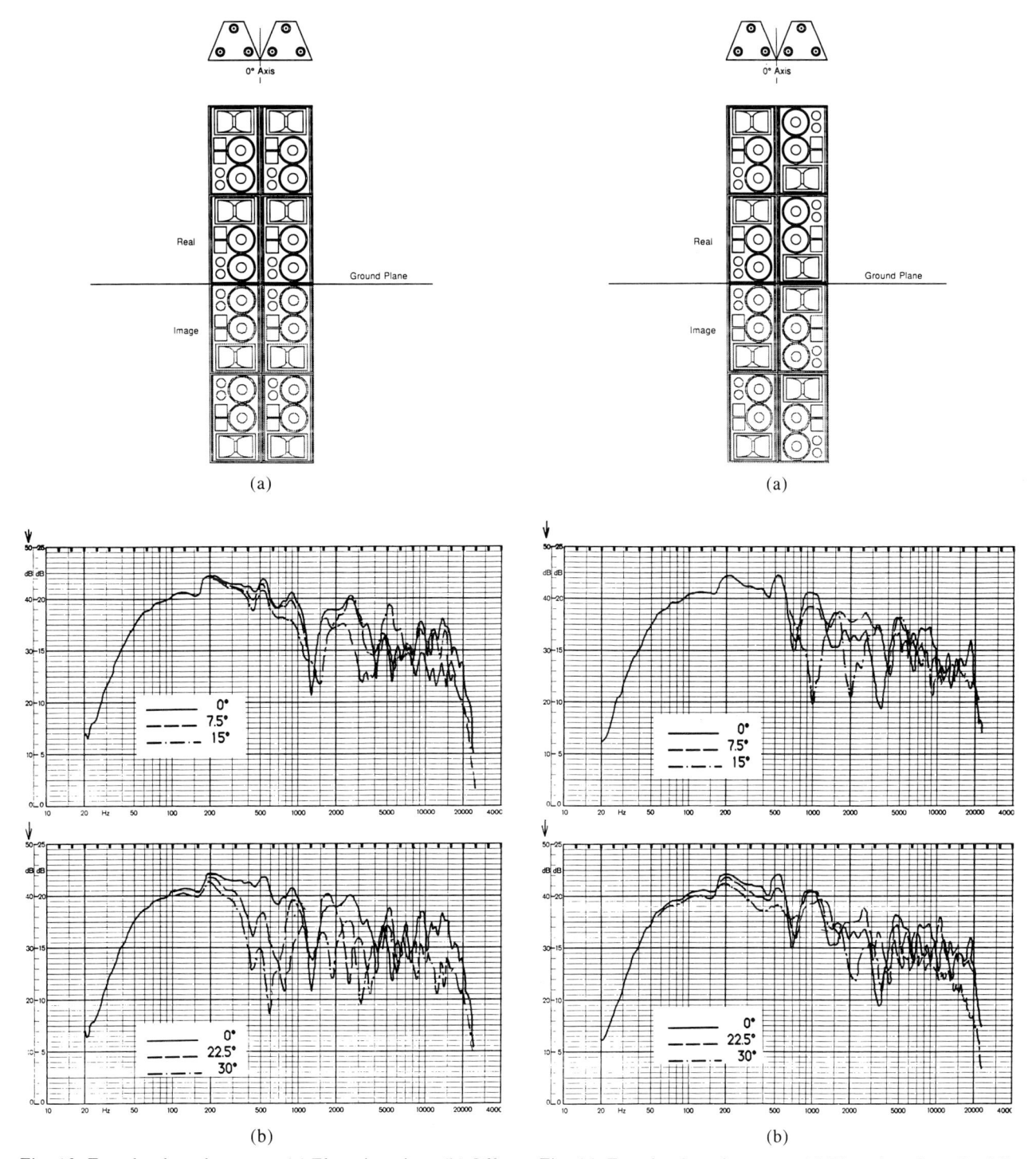

Fig. 13. Four-loudspeaker array. (a) Elevation view. (b) Off-axis response.

Fig. 14. Four-loudspeaker array. (a) Elevation view. (b) Off-axis response.

Fig. 21 shows another orientation of the loudspeakers in which the normal 45° splay between loudspeakers has been changed to 22.5°. There are only slight differences in response between this orientation and that in Figs. 18 and 20.

2.3 Corner Transitions

Two sets of corner transitions are shown in Figs. 22 and 23. Corner transitions would normally be used at the end of a planar section, wrapping around to

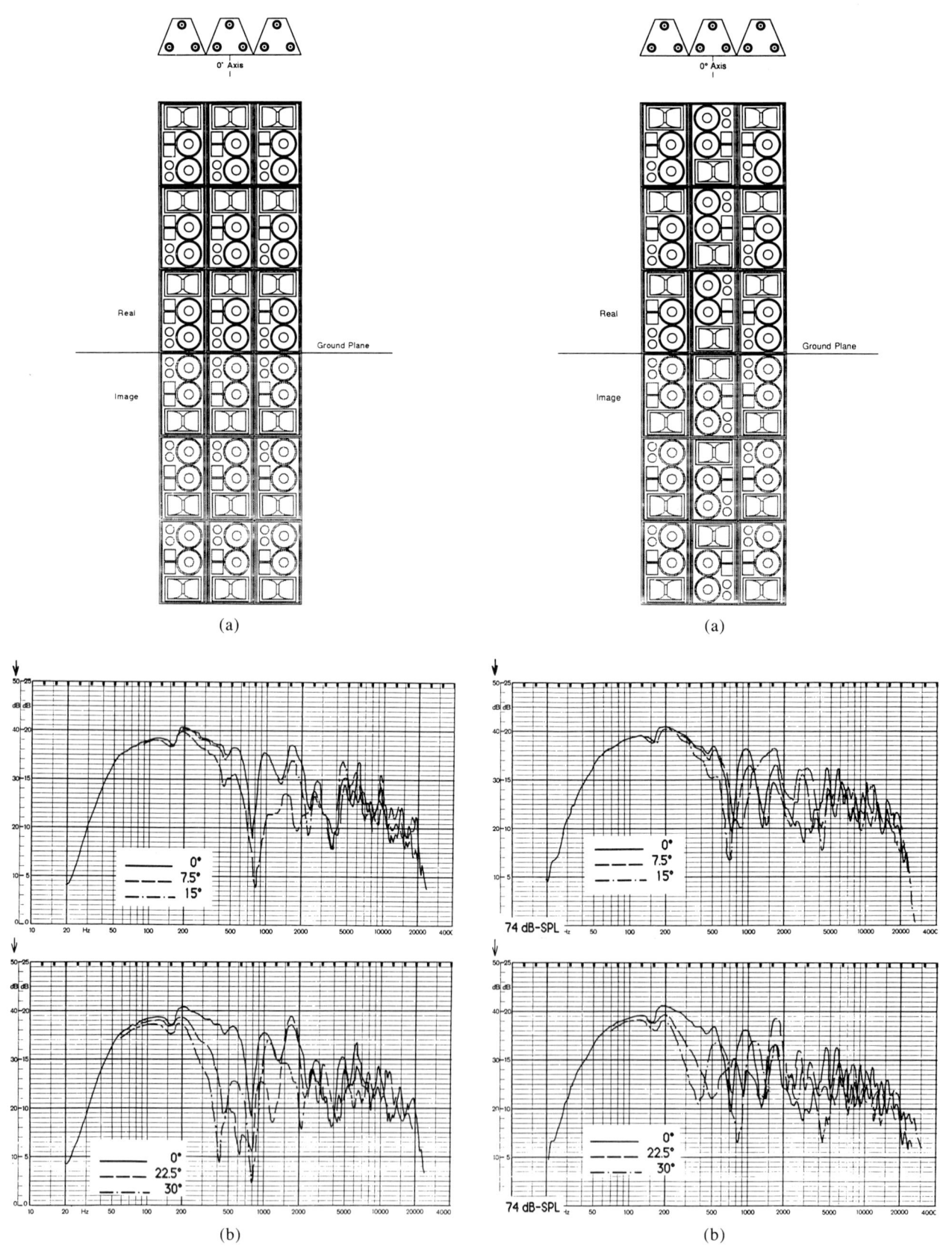

Fig. 15. Nine-loudspeaker array. (a) Elevation view. (b) Off-axis response.

Fig. 16. Nine-loudspeaker array. (a) Elevation view. (b) Off-axis response.

provide side coverage. Both the abrupt (45°) and the gentle (22.5°) splays provide a response that exhibits significant lobing up to 5 kHz. With more side elements, the response would probably be smoother, however.

2.4 Natural Splay Angle with Enclosure Separation

Fig. 24 shows details of an array in which adjacent enclosures are splayed at an angle of 60°, which is the

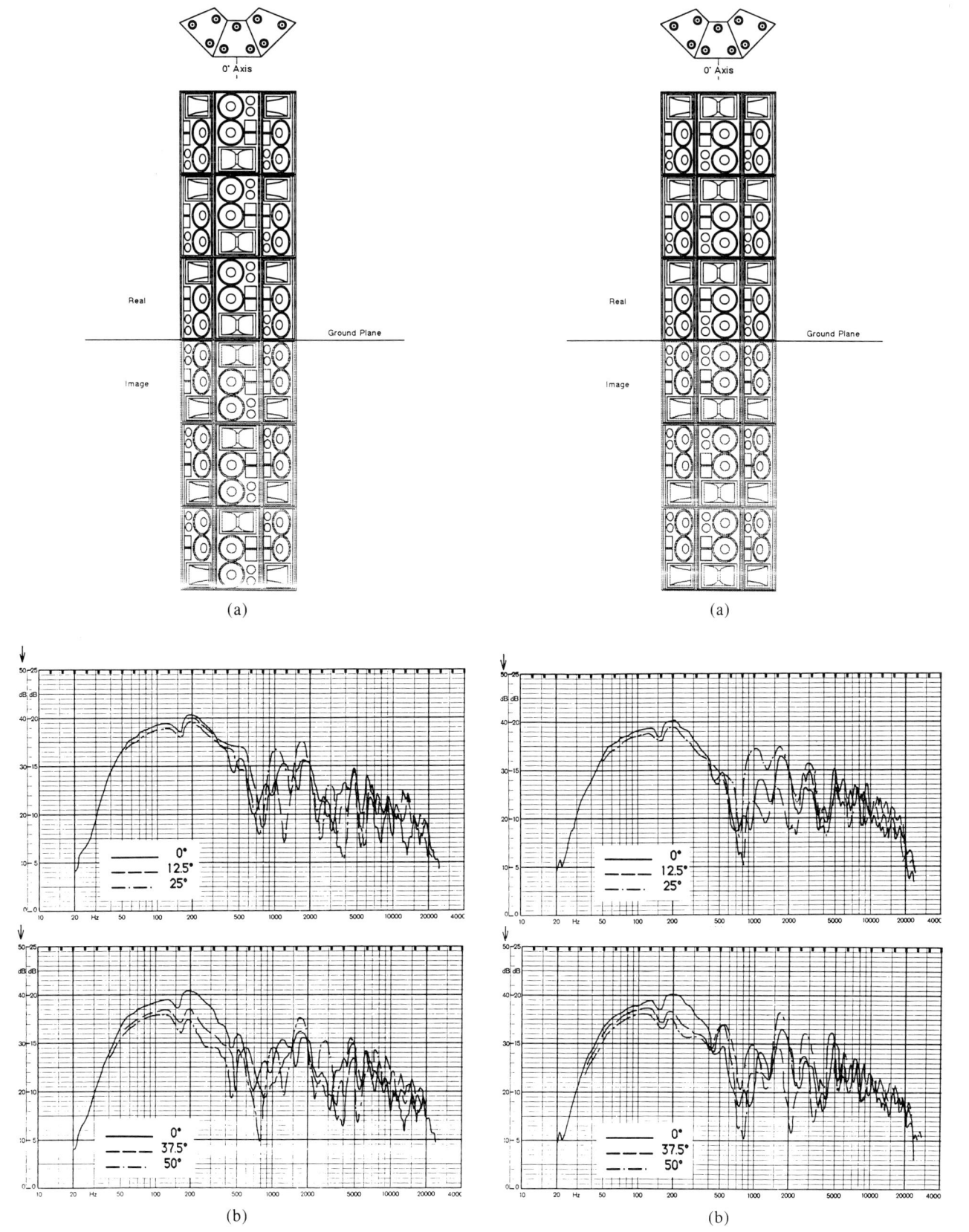

Fig. 17. Nine-loudspeaker array. (a) Elevation view. (b) Off-axis response.

Fig. 18. Nine-loudspeaker array. (a) Elevation view. (b) Off-axis response.

normal splay angle for 60° horns. Note that there are wide gaps between the adjacent enclosures and that they result in severe lobing out to the highest frequencies.

When solid wedges, as shown in Fig. 25, are used to fill in the gaps between adjacent loudspeakers, the response improves somewhat, with lobing reduced below 500 Hz. Cavity resonance at 250 Hz is also eliminated by the wedges [2]. However, the increased distance between horn mouths diminishes the theoretical advantage of splaying to the design horizontal coverage angle of the horns.

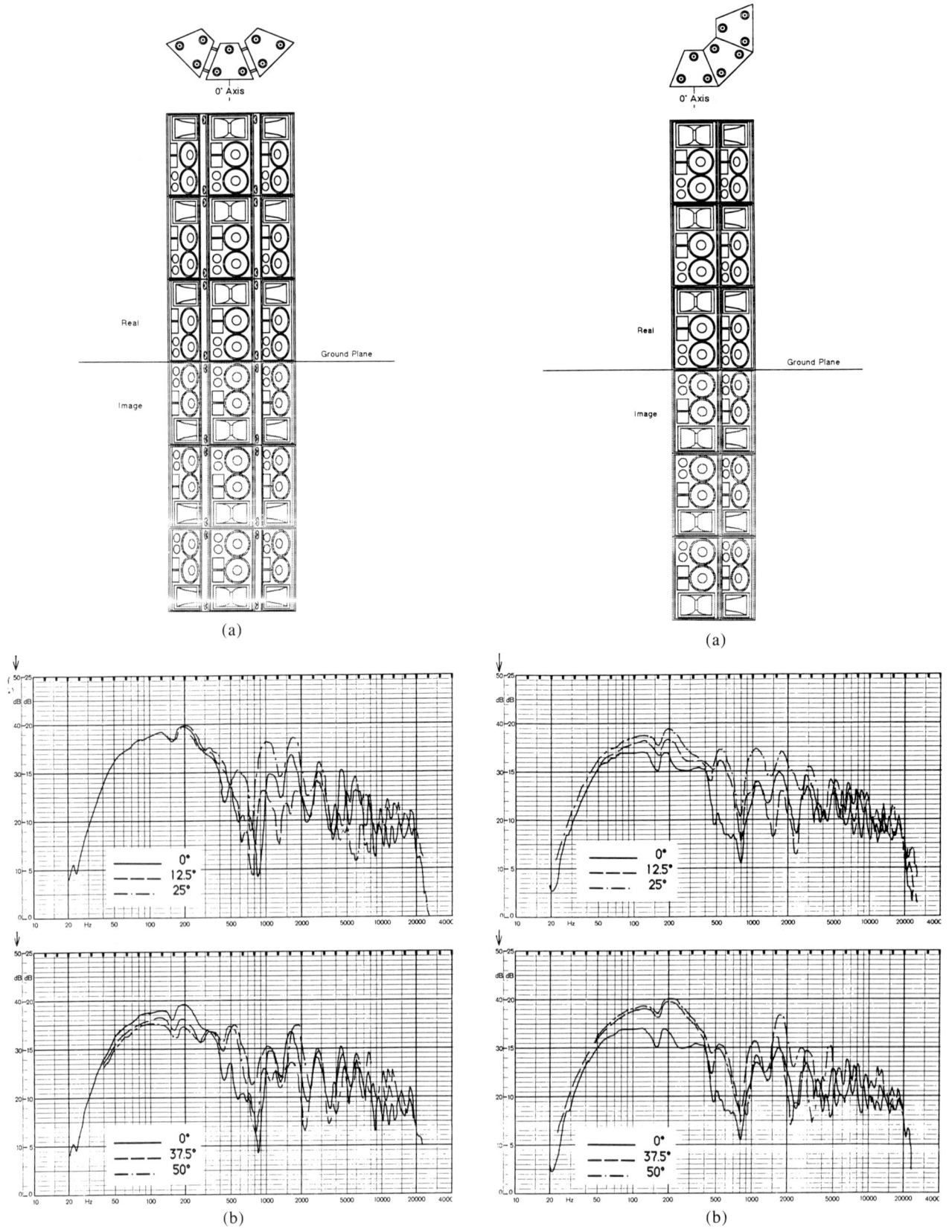

Fig. 19. Nine-loudspeaker array. (a) Elevation view. (b) Off-axis response.

Fig. 20. Nine-loudspeaker array. (a) Elevation view. (b) Off-axis response.

3 MEASUREMENTS ON SUBWOOFERS

The low-frequency systems used in this study each consisted of two 480-mm (18-in)-diameter transducers mounted in a ported enclosure. Frontal dimensions are 0.75 m (30 in) by 1.25 m (49 in). There were eight such systems, making a total of 16 transducers.

An interesting aspect of multiple-subwoofer performance is the downward shift in system resonance as the number of elements is increased [3]. Fig. 26

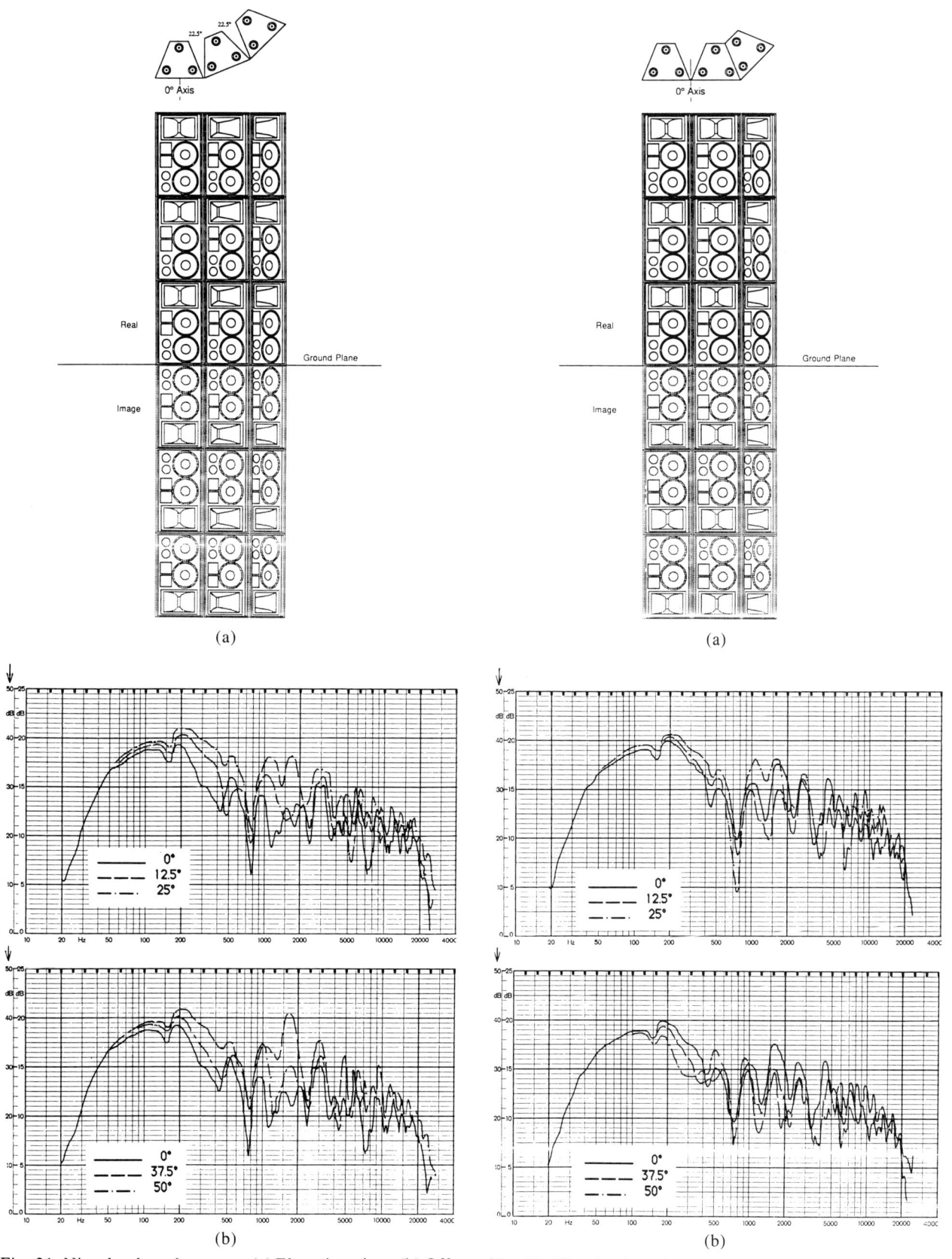

Fig. 21. Nine-loudspeaker array. (a) Elevation view. (b) Off-axis response.

Fig. 22. Nine-loudspeaker array. (a) Elevation view. (b) Off-axis response.

shows the shift as measured for a single unit, two units, and four units. The downward shift is caused by mutual coupling increasing the radiation reactance [4]. The radiation resistance portion of the complex radiation impedance is similarly increased for both the port and the diaphragm, increasing the efficiency of the array at low frequencies.

The net result of these effects is that less power may be required to generate a desired low-frequency output level. Those system designers familiar with Thiele–

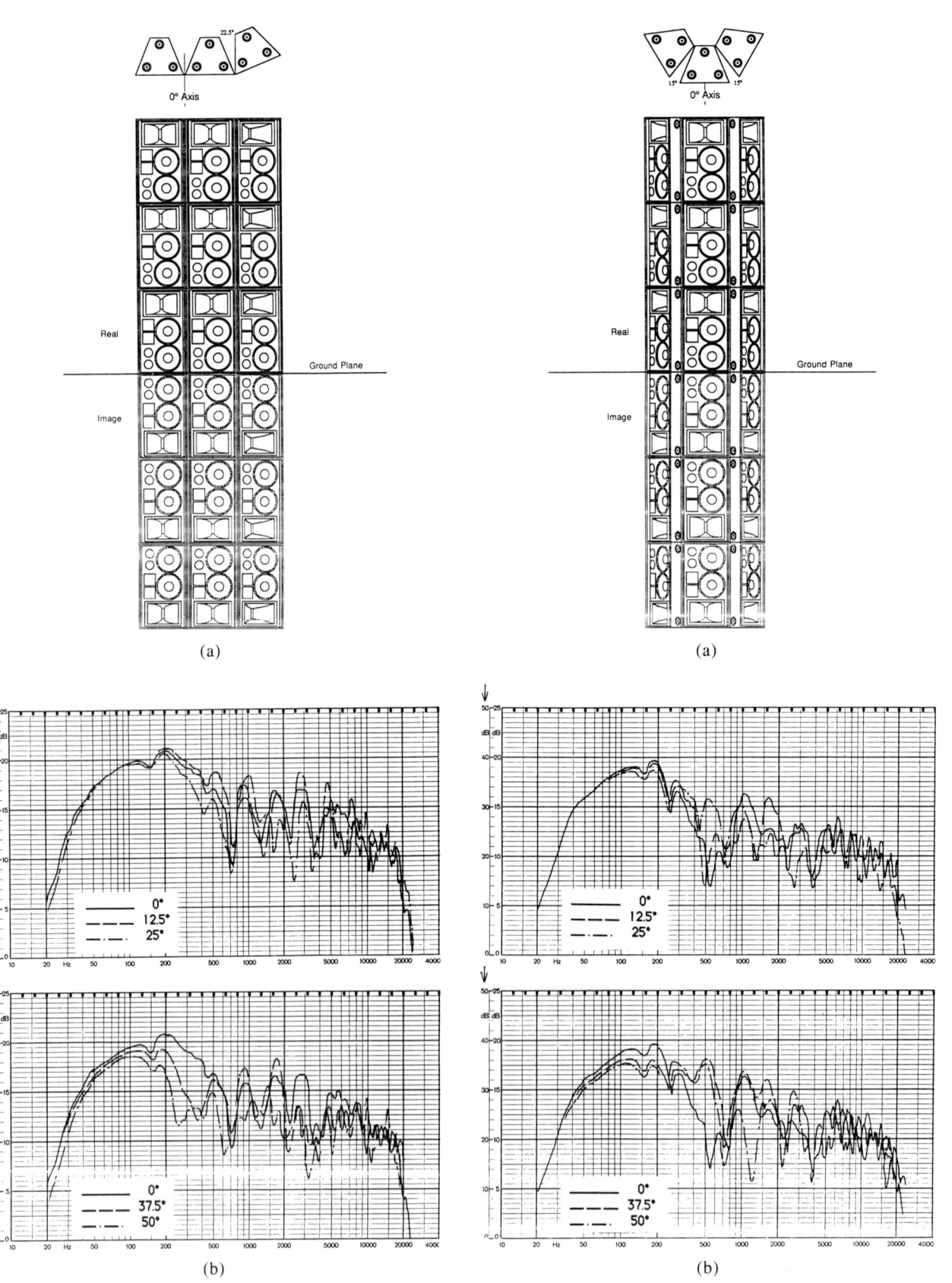

Fig. 23. Nine-loudspeaker array. (a) Elevation view. (b) Off-axis response.

Fig. 24. Nine-loudspeaker array. (a) Elevation view. (b) Off-axis response.

PAPERS

Small parameters will note that the low-frequency alignment itself may be profoundly altered, producing a bass-heavy or even "boomy" response. The system designer must take note and, if need be, design overdamped subwoofer modules, with the knowledge that in operation the ensemble will attain some desired alignment. Power compression effects can similarly affect alignment choice and shift with application [5].

Large subwoofer arrays are no less susceptible to lobing than full-range systems, as we can see in Fig. 27. A family of off-axis curves at 12.5° increments for a single subwoofer module is shown in Fig. 27(a), while data for a group of eight is shown in Fig. 27(b). For the group of eight subwoofer modules, lobing can be seen as low as 100 Hz.

4 ELECTRICAL INPUT POWER REQUIREMENTS AND ACOUSTICAL OUTPUT LEVELS

4.1 Full-Range Units

The basic full-range unit measured in these tests is intended for biamplification, and its voltage sensitivity (for flat output) is determined by its low-frequency section. The sensitivity is 98 dB, 1-W input referred to a distance of 1 m. At a distance of 10 m, as used in

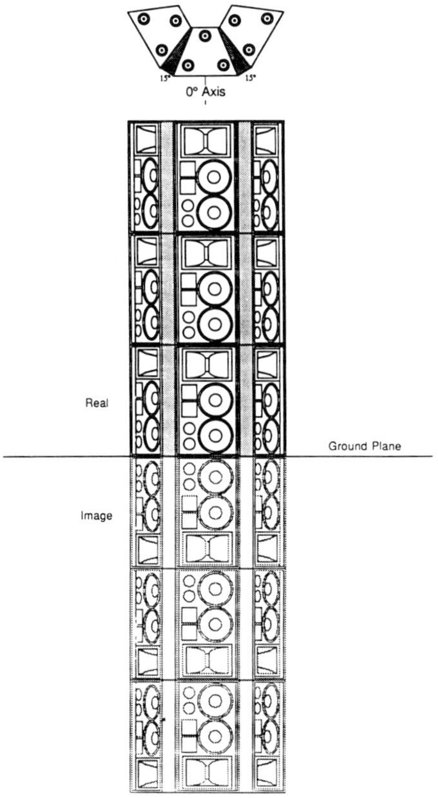

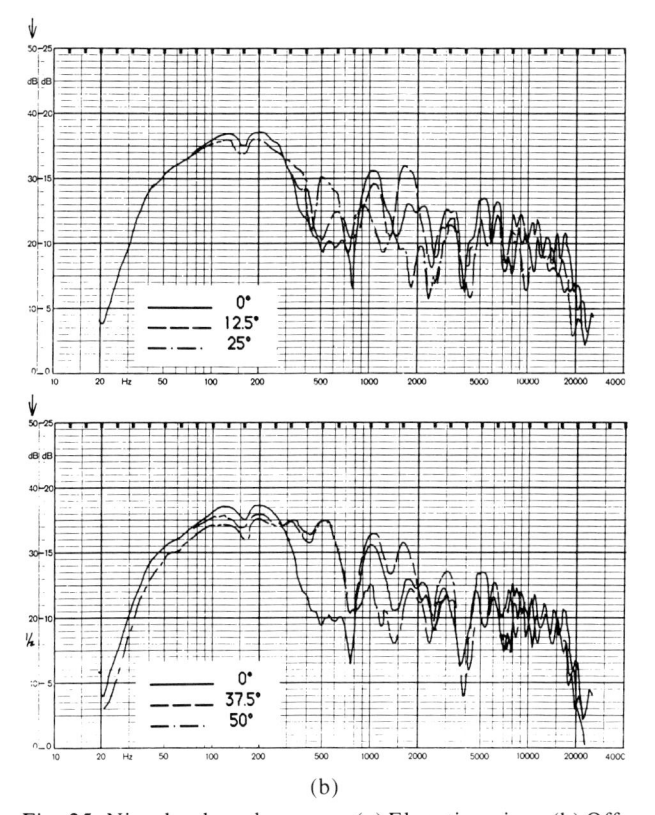

Fig. 25. Nine-loudspeaker array. (a) Elevation view. (b) Off-axis response.

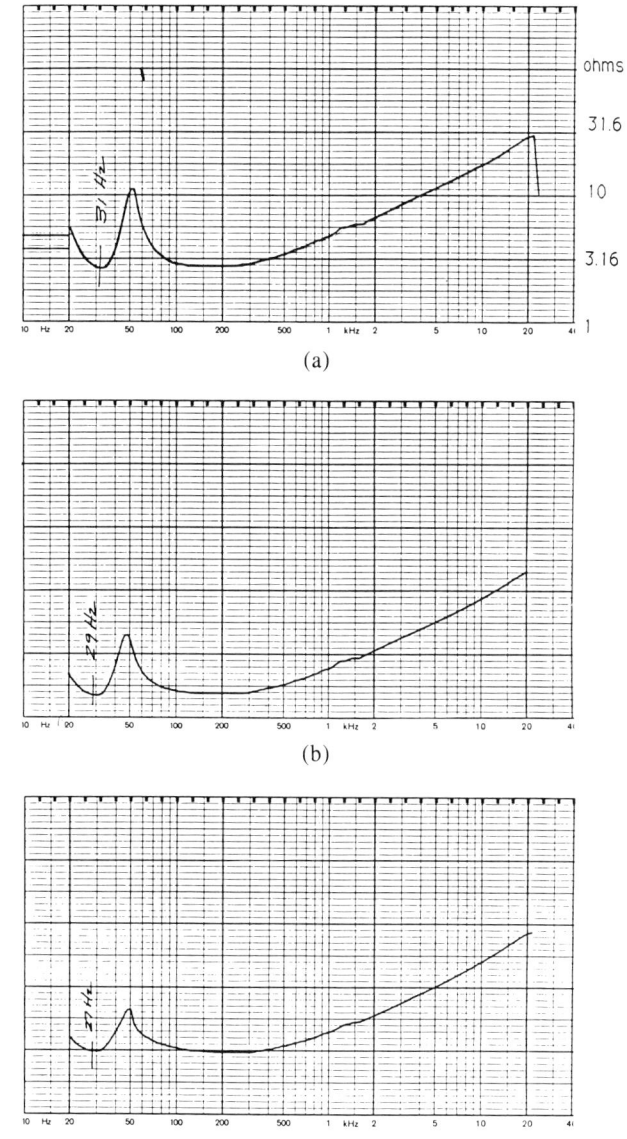

Fig. 26. Impedance shift of multiple subwoofer modules. (a) Single unit. (b) Two units. (c) Four units.

these tests, the output pressure level for a single unit, with a drive input of 1 W, would be 78 dB SPL, measured under free-space conditions. With ground plane conditions, the measured value would be 84 dB because of ground plane signal doubling.

The data shown in Fig. 16(b), for example, resulted from voltage input to each of the nine loudspeakers of 10 V rms. With a nominal input impedance per low-frequency system of 4 Ω, 10-V input would produce a power of 25 W. The pressure level output for each unit at a distance of 1 m would then be 118 dB SPL, measured on the ground plane. Nine such units operating independently, but with their powers summed, would produce a net level referred to 1 m of 127.5 dB SPL. Taking into account the measurement distance of 10 m, this would produce a level of 107.5 dB, measured on the ground plane.

However, if we refer to the data of Fig. 16, we see that the reference level for the bottom line on the graph is 74 dB SPL and that the level at 200 Hz is 115 dB SPL, a 7.5-dB difference. The difference has resulted from an increase in the forward directivity index of the array, as compared with the directivity of a single element [6].

4.2 Subwoofers

For the data shown in Fig. 27(a) the subwoofer was driven with 5 V and measured at a distance of 5 m. For the data shown in Fig. 27(b) the ensemble of eight modules was driven with 20 V and measured at a distance of 20 m. Since the drive level was 12 dB greater and the measuring distance accounted for 12 dB of inverse square loss, we can compare the two sets of curves directly.

At 200 Hz there is an on-axis measured difference of 20 dB. With eight modules as compared to a single module, we would expect a difference in power level of 9 dB. Taking the difference between 20 and 9 dB, we are left with an increase in output of 11 dB. This is due primarily to an increase in the directivity index of the array along its primary axis. This increase is consistent with the size of the array, relative to the wavelength at the frequency of measurement [6].

At 30 Hz we note a difference of 21 dB as we go from one module to eight. Again, 9 dB of this difference is due to the eightfold increase in power, leaving us with a net difference of 12 dB. Most of this increase is due to mutual coupling between the subwoofer modules, and the calculated maximum value of mutual coupling increase would be a 3-dB increase per doubling of units, or 9 dB total [7]. The remaining 3 dB can be accounted for by a slight increase in the array directivity index at 30 Hz.

From these two exercises we can see that the combination of mutual coupling and array directivity can be taken as a significant factor in determining both hardware and power requirements at very low frequencies. In very large arrays, both factors can be quite substantial, with directivity increases predominant at higher frequencies and mutual coupling increases predominant at lower frequencies.

The benefit of mutual coupling does not extend indefinitely as the number of radiators is increased. In general, the net half-space efficiency of a large array cannot be increased beyond 25% as a result of mutual coupling, and any apparent increase over this efficiency figure would reflect an increase in directivity of the array.

5 ATTENUATION WITH DISTANCE FROM LARGE ARRAYS

Inverse square relationships hold only in the far field. For closer distances, the attenuation law is different. Rathe [8] has shown that the attenuation with distance from a line array is 3 dB per doubling of distances, but only out to a distance of A/π, where A is the length of the line array. Beyond this value the attenuation becomes 6 dB per doubling of distance. The transition of course is gradual.

With a plane array, Rathe states that there is no attenuation with distance, out to a distance of A/π, where A is the short dimension of the plane. At that point the attenuation becomes 3 dB per doubling of distance, extending out to a point B/π, where B is the long dimension of the plane array. Beyond that point the attenuation approaches 6 dB per doubling of distance.

The size of the subwoofer array was such that the value of A/π was 0.5 m and that of B/π was 1.5 m. Fig. 28(a) is an elevation view of the array, and Fig. 28(b)

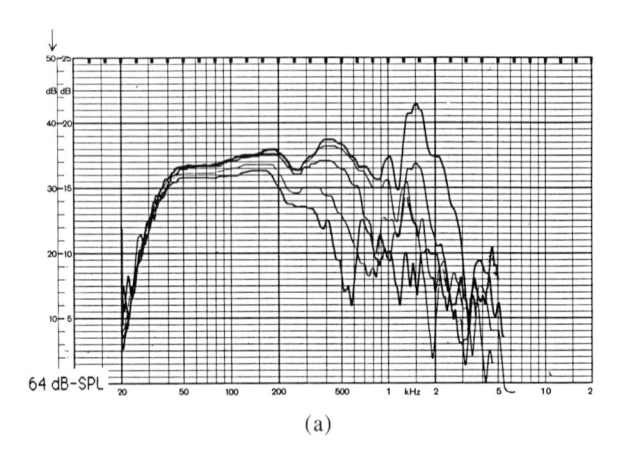

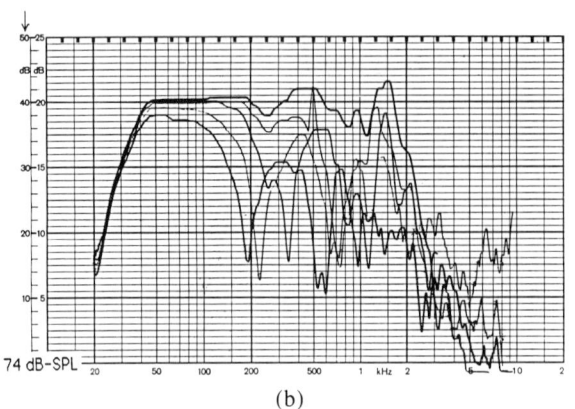

Fig. 27. Off-axis response of subwoofers at 12.5° increments. (a) Single unit. (b) Eight units.

shows a set of curves measured at intervals of 2 m, from a distance of 2 m out to 20 m along the primary axis of the array.

Rathe's observations apply to plane and line arrays which are evenly "illuminated" across their expanse. Thus it is difficult to detect clearly the presence of Rathe's theoretical break points in the attenuation data gathered here. The general trend, however, does follow Rathe's data, and it would be prudent to use these data in estimating the far- and near-field performance of arrays in the absence of measured data.

6 COMPUTER MODELING OF LOUDSPEAKER COVERAGE

JBL's central array design program (CADP) for estimating loudspeaker coverage can be used to observe the complex summation of multiple radiating elements [9]. Fig. 29 shows the modeled coverage of a nine-element array such as measured in Fig. 16. The CADP plots show in plan view a portion of the area covered by the loudspeaker array. Vertical and horizontal markers are spaced at 5-m intervals (as indicated by the legend "5 m" in the displays), so the entire area covered by the plots is about 14 by 12 m. The three stacks of loudspeakers are shown by small circles in the middle of the left sides of the plots. The numbers shown in the plots represent calculated relative sound pressure levels at each readout coordinate, normalized to 0 dB as the maximum value. Phasor summation of individual pressure contributions is made at frequencies of 7.9, 8, and 8.1 kHz, showing the effects of reinforcement and cancellation at the respective wavelengths. Note that the response changes markedly within

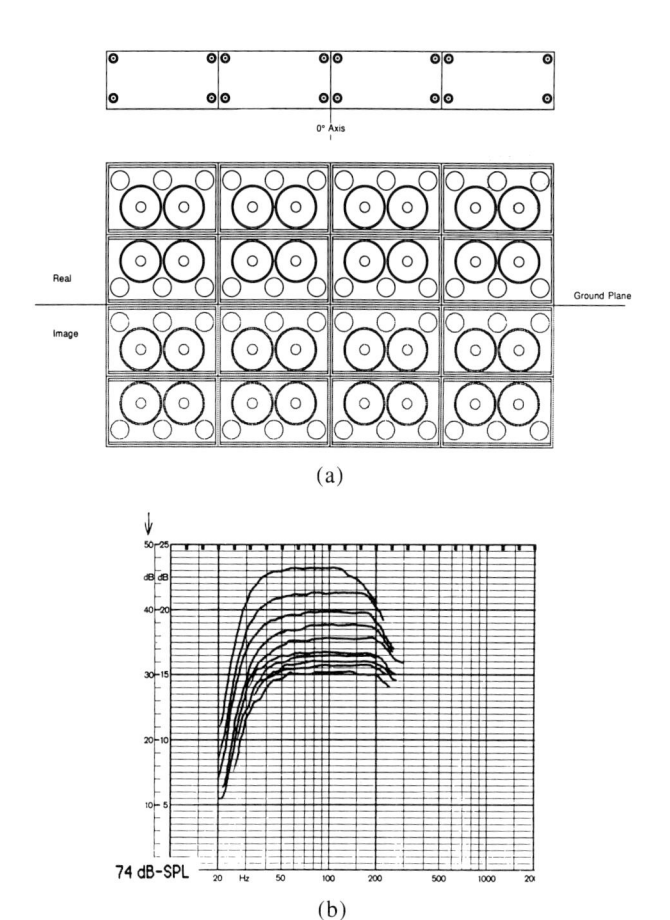

Fig. 28. Subwoofer array. (a) Elevation view. (b) Response at 2, 4, 6, 8, 10, 12, 14, 16, 18, and 20 m.

Fig. 29. Nine-element-array CADP phasor pattern. (a) Merging at 7.9 kHz. (b) Merging at 8 kHz. (c) Merging at 8.1 kHz.

this small range of frequencies, which is consistent with the data shown in Fig. 16.

The program can also model attenuation with distance, making use of the power summation of elements and ignoring phasor information. In the CADP example illustrated in Fig. 30, eight double-stacked subwoofers are shown as circles spaced along the left edge of the plot. The pattern-merging strategy used here ignored phasor data, presenting the rms pressure summation of all radiating elements at each readout coordinate. Normalization is again with respect to a maximum value of 0 dB. The attenuation along a line perpendicular to the array is shown in Fig. 31, along with actual measured values plotted from Fig. 28(b). Note that the CADP estimates and the actual measurements match quite well. Both curves follow Rathe's data fairly well. All of the curves presented in Fig. 31 were normalized at a distance of 8 m.

7 CONCLUSIONS

We can draw the following conclusions from our measurements.

1) Sine wave input signals seldom result in smooth frequency response curves with large arrays. However, sine wave signals are preferable to noise signals, which mask much detail.

2) Fine combing (less than one-third-octave bandwidth) on the order of ±6 dB is observed at high frequencies in all array measurements. Considering critical band theory, it is likely that the ear finds this acceptable [10].

3) The combination of high-frequency losses due to interference effects and the boost of low frequencies due to mutual coupling and directivity increase results in an overall skewed frequency response, even when the individual elements in the system may be adjusted for flat on-axis response. In general, the response falls off approximately 6 dB per decade, but this may be a function of overall array size. The rolloff will require a careful assessment of the hardware specification as well as the power allotment to the system.

4) Staggered, or "checkerboard," positioning of systems in flat arrays is generally beneficial, but is of no apparent benefit in curved arrays.

5) Any degree of excess spacing between adjacent elements in the array should be kept to a minimum. In addition, gaps between splayed elements in an array will benefit from being filled in, if practicable.

6) When large horn spacings are necessitated by enclosure size and mounting constraints, splaying of high-frequency horns along their nominal −6-dB zones is of no apparent benefit.

7) The effect of directivity increase in large arrays is very significant, but it is important to remember that the effect will be most apparent in the far field.

8) Mutual coupling is a significant factor at frequencies below 100 Hz. As the array size increases, the frequency below which it remains effective moves downward, inversely proportional to the square root of the array area [11].

The specification of full-range units in constructing large arrays for music performance is based on the goal of flat power bandwidth. As we have seen, there can be as much as 10-dB skew in the overall frequency response, even in arrays of limited size. This suggests some rethinking of the basic design philosophy and clearly leads to the conclusion that relatively more high-frequency hardware is needed. Stated differently, it is easier to "turn down the woofers" than it is to "turn up the tweeters."

Finally we must state that our knowledge of large arrays is still quite limited and that more data are needed by way of both modeling and actual measurement.

Fig. 30. Eight subwoofer modules; CADP pattern merging at 100 Hz, power summation.

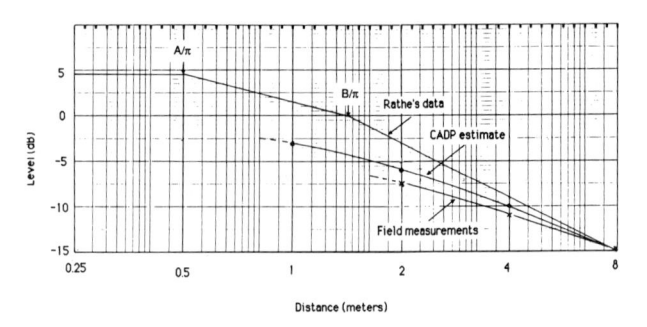

Fig. 31. Attenuation with distance from an array of eight subwoofer modules.

8 ACKNOWLEDGMENT

The authors wish to thank Mark Engebretson of Summit Laboratories, Warner Springs, CA, and Charlie Morgan of Morgan Sound, Lynnwood, WA, for their contributions to the work presented here.

9 REFERENCES

[1] M. R. Gander, "Ground-Plane Acoustic Measurement of Loudspeaker Systems," *J. Audio Eng. Soc.* (*Engineering Reports*), vol. 30, pp. 723–731 (1982

Oct.).

[2] R. Wickersham and D. Davis, "Experiments in the Enhancement of the Artist's Ability to Control His Interface with the Acoustic Environment in Large Halls," presented at the 51st Convention of the Audio Engineering Society, *J. Audio Eng. Soc. (Abstracts)*, vol. 23, p. 488 (1975 July/Aug.), preprint 1033.

[3] J. Eargle, J. Bonner, and D. Ross, "The Academy's New State of the Art Loudspeaker System," *J. Soc. Motion Picture Telev. Eng.*, vol. 94, no. 11 (1982).

[4] I. Wolff and L. Malter, "Sound Radiation from a System of Vibrating Circular Diaphragms," *Phys. Rev.*, vol. 33 (1929 June).

[5] M. R. Gander, "Dynamic Linearity and Power Compression in Moving-Coil Loudspeakers," *J. Audio Eng. Soc.*, vol. 34, pp. 627–646 (1986 Sept.).

[6] C. Molloy, "Calculations of Directivity Index for Various Types of Radiators," *J. Acoust. Soc. Am.*, vol. 20, no. 4 (1948).

[7] M. Engebretson, "Low-Frequency Sound Reproduction," *J. Audio Eng. Soc. (Engineering Reports)*, vol. 32, pp. 340–346 (1984 May).

[8] E. Rathe, "Notes on Two Common Problems of Sound Propagation," *J. Sound Vibration*, vol. 10, pp. 472–479 (1969).

[9] D. Albertz, J. Eargle, D. B. Keele, Jr., and R. Means, "A Microcomputer Program for Central Loudspeaker Array Design," presented at the 74th Convention of the Audio Engineering Society, *J. Audio Eng. Soc. (Abstracts)*, vol. 31, p. 964 (1983 Dec.), preprint 2028.

[10] J. Roederer, *Introduction to the Physics and Psychophysics of Music* (Springer, New York, 1975).

[11] D. B. Keele, Jr., "An Efficiency Constant Comparison between Low-Frequency Horns and Direct Radiators," presented at the 54th Convention of the Audio Engineering Society, *J. Audio Eng. Soc. (Abstracts)*, vol. 24, p. 498 (1976 July/Aug.), preprint 1127.

9.1 Additional Reading

[12] L. Beranek, *Acoustics* (McGraw-Hill, New York, 1954, and Am. Inst. Phys., New York, 1986), chap. 4, "Radiation of Sound."

[13] H. Knowles, "Loudspeakers and Room Acoustics," in Henney, *Radio Engineering Handbook*, 5th ed. (McGraw-Hill, New York, 1959).

[14] H. Olson, *Acoustical Engineering* (Van Nostrand, Princeton, NJ, 1957), chap. 2, "Acoustical Radiating Systems."

[15] C. N. Strahm, "Complete Analysis of Single and Multiple Loudspeaker Enclosures," presented at the 81st Convention of the Audio Engineering Society, *J. Audio Eng. Soc. (Abstracts)*, vol. 34, p. 1032 (1986 Dec.), preprint 2419.

THE AUTHORS

M. R. Gander

J. M. Eargle

Mark R. Gander was born in 1952 and was raised in rural Whitehouse, New Jersey, where he received an extensive musical education. He earned a B.S. degree from Syracuse University in 1974 and worked as a sound engineer and audio systems designer in broadcasting, studio recording, and concert reinforcement.

In 1976 he received an M.S.E.E. from The Georgia Institute of Technology, specializing in audio electronics and instrumentation, and including the multidisciplinary program in acoustical engineering. That same year he joined James B. Lansing Sound, Inc. as a transducer engineer, where he had design responsibility for various consumer and professional loudspeaker products, including the Cabaret Series and E Series musical instrument loudspeakers. In 1980 he moved from engineering into marketing as applications engineer for the JBL Professional Division and later became product manager for JBL Professional Products. He is currently vice president, marketing, for JBL Professional.

Mr. Gander has published in the *Journal of the Audio Engineering Society* as well as in the trade and popular press. He holds a First Class Radiotelephone Operator License and is a member of the AES, The Acoustical Society of America, and the Institute of Electrical and Electronics Engineers.

●

John M. Eargle holds degrees in music and electrical engineering from the Eastman School of Music (B.M.), University of Michigan (M.M.), University of Texas (B.S.E.E.), and Cooper Union (M.E.). He is a member of the Acoustical Society of America, the SMPTE, the National Academy of Recording Arts and Sciences, a senior member of the IEEE, and a Fellow and honorary member of the AES. He is a past president of the AES and is a reviewer for the *Journal*. He worked for RCA Record and Mercury Records during the 1960s. During the 70s he worked for the Altec Corporation and formed the firm JME Consulting Corporation. Since 1976 he has been associated with JBL Incorporated, Northridge, California, in the areas of product development and application.

Since 1982 he has expanded his activities in recording engineering and producing, and he has engineered and/or produced over 60 Compact Discs. He has published more than 60 technical articles and reviews. His books, *Handbook of Recording Engineering*, *The Microphone Handbook*, and *Handbook of Sound System Design*, are used as texts at the university level.

In 1984 Mr. Eargle was awarded the AES Bronze Medal.

Multiple-Beam, Electronically Steered Line-Source Arrays for Sound-Reinforcement Applications*

DAVID G. MEYER

School of Electrical Engineering, Purdue University, West Lafayette, IN 47907, USA

Long, narrow rooms with high ceilings are particularly difficult spaces for which to design sound-reinforcement systems. Rooms such as these typically require a combination of short-throw (low Q), medium-throw, and long-throw (high Q) loudspeaker components configured as a large, central cluster to provide uniform coverage over the seating space. Unfortunately many such rooms, in particular those of A-frame design, do not have adequate provisions for mounting a large cluster of loudspeaker components in an aesthetically pleasing fashion. A possible alternative for such a venue is described. In place of a central cluster with a large vertical dimension, we propose using an ensemble of horizontally mounted, electronically (phase-delay) steered line-source arrays. Each line source has a Q commensurate with the distance it must throw and the area it must cover. Shading is employed to minimize side-lobe energy as well as to help stabilize coverage patterns. Many desirable properties associated with a central cluster system, such as minimum energy directed toward the speaking position, stable coverage patterns, and preservation of locality of reference, apply to the proposed alternative described. Advances in signal-processing hardware technology help make a sound-reinforcement system design based on this approach both technically and economically feasible.

0 INTRODUCTION

It is generally agreed that, for most permanently installed sound-reinforcement applications, a central cluster design is preferred. Such a system, when designed using high-quality constant-directivity horns, offers the inherent advantages of stable directivity, uniform seating area coverage, minimum energy directed toward the ceiling and sidewalls, high efficiency, smooth frequency response, preservation of locality of reference, and generally straightforward design [1], [2]. Long, narrow, reverberant rooms, which typically require central clusters using a combination of low-, medium, and high Q loudspeaker components to cover the entire seating space with high intelligibility, rarely have adequate provisions for concealing such a cluster in an aesthetically pleasing fashion. This is particularly true for A-frame type structures, typical of many places of worship.

An alternative to the central cluster approach, particularly amenable to A-frame structures, is the subject of this paper. Instead of using a cluster of large vertical dimension (that is, hanging downward from the ceiling, typically from the central arch), we propose using an ensemble of *horizontally* mounted, electronically (phase-delay) steered line-source arrays. The perspective and side views of such an installation appear in Figs. 1 and 2, respectively. Each line source is designed to have a Q commensurate with the distance it must throw and the area it must cover—a task facilitated by computer-aided design software which has been specially written for this purpose. Separate mid- and high-frequency arrays can be provided for each coverage zone. Optionally, a separate array of large drivers can

---

* Presented at the 87th Convention of the Audio Engineering Society, New York, 1989 October 18–21; revised 1990 January 22.

be used to augment low-frequency reinforcement. To minimize side-lobe energy as well as to help stabilize the coverage pattern of each array, amplitude shading is used. A standard digital signal-processing (DSP) microprocessor can be used to implement the filters needed to shade the signal applied to each element, to provide an active crossover among the various low–mid–high frequency arrays, and to implement the tapped delay lines necessary for beam steering.

The overall goal of this work, then, is to demonstrate the feasibility of a viable alternative to a central cluster system for a particular class of venues. The primary objective is to maintain the desirable properties associated with central cluster designs, such as minimum energy directed toward the speaking position, relatively stable coverage patterns, and preservation of locality of reference, while dramatically improving the reinforcement system aesthetics.

We first describe the mathematical basis of phase-delay-steered line-source arrays. Included in this section is an analysis of the signal-processing requirements for control of the arrays. We then briefly describe the computer-aided design software specially written for multizone line array coverage and present the results of a case study.

1 ANALYSIS

The fundamental basis of the analysis presented in this section appears in [3]–[13]. Here we extend these results to accommodate multiple, electronically steered arrays consisting of elements possessing frequency-dependent amplitude, phase, and polar directivity characteristics. Also shown is how the array ensemble directivity pattern is mapped onto the "ear plane" of the audience seating area. The resulting mathematical model is incorporated into a computer-aided design tool, described in Sec. 2.

1.1 Pressure Intensity Mapping onto Seating Plane for Multiple Arrays

For this analysis, the coordinate system illustrated in Fig. 3 is assumed. The origin of the coordinate system is located at the central position of the seating space, with positive Z axis toward the front. The positive X axis is toward the right (facing front), and the positive Y axis is pointing up. The seating depth, along the Z axis, is S; the front and back widths of the seating space, parallel to the X axis, are W_F and W_B, respectively. A rake angle γ can also be specified for the seating space. An offset of Y_E relative to the floor can be specified to denote the ear plane of the listeners.

For the application domain considered in this paper, we assume that the sound-reinforcement loudspeaker arrays are mounted overhead, in a plane parallel to the X–Z plane. The arrays may either be line sources ($n \times 1$) or rectangular ($n \times m$) arrays. Each array j has central coordinates (xc_j, yc_j, zc_j) at which it is mounted; the coordinates of each array element i of array j, denoted by (x_{ij}, y_{ij}, z_{ij}), are relative to the array central coordinates. The frequency of analysis is f, typically chosen on an octave or one-third-octave basis. Element spacing for array j is b_j. The (frequency-independent) signal gain applied to array j is g_j, while the (frequency-dependent) amplitude shading values for each element i of array j are denoted by $a_{ij}(f)$; all gain parameters are in decibels. The (frequency-dependent) delay characteristic of the signal applied to each element i of array j is $d_{ij}(f)$. Each element i of array j can have an amplitude response $A_{ij}(f)$ and a time response $D_{ij}(f)$.

All array elements analyzed in this study are assumed

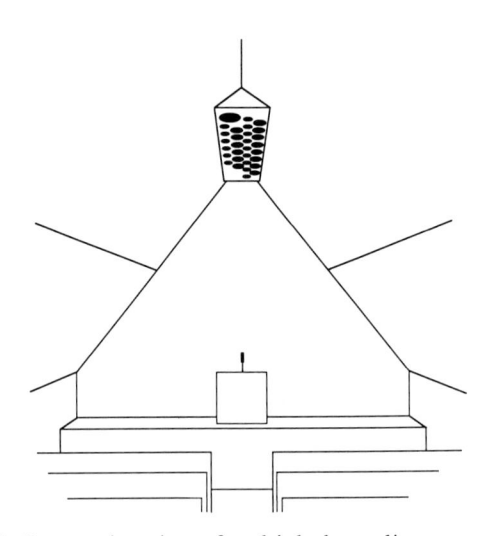

Fig. 1. Perspective view of multiple-beam line array installation.

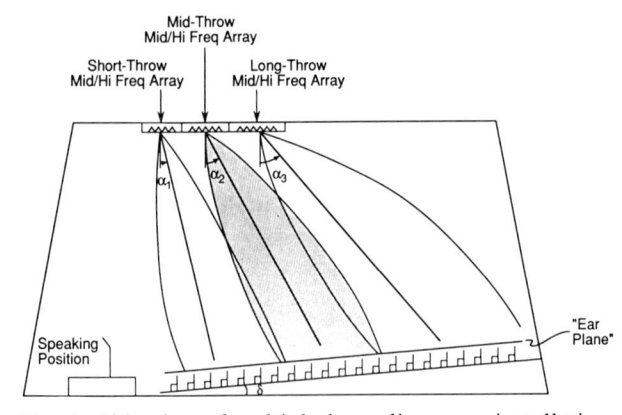

Fig. 2. Side view of multiple-beam line array installation.

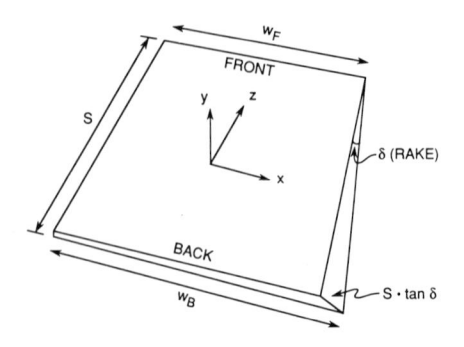

Fig. 3. Coordinate system used for analysis.

to be of the circular or elliptical cone variety.[1] Directivity characteristics are approximated by assuming that each transducer acts as a piston; detailed simulations of various transducer configurations based on this model appear in [9]. If v_k is the vertical radius of an elliptical array element of type k and h_k is its horizontal radius, the vertical and horizontal directivity functions are given by

$$V_k(\theta, f) = \frac{2J_1[(2\pi v_k f/c) \sin \theta]}{(2\pi v_k f/c) \sin \theta} \quad (1)$$

and

$$H_k(\theta, f) = \frac{2J_1[(2\pi h_k f/c) \sin \theta]}{(2\pi h_k f/c) \sin \theta} \quad (2)$$

where θ is the angle between the axis of the ellipse and the line joining the center of the ellipse, J_1 is the Bessel function of first order, f is the frequency of excitation, and c is the speed of sound at ambient conditions. For circular cone transducers $h_k = v_k$, and therefore $H_k(\theta, f) = V_k(\theta, f)$.

Each array element is viewed as a set of observation points (x_0, y_0, z_0) that span the ear plane of the entire seating space. At the rear of the seating space, x_0 ranges from $-W_B/2$ to $+W_B/2$; at the front of the seating space, x_0 ranges from $-W_F/2$ to $+W_F/2$. The range of z_0 is from $-S/2$ (back) to $+S/2$ (front), while the range of y_0 is from Y_E (ear plane height) to $Y_E + (S/2 - z_0) \tan \gamma$.

The (physical) distance from which each element i of array j is viewed at a given observation point (x_0, y_0, z_0), denoted by r_{ij}, is

$$r_{ij} = [(x_{ij} + xc_j - x_0)^2 + (y_{ij} + yc_j - y_0)^2 + (z_{ij} + zc_j - z_0)^2]^{1/2} . \quad (3)$$

The effect of delaying the signal applied to each element i of array j is to make it appear further away from the plane of observation. Here, electronically delaying the signal applied to element i of array j by d_{ij} seconds corresponds to moving element i a distance $\Delta y_{ij} = cd_{ij}$ behind the X–Z_j plane of the array (again, c denotes the speed of sound at ambient conditions). Incorporating the time response [14], [15] of element i of array j, denoted by $D_{ij}(f)$, the effective distance becomes $\Delta y_{ij}(f) = cd_{ij} + cD_{ij}(f)$. The *virtual* distance from which each element i of array j is viewed at a given observation point (x_0, y_0, z_0), denoted by r'_{ij}, is then

$$r'_{ij}(f) = \{(x_{ij} + xc_j - x_0)^2 + [y_{ij} + yc_j - y_0 - \Delta y_{ij}(f)]^2 + (z_{ij} + zc_j - z_0)^2\}^{1/2} . \quad (4)$$

The angles with which each element i of array j is observed may be determined as follows. Noting that the nominal orientation of each array element is "facing down" (that is, aimed in the $-Y$ direction), let θ_{ij} denote the angle of observation with respect to the central axis of element i of array j, and ψ_{ij} the angle of rotation in the X–Z_j plane around the element's central axis. Then

$$\theta_{ij} = \frac{\pi}{2} - \tan^{-1}\left\{\frac{y_{ij} + yc_j - y_0}{[(x_{ij} + xc_j - x_0)^2 + (z_{ij} + zc_j - z_0)^2]^{1/2}}\right\} \quad (5)$$

and

$$\psi_{ij} = \tan^{-1}\left(\frac{z_{ij} + zc_j - z_0}{x_{ij} + xc_j - x_0}\right) . \quad (6)$$

Each array element may optionally be mounted at an azimuth angle Θ_{ij} and a skew angle Ψ_{ij}. The effective angles of observation with respect to each element i of array j then become

$$\theta'_{ij} = \theta_{ij} - \Theta_{ij} \quad (7)$$

and

$$\psi'_{ij} = \psi_{ij} - \Psi_{ij} . \quad (8)$$

The observed directivity function, denoted by $R^k_{ij}(\theta_{ij}, \psi_{ij}, f)$, for type k element i of array j is given by

$$R^k_{ij}(\theta_{ij}, \psi_{ij}, f) = [\sin(\psi'_{ij})V_k(\theta'_{ij}, f) + \cos(\psi'_{ij})H_k(\theta'_{ij}, f)]^{1/2} . \quad (9)$$

Note that in the X–Z plane the directivity function of Eq. (9) produces elliptical isobars for cases in which $v_k \neq h_k$ and circular isobars for cases in which $v_k = h_k$.

Using the results of Eqs. (3), (4), and (9), the vector representation for pressure intensity contributions from a type k element i of array j can be expressed as

$$\overline{P}^k_{ij}(\theta_{ij}, \psi_{ij}, f) = \left[\frac{R^k_{ij}(\theta_{ij}, \psi_{ij}, f) \times 10^{[g_j + a_{ij}(f) + A_{ij}(f)]/10}}{r_{ij}}\right] e^{j2\pi ft} e^{-j[2\pi f r'_{ij}(f)/c]} . \quad (10)$$

Recalling the relation between the observation point (x_0, y_0, z_0) and the angles of observation θ_{ij} and ψ_{ij}, given in Eqs. (5) and (6), the vector sum of the individual pressure contributions from l line arrays, each consisting

---

[1] This is not a restriction, however. The model is designed such that array elements possessing *arbitrary* horizontal and vertical directivity functions may be utilized.

of $n_j \times m_j$ elements of type k, is then

$$\overline{P}(x_0, y_0, z_0, f) = \sum_{j=1}^{l} \sum_{i=1}^{n_j m_j} \overline{P}_{ij}^{k_j}(\theta_{ij}, \psi_{ij}, f)$$

$$= \sum_{j=1}^{l} \sum_{i=1}^{n_j m_j} \left[\frac{R_{ij}^{k_j}(\theta_{ij}, \psi_{ij}, f) \times 10^{[g_j + a_{ij}(f) + A_{ij}(f)]/10}}{r_{ij}} \right] e^{j2\pi f t} e^{-j[2\pi f r'_{ij}(f)/c]} . \quad (11)$$

Finally, the magnitude of this vector sum at each observation point (x_0, y_0, z_0) can be expressed as

$$|\overline{P}(x_0, y_0, z_0, f)| = \left| \sum_{j=1}^{l} \sum_{i=1}^{n_j m_j} \left[\frac{R_{ij}^{k_j}(\theta_{ij}, \psi_{ij}, f) \times 10^{[g_j + a_{ij}(f) + A_{ij}(f)]/10}}{r_{ij}} \right] e^{-j[2\pi f r'_{ij}(f)/c]} \right|$$

$$= \left\{ \left[\sum_{j=1}^{l} \sum_{i=1}^{n_j m_j} \frac{R_{ij}^{k_j}(\theta_{ij}, \psi_{ij}, f) \times 10^{[g_j + a_{ij}(f) + A_{ij}(f)]/10}}{r_{ij}} \cos\left(\frac{2\pi f r'_{ij}(f)}{c}\right) \right]^2 \right.$$

$$\left. + \left[\sum_{j=1}^{l} \sum_{i=1}^{n_j m_j} \frac{R_{ij}^{k_j}(\theta_{ij}, \psi_{ij}, f) \times 10^{[g_j + a_{ij}(f) + A_{ij}(f)]/10}}{r_{ij}} \sin\left(\frac{2\pi f r'_{ij}(f)}{c}\right) \right]^2 \right\}^{1/2} . \quad (12)$$

1.2 Beam Steering Resolution and Sampling Considerations

One of the design goals of the system described here is implementation of the tapped delay lines for beam steering using a commercially available DSP microprocessor (such as a Motorola 56000 or a Texas Instruments TMS32025). In this section we wish to analyze the trade-offs between sampling rate and beam steering resolution for various size arrays and element types. We also wish to determine memory requirements for the implementation of the tapped delay lines. The basic design constraints are as follows:

1) The sampling frequency must be low enough to:
   a) Be "comfortably" supported by a commercially available DSP microprocessor
   b) Minimize the tapped delay line memory requirements (implemented as circular buffers in software).
2) The sampling frequency must be high enough to:
   a) Provide sufficient audio bandwidth (such as 20 kHz) with nominal anti-image low-pass filter complexity
   b) Provide sufficient steering resolution for the driver–spacing combinations of interest.
3) The sampling frequency as well as delay tap spacing should be the same for each array in the system, to minimize hardware requirements and simplify the overall design.

To steer the main lobe of array j at an angle α_j requires the delay tap spacing δ_j to be

$$\delta_j = \frac{b_j \tan \alpha_j}{c} \quad (13)$$

where b_j is the center-to-center element spacing and c is the speed of sound at ambient conditions. (Note that the sampling frequency $f_s = 1/\delta_j$.) Of interest are the minimum steering angle and the steering angle resolution obtainable at different sampling frequencies for various driver–center-to-center spacing combinations. Fig. 4 illustrates how the minimum steering angle obtainable for various driver–spacing combinations over a range of sampling frequencies may be determined. At a sampling frequency of 50 kHz, for example, an array of 2-in (51-mm) drivers spaced 2.5 in (63.5 mm) center to center would have a minimum steering angle of about 6°, while an array of 8-in (203.2-mm) drivers spaced 10 in (254 mm) center to center would have a minimum steering angle of about 2°. (The center-to-center spacings chosen are based on recommendations that appear in [12].) Note that these represent the steer-

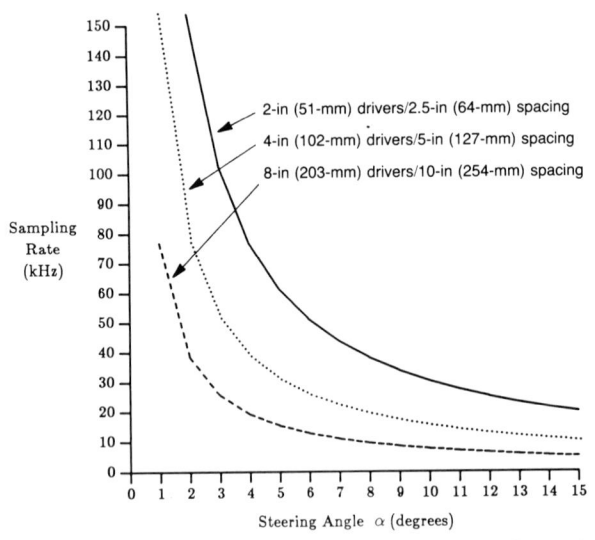

Fig. 4. Determination of minimum steering angle for various sampling frequencies.

ing angles associated with a single delay tap at the indicated sampling frequency. To obtain high steering resolution at small steering angles, then, requires relatively high sampling frequencies.

Fortunately, however, the steering resolution increases as the steering angle increases. The range of steering angles obtainable at a 50-kHz sampling rate for various driver–spacing combinations is illustrated in Fig. 5. Here it can be seen that, over the range of steering angles germane to the application described in this paper (generally 20–70°) a steering resolution of less than 1° per delay tap can typically be obtained. Sufficient steering resolution for most steering angles of interest (that is, greater than 20°) may therefore be obtained at relatively modest sampling frequencies (such as 50 kHz).

Another consideration is the amount of memory required to implement tapped delay lines for n-element line arrays. Fig. 6 illustrates the memory requirements for various-length line arrays at a sampling rate of 50 kHz. As is evident from this graph, even for relatively long arrays (such as 20 elements) steered at relatively large angles (such as 80°), the circular buffer memory requirements for implementation of the tapped delay lines are fairly modest (several thousand locations of RAM).

Based on the data presented in this section, it appears that a sampling frequency of 50 kHz satisfies most of the basic design constraints for multiple-beam electronically steered line array systems cited earlier: "comfortable" support by a commercially available DSP microprocessor, modest memory requirements, sufficient audio bandwidth, and sufficient steering resolution for all driver–spacing combinations of interest. The primary trade-off in using a 50-kHz sampling frequency, however, is the complexity of the requisite anti-image low-pass filter. Since, based on current state-of-the-art DSP microprocessor technology, the advantages far outweigh the disadvantages, 50 kHz will be the sampling frequency of choice for the case study presented in Sec. 3.

1.3 Amplitude Shading Considerations

A significant problem in designing transducer arrays is dealing with side lobes. If all array elements are driven with signals of equal amplitude and phase, the directivity pattern will exhibit side lobes of significant magnitude which shift with frequency [7], [11], [13]. To help reduce side-lobe energy, the amplitude of the signal applied to each element of array j is generally "tapered" proportional to its distance from the array central axis (xc_j, yc_j, zc_j). This technique, referred to as *shading*, can be performed using a variety of different tapering functions.

In sonar applications, use of shading to reduce side lobes is well understood [4], [6]. Significant work has also been done on the shading of steerable microphone arrays designed for transduction of speech in large rooms [5], [16]. Early attempts to use shading in line arrays designed for sound-reinforcement applications are documented in [8]. Here passive mechanical filters or passive crossover networks were used to reduce the high-frequency energy of the end-firing array elements, thus reducing the side-lobe energy as well as the otherwise narrow directivity at high frequencies.

Various array shading patterns (or tapering functions) have been utilized in practical sonar and delay-steered microphone arrays. The speech-band steerable microphone array described in [5] utilizes Dolph–Chebyshev shading, while the microphone array described in [16] utilizes a Hamming window function to determine the shading parameters. Due to the ease with which Hamming shading parameters can be calculated for different driver size–spacing combinations, as well as the demonstrated capability to reduce side-lobe energy significantly [16], we have chosen to use a Hamming window with $\beta = 0.3$ for the study presented here. For element i of an $n_j \times m_j$ array, the shading parameter (or element

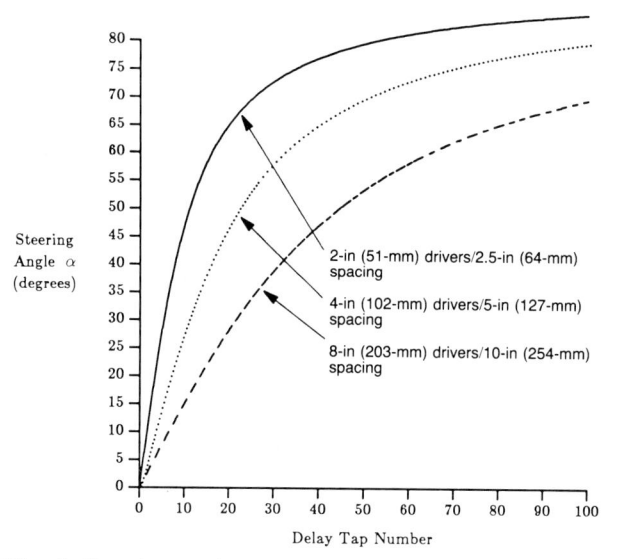

Fig. 5. Steering angle versus delay tap number at 50-kHz sampling rate.

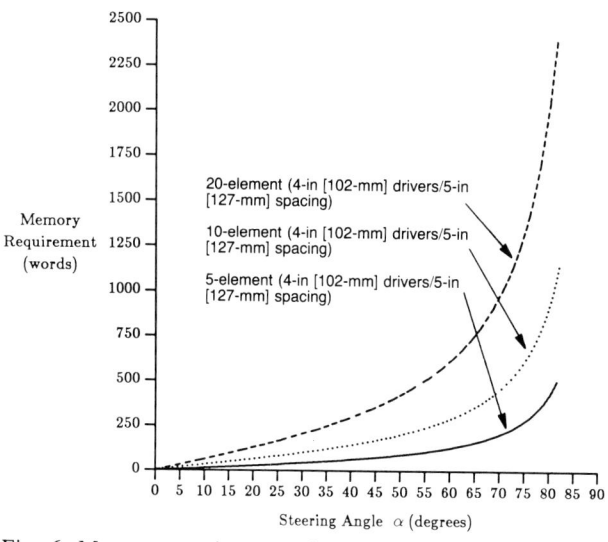

Fig. 6. Memory requirements for various-length line arrays at 50-kHz sampling rate.

weight) σ_{ij} is given by

$$\sigma_{ij} = \left[\frac{(1 + \beta) - (1 - \beta)\cos[2\pi N(i)(n_j + 1)]}{2}\right]\left[\frac{(1 + \beta) - (1 - \beta)\cos[2\pi M(i)(m_j + 1)]}{2}\right] \quad (14)$$

where $N(i)$ is the vertical index number of element i (range 1 to n_j) and $M(i)$ is the horizontal index number (range 1 to m_j). The effect of shading is demonstrated in the next section.

2 COMPUTER-AIDED DESIGN SOFTWARE

The design of a multiple-beam electronically steered sound-reinforcement array ensemble requires numerous iterations of trying various-length arrays of different driver type–spacing combinations steered toward different portions of the seating space at all frequency bands of interest. To help automate this task, two computer-aided design tools have been created: 1) a front-end program that automatically creates line arrays of arbitrary dimension, and 2) a back-end program that uses the data base created by the front-end program to create ear-plane SPL distribution maps (or *seatmap* plots).

The front-end program begins by prompting the system designer for the following room parameters: the room rear width W_B, the room front width W_F, the depth of the audience seating space S, the rake angle of the seating space γ, and the ear plane height Y_E. After prompting the designer for the frequency band of analysis f, data pertaining to each array j in the ensemble are requested. This includes a four-character driver type k (driver types currently available are listed in Table 1), the number of elements along the length (Z axis) of the array n_j, the number of elements along the width (X axis) of the array m_j, the element spacing b_j, the steering angle α_j, and the central mounting coordinates (xc_j, yc_j, zc_j). The relative mounting coordinates for each array element i of array j, (x_{ij}, y_{ij}, z_{ij}), are automatically generated. Also automatically generated are the signal delay parameters d_{ij} for each element i of array j to effect a steering angle of α_j, and the amplitude shading parameters σ_{ij} based on Eq. (14).

The back-end program generates direct field SPL distribution maps based on the parameters generated by the front-end program. A grid of fixed X dimension is generated independent of the actual room size; the Z dimension varies as a function of S. The SPL distribution at a distance Y_E from the floor of the space (the ear plane) is calculated at each grid location and subsequently normalized to 0 dB. The digits 0 through 9 are plotted at each grid location, corresponding to the number of decibels down the SPL is relative to 0 dB. If the SPL at a particular grid point is 10–20 dB down, the character * is plotted; if the SPL is 20–30 dB down, the character = is plotted; and if the SPL is more than 30 dB down, the character − is plotted.

An illustrative SPL distribution map is depicted in Fig. 7. Here $W_B = W_F = S = 10$ m, $Y_E = 1$ m, and $\gamma = 0$. A single 08CF driver (see Table 1) is placed at location (0 m, 10 m, 0 m), that is, 10 m up from the center of the square seating space area; the frequency of analysis is 2 kHz. As can be seen, the direct field SPL distribution map obtained is very close to what one might expect to measure in practice. The high resolution of the directivity data base for each

Table 1. Loudspeaker types available in data base.*

Loudspeaker type	Diameter (in)	Octave-band frequency response (dB)									
		31 Hz	63 Hz	125 Hz	250 Hz	500 Hz	1 kHz	2 kHz	4 kHz	8 kHz	16 kHz
01CA	1	−30	−30	−30	−30	−24	−18	−12	−6	0	0
02CA	2	−30	−24	−18	−12	−6	0	0	0	0	−6
04CA	4	−24	−18	−12	−6	0	0	0	0	−6	−12
08CA	8	−12	−6	0	0	0	0	−6	−12	−18	−24
15CA	15	0	0	0	0	−6	−12	−18	−24	−30	−30
02CB	2	−21	−18	−15	−12	−9	−6	−3	0	0	0
04CB	4	−18	−12	−9	−6	−3	0	−3	−6	−9	−12
08CB	8	0	0	0	0	−3	−6	−9	−12	−15	−18
01CF	1	0	0	0	0	0	0	0	0	0	0
02CF	2	0	0	0	0	0	0	0	0	0	0
03CF	3	0	0	0	0	0	0	0	0	0	0
04CF	4	0	0	0	0	0	0	0	0	0	0
05CF	5	0	0	0	0	0	0	0	0	0	0
06CF	6	0	0	0	0	0	0	0	0	0	0
08CF	8	0	0	0	0	0	0	0	0	0	0
10CF	10	0	0	0	0	0	0	0	0	0	0
12CF	12	0	0	0	0	0	0	0	0	0	0
15CF	15	0	0	0	0	0	0	0	0	0	0
2X4F	2 × 4	0	0	0	0	0	0	0	0	0	0
4X8F	4 × 8	0	0	0	0	0	0	0	0	0	0
6X9F	6 × 9	0	0	0	0	0	0	0	0	0	0

* All types listed have constant phase response.

driver (data points are stored at increments of 1°) helps provide accurate computer simulations.

The effect of substituting an elliptical 4X8F driver, with the larger driver dimension along the Z axis, is illustrated in Fig. 8. Note that the vertical (front-to-back) SPL distribution is similar to that of the 08CF driver illustrated in Fig. 7, while the horizontal (side-to-side) SPL distribution is similar to what would be obtained for a 04CF driver.

The SPL distribution for a simple eight-element line source, constructed using 04CF drivers spaced 5 in (127 mm) center to center, is illustrated in Fig. 9. Here the array is mounted directly overhead of the 10×10 m seating space, at location (0 m, 10 m, 0 m); the larger dimension of the array is along the Z axis and the frequency of analysis is again 2 kHz. Note that without amplitude shading (that is, all elements driven at equal amplitude), significant side-lobe energy is scattered over the seating area.

Upon shading the amplitude of the signal applied to each element according to a Hamming window (0.3), per Eq. (14), the SPL distribution map illustrated in Fig. 10 is obtained. Here a significant reduction in side-lobe energy can be observed, with very little "spreading" of the array main lobe.

Finally we wish to demonstrate the effect of electronically steering the simple eight-element line array discussed. The same 10×10 m seating area is used for the purpose of illustration, but now the array is moved to the front of the room (still facing in the $-Y$ direction) and mounted at location (0 m, 10 m, 5 m). Derivation of the steering angle α necessary to "hit" grid point (0 m, 1 m, 0 m), the ear plane of the room's central seating location (recall $Y_E = 1$ m), is illustrated in Fig. 11. As indicated, a steering angle of approximately 29° is required for this purpose. SPL distribution

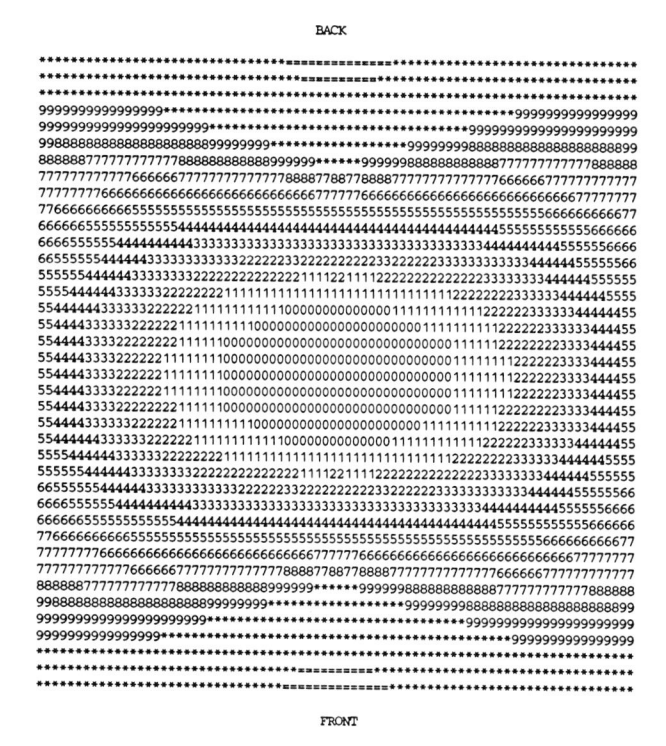

Fig. 8. Illustrative SPL distribution map for 10×10 m seating space with a single 4X8F driver located at (0 m, 10 m, 0 m) and operated at 2 kHz.

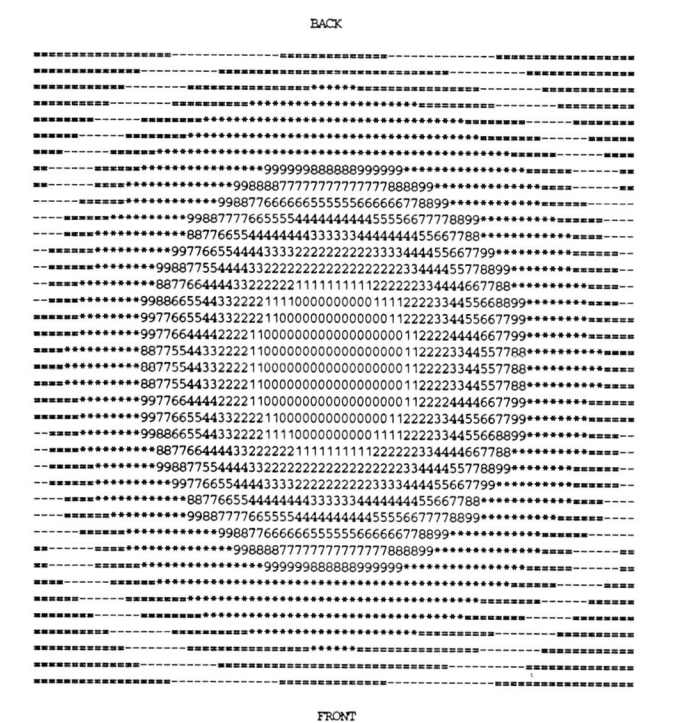

Fig. 7. Illustrative SPL distribution map for 10×10 m seating space with a single 08CF driver located at (0 m, 10 m, 0 m) and operated at 2 kHz.

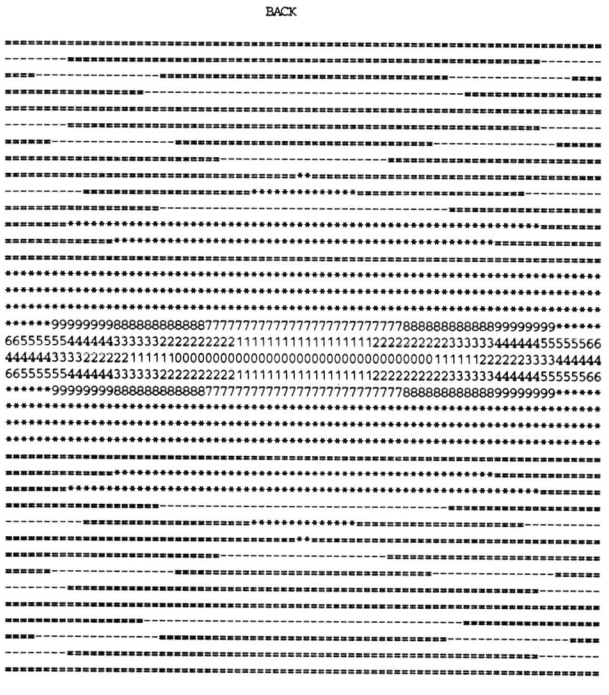

Fig. 9. Same 10×10 m seating space as in Figs. 7 and 8 with eight-element array of 04CF drivers spaced 5 in (127 mm) center to center, mounted at (0 m, 10 m, 0 m), operated at 2 kHz, and unshaded (each element at equal amplitude).

maps for the delay-steered unshaded array and the Hamming window (0.3) shaded array are illustrated in Figs. 12 and 13, respectively. Again, note the reduction in side-lobe energy attributed to shading.

3 CASE STUDY

As a practical illustration of the principles presented in this paper, the design of a multiple-beam delay-steered array ensemble is presented in this section. The room is characterized as follows: $W_B = 13.7$ m, $W_F = 12.2$ m, $\gamma = 3°$, and $S = 22$ m (actual seating depth is 18.3 m, with the speaking position 3.7 m in front of the seating area). For the purpose of illustration a two-way system is designed, consisting of three high-frequency arrays (short-, medium-, and long-throw) and a single low-frequency array. (It should be noted, however, that many different solutions are possible—what we wish to present here is simply a representative

Fig. 10. Same conditions as Fig. 9, but shaded using Hamming window (0.3).

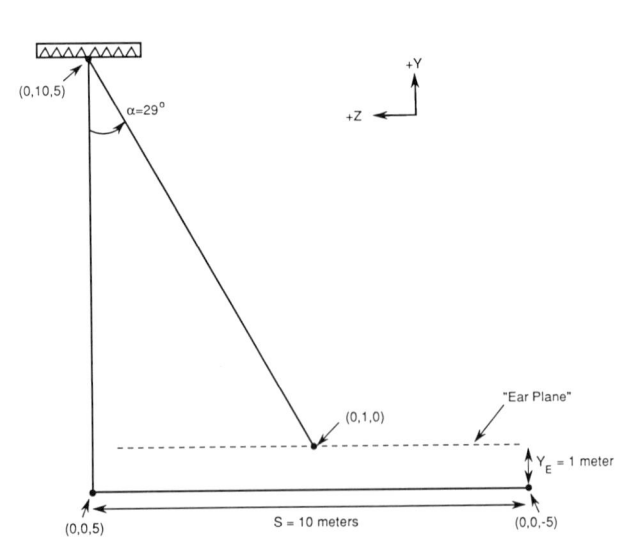

Fig. 11. Derivation of steering angle α necessary to "hit" ear plane of central seating location.

Fig. 12. Eight-element line array of 04CF drivers located at (0 m, 10 m, 5 m) steered 29°, unshaded, and operated at 2 kHz.

Fig. 13. Same conditions as Fig. 12, but shaded using Hamming window (0.3).

one.) A sketch of the array ensemble appears in Fig. 14. The high-frequency drivers are all 2 in. (51 mm) in diameter, spaced 635 mm center to center; the low-frequency drivers are 12 in (305 mm) in diameter, spaced 381 mm center to center. The directivity patterns for each driver are simulated based on the pistonic approximation model described in Eqs. (1) and (2).[2] For the high-frequency arrays, 1 kHz is used as the frequency of analysis; for the low-frequency array, 125 Hz is the frequency of analysis. Finally, the sampling frequency is 50 kHz. Table 2 lists the steering angles available for the high-frequency arrays at this sampling frequency, and Table 3 lists the steering angles available for the low-frequency array.

For the short-throw array (designated $j = 1$), a 5×1 configuration mounted at $(-0.18$ m, 9.7 m, 7.3 m$)$ is chosen. Based on a construction similar to that of Fig. 11, it can be determined that a steering angle α_1 of approximately 20° is needed to cover the front third of the seating space. Based on Table 2, the closest steering angle available is 17.9°. The SPL distribution map for the short-throw array thus aimed is given in Fig. 15. In examining this plot, note that the SPL at the speaking position is over 30 dB down from that at the center of the coverage zone.

Since a higher Q in both the horizontal and the vertical directivity patterns is required for the medium-throw array (designated $j = 2$), an 8×2 configuration mounted at $(0.18$ m, 9.7 m, 7.3 m$)$ is chosen. Using techniques similar to those described, it can be determined that a steering angle α_2 of approximately 50° is needed to cover the middle third of the seating space; based on Table 2, a steering angle of 49.8° is available. The SPL distribution map for the medium-throw array thus aimed is given in Fig. 16.

The long-throw array (designated $j = 3$) needs to be of somewhat higher Q than the medium-throw device described, so a 10×2 configuration mounted at $(-0.18$ m, 9.7 m, 6.5 m$)$ is chosen. Here it can be determined

---

[2] Actual designs, however, should be based on measured one-third-octave polar directivity data for the specific drivers being utilized. The software tools developed for this study allow use of either measured or simulated directivity data.

that a steering angle of about 65° will aim the main lobe appropriately to cover the back third of the seating space. Table 2 indicates that α_3 may be obtained very close to this value. The SPL distribution map for the long-throw array is illustrated in Fig. 17.

Next comes the critical step of combining the SPL

Table 2. Steering angles available at 50-kHz sampling frequency for 63.5-mm element spacing.

Delay tap	Steering angle
1	6.13
2	12.12
3	17.86
4	23.25
5	28.24
6	32.80
7	36.94
8	40.67
9	44.03
10	47.04
11	49.75
12	52.19
13	54.39
14	56.37
15	58.17
16	59.80
17	61.29
18	62.65
19	63.89
20	65.04
21	66.09
22	67.06
23	67.96
24	68.80
25	69.57
26	70.30
27	70.97
28	71.61
29	72.20
30	72.76

Table 3. Steering angles available at 50-kHz sampling frequency for 381-mm element spacing.

Delay tap	Steering angle
1	1.03
2	2.05
3	3.07
4	4.10
5	5.11
6	6.13
7	7.14
8	8.15
9	9.15
⋮	⋮
56	45.07
57	45.58
58	46.07
59	46.56
60	47.04
61	47.52
62	47.98
63	48.44
64	48.88
65	49.32
66	49.75
67	50.18
68	50.60
69	51.00
70	51.41

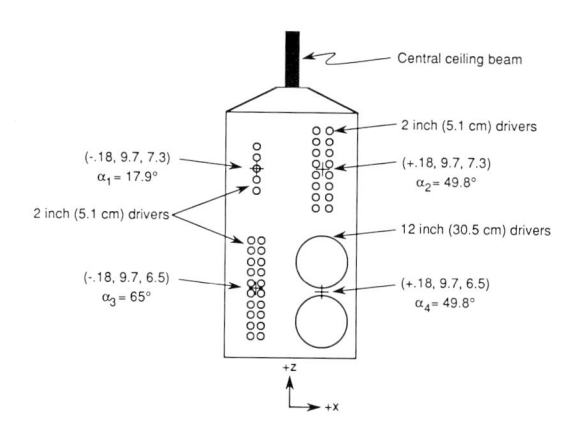

Fig. 14. Multiple-beam delay-steered array ensemble designed for case study.

distributions generated by each of the three high-frequency arrays into a composite map. Here the signal level of each array must be adjusted such that the maximum level over the entire seating space does not exceed 0 dB. The back-end program automatically normalizes the signal level of each array in generating the composite map and prints the result. For the composite high-frequency SPL distribution map illustrated in Fig. 18, $g_1 = -3.26$ dB (short throw), $g_2 = 0$ dB (medium throw), and $g_3 = -0.69$ dB (long throw).

The final step is to design a (single) low-frequency array which covers as much of the seating space as possible. (Note that multiple low-frequency arrays could be used if necessary, just as was done for the high-frequency case described.) Here a 2×1 array (designated $j = 4$) is chosen, aimed at 49.8° to hit the center of the seating space. The resulting SPL distribution map is illustrated in Fig. 19. Despite its simplicity, low-frequency coverage at the given frequency of analysis varies no more than 6 dB over the entire seating space.

4 DISCUSSION

An interesting question that comes to mind in considering array designs based on the principles presented in this paper concerns the possibility of generating multiple beams, steered in different directions, using a *single* array. While it would be nice if this were possible—considerably fewer drivers would be used, and the system would be much more compact—it is inherently not feasible: since *correlated* signals, delayed different amounts, would be fed to each driver, a severe comb-filtering effect would result. Use of array ensembles is therefore necessitated.

Some "extra benefits" of implementing the type of sound-reinforcement array ensemble described in this paper include 1) the ability to electronically "re-aim" the main lobes to compensate for partial room fill and 2) the ability to compensate for "beam bending," which occurs due to changes in temperature gradients. (This could potentially be done automatically, under computer control, using data from an array of temperature sen-

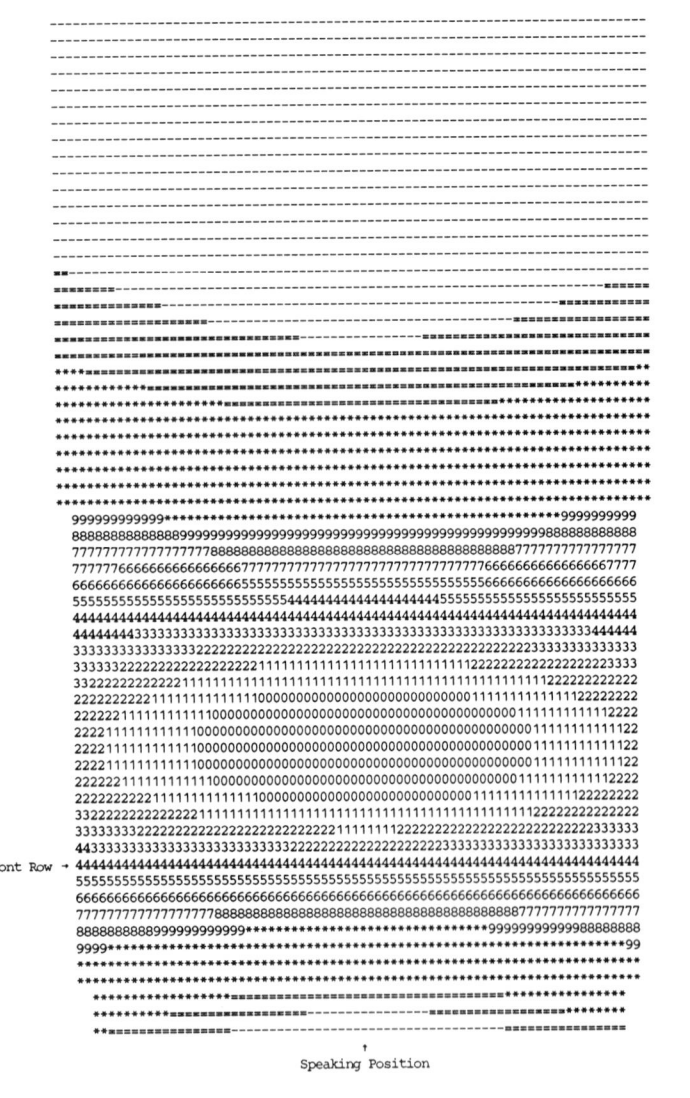

Fig. 15. SPL distribution map for 5×1 short-throw array of 02CF drivers spaced 2.5 in (635 mm) center to center, mounted at (−0.18 m, 9.7 m, 7.3 m), aimed at 17.9° and operated at 1 kHz.

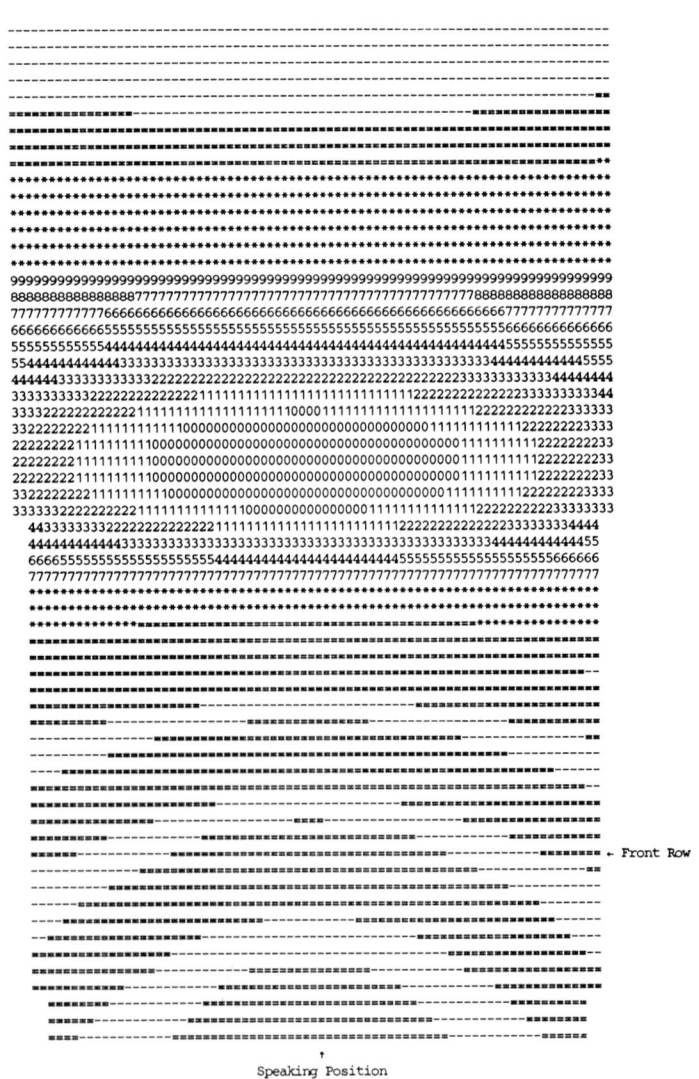

Fig. 16. SPL distribution map for 8×2 medium-throw array of 02CF drivers spaced 2.5 in (635 mm) center to center, mounted at (0.18 m, 9.7 m, 7.3 m), aimed at 49.8°, and operated at 1 kHz.

sors.)

One of the primary difficulties in implementing a delay-steered sound-reinforcement array using commercially available components is the virtual nonexistence of modular amplifier components: 10 20-W amplifier modules, for example, would be much more useful than a single 200-W amplifier. Also helpful would be the availability of a standard DSP box, configurable via a standard personal computer, that implements the tapped delay lines as well as the digital filters for the crossover networks and amplitude shading.

Areas in which more research is needed include 1) sensitivity of directivity patterns to perturbations in control parameters as well as both amplifier and driver characteristics, 2) algorithms for dynamic array tapering that help stabilize the array directivity patterns as the frequency of excitation varies (that is, a "digital" version of what was attempted in [8] and [13]), and 3) amplitude shading patterns which minimize side-lobe energy while optimizing both efficiency and broad-band directivity stability.

5 SUMMARY

In this paper we have presented both theoretical and practical aspects of multiple-beam delay-steered array ensemble design for sound-reinforcement applications. Such an approach offers many of the desirable properties associated with a central cluster design, with the benefit of being much more aesthetically pleasing from an architectural point of view, in particular for A-frame style structures. Relatively recent advances in signal-processing hardware technology have helped make a sound-reinforcement design based on this approach both technically and economically feasible.

6 REFERENCES

[1] D. L. Klepper, "Sound Systems in Reverberant Rooms for Worship," *J. Audio Eng. Soc.*, vol. 18, pp. 391–401 (1970 Aug.).

[2] D. L. Klepper, "Requirements for Theatre Sound and Communication Systems," *J. Audio Eng. Soc.*,

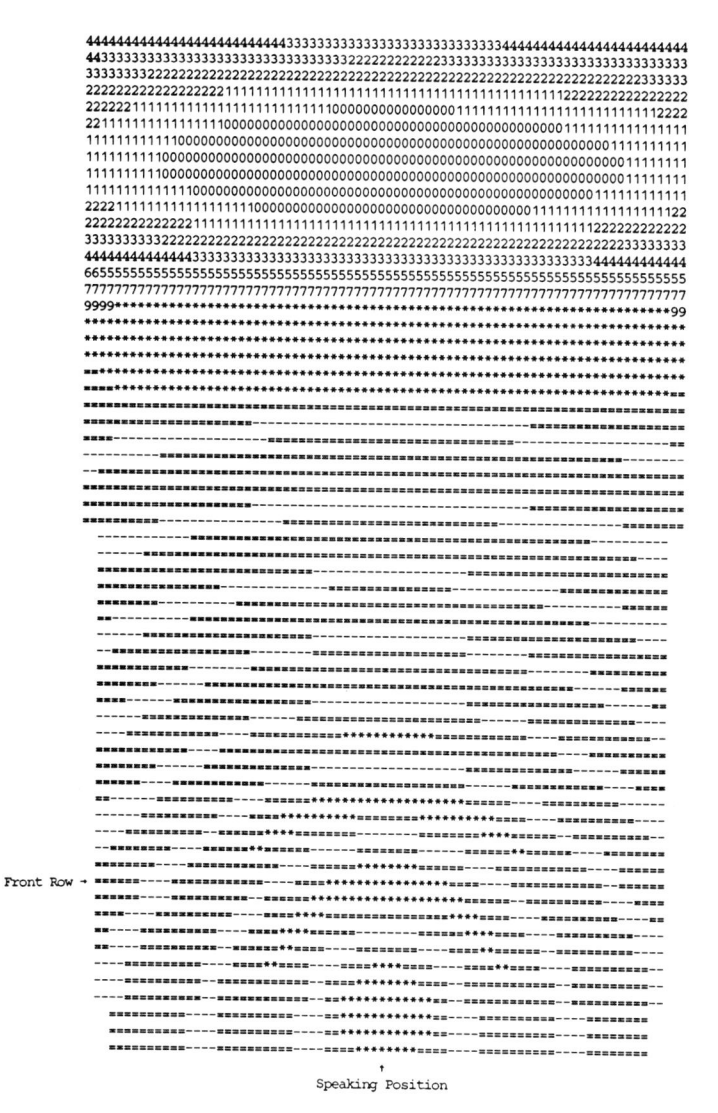

Fig. 17. SPL distribution map for 10 × 2 long-throw array of 02CF drivers spaced 2.5 in (635 mm) center to center, mounted at (−0.18 m, 9.7 m, 7.3 m), aimed at 65°, and operated at 1 kHz.

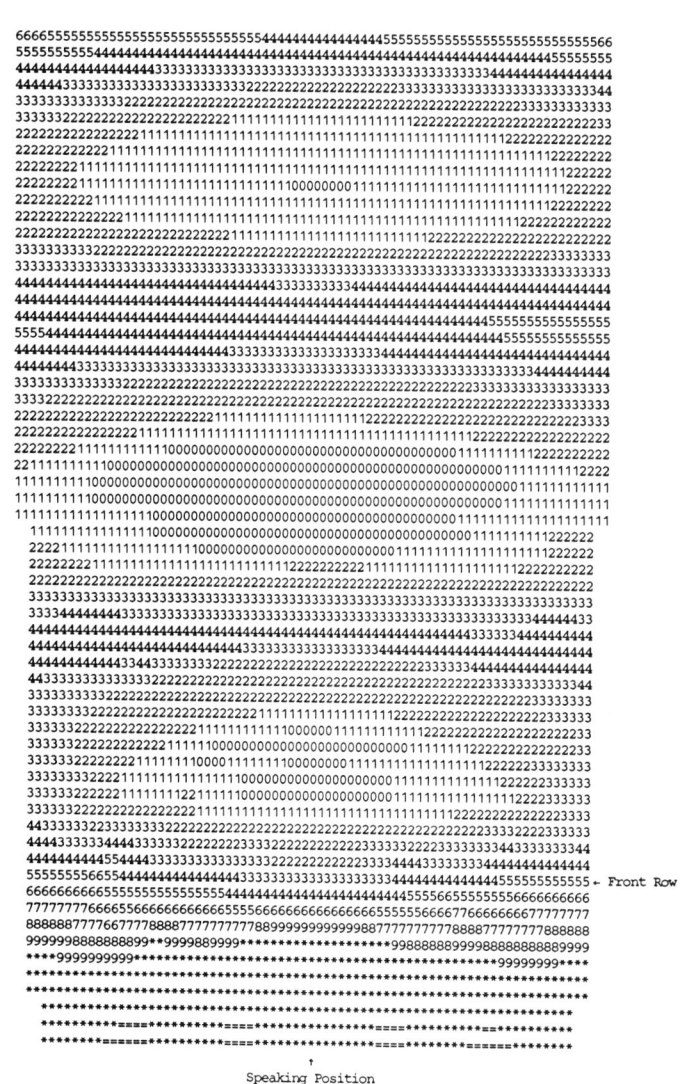

Fig. 18. Normalized composite SPL distribution map for arrays of Figs. 15–17 (high-frequency array ensemble), operated at 1 kHz.

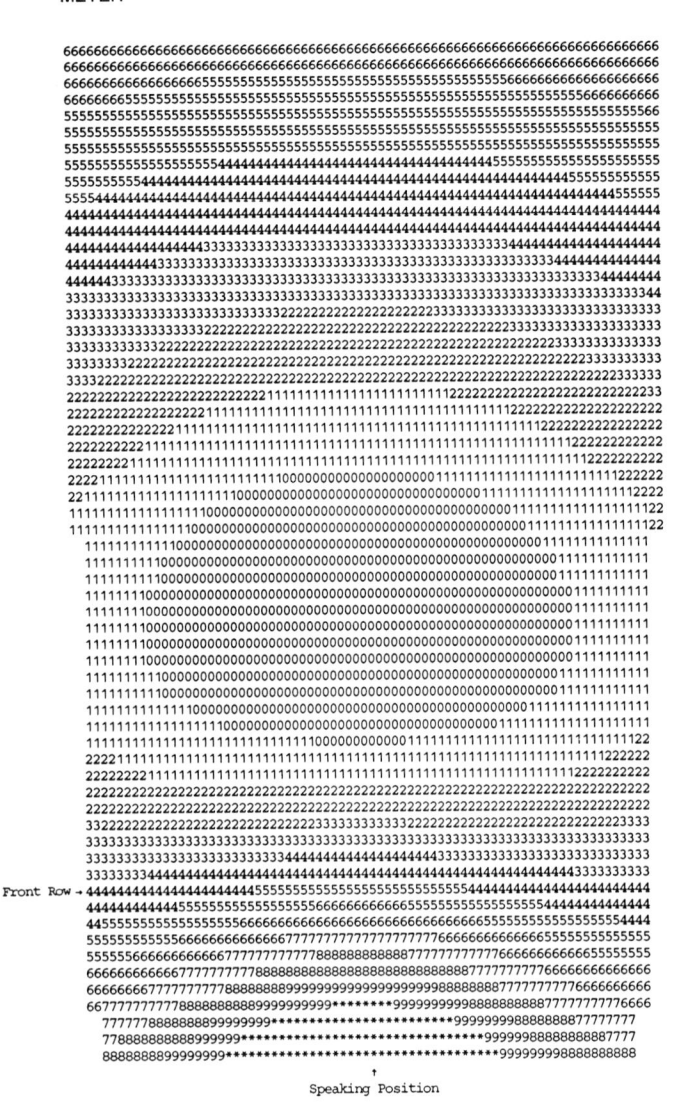

```
6666666666666666666666666666666666666666666666666666666666666666
6666666666666666666666666666666666666666666666666666666666666666
666666666666666666655555555555555555555555555555555566666666666666
6666666655555555555555555555555555555555555555555555555566666666
5555555555555555555555555555555555555555555555555555555555555566
5555555555555555555555555555555555555555555555555555555555555555
5555555555555555555555555555555555555555555555555555555555555555
5555555555555555554444444444444444444444444444444455555555555555
5555555555554444444444444444444444444444444444444444444455555555
55554444444444444444444444444444444444444444444444444444444455555
4444444444444444444444444444444444444444444444444444444444444444
4444444444444444444444444444444444444444444444444444444444444444
44444444444444444333333333333333333333333333333333344444444444444
444444444444443333333333333333333333333333333333333334444444444444
4444443333333333333333333333333333333333333333333333333344444444
333333333333333333333333333333333333333333333333333333333333344
333333333333333333333333333332222222222222222333333333333333333
3333333333333333322222222222222222222222222222223333333333333333
33333333333322222222222222222222222222222222222222223333333333333
33333333222222222222222222222222222222222222222222222223333333333
3333222222222222222222222222222222222222222222222222222223333333
222222222222222222222221111111111111111111122222222222222222233
2222222222222222222211111111111111111111111111112222222222222222
22222222222222221111111111111111111111111111111112222222222222222
222222222111111111111111111111111111111111111111111112222222222
2222222211111111111111111111111111111111111111111111111122222222
22221111111111111111110000000000000000111111111111111111112222
2211111111111111111100000000000000000000001111111111111111112222
1111111111111111100000000000000000000000000001111111111111122
111111111111111100000000000000000000000000000000011111111111122
1111111111110000000000000000000000000000000000000011111111111
111111111000000000000000000000000000000000000000000001111111111
11111111100000000000000000000000000000000000000000000001111111111
1111111100000000000000000000000000000000000000000000000001111111
1111111100000000000000000000000000000000000000000000000001111111
1111111100000000000000000000000000000000000000000000000001111111
1111111100000000000000000000000000000000000000000000000001111111
1111111100000000000000000000000000000000000000000000000001111111
111111110000000000000000000000000000000000000000000000001111111
1111111111000000000000000000000000000000000000000000000001111111
11111111111110000000000000000000000000000000000000000001111111
1111111111111111000000000000000000001111111111111111111122
222211111111111111111111111111111111111111111111111111111122
22222222111111111111111111111111111111111111111111111111111112222222
2222222222111111111111111111111122222222222222222222222222222222
2222222222222222222222222222222222222222222222222222222222222222
3322222222222222222222222223333333333333333322222222222222222222
3333333333333333333333333333333333333333333333333333333333333333
3333333333333333334444444444444444433333333333333333333333333333
3333333344444444444444444444455555555555555555555544444444443333333
Front Row → 4444444444444444445555555555555555555555554444444444444444444444
44444444444455555555555555566666666666666666666555555555555555544
4455555555555555555566666666666666666666666665555555555555555544
555555555556666666666677777777777777777777666666666666666655555
5555556666666666677777777778888888888887777777777666666666666666
6666666666677777777788888888888888888888887777777766666666666666
666666667777777777888888888999999999999888888887777777776666666
667777777777888888888999999999999999999888888887777777776666
777777788888888899999999 ••••••••••• 99999999988888888887777777
778888888888999999 ••••••••••••••••••••••• 9999998888888888887777
88888888899999999 ••••••••••••••••••••••••• 99999998888888888
                                ↑
                        Speaking Position
```

Fig. 19. SPL distribution map for 2 × 1 low-frequency array of 12CF driver spaced 15 in (381 mm) center to center, mounted at (0.18 m, 9.7 m, 6.5 m), aimed at 49.8°, and operated at 125 Hz.

vol. 20, pp. 642–649 (1972 Oct.).

[3] L. Beranek, *Acoustics* (McGraw-Hill, New York, 1954).

[4] N. Davids, E. Thurston, and R. Mueser, "The Design of Optimum Directional Acoustic Arrays," *J. Acoust. Soc. Am.*, vol. 24, pp. 50–56 (1952 Jan.).

[5] J. Flanagan, J. Johnston, R. Zahn, and G. Elko, "Computer-Steered Arrays for Sound Transduction in Large Rooms," *J. Acoust. Soc. Am.*, vol. 78, pp. 1508–1518 (1985 Nov.).

[6] J. Jarzynski and W. Trott, "Array Shading for a Broadband Constant Directivity Transducer," *J. Acoust. Soc. Am.*, vol. 64, pp. 1266–1269 (1978 Nov.).

[7] L. Kinsler, A. Frey, A. Coppens, and A. Sanders, *Fundamentals of Acoustics*, 3rd ed. (Wiley, New York, 1980).

[8] D. L. Klepper and D. W. Steel, "Constant Directional Characteristics from a Line Source Array," *J. Audio Eng. Soc.*, vol. 11, pp. 198–202 (1963 July).

[9] D. G. Meyer, "Computer Simulation of Loudspeaker Directivity," *J. Audio Eng. Soc.*, vol. 32, pp. 294–315 (1984 May).

[10] H. Olson, *Acoustical Engineering* (Van Nostrand, Princeton, NJ, 1967).

[11] L. H. Schaudinischky, A. Schwartz, and S. Mashiah, "Sound Columns—Practical Design and Applications," *J. Audio Eng. Soc.*, vol. 19, pp. 36–40 (1971 Jan.).

[12] A. P. Smith, "A Three-Way Columnar Loudspeaker for Reinforcement of the Performing Arts," *J. Audio Eng. Soc.*, vol. 19, pp. 213–219 (1971 Mar.).

[13] I. Wolff and L. Malter, "Directional Radiation of Sound," *J. Acoust. Soc. Am.*, vol. 2, pp. 201–241 (1930 Feb.).

[14] R. C. Heyser, "Loudspeaker Phase Characteristics and Time Delay Distortion, Parts I and II," *J. Audio Eng. Soc.*, vol. 17, pp. 30–41 (1969 Jan.); pp. 130–137 (1969 Apr.).

[15] R. C. Heyser, "Determination of Loudspeaker Signal Arrival Times, Parts I–III," *J. Audio Eng. Soc.*, vol. 19, pp. 734–743 (1971 Oct.); pp. 829–834 (1971 Nov.); pp. 902–905 (1971 Dec.).

[16] H. Abe, N. Miyaji, M. Iwahara, A. Sakamoto, and L. Boden, "Practical Applications and Digital Control of Microphone Arrays," presented at the 76th Convention of the Audio Engineering Society. *J. Audio Eng. Soc. (Abstracts)*, vol. 32, p. 1006 (1984 Dec.), preprint 2116.

THE AUTHOR

David George Meyer received the B.S. degree in electrical engineering in 1973, the M.S.E. degree in electrical engineering in 1975, the M.S. degree in computer science in 1979, and the Ph.D. degree in electrical engineering in 1981, all from Purdue University, West Lafayette, Indiana.

In 1982 he joined the School of Electrical Engineering at Purdue University, where he is currently an Associate Professor specializing in advanced architecture microprocessors, computer systems engineering, acoustics, digital signal processing, and educational delivery systems.

He has published over 30 technical papers on acoustics, computer system design, audio system engineering, and educational delivery systems in addition to authoring book chapters on parallel processing and advanced architecture microprocessors. He has done consulting in acoustics for both AT&T Bell Laboratories and Electro-Voice, Inc., as well as designed and analyzed sound-system installations for various clients.

Dr. Meyer is a member of the Institute of Electrical and Electronics Engineers, the Audio Engineering Society, the Association of Computing Machinery, and the American Society for Engineering Education.

Prediction of the Full-Space Directivity Characteristics of Loudspeaker Arrays*

KENNETH D. JACOB AND THOMAS K. BIRKLE

Bose Corporation, Framingham, MA 01701, USA

Traditionally the problem of predicting the behavior of loudspeaker arrays has been approached by making simplifying assumptions about the individual array elements. Assuming omnidirectional or piston behavior, or simplifying or disregarding phase response can lead at best to good approximation of actual array behavior, and at worst to serious error. A new graphics-based array simulation program has been developed which allows four traditional techniques as well as a new hybrid technique to be used in predicting the behavior of arbitrarily configured arrays. The correlation between actual and predicted behavior of three test arrays is presented. Results show that the hybrid technique, based on measurements of an array element's magnitude and phase response, is the most accurate predictor. A second technique, based on assumed phase response, is shown to be accurate in cases where the element's acoustic origin is fixed. Three other traditional prediction techniques are shown to have significant limitations which can lead to serious errors.

0 INTRODUCTION

New emphasis has been placed on the role that loudspeaker directivity plays in the success or failure of sound systems designed for large audiences. The directivity of a sound source can be responsible for poor speech intelligibility if it causes echoes or too many late arriving reflections. Similarly, erratic directivity can be the cause of uneven perceived frequency response over an audience area. It seems clear that the goal of even sound distribution and good intelligibility over a large audience area depends in part on the directivity characteristics of the sound sources.

If source directivity is known, a variety of microcomputer-based software programs can be used to predict the sound arriving at a listener location.[1] The usefulness of these programs is limited, however, by the number of sources whose full-space directivity is known. Unfortunately these directivity characteristics are rarely known in the case of loudspeaker arrays or clusters.

Although the need to know the full-space directivity characteristics of multiple-element arrays is compelling, acquiring the data involves difficult, unproven, or inaccurate techniques. Careful measurements guarantee accurate information, but are time consuming and require specialized equipment and personnel. Prediction techniques, if they can be proven accurate, offer clear advantages over measurement techniques since array behavior can be explored on the desktop as opposed to in the laboratory.

In this study the various techniques currently used to predict array behavior are reviewed. In addition, a new hybrid prediction technique is presented. An experiment has been designed to test the accuracy of the various techniques by comparing predictions to the actual behavior of three arrays. The overall accuracy of the various predictors is computed by averaging the errors between actual and predicted directivities. Examples illustrating the major sources of error are presented along with a discussion of the results.

1 ARRAY PREDICTION TECHNIQUES

1.1 Basic Mathematics

Array prediction techniques require summation of the contributions from each of the individual elements. This summing function is modified in important ways by the assumptions inherent in the various techniques.

---

* Presented at the 87th Convention of the Audio Engineering Society, New York, 1989 October 18–21; revised 1990 January 24.
[1] For a description of some of these programs see [1].

In this section the basic mathematics for complex pressure summation from multiple sources is presented first, followed by the specific summation formulas of the various predictions techniques.

The coordinate system used to define an arbitrary array of elements is shown in Fig. 1. Henceforth the sphere centered at the array center and on which predictions are calculated will be called the *polar sphere*. The term *polar circle* will be used to describe the intersection between the polar sphere and any plane that contains the array center and the sphere's pole. This includes the circular paths used in traditional polar plots. The general equation for computing the composite pressure at a point on the polar sphere is

$$P(\theta, \varphi, f, R) = \sum_{n=1,N} p_n(\theta, \varphi, f, r_{n,\theta,\varphi}) \quad (1)$$

where

$P(\theta, \varphi, f, R)$ = total complex pressure at frequency f, at distance R from array center, and at polar angle defined by θ and φ
R = radius of polar sphere
N = number of sources
$p_n(\theta, \varphi, f, R)$ = complex pressure contribution from nth source, which can be expanded to

$$p_n(\theta, \varphi, f, r_{n,\theta,\varphi}) = \frac{a_{n,\theta,\varphi}}{r_{n,\theta,\varphi}} \times \exp[-j(kr_{n,\theta,\varphi} + \phi_{n,\theta,\varphi})] \quad (2)$$

with

$a_{n,\theta,\varphi}$ = pressure amplitude of nth source at polar angle θ and φ and a reference distance (usually 1 m)
$\phi_{n,\theta,\varphi}$ = phase, in radians, of nth source at θ and φ and the reference distance
k = wave number, $= 2\pi f/c$, c being wave speed
$r_{n,\theta,\varphi}$ = distance from point on polar sphere to reference mark on nth source.

1.2 Simple Source Prediction Technique

The simple source technique assumes that each source is omnidirectional and has as phase response the result of propagation delay only. The assumptions of this model are realistic when the source is much smaller than a wavelength of sound. The simple source technique has been extended by several investigators to regions of higher frequency with various degrees of success [2]–[7]. The technique has been shown to be useful in predicting major directivity features, but has not been used to compare predictions with actual measurements to verify its accuracy. Because loudspeakers typically operate in regions where wavelengths are on the same order as or smaller than the sources, it can be expected that actual array behavior will deviate from predictions made using this technique.

The equation used to calculate directivity using the simple source model is

$$P(\theta, \varphi, f, R) = \sum_{n=1,N} p_n(\theta, \varphi, f, r_{n,\theta,\varphi}) \quad (3)$$

where

$$p_n(\theta, \varphi, f, r_{n,\theta,\varphi}) = \frac{K_n}{r_{n,\theta,\varphi}} \times \exp[-jkr_{n,\theta,\varphi}] \quad (4)$$

with K_n being the pressure amplitude of the nth source.

1.3 Power Sum Technique

The power sum technique assumes that phase interaction between elements can be disregarded. In this

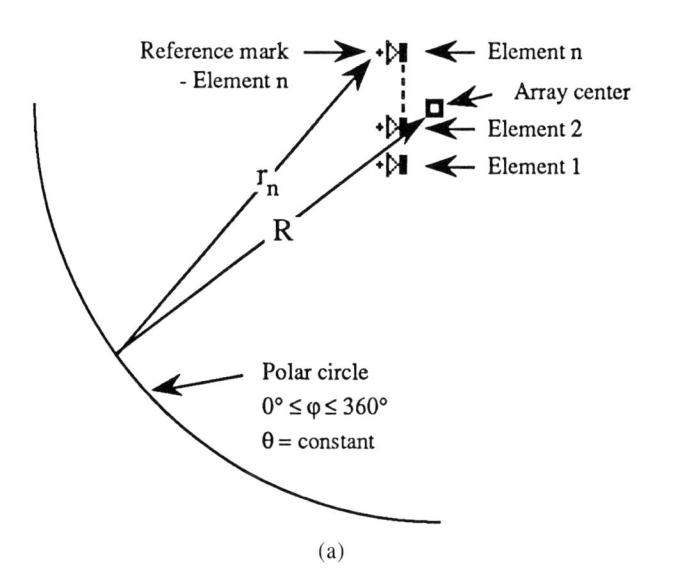

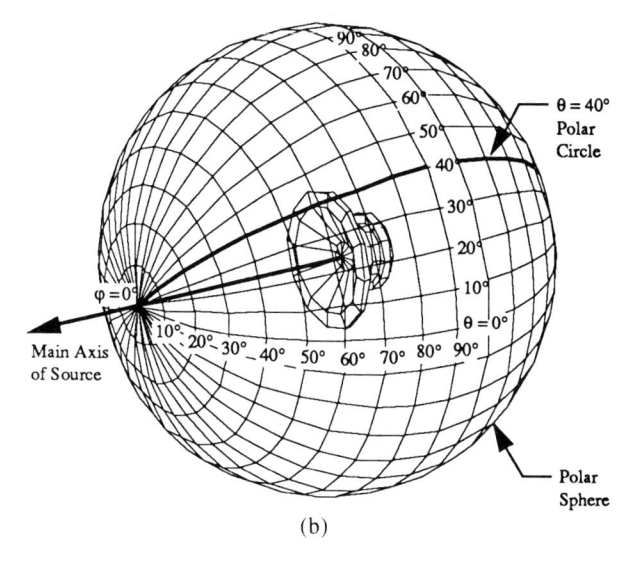

Fig. 1. (a) Coordinate system for array of n elements. (b) Schematic showing polar sphere, polar circles, and main source axis. Measurements are made at intersection points of longitude and latitude lines. Predictions are made at same or higher resolution.

model only the magnitudes of the energy contributions from the various elements are added. While phase interaction is responsible for many forms of array behavior—a simple dipole is an example—there are situations where the interaction between elements is assumed to be negligible. This is an assumption usually made in predicting horn cluster behavior, for example.

The equation used to calculate array behavior using the power sum model is

$$P^2(\theta, \varphi, f, R) = \sum_{n=1,N} |p_n(\theta, \varphi, f, r_{n,\theta,\varphi})|^2 \qquad (5)$$

where

$$|p_n(\theta, \varphi, f, r_{n,\theta,\varphi})|^2 = \left(\frac{a_{n,\theta,\varphi}}{r_{n,\theta,\varphi}}\right)^2 \qquad (6)$$

where $P^2(\theta, \varphi, f, r)$ is the total sound energy at frequency f and radius R from the array center, and at a polar angle defined by θ and φ.

1.4 Piston Source Technique

Another technique assumes that an array element can be modeled as a piston in an infinite baffle.[2] This technique is particularly attractive since the pressure can be computed from a closed-form equation. Many authors have used the piston model to predict array behavior, but to these authors' knowledge none has correlated prediction with actual measured behavior.

The equation used to calculate dispersion using the piston source model [9] is

$$P(\theta, \varphi, f, R) = \sum_{n=1,N} p_n(\theta, f, r_{n,\theta,\varphi}) \qquad (7)$$

where

$$p_n = \frac{j\rho_0 c k l_n U_n}{r_{n,\theta,\varphi}} \times \frac{J_1(kl_n \sin \theta_n)}{kl_n \sin \theta_n} \times \exp[-jkl_n] \qquad (8)$$

where

- ρ_0 = density of air
- l_n = radius of nth piston source
- U_n = velocity amplitude of nth source
- J_1 = Bessel function of the first kind
- θ_n = angle between line from point on polar circle and main axis of nth source, bounded by $-90° < \varphi < +90°$.

The assumption that the piston is located in an infinite baffle restricts predictions to the region $-90° < \varphi < +90°$; rear radiation is not predicted. The piston model is useful for cone-type loudspeakers operating in frequency regions where each part of the cone is moving with the same velocity and phase. The piston model is not useful for horn sources or other sources that have a behavior fundamentally different from that of pistons.

1.5 Phasor Sum Technique

In the phasor sum technique the complex pressure contribution of an array of sources is computed, but it is assumed that the phase response is due solely to propagation delay. This method is identical to the simple source model with the exception that it does not assume omnidirectionality. Advocates of this technique [10] state that the phasor sum should be restricted to arrays of like devices. Furthermore they state that the technique is not recommended for predicting the dispersion characteristics of any cluster made up of different devices; in these cases, the power sum model is recommended.

The equation used to calculate dispersion using the phasor sum technique is

$$P(\theta, \varphi, f, R) = \sum_{n=1,N} p_n(\theta, \varphi, f, r_{n,\theta,\varphi}) \qquad (9)$$

where

$$p_n(\theta, \varphi, f, r_{n,\theta,\varphi}) = \frac{a_{n,\theta,\varphi}}{r_{n,\theta,\varphi}} \times \exp[-jkr_{n,\theta,\varphi}]. \qquad (10)$$

1.6 Other Models

Other investigators have worked to predict the dispersion characteristics of the individual elements theoretically [11]–[13]. These predictions are usually based on a finite-element model of the individual elements. While some of these models appear promising, none has emerged that is suitable for predicting the phase and magnitude behavior of the broad range of transducer types used in sound reinforcement.

2 NEW HYBRID TECHNIQUE

The limitations of existing techniques led us to a new approach. In this approach the simplifying assumptions made by the simple source, piston, power sum, and phasor sum models are avoided through the use of actual measurements of array elements. Prediction of array behavior is carried out on a new software program, and calculation of array behavior is made using the complex pressure summation scheme described in Sec. 1.1. This hybrid approach is suitable for use in predicting the behavior of most types of modern arrays, from horn clusters to arrays of wide-bandwidth drivers.

2.1 Measurement of Array Elements

The measurement of individual array elements is based on the need to capture both phase and magnitude information over the full polar sphere. The measurement system used is shown in Fig. 2. In this setup a computer-based dual-channel fast Fourier transform system is used. The exact phase relationship between the source and the microphone is preserved by ensuring that a

---

[2] A less well-known closed-form solution exists for the radiation of a piston in a rigid sphere but has not been tested in this study see [8].

single reference point marked on the source is always directly over the center of turntable rotation.

At each location on the polar sphere a 4096-point frequency response curve was obtained over a bandwidth of 0–10 kHz. There were 64 separate files generated for a 360° rotation of the turntable, corresponding to a polar resolution of 5.625°. A signal-to-noise ratio of at least 40 dB and power compression of less than 0.5 dB were ensured over the passband of the loudspeaker for all measurements.

2.2 Data Reduction

Data reduction is possible in the spatial and spectral domains. In general a source whose spectral or spatial response is relatively smooth and gradual can tolerate more data reduction than a source that has an erratic or rapidly changing response. For the three transducers used in this study it was found that spatial resolution could be reduced to 10° crossings of longitude and latitude lines, and spectral resolution could be reduced to one-third-octave bands. (As is shown later, the reduction of *measured* data to 10° and one-third-octave resolution does not preclude the ability to predict higher resolution spectral or spatial behavior.)

2.3 Program for Predicting Array Behavior

The software program written for the purpose of predicting array behavior is called ArrayCAD[3] prediction program and runs on the Apple[4] Macintosh family of computers. The program is designed to allow the rapid construction and modification of array models through the use of an advanced user interface.

The physical shape of each sound source is represented by a wire-frame model. Source acoustics are represented by full-space magnitude and phase response and on-axis sensitivity. Individual array elements can be placed and aimed into an array using Cartesian or spherical coordinate systems. Each array element can be assigned an arbitrary electrical driving signal.

---

[3] ArrayCAD is a trademark of Bose Corporation, Framingham, MA, USA.

[4] Product names are trademarks of their respective makers.

In the program the user can choose between simple source, power sum, piston source, and phasor sum techniques or the new hybrid technique. Sine wave or constant-bandwidth array response can be predicted; for example, predictions can be made at 2000 Hz or in the one-third-octave band whose center frequency is 2000 Hz.

3 TEST ARRAYS

In order to test the accuracy of the various prediction techniques it was necessary to construct and measure real arrays. Three arrays employing different transducers were used. Measurements were made on the setup shown in Fig. 2. For the two arrays composed of cone loudspeakers, measurements were made with 5.625° resolution. For the array composed of horns, where narrow-angle interference effects are known to occur [14]–[17], a much higher measurement resolution of 0.5° was used.

3.1 Twiddler Array[5]

The driver used in this array is a wide-bandwidth 2.25-in (57-mm) diameter loudspeaker. The driver is housed in a 45-in³ (730-cm³) sealed enclosure. Two of these units were used to create a simple array where the centers of the drivers were separated by 4.7 in (120 mm), as shown in Fig. 3.

3.2 4.5-in (114-mm) Driver Array

The driver used in the second array is a very wide-bandwidth 4.5-in (114-mm)-diameter loudspeaker. The

---

[5] Twiddler is a registered trademark of Bose Corporation, Framingham, MA, USA.

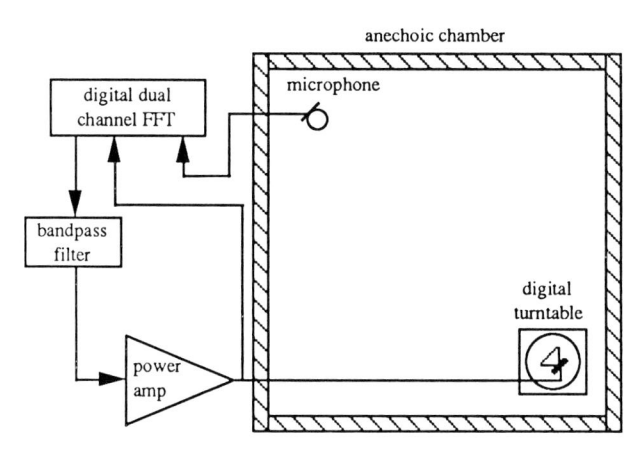

Fig. 2. Setup for measuring full-sphere magnitude and phase polar response of an array element.

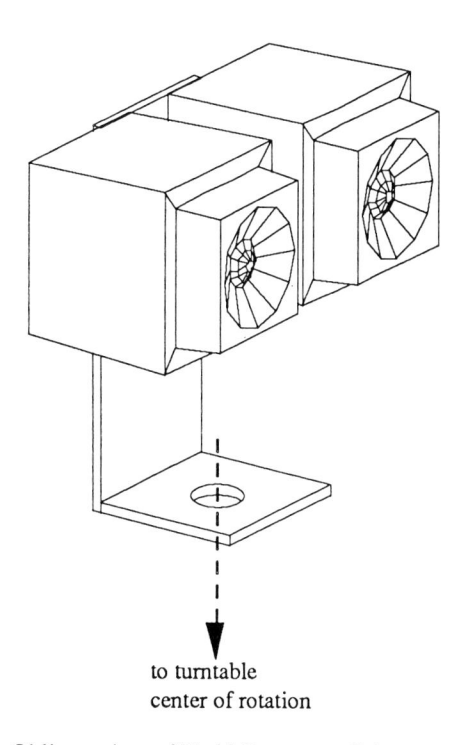

Fig. 3. Oblique view of Twiddler array. Driver centers are separated by 4.7 in (120 mm).

array consisted of four drivers[6] located in a line and angled as shown in Fig. 4.

3.3 Horn Array

The third test array consisted of two Electro-Voice HP-6040A constant-directivity horns driven by DH2 compression drivers. The horns were aimed such that their 2-kHz patterns overlapped at their −6-dB points. This corresponded to aim angles of ±25°. The horns were placed so as to ensure that their drivers lay on the surface of a sphere and were as close together as possible, as shown in Fig. 5.

3.4 Models of the Three Test Arrays

For each of the three test arrays a model was specified. Predictions were computed over the 10 one-third-octave-band frequencies ranging from 500 to 4000 Hz and in the horizontal polar plane. Each one-third-octave band was represented by 10 frequencies. Array predictions were made from an average of the 10 frequencies. In the case of the Twiddler array and the 4.5-in (114-mm), array, predictions were made using a resolution of 10°. In the case of the horn array, 1° resolution was used in an attempt to predict any narrow-angle interference effects.

4 RESULTS

4.1 Overall Accuracy of Prediction Techniques

For each array and for each prediction technique, a series of 10 polar plots corresponding to the 10 one-third-octave bands were generated and compared to the polar plots from the actual arrays. Only polar data within 12 dB of the axial sound pressure level were considered. If both the actual and the predicted polar data were more than 12 dB down compared to the on-axis level, the data were discarded. This choice was made in order to compare the accuracies of the predictors in the high-energy region of the arrays' directivity patterns.

For the Twiddler and 4.5-in (114-mm) driver arrays the analysis technique described was used over the entire 360° polar circle. In the case of the horn array, the technique was only used over the first 30° of the polar circle since each of the predictive techniques tested predicted near-identical results outside of the 30° region.

Errors between predicted and actual polar responses at each point around the 10 polar plots were computed and averaged to yield a single standard error repre-

---

[6] While the baffle used in this experiment is the same as that employed in the Bose model 402 loudspeaker, the electronic shading circuit used in the 402 to control dispersion at high frequencies was removed. Therefore the polar data presented for this 4.5-in (114-mm) driver array does not resemble that of the 402 loudspeaker.

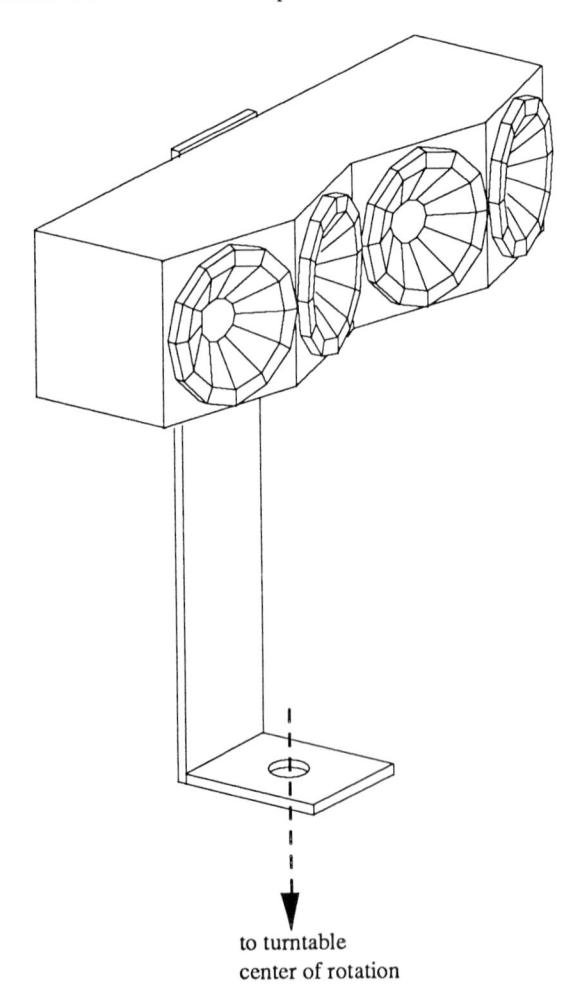

Fig. 4. Oblique view of 4.5-in (114-mm) array. Driver centers are separated by 4.8 in (122 mm) and are aimed at ±15° from center.

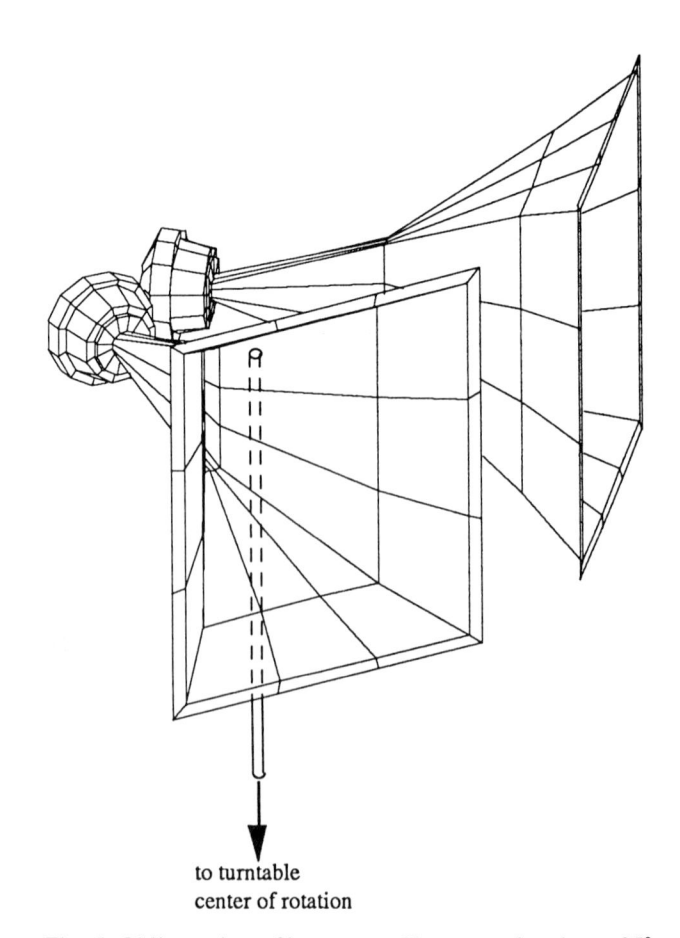

Fig. 5. Oblique view of horn array. Horns are aimed at ±25° and are aligned so that the backs of the compression drivers fall on a sphere of radius ≈ 13.3 in (350 mm).

sentative of the accuracy of the prediction technique. Thus the higher the standard error, the less accurate the predictor. The overall results are shown in Figs. 6 and 7.

From the overall results it can be seen that the new hybrid technique has low overall error for each of the three different arrays. The phasor sum technique is equally accurate except in the case of the horn array, where it is significantly less accurate than the hybrid technique. The piston source technique is accurate in the spatial region and for the transducer type to which it is restricted. The power sum and simple source techniques are much less accurate.

4.2 Illustrative Examples

While the technique described in Sec. 4.1 indicates the overall accuracy of the various prediction techniques, it does not show the major sources of error. Closer examination of the polar data can be used to expose these sources.

4.2.1 Twiddler and 4.5-in (114-mm) Driver Arrays

In the case of the Twiddler and 4.5-in (114-mm) driver arrays the driver size results in wide dispersion, even up to 4 kHz. Thus significant interdriver interference can occur. The power sum technique, with its neglect of phase interaction, cannot predict the strong interaction frequencies where interference occurs. The simple source technique, while accurately predicting the major interference behavior of the spaced sources, results in serious errors due to its assumption that the sources are omnidirectional.

Both the piston technique and the phasor technique can be expected to predict the Twiddler and 4.5-in (114-mm) array behavior accurately since their basic assumptions are mostly valid in the frequency range used here. The modest error observed in the piston technique relative to the phasor sum technique may be due to the fact that at the higher frequencies the driver does not behave like a simple piston, and second, to the incorrect assumption that the driver is mounted in an infinite baffle. Side-by-side polar plots illustrating these effects are shown in Fig. 8.

It is interesting to note that as the number of drivers in the array increases from the two-driver Twiddler to the four-driver 4.5-in (114-mm) array, the overall error increases by the same factor of 2. This result is not surprising considering that the complexity of the directivity pattern increases with the number of elements.

4.2.2 Horn Array

In the case of the horn array, both the hybrid and the phasor sum techniques were able to show basic narrow-angle interference effects measured in the horn overlap region. The power sum technique is incapable of predicting these phase-based interference effects. (The simple and piston source techniques were not used since their assumptions are violated by basic horn behavior.)

Although the overall error of the power sum technique is lower than that of the phasor technique, it must be stressed that qualitatively, the power sum technique should be considered much less accurate since it is unable to predict the narrow-angle interference effects. The phasor technique, while being quantitatively least accurate, does indicate the presence of these interference effects, but errs in their exact spatial and spectral locations.

The hybrid technique is most accurate, due primarily to its ability to predict more exactly the angle of the interference nulls. Because the new technique relies on measured phase information while the phasor sum

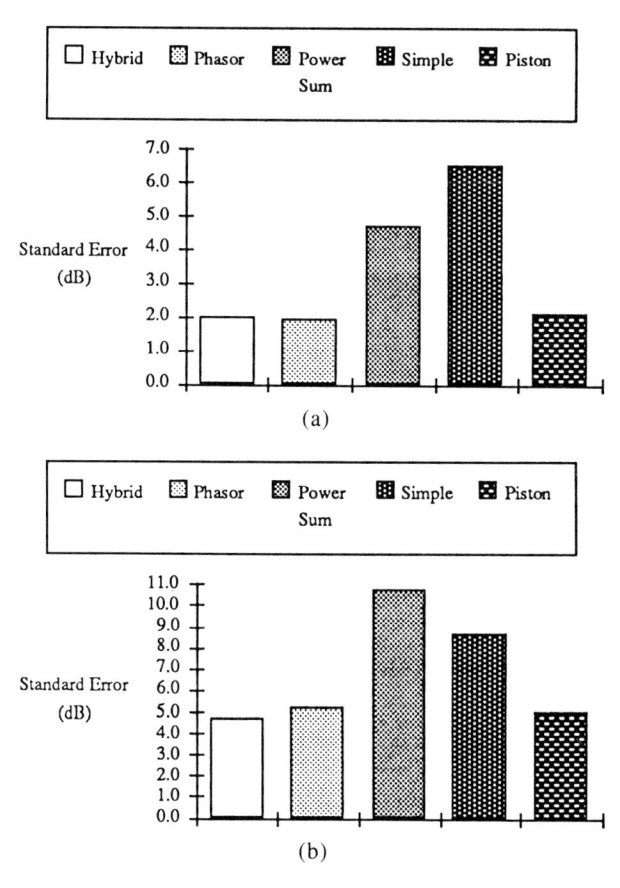

Fig. 6. Overall standard errors for five different array prediction techniques. (a) Twiddler array. (b) 4.5-in (114-mm) driver array.

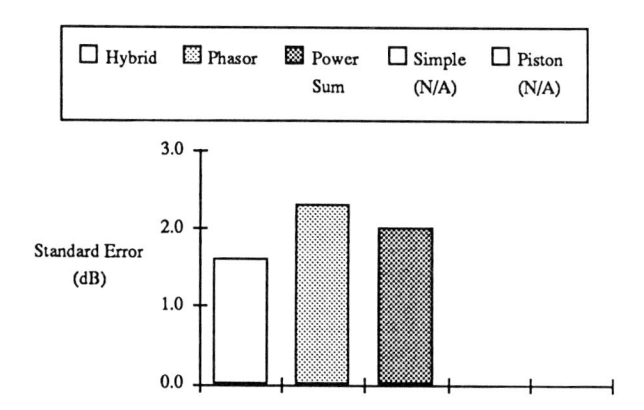

Fig. 7. Overall standard errors for three different array prediction techniques; horn array.

technique relies on assumed phase response, the location of these nulls can be expected to be predicted more accurately by the new technique. Fig. 9 shows measurements and predictions illustrating these trends.

5 DISCUSSION

5.1 Stability of Acoustic Origin[7] and Prediction Accuracy

Results comparing the accuracy of the different array behavior prediction techniques indicate that the hybrid and phasor techniques are similar in accuracy in the case of the Twiddler and 4.5-in (114-mm) driver arrays. The only difference between the two techniques is the use of measured versus assumed phase response. The phasor technique assumes that the phase response is constant and a result of propagation delay only. If the actual phase response of an array element can be shown to be due to propagation delay only, then the primary assumption of the phasor model will have been proven.

The conditions for propagation-time only phase response are shown in Fig. 10, where it is assumed that the acoustic origin is placed somewhere near to, but not exactly coincident with, the turntable axis of rotation. Therefore,

$$\phi_T = \phi_R + \phi_\varphi = K - kr\cos\varphi$$

where

ϕ_T = total measured phase
ϕ_R = phase due to delay from center of rotation to microphone
ϕ_φ = phase due to acoustic origin not being exactly aligned to center of rotation
φ = angle of rotation of turntable
K = constant
k = wave number; = $2\pi f/c$.

Under these conditions, the acoustic origin simply rotates about the turntable axis of rotation, tracing out a small circle, and thus giving rise to a sinusoidally varying phase from the acoustic origin to the measurement microphone. One complete revolution of the turntable corresponds to one sinusoidal period in the phase versus angle response.

The Twiddler and the 4.5-in (114-mm) drivers showed measured phase responses that did not significantly deviate from the sinusoidal, although this result was not expected. The horn showed significant phase deviations, demonstrating a nonstable acoustic origin. Fig. 11 illustrates the difference between the Twiddler

---

[7] Acoustic origin is defined as the point in space from which all sound is assumed to radiate.

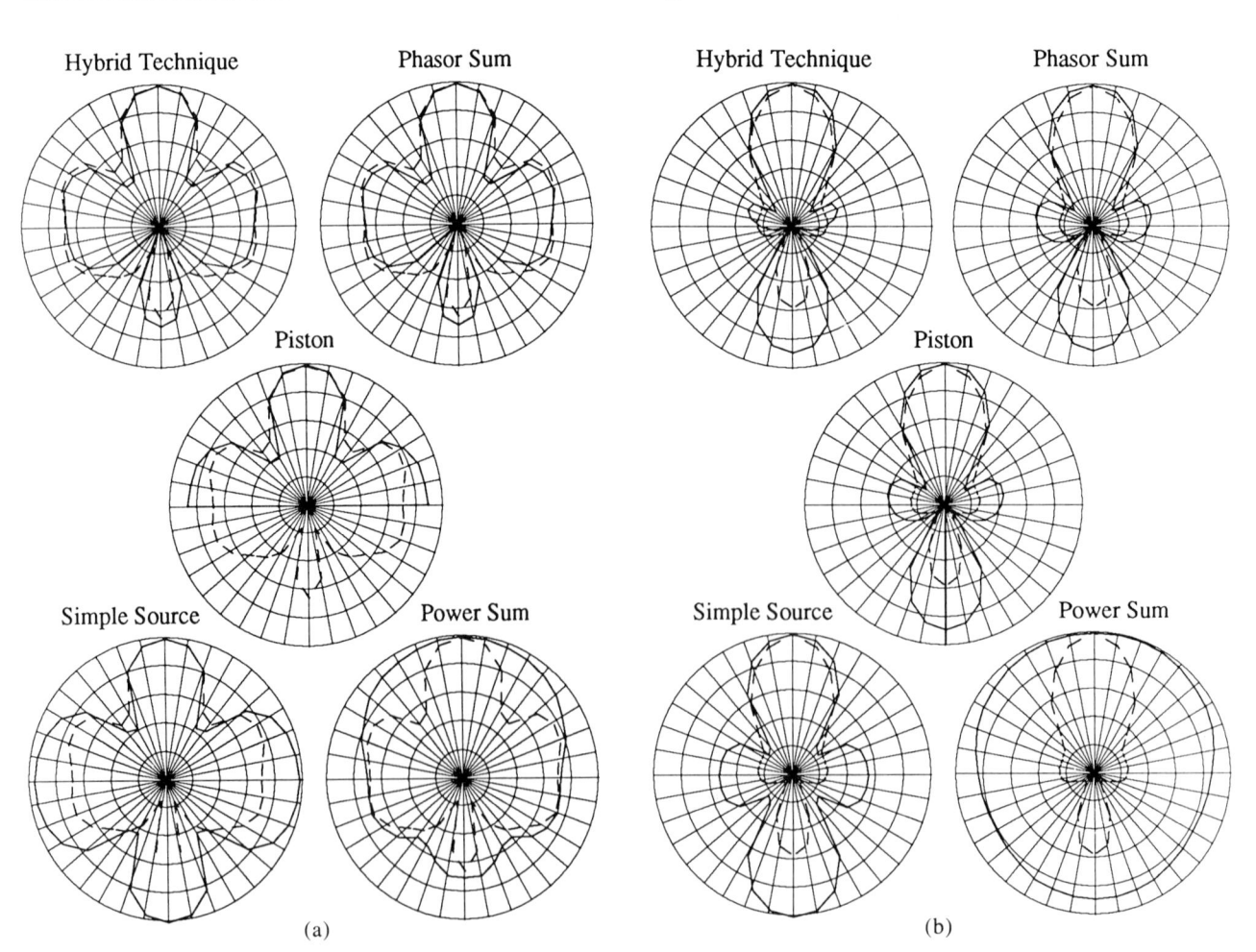

Fig. 8. Horizontal polar plots for five different array prediction techniques. 5 dB/div. Solid lines—predicted; dashed lines —measured. (a) Twiddler array, 2500-Hz band. (b) 4.5-in (114-mm) driver array, 1000-Hz band.

and the horn elements with regard to acoustic origin stability. (No preference is intended or implied here concerning the advantages or disadvantages of acoustic origin stability.)

A stable acoustic origin, therefore, is thought to be primarily responsible for the similarities in accuracy of the hybrid and phasor techniques in the case of the Twiddler and 4.5-in (114-mm) driver arrays. In the one case where an element was used with the acoustic origin not fixed, the hybrid prediction technique remains accurate because it includes the nonsinusoidal phase response, while the phasor technique becomes less accurate.

5.2 Extension of the Phasor Technique

The usefulness of the phasor technique has been extended in this study. The authors who introduced the technique stated that it should be restricted to arrays of like devices and to single frequencies [10]. In this study the phasor technique has been shown to be useful for bandwidths of at least one-third octave and for a variety of different transducers. These results indicate that the phasor model is useful for arrays composed of elements whose acoustic origins are known to be fixed.

5.3 Effectiveness of Data Reduction

The data base used to represent array elements consisted of a single complex number for each 10° crossing of longitude and latitude lines on the measurement sphere. This corresponds to a data-base size of only $36 \times 18 = 648$ complex numbers for each one-third-octave band. What is remarkable is that narrow-angle interference effects can be predicted with excellent accuracy using only low-resolution array element data bases. The low-resolution data base was sufficient to predict narrow-angle effects simply by high-resolution sampling in the array prediction program. This indicates that high-resolution measurements of array elements are not needed to predict complex yet highly significant array behavior patterns.

5.4 Effectiveness of Other Techniques Tested

The piston model, while shown to be useful for arrays using cone loudspeakers, is not applicable to horn arrays or any other array element that differs fundamentally from a piston, and it cannot be used to predict rear radiation in its simple form. It is possible that the accuracy of the piston model could be improved if a more exact solution for a piston in a rigid sphere [8] were exploited. This study shows that the piston model is a good choice when neither magnitude nor phase data are available for array elements.

The simple source model is of less value as a predictive technique, but nevertheless is capable of exposing the major front-radiation interference effects of spaced sources. The simple source model is probably of most value for predicting the behavior of low-frequency arrays. When the individual elements become directional, or have nonconstant phase response as a function of angle, the model will produce serious errors. In this study most of the error associated with the simple source method is in the rear radiation region. The simple

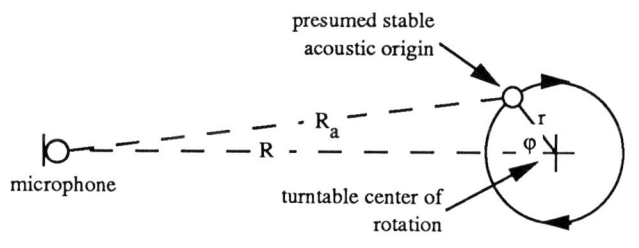

Fig. 10. Schematic diagram showing a fixed acoustic origin traveling about turntable axis of rotation. Phase versus φ response of such a system will be sinusoidal in shape. $R_a \approx R - r \times \cos \varphi$ when $R \gg r$.

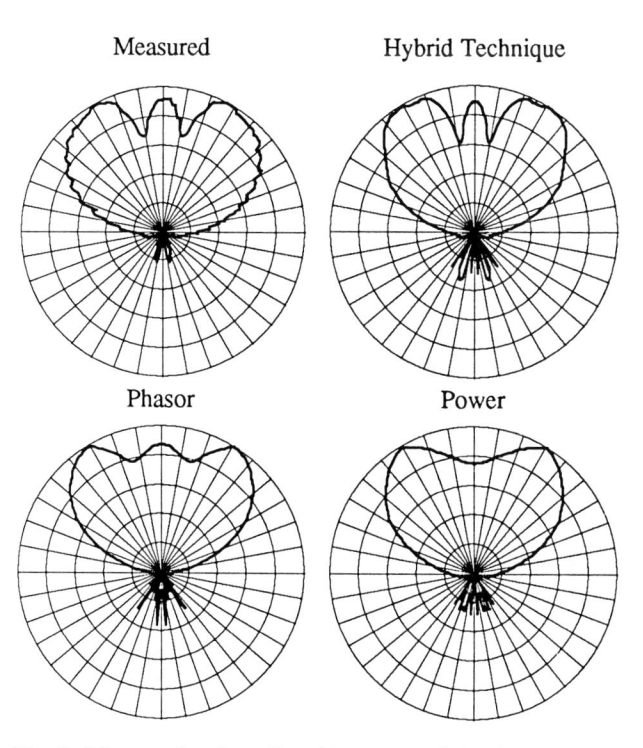

Fig. 9. Measured and predicted horn array behavior in 1000-Hz band for three different array prediction techniques. 5 dB/div. Notice that angle of dip near ±10° and off-axis region between 10 and 20° are better predicted by hybrid technique.

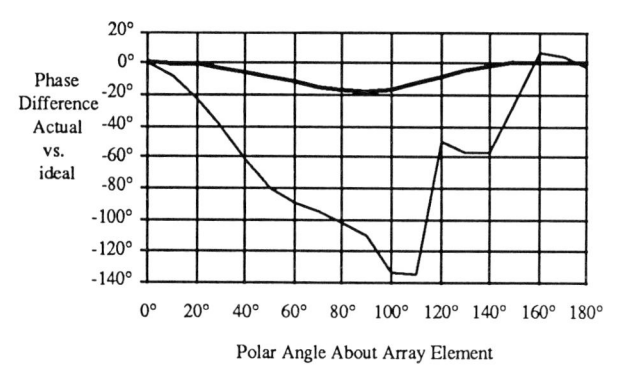

Fig. 11. Phase difference between actual measured phase and phase if acoustic origin is assumed fixed. Twiddler driver exhibits minor phase difference when compared to horn at 1000 Hz. Instability of acoustic origin is accounted for by hybrid prediction technique, but cannot be accounted for by phasor technique. Heavy line—twiddler; light line—horn.

source method is not an appropriate method for predicting horn array behavior.

The power sum technique is only useful in arrays where the individual elements have minimal interaction. In the broad range of frequencies tested here, both the Twiddler and the 4.5-in (114-mm) 4.5 drivers interact strongly, and therefore the power sum model is seriously flawed. In the case of horn arrays, the success of the power sum model is dependent on the aiming of the individual horns. If the horns are aimed so that their −6-dB points are aligned, significant interference results, and the power sum model will be unable to account for the interference effects. If horn elements are angled wider than their −6-dB points, then the power sum model can be expected to be more accurate. However, this increased angling may not be practical in cluster design since it is likely to create a hole or a dark spot in the horn array's polar response. The power sum technique has the undesirable effect of masking the interference effects that occur in the proper aiming of constant-directivity horns.

6 CONCLUSION

The relative accuracies of five different methods used to predict the dispersion behavior of loudspeaker arrays have been tested against the actual behavior of three arrays. Results indicate that a hybrid technique proposed here, based on the measurement of both phase and magnitude over the entire polar sphere of an array element is most promising as being applicable to most transducer and array types. The phasor sum technique, based on measured magnitude and assumed phase response, was shown to be accurate except in situations where array elements do not meet the requirement of phase response due to propagation time only, such as in the case of a constant-directivity horn.

Other traditional prediction techniques were shown to have limitations and restrictions, but nevertheless were useful for predicting array behavior in some circumstances. The piston model is limited to array elements whose behavior resembles a piston, such as small cone drivers, but cannot be used to predict rear radiation from arrays, or for predicting the behavior of horn arrays. The simple source model is limited to low frequencies or, at higher frequencies, to predicting general front-radiation interference effects. And the power sum model is only useful in the case where array elements have negligible interaction.

Significantly, it was shown that only low-resolution, full-sphere polar response was required to predict narrow-angle interference effects. These predictions were made by implementing high-resolution sampling about the array model polar circle. The significance of this finding is that high-resolution measurements of actual array elements were not required for an accurate prediction of detailed array behavior.

These results indicate that significantly increased accuracy in the prediction of array behavior requires the inclusion of phase information. If the new hybrid prediction technique described in this study can be generalized, manufacturers will need to publish both magnitude and phase information for their arrayable loudspeakers, but they will only need to do this at moderate measurement resolution. At the very least, this study indicates the need for manufacturers to publish full-sphere polar responses for their most common arrays.

7 ACKNOWLEDGMENT

The authors gratefully acknowledge Chris Ickler, staff engineer, Bose Corporation, for many valuable conversations relevant to this study.

8 REFERENCES

[1] *Proc. 6th AES Int. Conf. on Sound Reinforcement* (1988), sec. 4.

[2] B. Zegada and E. Hixson, "Three Octave Constant Beamwidth End-Fire Line Array," Electro Acoustics Res. Lab., Dept. of Elec. Comput. Eng., University of Texas, Austin, Rep. (1985 May).

[3] J. Lardies and J. P. Guilhot, "A Very Wide Bandwidth Constant Beamwidth Acoustical End-Fire Line Array without Side Lobes," *J. Sound Vibration*, vol. 120, no. 3 (1988).

[4] D. G. Meyer, "Computer Simulation of Loudspeaker Directivity," *J. Audio Eng. Soc.*, vol. 32, pp. 294–315 (1984 May).

[5] D. G. Meyer, "Digital Control of Loudspeaker Array Directivity," *J. Audio Eng. Soc. (Engineering Reports)*, vol. 32, pp. 747–754 (1984 Oct.).

[6] G. L. Augspurger and J. S. Brawley, "An Improved Colinear Array," presented at the 74th Convention of the Audio Engineering Society, *J. Audio Eng. Soc. (Abstracts)*, vol. 31, p. 964 (1983 Dec.), preprint 2047.

[7] G. L. Augspurger, "Near-Field and Far-Field Performance of Large Woofer Arrays," *J. Audio Eng. Soc.*, vol. 38, this issue (1990 April).

[8] P. Morse and K. Ingard, *Theoretical Acoustics* (McGraw-Hill, New York, 1968).

[9] See, for example, L. Kinsler and A. Frey, *Fundamentals of Acoustics*, 2nd ed. (Wiley, New York, 1962), p. 165.

[10] D. Albertz, J. Eargle, D. B. Keele, Jr., and R. Means, "A Microcomputer Program for Central Loudspeaker Array Design," presented at the 74th Convention of the Audio Engineering Society, *J. Audio Eng. Soc. (Abstracts)*, vol. 31, p. 964 (1983 Dec.), preprint 2028.

[11] I. C. Shepard and R. J. Alfredson, "An Improved Computer Model of Direct-Radiator Loudspeakers," *J. Audio Eng. Soc.*, vol. 33, pp. 322–329 (1985 May).

[12] Y. Kagawa, T. Yamabuchi, and K. Sugihara, "A Finite Element Approach to a Coupled Structural-Acoustic Radiation System with Application to Loudspeaker Characteristic Calculation," *J. Sound Vibration*, vol. 69, no. 2 (1980).

[13] A. J. M. Kaizer and A. Leeuwestein, "Calculation of the Sound Radiation of a Nonrigid Loudspeaker Diaphragm Using the Finite-Element Method," *J. Audio Eng. Soc.*, vol. 36, pp. 539–551 (1988 July/Aug.).

[14] See, for example, W. J. W. Kitzen, "Multiple Loudspeaker Arrays Using Bessel Coefficients," *Elec. Components Appl.*, vol. 5 (1983 Sept.).

[15] C. Foreman, "Applications for the Altec Lansing Mantaray Constant Directivity Horns," Altec Eng. Notes, Tech. Letter 241.

[16] Electro-Voice, Inc., "Pro Sound Facts," no. 4 (1976).

[17] R. Sinclair, "Stacked and Splayed Acoustical Sources," presented at the 61st Convention of the Audio Engineering Society, *J. Audio Eng. Soc.* (Abstracts), vol. 26, p. 994 (1978 Dec.), preprint 1389.

THE AUTHORS

K. Jacob

T. Birkle

Ken Jacob is a staff engineer of Bose Corporation, Framingham, Massachusetts. He received his master's degree from the Massachusetts Institute of Technology and his bachelor's degree from the University of Minnesota, both with specializations in acoustics.

In addition to research in speech intelligibility, Mr. Jacob is the acoustic engineer of the Bose 102[1] commercial sound system and the Acoustic Wave[1] Cannon system, and is the project engineer of the Sound System[1] software family including the Modeler[1] design program.

Tom Birkle is a staff engineer for Bose Corporation, Framingham, Massachusetts. He received a B.S.E.E. degree from the University of Colorado at Boulder in 1982. From 1983 until 1987 he worked as a sound system designer with the acoustical consultants David L. Adams Associates, Inc. in Denver, Colorado. He then joined Bose Corporation as the programmer for the Sound System[1] software family.

[1] Trademarks of Bose Corporation.

An Array Filtering Implementation of a Constant-Beam-Width Acoustic Source*

JEFFERSON A. HARRELL AND ELMER L. HIXSON

University of Texas, Austin, TX 78712, USA

For acoustic arrays (microphones or loudspeakers) the main beam width of the radiation pattern is determined by the spacing between the elements in terms of the wavelength. Thus the beam pattern depends on the frequency. It is demonstrated that by controlling the spacing and the frequency response of a pair of superposed arrays, constant directivity of the main beam lobe over a wide frequency range can be achieved. Large powers can be achieved with relatively low energy densities at each element. A procedure is described by which line arrays of loudspeakers can be superposed with appropriate signal conditioning to control the main beam coverage with frequency. This methodology is extended to an X-shaped planar array for three-dimensional beam control with frequency. Computer models and experiments verify the process for one octave, and models are used to extend the method to wider frequency ranges and more optimal beam synthesis.

0 INTRODUCTION

Broad-band acoustic signals (such as voice and music) intended for entertainment require uniform coverage over an area (such as the audience in a theater) so that the spectrum distribution in the area remains constant. This ensures, for instance, that everyone in the audience hears the same relative amplitudes across the frequency range of the sound track. Thus the effective beam width of the acoustic source needs to be constant with frequency.

An effort to solve this problem is included in the work of Keele [1]. His work in constant-directivity horns demonstrates an ability to cover an area uniformly with broad-band characteristics. However, the sound intensities achieved in the throat of the horn threaten to enter the nonlinear region at lower frequencies, thus introducing the possiblity of unacceptable distortion for high power levels in theater applications. The array implementation discussed here suggests a natural low-frequency complement to constant-directivity high-frequency horns applied to theater sound systems.

Linear arrays of transducers have a beam width that is proportional to the wavelength of the signal, thus failing to provide a consistent sound field with frequency. This phenomenon is referred to as spectral distortion and is illustrated in Fig. 1.

Another solution is to use one or more arrays, with each transducer receiving individually conditioned signals, effectively shading the array. This makes the array seem smaller at higher frequencies to approximate a constant beam width. Unfortunately this also requires complex filters to achieve linear or constant-phase response and greatly increases the component count. In addition, as fewer elements are energized in the array, the overall sound level goes down with increasing frequency, even though the relative beam width stays constant with frequency. This effect requires signal equalization, adding further to the component count.

Linear array superposition minimizes these problems

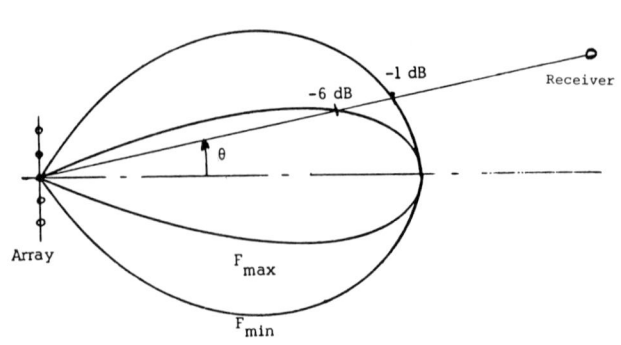

Fig. 1. Spectral distortion of the main beam of an array.

---

* Presented at the 87th Convention of the Audio Engineering Society, New York, 1989 October 18–21; revised 1990 January 5.

by combining two arrays having equally spaced elements, each having the desired beam width at a single frequency corresponding to the half-wavelength of the array. Then by using appropriate filters, the arrays are smoothly combined over the bandwidth of the system such that the 0-dB and −3-dB points on the main lobe remain constant in angle and space. The advantage of this approach is that the on-axis amplitude remains constant with frequency in addition to the main beam width, thus not requiring signal equalization.

1 THEORY

Consider a three-element array with transducer elements spaced a half-wavelength apart, the polar beam pattern of which is illustrated in Fig. 2. This pattern arises as the result of the superposition of three ideal acoustic point sources. The maxima occur at angles where the sources appear in phase to a far-field observation point, and the minima (or nulls) at angles where the sources are out of phase to some degree. The plot for $f = f_0$ is for 850 Hz where the spacing is a half-wavelength, and the plot for $f = 2f_0$ is for 1700 Hz where the spacing is then one wavelength.

This polar pattern is relative, with the on-axis response normalized to 0 dB. If this pattern (Fig. 2) represents an array at 850 Hz, then the same pattern would exist for an array with half the spacing and half the size at 1700 Hz. By appropriate superposition, the main lobe can be held approximately constant over an octave with these two arrays.

Each array has a frequency-dependent directivity function associated with it. By applying two constraints on the functions, two relative weighting functions can be derived in terms of the directivity functions. The constraints concern amplitude and angle and reduce to the requirement that the on-axis amplitude and half power angle amplitudes are held constant over the octave. This translates to require that the combination of directivity functions stay at 0 dB at 0° and at −3 dB at 18°. (This is the half power angle for the array of Fig. 2.) Mathematically, this becomes

$$R_1(f)D_1(f, 0°) + R_2(f)D_2(f, 0°) = 1.0 \quad (1)$$

$$R_1(f)D_1(f, 18°) + R_2(f)D_2(f, 18°) = 0.707 \quad (2)$$

where the functions $D_1(f, \theta)$ and $D_2(f, \theta)$ are the individual array directivity functions, given by

$$D_1(f, \theta) = \frac{\sin 3\phi_1}{3 \sin \phi_1} \quad (3)$$

where

$$\phi_1 = \frac{kd_1 \sin \theta}{2} \quad (4)$$

with k the propagation constant and d_1 the spacing between array 1 elements. The directivity for array 2 has a similar expression.

R_1 and R_2 are the frequency-dependent array weighting functions. With these determined, we now have the required amplitude response of the filters necessary to direct the signal to the two arrays and produce the constant main lobe (between 0 and 18°). Note that only two filters are required; all elements in either array receive the same signal. This method of filter definition is from work done with line arrays by Hixson and Au [2].

Fig. 3 illustrates the simulated effect of this array combination as a linear function of angle and frequency. The angular axis is from 0 to 90° with 2.25° increments. At f_0 the element spacing of the array pair is one-half and one-quarter wavelengths. Note the high degree of control up to 18°, after which some significant deviations occur with frequency.

This concept can be extended from a line array to a two-dimensional array by superposition again. By using four superposed arrays, two for vertical and two for horizontal control, the now three-dimensional shape of the main lobe can be adjusted to audience requirements. For example, a main beam would be desired to be wide in the horizontal dimension but narrow in the vertical for most efficient coverage of a theater audience.

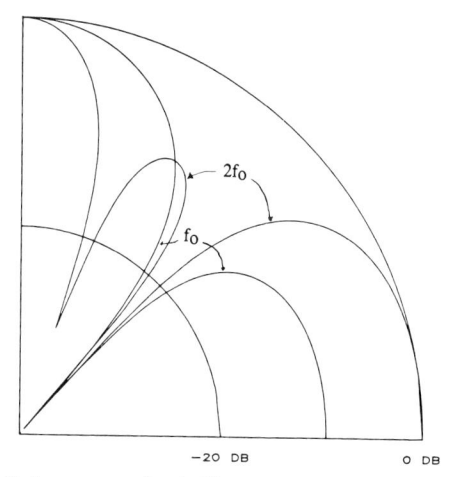

Fig. 2. Polar pattern for half-wavelength, $N = 3$ array; $f = f_0$ and $f = 2f_0$.

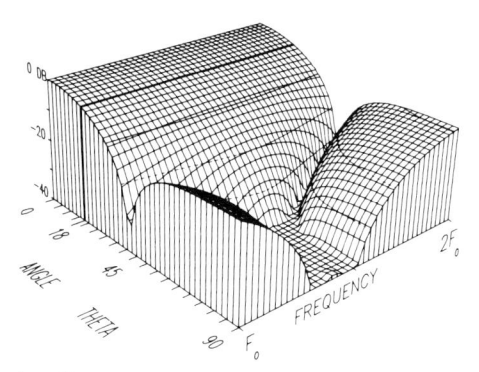

Fig. 3. Amplitude versus angle versus frequency for superposed arrays.

2 IMPLEMENTATION

Our work with three-element basic arrays used elements spaced one-half wavelength for a half power pattern, which is 36° vertically between −3-dB points, and three-eighth wavelength for a half power pattern totaling 50° horizontally between −3-dB points. Our X-shaped array totaled nine elements for the four arrays.

We decided to use an X-shaped array as illustrated in Fig. 4 instead of a square array, since the X shape requires fewer elements to demonstrate the concept. Fig. 4(a) shows a two-dimensional view of the array plane to clarify the driver layout and tilt angle. Fig. 4(b) illustrates the complete geometry with axis orientations and angle definitions. Although the square array would have a lower power density per element, computer modeling showed the square array pattern to have many more side lobes and a narrow main lobe, thus being less controllable and desirable for our application.

The required filter functions were implemented with a combination of analog "twin tee" notch and first-order low-pass filters. The resultant phase difference between the functions was very nearly constant at 173° over one octave. Since the functions must be in phase to match the requirements of their derivation, an inverter was added to the signal path of the function that was effectively a low-pass filter.

However, these analog filters required to do the weighting of the signal can only approximate the function required, thus providing considerably less than ideal coverage for some applications. Fortunately, for our application the predicted response of our filters was within 12% of the ideal. Fig. 5 shows the ideal filter functions along with the predicted response of the analog filters.

The best way to implement the filter functions would be with linear-phase digital filters. This would provide a much closer match to the ideal amplitude response, and it could become a critical difference if derived filter responses presented shapes that were not smooth with frequency. Ultimately, digital implementation is the clear choice for more complicated systems.

3 RESULTS

The loudspeakers used in our array were 3.0-in (76-mm) Bose drivers. Theoretically, in the octave we used for our measurements (844–1688 Hz), these sources should have approximated point sources individually, even at the range of measurement we were confined to in our anechoic chamber. However, the directivity plot in Fig. 6 taken for the center element of the array driven individually shows a considerable deviation from an ideal point source. Over the octave of interest, the amplitude variations within 45° of the on-axis response are as much as 6 dB. After considerable investigation it was determined that the cause was due to diffraction about the baffle, although the baffle extended to almost a full wavelength beyond the furthermost element.

The diffraction was verified with considerable woodwork. Initially the drivers were mounted in the baffle with no backing foam or wood. This results in a baffle with drivers mounted in holes in a sheet of plywood. Our first response to the nonomnidirectional results was to presume that the loudspeaker was acting as a dipole and diffracting around the baffle to interact with its effective backside source. Accordingly, the

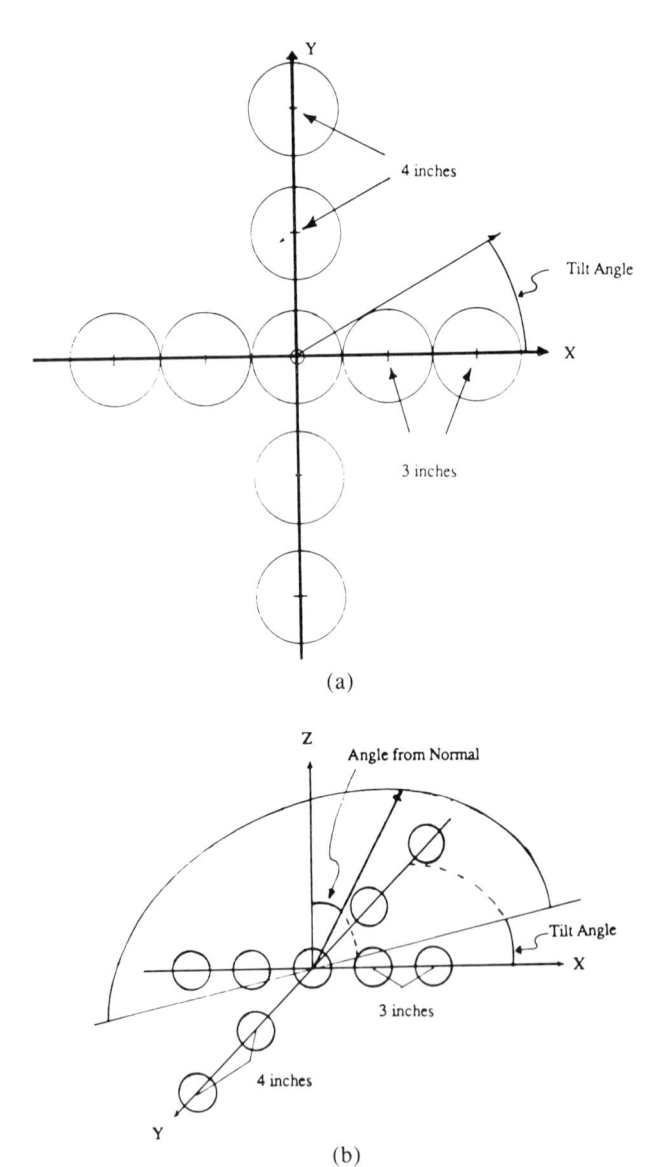

Fig. 4. (a) Frontal view of array plane. (b) Geometry and angle definitions for X array.

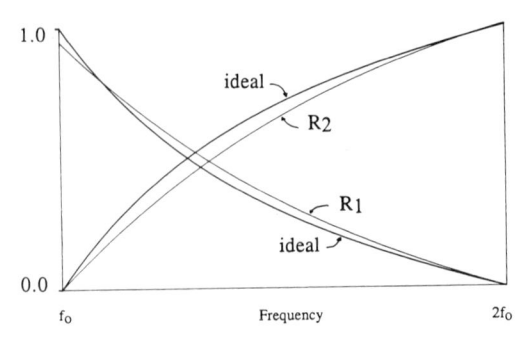

Fig. 5. Ideal and analog amplitude filter functions.

solution seemed to be to build boxes of identical dimensions around the back of every driver to isolate them from themselves and each other. After considerable carpentry, the boxes were mounted and no significant improvement was observed. Polar plots revealed that the power from the back side was reduced dramatically, but the forward pattern was as nonideal as before. The next consideration was that the elements might be interacting even with only one being driven. This was investigated by mounting an additional baffle over the original so as to cover all elements but the center one. The polar response of this arrangement indicated the same lack of omnidirectional coverage we had experienced all along. The only effect left to attribute this result to was diffraction of the forward pattern from the edge of the baffle, with negligible interaction between elements.

Before the array was assembled, the drivers were mounted in a baffle 21 in (533 mm) square, one at a time, to determine their frequency responses on axis. Data were taken from 200 Hz to 15.0 kHz. Since we had more drivers than were necessary, we could select the nine best matched drivers for the array. After examining the responses, the drivers selected were matched within 1.5 dB of each other and no pair of drivers destined for symmetric placement differed by more than 1 dB. In order to achieve our desired spacing of one-half and three-eighth wavelengths, the drivers were ultimately mounted 4.0 and 3.0 in (102 and 76 mm) apart, respectively.

In order to compare measured results with theoretical predictions, a computer program was written to calculate the polar responses of an array of ideal point sources. The program used distances that are referenced to the wavelength of the lowest frequency of the octave for which calculations were made, and the radius of observation was in terms of this wavelength. The program was used to calculate far-field and near-field responses. The latter were necessary for comparison with our data, which were taken at near field due to the limited size of our anechoic chamber.

Our array was verified initially by energizing line array pairs only to confirm expected unfiltered beam patterns. Then the four arrays were connected together with filters and patterns were measured at tilt angle intervals of 15°, with each orientation measured at five frequencies in the octave (that is, every quarter-octave inclusive).

Figs. 7–9 give results for tilt angles of 0, 45, and 90°, respectively. Parts (a) show the predicted far-field responses of the array, parts (b) the predicted near-field responses at a distance corresponding to our measurements, and parts (c) the polar plots made from the data taken in the chamber. In addition, calculations and measurements were taken for 15, 30, 60, and 75°. These plots are omitted since they are similar in shape to the figures presented.

The first observation made on the results is that the shapes of the patterns show a remarkable tolerance for the nonideal coverage of the individual drivers (recall their considerable deviation from omnidirectional). This indicates a significant averaging effect of the array with regard to how the sound fields from each driver are superposed to create the array beam pattern. Apparently 6-dB variations in supposedly omnidirectional coverage from drivers are averaged out to provide an almost ideal main beam shape.

The figures indicate that the main lobes were close in shape to the near-field calculations and the beam widths meet or exceed far-field calculations. Note that the near-field condition seems to show up mostly as a reduction of the depth of the nulls.

Beam patterns for a total of seven tilt angles were calculated. While the 3-dB angles did vary with frequency, they stayed quite close to the values predicted by the computer near-field model. The only notable exception was for the plots taken for 1010 Hz. The average 3-dB angle over frequency is quite close to that predicted, tending to be on the low side. Fig. 10 is a plot of the measured 3-dB angles versus frequency for the seven tilt angles. Deviations are from the computer-calculated values listed on the left.

4 TWO-DIMENSIONAL EXTENSIONS

The array superposition concept shows promise for more than one octave. The computer-modeled results for a pair of five-element arrays (superposed for a total of nine elements in line) are shown in Fig. 11. The individual arrays were spaced at a half-wavelength but apart in base frequencies by three octaves rather than one, as in previous cases. Some deviation between the constraint angles becomes apparent in the third octave, but the result is still within 3 dB of the on-axis response (with one small region just over 3 dB down) and well behaved.

However, for situations where side-lobe suppression is important, the previous results are not adequate. In order to suppress side lobes over an octave, array element tapering must be considered. This is amplitude variation within a base array, in contrast to the constant-

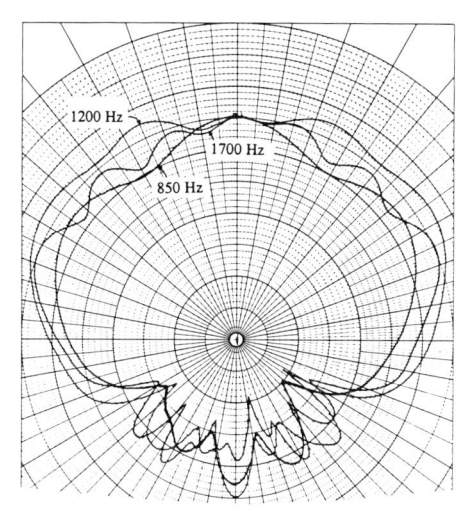

Fig. 6. Polar response of center array element driven individually.

amplitude signal applied to all elements of a base array in previous cases. One example of such a taper are the binomial coefficients. As discussed in [3], the effect of the binomial taper is to take the pattern of a dipole and raise it to a power proportional to the number of elements in the array. This can take the familiar figure-eight pattern and multiply it by itself to create a narrower main lobe with absolutely no side lobes for half-wavelength spacing.

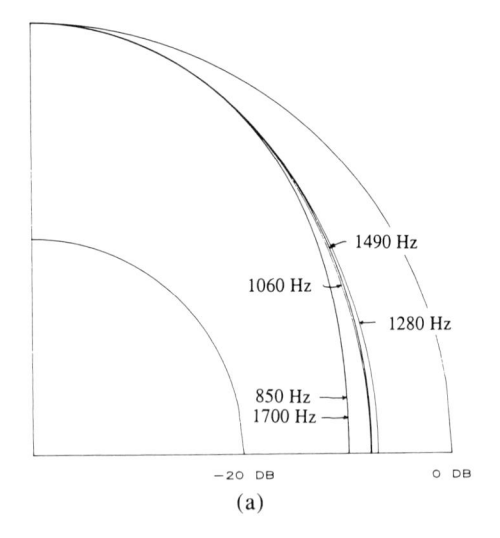

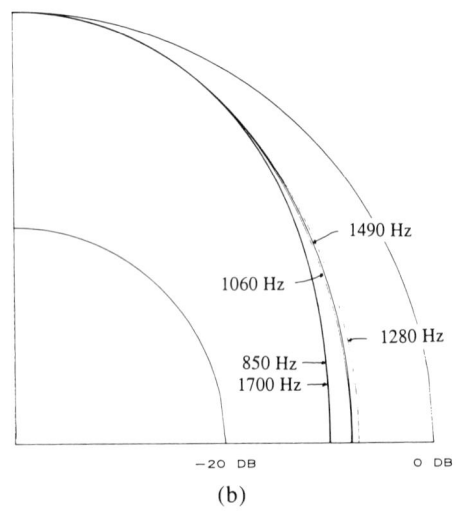

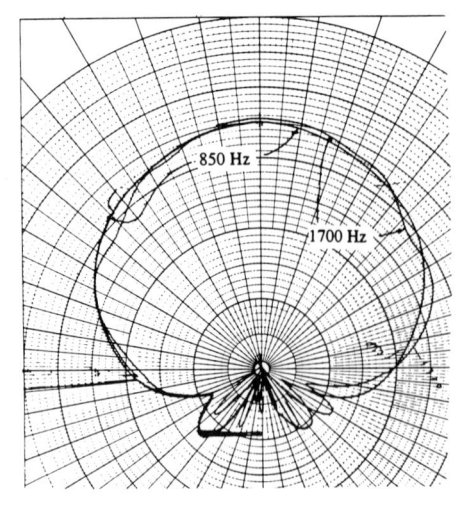

The application of binomial tapering with array superposition has been studied for one-octave array spacing (in frequency). In [4] the base arrays were configured as end fire with binomial tapering. End fire refers to the addition of a linear phase delay between elements so that for the base frequency the acoustic travel time between the elements is matched by a signal delay, creating an in-phase signal traveling along the forward direction of the line array. Now the array is no longer broadside in pattern, but fires from one end and is symmetric about its axis rather than perpendicular, as for the previous broadside array.

The application in [4] has been simulated for a pair of five-element base arrays. For the base arrays with binomial taper and end-fire phasing, the directivity becomes

$$D(\theta, f) = \cos^4 \phi \qquad (5)$$

where

$$\phi = \frac{kd(\cos \theta - 1.0)}{2} \qquad (6)$$

with k being the propagation constant and d the spacing between elements.

With two such arrays superposed and with array weighting functions similar to the previously discussed R_1 and R_2 fading array 1 out and array 2 in, respectively, with increasing frequency, a relatively constant beam pattern over the octave can be generated. Computer-simulated results are shown in Fig. 12, where Fig. 12(a) shows the amplitude in decibels and Fig. 12(b) illustrates the same with a linear amplitude scale. Note that the largest side-lobe deviation along the right side of the pattern is down approximately 26 dB from the on-axis amplitude.

If one wants to distribute the arrays more sparsely in frequency, that is, with the base arrays more than an octave apart, side-lobe problems arise. Simply trying

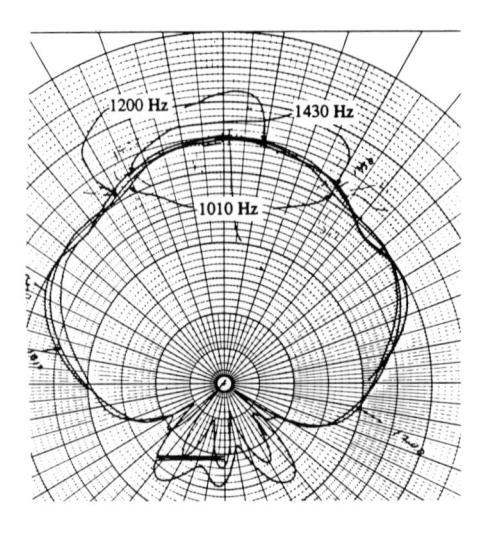

Fig. 7. (a) Far-field calculation. (b) Near-field calculation. (c) Chamber measurement for 0° tilt.

to stretch the binomial arrays with rederived R_1 and R_2 proves a poor solution, as illustrated in the linear plots of Fig. 13. For a two-octave stretch, as shown in Fig. 13(a), a significant lobe is clear at the front of the right side, hardly 6 dB down from the on-axis response. Even for the modest 1.5-octave stretch shown in Fig. 13(b), the same lobe is only supressed by approximately 11 dB.

Trying some other classical tapers on the five-element base arrays indicated some promise, however. Using our computer model, we experimented with side-lobe suppression for the following tapers: triangle and Dolph–Chebyshev for 20- and 30-dB suppression. The optimum solution of the three turns out to be the Dolph–Chebyshev 20-dB suppression taper for a two-octave stretch, as illustrated in the linear plot of Fig. 14. While the lobe suppression is at worst only 12 dB, this was the best result achieved for a two-octave stretch.

The preceding seems to indicate that in order to "stretch" the base arrays for superposition and suppress side lobes, a mathematical compromise must be reached. If one insists on infinite side-lobe suppression at the base array element spacing frequencies (such as the binomial taper), then one generates significant side lobes at frequencies in between. However, if one relaxes the side-lobe constraint (to a degree) at the base array element spacing frequencies (such as the Dolph–Chebyshev taper permitting side lobes 20 dB down), then side lobes at frequencies in between can be reduced significantly. The optimization of this compromise is an area that we intend to pursue.

5 THREE-DIMENSIONAL EXTENSIONS

Another promising area is that of three-dimensional beam control. Our measurements have indicated the feasibility of an X-shaped array with differing horizontal and vertical beam-width control. Now we would like to explore the utilization of end-fire line arrays to create the same effect.

Consider the superposed binomial tapered end-fire line array pair of the previous section. If one assembled an array of elements where each element was such an array pair, and arranged such that the array pairs were parallel to each other with a uniform spacing between them, then a rectangular array of drivers would be the geometric result. Now if another rectangular array is established with the spacing between parallel array pairs half that of the first rectangle, then this rectangle would have half the width of the first but the same length. Lastly, geometrically superpose the two rectangles along the axis parallel to the array pairs. Fig. 15 illustrates the result, with the base array elements themselves a binomial end-fire line array pair.

This is the three-dimensional equivalent of the element superposition of two line arrays discussed in Sec. 1, but with each element now an entire superposed binomial end-fire array pair. The same array filtering functions R_1 and R_2 discussed in Sec. 1 can also be applied to the superposed rectangles, permitting a similar degree of control over one octave. Over the octave, the vertical beam width (perpendicular to the plane of the rectangles) would be controlled by the characteristics of the end-fire elements, while the horizontal beam width (in the plane of the rectangles) would be controlled primarily by the broadside directivity created by the spacing between the end-fire array axes.

The resulting beam pattern of such a rectangle at the bottom or top of the octave of superposition is illustrated in Fig. 16 for half of the forward half-space. Fig. 16 is a three-dimensional polar plot, sometimes referred

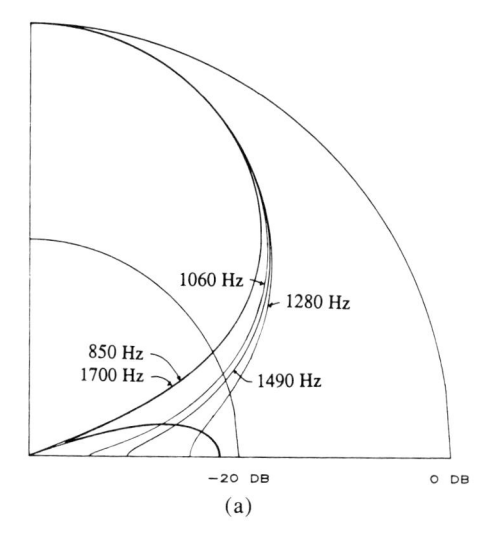

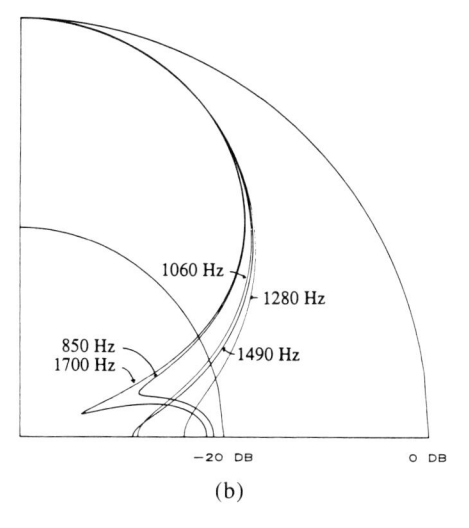

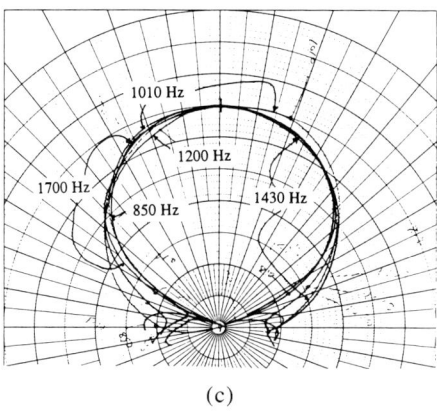

Fig. 8. (a) Far-field calculation. (b) Near-field calculation. (c) Chamber measurement for 45° tilt.

to as a balloon plot. The X, Y, and Z axes are shown for geometric orientation and do not represent independent or dependent variables. The maximum radius, along the Z axis, is 40 dB. The pattern is symmetric about the Z–Y plane, and thus the far half is omitted for clarity. Our computer model demonstrates that except for minor side-lobe variations over the octave, the main beam shape remains remarkably constant. The array pair lines are pointing up the Z axis and the rectangle is in the Z–Y plane. Fig. 17 shows two polar slice plots of the three-dimensional polar balloon previously shown, with Fig. 17(a) showing the Z–Y plane and Fig. 17(b) the Z–X plane.

These results are similar to those obtained by Meyer [5], but are realizable with less hardware since only two digital filters are required. (The end-fire arrays and

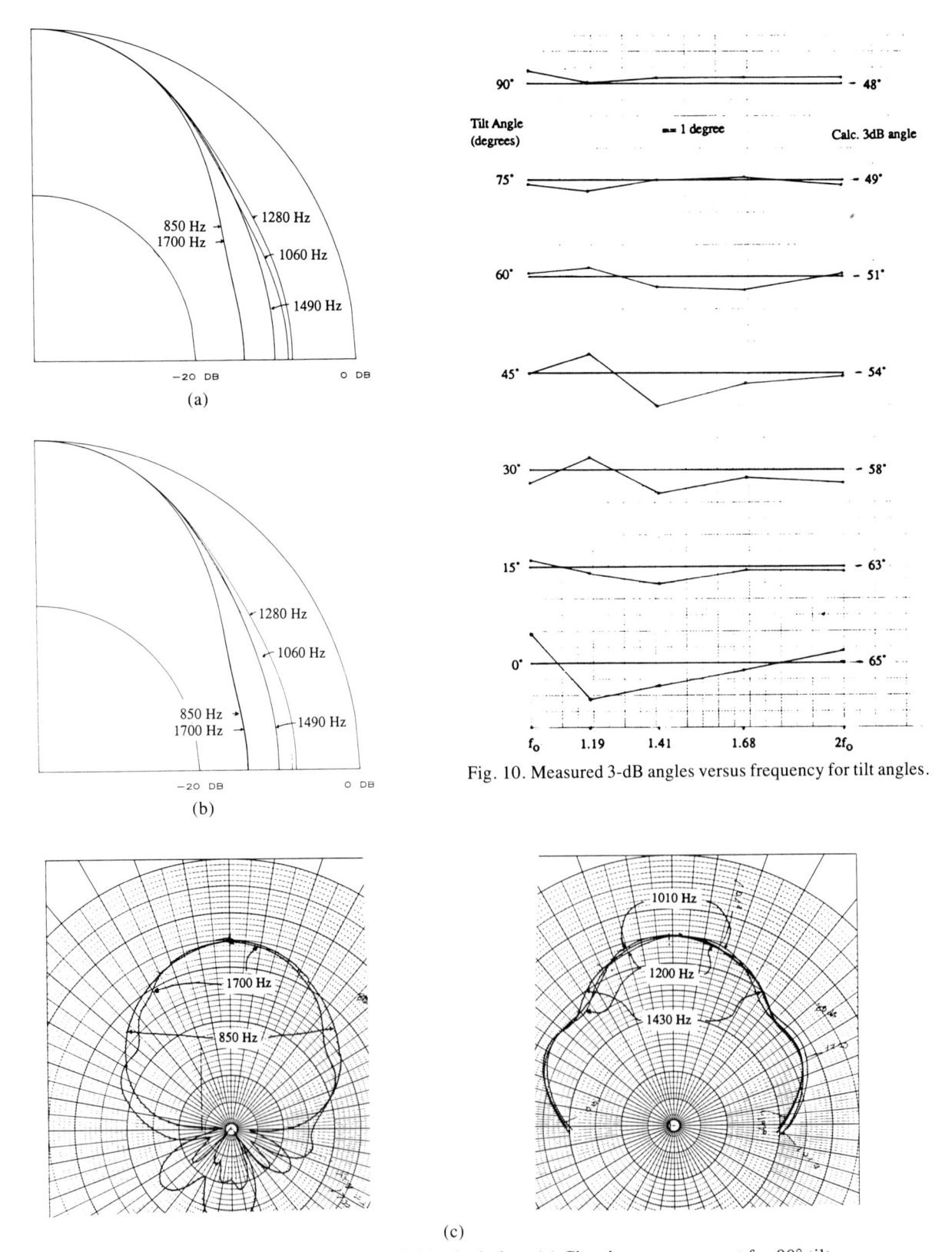

Fig. 10. Measured 3-dB angles versus frequency for tilt angles.

Fig. 9. (a) Far-field calculation. (b) Near-field calculation. (c) Chamber measurement for 90° tilt.

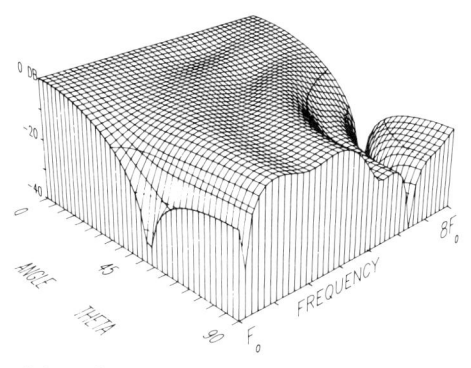

Fig. 11. Response of five-element superposed line arrays spanning three octaves.

(a)

(b)

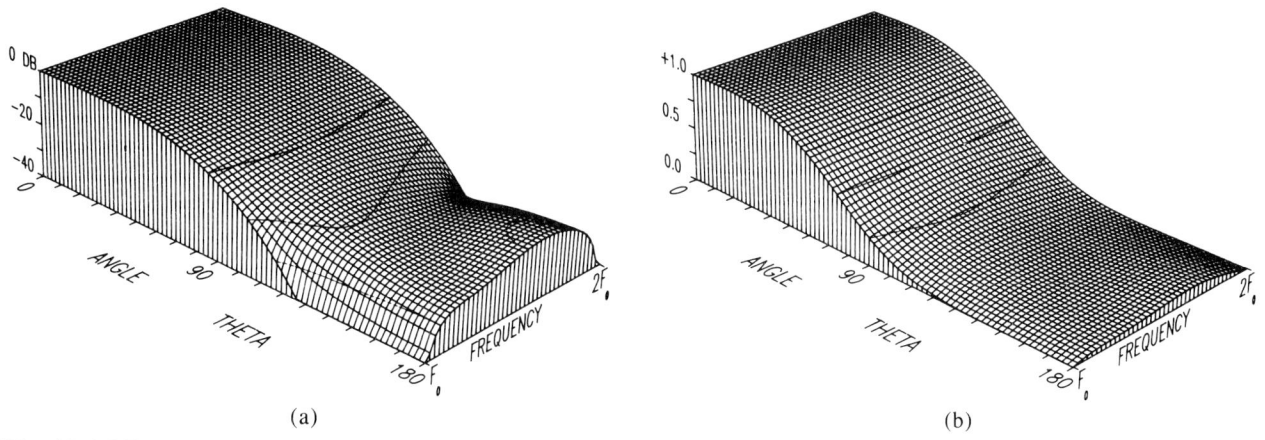

Fig. 12. (a) Beam pattern for binomial tapered array over an octave, plotted in decibels from 0 to 180°. (b) Linear amplitude plot.

(a)

(b)

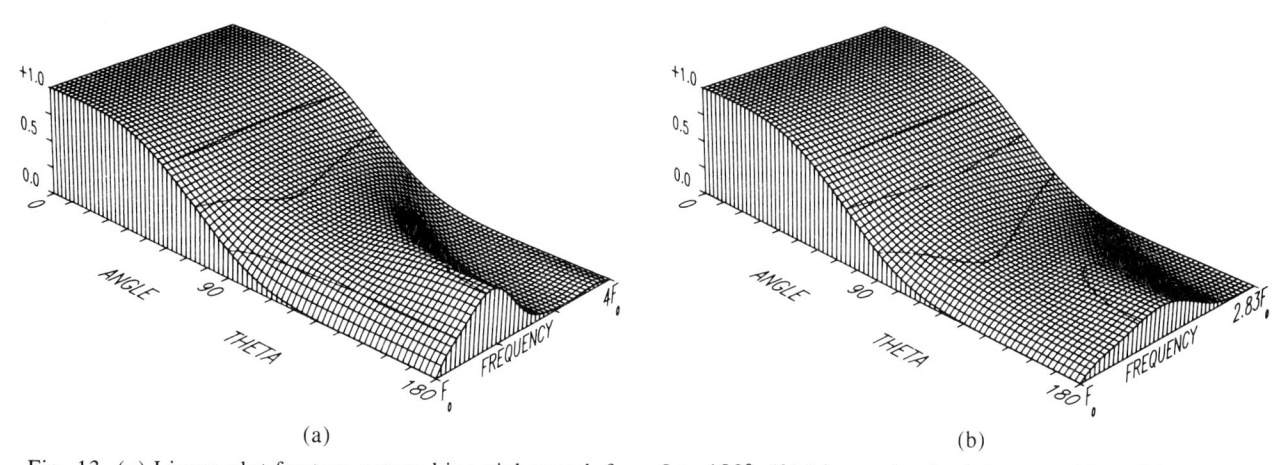

Fig. 13. (a) Linear plot for two-octave binomial stretch from 0 to 180°. (b) Linear plot for 1.5-octave binomial stretch.

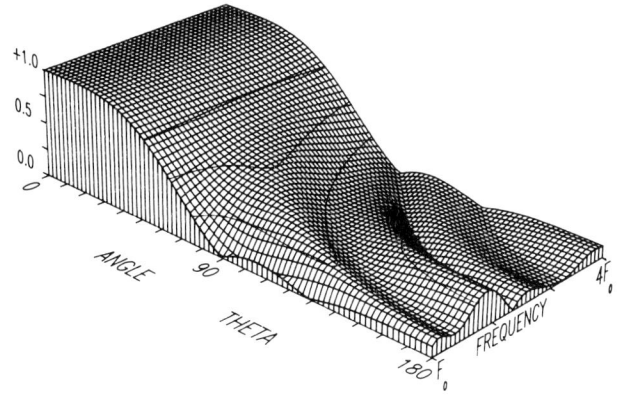

Fig. 14. Linear plot for Dolph–Chebyshev 20-dB suppression taper for 0–180°.

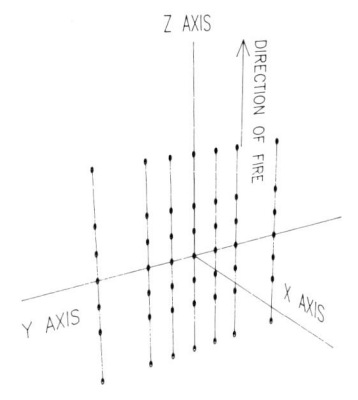

Fig. 15. Geometry of the array of arrays.

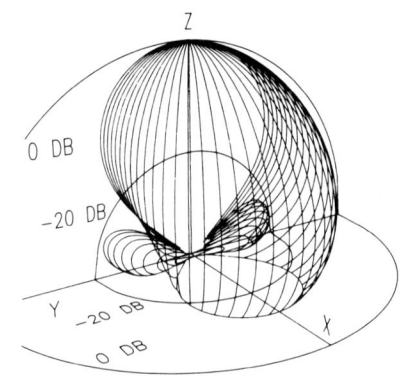

Fig. 16. Forward three-dimensional polar beam pattern for superposed rectangles.

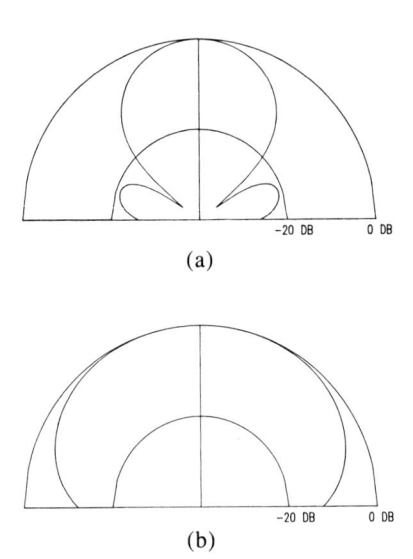

Fig. 17. (a) Polar slice of Z–Y plane. (b) Polar slice of Z–X plane.

the rectangles are superposed with the same filter weighting functions.) Signals derived for each individual driver require fewer individual manipulations (many drivers receive identical signals), but beam steering is fixed by the positioning of the drivers.

While the preceding is to be pursued, another application of the binomial array pairs would be in a Bessel array [6]. This Bessel weighting would fix the beam pattern such that the horizontal and vertical coverage would not be independent, but the beam pattern in the plane of the rectangle would be inherently fixed by the spacing between end-fire line arrays, and remain constant with frequency as long as the line arrays maintained their individual patterns. With the number of lines to be used known, one could determine the pattern ripple versus frequency, along with other relevant parameters, from evaluation methods [6], or start with a requirement and synthesize an array to meet it.

6 DISCUSSION AND CONCLUSIONS

We have implemented a filtering technique that has achieved a nearly constant main beam width of four superposed constant-spaced arrays over one octave. This method yields 3-dB angles very close to those predicted by our computer model. The technique seems to be forgiving in that it tolerates significant deviations from ideal coverage of the individual elements of the array.

The coverage pattern produced allows for a different beam width in the horizontal and vertical directions, controlled by the spacing of the arrays and their associated filter functions. The drivers used were small broad-band loudspeakers requiring a total of three low-power amplifiers.

The highest frequency that this technique can produce is limited by the spacing constraint of the size of the driver itself. However, the lower frequencies provide no constraint in a theater environment since the array size required for 100 Hz is approximately 12 ft (3.6 m).

The size of the main beam can be varied by using more elements and different filtering functions for the array. If the number of elements in the horizontal and vertical arrays is the same, then the filter functions are the same for each array. The most flexible implementation of the filter functions would be in the digital domain, since the functions could be derived with loudspeaker response corrections as part of the synthesis task, and phase linearity could be guaranteed.

From these results for a line array, this technique seems promising for X arrays and parallel end-fire line arrays spanning several octaves. We intend to investigate the limits of this technique and perhaps discover a more general approach of which ours is a subset.

7 ACKNOWLEDGMENT

The authors would like to thank Bose for the donation of 25 3-in drivers without which this experiment could not have been performed.

8 REFERENCES

[1] D. B. Keele, "Loudspeaker Horn," US Patent 4308932, 1982.

[2] E. L. Hixson and K. T. Au, "Wide-Bandwidth

Constant Beamwidth Acoustic Array," *J. Acoust. Soc. Am.*, vol. 48, p. 117 (1970 July).

[3] Stutzman and Thiele, *Antenna Theory and Design* (Wiley, New York, 1981).

[4] J. Lardies and J. P. Guilhot, "A Very Wide Bandwidth Constant Beamwidth End-Fire Line Array without Sidelobes," *J. Sound Vibration*, vol. 120, no. 3 (1988).

[5] D. G. Meyer, "Digital Control of Loudspeaker Array Directivity," *J. Audio Eng. Soc. (Engineering Reports)*, vol. 32, pp. 747–754 (1984 Oct.).

[6] D. B. Keele, Jr., "Effective Performance of Bessel Arrays," presented at the 87th Convention of the Audio Engineering Society, *J. Audio Eng. Soc. (Abstracts)*, vol. 37, p. 1068 (1989 Dec.), preprint 2846.

THE AUTHORS

J. Harrell

E. Hixson

Jeff Harrell was born in Memphis, Tennessee, in 1956. He received the B. S. degree from the University of Missouri at Rolla in 1977 and the M. S. degree from the University of Illinois at Urbana in 1978, both in nuclear engineering.

After working for Combustion Engineering as a reactor physicist, he returned to school and received the M. S. degree in electrical engineering from the University of Arkansas at Fayetteville in 1981.

From 1981 to 1985 he worked at Sandia National Laboratories in Albuquerque, New Mexico, dealing with the effects of radiation in various types of circuits. He is presently a Ph.D. candidate in the Electrical Engineering Department at the University of Texas, Austin. His research interests include acoustic systems, wave phenomenon, signal processing, and analog circuits.

Elmer Hixson received the B.S., M.S., and Ph.D. degrees in electrical engineering from the University of Texas, Austin, the latter in 1960.

From 1948 to 1954 he worked in atmospheric acoustics and passive submarine sonar at the U.S. Navy Electronics Laboratory in San Diego, CA. He joined the faculty of the Electrical Engineering Department of the University of Texas in 1954 and is at present a Professor of Electrical Engineering. In 1962 he started the Engineering Acoustics Program, which now has four full-time faculty members. His major research interest is in electroacoustics and he consults in acoustics and noise control.

Dr. Hixson is a Fellow of the ASA, a senior member of the IEEE, a member of the American Society of Engineering Education, and a founding member of the Institute of Noise Control Engineering.

Low-Frequency Sound Reproduction*

MARK E. ENGEBRETSON

AB Systems Design, Inc., Folsom, CA 95630, USA

The usual procedure for selecting loudspeaker systems in low-frequency commercial applications is to consult a manufacturer's catalog sheets and select a device, or a group of devices, which appear to satisfy the criteria for the desired sound-pressure levels in the installation environment. The criteria normally used by sound consultants and contractors are pressure sensitivity, frequency response, power capacity, and attenuation-with-distance formulas. As a consequence, the design of low-frequency loudspeaker systems is largely a trial-and-error process based upon limited available data, and the results are often disappointing. Recently, however, cognizant manufacturers have made Thiele–Small parameters available for commercial drivers, greatly simplifying the system design process and enabling predictable results to be obtained.

0 INTRODUCTION

The proper design of systems for low-frequency sound reproduction seeks answers to the following questions:
1) Will the low-frequency system deliver the desired acoustic levels?
2) Will the system meet the lower f_3 requirements?
3) Will the system meet the upper f_3 requirements?
4) How much electric power will be required?

It will be shown that the answers to these four propositions take into account architectural acoustics, loudspeaker directivity, loudspeaker mechanical displacement limits, loudspeaker thermal power limits, and the effects of mutual radiation for multiple-loudspeaker systems.

Virtually all commercial sound systems of any consequence are equipped with detailed amplitude response equalizers. Standard practice is to use a real-time audio spectrum analyzer and a calibrated microphone to smooth the system's reverberant-field response to some predetermined characteristic by adjustment of the equalizer.

It is worth noting that one-third-octave equalization devices are very powerful tools in competent hands, and may be employed to correct rather severe system response defects. However powerful, the use of detailed equalization also carries certain liabilities in application. Because a real-time analyzer cannot differentiate between loudspeaker response characteristics and so-called room-gain effects, many sound systems employ loudspeaker components that demonstrate gross response deficiencies. More often than not these response deficiencies are mistakenly identified as acoustical anomalies, which are then "corrected" through equalization. By failing to identify the *real* source of the response deficiencies, expensive and ineffective loudspeaker systems continue to be specified, requiring costly equalization devices and practices to rectify.

Of the three system types (closed box, horn, and vented box), the vented-box system emerges as clearly superior to the others in terms of physical size, cost, and usable power bandwidth, given the commercially available drivers around which systems are designed.

1 GLOSSARY OF SYMBOLS

c	=	velocity of sound in air, = 345 m/s
f	=	frequency, hertz
f_B	=	resonance frequency of vented enclosure
f_S	=	resonance frequency of unenclosed driver
f_3	=	frequency at which the system response is 3 dB below passband level
P_A	=	acoustic power, watts
P_{AR}	=	displacement-limited acoustic power output rating
P_{ER}	=	displacement-limited electric power rating
$P_{E(max)}$	=	thermally limited maximum input power
Q_{ES}	=	Q of driver at f_S, considering electromagnetic damping only
Q_{TS}	=	total Q of driver at f_S, considering all driver resistances
S_D	=	effective surface area of driver diaphragm
S_V	=	effective surface area of vent opening
V_B	=	net internal volume of enclosure

---

* Presented at the 72nd Convention of the Audio Engineering Society, Anaheim, CA, 1982 October 23–27.

V_D = peak displacement volume of driver diaphragm, = $S_D x_{max}$
x_{max} = peak linear displacement of driver diaphragm
x_{peak} = displacement of driver diaphragm
η = reference efficiency of system, percent (half-space)
ρ_0 = density of air, = 1.21 kg/m^3

2 CLOSED-BOX SYSTEMS

Closed-box systems offer an advantage in that they are the simplest of all types to design and construct. The minimum low-frequency cutoff is easily predicted from published driver parameters [1] and may be expressed as

$$f_3(\text{min}) = 0.8 \frac{f_S}{Q_{ES}}.$$

For a low f_3 frequency this directly implies some combination of large mass and low motor strength, neither characteristic being conducive to the realization of high midband efficiency.

Standard practice for sealed-box systems has been to employ drivers that are more suited for vented-system applications. If midband efficiency characteristics are to equal those of a vented system, the trade-off is an increase in the lower f_3 frequency. Conversely, midband efficiency may be traded for a decrease in lower f_3 frequency. Considering the almost universal requirement for high efficiency in commercial applications, this explains why closed-box systems have not been widely accepted and specified.

3 HORN SYSTEMS

Horn systems have enjoyed commercial popularity due to their inherent potential for high efficiency. The lower cutoff frequency of a horn system is a function of the flare rate and the horn-mouth perimeter. The upper frequency limits are totally dependent upon driver selection [1]–[3], and, as we shall see, this presents a redoubtable case for the selection of vented systems in a wide range of commercial applications.

Horn systems must be large for low-frequency reproduction. The lower the cutoff frequency desired, the longer the horn must be, with an attendant increase in mouth perimeter. Horns have been folded in an effort to contain one of the three dimensions. However, the mouth perimeter must remain large for proper loading at low frequencies.

A compromise was reached in the 1940s through the development of the combination-type enclosure, which married a short front-loading horn to the bass reflex cabinet. The horn provided a needed efficiency advantage in the lower speech frequency region (powerful amplifiers were very expensive in the forties), and the bass reflex gave useful, albeit diminished, output to the lowest frequencies that could then be printed on a photographic-film sound track. The combination enclosure afforded the added advantage of allowing a coplanar mounting arrangement of the low- and high-frequency drivers in two-way systems. Precise time alignment of low- and high-frequency driver units at the crossover frequency could be obtained, a significant improvement over folded horn systems. The combination-type low-frequency system remains a frequently specified product in commercial projects.

Keele [3] has shown that the moving-mass corner, which is controlled by motor strength and mechanical mass, can be expressed as (Fig. 1)

$$f_{HM} = \frac{2 f_S}{Q_{TS}}.$$

This permits simple calculation of power response break frequencies for piston-band performance. Similarly, if a horn driver behaves in a pistonlike manner, the same relationship can be used to approximate the frequency at which $ka = 1$. Above this frequency the directivity index will increase at approximately 6 dB per octave, while the power response will fall off at the same rate. (The difference between axial frequency response and relative power response yields the directivity index in decibels versus frequency.) These relationships are shown in Figs. 2 and 3. Fig. 4 illustrates on- and off-axis responses for 0, 15, 30, and 45° for a single-driver combination-type system. The frequency at which directional effects begin to appear corresponds remarkably to the calculated f_{HM}.

It is interesting to observe that the drivers originally designed for theater horn use incorporated magnetic fields of high flux (due to focused air gaps), extremely low unmounted moving mass, and an attendant high resonant frequency f_S. Calculating f_{HM} for one such device, we have

$$f_{HM} = 2\left(\frac{45 \text{ Hz}}{0.17}\right) = 530 \text{ Hz}.$$

This driver is well suited for horn use up to a crossover frequency of 500 Hz.

Turning our attention to today's products, few (if any) similar items exist in the inventory of driver loudspeakers specified for horn applications. One product often selected for horn cabinet use yields

$$f_{HM} = 2\left(\frac{24 \text{ Hz}}{0.254}\right) = 189 \text{ Hz}$$

where f_S and Q_{TS} are supplied by the manufacturer [4].

This is consistent with the reverberant-field response measurement results depicted in Fig. 5. In commercial practice most combination-type systems afford fourth-order f_3 frequencies between 40 and 70 Hz, with horn loading beginning to take effect approximately one octave above the f_3 frequency. In the example selected for Fig. 4 the horn loading is fully developed at 200 Hz, followed almost immediately by the moving-mass corner, which results in a 6-dB-per-octave rolloff in power response.

These factors indicate that combination-type and horn systems, when fitted with contemporary driver units, afford 25–40% efficiency, but this high efficiency is concentrated over a very narrow bandwidth. Above the moving-mass corner frequency, horn cabinets roll off in power response at 6 dB per octave until additional losses due to voice-coil inductance and front-cavity compliance are encountered (Fig. 1). This effectively negates any efficiency advantages that are afforded by horn loading due to the need for electrical equalization, the inherent complexity of horn cabinets, and their physical size requirements.

4 VENTED SYSTEMS

It has been shown from the preceding and [4] that vented systems afford the most satisfactory trade-off between bandwidth and efficiency, given the available drivers for commercial applications. A further advantage is that of physical size. For a given bandwidth and efficiency, vented systems can be designed that are considerably smaller than other system types.

5 INFLUENCE OF ARCHITECTURAL ACOUSTICS

It is assumed that readers are familiar with system design criteria for free-field (outdoor) applications. (It should be noted, however, that for long wavelengths out of doors, half-space conditions prevail.)

In enclosed spaces two factors influence the reverberant-field sound-pressure level: the power level introduced into the space and the power that is absorbed by the space. The average absorption coefficient may be calculated if volume, surface area, and reverberation time are known:

$$\bar{a} = 1 - e\exp\left[-\frac{0.161V}{ST_{60}}\right] \quad (1)$$

where $\bar{a}$ and S are in square meters, V is in cubic meters, and the room constant R may be expressed in square meters and calculated by

$$R = \frac{S\bar{a}}{1 - \bar{a}} . \quad (2)$$

The reverberant-field pressure in an enclosure may be calculated from a known acoustic power by

$$P = \left(\frac{4\rho_0 c P_A}{R}\right)^{1/2} \quad (3)$$

where
P_A = acoustic power, watts
P = rms pressure, newtons per square meter
R = room constant, square meters
$\rho_0 c$ = characteristic impedance of air, 415 N · s/m^3, at 22°C and 760 mm Hg.

Since pressure is in newtons per square meter, we can solve for dB sound-pressure level dB-SPL by

$$\text{dB-SPL} = 20 \log p + 94 \text{ dB} . \quad (4)$$

If we let dB-SPL equal the desired sound-pressure level, then by substitution

$$P_A = \frac{R \times 10 \exp[(\text{desired SPL} - 94 \text{ dB})/10]}{1660} . \quad (5)$$

Knowing the acoustic power required to achieve a desired sound-pressure level in the reverberant field, we can design the loudspeaker system to meet the acoustic power objective.

6 DISPLACEMENT AND POWER

All sound systems should be sufficiently rugged to preclude failures. Loudspeaker failures take two forms, failures resulting from exceeding displacement limits and failures resulting from exceeding the thermal ca-

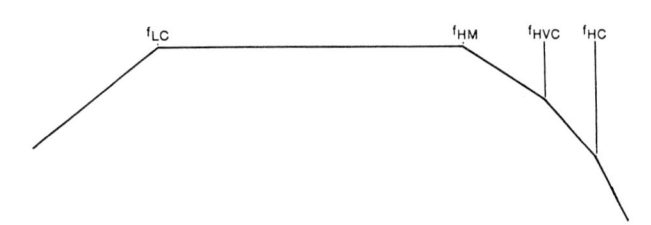

Fig. 1. Driver parameters and horn loading. $f_{HS} = f_S/Q_{TS}$ and $f_{HM} = 2 f_{HS}$ where f_{HS} is the upper bound of resistance-controlled operation, f_{HM} the driver moving-mass corner, f_{LC} the driver suspension compliance, f_{HVC} the driver voice-coil inductance, and f_{HC} the front cavity compliance.

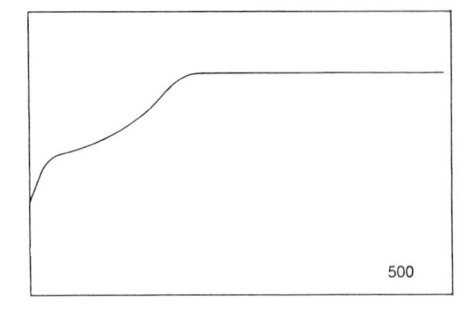

Fig. 2. Axial frequency response.

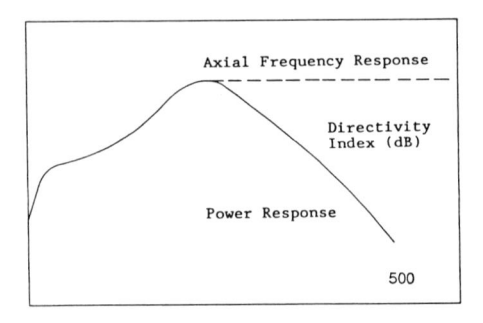

Fig. 3. Axial, directivity and power response.

pacity of the driving voice coil. Good engineering practice suggests that we design systems to eliminate the possibility of mechanical (displacement) failures by limiting the band-pass to low-frequency devices and by restricting the available electric power to values below the thermal limits of the voice coils.

We therefore let $P_{ER} = P_{E(max)}$ and design the total system to operate within the thermal power capacity of the voice coil or coils.

The quantity x_{peak} may be solved for dB-SPL desired at 1 m [5] by

$$x_{peak} = \frac{(1.18 \times 10^3)10^{SPL/20}}{f^2 a^2} \quad (6)$$

where x_{peak} is the displacement and a the piston radius, both in millimeters.

If the quantity x_{max} is given, we can transpose Eq. (6) and solve for dB-SPL at 1 m:

$$dB\text{-}SPL = 20 \log\left(\frac{x_{max} f^2 a^2}{1180}\right) . \quad (7)$$

For values of $f << ka = 1$ in half-space conditions at 1 m,

$$db\text{-}SPL = 10 \log P_A + 112.2 \text{ dB} \quad (8)$$

and, by substitution,

$$P_A = 10 \exp\left[\frac{20 \log(x_{max} f^2 a^2/1180) - 112.2}{10}\right] . \quad (9)$$

Similarly we may solve for f if x_{max}, P_A, and a are known,

$$f = \left\{\frac{(1.18 \times 10^3) 10 \exp[(10 \log P_A + 112.2 \text{ dB})/20]}{x_{max} a^2}\right\}^{1/2} \quad (10)$$

and transposing for piston radius, we have

$$a = \left\{\frac{(1.18 \times 10^3) 10 \exp[(10 \log P_A + 112.2 \text{ dB})/20]}{f^2 x_{max}}\right\}^{1/2} . \quad (11)$$

The preceding displacement equations hold for closed-box design. Fig. 6 illustrates displacement versus frequency for constant power output from a typical 12-inch vented system.

A child's rubber ball connected to an elastic string provides a suitable analogy to understand how a vented box works. The ball can be set into motion suspended from the elastic such that the hand motion required to prolong large displacements of the ball is minimal. The motion of the ball is analogous to the power radiated by the vent, while the driving force, the hand, is analogous to the loudspeaker. The elastic string corresponds to the air volume of the cabinet. At the box alignment frequency, cone displacement is minimal, whereas for all other frequencies the cone peak displacement will increase by a factor of four for each octave *decrease* in frequency for constant power output. Below the box tuning frequency the displacement will increase dramatically, but no useful output will result, leading to the conclusion that *all vented systems require high-pass filters at, or above, the box tuning frequency.*

The *only* exception to this rule is when a system is operated well below power levels that might present a displacement hazard. Such conditions indicate that perhaps a smaller system and less amplifier power is warranted. (Although it is negligence to underdesign, it is also wasteful to overdesign systems.)

In vented systems the minimum frequency for displacement-limited operation will be lower than the predicted x_{max} frequency and may be approximated by

$$f_{min} = \frac{f x_{max}}{2^{1/2}} . \quad (12)$$

7 MINIMUM VENT SIZE

At the box tuning frequency f_B virtually all of the acoustic power is radiated by the vent. Many manufacturers apply filter theory techniques successfully to the design of vented systems for small signals, but few products marketed today perform satisfactorily for large signals due to power compression resulting from inadequate vent areas.

Minimum vent size is a function of the peak volume displacement of the driver V_D and the box tuning frequency. The minimum effective vent surface area S_V is determined as

$$S_V = \frac{3.23 V_D}{f_B^{1/2}} \quad (13)$$

where S_V and V_D are in centimeters. An 18-in (460-mm) loudspeaker with $x_{max} = 9.5$ mm requires a duct area of 706 cm^2 (109 in^2) for a 32-Hz alignment if maximum volume displacement is to be utilized without power compression.

8 MUTUAL RADIATION

When sound is radiated from one diaphragm that is in close proximity to another diaphragm being similarly driven, the diaphragms are said to be coupled [6].

The radiation impedance into which a diaphragm looks is the sum of the self-radiated and the mutual-radiation impedance from the neighboring diaphragms.

The acoustic power radiated by the coupled array is

$$P_A = (r_s + r_m)v^2 \times 10^{-7} \text{W}$$

where

- r_s = real part of self-radiated impedance
- r_m = real part of mutual-radiation impedance
- v = rms diaphragm velocity, cm/s.

The radiated power will be proportional to the square of the number of diaphragms for constant velocity. Half of the gain in acoustic power is due to the increase in applied electric power, the remaining half is an increase in efficiency. The increase in efficiency will hold to a frequency where the array becomes large with respect to the wavelength above which the diaphragms no longer "couple." At intermediate frequencies and above, the diaphragms will begin to interfere with one another, resulting in an actual reduction in radiation resistance.

It is assumed that the array is configured to realize the tightest possible density of drivers. The larger the array, the lower the frequency at which interference effects take place. The upper frequency at which the array will be usable may be approximated by

$$f_{max} = k \frac{c}{\pi D(N^{1/2})} \qquad (14)$$

where

- f_{max} = upper bounds of mutual (nondirectional) radiation
- c = velocity of sound
- D = nominal piston diameter
- k = approximate ratio of piston diameter to driver spacing
- N = number of coupled drivers.

Table 1 lists for different numbers of mutually coupled drivers the system reference efficiencies, the system thermal power capacities, the system power output capabilities, and the upper passband limits for mutual radiation. [Individual units are 18-in (460-mm) devices.]

9 DESIGN EXAMPLE

Design a vented subwoofer system to deliver 115 dB-SPL to 32 Hz for an auditorium with a volume of 14 000 m³, a surface area of 3700 m², and a measured reverberation time of 2.0 s.

From Eq. (1),

$$\bar{a} = 1 - \exp\left[-\frac{0.161 \times 14\,000}{3700 \times 2.0}\right] = 0.26$$

and from Eq. (2),

$$R = \frac{3700 \times 0.26}{1 - 0.26} = 1300 \text{ m}^2 .$$

Knowing the room constant R, we can calculate the required acoustic power for 115 dB-SPL from Eq. (5),

$$P_A = \frac{1300 \times 10 \exp[(115 \text{ dB} - 94 \text{ dB})/10]}{1660}$$

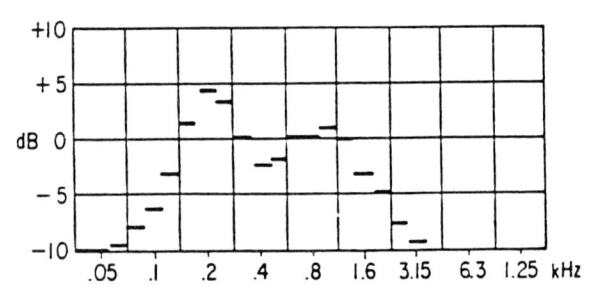

$$= 99 \text{ W} .$$

Fig. 5. Reverberant-field response of vented horn.

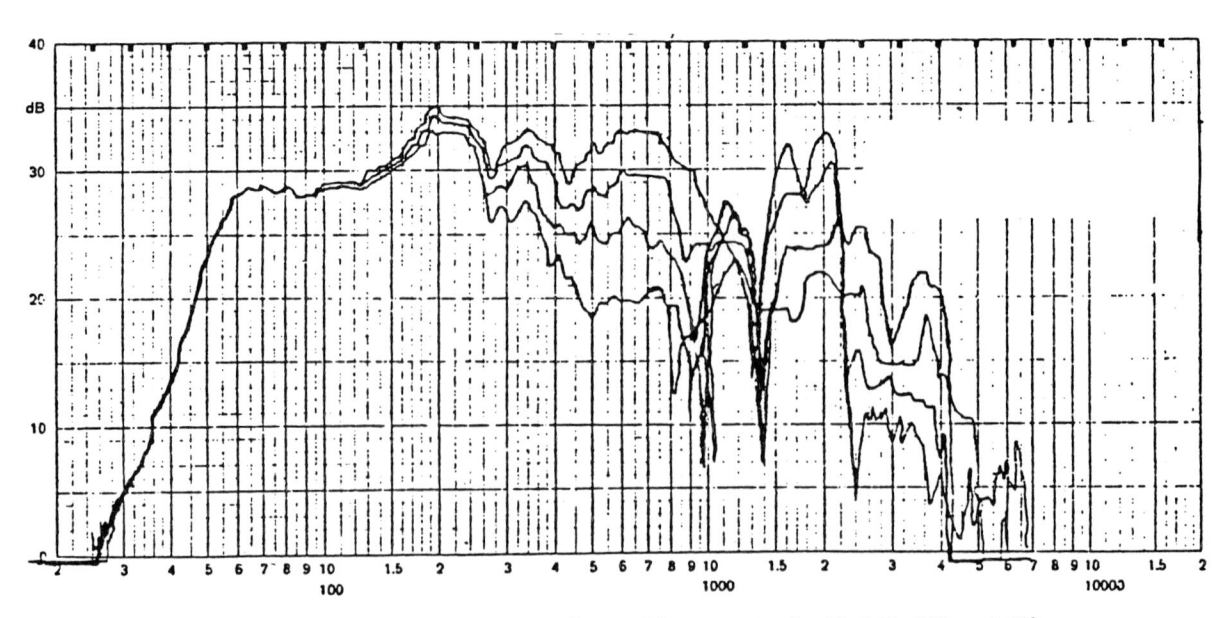

Fig. 4. On- and off-axis responses of vented horn system for 0°, 15°, 30°, and 45°.

We have elected to use a commercial 18-in (460-mm) driver having the following characteristics:

$V_D = 1230$ mm^3

$x_{max} = 9.5$ mm

$\eta = 2.1\%$

$P_{E(max)} = 200$ W .

Allowing f_{min} to be 32 Hz, we have from Eq. (12)

$fx_{max} = 32 \times 2^{1/2} = 45$ Hz .

Using 45 Hz for f, from Eq. (9) a vented system using a single driver will be capable of

$$P_A = 10 \exp\left[\frac{20 \log(9.5 \times 45^2 \times 203^2/1180) - 112.2}{10}\right] = 2.7 \text{ W} .$$

Because radiated power in a mutually coupled array will be proportional to the square of the number of diaphragms for constant velocity, we have

$$N = \left(\frac{P_{An}}{P_A}\right)^{1/2} = \left(\frac{99}{2.7}\right)^{1/2} = 6 \text{ drivers}$$

where P_{An} = required acoustic power.

Practical combinations of impedance and power indicate that an array of eight units would be required to meet the prescribed criteria. Given $P_E = 200$ W per driver (1600 W total), the acoustic power usable from the array would be 269 W, and the resulting sound-pressure level in the reverberant field may be calculated as

$$\text{dB-SPL} = 10 \log\left(\frac{269}{99}\right) + 115 \text{ dB} = 119 \text{ dB} .$$

The minimum vent opening for the array may be found from Eq. (13),

$$S_V = \frac{3.23 \times 8 \times 1230}{32^{1/2}} = 0.562 \text{ m}^2 .$$

The maximum frequency to which the array will provide mutual coupling may be computed from Eq. (14),

$$f_{max} = 0.8\left(\frac{345}{0.41\pi \times 8^{1/2}}\right) = 76 \text{ Hz} .$$

Box construction may divide the array (and the minimum vent opening) into smaller units that are more easily handled, and traditional vented-box design methods can be followed.

10 COMMENTS

The preceding relationships assume that a reverberant field exists in the installation environment, a highly reasonable assumption for theaters, auditoriums, and churches. Note that loudspeakers behave as pistons at low frequencies, and the displacement-limited acoustic power that can be realized operates independently of efficiency and the brand of loudspeaker selected (other than to establish S_D and x_{max} criteria). It should be apparent that an electrical high-pass filter will be required at or above the box tuning frequency in order to protect the system from exceeding x_{max} limits on a program that is below the design passband. The calculated upper passband limit f_{max} indicates satisfactory operation to an f_3 of 80 Hz, but does not consider the baffle surface area that is required for ducts. For passband programs the acoustic power output has been restricted to values within the thermal limits of the drivers.

The practical upper frequency limit f_{max} for single-driver systems is reached at $ka = 2$, assuming flat axial response characteristics. This will result in a 3-dB drop in low-frequency power response at the transition frequency, which is within acceptable limits. This directly implies f_{max} frequencies for 18-in (460-mm), 15-in (380-mm), and 12-in (300-mm) direct-radiator single-driver systems of 500, 630, and 800 Hz, respectively.

11 CONCLUSION

The use of Thiele–Small driver parameters in the design of systems for low-frequency sound reproduction greatly aids the system designer in the accurate pre-

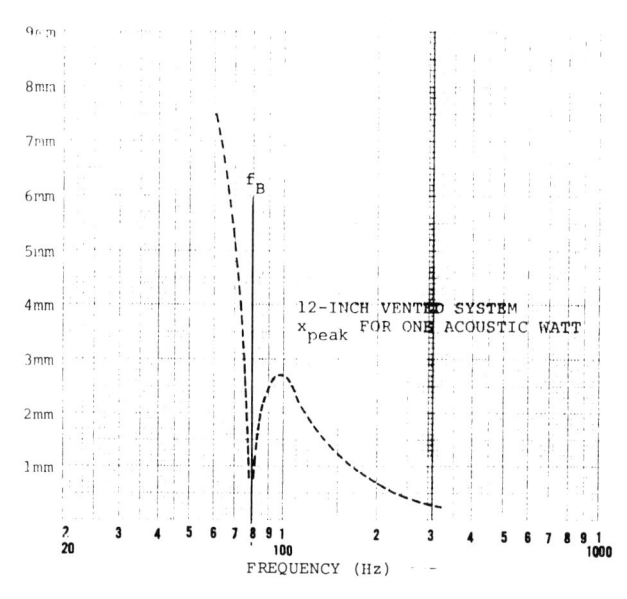

Fig. 6. Displacement versus frequency for vented systems.

Table 1.

Number of Drivers	Efficiency (%)	$P_{E(max)}$ (W)	P_A (W)	f_{max} (Hz)
1	2.1	200	4	270
2	4.2	400	17	150
4	8.4	800	67	100
8	16.8	1600	269	75
16	25.0*	3200	800	50

* 25% is assumed as the practical limit of efficiency for mutual radiation.

diction of performance possibilities. By contrast, traditional design methods that rely upon pressure sensitivity, axial frequency-response characteristics, and no knowledge of loudspeaker mechanical displacement limits often fall woefully short of the design objectives and have little merit, given the quality of information available today.

An added benefit from this design approach is in system reliability. When loudspeaker components are operated within safe mechanical and thermal power limits, failures for any reason become highly unlikely and become matters to take up with loudspeaker or amplifier manufacturers' warranty departments.

12 ACKNOWLEDGMENT

The author would like to thank Mark Gander and the engineering staff of James B. Lansing Sound, Inc., for providing technical support and a "sounding board" for the insights presented in this paper.

13 REFERENCES

[1] R. Small, "Suitability of Low-Frequency Drivers for Horn-Loaded Loudspeaker Systems," presented at the 57th Convention of the Audio Engineering Society, *J. Audio Eng. Soc. (Abstracts)*, vol. 25, p. 526 (1977 July/Aug.), preprint 1251.

[2] D. B. Keele, Jr., "An Efficiency Constant Comparison between Low-Frequency Horns and Direct Radiators," presented at the 54th Convention of the Audio Engineering Society, *J. Audio Eng. Soc. (Abstracts)*, vol. 24, p. 498 (1976 July/Aug.), preprint 1127.

[3] D. B. Keele, Jr., "Low-Frequency Horn Design Using Thiele/Small Driver Parameters," presented at the 57th Convention of the Audio Engineering Society, *J. Audio Eng. Soc. (Abstracts)*, vol. 25, p. 526 (1977 July/Aug.), preprint 1250.

[4] D. B. Keele, Jr., "Direct Low-Frequency Driver Synthesis from System Specifications," presented at the 69th Convention of the Audio Engineering Society, *J. Audio Eng. Soc. (Abstracts)*, vol. 29, p. 548 (1981 July/Aug.), preprint 1797.

[5] M. R. Gander, "Moving-Coil Loudspeaker Topology as an Indicator of Linear Excursion Capability," *J. Audio Eng. Soc.*, vol. 29, pp. 10–26 (1981 Jan./Feb.)

[6] I. Wolff and L. Malter, "Sound Radiation from a System of Circular Diaphragms," *Phys. Rev.*, vol. 13, p. 1061 (1929 June).

[7] L. L. Beranek, *Acoustics* (McGraw-Hill, New York, 1954).

[8] M. Engebretson, "Product Applications Information," Tech. Note, AB Systems Design, Inc., 1981 June.

[9] M. Engebretson and J. Eargle, "The State-of-the-Art Cinema Sound Reproduction System—Technology Advances and System Design Considerations," presented at the 69th Convention of the Audio Engineering Society, *J. Audio Eng. Soc. (Abstracts)*, vol. 29, p. 545 (1981 July/Aug.), preprint 1799.

[10] C. A. Hendricksen, "Directivity Response of Single Direct Radiation Loudspeakers in Enclosures," Altec Corp. Tech. Lett. 237 (1977 Sept.).

[11] C. A. Hendricksen, "Vented Box Design Method for Altec Low Frequency Loudspeakers," Altec Corp. Tech. Lett. 245, 1980 Apr.

[12] D. B. Keele, Jr., "Optimum Horn Mouth Size," presented at the 46th Convention of the Audio Engineering Society, *J. Audio Eng. Soc. (Abstracts)*, vol. 21, p. 748 (1973 Nov.), preprint 933.

[13] A. N. Thiele, "Loudspeakers in Vented Boxes," *J. Audio Eng. Soc.*, vol. 19, pp. 382–392 (1971 May); pp. 471–483 (1971 June).

THE AUTHOR

Mark Engebretson was born in Minneapolis, Minnesota, and attended the University of Minnesota from 1960 to 1964. From 1965–1970 he worked as an audio engineer and sound systems designer in sound reinforcement, corporate boardroom, and audio-visual communications.

In 1970 he joined the Altec Lansing Corporation as a field sales engineer, holding various positions with the company culminating in that of vice president, new product development, in 1976. That same year he joined Paramount Pictures to develop a new motion picture sound playback system, and held the position of vice president, research and development. His work at Paramount resulted in his patent 4,256,389, "Method and System of Controlling Sound and Effects Devices by a Film Strip." In 1979 he left Paramount to form Advanced Technology Design Corp., developing a family of complementary-bandpass power amplifiers and loudspeaker products. In 1982, AB Systems Design, Inc. acquired assets of Advanced Technology Design, with Mr. Engebretson joining AB as director of marketing.

In 1976, Mr. Engebretson served as chairman of the West Coast convention of the Audio Engineering Society. He has written numerous articles, and has been published by the Audio Engineering Society, The Society of Motion Picture and Television Engineers, and in *Modern Recording* magazine. He is a member of the AES and SMPTE. In addition to his activities at AB Systems, Mr. Engebretson has an active consulting and design practice, dealing in a broad range of electroacoustic technologies.

Near-Field and Far-Field Performance of Large Woofer Arrays*

G. L. AUGSPURGER

Perception Inc., Los Angeles, CA 90039, USA

In recent years it has become practical to predict coverage patterns of loudspeaker arrays using computer simulations. Comparisons of low-frequency array performance with computer analysis reveal three interesting truths: (1) Some favorite array geometries perform better in practice than they should. (2) Conventional calculations of polar patterns are inadequate to predict the performance of large arrays. (3) Computer optimization of array geometry can yield substantially better performance than what we have come to accept.

1 MODELING TECHNIQUE

In essence, the distance from each element in an array to a given listening location is calculated. Then, at any desired number of frequencies, the distances are translated to equivalent phase angles and the sources are summed vectorially.

Inverse-square losses are included in the summation. For large arrays, a number of listeners may be in the near field, and this factor becomes important. Indeed, for some array simulations (distributed ceiling loudspeakers, for example), inverse-square effects are an essential performance element.

Phase response is no more difficult to calculate than amplitude response. Since a low-frequency array must be combined with some kind of high-frequency reproducers, this becomes a key element in synchronizing the complete system.

A beneficial side effect of this array model is that first-order room reflections are easily included as virtual sources. On the other hand, it is very difficult to predict the effect of irregular surfaces such as auditorium seats. This is a shortcoming of all architectural ray tracing schemes.

Another drawback of the computer model is that there is no easy way to predict the contribution of hidden elements in an array. This problem is discussed in more detail in Sec. 3

Ignoring the directivity of individual array elements would seem to be a dangerous simplification. This turns out not to be the case. Any array analysis assumes that amplitude and phase responses of all elements are identical, or at least predictable. In my experience, typical cone loudspeakers maintain well-matched amplitude and phase characteristics only over the range in which they are essentially nondirectional. For example, commercial 300-mm (12-in) loudspeakers may show appreciable deviations above 500 Hz or so. At this frequency, the effective piston diameter is about $3/8$ wavelength, and at 90° off-axis the error is usually less than 3 dB.

2 GRAPHICS PRESENTATION

In this study, array performance is presented graphically in two basic formats. The first is the familiar far-field polar response curve. Typically, 10 curves are overlaid, covering the range from 63 to 500 Hz on standard one-third-octave center frequencies.

The second format overlays four frequency response curves at typical seating locations. The theoretical array is centered 8 m above ear level. Responses are then computed at a point directly under the array and for listeners 5, 15, and 25 m from the stage (Fig. 1). This corresponds to the general observation that a directional loudspeaker should be able to compensate for inverse-square effects out to a distance about three times its height above the ear plane. If the distance to the farthest seat is only twice the height of the loudspeaker, then an omnidirectional coverage pattern is probably adequate. On the other hand, if the far-throw distance is 10 times that of the short throw, your computer will

---

* Presented at the 87th Convention of the Audio Engineering Society, New York, 1989 October 18–21, revised 1990 January 4.

get a thorough workout.

The frequency response curves assume that individual array elements have flat response. For most arrays this means that their combined output builds up at low frequencies. What is important is not the flatness of the response, but how closely the curves track each other. The overall response can be equalized electronically or by using more efficient loudspeakers that behave as constant-velocity sources.

3 DIFFRACTION AND SCATTERING

Consider a large 360° "wedding cake" array with four tiers of loudspeakers set on 30° wedges. If this is an open array with individual woofer boxes separated from each other, its performance will obviously be different from that of a large barrel in which loudspeakers are mounted on a continuous baffle surface.

The computer model used here assumes line-of-sight contributions from all array elements. The most practical way to approximate the amplitude and phase responses of hidden elements seems to be a correction factor in the form of an overdamped second-order low-pass filter. Fig. 2 compares the responses of the two configurations at a typical seating location. For this kind of array, neither simulation is terribly accurate, but general trends can easily be seen.

In most situtations the coverage pattern of an enclosed array is superior to that of its open counterpart. This is one reason why some arrays that simulate cylindrical or conical sources perform much better than one would expect.

Another factor that is generally ignored is the effect of nearby scattering surfaces. Early reflections are provided by surfaces near the loudspeaker *or* the listener. If the listener is surrounded by chairs, benches, or other listeners, then the response curve of the loudspeaker will be scrambled in some unpredictable way.

If one is lucky, such early reflections may compensate for deficiencies in array performance. As an example, one of the woofer arrays tested at the Hollywood Bowl was predicted to have a huge 500-Hz dip in a certain seating area. Yet, when the array was driven with pink noise, the dip was barely perceptible—it disappeared into the fine comb structure produced by scattering from the seats.

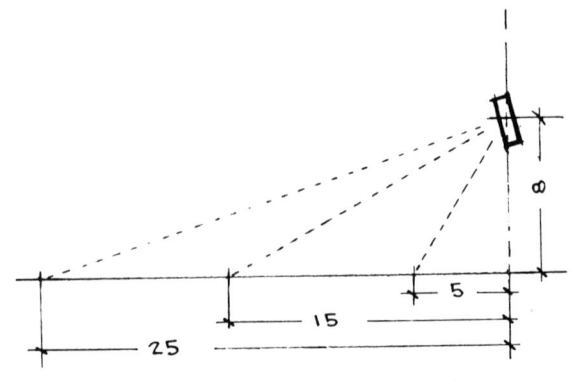

Fig. 1. Basic geometry for analysis.

4 ARE LARGE ARRAYS NECESSARY?

Having analyzed and measured a variety of woofer array geometries, it seems to me that there is really little to choose between most of the stacks, bananas, baskets, and barrels that have been my favorites at various times. Coverage patterns too often take the form of random pincushions. Off-axis lobing into microphones can be more intense than on-axis sound to the audience.

Fig. 3 shows the predicted responses of a simple woofer stack at several seating locations. The stack is two woofers wide by six woofers high. In Fig. 3(a) it is set vertically on the stage; in Fig. 3(b) it is suspended above the stage and tilted. As is confirmed by practical experience, one location is about as good as the other. In fact, if stage floor reflections are included in the simulation, a microphone 8 m below the overhead array is more apt to go into feedback than one 3 m beside the stage stack.

A single woofer mounted at ceiling level has a directivity index of 3 dB. If mounted at the intersection of a wall and a ceiling, its directivity index is 6 dB. Even with very large, very sophisticated arrays, it is hard to do better than this. Moreover, a single loudspeaker has a precisely defined acoustic origin and can be synchronized with a dozen high-frequency horns if need be. For medium-size theaters, auditoriums, and churches, a single box housing one or two woofers may be the ideal array.

Where greater sound power is needed, woofers can be packed into a single sound chamber, as in the "manifold" systems produced by Electro-Voice [1]. Or one may be tempted to make use of the omnidirectional characteristic of a Bessel line array. In one popular implementation patented by Philips [2], five identical equally spaced loudspeakers are driven as follows:

Level, dB: -6 0 0 0 -6
Polarity: $+$ $+$ $+$ $-$ $+$

Performance seems almost too good to be true: the

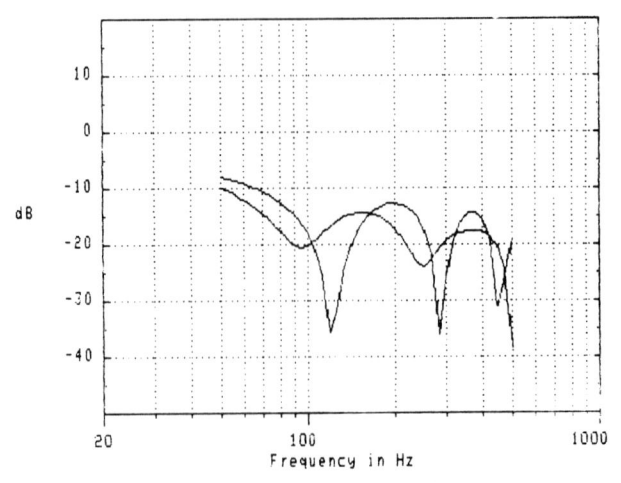

Fig. 2. Open versus baffled cylindrical array.

acoustic power of three and a half woofers with the directional pattern of one. However, the array acts as an unusual acoustical all-pass filter. If wide coverage is needed in both the horizontal and the vertical planes, it may be impossible to synchronize the array with high-frequency transducers for all listener locations.

Arranging a dozen or so woofers on the surface of a large sphere does not simulate a point source except (as is true for any array) for wavelengths several times the diameter of the sphere. At higher frequencies the pattern develops strong frequency-dependent lobes.

5 SHADED VERTICAL LINE ARRAYS

The classical array for mathematical analysis is a line of uniformly spaced elements, all driven at the same level. If the usable frequency range of such an array is to extend up to 500 Hz or so along any axis, then the center-to-center spacing of elements can be no greater than 0.34 m. For a practical array this is a convenient spacing for 300-mm (12-in) woofers. If good directional control is to be maintained down into the 65-Hz region, then the height of the array must be at least 5 m.

Ideally, such an array would produce a plate-shaped pattern with most of its energy projected in the plane of the equator and very little at the poles. In practice, the directivity of the main lobe is proportional to the frequency, and as the main lobe becomes more directional, additional upper and lower lobes develop. These unwanted lobes can be suppressed to almost any degree desired by tapering or shading the array. In this paper I will use the terms "tapered" to mean some kind of frequency-dependent processing of drive signals and "shaded" to mean a simple gradation in level.

Fig. 4(a) shows the vertical polar response of a uniformly driven 21-element array at octave-spaced frequencies from 63 to 500 Hz. Fig. 4(b) illustrates the secondary lobe suppression achieved with shading. Drive levels decrease from the center to the top and bottom elements according to a raised cosine window function, just as if we were designing a transversal filter (which is exactly what we are doing). In this frequency range the pattern is about as close to a lighthouse beam as anyone is apt to get.

The array, while a great improvement over an off-the-shelf column loudspeaker, has two obvious defi-

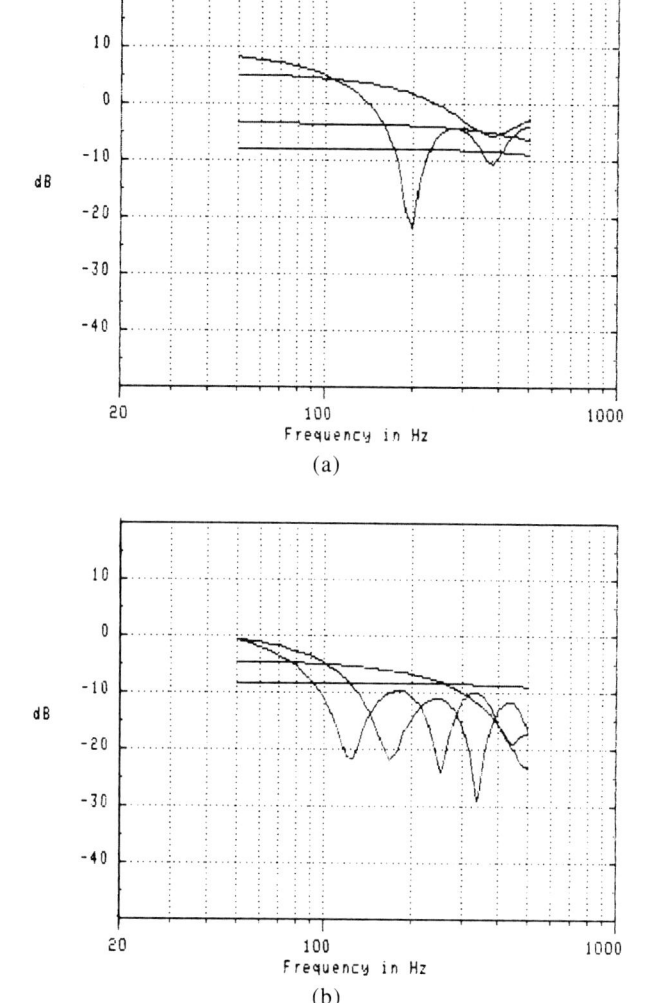

Fig. 3. (a) Stage woofer stack response. (b) Suspended woofer stack response.

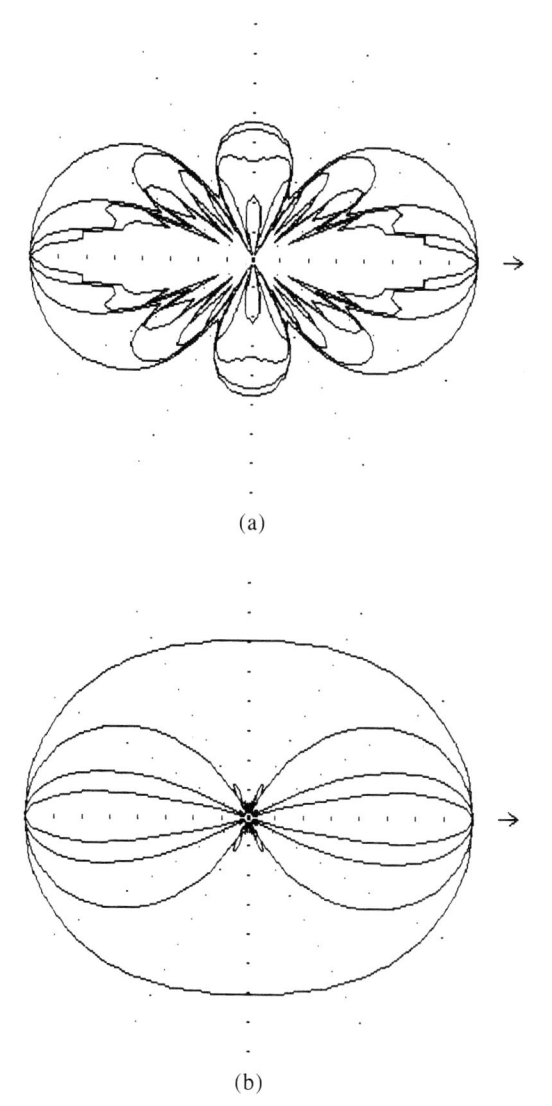

Fig. 4. (a) Unshaded vertical line array. (b) Array of (a) with shading. Symmetrical 21-loudspeaker column; no taper; one-octave centers: 63–500 Hz.

ciencies for many applications. First, the pattern is still frequency dependent, and, second, the main lobe is at right angles to the axis of the array.

We might attack the second problem with electronic beam steering. By progressively delaying drive signals from top to bottom, the main lobe becomes more of a conical lid than a flat plate. Unfortunately, from the standpoint of an observer on one of the poles, the effective spacing between elements has changed. As a result, unwanted lobes appear again.

6 TAPERED VERTICAL LINE ARRAYS

For a line source to produce constant directivity its effective length must vary inversely with frequency. Tapering a line array by driving individual elements through low-pass filters is an established technique. At one time at least three U.S. manufacturers produced tapered arrays, but these were relatively small devices with only a few elements. Performance was not sufficiently better than that of other column loudspeakers to justify the cost. Klepper and Steele designed an interesting array with acoustical low-pass tapering [3]. This exhibited good pattern control over a vertical angle of about 30°, but the acoustical filter was not well enough controlled to realize the full potential of a tapered array.

Computer simulation allows the concept of low-pass tapering to be explored in detail and confirms the soundness of the basic theory. It also reveals some unexpected quirks.

In Sec. 5 we examined the performance of a 21-element vertical array. By tapering its drive signals from wide-band (top) to 50-Hz low-pass (bottom) we might hope for a constant coverage pattern over the range from 50 to 500 Hz. Moreover, because of the progressive delay introduced by the low-pass filters we would expect the beam to be tilted down 10° or so. The good news is that the pattern really is independent of frequency and the beam really does tilt down. The bad news is that the lower hemisphere is almost nondirectional. This might make a good low-frequency array for a boxing arena but not for a 3:1 throw distance range.

By symmetrically tapering the array from its center the pattern tightens substantially and inverse-square correction is almost perfect. Fig. 5(a) shows the frequency responses at the four designated locations. Fig. 5(b) overlays polar response curves from 63 to 500 Hz. This is not only a true constant-directivity array but a useful one.

The basic concept can be modified to optimize vertical arrays for a variety of real-life situations. How about a woofer stack on the stage? Fig. 6 shows the frequency response of an array designed for this near-field application. There are nine elements in the array, spaced logarithmically to a height of about 7 m.

In this case the curves are for the usual three audience locations plus a microphone 3 m to one side of the array. Although most of the curves represent locations well into the near field, the acoustic origin of the array remains sharply defined. In the 500-Hz crossover region, phase response lies within a 45° window for the locations plotted.

Low-pass tapering is a powerful technique. Constant directivity can be achieved with relatively few elements. The acoustic origin is stable, allowing accurate synchronization with high-frequency transducers. Multiple power amplifiers with low-pass filters at their inputs are not that much more complex than the two or more channels needed to power a conventional woofer stack.

7 TAPERED END-FIRE LINE ARRAY

Conventional line arrays are usually wall mounted to provide a horizontal pattern that is uniform through 180°. When hung in free space a 360° pattern results. But what if we need pattern control horizontally and vertically? The benefits of low-pass tapering can be extended to large two- and three-dimensional arrays. However, where huge amounts of acoustic power are not required, more elegant solutions exist.

One of these is the end-fire line array, analogous to

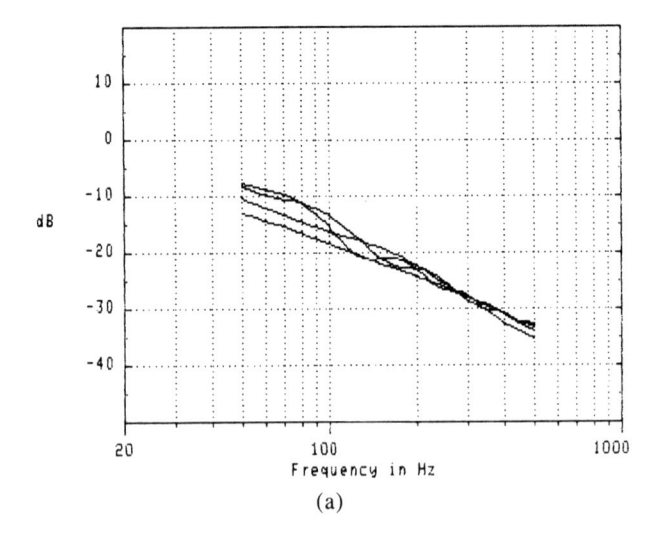

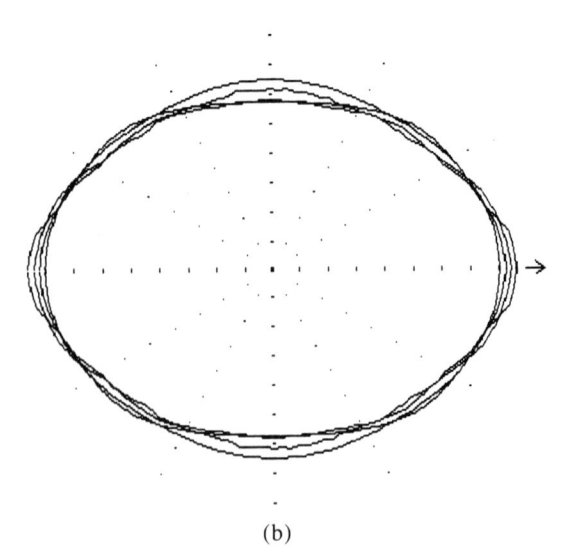

Fig. 5. (a) Vertical line array with low-pass taper. (b) Polar response of (a). Symmetrical 21-loudspeaker column; low-pass taper; one-octave centers: 63–500 Hz. 5 dB/div.

a shotgun microphone. It can also be thought of as a line array with 90° electronic beam tilting. For well-behaved low-frequency arrays, a tapped delay line is needed in addition to low-pass tapering. Second-order or even fourth-order low-pass filters do not have enough delay to properly synchronize the array elements.

Fig. 7 shows the polar response of a 13-element end-fire array 5 m long. The pattern is fairly broad, but since it now takes the form of an ice cream cone rather than a doughnut, the directivity index along a preferred axis is much greater than that of a comparable vertical array. A horizontal end-fire array could be mounted on or flushed into the ceiling to increase its directivity index by another 3 dB.

Again, computer simulation suggests that sophisticated pattern control, independent of frequency, can be achieved with signal-processed line arrays. I am not aware of any commercial or experimental end-fire loudspeaker array, even though the concept was described by Olson more than 30 years ago [4]. With the addition of low-pass tapering it appears to be a practical method of achieving good low-frequency pattern control.

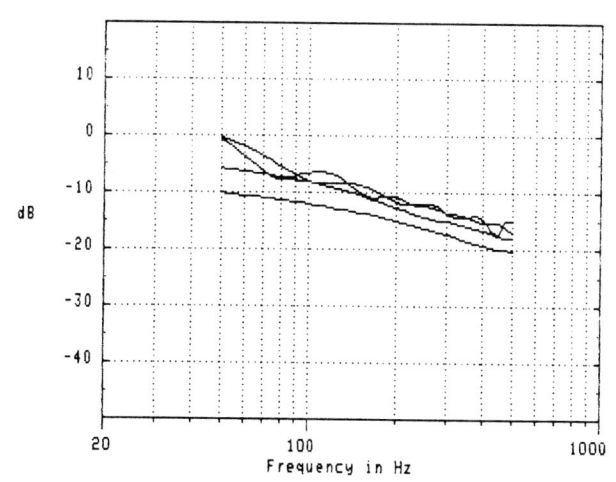

Fig. 6. Vertical near-field array with low-pass taper.

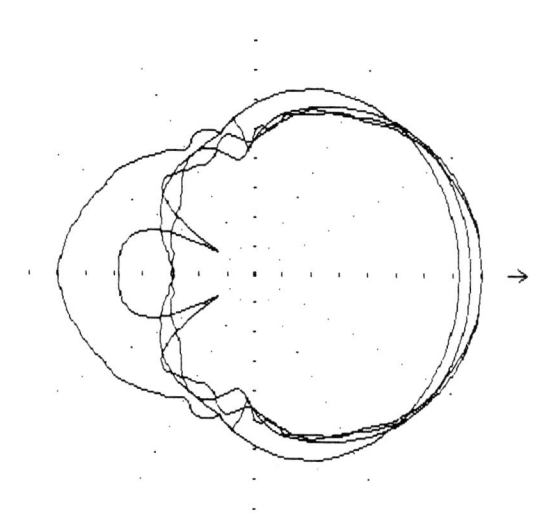

Fig. 7. End-fire line array with low-pass taper; one-octave centers: 63–500 Hz. 5 dB/div.

8 DOUBLE-BARREL CARDIOID ARRAY

Arrays of unbaffled cone loudspeakers have occasionally been used for sound reinforcement applications. If a simple acoustic low-pass filter is added at the rear, the pattern approaches that of a true cardioid [5].

A cardioid loudspeaker array can be very small, and there is no low-frequency limit to its unidirectional pattern. Unfortunately it is an inefficient way to pump air. Low-frequency response rolls off at 6 dB per octave as compared with an equivalent array of baffled loudspeakers. Cone excursions of 25 mm or more may be required to produce appreciable output in the 63-Hz band.

Instead of using an unbaffled loudspeaker as the basic element, let us go back to the textbook doublet source: two point sources separated by a small distance and driven in opposite polarity. For maximum efficiency below 125 Hz or so, the small distance turns out to be about 0.7 m. By delaying the signal to the rear loudspeaker we can produce any pattern from bipolar through hypercardioid to true cardioid.

The useful frequency range can be extended upward to 500 Hz by crossing over to a second element with smaller spacing between loudspeakers and shorter delay, just as in some two-way cardioid microphones. The geometry of this four-loudspeaker array is shown in Fig. 8.

When I first began toying with this scheme, my main concern was pattern control through the crossover region. Computer simulation shows that this is not a problem. A slightly asymmetrical 24-dB per octave crossover network maintains a frequency-independent cardioid pattern and at the same time equalizes pressure response. Other directional patterns seem to be equally robust. For most sound reinforcement applications, a supercardioid is probably the best compromise between directionality and suppression of rear lobes.

Predicted response is shown in Fig. 9. The main lobe is not tight enough to compensate for a 3:1 throw distance ratio. However, the well-defined null zone and the unchanging pattern over a range of more than three octaves makes this a definite candidate for real-world testing.

How much efficiency has been sacrificed? In comparison with the same four loudspeakers in a conventional array, on-axis sensitivty is only about 3 dB less, but total energy output is about 9 dB less. Used in a

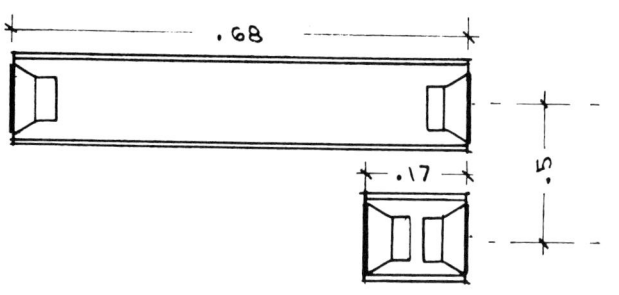

Fig. 8. Double-barrel array geometry.

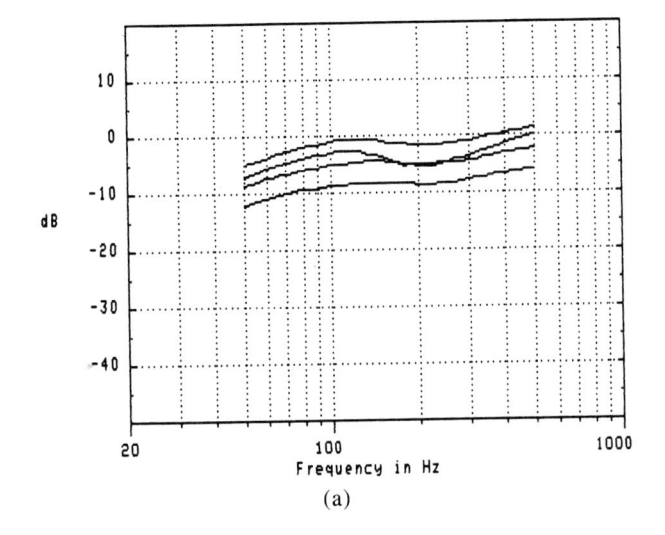

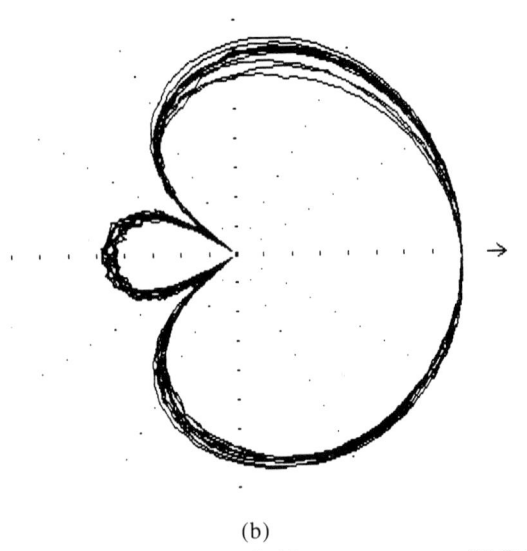

Fig. 9. (a) Double-barrel cardioid array response. (b) Polar response of (a). 150-Hz crossover; one-third-octave centers: 50–500 Hz.

semireverberant hall, the comparative loss in efficiency would be about 6 dB.

9 CONCLUSION

For many applications a conventional low-frequency array with its uncontrolled directional characteristics may indeed be good enough. However, loudspeaker design and signal processing technology have progressed to the point where constant-directivity low-frequency arrays can be realized. As these become available, we will learn how to exploit the benefits of low-frequency pattern control.

The computer simulations presented in this paper illustrate only a few of the promising array geometries that deserve investigation. In my view, the tapered end-fire array, the near-field woofer stack, and the double-barrel cardioid array are not only novel but offer promising solutions to practical sound-reinforcement problems.

10 REFERENCES

[1] N. V. Franssen, "Direction and Frequency Independent Column of Electro-Acoustic Transducers," U.S. Patent 4,399,328 (1983 Aug. 16).

[2] R. J. Newman and D. E. Carlson, "High Output Loudspeaker for Low Frequency Reproduction," U.S. Patent 4,733,749 (1988 Mar. 29).

[3] D. L. Klepper and D. W. Steele, "Constant Directional Characteristics from a Line Source Array," *J. Audio Eng. Soc.*, vol. 11, p. 198 (1963 July).

[4] H. F. Olson, *Acoustical Engineering* (Van Nostrand, New York, 1957).

[5] J. K. Hilliard, "Unbaffled Loudspeaker Column Arrays," *J. Audio Eng. Soc.* (*Project Notes/Engineering Briefs*), vol. 18, pp. 672–673 (1970 December).

THE AUTHOR

George L. Augspurger received his B.A. degree from Arizona State University at Tempe, M.A. from UCLA, and has done additional postgraduate work at Northwestern University. After working in sound contracting and television broadcasting, he joined James B. Lansing Sound, Inc. in 1958 where he served as Technical Service Manager and, later, as Manager of the company's newly-formed Professional Products Department. In 1968 he was appointed JBL's Technical Director.

In October, 1970 he left JBL to devote full time to Perception Inc., a consulting group specializing in architectural acoustics and audio system design. Mr. Augspurger is a member of the Audio Engineering Society, Acoustical Society of America, and United States Institute of Theatre Technology.

Large Arrays: Measured Free-Field Polar Patterns Compared to a Theoretical Model of a Curved Surface Source*

JOHN MEYER AND FELICITY SEIDEL

Meyer Sound Laboratories, Inc., Berkeley, CA 94941, USA

A method used to measure horizontal and vertical polar patterns of large arrays under actual free-field conditions is described. Ways in which the polar patterns of loudspeaker arrays deviate from a theoretical model of a curved-surface source are discussed; thus the concept of arrayability is introduced. Criteria for arrayability are suggested.

0 INTRODUCTION

Sound-reinforcement loudspeakers are commonly used in arrays for acoustical output power and directivity. While professional practice has historically given rise to varied geometrical arrangements, the predominant tendency (especially in recent years) has been to array loudspeaker components in a convex arc pursuant to the ideal of a curved radiating surface.

A general theoretical foundation for predicting curved array behavior exists [1, pp. 40–55], but has rarely been subjected to precise experimental verification in specific applications. Given the broad variations in contemporary loudspeaker component designs, cabinet structures, and crossover arrangements, one might reasonably expect that the interactions among elements in a physical array could produce deviations from the theoretical ideal.

It should not be taken for granted, therefore, that all loudspeaker arrays form polar patterns that are easily predictable. Yet the polar response of an array is one substantial determinant of its applicability in specific environments. Thus arises the need to study the behavior of arrayed loudspeaker systems.

In this paper an apparatus and a method for measuring free-field polar patterns of loudspeaker arrays are described. The technique has been applied, over a two-year period, to the detailed measurement and characterization of a large number of arrays. The aims of this research have been twofold: one, to compare the polar patterns of curved arrays with those of other array geometries; and two, to quantify the performance of specific curved arrays in an effort to test for predictability.

1 LOUDSPEAKER ARRAYS

Olson describes the ideal sound source as a curved radiating surface [1, p. 50]: "A sphere vibrating radially radiates sound uniformly outward in all directions. A portion of a spherical surface, large compared to the wavelength and vibrating radially, emits uniform sound radiation over a solid angle subtended by the surface at the center of curvature. To obtain uniform sound distribution over a certain solid angle, the radial air motion must have the same phase and amplitude over the spherical surface intercepted by the angle having its center of curvature at the vertex and the dimensions of the surface must be large compared to the wavelength. When these conditions are satisfied for all frequencies, the response characteristic will be independent of the position within the solid angle." (See also Appendix 1.)

---

* Presented at the 87th Convention of the Audio Engineering Society, New York, 1989 October 18–21; revised 1990 January 6.

This ideal sound source may be modeled as an infinite number of point sources distributed along a spherical section, which in turn may be approximated by a finite number of loudspeaker elements arrayed in an arc. Theoretically, then, for a particular loudspeaker with its given frequency response, phase response, and coverage angle, there should be some ideal number and arrangement of cabinets such that the array has a single focal point located at some determinable distance behind the loudspeaker cabinets. Where the array's surface dimensions are large compared with the wavelength, its polar pattern should be uniform within the angle subtended by the array at the focal point.

In practice, however, we can only approximate the ideal of uniform sound distribution within a specified solid angle. For example, the requirement that the arc dimensions be large with respect to wavelength is most often violated at low frequencies in acoustical applications, and this has a fundamental effect on the distribution patterns of loudspeaker arrays. Olson shows that the sound distribution narrows as the wavelength approaches the dimensions of the arc, then widens again with increasing wavelength. Where the arc dimensions are small compared with the wavelength, the distribution pattern becomes omnidirectional [1, pp. 50–52]. (The distribution patterns of radial, constant-directivity and multicellular horns conform to this model, narrowing where the wavelength approaches the horn dimensions and becoming omnidirectional at lower frequencies.)

Moreover, in order for the array to behave exactly as if it were the radiating arc depicted by Olson, the individual loudspeakers must emit sound radially as though from a spherical arc having a single center of curvature, with uniform phase and amplitude at all frequencies. All actual loudspeakers, however, exhibit characteristics (nonuniform amplitude and phase response, and directional characteristics that vary with frequency) which distinguish them to some degree from this ideal and from each other. It is reasonable to expect that these characteristics would affect the polar patterns of arrays.

Accordingly, there is a need for some mechanism to predict the sound distribution of actual arrays from the characteristics of the individual loudspeaker elements and the array geometry. The extent to which a given actual array deviates from a theoretical model can be determined by careful free-field measurement, and may serve as an indication of the accuracy of the model.

2 MEASUREMENT APPARATUS

In order to study the behavior of arrayed loudspeaker systems carefully and analytically, we had to ensure that what we measured was attributable only to the loudspeaker array by eliminating complex interactions with the surrounding environment.

To accomplish this, the measurements were conducted in a large outdoor field with the aid of a device effectively capable of moving a microphone along a full 360° arc in both the horizontal and the vertical planes (Figs. 1 and 2). The device consists of an octagonal scaffold, an open grating floor, and a microphone apparatus attached to a rotating ring mounted atop the scaffold.

Fig. 1. Scaffolding with microphone apparatus in horizontal plane.

2.1 The Scaffold

The scaffold was 18 ft (5.49 m) in diameter and octagonal. Its 21-ft (6.4-m) height was determined both by manufacturer specification and by earlier experiments, in which we found that a loudspeaker suspended from 20 ft (6.1 m) to 60 ft (18.3 m) above the earth did not vary substantially in its interaction with the ground surface. The shape allowed us to mount rollers such that a circular ring with a 15-ft (4.6-m) diameter can rotate through a full 360° about the center of the scaffold, while minimizing the amount of reflective steel tubing outside of the ring.

A steel floor grating rested upon steel beams hung between the radial arms at the top of the scaffold. The floor and the beams provided a strong, stable surface on which to place arrays, while the 70% open grating reduced reflections.

2.2 The Microphone Apparatus

The microphone apparatus (see Figs. 1 and 2) mounted on two vertical steel tubes that were fastened, 180° apart, to the circular ring mentioned. Two aluminum tubes were mounted to the vertical steel members with aluminum couplers, such that the height of attachment was adjustable. (Because of the counterweight system, the minimum height requirement was about 4 ft (1.2 m) above the floor.) These tubes were arranged to form a U shape with right angles between either side and the bottom of the U, which was connected to the side tubes with aluminum couplers so that the point of attachment was adjustable. Between each of the aluminum couplers on the vertical tubes and the sides of the U there were gear mechanisms which allowed the entire assembly to rotate 180° over the center of the scaffold (Fig. 2).

A measurement microphone was mounted to the center of the bottom leg of the U so that, as the U was rotated, the microphone traveled 180° along the arc of a circle in the vertical plane. By fixing the vertical axial position of the moving microphone and rotating the ring to which the vertical members were mounted, the microphone could be made to traverse the arc of a complete circle in the horizontal plane at a specified height and radius.

All of the tubes on this apparatus, as well as the scaffolding and floor in front of the array, were covered with acoustic foam to minimize reflections during the measurement process.

3 DATA ACQUISITION

The measurement data were compiled using a Hewlett-Packard model 3582A dual-channel Fast Fourier transform (FFT) analyzer driven by a Hewlett-Packard model 9807A integral personal computer running our proprietary Polar Cad software.

The analyzer was set for dual-channel operation in the transfer function mode, using two microphone inputs: one (the moving microphone) mounted on the preceding microphone apparatus, and the other (the reference microphone) fixed on axis of the system under test. (Information on the general theory of FFT measurements may be found in [2]–[4].) For these mea-

Fig. 2. Scaffolding with microphone apparatus in vertical plane.

surements we used a pair of calibrated Brüel & Kjaer model 4133 half-inch (13-mm) microphones which were matched by the manufacturer. We chose this model both because it exhibits its flattest response at 0° of incidence and because of its low distortion characteristics at high sound pressure levels [5].

To obtain one-twelfth-octave resolution over the entire audio spectrum, we measured over five frequency spans, selecting shorter capture times for high-frequency ranges and longer times for low frequencies. Periodic noise, generated by the analyzer, was the test signal. Time averaging was used, with a choice of four, eight, or sixteen averages selected according to weather or noise conditions. The computer, in conjunction with the Polar Cad software, controlled the analyzer and wrote the data to disk at each measurement point.

3.1 Loudspeaker Positioning

The loudspeaker array to be tested was assembled on the open grating floor around the center of the scaffold, and was placed so that the reference microphone lay on the central axis of the array. To obtain accurate polar data, the moving microphone must traverse a circle whose center is coincident with the focal point of the array. Accordingly, each array was tested to ascertain its focal point by application of the inverse square radiation attenuation law.

First we took the frequency response of the array at an arbitrary, measured distance on axis. Then we moved the microphone farther from the array along the primary axis while continuously monitoring the frequency response as measured at the microphone position. If the response drops evenly in level across a significant portion of the audio band, then the array exhibits a single definable focal point. In the converse case (that is, if the decrease in level varies with frequency), the focal point of the array is frequency dependent.

Given that the array exhibited a single focal point, we continued increasing the distance between the microphone and the array until the level had dropped by 6 dB relative to the initial microphone position. We then measured and recorded the distance.

Repeating the process, we measured and recorded the distance at which the overall level had dropped by 12 dB relative to the initial microphone position. Finally we extrapolated from these data, using the inverse square law, to find the focal point by successive approximation, determined the appropriate radial displacement of the array from the scaffold center, and offset the array accordingly.

3.2 Measurement Procedure

The microphone apparatus was adjusted such that the moving microphone measured along a circle with a 17-ft (5.2-m) radius and was initially positioned so that the moving microphone was as close to the reference microphone as possible [in practice, approximately 1 in (25.4 mm) above it]. At this point we took the on-axis response. The moving microphone then was rotated in 10° increments about the array, first in the horizontal plane and then in the vertical plane, with data taken at each increment. This procedure gave 36 measurement points per plane, for a total of 72 points per array. (In the horizontal plane, the microphone can move about the array through a full 360° arc, but in the vertical plane its path is restricted by the scaffolding beneath the array. Therefore when measuring on the vertical axis, we rotated the microphone 180° over the array and then placed the array upside down, sweeping back through the same 180° arc to complete a full circle.)

For each measurement point three classes of information were gathered: the on-axis frequency response (the reference microphone), the transfer function between the moving microphone and the reference microphone, and the coherence information.

Storing the on-axis frequency response for each measurement point allowed us to monitor the consistency of the loudspeaker system itself: if a problem were to arise within the system during the measurement process, it would most likely be reflected by the on-axis frequency response.

We obtained two major benefits by operating the analyzer in the transfer function mode. First, each measurement point was normalized, in real time, to the on-axis position. Second, the effects of extraneous noise during the measurement process were minimized, because we were measuring the difference between the two microphones.

The coherence information allowed us to determine the quality of the data collected at each measurement point. The coherence tends to be quite good in an undisturbed free-field environment. It is nonetheless another means of monitoring our test conditions.

4 ARRAY MEASUREMENTS

In this section we present and analyze the data collected for the horizontal radiation pattern of two simple two-cabinet arrays.

4.1 The Objects of Measurement

The loudspeaker system used for these experiments was a two-way full-range unit of compact dimensions, comprising a 12-in (0.3-m) cone low-frequency driver in a vented enclosure with a 90° modified radial high-frequency horn and a driver having a 1.4-in (36-mm) throat. The enclosure dimensions and high-frequency horn geometry are shown in Fig. 3. An external line-level signal-processing unit provided an optimized active crossover for biamplification, along with correction to yield nominally flat amplitude and phase response from a single cabinet in half-space. The crossover networks provided for an acoustical crossover between the two drivers at approximately 1.2 kHz. The acoustic centers of the low-frequency and high-frequency drivers are aligned to the same plane by a delay circuit in the low-frequency path. Fig. 4 shows two simple two-cabinet arrays formed using this loudspeaker.

The configuration of Fig. 4(a) is a curved array formed by placing the cabinets such that the rear corners touch and the angle between their respective main axes is 60°. This arrangement is specified by the manufacturer to exhibit a 120° directivity pattern between −6-dB points; we refer to it as the wide-throw configuration.

The array shown in Fig. 4(b) is referred to as the parallel configuration. It is formed by placing the cabinets side by side such that their main axes are parallel and their front faces are flush with one another.

The wide-throw configuration of Fig. 4(a) was found to exhibit a clearly defined focal point at a radial distance of approximately 27 in. (686 mm) from the array surface. By contrast, the focal point for the parallel configuration of Fig. 4(b) was ill defined and frequency dependent.

4.2 Data Analysis

Fig. 5 shows the pressure magnitude data collected on a single loudspeaker (D), the array of Fig. 4(a) (E), and the array of Fig. 4(b) (F). We present our data in one-third-octave resolution, since we found this to be the minimum resolution at which broad sound distribution characteristics remain clearly defined. To achieve this, we developed a processing routine that combines our original one-twelfth-octave bins to give one-third-octave resolution. These data were then collated and plotted in a polar coordinate system.

For purposes of comparison, we also included Olson's predicted polar patterns for a 90° radiating arc (A), a 120° radiating arc (B), and a line radiator (C) of dimensions similar to those of the test arrays [that is, each arc having a radius of 27 in (686 mm) and the line array being 29 in (737 mm) long; this corresponds to the wavelength of sound in air at approximately 500 Hz] [1, p. 42]. We transposed Olson's data to the same logarithmic scale used in plotting our data in order to facilitate comparisons; data that Olson does not give are omitted in the chart.

A number of observations may be made concerning this figure.

1) Both the 90° arc (A) and the 120° arc (B) exhibit

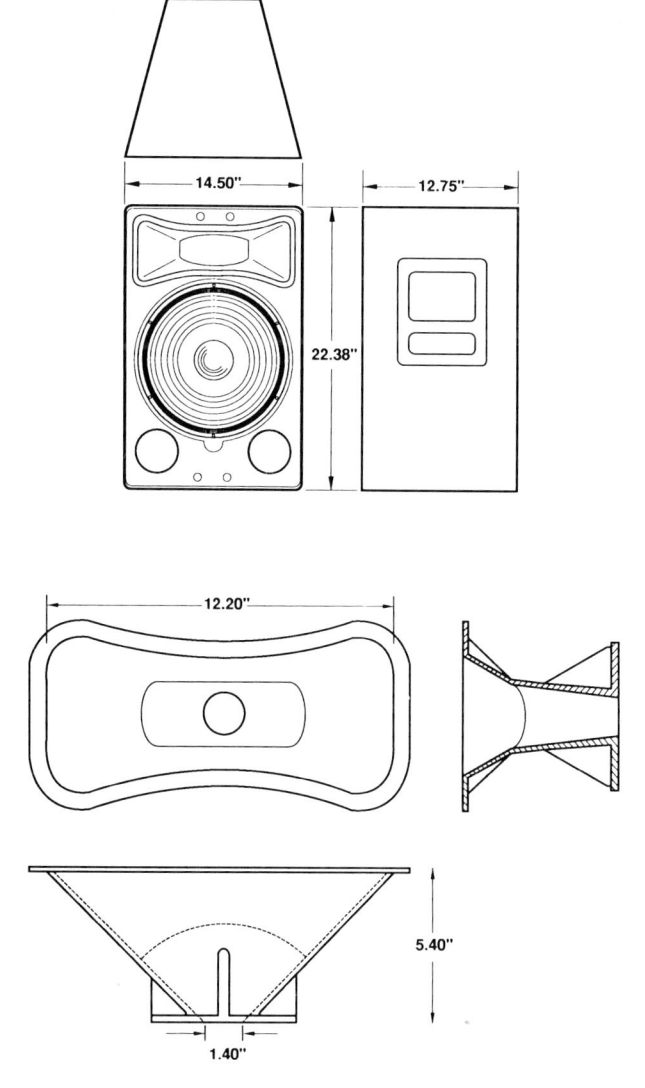

Fig. 3. Loudspeaker system and its horn dimensions.

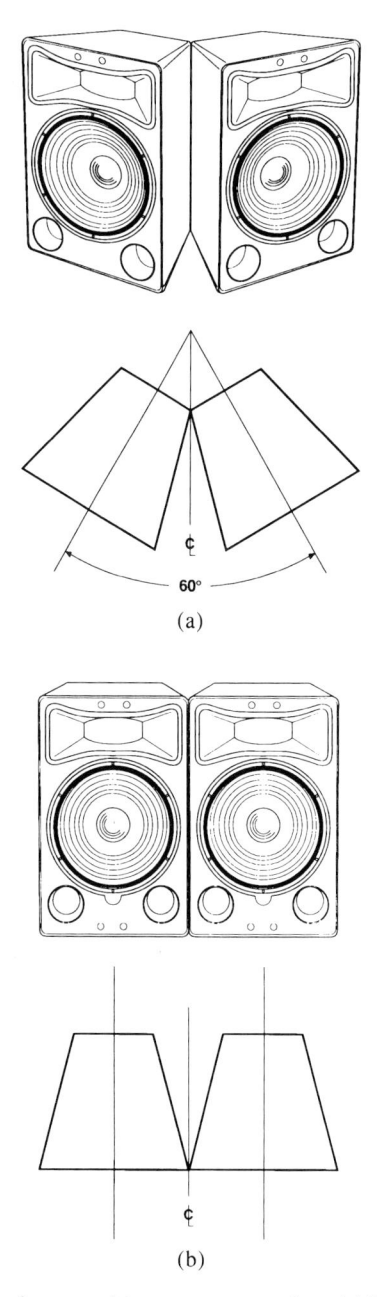

Fig. 4. Simple two-cabinet array examples. (a) Wide-throw configuration. (b) Parallel configuration.

the characteristics described in Sec. 1. Above 1 kHz, where the arc dimensions are large in comparison to the wavelength, each maintains relatively uniform distribution within the angle subtended by the arc. The number of lobes increases with increasing frequency, and the amplitude range of the lobing appears simultaneously to decrease. Where the wavelength equals the arc dimensions, each distribution pattern narrows substantially; for the 90° arc this occurs at 500 Hz, while the narrowing for the 120° arc is most pronounced at 250 Hz. At 125 Hz and below, both approach omnidirectionality.

2) The line array (C) is essentially omnidirectional at 125 Hz and narrows progressively with increasing frequency.

3) The behavior of the single loudspeaker (D) conforms in a very general way to that of a 90° arc (A), particularly at 125 Hz, 2 kHz, and 4 kHz. This confirms the specified 90° distribution pattern for the unit. [The narrowing at 250 and 500 Hz is considerably less severe than is the case for the 90° arc, however, due to the fact that these frequencies are reproduced by the 12-in (0.3-m) low-frequency driver.]

4) Similarly, the behavior of the wide-throw configuration (E) shows good general agreement with that predicted for a 120° arc (B): the distribution pattern is essentially omnidirectional at 125 Hz and below, narrows at 250 through 500 Hz, then widens again with increasing frequency. We therefore can conclude that the array is acting as an approximation of a 120° radiating arc whose focal length (radius) is 27 in (686 mm).

5) Such general agreement aside, specific features of the measurement data for the single loudspeaker (D) and the wide-throw configuration (E) differ from the arc models (A and B), and the distinctions are particularly apparent at 500 Hz, 1 kHz, and 4 kHz. As a rule, both the single loudspeaker and the wide-throw configuration exhibit smoother and wider coverage than the corresponding models would indicate.

6) The parallel configuration (F) exhibits generally "poor" characteristics. Rear lobing appears as low as 63 Hz. At 500 Hz and above the distribution pattern is uneven, with pronounced lobing over a 10-dB range from −30 to +30°.

7) While the differences between the two array configurations (E and F) are most pronounced at 1 kHz, the wide-throw configuration consistently maintains a wider coverage angle with a more even distribution pattern than does the parallel configuration.

8) Below 250 Hz (where the wavelength exceeds the array dimensions) there is little difference between the distribution patterns of the two arrays (E and F). This would indicate that where the array dimensions are small relative to the wavelength, array geometry appears to have little effect.

9) The behavior of the parallel configuration is not directly predictable from Olson's data. Neither radiating arc nor line array, it appears in broad terms to function as a kind of hybrid between the two.

5 CONCLUSION

Clearly, loudspeakers do not act independently of one another in an array. On the contrary, the polar patterns of arrayed loudspeakers seem to be fundamentally dependent upon not only the attributes of the individual loudspeakers, but also the geometry of the array. These variables appear to interact critically in ways that are far more complicated than is popularly believed

For example, we see nothing in the pressure magnitude polar patterns of the single loudspeaker which, taken alone, would predict that loudspeaker's measured behavior in either the parallel or the wide-throw geometry that we discussed. If simple scalar summation of the pressure responses of each loudspeaker primarily determines array behavior, then the parallel configuration in our example could be expected to yield polar patterns as consistent and smooth as those of the wide-throw configuration. As the measurement data clearly show, this is not the case.

We summarize that, in addition to the pressure magnitude polar patterns of the individual loudspeaker elements, the following factors must be included in any predictive calculation of array polar patterns: the phase response and positions of the acoustical centers of the individual loudspeakers, the array geometry, and the propagation vector.

Our future research will be devoted to predicting array polar patterns from the data that we have gathered. (We have calculated a rough estimation of the nulls in the polar pattern of the parallel configuration from the pressure magnitude data; we give these calculations in Appendix 2. Because of the complexity, no similar calculations have yet been made for the wide-throw configuration.) For researchers who are interested in pursuing similar work, we will make our raw measurement data available.

5.1 Arrayability

We have shown that, for a given loudspeaker element, some array geometries have associated frequency-dependent polar patterns whose point-by-point amplitude change over a specified coverage angle approaches 10 dB, while others are well within a 6-dB amplitude range over a similar angle.

To be certain, the human ear, trained or untrained, can detect a 6-dB change in amplitude over most of the audible range [6]. Accordingly, we propose the term "arrayability" as a reference to the fact that different loudspeaker arrays have measurably different polar patterns. Arrayability implies that some loudspeakers in certain array configurations behave better, that is, generate more even polar patterns, than others.

As criteria for arrayability, we suggest that if the measured one-third-octave polar plots of a given loudspeaker system in a given array configuration are even, within a 6-dB tolerance over a specified coverage angle, then the loudspeaker system in question may be said to be arrayable in that configuration. For example, from

A.

B.

C.

D.

E.

F.

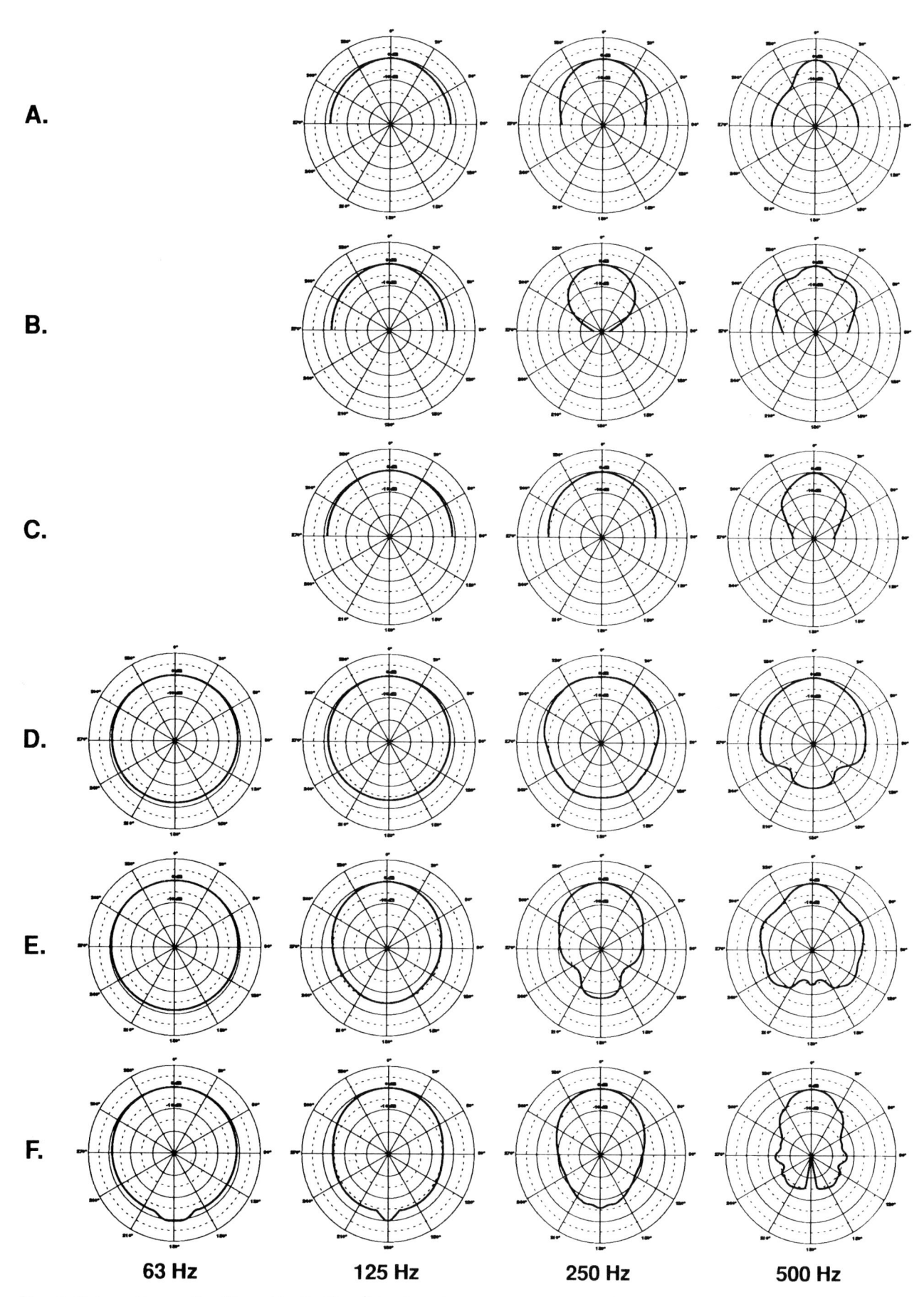

63 Hz   **125 Hz**   **250 Hz**   **500 Hz**

Fig. 5. Polar patterns after Olson (A, B, C) and loudspeaker array measurement data (D, E, F) on octave centers. (a) 63–500 Hz. (b) 1–8 kHz.

PAPERS LARGE ARRAY MEASUREMENT

A.

B.

C.

D.

E.

F.

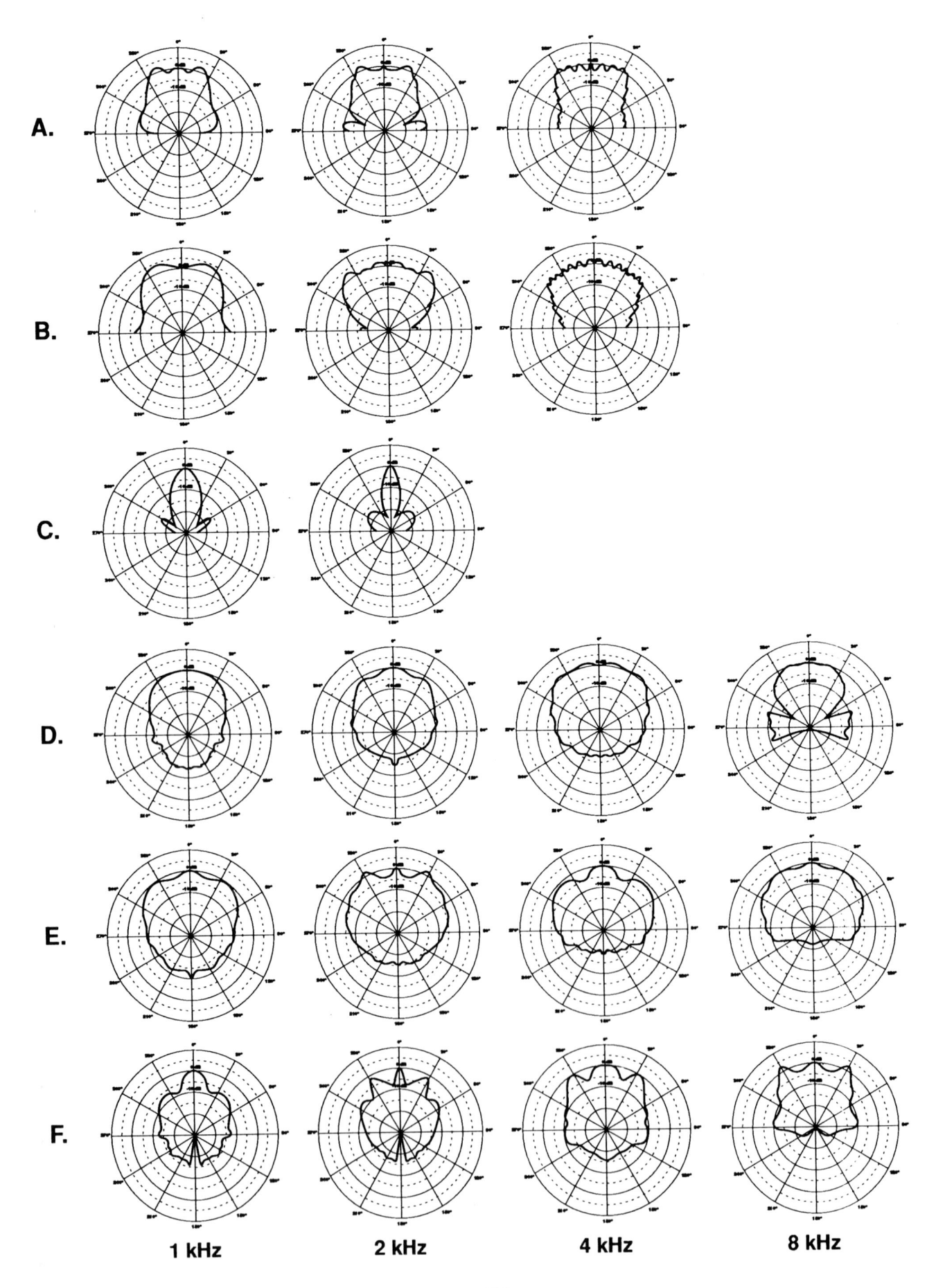

        1 kHz   **2 kHz**   **4 kHz**   **8 kHz**

Fig. 5(b).

the preceding data analysis we would say that the loudspeaker of Fig. 3 is arrayable in the wide-throw configuration of Fig. 4(a), but is nonarrayable in the parallel configuration of Fig. 4(b).

Distinctions between arrayable and nonarrayable loudspeaker systems are important, and will become increasingly useful as more data are gathered. As the necessary data are collected, arrayability, as described in the preceding examples, should be given as an additional loudspeaker specification. This is an inevitable step, given that professional loudspeakers are almost always used in arrays.

6 ACKNOWLEDGMENT

The authors would like to thank Kyle Takemori and Bob McCarthy for developing the Polar Cad software and routines to convert and plot the data at one-third-octave resolution in polar coordinates; Jean-Pierre Mamin for the mechanical design of the measurement apparatus; Ralph Jones, Chuck McDowell, and Tom Paddock for their assistance in the preparation of this manuscript; Betsy Cohen for technical editing; Tien Lee for his advice on this manuscript and for the calculations presented in Appendix 2; and Tracy Korby and Brandon Lortz for their assistance in the field.

7 REFERENCES

[1] H. F. Olson, *Acoustical Engineering* (Van Nostrand, New York, 1957).

[2] R. W. Ramirez, *The FFT Fundamentals and Concepts* (Prentice-Hall, Englewood Cliffs, NJ, 1985).

[3] R. B. Randall, *Frequency Analysis* (K. Larson and Søn; Brüel and Kjaer Instruments, 1987).

[4] J. Meyer, "Equalization Using Voice and Music as the Source," presented at the 76th Convention of the Audio Engineering Convention, *J. Audio Eng. Soc. (Abstracts)*, vol. 32, p. 1014 (1984 Dec.), preprint 2150.

[5] *Condenser Microphones and Microphone Preamplifiers: Theory and Application Handbook* (K. Larson and Søn; Brüel and Kjaer Instruments, 1976).

[6] J. R. Pierce, *The Science of Musical Sound* (Scientific American Books, New York, 1983), p. 131.

[7] L. Kinsler and A. Frey, "Fundamentals of Acoustics (Wiley, New York, 1950), pp. 155–164.

[8] T. A. Milligan, *Modern Antenna Design* (McGraw-Hill, New York, 1985).

APPENDIX 1

THEORETICAL MODEL OF A SPHERICAL RADIATOR

Using the derivation of spherical wave properties given in Kinsler and Frey [7], we begin with the general acoustic wave equation

$$\frac{\partial^2 p}{\partial t^2} = c^2 \nabla^2 p \ . \tag{1}$$

When expressed in spherical coordinates, the Laplacian operator is given by

$$\nabla^2 = \frac{\partial^2}{\partial r^2} + \frac{2}{r}\frac{\partial}{\partial r}$$
$$+ \frac{1}{r^2 \sin\theta}\frac{\partial}{\partial\theta}\left(\sin\theta\,\frac{\partial}{\partial\theta}\right) \tag{2}$$
$$+ \frac{1}{r^2 \sin^2\theta}\frac{\partial^2}{\partial\psi^2}$$

where

$x = r \sin\theta \cos\psi$
$y = r \sin\theta \sin\psi$
$z = r \cos\theta$.

If the waves have spherical symmetry, that is, if the acoustic pressure $p = p(r, t)$ is a function of radial distance and time but not a function of the angular coordinates θ and ψ, Eq. (2) can be substituted into Eq. (1) to give

$$\frac{\partial^2 p}{\partial t^2} = c^2 \left(\frac{\partial^2 p}{\partial r^2} + \frac{2}{r}\frac{\partial p}{\partial r}\right) \tag{3}$$

or

$$\frac{\partial^2 p}{\partial t^2} = c^2 \left(\frac{1}{r}\frac{\partial^2 (rp)}{\partial r^2}\right) \ . \tag{3b}$$

Since the spacial coordinate r is an independent variable which is not a function of time, the partial derivative $\partial^2 p/\partial t^2$ can be written as

$$\frac{\partial^2 p}{\partial t^2} = \frac{1}{r}\frac{\partial^2 (rp)}{\partial t^2} \ . \tag{4}$$

Substituting Eq. (4) into Eq. (3b) gives

$$\frac{\partial^2 (rp)}{\partial t^2} = c^2 \frac{\partial^2 (rp)}{\partial r^2} \ . \tag{5}$$

If the product rp in this equation is considered a single variable, Eq. (5) has the general solution

$$rp = f_1(ct - r) + f_2(ct + r) \tag{6a}$$

or

$$p = \frac{1}{r}f_1(ct - r) + \frac{1}{r}f_2(ct + r) \ . \tag{6b}$$

The first term of Eq. (6) represents a spherical wave diverging from the origin of coordinates with a velocity c while the second term represents a similar wave converging on the origin. Note that since $r = 0$ at the origin, the equation predicts infinite values for the acoustic pressure at the focal point of a converging wave. In actual practice these values remain finite but become so large that many of the assumptions made in deriving the wave equation are no longer valid.

By combining the three force equations describing the pressure gradient in each of the x, y, and z directions, we can show that for spherical waves, the radial pressure gradient $\partial p/\partial r$ is related to the radial acceleration by

$$-\frac{\partial p}{\partial r} = \rho_0 \frac{\partial^2 \xi}{\partial t^2} \qquad (7)$$

where ξ represents a radial particle displacement.

We can integrate Eq. (7) with respect to time to obtain

$$u = \frac{\partial \xi}{\partial t} = -\frac{1}{\rho_0} \int \frac{\partial p}{\partial r} \, dt \qquad (8)$$

where u represents a radial particle velocity. This is a general equation relating acoustic pressure and radial particle velocity.

Now consider the case of a pulsating sphere. Let a be the average radius of the sphere and assume the radial velocity u_s of any point on the surface to be given by

$$u_s = U_0 \cos \omega t \qquad (9)$$

where U_0 is the velocity amplitude of the surface. The complex form of Eq. (9) is

$$\boldsymbol{u}_s = U_0 \, e^{j\omega t} \, . \qquad (10)$$

At other than those large velocity amplitudes where the associated acoustic pressure amplitudes approach or exceed the mean equilibrium pressure in the fluid, the fluid medium surrounding the sphere must remain in contact with the surface of the sphere at all times, and consequently the partical velocity of an acoustic wave of radius a must be equal to the surface velocity of the sphere. This boundary condition can be expressed as

$$\frac{A}{az_a} e^{j(\omega t - ka)} = U_0 e^{j\omega t} \qquad (11)$$

where z_a is the specific acoustic impedance of a spherical wave evaluated at $r = a$. So

$$A = aU_0 z_a e^{jka} \, . \qquad (12)$$

If the radius of the pulsating sphere is small such that $ka \ll 1$ at all operating frequencies,

$$A = aU_0 \frac{\rho_0 ck a(ka + j)}{1 + k^2 a^2} (\cos ka + j \sin ka)$$

$$\approx a^2 U_0 \rho_0 ck \frac{(ka + j)(1 + jka)}{1 + k^2 a^2} \qquad (13)$$

or

$$A \approx j\rho_0 cka^2 U_0 \, . \qquad (14)$$

The equation for the acoustic pressure is therefore

$$\boldsymbol{p} = \frac{j\rho_0 cka^2 U_0}{r} e^{j(\omega t - kr)} \, . \qquad (15)$$

The real part of this expression,

$$p = \frac{-\rho_0 cka^2 U_0}{r} \sin (\omega t - kr) \qquad (16)$$

represents the actual pressure in the wave.

APPENDIX 2

PREDICTION OF NULLS IN THE PARALLEL-CONFIGURATION POLAR PATTERN

It is well known in electromagnetic antenna design theory that an array of elements (as shown in Fig. 6) has a radiation polar pattern which is the product of the polar pattern of each element and the array factor $A(\psi)$ given generally in the form [8]

$$A(\psi) = \frac{\sin (N\psi/2)}{N \sin (\psi/2)} \qquad (17)$$

where

N = number of elements in array in one dimension
ψ = $(2\pi d \sin \theta)/\lambda$, in radians
λ = wavelength
d = distance between elements
θ = angle between normal vector to array plane and vector from array center to observer.

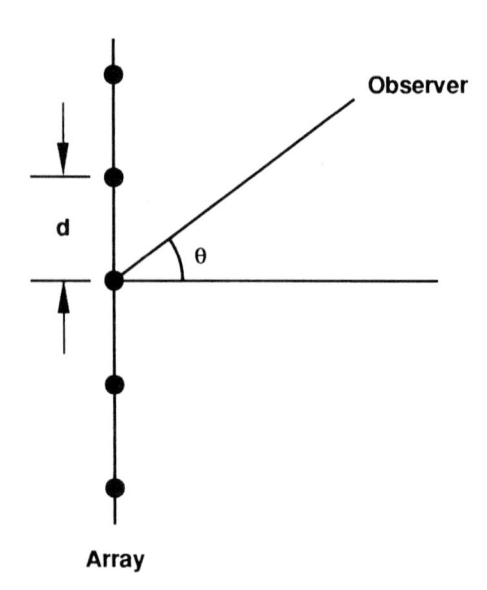

Fig. 6. Line array of N elements at spacing d.

To predict the polar pattern of two loudspeakers parallel to each other, we calculate nulls of the array factor $A(\psi)$ using $d = 14.5$ in $= 368.3$ mm (the width of the loudspeaker cabinet in our example), assuming the acoustic center of the loudspeaker is located along the centerline of the cabinet. Here $N = 2$ and the nulls of $A(\psi)$ exist at θ_n such that

$$\frac{N\pi d \sin \theta_n}{\lambda} = n\pi, \qquad n = 1, 3, 5 \ldots . \qquad (18)$$

For values at

250 Hz	no null				
500 Hz	$\theta_1 = 64°$				
1 kHz	$\theta_1 = 27°$				
2 kHz	$\theta_1 = 13°$		$\theta_3 = 42°$		
4 kHz	$\theta_1 = 6°$		$\theta_3 = 20°$		$\theta_5 = 34°$

The nulls for the polar pattern of the parallel configuration in our example are recognizable as calculated at 500 Hz, 1 kHz, and 2 kHz, but smear at 4 kHz and above.

THE AUTHORS

J. Meyer

F. Seidel

John Meyer was born in 1943 in Oakland, California. Following engineering studies at Heald Institute of Technology, he worked for Berkeley Custom Electronics before joining the staff of Harry McCune Sound Service, a San Francisco multimedia rental firm for which he designed several loudspeaker systems. Invited to Montreux, Switzerland, in 1973 to consult for the Institute for Advanced Musical Studies, Meyer embarked on an extensive program of research in sources of nonlinearity in audio transducers. Among the products of this work were a body of data on distortion sources in microphones and an ultra-low distortion horn driver design (U.S. Patent No. 4,152,552). Upon his return to the U.S. in 1975, he worked as a consultant in sound contracting and served as technical director for a direct-to-disk audiophile record company. In 1979, he founded Meyer Sound Laboratories, Inc., the professional loudspeaker manufacturing company of which he is president. Meyer is a member of the Audio Engineering Society.

●

Felicity Seidel graduated magna cum laude with a Bachelor's degree in physics from San Francisco State University in 1987. Two months after graduating she began work at Meyer Sound Laboratories in Berkeley, California. The majority of her work there has been devoted to the study of arrayed loudspeaker systems.

Comparative Performance of Three Types of Directional Devices Used as Concert-Sound Loudspeaker Array Elements*

PAUL F. FIDLIN AND DAVID E. CARLSON

Electro-Voice, Inc., Buchanan, MI 49107, USA

A number of approaches have been used in the design of concert-sound loudspeaker arrays. The basic requirements for large-scale sound reinforcement are introduced, and the acoustic power summation of multiple sources is investigated in terms of amplitude response, directivity, and time delay. Experimental data are presented for a variety of multidimensional horn arrays, demonstrating the effect of spacing and angular displacement between source elements as a function of the directional characteristics of the individual sources. This information is used to develop guidelines for the design of large-scale concert arrays.

0 INTRODUCTION

The touring concert sound industry has a unique set of problems to deal with:

1) The sound system is always on the move from venue to venue, each of which is different.
2) The system needs to be modular and relatively straightforward to assemble and take down.
3) Truck packing, hanging requirements, and a multitude of other constraints limit the physical format of a concert system.
4) Due to the high sound-pressure levels required and the large venues, multiple-loudspeaker systems (as many as 100 boxes) are needed.

To meet requirements 1, 2, and 3, identical boxes with identical radiation patterns are generally used. This is typically accomplished with one-box systems (all-in-one full-range boxes) or two-box systems (where the high-frequency box operates from a lower limit of about 80–160 Hz up to 20 kHz). Requirement 4 results in the use of many boxes pointed in the same direction with overlaping coverage.

A limited body of work has been directed toward the touring concert-sound industry with regard to the performance of multiple-loudspeaker systems. Strahm [1], Woolley [2], and Augspurger [3] focused on the mutual coupling effects with low-frequency drivers in multiple-cabinet arrays, but did not touch upon directivity control. Carlson and Gunness [4] discussed manifolding techniques for achieving high sound-pressure levels while minimizing lobing from multiple sources.

This paper is intended to be the first in a series of publications that will specifically address the problems involved in designing arrays for touring concert sound. This installment discusses and shows experimentally the directional and amplitude responses of both linear and curved arrays. The focus is on the comparison of three different array elements and how they combine as multiple sources in straight-line and curved arrays

---

* Presented at the 86th Convention of the Audio Engineering Society, Hamburg, West Germany, 1989 March 7–10, under the title "The Basic Concepts and Problems Associated with Large-Scale Concert-Sound Loudspeaker Arrays"; revised 1990 January 16.

with various spacings and splay angles. Three different types of devices are considered.

1) Constant-directivity horn-type devices. (The horn has multiple flare rates in both the horizontal and the vertical directions. It maintains a constant coverage angle independent of frequency.)

2) Radial-horn-type devices. (The horn has straight sides defining a radial arc and vertical dimensions defined by an exponential area expansion rate. The horizontal pattern maintains a relatively constant coverage pattern, as defined by the radial arc with a midband-narrowing deviation. The vertical pattern varies with frequency—the coverage pattern narrowing with increasing frequency.)

3) Collapsing-polar devices. (These devices are also known as beaming devices. Their coverage pattern narrows with increasing frequency in both the horizontal and the vertical directions. It may be a horn device or a direct-radiating cone loudspeaker.)

1 DISCUSSION

Wolff and Malter [5] laid the basic ground work for the directional behavior of loudspeaker arrays used today by adapting antenna radiation and optical diffraction theories to acoustics. Their theoretical analysis included point sources, line sources, and piston sources for straight-line arrays, curved arrays, and circular arrays. The theories were verified with experimental data using various cone loudspeakers. Wolff and Malter also presented experimental directional data for various horn and lens devices but did not attempt to explain the results in any detail theoretically.

While Wolff and Malter limited their work to the directional behavior of arrays with all of the elements being driven with identical amplitude and phase, Pritchard [6] expanded array theory by allowing the level, polarity, and relative phase to be manipulated to modify and control the directivity of an array. Pritchard's work was a theoretical work based on point sources, offering means of predicting and controlling beamwidths and steering the array.

Beranek [7] presented a brief summary of directivity theory for point sources, line sources, piston sources, and arrays, as well as introducing some experimental data. Olson [8] followed with a comprehensive collection of modern theoretical array analysis. Olson also presented a large body of experimental data for a variety of horns and cone loudspeakers.

More recently a number of individuals [9]–[12] have investigated the benefits of varying amplitude, phase, and time delay to the individual array elements to produce particular directivity. For example, electronic manipulation can allow the coverage pattern to be adjusted, superdirectivity to be achieved, and the directivity pattern to be steered.

To date, all of the studies that seek to describe and predict the directional performance of arrays theoretically assume ideal sources (point sources, line sources, and pistons) as the individual array elements. The theory is then confirmed with actual loudspeakers, where the loudspeakers are operated only in the frequency range in which they behave as ideal sources. In real-life concert situations, however, the loudspeakers in the array are operating well out of the range of ideal source approximations, that is, acting as point sources or maintaining true piston motion. With this paper we intend to lay the initial groundwork for full-beamwidth analysis of the directional performance of concert-type arrays.

We first compare the directivity and the amplitude response of three different devices: a constant-directivity horn, a radial horn, and a collapsing-polar (or beaming) horn. We compare them individually, then in dual arrays, both pointed straight ahead and splayed at two different angles—the first at one half of the horn nominal coverage angle (so that half of the patterns overlap) and the second equal to the horn nominal coverage angle (so that the edges of patterns are adjacent without overlap). Next we investigate quad arrays with all four horns in a straight line facing straight ahead and then in a curved array. The horns are spaced at distances typical for large concert arrays. Finally we repeat the dual and quad arrays with the horns in a line facing straight ahead with no space between them (adjacent horn flanges touching) to investigate the effects of array element spacing.

We concern ourselves, for the time being, with the specific case of equal amplitude and phase signals being sent to each driver in an array. It is interesting to speculate about generating specific directivity patterns, or steering the patterns from arrays by manipulating the amplitude, polarity, and phase of the signals fed to the individual elements. However, the consequence of using such techniques results in reduced efficiency of the overall array. Due to the relatively large amounts of acoustic output required within this industry, and the size and weight of the loudspeaker systems and amplifiers (and the resulting transportation costs), the benefits do not outweigh the tradeoffs at the present time.

Accurate polar information can be obtained only when the distance from the measurement microphone to the array is much greater than the frontal dimensions of the array. For most accurate data, the measurement distance should be a factor of 10 times the largest array dimension; however, reasonable results can be obtained with a factor of 4 or 5. In practice, large concert arrays often reach widths of 20 ft (6 m). This would result in a measurement distance requirement of about 100 ft (30 m), which is not only impractical, but impossible since the furthest distance we can get from an array in our anechoic chamber is 20 ft (6 m). We can get around this problem by frequency scaling. For example, if we had a scale model horn that was exactly one-third the size of a horn that we wanted to evaluate in an array situation, we could construct the array with the smaller horns positioned at one-third the spacing and measure at three times that frequency, and we would get the same results as for the full-scale sized array. One-third scaling would allow us to get effectively 60 ft (18 m) from an array in our chamber situation, which, as you will see, is sufficient for our experiments.

We were able to obtain three different horns, a constant-directivity type, a radial type, and a collapsing-polar type, which had the same nominal horizontal patterns, nearly the same mouth dimensions, and the same driver-mounting schemes (Table 1). Horn drivers were chosen to represent typical production tolerances with amplitude response held to ±1.5 dB between drivers. Angular coverage variation between production horns was held to within ±2°.

Since the horn parameters are nearly identical in the horizontal plane and they can accept identical drivers, convenient comparisons can be made between the three different types of horns, with the only variable being the type of directional radiation pattern: constant directivity, radial, or collapsing polar. There are substantial differences in the vertical radiation patterns between the three horns. However, this will not affect our analysis in the horizontal plane. Consequently, we limit our discussion primarily to horizontal directivity responses.

Our intent is to approximate an entire concert box with a single element through the use of frequency scaling. If we use a scaling factor of one-third, the effective size of the three different horn mouths becomes approximately 12.5 × 12.5 in (318 × 318 mm), with the usable frequency range being approximately 400 Hz to 7 kHz.

In typical concert situations the bulk of the musical information defining clarity, intelligibility, and articulation falls in the 400-Hz-to-7-kHz frequency region. The effective horn size is typical of what is used in the midfrequency range of many concert loudspeaker systems—a little bit smaller than what might typically be used in the midbass region and a little bit larger than what might be used for tweeter horns. Overall, however, it is reasonable as a single-element approximation of a typical concert box in the 400-Hz-to-7-kHz frequency range.

To determine spacing between array elements we must make an assumption about the loudspeaker enclosure. We arbitrarily chose a rectangular enclosure 36 in (914 mm) wide. In addition, we allowed 9 in (229 mm) on either side of the enclosure to accommodate horizontal splaying, resulting in an effective front/baffle dimension of 54 in (1.37 m) as shown in Fig. 1. Applying our one-third scaling factor, the effective measurement baffle becomes 18 in (457 mm). The majority of measurements were made with the horns mounted in the center of an 18-in (457-mm) square baffle. The baffles were then hinged together to accommodate straight, angled, or curved arrays.

All measurements were made with one-third-octave-band pink noise from 1.25 to 20 kHz centered at standard ISO frequencies. As a reference, Table 2 converts the actual measurement frequency to the effective scaled frequency.

2 EXPERIMENTAL SETUP

2.1 Directional Data

All directional information was gathered using a fully automated rotational system (Figs. 2–5) capable of gathering complete three-dimensional directional data.

Table 1.

Horn type	Constant directivity (CD)	Radial (RAD)	Collapsing polar (COL)
Nominal pattern ($W \times H$)	90 × 40°	90 × 60°	90 × 90°
Mouth dimension ($W \times H$)	4.20 × 4.20 in (107 × 107 mm)	4.15 × 4.35 in (105 × 110 mm)	4.05 × 4.05 in (103 × 103 mm)
Mounting flange dimension ($W \times H$)	5.25 × 5.25 in (133 × 133 mm)	6.00 × 5.00 in (152 × 127 mm)	4.70 × 4.70 in (119 × 119 mm)
Driver mount (screw on)	1.38 in (35 mm)	1.38 in (35 mm)	1.38 in (35 mm)
Frequency range	1.25–20 kHz	1.25–20 kHz	1.25–20 kHz

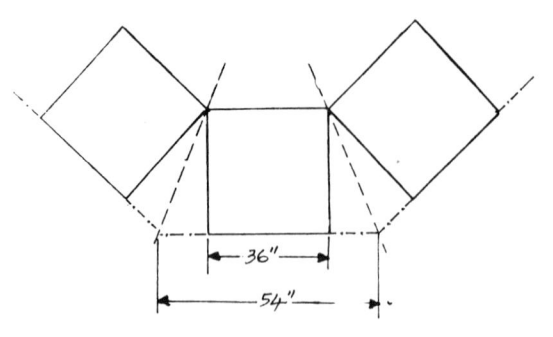

Fig. 1. Curved array of concert enclosures, showing effective baffle dimensions.

Table 2. Frequency scaling conversion table.

Actual measured frequency (Hz)	Effective scaled frequency (Hz)
1 250	417
1 600	533
2 000	667
2 500	833
3 150	1 050
4 000	1 333
5 000	1 667
6,300	2 100
8 000	2 667
10 000	3 333
12 500	4 167
16 000	5 333
20 000	6 667

The system is supported by a 80286-based personal computer that controls platform and crossbar angular positioning, a digital filter set, and a disk mass storage system. It interfaces with the operator through menu-driven software.

The array under test was attached to the positioning platform in the anechoic chamber and fed an amplified signal from a pink noise generator. Equal voltage was fed into each array, with all drivers connected in parallel.

The acoustic radiation from the array under test was gathered by a stationary B&K 4135 microphone placed 20 ft (6 m) away in the opposite corner. A calibrated preamplifier was used to amplify the signal and send it to the "front end" of the data collection/positioning control system. A digital signal-processing chip synthesizes the required one-third-octave-band digital filters for each band of data stored. The stored data are later played back to an $X-Y$ plotter for hard copy.

The following outputs are available from the system: 1) Q and directivity index; 2) 6 dB beamwidth, 3) polar directivity plots—arbitrary, vertical or horizontal; and 4) three-dimensional visualization of the directivity pattern. The system provides two output forms for the

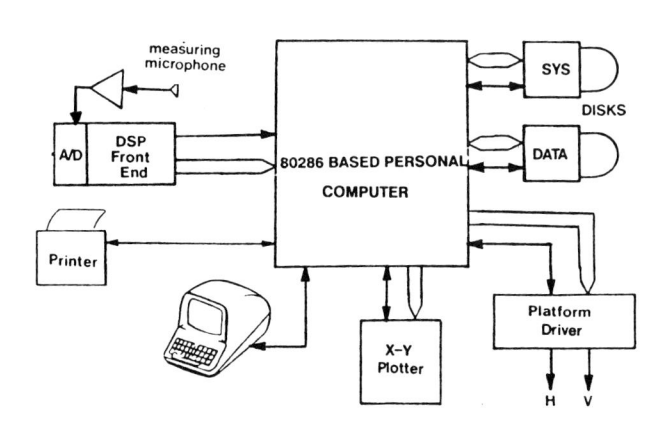

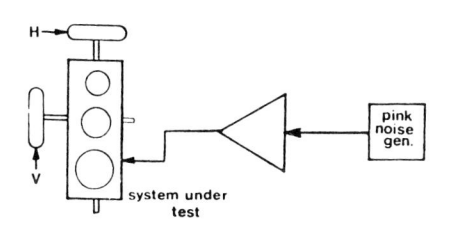

Fig. 2. Automated directional information-gathering system.

Fig. 3. Control terminal and visual output from system.

Fig. 4. Quad array mounted on rotational platform.

Fig. 5. Dual array mounted on rotational platform.

polar data:

1) *Raw.* In this form the data are plotted, in 2° increments, directly onto paper (Fig. 6).

2) *Smooth.* A convolution window is applied to eliminate noise jitter, and in addition the polars are normalized by increasing levels until some portion of the curve hits the 0-dB line (Fig. 7).

2.2 Amplitude Response

To obtain a complete and meaningful picture of array behavior it is necessary, in addition to considering directional information, to look at the amplitude response and its variation off axis. Amplitude response data were gathered with the same microphone used to collect polar data (that is, it was always left in a fixed position). The same pink noise signal was fed to the loudspeaker array. However, this time the measuring system was an Ivie IE-30A one-third-octave real-time analyzer with storage capabilities. The stored response was plotted out using an $X-Y$ plotter. In addition a sine wave response curve was obtained by using a B&K 2010 sine wave generator/measuring amplifier. This too was plotted with an $X-Y$ plotter.

A high degree of correspondence was found to exist between sine and real-time analyzer curves. Typical sine wave responses with the corresponding one-third-octave responses are shown in Fig. 8. Only one-third-octave data are presented here.

Amplitude responses were gathered in 22.5° increments up to and beyond the nominal coverage angle of a given array. For each array a second set of amplitude responses was generated by equalizing flat the 0° amplitude response and adjusting accordingly each additional off-axis amplitude response. (This would correspond to the array being equalized flat on axis.)

Data were generated on the following combinations of devices and arrays: horizontal and vertical polars, beamwidth, amplitude, and adjusted amplitude in each case. Horizontal results for arrays 1–20 are shown in Appendix 1 (Figs. 22–41) and vertical results for Arrays 1, 2, 7, 9, 13, 14, and 17–20 in Appendix 2 (Figs. 42–51). Fig. 9 is a pictorial representation of the arrays.

A. *Singles (horizontal, Figs. 22–24; vertical, Figs. 42, 43)*
  Array 1: CD horn (CD)
  Array 2: Radial horn (RAD)
  Array 3: Collapsing horn (COL)

B. *Dual arrays, 18-in (457-mm) baffles at 0° (horizontal, Figs. 25–27)*
  Array 4: Two CD horns
  Array 5: Two radial horns
  Array 6: Two collapsing horns

C. *Dual arrays, 18-in (457-mm) baffles at 45° (horizontal, Figs. 28–30; vertical, Figs. 44, 45)*
  Array 7: Two CD horns
  Array 8 : Two radial horns
  Array 9: Two collapsing horns

D. *Dual arrays, 18-in (457-mm) at 90° (horizontal, Figs. 31–33)*
  Array 10: Two CD horns
  Array 11: Two radial horns
  Array 12: Two collapsing horns

E. *Quad arrays, 18-in (457-mm) at 0° (horizontal, Figs. 34, 35; vertical, Figs. 46, 47)*
  Array 13: Four CD horns
  Array 14: Four collapsing horns

F. *Quad arrays, 18-in (457-mm) baffles at 30° (horizontal, Figs. 36, 37)*
  Array 15: Four CD horns
  Array 16: Four collapsing horns

G. *Dual arrays, no baffles at 0° (horizontal, Figs. 38, 39; vertical, Figs. 48, 49)*
  Array 17: Two CD horns
  Array 18: Two collapsing horns

H. *Quad arrays, no baffles at 0° (horizontal, Figs. 40, 41; vertical, Figs. 50, 51)*
  Array 19: Four CD horns
  Array 20: Four collapsing horns

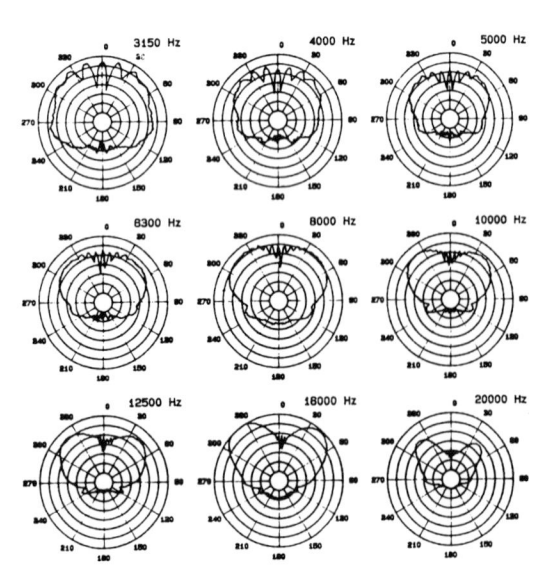

Fig. 6. Raw polar plot.

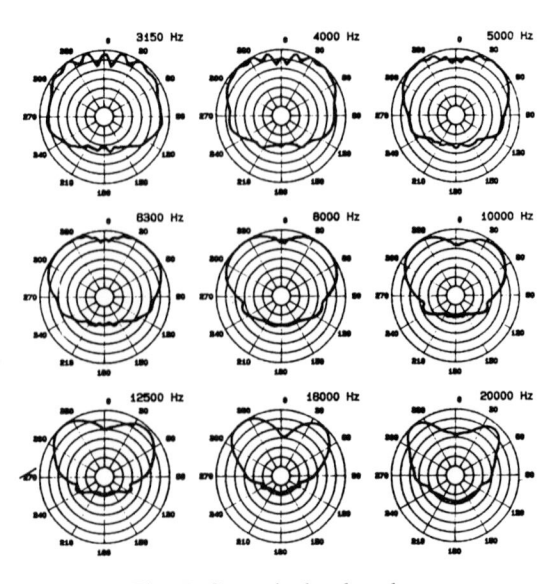

Fig. 7. Smoothed polar plot.

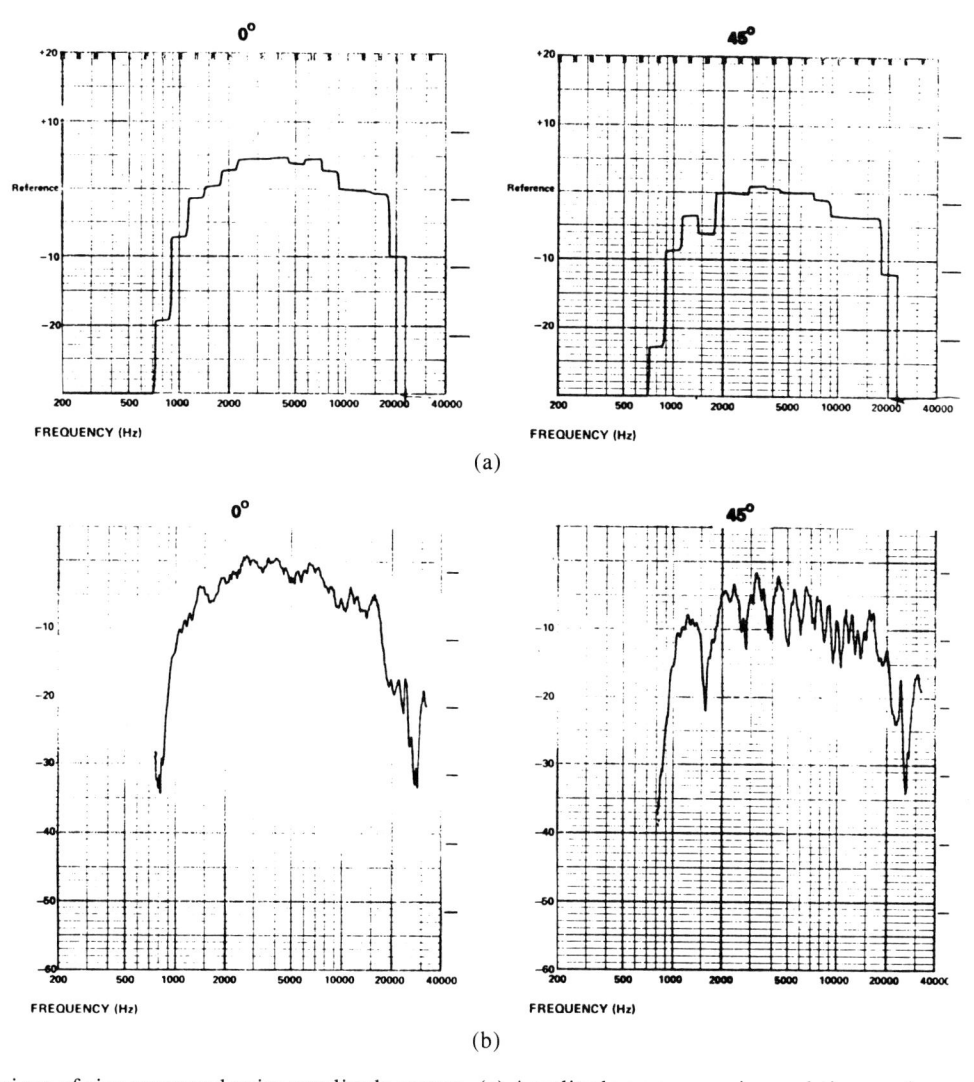

Fig. 8. Comparison of sine wave and noise amplitude curves. (a) Amplitude response using real-time analyzer. (b) Sine wave amplitude response.

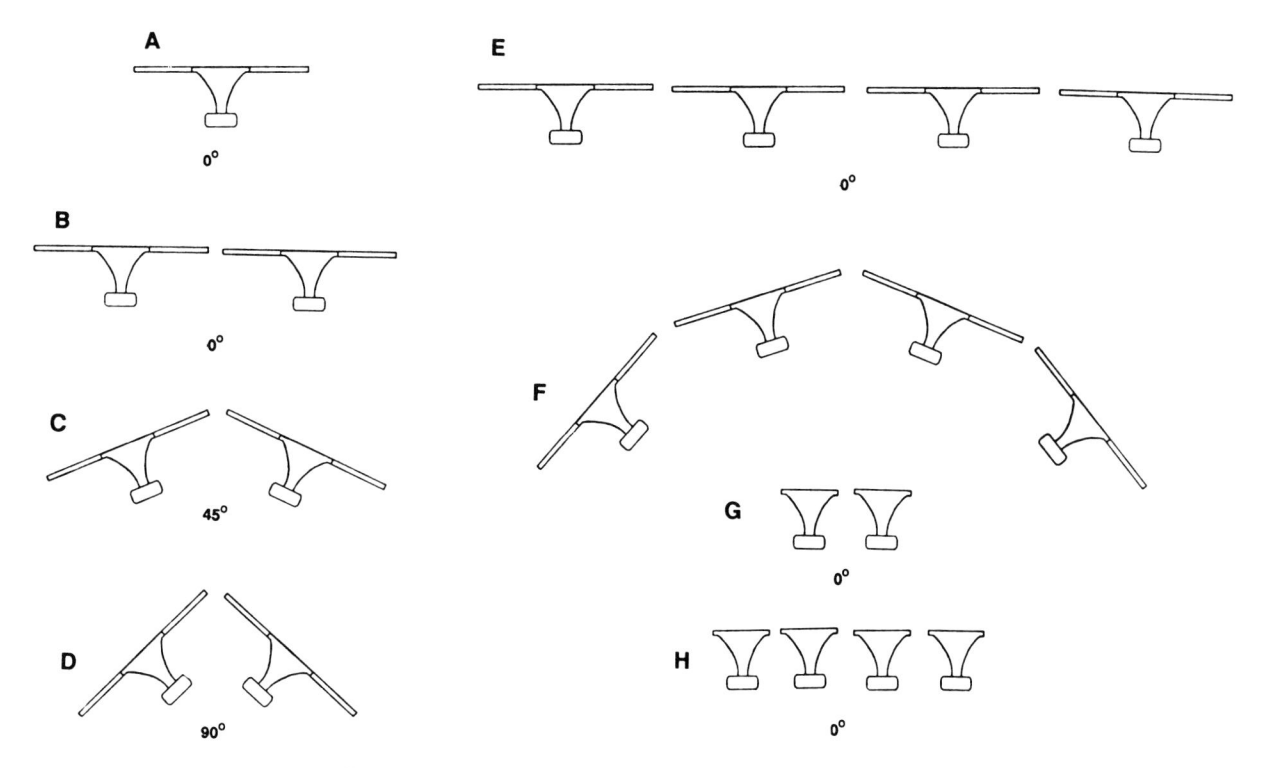

Fig. 9. Pictorial representation of the arrays measured.

3 EXPERIMENTAL RESULTS

The following observations were made from the volumes of data gathered.

3.1 Directivity Related

1) The single-element coverage patterns vary drastically (Fig. 10).

2) Horizontal coverage angles are the same for single, dual, and quad 0° arrays of the same horn type (Fig. 11).

3) At higher frequencies the nominal horizontal coverage angle for dual and quad closed-spaced arrays remains the same as for a single horn (Fig. 12). However, at lower frequencies the pattern is cut in half for duals, exhibiting large main central lobes (Fig. 13). These would be very problematic in a concert situation. It is the classic line-radiator effect, or beamwidth narrowing, which occurs when loudspeakers or horns are stacked or lined up in a row.

4) As the spacing between sources increases, the number of lobes in the polar response increases, with the lobes becoming narrower and the gaps between lobes decreasing. Wolff and Malter [5] predicted this for point sources. Fig. 14(a) compares the polar responses of the dual CD horns facing straight ahead for

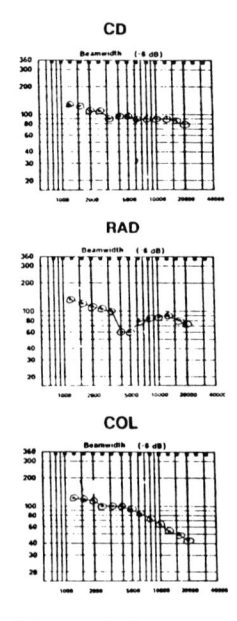

Fig. 10. Horizontal beamwidth. Comparison of single elements.

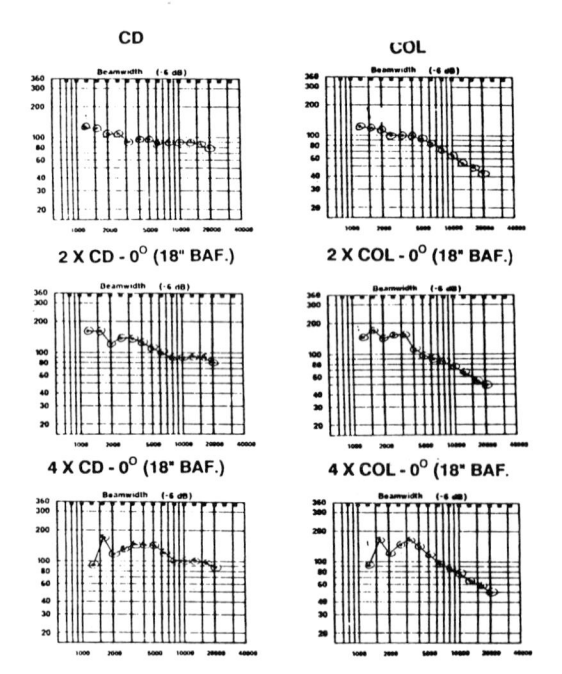

Fig. 11. Horizontal beamwidth. Comparison of single, dual, and quad systems, 18-in (457-mm) baffles.

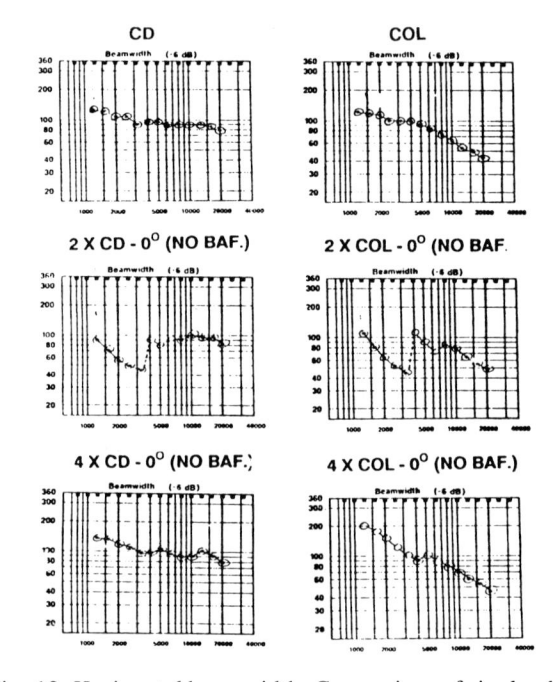

Fig. 12. Horizontal beamwidth. Comparison of single, dual, and quad systems; close spaced.

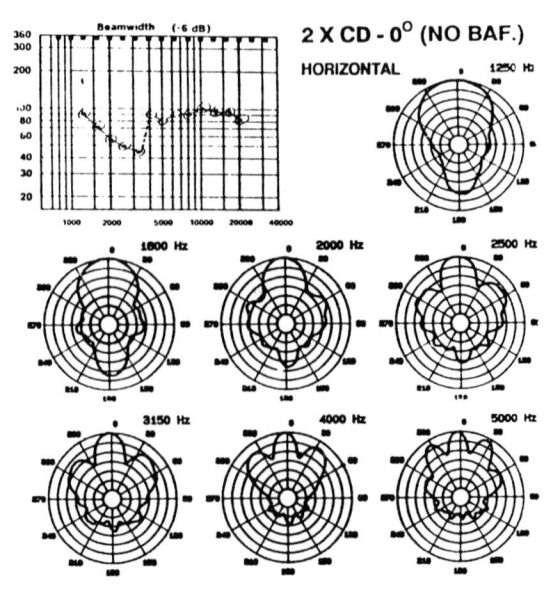

Fig. 13. Line radiator effect.

the 18-in (457-mm)-baffle and the no-baffle (5.25-in (133-mm) spacing) conditions. Fig. 14(b) shows the theoretical polar responses for two point sources at the same spacings as the CD horn measurements of (a). (See Appendix 3 for calculations.) Note that there is remarkable agreement between the point-source theory and the actual horn measurements in the front hemisphere where the two individual horn patterns overlap (±45°). By separating the sources an appropriate distance, the problem of beamwidth narrowing discussed can be minimized.

5) Vertical coverage angles are unaffected by straight-line horizontal arrays, and the coverage angle remains as for a single horn (Fig. 15).

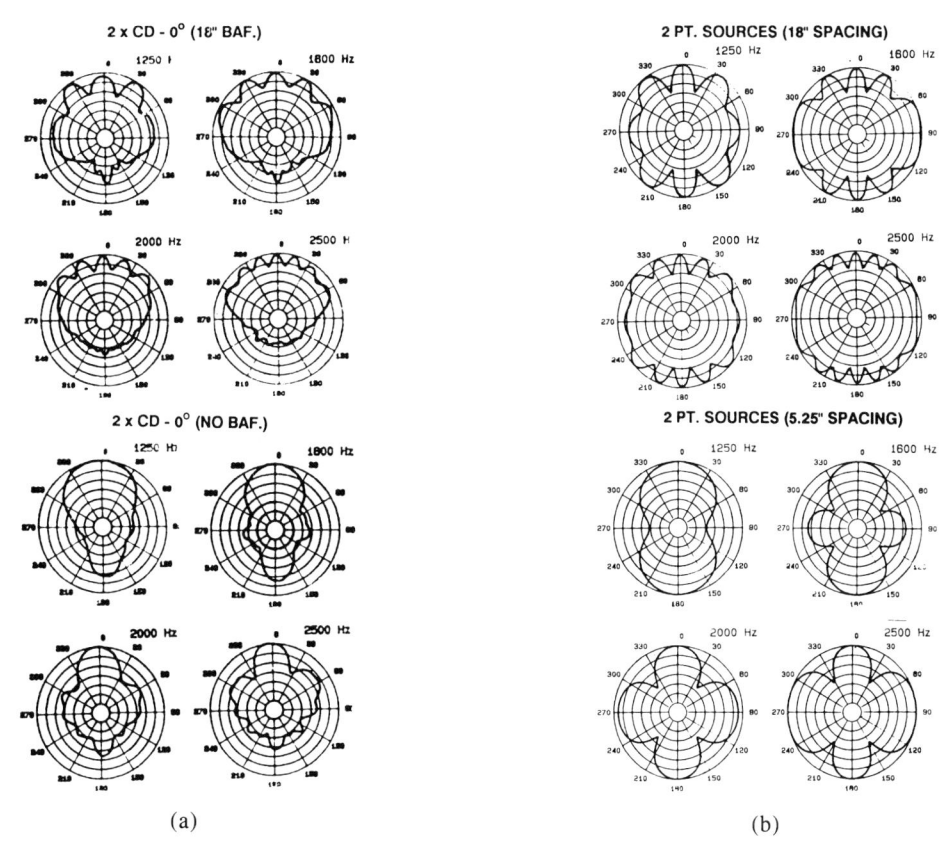

Fig. 14. Comparison of spacings with dual arrays. (a) Measured data. (b) Theoretical data.

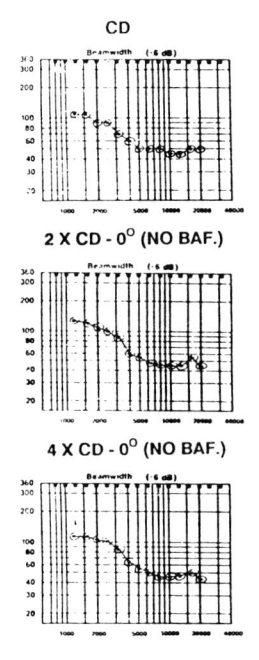

Fig. 15. Vertical beamwidth. Comparison of single, dual, and quad systems, close spaced.

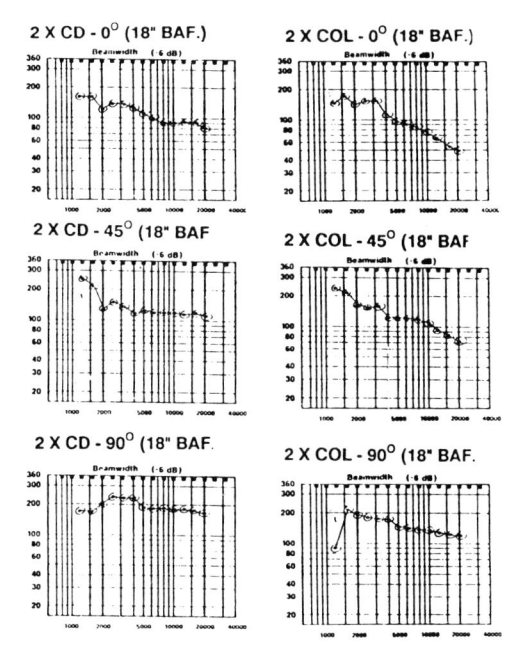

Fig. 16. Increased horizontal coverage pattern with increased angle between elements. 0, 45, and 90°.

6) Curved arrays result in an increase in the effective horizontal coverage pattern due to the addition of the splay angle (Fig. 16). However, larger splay angles result in drastic gaps in the polar plots at the junction between two elements in the case of the COL and RAD horns (Fig. 17).

7) Low-frequency polars (up to 2500 Hz) are very similar for all three horn styles. In each case polars show lobes at approximately the same angles for corresponding arrays. This is because the width of the horn mouth is the controlling factor up to 2500 Hz, and all three devices have nearly identical dimensions (see Appendix 1).

8) When the number of sources increases with the spacing between them remaining constant, the number of major lobes and their polar positions remain constant. This result can be predicted from the theory developed by Wolff and Malter [5] for point sources. Fig. 18(a) compares the polar responses of the dual CD horn array and the quad array with the horns facing straight ahead on the 18-in (457-mm) baffles. Fig. 18(b) shows the theoretical polar responses for two- and four-point sources at the same spacings as the CD horn measurements. (See Appendix 3 for calculations.) Note that there is remarkable agreement between the point-source theory and the actual horn measurements in the front hemisphere, where the individual horn patterns overlap ($\pm 45°$).

3.2 Amplitude Related

When a second source is added in parallel to a single source, the impedance halves, and the power taken on by the array doubles (+3 dB). In addition the two elements can be expected to provide pressure fields that reinforce constructively at the microphone, resulting in an additional 3 dB. We would therefore expect to see a 6-dB increase in level for every doubling of the number of elements.

1) Looking at the close-spaced 0° arrays in Fig. 19, we can see that the amplitude increases +5 dB with each doubling. This is very close to the theoretical maximum level increase discussed. Note that the CD devices fare significantly better at frequencies above 5 kHz than do the COL devices.

2) However, comparing the 0° 18-in (457-mm) baffle arrays in Fig. 20, single to dual, we notice that any increase in amplitude is frequency dependent due to imperfect geometry.

3) Comparing dual to quad 18-in (457-mm) spaced arrays, we see no increase in amplitude. The path-length difference between the central pair and the outer pair in the quad array is significant enough that we are getting phase cancellations and not achieving pure pressure summation (Fig. 20).

4) In all array comparisons the CD horn arrays show flatter and more consistent off-axis amplitude curves than either the RAD or the COL horn arrays (Fig. 21, and Appendix 1).

4 CONCLUSIONS

Translating our experimental results into the full-scale-sized concert-sound domain requires a 3:1 down scaling in frequency. Consequently our conclusions refer to the 400-Hz-to-7-kHz region where the vast majority of musical information is contained. (Varying

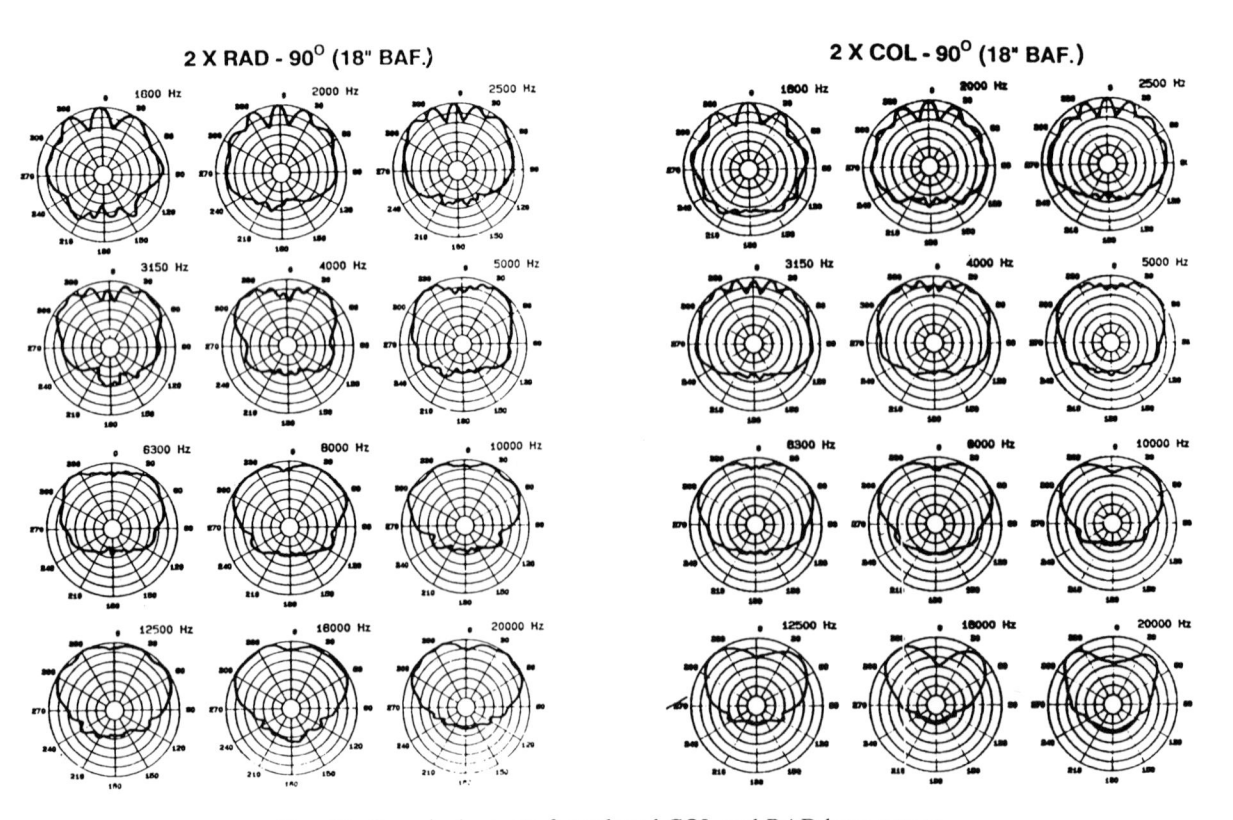

Fig. 17. On-axis dropouts for splayed COL and RAD horn arrays.

the spacing between elements, with appropriate changes in the scaling factor, would allow us to look in more detail at higher and lower frequency ranges.) Our most important findings follow.

1) Lobing for concert-spaced arrays will vary drastically with the geometry of the array and the directional characteristics of the individual array elements.

2) At low frequencies (where the wavelength is greater than the dimensions of the array element) the directional responses of constant-directivity, radial, and collapsing-polar devices will be the same both individually and in array configurations if their radiating areas are the same.

3) Constant-directivity devices have superior directional performance over radial or collapsing polar devices, maintaining more even energy distribution spatially throughout the nominal coverage pattern.

4) Arrays constructed with constant-directivity horns

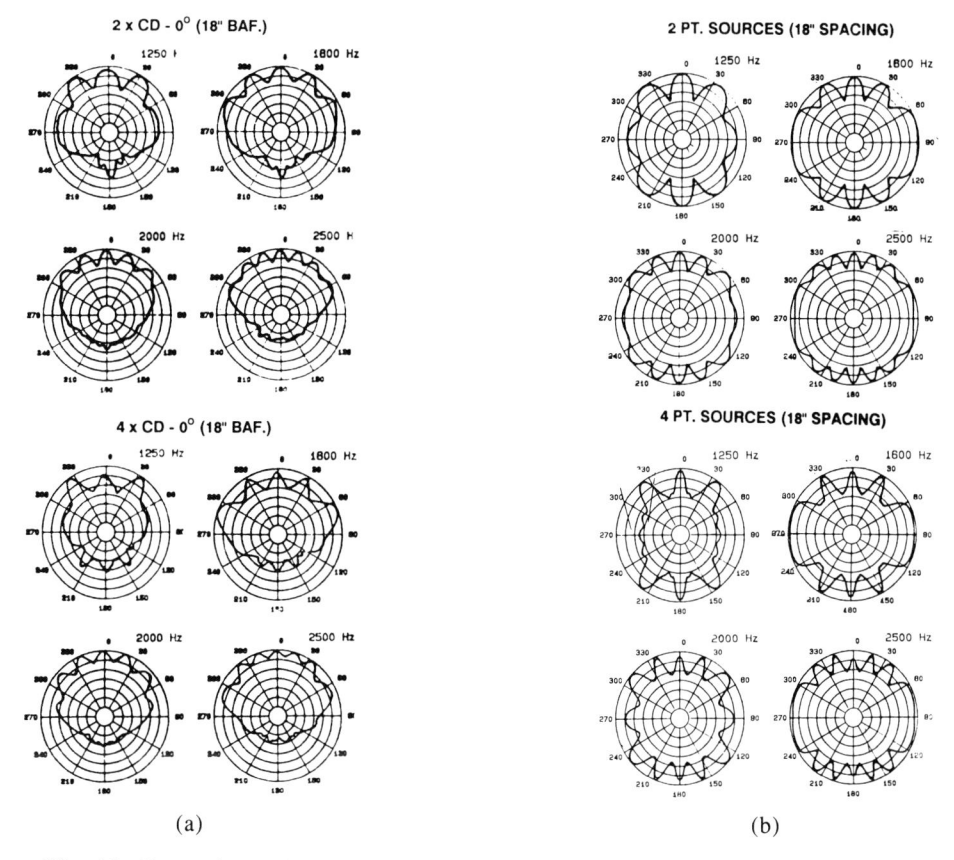

Fig. 18. Comparison of dual and quad arrays. (a) Measured data. (b) Theoretical data.

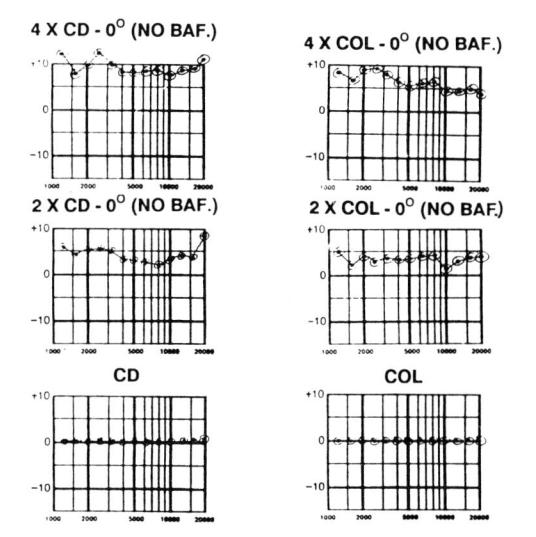

Fig. 19. Increase in amplitude with increasing number of elements in an array; close spaced.

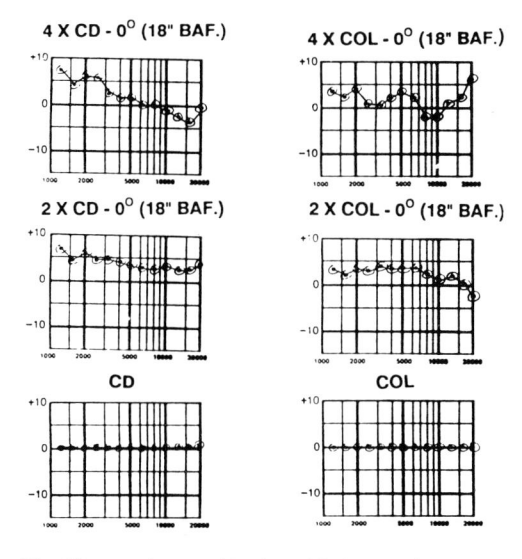

Fig. 20. Change in amplitude with increasing number of elements in an array; 18-in (457-mm) baffles.

consistently provided superior performance, with both smoother amplitude response and more even angular coverage.

5) There is evidence that the frontal lobing of arrays can be approximately calculated using point-source theory in spatial regions where the coverage patterns of the individual array elements overlap.

5 SUMMARY

The experimental data shown and the conclusions drawn relate specifically to one frequency band (400 Hz to 7 kHz) of a multiway system. We have seen that constant-directivity devices should always be used for constructing arrays because they provide superior results to other directivity-type devices.

6 ACKNOWLEDGMENT

Special thanks are due to David Gunness for establishing the data collection operating system and formatting the output data. Many thanks are also due to Mark Blanchard for running all the data and laboriously mounting each array, and finally to Kathy Reed for her secretarial skills.

7 REFERENCES

[1] C. N. Strahm, "Complete Analysis of Single and Multiple Loudspeaker Enclosures," presented at the 81st Convention of the Audio Engineering Society *J. Audio Eng. Soc. (Abstracts)*, vol. 34, p. 1032 (1986 Dec.), preprint 2419.

[2] S. J. Woolley, "Miniature Direct-Radiator Subwoofer Modules," presented at the 6th Int. Conf. of Sound Reinforcement of the Audio Engineering Society (1988 May).

[3] G. L. Augspurger, "Performance of Woofer Arrays," presented at the 85th Convention of the Audio Engineering Society, *J. Audio Eng. Soc. (Abstracts)*, vol. 36, p. 1030 (1988 Dec.).

[4] D. E. Carlson and D. W. Gunness, "Loudspeaker Manifolds for High-Level Concert Sound Reinforcement," presented at the 81st Convention of the Audio Engineering Society, *J. Audio Eng. Soc. (Abstracts)*, vol. 34, p. 1034 (1986 Dec.), preprint 2387.

[5] I. Wolff and L. Malter, "Directional Radiation of Sound," *J. Acoust. Soc. Am.*, vol. 2, pp. 201–241 (1930 Oct.).

[6] R. L. Pritchard, "Optimum Directivity Patterns for Point Arrays," *J. Acoust. Soc. Am.*, vol. 25, pp. 879–891 (1950 Sept.).

[7] L. L. Beranek, *Acoustics* (McGraw-Hill, New York, 1954).

[8] H. F. Olson, *Elements of Acoustical Engineering* (Van Nostrand, New York, 1957).

[9] W. J. W. Kitzen, "Multiple Loudspeaker Arrays Using Bessel Coefficients," Philips (reference incomplete.

[10] D. G. Meyer, "Digital Control of Loudspeaker Array Directivity," *J. Audio Eng. Soc.*, vol. 32, pp. 747–754 (1984 Oct.).

[11] D. G. Meyer, "Development of a Model for Loudspeaker Dispersion Simulation," presented at the 72nd Convention of the Audio Engineering Society *J. Audio Eng. Soc. (Abstracts)*, vol. 30, p. 946 (1982 Dec.), preprint 1912.

[12] V. M. Moser, "Amplitude and Phase-Controlled Acoustic Linear Arrays with Uniform Horizontal Directivity," *Acustica*, vol. 60, pp. 91–104 (1986 Apr.).

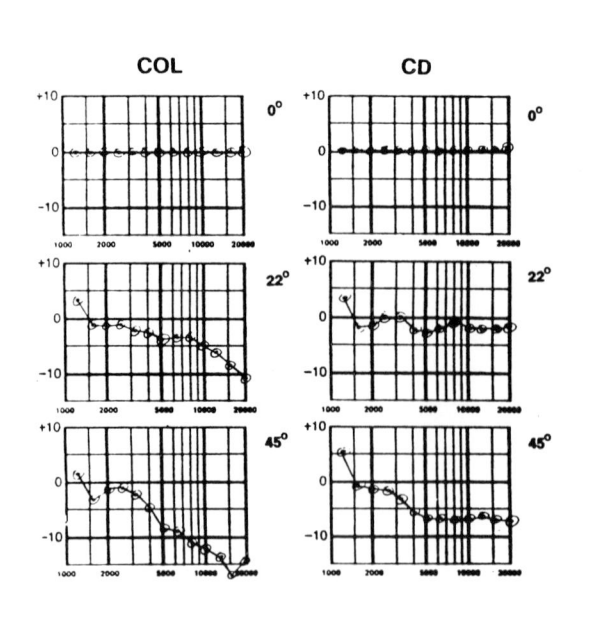

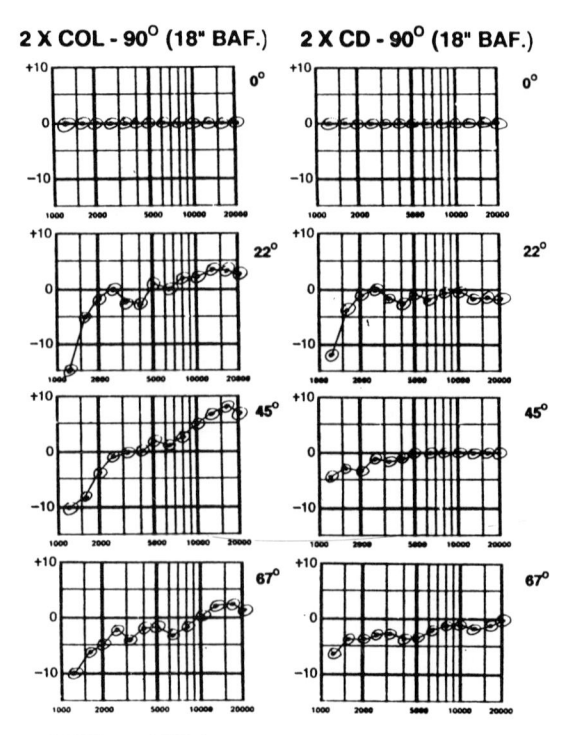

Fig. 21. Off-axis amplitude comparison of COL and CD horn arrays.

APPENDIX 1

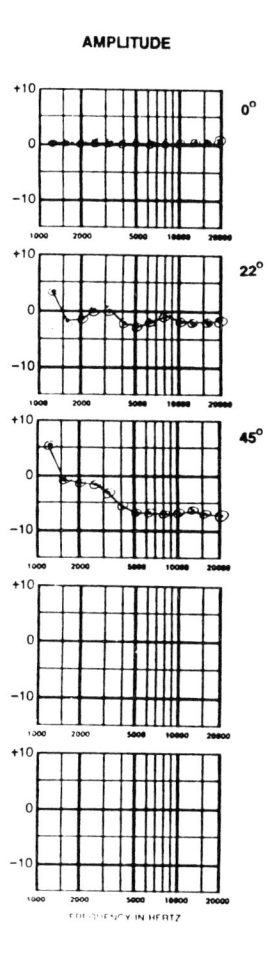

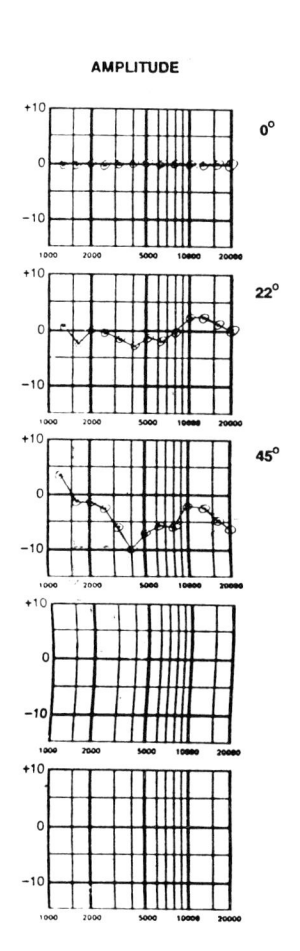

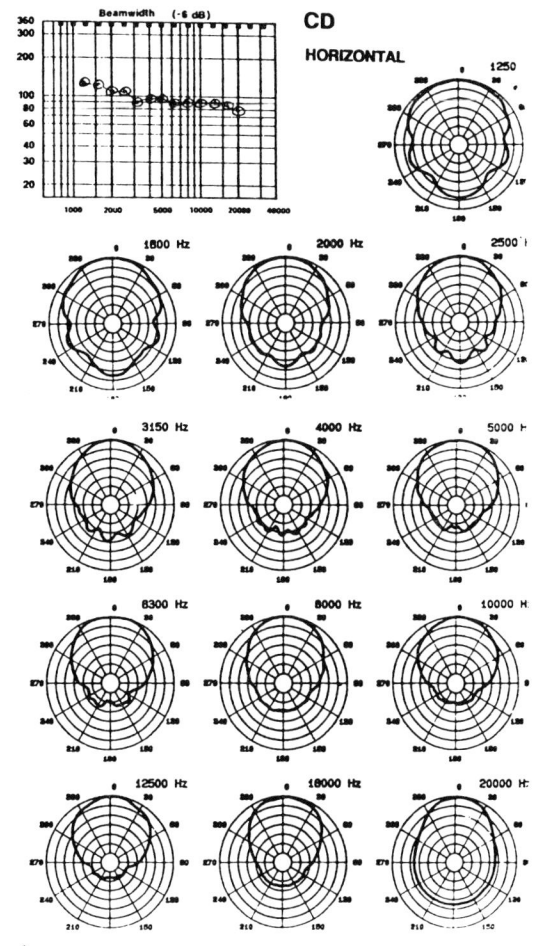

Fig. 22. Array 1.

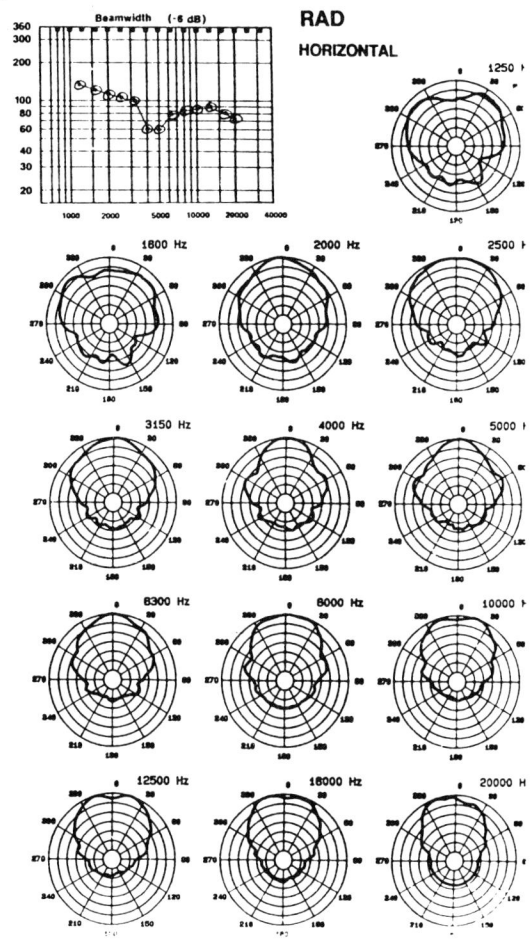

Fig. 23. Array 2.

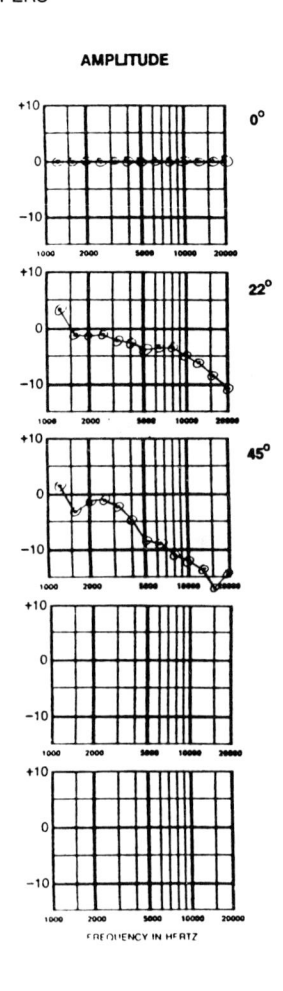

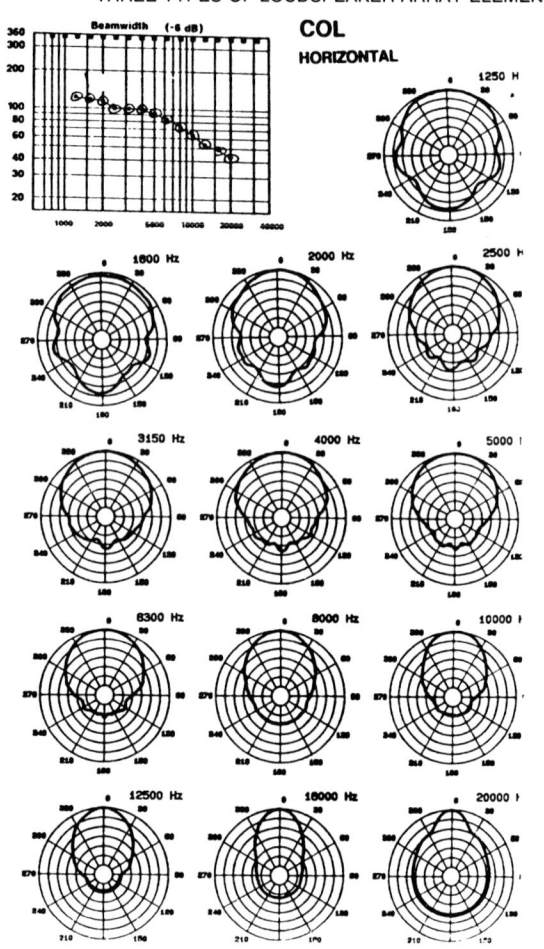

Fig. 24. Array 3.

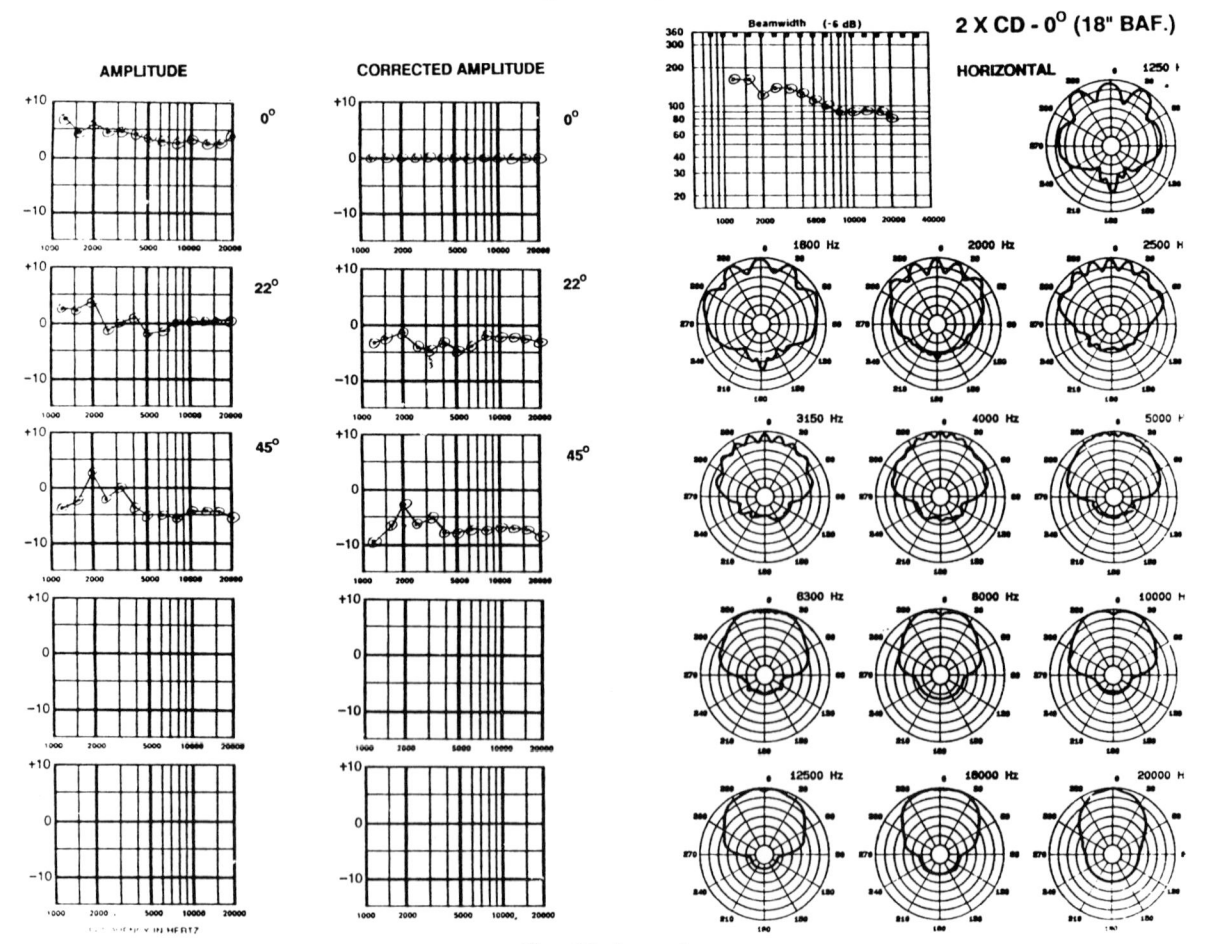

Fig. 25. Array 4.

Fig. 26. Array 5.

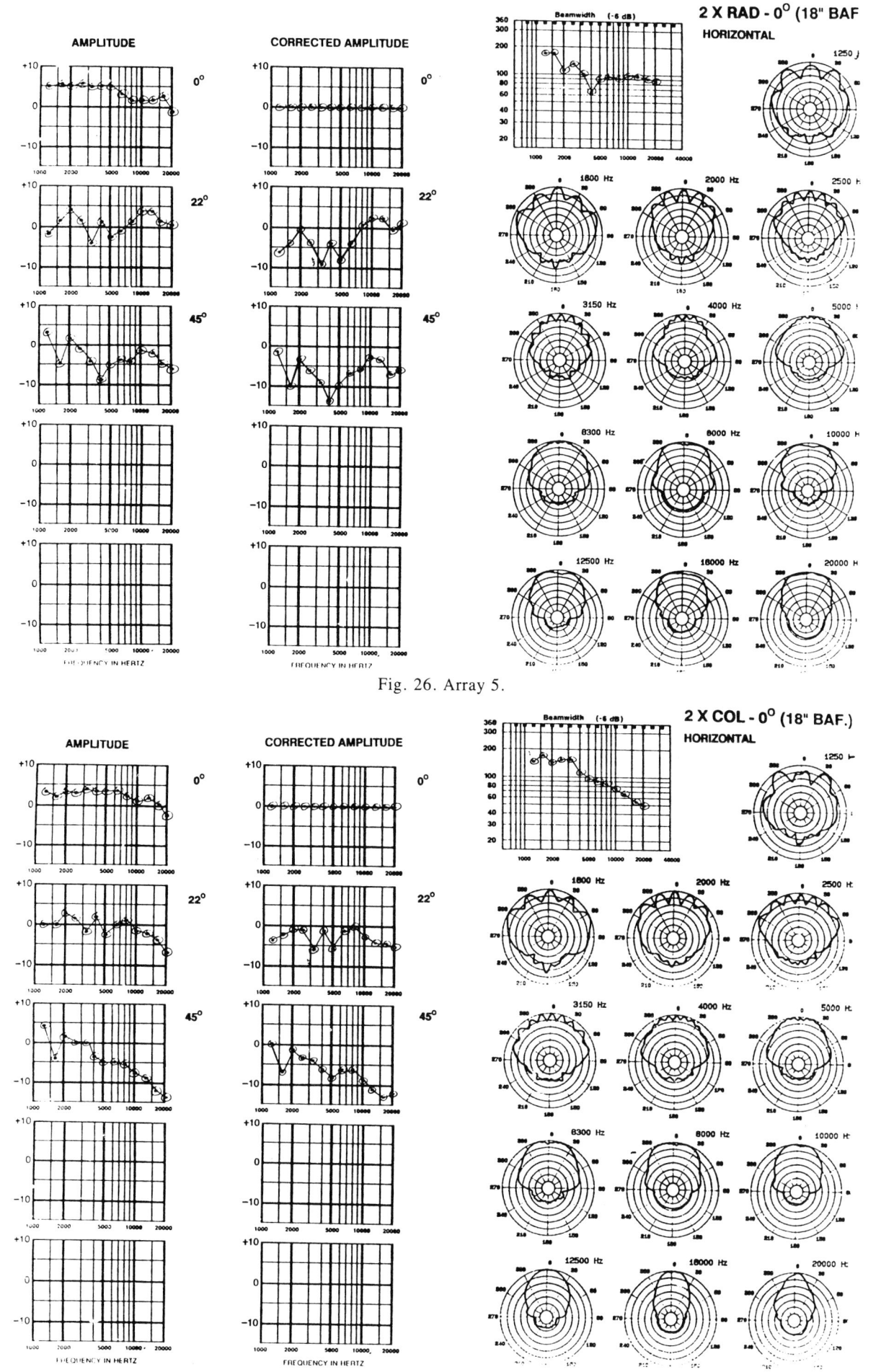

Fig. 27. Array 6.

Fig. 28. Array 7.

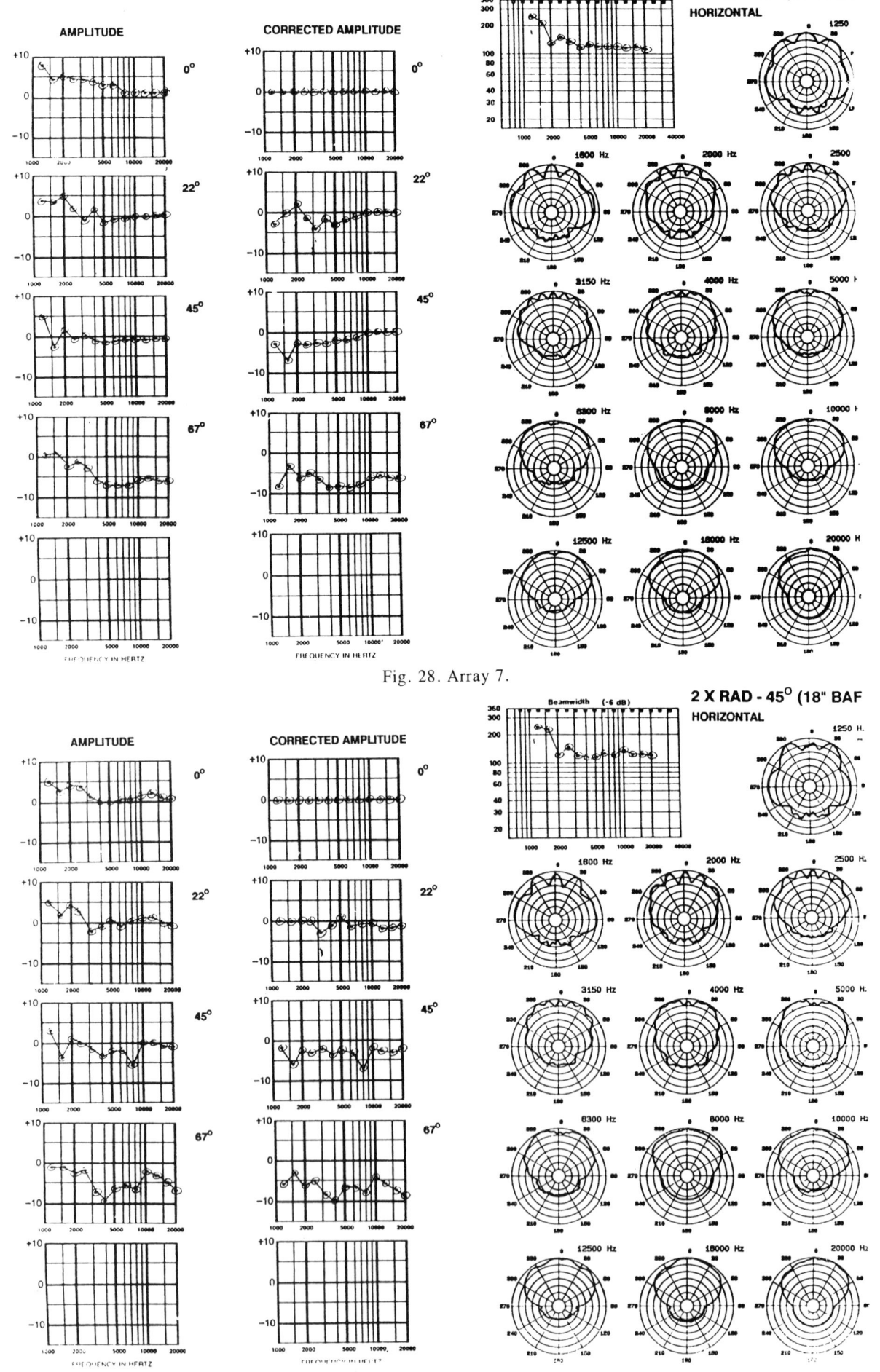

Fig. 29. Array 8.

Fig. 30. Array 9.

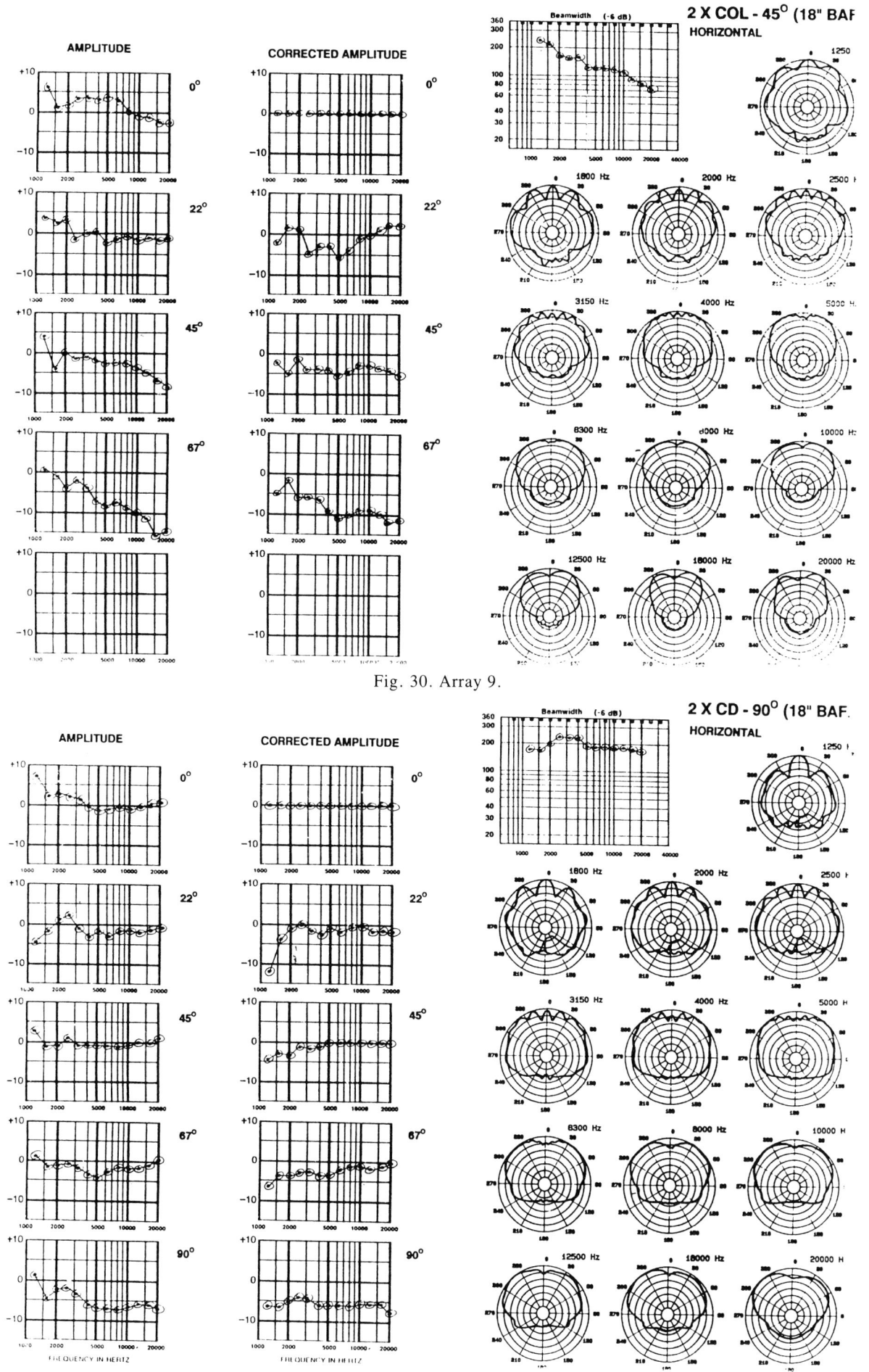

Fig. 31. Array 10.

Fig. 32. Array 11.

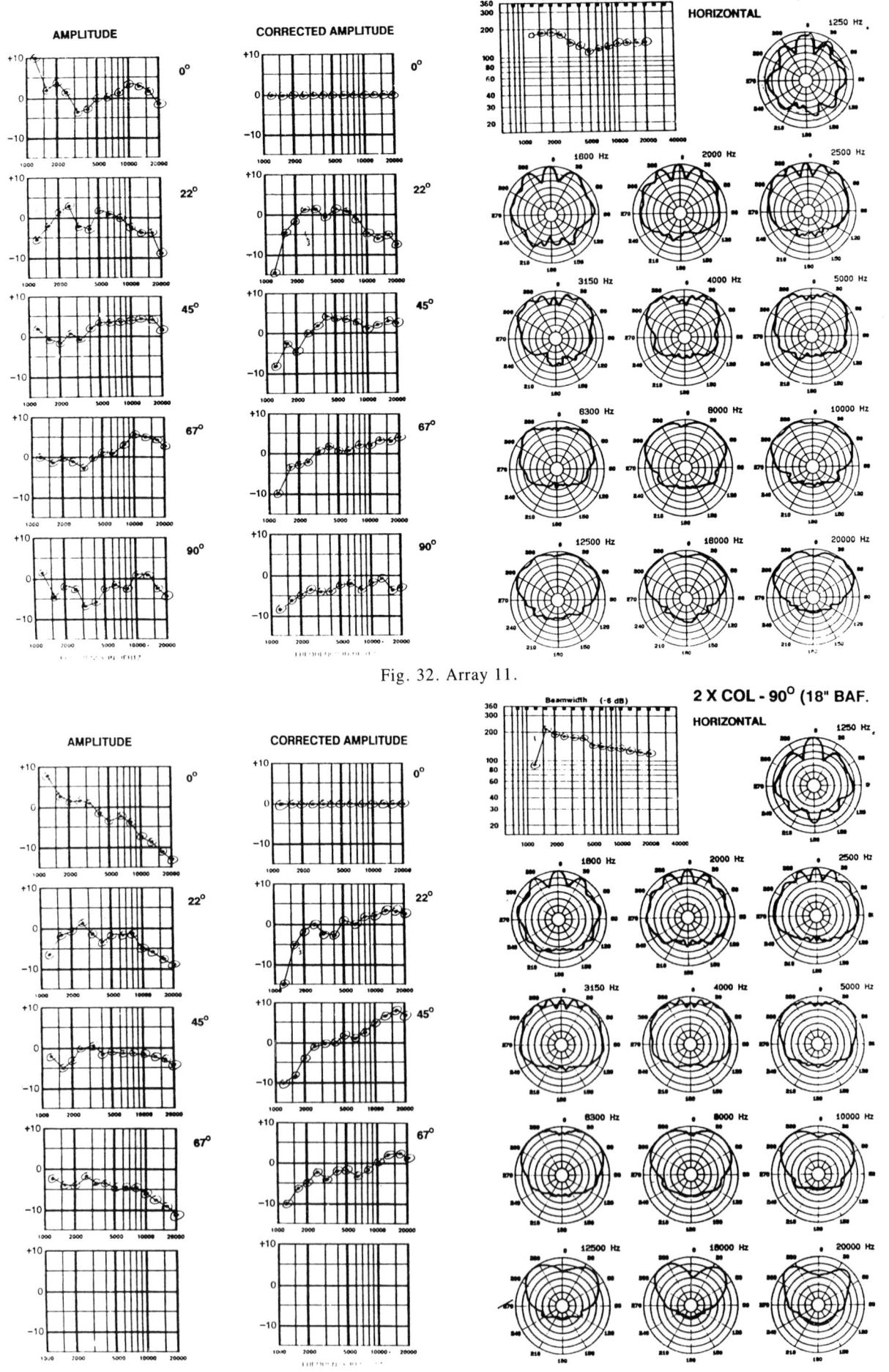

Fig. 33. Array 12.

Fig. 34. Array 13.

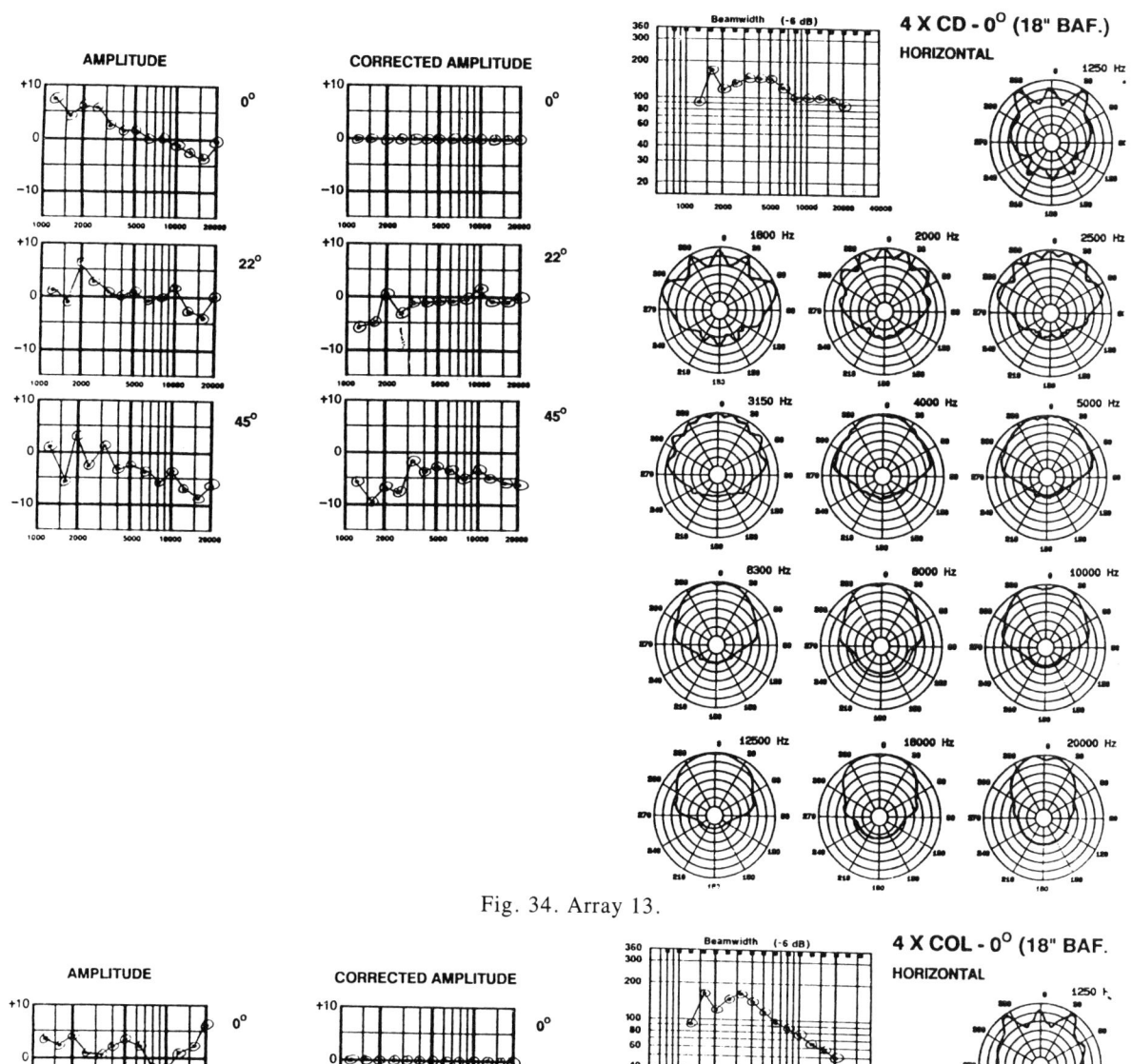

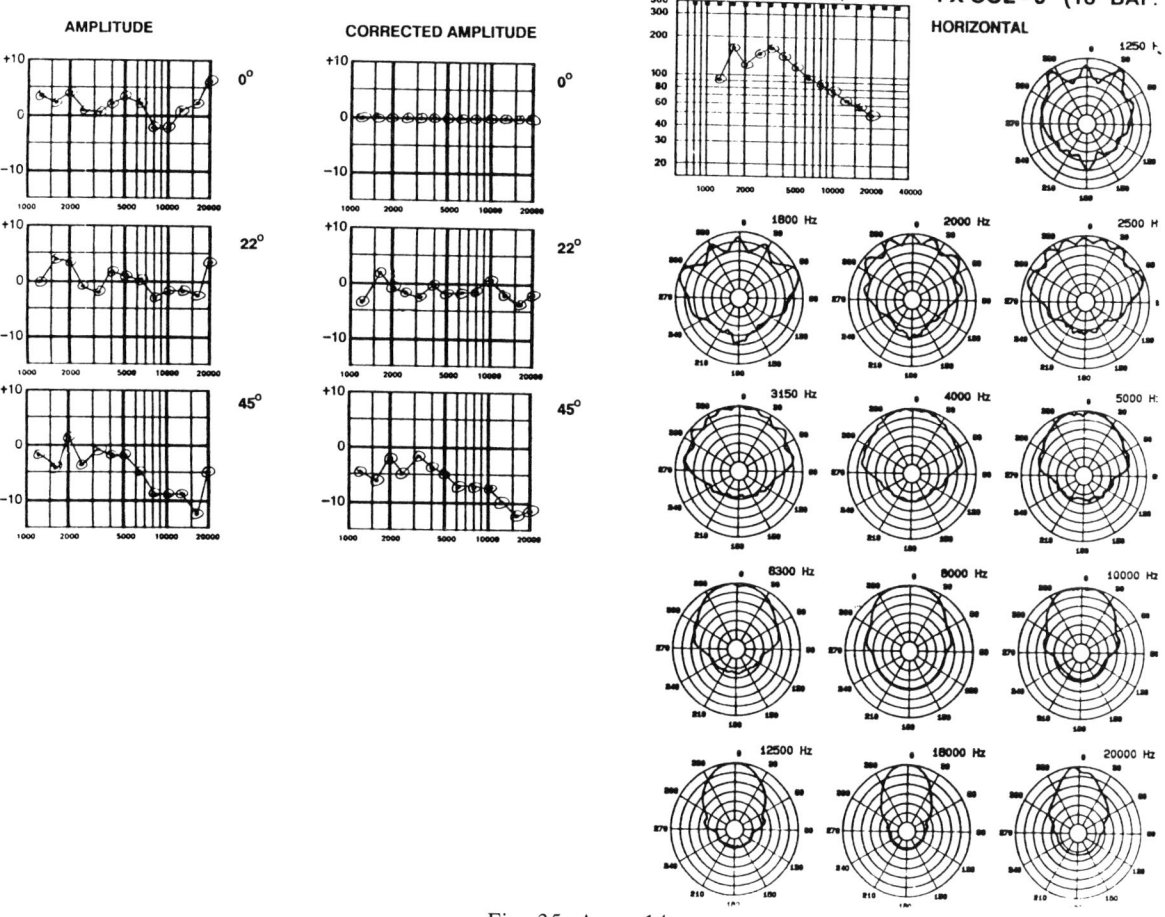

Fig. 35. Array 14.

Fig. 36. Array 15.

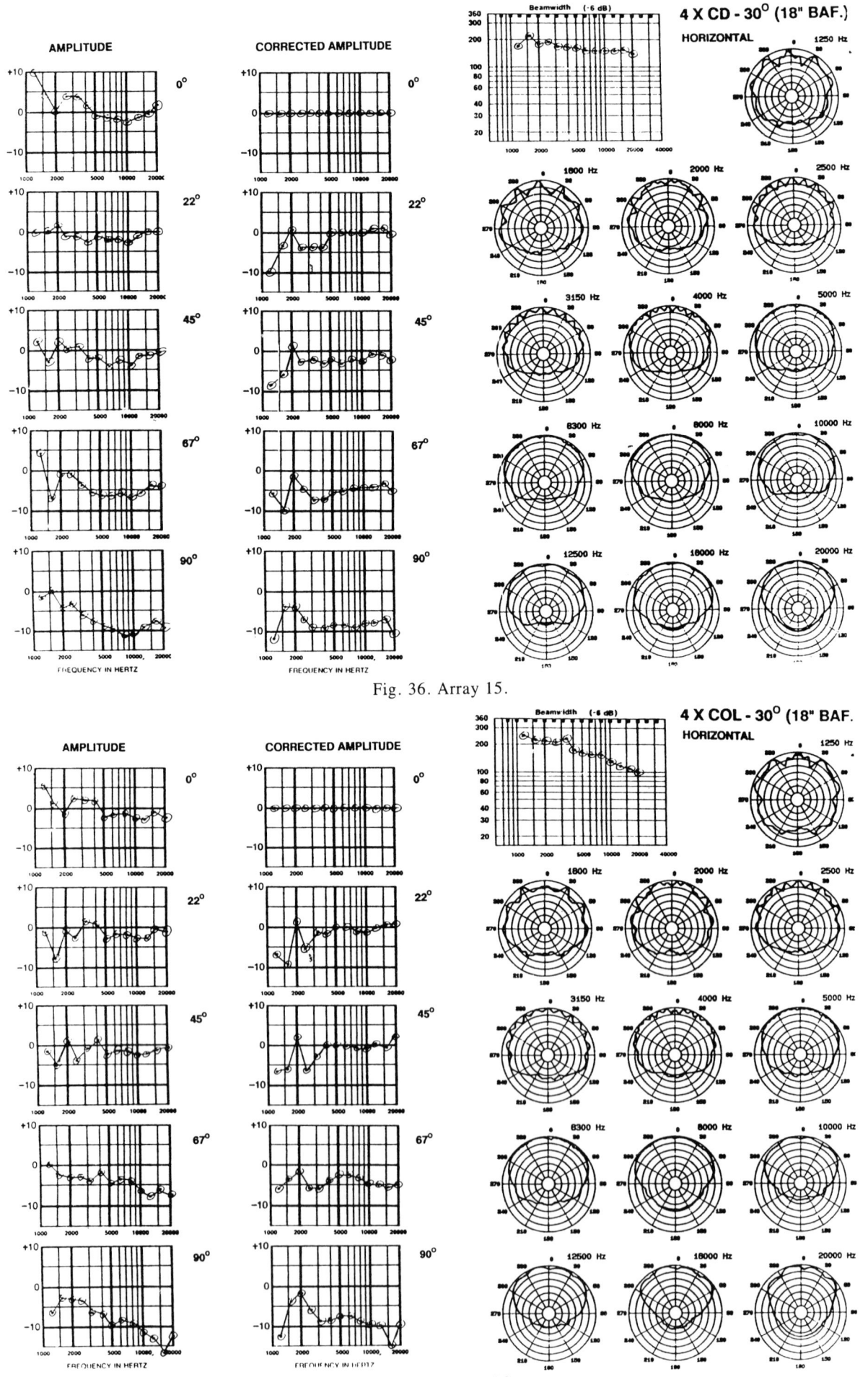

Fig. 37. Array 16.

Fig. 38. Array 17.

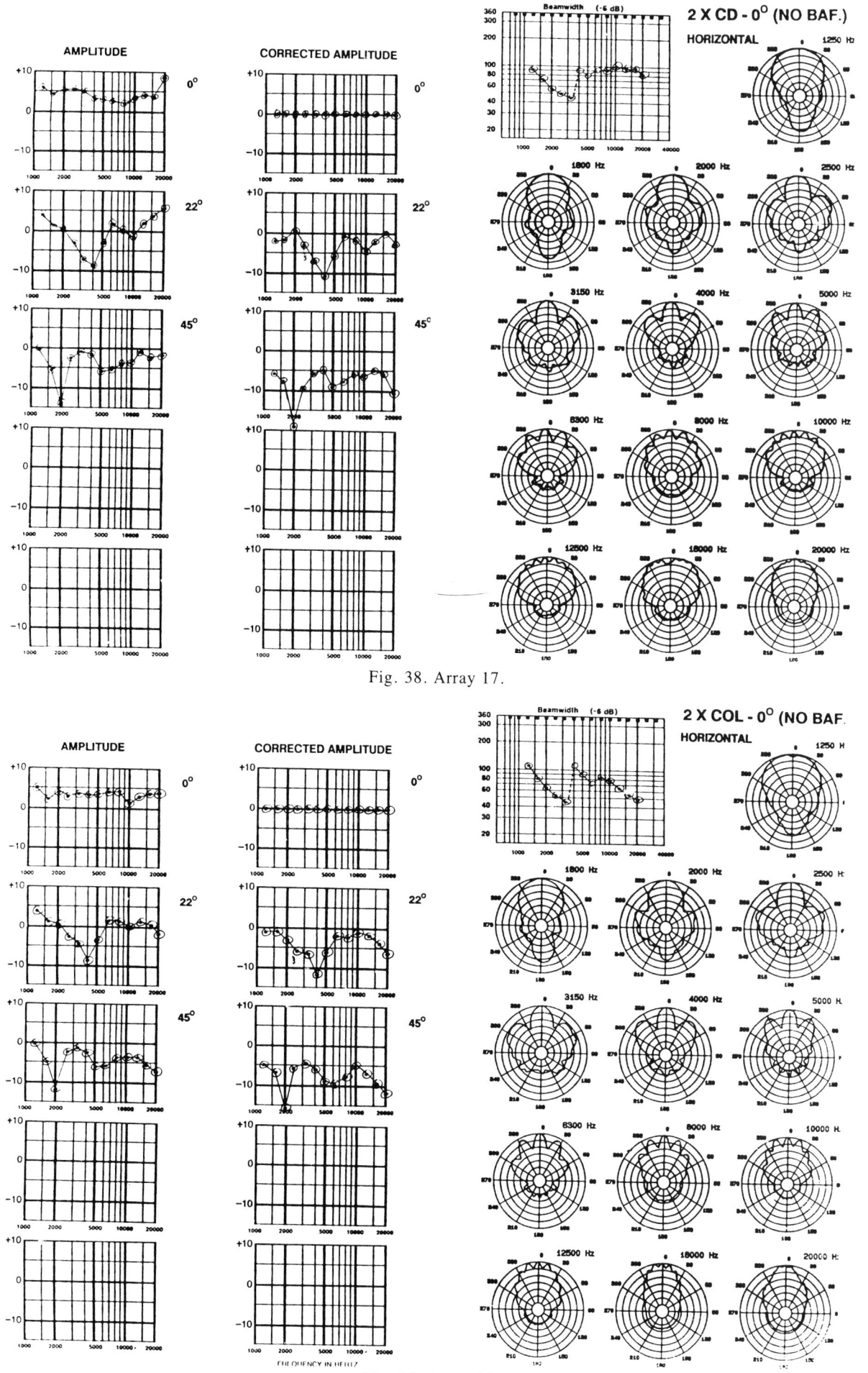

Fig. 39. Array 18.

Fig. 40. Array 19.

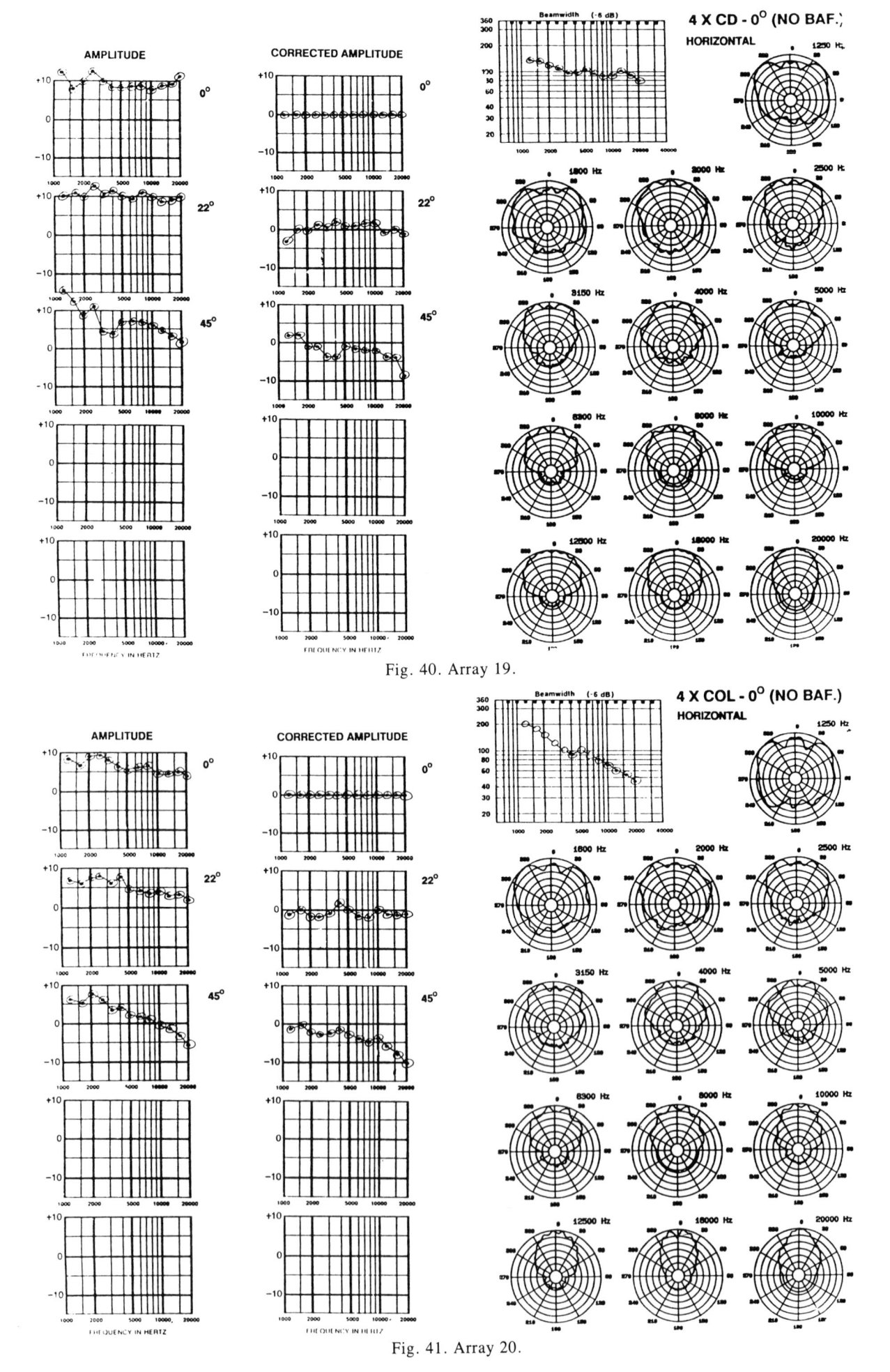

Fig. 41. Array 20.

Fig. 42. Array 1.

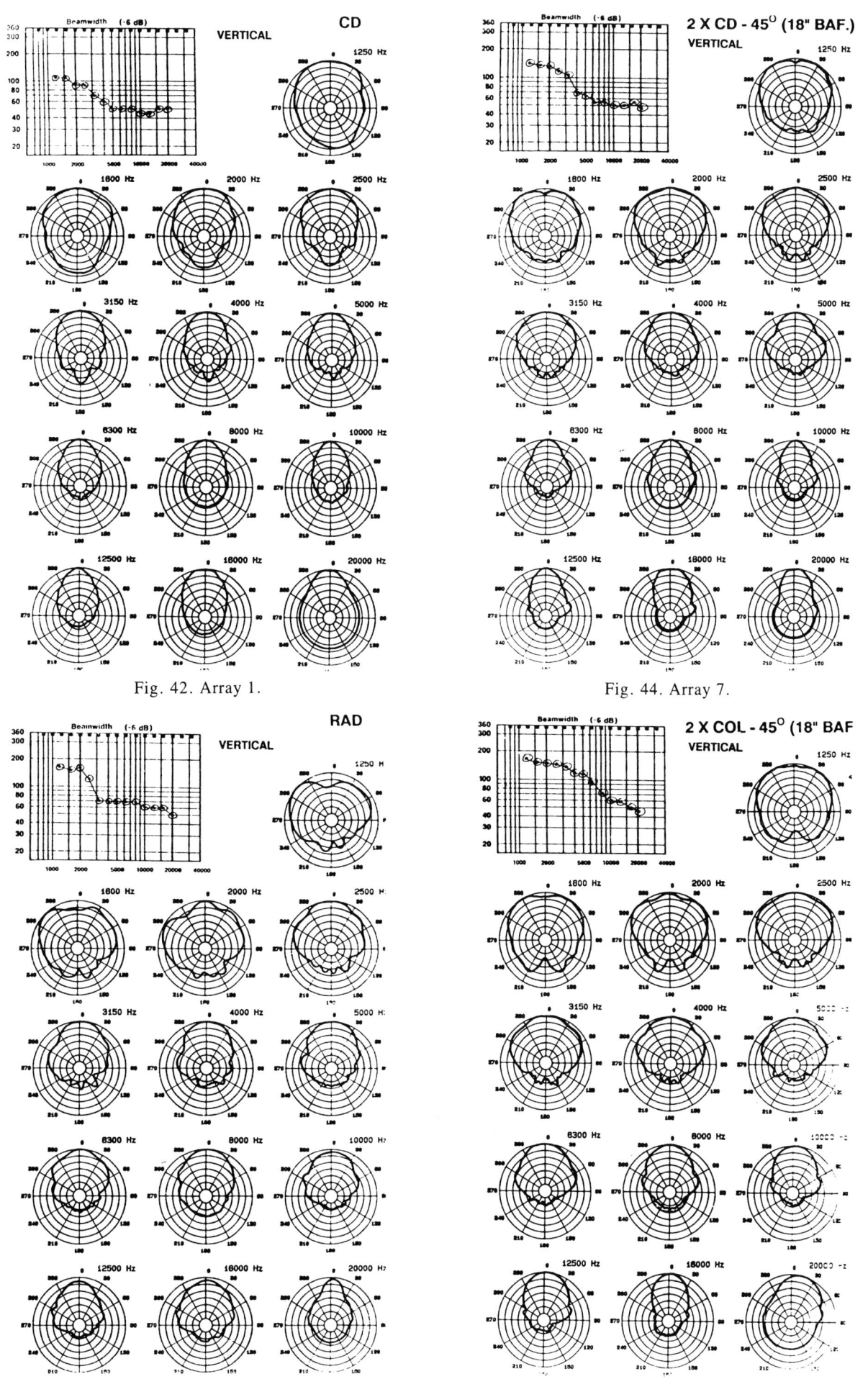

Fig. 44. Array 7.

Fig. 43. Array 2.

Fig. 45. Array 9.

Fig. 46. Array 13.

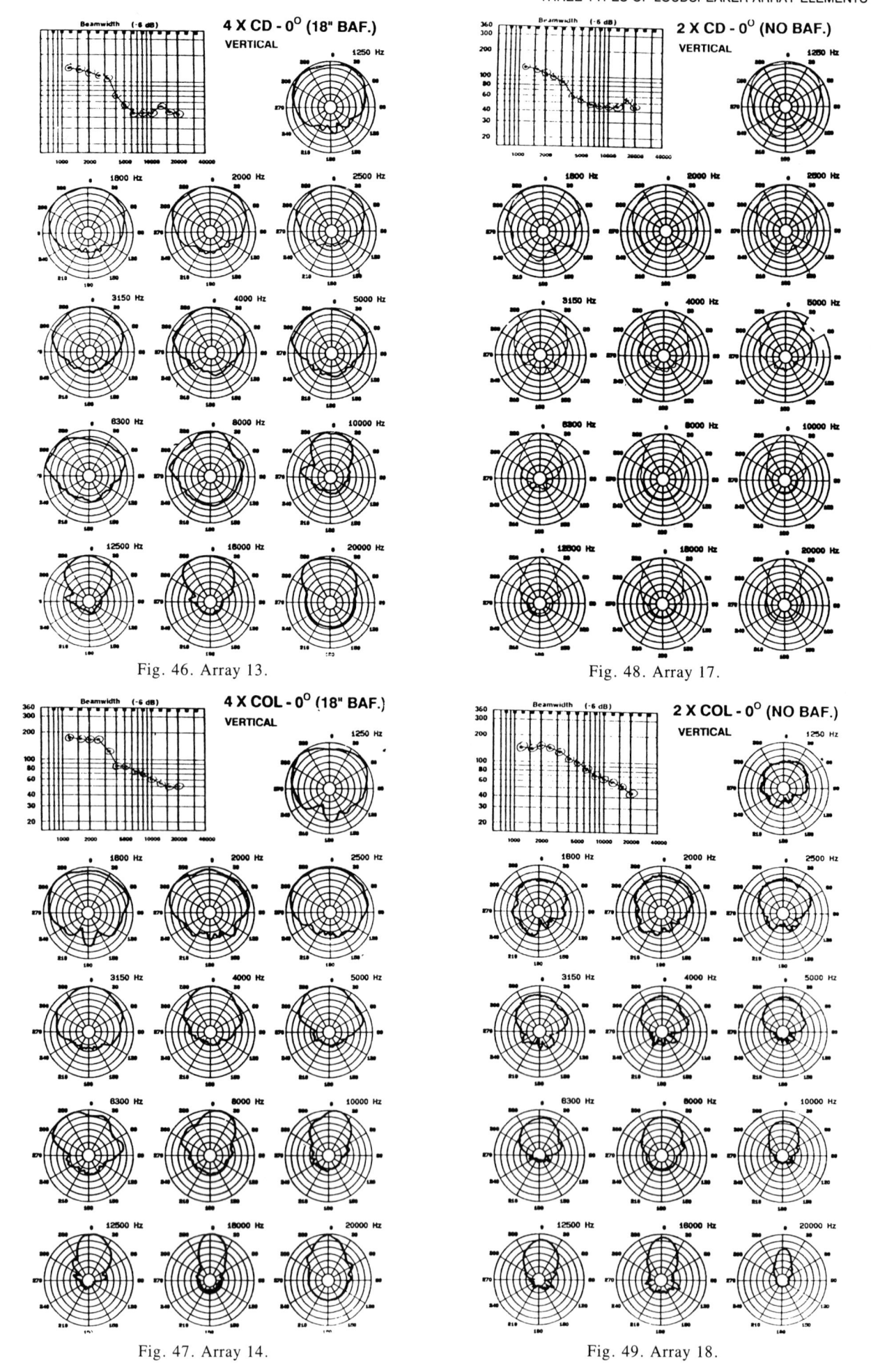

Fig. 47. Array 14.

Fig. 48. Array 17.

Fig. 49. Array 18.

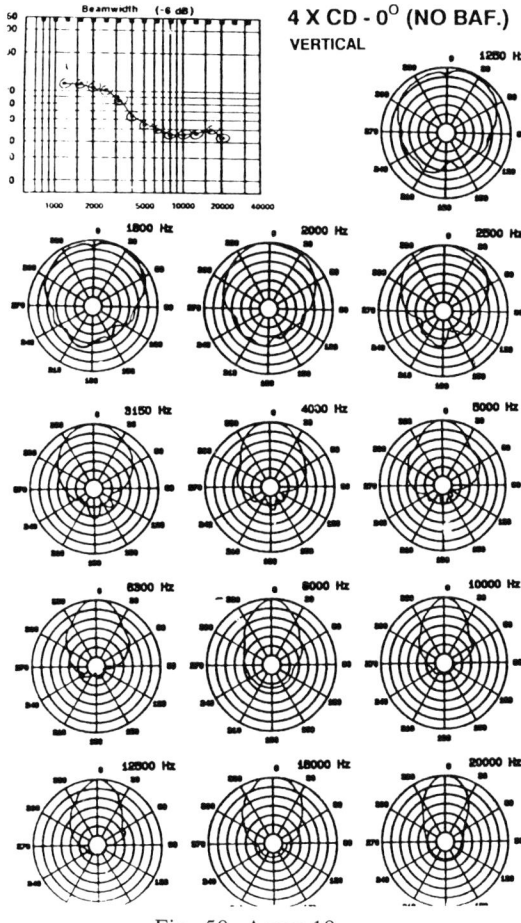

Fig. 50. Array 19.

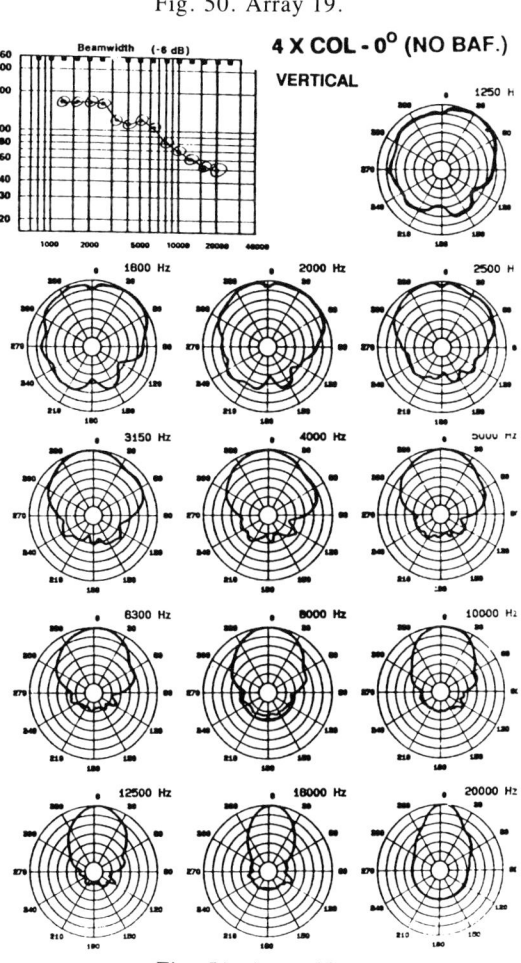

Fig. 51. Array 20.

APPENDIX 3
CALCULATION OF POLAR RESPONSE

Wolff and Malter [5] combined the theories of antennas and optics to predict the directional behavior of arrays of acoustic point sources. For a line array of n equally spaced sources, they developed a mathematical relationship for the acoustic pressure at any location on a circle in a plane passing through the line of points as

$$R_\alpha = \frac{\sin\left(\frac{n\pi d}{\lambda}\sin\alpha\right)}{n\sin\left(\frac{\pi d}{\lambda}\sin\alpha\right)} \quad (1)$$

where

R_α = ratio of pressure at an angle α to pressure at $\alpha = 0$,
n = number of sources
d = distance between sources
λ = wavelength
α = angle that the line from the center of the line array to the point of interest makes with the normal to the line array.

This is illustrated in Fig. 52.

Using Eq. (1) we calculated the relative pressure every 2° in one-ninth-octave frequency increments, and averaged the pressures within the standard ISO one-third-octave frequency bands. We then converted to a logarithmic scale:

$$R_{\log} = 20\log|R_\alpha| \quad (2)$$

and shifted the reference level for each plot so that the maximum pressure was full scale (0 dB).

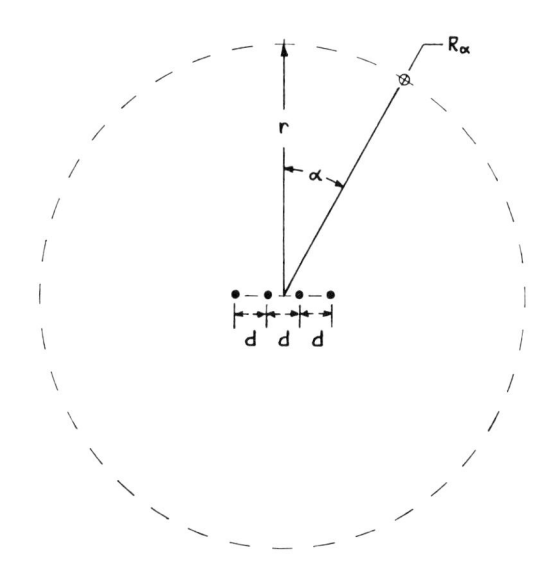

Fig. 52. Geometry for calculating polar response of a line source. $r \gg (n - 1)d$.

THE AUTHORS

P. F. Fidlin

Paul F. Fidlin was born in Southampton, England, in 1955. He received the B.S.C. degree, with honors, in electronics from Southampton University in 1974.

From 1974 to 1976 he worked for Plessey as an analog/digital circuit designer. He spent the next 5 years as an engineer and later as senior engineer with Celestion International based in Ipswich, England. His responsibilities included the design and development of both guitar loudspeakers and hi-fi loudspeaker systems. In 1983 he joined Electro-Voice, Inc., Buchanan, Michigan, as a loudspeaker design engineer and since 1987 has been a group leader of the loudspeaker components group. His responsibilities include the design and development of horns, compression drivers, and woofers. This includes products such as the DL series woofers, HP series horns, and DH1A and N/DYM-1 compression drivers.

Mr. Fidlin is a multi-instrumentalist and heavily involved in both composing and home recording. He is a member of the AES and has presented a number of technical papers. He is also a member of the IEE (London).

•

David E. Carlson was born in Edinboro, Pennsylvania, in 1952. He received the B.S. degree in electrical engineering and the M.S. degree in acoustics from the Pennsylvania State University in 1975 and 1979, respectively.

From 1972 to 1974 he was employed by the Public Broadcasting System affiliate WQLN in Erie, Pennsylvania, as a cameraman, audio engineer, and lighting director for both film and video productions. From 1975 to 1977 he worked as a research assistant at the Noise Control Laboratory at Pennsylvania State University, where he conducted research on the use of trees and other vegetative barriers for noise reduction outdoors. He was employed by Custom Audio Services in State College, Pennsylvania, from 1977 to 1979 as a senior engineer directing the development and design of touring concert-sound systems. In 1980 he formed CSS Systems, Inc., in State College, where he served as vice president until 1985. There his responsibilities included the development, design, and manufacture of loudspeaker systems and electronics for professional sound-reinforcement applications. In addition, he directed the operations of the touring-sound rental division and the permanent-installation contracting division. During the time period of 1977 to 1985, Mr. Carlson also served as a live-mix sound engineer for numerous touring artists. Since 1985 he has been employed by Electro-Voice, Inc., in Buchanan, Michigan, as a loudspeaker design engineer. His responsibilities include the development and design of components and systems for high-level sound-reinforcement applications. He played an instrumental role in the development of the Manifold Technology® and the DeltaMax™ series of loudspeaker systems.

Mr. Carlson holds several patents on loudspeaker design and is a member of the AES and the IEEE.

Effective Performance of Bessel Arrays*

D. (DON) B. KEELE, JR.

Audio Magazine, Diamandis Communications, Inc., New York, NY 10036, and Techron, Division of Crown International Inc., Elkhart, IN 46517, USA

The Bessel array is a configuration of five, seven, or nine identical loudspeakers in an equal-spaced line array that provides the same overall polar pattern as a single loudspeaker of the array. The results of a computer simulation are described, which uses point sources to determine the effective operating frequency range, working distance, efficiency, power handling, maximum acoustic output, efficiency–bandwidth product, and power–bandwidth product of the array. The various Bessel configurations are compared to one-, two-, and five-source equal-spaced equal-level equal-polarity line arrays.

As compared to a single source, a five source Bessel array is 14% (0.6 dB) more efficient, can handle 3.5 (+5.4 dB) more power, and has 4 times (+6 dB) the maximum midband acoustic output power, and is usable for omnidirectional radiation up to the frequency where the overall length is 11 wavelengths long. As compared to a two-source equal-level in-phase array, a five-source Bessel array is 43% (2.4 dB) less efficient, can handle 1.75 (+2.4 dB) more power, has the same maximum midband acoustic output power, and is usable for omnidirectional radiation 10 times higher in frequency. A working distance of 20 times the length of the Bessel array was assumed, with the length of the Bessel array (center-to-center distance of outside sources) being four times that of the two-source array. Analysis reveals that the three Bessel arrays have equal maximum acoustic output, but that the five-element Bessel array has the highest efficiency and power–bandwidth product. The seven- and nine-source Bessel arrays are found to be effectively unusable, as compared to the five-source array, due to much lower efficiency, requirement for more sources, and poor high-frequency performance. Judging polar peak-to-peak ripple and high-frequency response, the performance of the Bessel array is found to improve in direct proportion to the working distance away from the array. Unfortunately the phase versus direction and phase versus frequency characteristics of the Bessel array are very nonlinear and make it difficult to use with other sources.

0 INTRODUCTION

The Bessel array is a patented configuration [1] of equally spaced identical transducers, which is said to provide the same overall polar pattern as the polar pattern of a single transducer of the configuration. It is a method that extends the directional operating bandwidth of an array of transducers up into the region where the length of the array is a large number of wavelengths [2], [3]. The introduction to the Philips paper describes the justification for the Bessel configuration [2]:

An array of N loudspeakers connected in parallel and in phase can radiate N^2 times as much power as a single loudspeaker at very low frequencies, but only N times as much at high frequencies. The power response of the array is therefore quite different from that of the single loudspeakers that compose it. This is due to the increased directivity of the array; whereas the radiation pattern of a single loudspeaker is reasonable omnidirectional, usually up to at least a few kilohertz, that of an array is so only at low frequencies. At high frequencies it becomes much more directive; moreover, the directivity varies considerably with frequency. . . . These shortcomings can be remedied, at some expense to power radiation, by correctly proportioning the drive to the individual speakers of the array. The required proportioning coefficients are based on the Bessel functions.

The configuration normally takes the form of a five-, seven-, or nine-element line array or a 25 (5 × 5)-element symmetrical planar (panel) source. Only the line array Bessel configurations are analyzed in this study. The method used to set the drive levels of the

---

* Presented at the 87th Convention of the Audio Engineering Society, New York, 1989 October 18–21.
** Now also with DBK Associates, Elkhart, IN 46517, USA.

transducers in the array essentially randomizes the polarity of each of the elements [4], [5]. These polarity reversals reduce the sensitivity and the efficiency of the resultant array dramatically as compared to an equal-drive-level equal-polarity array. However, the chosen drive levels do extend the directional operating bandwidth of the array up into the region where the array is many wavelengths long.

To my knowledge, most (if not all) of the available references to the Bessel array contain hardly any information on the effective operation of the configuration. Some questions that immediately come to mind are: How high in frequency does the array operate? How far away from the array must you be? How do the efficiency, power handling, and maximum acoustic output compare to those of other array configurations? Which of the three array types, five-, seven-, or nine-element Bessel, is the best?

These and other questions are answered in this paper by analyzing the Bessel configuration using simulations based on arrays of point sources. The point source, being omnidirectional, should provide omnidirectional radiation when arrayed in a Bessel configuration. The degree to which the analyzed configurations provide omnidirectional coverage is the basis for evaluating their effective performance.

1 REVIEW OF BESSEL-DERIVED SOURCE LEVELS

Quoting again from [2]:
Consider an array of $2N + 1$ speakers equidistantly spaced in a straight line and driven by a common signal multiplied by coefficients ($a_{-N}, a_{-N+1}, \ldots, a_0, \ldots, a_{N-1}, a_N$) peculiar to each speaker. Assume that
- The point of observation P is in the far-field region of each speaker.
- The radiation of each speaker is not influenced by the others.
- All speakers have the same frequency and directional response $A(\omega, \theta)$.

The required proportioning of the drive levels (both level and polarity) of each of the transducers of the configuration is based on numbers derived from the Bessel function of the first kind and order n [2], [6]:

$$J_n(z) = \left(\frac{z}{2}\right)^n \sum_{k=0}^{\infty} \frac{(-z^2/4)^k}{k!\,(n+k)!} \,. \quad (1)$$

The method relies on a mathematical property of the Bessel function, which is

$$\left|\sum_{n=-\infty}^{\infty} J_n(z)\right| = \left|\sum_{n=-\infty}^{\infty} J_n(z)\, e^{(jnx)}\right|$$

$$= \left|e^{(jz\,\sin x)}\right| = 1 \,. \quad (2)$$

This property, combined with the equation that gives the pressure magnitude and phase at point P, at a particular frequency and angle for an array of sources (at an infinite distance),

$$p(\omega, \theta) = A(\omega, \theta) \sum_{n=-N}^{N} a_n\, e^{(-jnx)} \quad (3)$$

where

x	$= \omega l \sin \theta / c$ (assumes sample point at infinite distance)
ω	= frequency, rad/s, $= 2\pi f$
c	= velocity of sound
l	= distance between loudspeakers
θ	= angle between sample point vector and array axis
$A(\omega, \theta)$	= amplitude–phase function giving directional characteristics of a single source
a_n	= drive level of source n giving strength and polarity

yields a function that makes the dependence of the magnitude of p on direction and frequency the same for the array as for a single loudspeaker that makes up the array:

$$p(\omega, \theta) = A(\omega, \theta) \sum_{n=-N}^{N} J_n(z)\, e^{(-jnx)}$$

$$= A(\omega, \theta)\, e^{(-jz\,\sin x)}, \quad N \to \infty \quad (4)$$

or

$$|p(\omega, \theta)| = |A(\omega, \theta)| \,.$$

Eq. (4) clearly shows that the polar pattern of the array will be the same as that of one of the sources that make up the array. This function only works exactly, of course, for an infinite array of sources and a sample point at an infinite distance from the array. A finite-sized array of five, seven, or nine sources is also found to work quite well even if the drive levels are restricted to approximate values limited to the integer ratios ± 1 and ± 2 (± 0.5 and ± 1 in practice). These approximate values allow the drive levels of the array to be set by simple series–parallel connections of the drivers.

The approximate coefficient values are derived from the Bessel function by searching for arguments (both integer and noninteger values of z are allowed) that yield a coefficient ratio series that can be approximated by ± 1 and ± 2. An argument value of $z = 1.5$ is found to be a good choice for the five-element array coefficients. Fig. 1 shows the resultant coefficient values of $J_n(1.5)$ over the range $-10 \leq n \leq +10$, plotted in bar graph form. Both the actual values and the absolute values are plotted for comparison purposes. The plotted values show that the function decreases very rapidly to very small values for n beyond ± 3. Truncating the series beyond these values eliminates relatively little from the sum.

Choosing the range $-2 \leq n \leq 2$ yields the series

$J_{-2}(1.5), J_{-1}(1.5), J_0(1.5), J_2(1.5), J_3(1.5)$

$= 0.232, -0.558, 0.512, 0.558, 0.232$

the ratios of which, can be approximated as follows:

five-element Bessel ratios

$$= +0.5 : -1 : +1 : +1 : +0.5$$

configuration:

⊕⊕⊕⊕⊕
|— 1.0 —|

Likewise, the corresponding series for $z = 2.405$ and $z = 3.83$ yield the approximate drive ratios for the seven- and nine-element Bessel arrays:

seven-element Bessel ratios

$$= -0.5 : +1 : -1 : 0 : +1 : +1 : +0.5$$

configuration:

⊕⊕⊕ ⊕⊕⊕
|—— 1.5 ——|

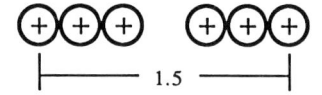

nine-element Bessel ratios

$$= +0.5 : -1 : +1 : 0 : -1 : 0 : +1 : +1 : +0.5$$

configuration:

⊕⊕⊕ ⊕ ⊕⊕⊕
|———— 2.0 ————|

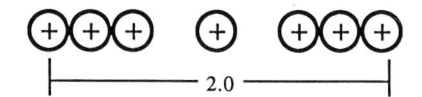

Note that sources that have zero drive levels can be eliminated from their respective arrays, making the seven-element Bessel array have six actual sources and the nine-element Bessel array have seven actual sources. Note also that the spaces for the removed sources must be preserved.

These drive ratios can be implemented by simple series–parallel hookups for each of the three configurations. Each ratio combination can be connected in a mostly parallel or a mostly series hookup. Only the more practical mostly parallel connection is analyzed here.

2 SIMULATION METHODS

The polar and frequency response simulations were accomplished by evaluating a more complete version of Eq. (3), which takes proper account of the actual distance from each source to the sample point, with no approximations. This more complete equation allows a proper evaluation of the effective working distances from the array. The equation appears as

$$p(\omega, \theta) = A(\omega, \theta) \sum_{n=-N}^{N} \frac{a_n}{r_n} e^{(-jkr_n)} \quad (5)$$

where

k = wavenumber, $= \omega/c = 2\pi f/c = 2\pi/\lambda$
ω = frequency, rad/s, $= 2\pi f$
c = velocity of sound
λ = wavelength, $= c/f$
f = frequency, Hz
r_n = distance from source n to sample point, $= |r_n|$
a_n = strength and polarity of source n.

All distances in this paper are referenced to a system that has a unit velocity of transmission. This means that at a frequency of 1 Hz a unit distance is 1 wavelength

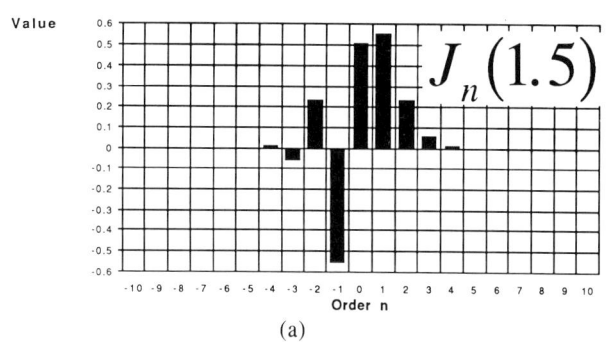

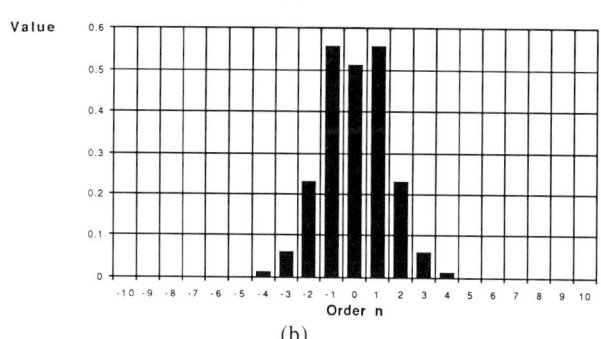

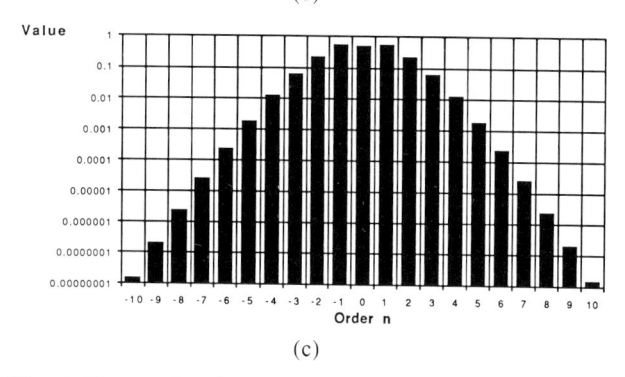

Fig. 1. Bar graphs of Bessel function of first kind and order n for argument of 1.5 [$J_n(1.5)$]; $-10 \leq n \leq +10$. (a) Linear vertical scale. (b) Absolute values using linear vertical scale. (c) Absolute values using logarithmic vertical scale, ranging from 10^{-8} to 1.0. Note that values grow rapidly very small for n beyond ± 3. Values in the range of $|n| \leq 2$ are used to generate source drive levels and polarities for five-source Bessel array ($+0.5 : -1 : +1 : +1 : +1 : +0.5$).

long. All the five-element arrays have a unit overall source center-to-center length. A working distance of 20 units implies that the pressure sampling point is 20 times the length of the five-element array away from the center of the array.

The array was oriented so that its axis was along the y axis, with its center at the origin of the coordinate system. Rotation was always around the center of the array (the origin), with the zero angle in the direction of the positive x axis and positive angles in the counter-clockwise direction. For the Bessel arrays, the sources with higher number Bessel coefficients were above the x axis (positive angles).

A number of the analysis factors require the calculation of the peak-to-peak ripple, in decibels, for the polar directional pattern at a specific frequency. The equation to calculate this ripple factor [2] is

$$\text{peak-to-peak ripple [dB]} = R(\omega)$$
$$= 20 \log \left(\frac{|p(\omega, \theta)|_{\max}}{|p(\omega, \theta)|_{\min}} \right)_{-\pi \leq \theta \leq \pi} . \quad (6)$$

All calculations for this paper were done on a Macintosh II desktop computer using a combination of the Microsoft spreadsheet program Excel and the math analysis program Mathematica by Wolfram Research, Inc. (used for all graphic output).

3 ARRAY ANALYSIS FACTORS

The various Bessel configurations (five-, seven-, and nine-source line arrays) were compared to one-, two-, and five-source equal-spaced equal-level equal-polarity line arrays. A number of analysis factors were used in the comparison: voltage sensitivity, impedance, efficiency, maximum input power handling, maximum acoustic output power and SPL, maximum operating frequency, working distance, polar directional response, frequency response (both magnitude and phase), polar peak-to-peak ripple versus frequency, efficiency–bandwidth product, power–bandwidth product, and power–bandwidth product per unit. These analysis factors are now described individually.

3.1 Voltage Sensitivity

The voltage sensitivity of a system is the on-axis sound pressure level (SPL) generated at a specific distance for a particular applied voltage. In this paper all measurements are referenced or normalized to a point source that is assumed to have all unit specifications, that is, a sensitivity of 1- or 0-dB SPL for an applied unity voltage, at a 1-unit distance. The sensitivity of the analyzed arrays is simply the total of the individual drive levels.

3.2 Impedance

The input electrical impedance for each analyzed array was computed assuming a unity impedance (resistance) for each of the individual sources.

3.3 Efficiency

The electroacoustic efficiency (electric input power divided by the resultant acoustic output power) of each array was computed by direct comparison to a single point source, in the omnidirectional radiation region of the array's frequency range. An efficiency of unity was assigned to the point source. The efficiency of an array was computed by squaring its sensitivity and multiplying by its impedance:

$$\eta_0 = \frac{P_{\text{out}}}{P_{\text{in}}} = \text{Sens}^2 Z_{\text{in}} . \quad (7)$$

3.4 Power Handling

The maximum input electric power handling of an array was computed by summing the individual source powers computed by applying a unity input voltage and assuming unity impedances for all individual sources.

3.5 Maximum Acoustic Output Power and SPL

The maximum acoustic output power was computed by multiplying the array's efficiency by its maximum input electric power,

$$P_{\text{out}} = \eta_0 P_{\text{in}} \quad (8)$$

The sound pressure level, in decibels, was calculated by using $10 \log_{10}(P_{\text{out}}/P_{\text{ref}})$, where P_{ref} is unity (the power output of a single source).

3.6 Maximum Operating Frequency

The maximum operating frequency was assessed by simulating the polar response of the analyzed array using Eq. (5) and then finding the maximum frequency up to which the peak-to-peak polar magnitude ripple [Eq. (6)] did not exceed a specific amount, usually 3, 4, 6, or 9 dB. Note that a point source has a maximum operating frequency of infinity, using this definition.

3.7 Working Distance

The working distance was assessed similarly to the maximum operating frequency by polar simulations and then noting the minimum operating distance that provided a specific peak-to-peak polar ripple. Usually a specific working distance (in terms of array lengths of 5, 10, or 20 units) was chosen, and then all the relevant parameters were calculated.

3.8 Polar Response

Polar directional responses were computed using Eq. (5) at various angles and distances for each of the analyzed configurations. A set of polar responses at a fixed working distance (usually 20 units) at various frequencies were simulated along with a set at a fixed frequency (usually 10 Hz) at various working distances. Both magnitude and sometimes phase versus angle plots are displayed. The linear phase effects of transport delay between source and sample point were removed in all phase displays.

3.9 Frequency Response (Magnitude and Phase)

Magnitude and phase frequency responses were computed using Eq. (5) at various angles and distances for each of the analyzed configurations. A set of frequency responses at various angles at a fixed distance (usually 20 units) were simulated along with a set at a fixed angle (usually 45°) at various working distances. Both magnitude and sometimes phase (also group delay in one case) versus frequency plots are displayed. The linear phase effects of transport delay between source and sample point were removed in all phase displays.

3.10 Polar Peak-to-Peak Ripple versus Frequency

A plot of the polar peak-to-peak ripple, in decibels, versus frequency indicates the tradeoff of polar nonlinearities versus the high-frequency limit. In general, all arrays exhibit increasing polar ripple as the frequency is increased.

3.11 Efficiency–Bandwidth Product

The efficiency–bandwidth product was computed by forming the product of the efficiency and the maximum operating frequency. This number gives a comparative value that indicates how thrifty the analyzed array is in terms of its efficiency and operating frequency range.

3.12 Power–Bandwidth Product

The power–bandwidth product was computed by forming the product of the maximum acoustic output power and the maximum operating frequency. This number gives a comparative value, which indicates how well the analyzed array functions in terms of its output power and operating frequency range.

3.13 Power–Bandwidth Product per Unit

The power–bandwidth product per unit was computed by dividing the power–bandwidth product by the number of units in the array. This number can be thought of as a figure of merit for comparing the operating effectiveness of the analyzed arrays on a per-unit basis.

4 SIMULATION RESULTS

Several different point source configurations were analyzed and compared for this study. All configurations were analyzed in terms of the performance of a single point source. The following configurations were analyzed.

1) Two equal-level equal-polarity sources with center-to-center spacing of 0.25 unit (same center-to-center spacing as the individual spacing of the five-element arrays)

2) Two equal-level equal-polarity sources with center-to-center spacing of 1.0 unit (same overall center-to-center length as the five-element arrays)

3) Five equal-level equal-polarity equal-spaced sources with individual center-to-center spacing of 0.25 unit and overall center-to-center length of 1.0 unit

4) Five-source Bessel array with individual center-to-center spacing of 0.25 unit and overall center-to-center length of 1.0 unit

5) Seven-source Bessel array with individual center-to-center spacing of 0.25 unit and overall center-to-center length of 1.5 units

6) Nine-source Bessel array with individual center-to-center spacing of 0.25 unit and overall center-to-center length of 2.0 units.

Note that all the arrays have the same individual center-to-center spacings (except for the two-source configuration with center-to-center length of 1.0 unit). This means that the overall array length increases in direct proportion to the number of sources. This models the real-world situation of using the same-size transducers packed as close together as possible.

For each configuration, several possible analysis factors were calculated: voltage sensitivity, impedance, efficiency, maximum input power handling, maximum acoustic output power and SPL, maximum operating frequency, working distance, polar directional response, frequency response (magnitude, phase, and group delay), efficiency–bandwidth product, power–bandwidth product, and power–bandwidth product per unit. Further explanations of these factors are given in Sec. 2. The results of the simulations are described in the following sections and shown in Figs. 2 to 32.

4.1 Single Point Source

A single point source is the reference for all the following array configurations. The single point source is arbitrarily assigned all unit parameters and its characteristics are shown in Table 1. Note that all the frequency-dependent factors have infinite values because the point source by definition has no upper frequency limit.

The polar response of the point source (not shown) is a perfect circle, while its frequency and phase responses (not shown) are straight lines. The polar and frequency responses of the reference point source are not distance dependent. Note that table entries have been reserved for working distances of 5, 10, and 20 units at peak-to-peak ripple values of 3, 4, and 6 dB.

4.2 Two Sources, Equal Level, Equal Polarity, with 0.25- and 1.0-Unit Center-to-Center Spacings

The two-source array is the simplest configuration, one step above the single source, and is used quite frequently to increase the acoustic output as compared to a single source. Unfortunately, as the following simulations show, the maximum frequency of operation drops dramatically because of source interference and lobing. Two double-source configurations, with center-to-center spacing of 0.25 and 1.0 unit, were analyzed and are described in the following section.

The two-source array with 0.25-unit center-to-center spacing has the same center-to-center spacing as the individual spacing of the five-element arrays. This close side-by-side spacing is the logical configuration for getting the most performance (highest operating bandwidth) from a two-source array. All the characteristics and calculated parameters for the two-source array with 0.25-unit center-to-center spacing are shown in Table 2.

The 1.0-unit center-to-center spaced two-source array has the same center-to-center spacing as the overall center-to-center spacing (outside sources) of the five-element arrays. If you just simply remove the center three sources of the five-element array, you get this spacing. All the characteristics and calculated parameters for the two-source array with 1.0-unit center-to-center spacing are shown in Table 3.

All the responses and characteristics for the 1.0-unit center-to-center spacing array are the same as those for the 0.25-unit center-to-center spacing array, but shifted down in frequency by two octaves (frequency × ¼). The data on the 1.0-unit center-to-center spacing array have been included for comparing against the five-source arrays, which have the same length.

4.2.1 Polar Responses

The polar magnitude responses of the two-source array with 0.25-unit center-to-center spacing, at constant distance, are shown in Fig. 2. The polars are displayed at half-decade intervals from 0.316 to 31.6 Hz and at a working distance of 20 units. An additional polar at 2.0 Hz is also displayed. All the polar plots

Table 1. Array type: single source.

Number Units (N): 1
Overall Length (c-c): 0
Strengths: 1
Impedance (Z_{in}): 1 (0.0 dB)
Voltage Sensitivity: 1 (0.0 dB)
Efficiency (η_o): 1 (0.0 dB)
Maximum Input Power (P_{in}): 1 (0.0 dB)
Maximum Acoustic Output Power (P_{out}): 1 (0.0 dB)
Maximum Sound Pressure Level: 1 (0.0 dB)

Maximum Upper Frequency (F_{max}):

Distance ->	5	10	20
Ripple (dB): 3	Infinity	Infinity	Infinity
4	Infinity	Infinity	Infinity
6	Infinity	Infinity	Infinity

Efficiency-Bandwidth Product ($\eta_o \times F_{max}$):

Distance ->	5	10	20
Ripple (dB): 3	Infinity	Infinity	Infinity
4	Infinity	Infinity	Infinity
6	Infinity	Infinity	Infinity

Power-Bandwidth Product ($P_{out} \times F_{max}$):

Distance ->	5	10	20
Ripple (dB): 3	Infinity	Infinity	Infinity
4	Infinity	Infinity	Infinity
6	Infinity	Infinity	Infinity

Power-Bandwidth Product per Unit ($P_{out} \times F_{max} / N$):

Distance ->	5	10	20
Ripple (dB): 3	Infinity	Infinity	Infinity
4	Infinity	Infinity	Infinity
6	Infinity	Infinity	Infinity

Table 2. Array type: two sources ($L = 0.25$ unit), equal level, same polarity.

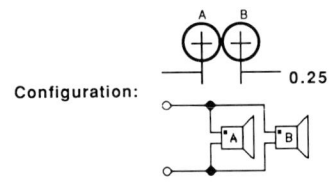

Number Units (N): 2
Overall Length (c-c): 0.25
Strengths: 1:1
Impedance (Z_{in}): 0.5 (-3.0 dB)
Voltage Sensitivity: 2 (+6.0 dB)
Efficiency (η_o): 2 (+3.0 dB)
Maximum Input Power (P_{in}): 2 (+3.0 dB)
Maximum Acoustic Output Power (P_{out}): 4 (+6.0 dB)
Maximum Sound Pressure Level: 2 (+6.0 dB)

Maximum Upper Frequency (F_{max}):

Distance ->	5, 10, 20
Ripple (dB): 3	1.00
4	1.10
6	1.30

Efficiency-Bandwidth Product ($\eta_o \times F_{max}$):

Distance ->	5, 10, 20
Ripple (dB): 3	2.00
4	2.20
6	2.60

Power-Bandwidth Product ($P_{out} \times F_{max}$):

Distance ->	5, 10, 20
Ripple (dB): 3	4.00
4	4.40
6	5.20

Power-Bandwidth Product per Unit ($P_{out} \times F_{max} / N$):

Distance ->	5, 10, 20
Ripple (dB): 3	2.00
4	2.20
6	2.60

Table 3. Array type: two sources ($L = 1.0$ unit), equal level, same polarity.

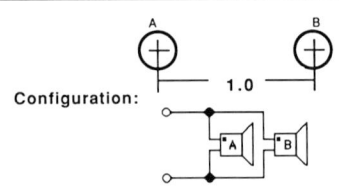

Number Units (N): 2
Overall Length (c-c): 1.0
Strengths: 1:1
Impedance (Z_{in}): 0.5 (-3.0 dB)
Voltage Sensitivity: 2 (+6.0 dB)
Efficiency (η_o): 2 (+3.0 dB)
Maximum Input Power (P_{in}): 2 (+3.0 dB)
Maximum Acoustic Output Power (P_{out}): 4 (+6.0 dB)
Maximum Sound Pressure Level: 2 (+6.0 dB)

Maximum Upper Frequency (F_{max}):

Distance ->	5, 10, 20
Ripple (dB): 3	0.25
4	0.28
6	0.33

Efficiency-Bandwidth Product ($\eta_o \times F_{max}$):

Distance ->	5, 10, 20
Ripple (dB): 3	0.50
4	0.55
6	0.65

Power-Bandwidth Product ($P_{out} \times F_{max}$):

Distance ->	5, 10, 20
Ripple (dB): 3	1.00
4	1.10
6	1.30

Power-Bandwidth Product per Unit ($P_{out} \times F_{max} / N$):

Distance ->	5, 10, 20
Ripple (dB): 3	0.50
4	0.55
6	0.65

displayed in this paper cover a range of 40 dB with +6 dB at the outer edge and −34 dB at the center. All polars are normalized so that the on-axis level is 0 dB. Note that at 2 Hz, where the sources are one-half wavelength apart, the first polar null at 90° off axis occurs. Note also that above about 1.8 Hz, the polar response is multilobed and hence unusable for omnidirectional response.

The polar responses for the 0.25-unit center-to-center spacing array, at a fixed frequency of 10 Hz and at different working distances, are shown in Fig. 3. Polars at distances of 1.25, 2.5, 5, 10, and 100 000 units are shown. Observe that the polar responses essentially exhibit no change with increasing working distances beyond about 2.5 units (10 times array length). Note that the 100 000-unit distance is extremely far from the array; essentially an effective infinity. If the overall length of the array were 2 ft (0.6 m), this distance would be about 38 mi (60 km) away.

Only a few polar responses were done on the two-source array with 1.0-unit center-to-center spacing, mainly to illustrate the variation of phase versus angle with frequency and working distance. The previous two-source array exhibits the same behavior but is four times higher in frequency. Note that the first polar null at 90° off axis occurs at a frequency of 0.5 Hz, where the sources are one-half wavelength apart (not shown).

Fig. 4 shows various magnitude and phase polar responses at different frequencies f and distances D for the two-source array with 1.0-unit center-to-center spacing. The following four combination are plotted:
1) $f = 1$ Hz, $D = 20$ units
2) $f = 1$ Hz, $D = 100\,000$ units
3) $f = 2$ Hz, $D = 20$ units
4) $f = 2$ Hz, $D = 100\,000$ units.

The phase versus direction plots show the phase of the pressure at the sample point versus the off-axis direction. The phase values are referenced to the input signal of the array. The effects of linear phase lag and delay due to sample distance have been removed in

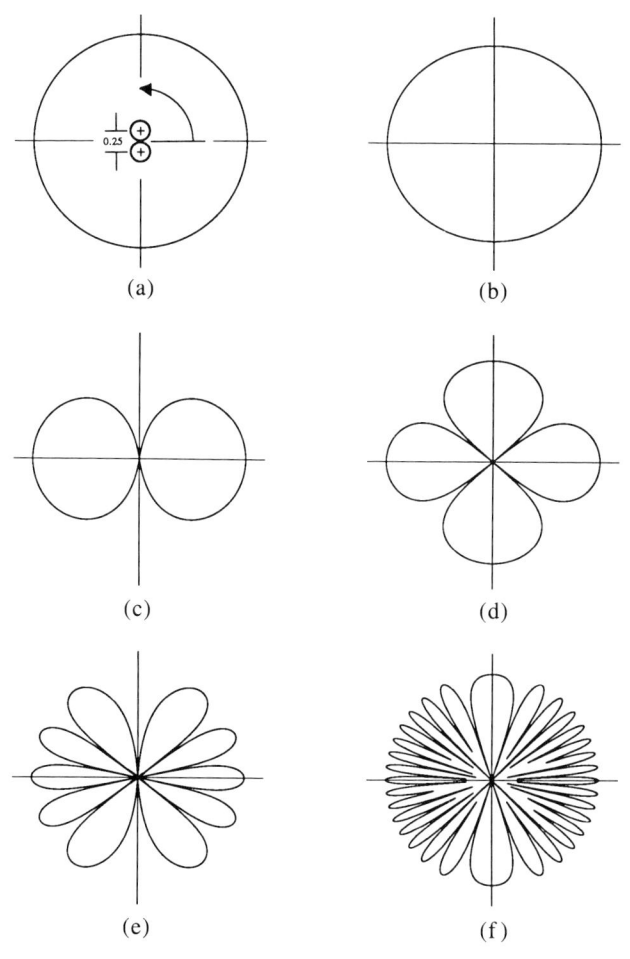

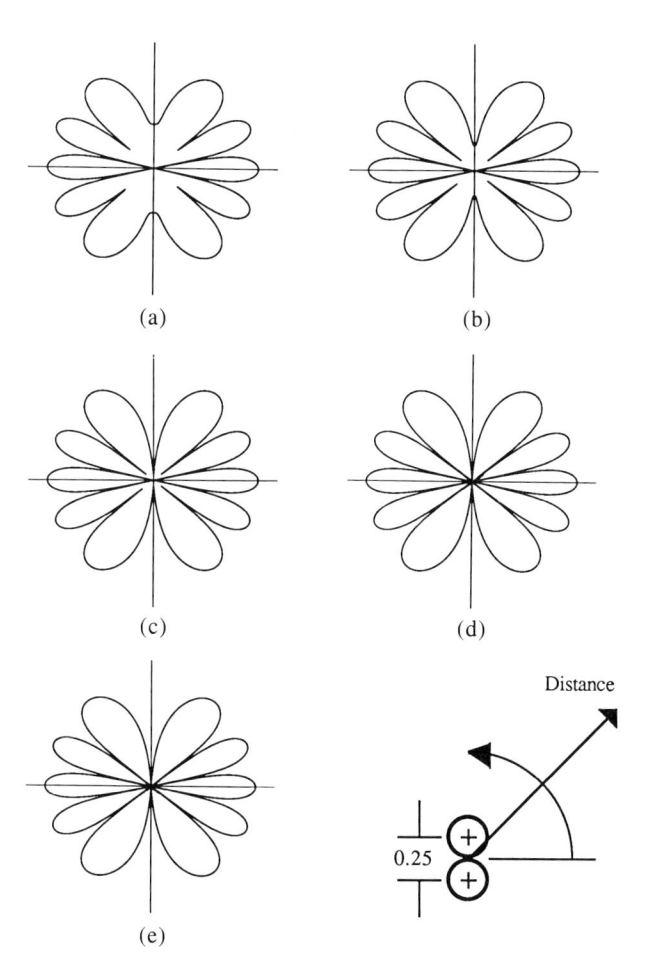

Fig. 2. Polar magnitude responses for two-source equal-level equal-polarity array with 0.25-unit center-to-center spacing at a constant working distance of 20 units. Spacing is the same as individual center-to-center spacing of five-source arrays. Polars are displayed at half-decade intervals from 0.316 to 31.6 Hz, with additional polar at 2.0 Hz. (a) 0.316 Hz. (b) 1 Hz. (c) 2 Hz. (d) 3.16 Hz. (e) 10 Hz. (f) 31.6 Hz. Polar plot covers a range of 40 dB with +6 dB at the outer edge and −34 dB at the center. All polars are normalized so that on-axis level is 0 dB. Sources are one-half wavelength apart at 2 Hz and exhibit a null at ±90° off axis (c). Note that polar response is mostly omnidirectional at and below 1 Hz, but gets progressively narrower and gains additional lobes as frequency increases.

Fig. 3. Polar magnitude responses for two-source equal-level equal-polarity array with 0.25-unit center-to-center spacing at fixed frequency of 10 Hz and different working distances. (a) 1.25 units. (b) 2.5 units. (c) 5 units. (d) 10 units. (e) 100 000 units. Observe that polar responses essentially exhibit no change with increasing working distance beyond about 2.5 units (10 times array length).

this plot and in all the polar and frequency response plots of this paper. Note that the phase values switch between 0 and ±180°, depending on the polar lobe on which the pressure sample point happens to be. The phase always starts out at 0° (on axis). At distances far from the array, the phase transitions occur very abruptly according to the angle, with no rounded corners.

4.2.2 Frequency Responses

The magnitude versus frequency responses of the two-source array with 1.0-unit center-to-center spacing at constant distance are shown in Fig. 5. The responses are shown at angles ranging from 0 to +90°, with steps of 15°, at a working distance of 20 units. Note that the response gets progressively rougher as the angle increases due to the nulls in the response moving down in frequency. The magnitude versus frequency responses at a fixed angle of 45° and at different working distances are given in Fig. 6. Distances of 1.25, 2.5, 5, 10, and 100 000 units are shown. Note that the frequency response changes very little with distance beyond 5 units.

The frequency range of the responses goes from 0.1 to 10 Hz with a log frequency scale. Note that the frequency scale is marked with decade number (log f) rather than frequency (−1.0 = 0.1 Hz, 0.0 = 1 Hz, and so on).

To illustrate the variation of phase versus frequency and working distance, several magnitude and phase responses were done on the two-source array with 1.0-unit center-to-center spacing. Fig. 7 shows these responses with a fixed angle of 45° and distances of 5 and 100 000 units. Observe that the phase again is either 0 or ±180°, depending on the polar lobe in which the sample point happens to be. This phase versus frequency behavior looks suspiciously nonminimum phase, but is actually minimum phase [7]. This comment only applies to the two-source array, however, where the response at the sample point is strictly due to a signal plus a single delayed signal of reduced amplitude.

4.2.3 Polar Peak-to-Peak Ripple versus Frequency

Fig. 8 shows a plot of the polar peak-to-peak ripple, in decibels, versus frequency for both two-source arrays

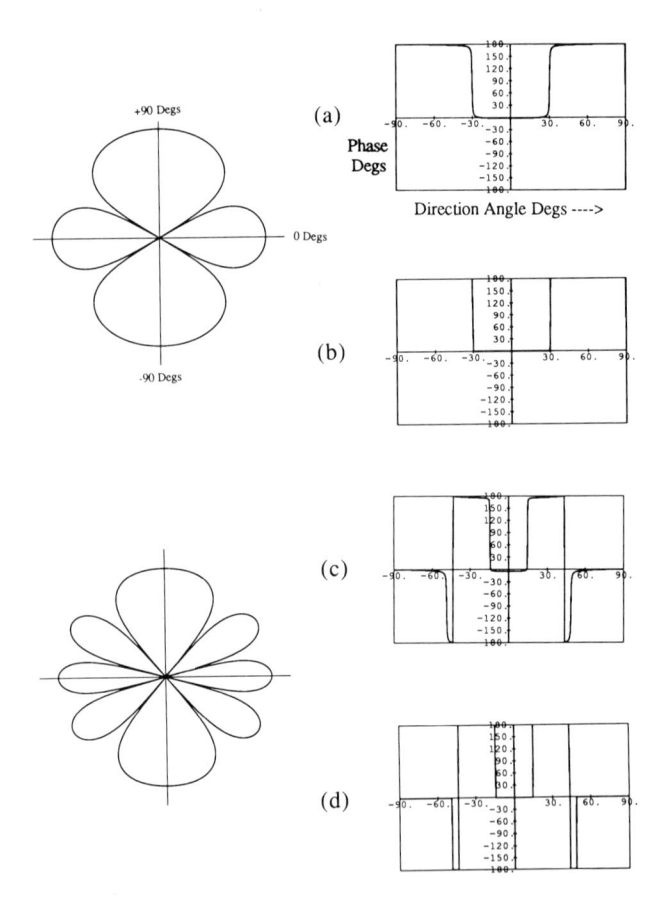

Fig. 4. Polar magnitude and phase responses at various frequencies and distances for two-source equal-level equal-polarity array with 1.0 unit center-to-center spacing. Phase versus direction plots show phase of pressure at sample point versus off-axis direction. (a) $f = 1$ Hz, $D = 20$ units. (b) $f = 1$ Hz, $D = 100\,000$ units. (c) $f = 2$ Hz, $D = 20$ units. (d) $f = 2$ Hz, $D = 100\,000$ units. Phase values are referenced to array input signal. Effects of linear phase lag and delay due to sample distance have been removed. Note how phase changes rapidly from 0 to ±180° as direction angle increases, as each separate lobe is transversed. Note also that this array is spaced one-half wavelength apart at 0.5 Hz.

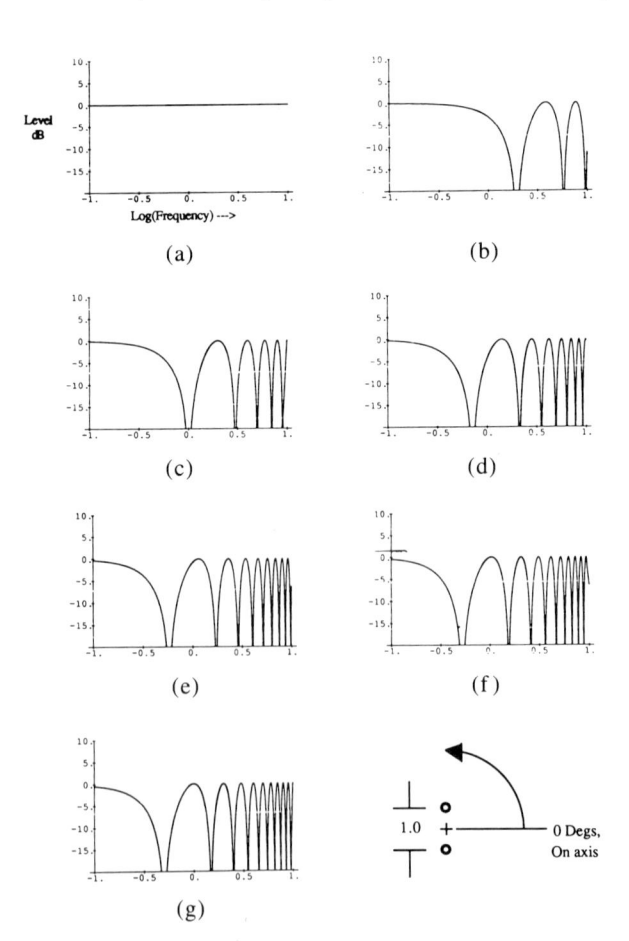

Fig. 5. Magnitude frequency responses of two-source equal-level equal-polarity array with 1.0-unit center-to-center spacing at constant distance of 20 units and frequency range of 0.1 to 10 Hz. Note that log of frequency is indicated (−1 = 0.1 Hz, 0 = 1 Hz, etc.). Responses are shown at angles ranging from 0 to +90°, with steps of 15°. (a) 0°. (b) 15°. (c) 30°. (d) 45°. (e) 60°. (f) 75°. (g) 90°. Note that response gets progressively rougher as angle increases.

at a working distance of 20 units. Note that the ripple increases very rapidly above 0.25 Hz for the 1.0-unit spacing and above 1 Hz for the 0.25-unit spacing.

Fig. 9 shows the polar peak-to-peak ripple versus frequency at several different working distances from 2.5 to 100 000 units for the 1.0-unit spaced two-source array. The graph exhibits essentially no change at distances beyond 5 units.

Fig. 10 shows a polar response of the 0.25-unit spaced array at a distance of 20 units and a frequency of 1.1 Hz, which corresponds to the frequency where the peak-to-peak ripple is 4 dB. Note that the polar response is very smooth, but squashed vertically, and exhibits its maximum deviation (-4 dB) at $\pm 90°$. As will be seen, this is a characteristic of all the equal-level equal-phase arrays.

The following approximate equations relate the maximum operating frequency for omnidirectional radiation f_{max} to the array length for the two-source equal-level equal-polarity arrays.

For the 0.25-unit spaced array,

$$f_{max} \approx 1.0 \frac{c}{L}, \text{ for 3-dB peak-to-peak polar ripple}$$

$$\approx 1.1 \frac{c}{L}, \text{ for 4-dB peak-to-peak polar ripple}$$

$$\approx 1.3 \frac{c}{L}, \text{ for 6-dB peak-to-peak polar ripple}$$

$$\approx 1.6 \frac{c}{L}, \text{ for 9-dB peak-to-peak polar ripple}$$

(9)

For the 1.0-unit spaced array,

$$f_{max} \approx 0.25 \frac{c}{L}, \text{ for 3-dB peak-to-peak polar ripple}$$

$$\approx 0.28 \frac{c}{L}, \text{ for 4-dB peak-to-peak polar ripple}$$

$$\approx 0.33 \frac{c}{L}, \text{ for 6-dB peak-to-peak polar ripple}$$

$$\approx 0.40 \frac{c}{L}, \text{ for 9-dB peak-to-peak polar ripple}$$

(10)

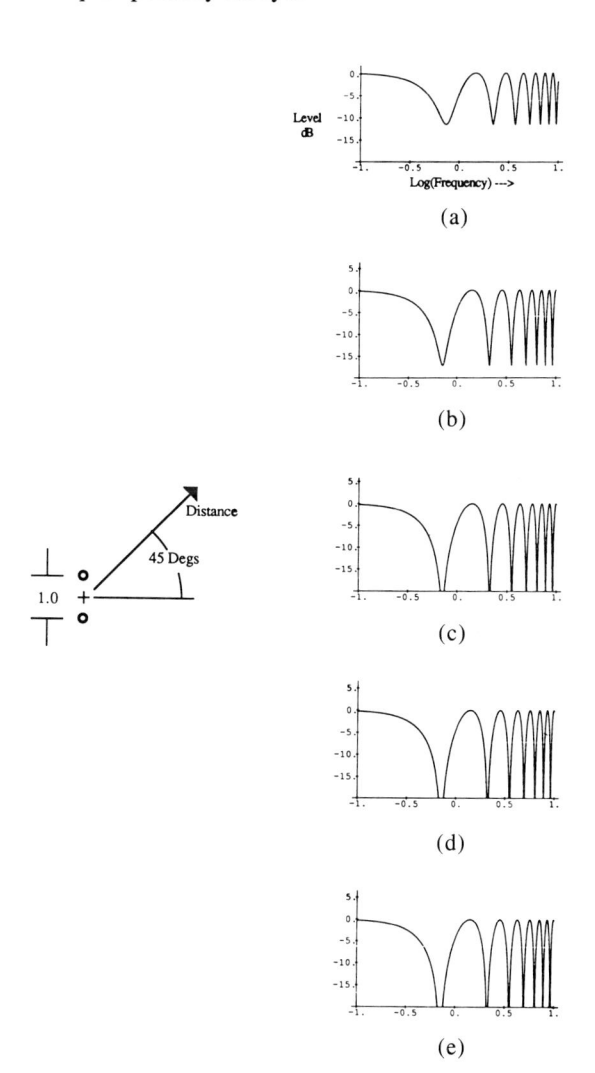

Fig. 6. Magnitude frequency responses at fixed angle of 45° and different working distances for two-source equal-level equal-polarity array with 1.0-unit center-to-center spacing and frequency range of 0.1 to 10 Hz. Note that log of frequency is indicated ($-1 = 0.1$ Hz, $0 = 1$ Hz, etc.). (a) 1.25 units. (b) 2.5 units. (c) 5 units. (d) 10 units. (e) 100 000 units. Observe that frequency response changes very little with distance beyond 5 units.

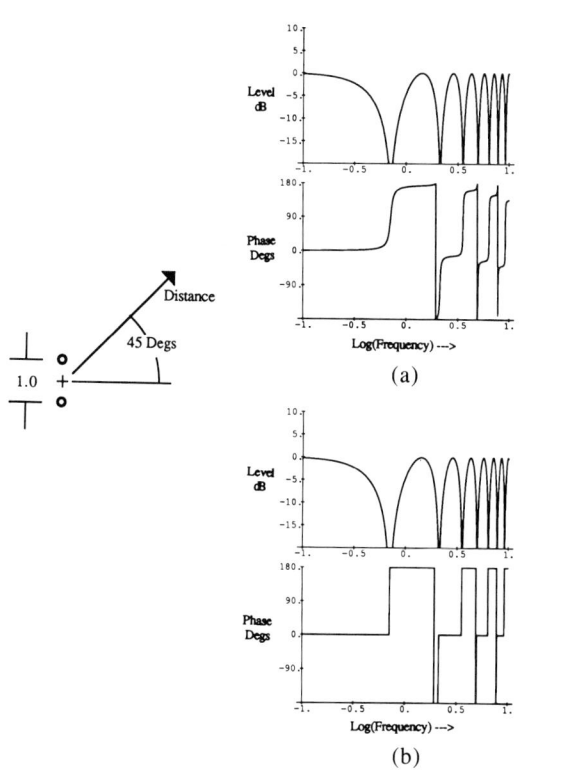

Fig. 7. Off-axis +45° magnitude and phase versus frequency responses for two-source equal-level equal-polarity array with 1.0-unit center-to-center spacing. (a) 5 units. (b) 100 000 units. Observe that phase is either 0 or $\pm 180°$, depending on the polar lobe in which the sample point happens to be. At the farther distance the phase switches very rapidly. Phase versus frequency behavior is near nonminimum phase but is actually minimum phase.

where c is the velocity of sound and L the length of the array (center-to-center distance of sources).

4.2.4 Discussion

At low frequencies, the two-source arrays exhibit mostly omnidirectional behavior below 0.25 Hz for the 1.0-unit spaced array and below 1.0 Hz for the 0.25-unit spaced array. The upper frequency limit for omnidirectional radiation occurs at the frequency where the sources are about one-quarter wavelength apart. Below this frequency, the efficiency is twice that of the single source, while the maximum output is four times that of the single source.

The directional characteristics essentially do not change with working distance beyond a point that is roughly 10 times the length of the array. The behavior of the 1.0-unit spaced two-source array exhibits the same activity as the 0.25-unit spaced two-source array, but at one-fourth the frequency.

The off-axis polar phase alternates between 0 and 180° depending on the polar lobe in which the sample point is. The off-axis phase versus frequency data exhibit the same switching behavior with increasing frequency, but are found to be minimum phase.

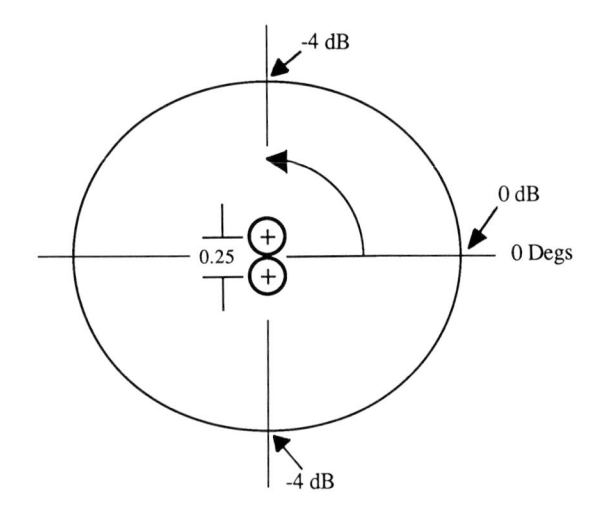

Fig. 10. Magnitude polar response of 0.25-unit center-to-center spaced two-source equal-level equal-polarity array at a distance of 20 units and a frequency of 1.1 Hz. This polar is at the frequency where the peak-to-peak ripple is 4 dB. Note that polar is very smooth, but squashed vertically, and exhibits its maximum deviation (-4 dB) at $\pm 90°$. Polar plot covers a range of 40 dB with $+6$ dB at the outer edge and -34 dB at the center. Polar is normalized so that on-axis level is 0 dB.

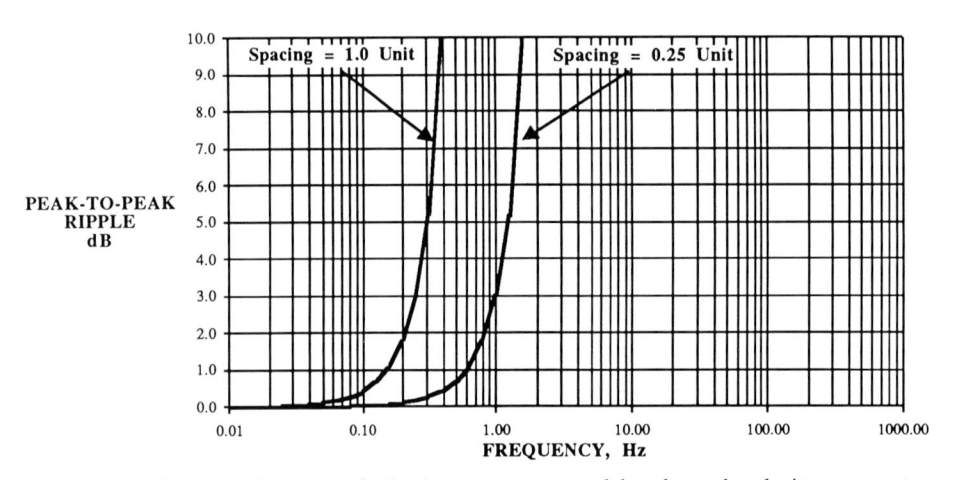

Fig. 8. Polar peak-to-peak ripple versus frequency for both two-source equal-level equal-polarity arrays at a working distance of 20 units. Note that ripple increases very rapidly above 1 Hz for 0.25-unit spaced array and above 0.25 Hz for 1.0-unit spaced array. Velocity of propagation 1 unit/s.

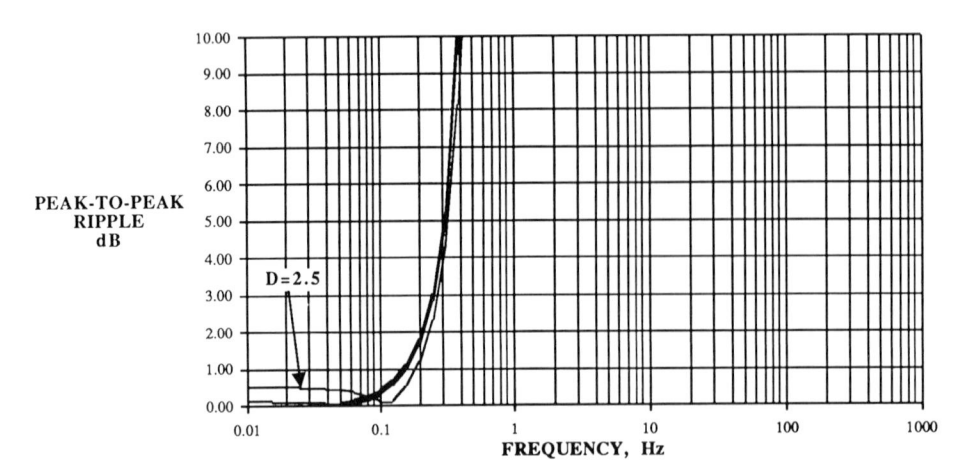

Fig. 9. Polar peak-to-peak ripple, versus frequency at working distances of 2.5, 5, 10, 20, 40, 80, 160, 1000, 10 000, and 100 000 units for 1.0-unit spaced two-source equal-level equal-polarity array. Graph exhibits essentially no change at distances beyond 5 units. Note close bunching of all curves.

4.3 Five Sources, Equal Level Equal Polarity Equal Spaced, with Overall Center-to-Center Length of 1.0 Unit

This array contains five sources equally spaced with equal levels and equal polarities. The overall length, measured from the centers of the outside sources, is 1 unit. The individual source center-to-center spacing is 0.25 unit. The characteristics and calculated parameters for this array are shown in Table 4.

This array provides 25 times the acoustic output power at low frequencies as compared to a single source. This array was included for direct comparison to the five-source Bessel array. The only difference between this array and the Bessel array is the amplitude and polarity of the source drive levels.

4.3.1 Polar Responses

The polar magnitude responses of the five-source equal-level array, at constant distance, are shown in Fig. 11. They are displayed at half-decade intervals from 0.1 to 10 Hz at a working distance of 20 units. The polar plot covers a range of 40 dB with +6 dB at the outer edge and −34 dB at the center. All polar plots are normalized so that the on-axis level is 0 dB. The polars of this array are much more complex and directive than the two-source polars (Fig. 2). For omnidirectional radiation, the five-source equal-level array ios unusable above about 0.6 Hz.

The polar magnitude responses for the five-source equal-level array, at a fixed frequency of 10 Hz and at different working distances, are shown in Fig. 12. Polars at distances of 1.25, 2.5, 5, 10, 20, and 100 000 units are shown. Note that the polar response changes very little with distance beyond roughly 10 units (10 array lengths).

Fig. 13 shows various magnitude and phase polar responses at different frequencies f and distances D for the five-source array. The following five combinations are plotted:

1) $f = 0.5$ Hz, $D = 20$ units
2) $f = 1$ Hz, $D = 20$ units
3) $f = 1$ Hz, $D = 100\ 000$ units
4) $f = 2$ Hz, $D = 20$ units
5) $f = 2$ Hz, $D = 100\ 000$ units.

The phase versus direction plots show the phase of the pressure at the sample point versus the off-axis direction. The effects of the linear phase delay due to sample distance have been eliminated. Note that the phase values switch between 0 and +180° depending on the polar lobe on which the pressure sample point happens to be. The phase always starts out at 0° (on

Table 4. Array type: five sources, equal level, equal spacing, same polarity.

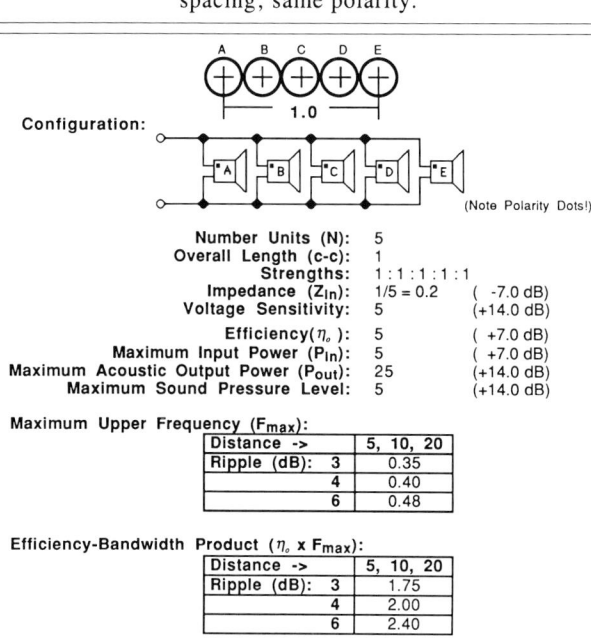

Configuration:

Number Units (N): 5
Overall Length (c-c): 1
Strengths: 1 : 1 : 1 : 1 : 1
Impedance (Z_{In}): 1/5 = 0.2 (−7.0 dB)
Voltage Sensitivity: 5 (+14.0 dB)
Efficiency (η_o): 5 (+7.0 dB)
Maximum Input Power (P_{In}): 5 (+7.0 dB)
Maximum Acoustic Output Power (P_{out}): 25 (+14.0 dB)
Maximum Sound Pressure Level: 5 (+14.0 dB)

Maximum Upper Frequency (F_{max}):

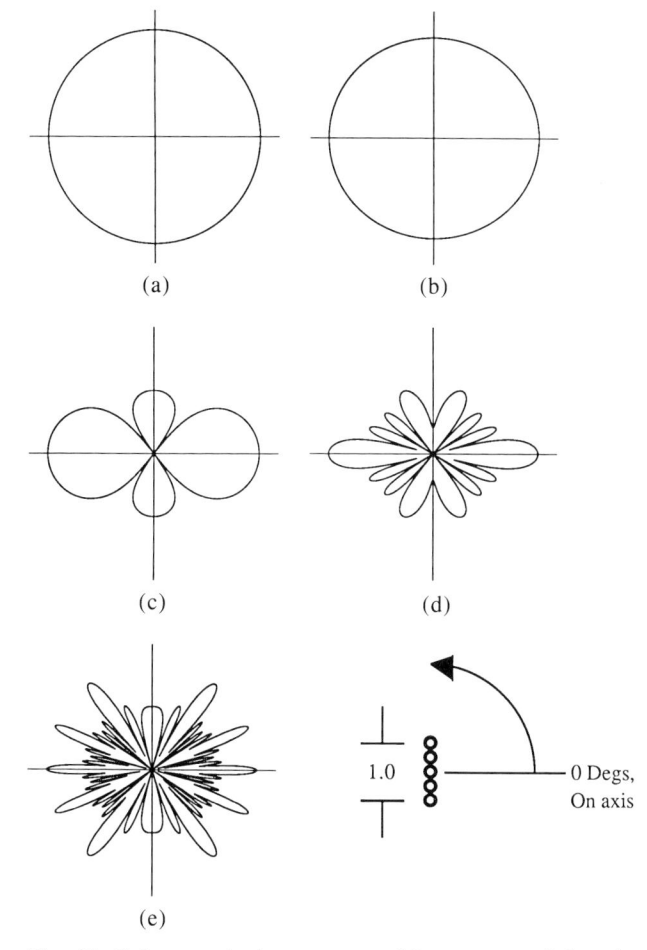

Distance →	5, 10, 20
Ripple (dB): 3	0.35
4	0.40
6	0.48

Efficiency-Bandwidth Product ($\eta_o \times F_{max}$):

Distance →	5, 10, 20
Ripple (dB): 3	1.75
4	2.00
6	2.40

Power-Bandwidth Product ($P_{out} \times F_{max}$):

Distance →	5, 10, 20
Ripple (dB): 3	8.75
4	10.00
6	12.00

Power-Bandwidth Product per Unit ($P_{out} \times F_{max} / N$):

Distance →	5, 10, 20
Ripple (dB): 3	1.75
4	2.00
6	2.40

Fig. 11. Polar magnitude responses of five-source unit-length equal-level equal-polarity equal-spaced array at a constant working distance of 20 units. Polars are displayed at half-decade intervals from 0.1 Hz to 10 Hz. (a) 0.1 Hz. (b) 0.316 Hz. (c) 1 Hz. (d) 3.16 Hz. (e) 10 Hz. Note how directive and complex polars get above 0.316 Hz. Polar plot covers a range of 40 dB with +6 dB at the outer edge and −34 dB at the center. All polars are normalized so that on-axis level is 0 dB.

axis). At distances far from the array, the phase changes occur more abruptly with angle. The phase variation with direction for the five-source array is very similar to the behavior of the two-source arrays.

4.3.2 Frequency Responses

The magnitude versus frequency responses of the five-source equal-level array at constant distance are shown in Fig. 14. The responses are plotted at angles ranging from 0 to 90°, with steps of 15° at a working distance of 20 units. Note that the response gets progressively rougher as the angle increases, similarly to the two-source arrays.

The magnitude versus frequency responses at a fixed angle of 45° and at different working distances are plotted in Fig. 15. Distances of 1.25, 2.5, 5, 10, and 100 000 units are shown. Note that the frequency response changes very little with distance beyond about 5 units.

The phase versus frequency behavior of the five-source array is shown in Fig. 16, where responses at 45° off axis at distances of 20 and 100 000 units are plotted. The effects of linear phase lag and delay due to sample distance have been eliminated. The phase activity versus frequency is very similar to that of the two-source arrays, but is highly likely to be nonminimum phase due to the existence of the additional sources. The phase toggles rapidly between 0 and ±180° as frequency increases.

4.3.3 Polar Peak-to-Peak Ripple versus Frequency

Fig. 17 exhibits a plot of polar peak-to-peak ripple, in decibels, versus frequency for the five-source equal-level array at a working distance of 20 units. Note that

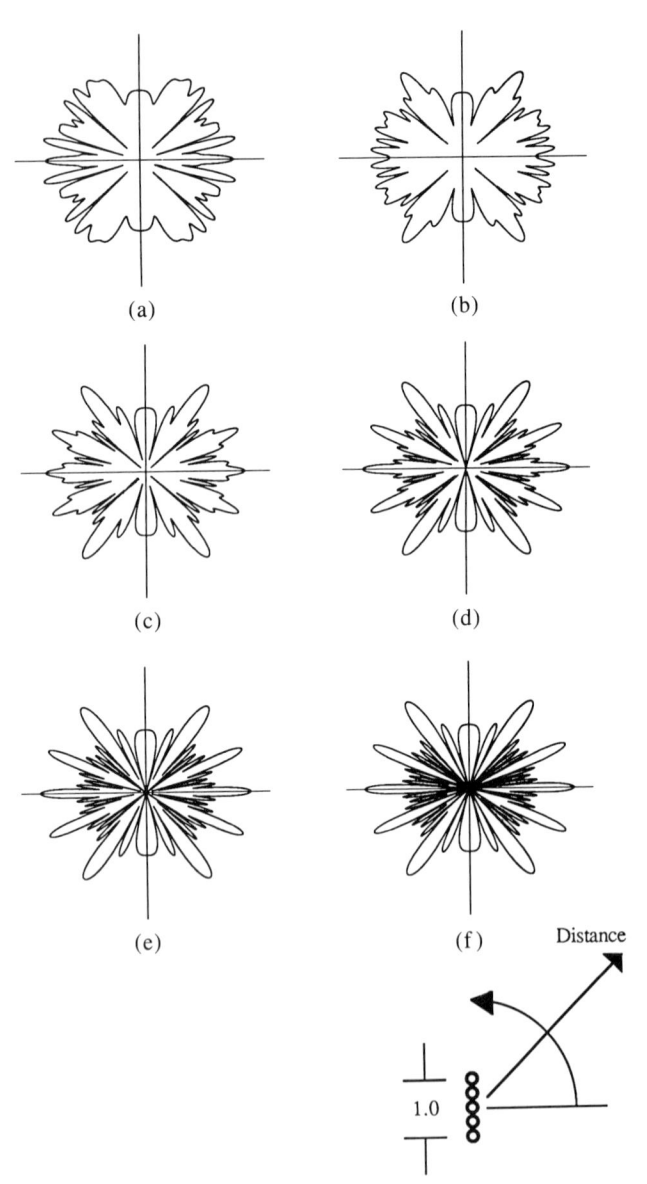

Fig. 12. Polar magnitude responses for five-source unit-length equal-level equal-polarity equal-spaced array at a fixed frequency of 10 Hz and different working distances. (a) 1.25 units. (b) 2.5 units. (c) 5 units. (d) 10 units. (e) 20 units. (f) 100 000 units. Note that polar response changes very little with distance beyond roughly 10 units (10 array lengths).

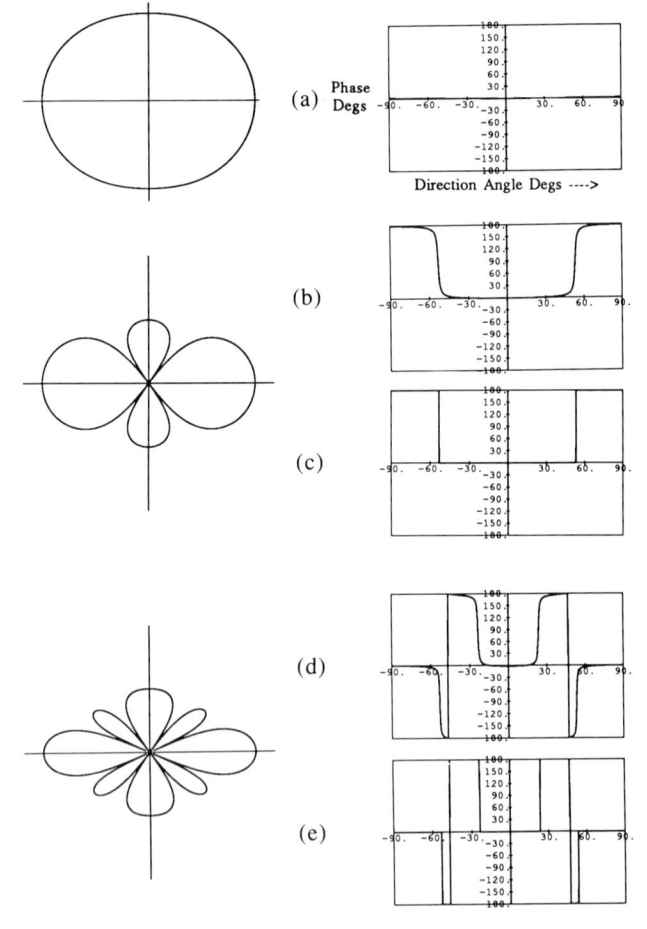

Fig. 13. Magnitude and phase polar responses at different frequencies and distances for five-source unit-length equal-level equal-polarity equal-spaced array. (a) $f = 0.5$ Hz, $D = 20$ units. (b) $f = 1$ Hz, $D = 20$ units. (c) $f = 1$ Hz, $D = 100\,000$ units. (d) $f = 2$ Hz, $D = 20$ units. (e) $f = 2$ Hz, $D = 100\,000$ units. Phase versus direction plots show phase of pressure at sample point versus off-axis direction. Effects of linear phase delay due to sample distance have been eliminated. Note that phase values switch between 0 and ±180°, depending on the polar lobe on which the pressure sample point happens to be. Phase always starts out at 0° (on axis). At distances far from array, phase changes occur more abruptly with angle.

the ripple increases very rapidly above 0.35 Hz. The five-source array has somewhat better performance than the two-source 1.0-unit array, but significantly lower performance than the 0.25-unit two-source array (see Fig. 8). Additional data (not shown) indicate that the polar peak-to-peak ripple essentially does not change with working distances beyond about 10 units. This behavior is similar to that of the two-source arrays (see Fig. 9).

The following approximate equations relate the maximum operating frequency for omnidirectional radiation f_{max} to the array length for the five-source equal-level equal-polarity array:

$$f_{max} \approx 0.35 \frac{c}{L}, \text{ for 3-dB peak-to-peak polar ripple}$$

$$\approx 0.40 \frac{c}{L}, \text{ for 4-dB peak-to-peak polar ripple}$$

$$\approx 0.48 \frac{c}{L}, \text{ for 6-dB peak-to-peak polar ripple}$$

$$\approx 0.56 \frac{c}{L}, \text{ for 3-dB peak-to-peak polar ripple}$$

(11)

where c is the velocity of sound and L the length of the array (center-to-center distance of outside sources). Compare these multipliers to the previous values for the two-source arrays given in Eqs. (9) and (10).

4.3.4 Discussion

At low frequencies, below about 0.35 Hz, the five-source array exhibits mostly omnidirectional behavior. The upper frequency limit for omnidirectional radiation occurs at the frequency where the length of the array is about one-third wavelength, which is somewhat higher than for the two-element array. Below this frequency, the efficiency is five times that of the single source, while the maximum output is 25 times higher.

The five-source equal-level array of 1.0-unit center-to-center length operates slightly higher in frequency than the two-source equal-level array with 1.0-unit center-to-center spacing, but significantly lower than the two-source equal-level array with 0.25-unit center-to-center spacing. The phase versus frequency curve is nonminimum phase. The phase versus angle and phase versus frequency curves alternate between 0 and ±180°.

The directional characteristics essentially do not

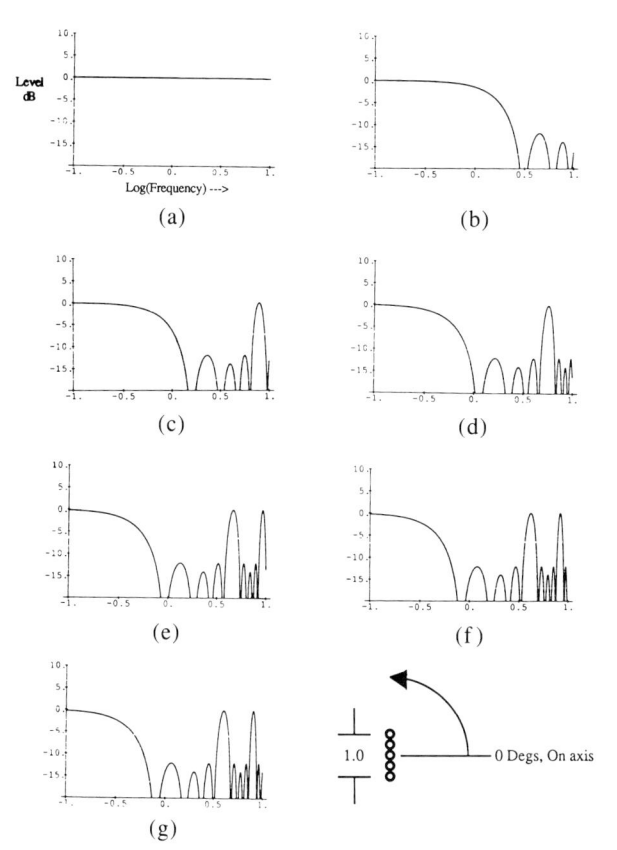

Fig. 14. Magnitude versus frequency responses for five-source unit-length equal-level equal-polarity equal-spaced array at a constant distance of 20 units and frequency range of 0.1 to 10 Hz. Note that log of frequency is indicated (−1 = 0.1 Hz, 0 = 1 Hz, etc.). Responses are shown at angles ranging from 0 to +90° with steps of 15°. (a) 0°. (b) 15°. (c) 30°. (d) 45°. (e) 60°. (f) 75°. (g) 90°. Note that response progressively gets rougher as angle increases, similarly to two-source arrays.

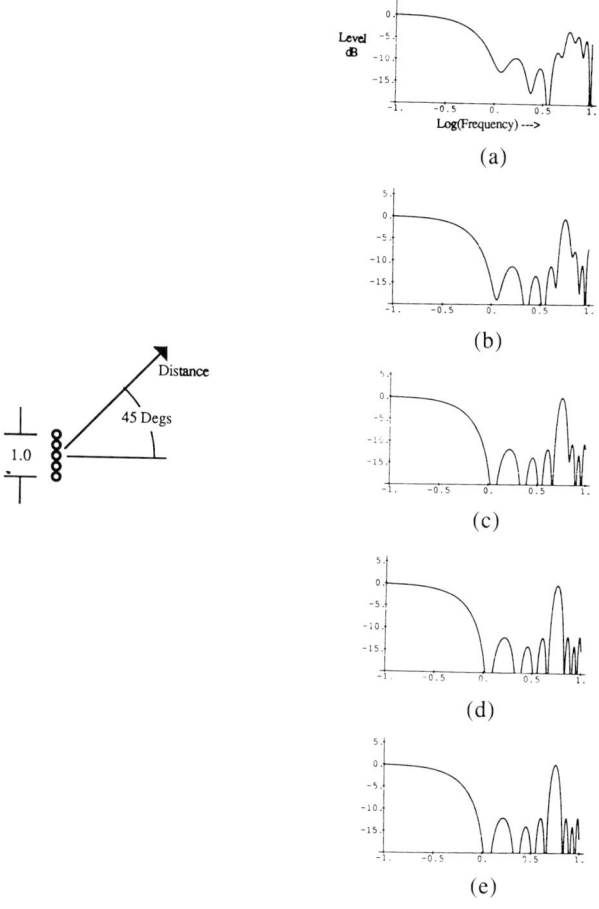

Fig. 15. Magnitude versus frequency responses at a fixed angle of 45° and different working distances for five-source unit-length equal-level equal-polarity equal-spaced array, with frequency range of 0.1 to 10 Hz. Note that log of frequency is indicated (−1 = 0.1 Hz, 0 = 1 Hz, etc.). (a) 1.25 units. (b) 2.5 units. (c) 5 units. (d) 10 units. (e) 100 000 units. Note that frequency response changes very little with distance beyond about 5 units.

change beyond a point that is roughly 10 times the length of the array (roughly the same as for the two-source equal-level array).

4.4 Five-Source Bessel Array with Overall Center-to-Center Length of 1.0 Unit

The Bessel configuration is used to gain increased acoustic output without the severe narrowing directional characteristics with frequency exhibited by the equal-level equal-polarity equal-spaced line arrays. The Bessel array is said to have the same overall directional pattern as one of the sources that make up the array.

The following simulation uses omnidirectional point sources to form the Bessel structure. The degree to which the overall polar response matches an omnidirectional pattern is used to judge the effectiveness of the Bessel array. The five-source Bessel array contains the fewest number of sources of the three analyzed Bessel configurations. The characteristics and calculated parameters for the five-source Bessel array with 1.0-unit overall length are shown in Table 5.

Because of the much greater upper frequency of the Bessel array, all the bandwidth product values are much higher than those for the previous arrays. However, the efficiency is only about 14% (+0.6 dB) greater than that of a single source. With the higher power handling of 3.5 times a single source, the maximum acoustic output of the Bessel array is the same as that of the two-source arrays.

4.4.1 Polar Responses

The polar magnitude responses of the five-source Bessel array, at constant distance, are shown in Fig. 18. The polars are displayed at half-decade intervals from 0.316 to 100 Hz at a working distance of 20 units. The polar plot covers a range of 40 dB with +6 dB at the outer edge and −34 dB at the center. All polars are normalized so that the on-axis level is 0 dB.

Note the much greater high-frequency range of operation as compared with the previous arrays. The polar ripple does not get significant until frequencies higher than about 10 Hz.

The polar magnitude responses for the five-source Bessel array, at a fixed frequency of 10 Hz and at different working distances, are shown in Fig. 19. Polars at distances of 1.25, 2.5, 5, 10, 20, 40, and 100 000 units are plotted. Note that, unlike for the previous arrays, the polar ripple appears to get smaller and smaller

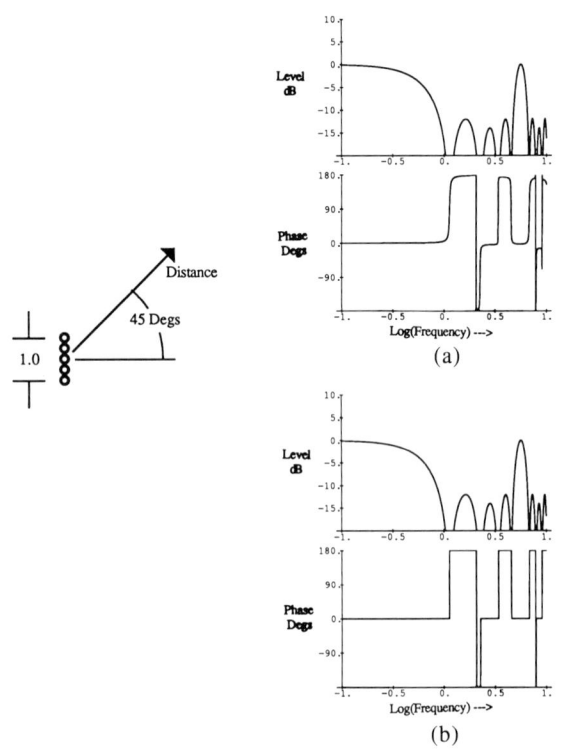

Fig. 16. Off-axis 45° magnitude and phase versus frequency plots over frequency range of 0.1 to 10 Hz for five-source unit-length equal-level equal-polarity equal-spaced array. (a) 20 units. (b) 100 000 units. Note that log of frequency is indicated (−1 = 0.1 Hz, 0 = 1 Hz, etc.). Effects of linear phase lag and delay due to sample distance have been eliminated. Phase activity versus frequency is very similar to two-source arrays, but is highly likely to be nonminimum phase due to existence of additional sources. Phase toggles rapidly between 0 and ±180° as frequency increases.

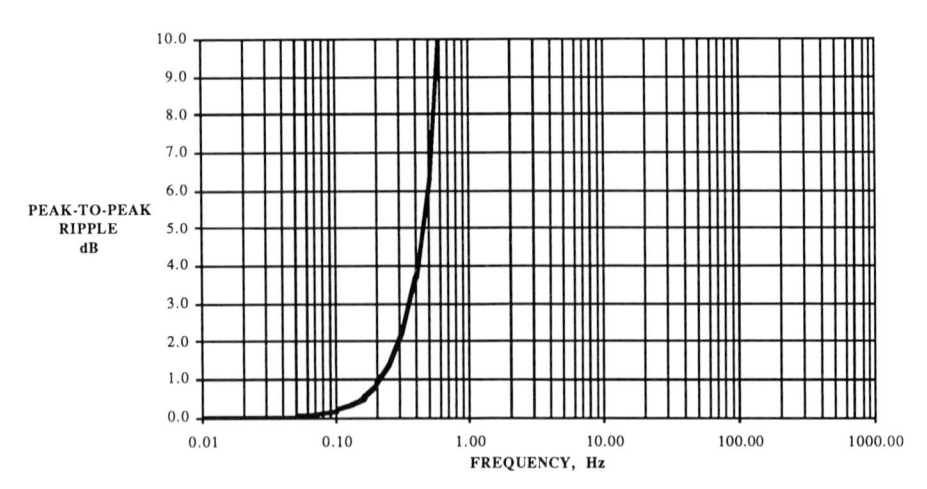

Fig. 17. Plot of polar peak-to-peak ripple versus frequency for five-source unit-length equal-level equal-polarity equal-spaced array at a working distance of 20 units. Note that ripple increases very rapidly above 0.35 Hz. Five-source unit-length array has somewhat better performance than two-source unit-length array, but significantly lower than 0.25-unit two-source array (see Fig. 8). Velocity of propagation 1 unit/s.

the farther away you get from the array. However, a limit of about 1.2 dB peak-to-peak ripple appears to exist even at the farthest distance. This figure is confirmed in [2, table 2].

For omnidirectional radiation, at a distance of 20 units, with no more than 6-dB peak-to-peak ripple, the five-source Bessel array is usable up to beyond 18 Hz. As compared to an equal-level equal-polarity two-source array with center-to-center spacing equal to the overall center-to-center spacing of the Bessel array (0.33 Hz from Table 3), this represents an increase in upper frequency of about 55 times ($\approx 18/0.33$).

A further study of the variation of polar ripple with distance was performed by simulating at the much higher frequency of 100 Hz (where the array length is 100 wavelengths) and then varying the working distance from 10 to 1000 units in three steps of one decade each. Fig. 20 shows the results of these simulations. The polar at 10-unit distance is unusable due to severe polar ripple (about 40 dB peak to peak). It settles down to about 2-dB peak-to-peak ripple at a distance of 1000 units (a long way away). It appears that there is no effective upper limit to the frequency of operation of the Bessel array if you can get far enough away. Practically, however, working distances in the range of 5–20 times the length of the array will define the operation of the array.

Fig. 21 shows a series of phase polar responses (phase versus direction angle) at a constant distance of 20 units and frequencies of 0, 0.1, 0.5, 1, 2, 4, 5, 10, and 20 Hz. The delay effects of the working distance have been compensated for, thus making the on-axis phase zero in every case. Also shown is a phase polar response at a distance of 100 000 units at 20 Hz. The phase curves exhibit a highly nonlinear sinusoidal-like variation of phase with angle with a peak-to-peak amplitude of $\pm 90°$. For a fixed angular increment, the number of oscillation cycles increases with frequency.

4.4.2 Frequency Responses

The magnitude versus frequency responses of the five-source Bessel array, at constant distance, are shown in Fig. 22. The responses cover the range from 0.1 to 10 Hz, and are plotted at angles ranging from 0 to $+90°$, with steps of $15°$, at a working distance of 20 units. Unlike the previous equal-level arrays, the ripple does not increase continually with angle.

Table 5. Array type: five-source Bessel array.

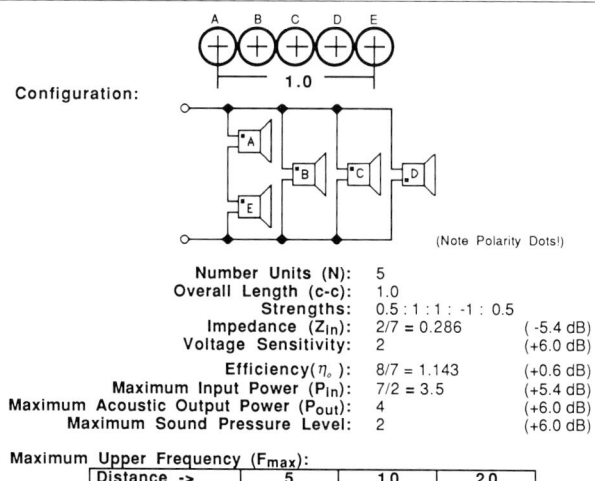

Configuration:

Number Units (N): 5
Overall Length (c-c): 1.0
Strengths: 0.5 : 1 : 1 : -1 : 0.5
Impedance (Z_{in}): 2/7 = 0.286 (-5.4 dB)
Voltage Sensitivity: 2 (+6.0 dB)
Efficiency(η_o): 8/7 = 1.143 (+0.6 dB)
Maximum Input Power (P_{in}): 7/2 = 3.5 (+5.4 dB)
Maximum Acoustic Output Power (P_{out}): 4 (+6.0 dB)
Maximum Sound Pressure Level: 2 (+6.0 dB)

Maximum Upper Frequency (F_{max}):

Distance ->	5	10	20
Ripple (dB): 3	2.05	4.00	8.00
4	3.00	6.00	11.00
6	4.50	8.80	18.00

Efficiency-Bandwidth Product ($\eta_o \times F_{max}$):

Distance ->	5	10	20
Ripple (dB): 3	2.34	4.57	9.14
4	3.43	6.86	12.57
6	5.14	10.06	20.57

Power-Bandwidth Product ($P_{out} \times F_{max}$):

Distance ->	5	10	20
Ripple (dB): 3	8.20	16.00	32.00
4	12.00	24.00	44.00
6	18.00	35.20	72.00

Power-Bandwidth Product per Unit ($P_{out} \times F_{max} / N$):

Distance ->	5	10	20
Ripple (dB): 3	1.64	3.20	6.40
4	2.40	4.80	8.80
6	3.60	7.04	14.40

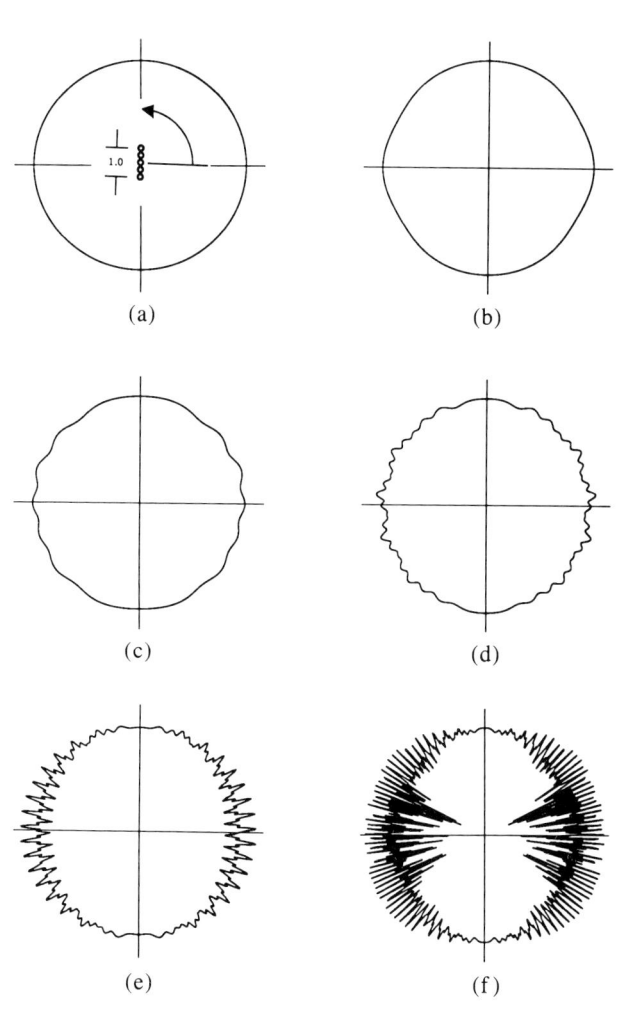

Fig. 18. Polar magnitude responses for five-source unit-length Bessel array at a constant working distance of 20 units. Polars are displayed at half-decade intervals from 0.316 to 100 Hz. (a) 0.316 Hz. (b) 1 Hz. (c) 3.16 Hz. (d) 10 Hz. (e) 31.6 Hz. (f) 100 Hz. Note much greater high-frequency range of operation as compared with previous arrays. Polar ripple does not get significant until frequencies higher than about 10 Hz, where line length is 10 wavelengths. Polar plots cover a range of 40 dB with +6 dB at the outer edge and −34 dB at the center. All polars are normalized so that on-axis level is 0 dB.

The magnitude versus frequency responses at a fixed angle of 45° and at different working distances are shown in Fig. 23. Note the wider frequency range of 0.1 to 100 Hz. Responses at distances from 1.25 to 160 units with 1:2 steps are simulated, in addition to one at 100 000 units. Note, however, that, unlike for the previous arrays, the frequency response ripple decreases continually with distance until about a 2-dB peak-to-peak ripple is attained. This again reinforces the observation that the Bessel array performance can reach any arbitrary upper frequency if you move far enough away from the array.

The phase versus frequency behavior of the five-source Bessel array is shown in Fig. 24, with magnitude, phase, and group delay responses at 45° off axis at distances of 20 units. Both log and linear frequency scale plots are shown, up to a frequency of 10 Hz. The phase varies nonlinearly, in a somewhat sinusoidal manner with frequency, oscillating between ±90°. The magnitude response is mostly flat, with peak-to-peak ripple, with more amplitude variations per unit frequency.

Because the magnitude is mostly flat and the phase varies dramatically with frequency, this magnitude–phase behavior versus frequency is highly nonlinear and nonminimum phase. The group delay plots of Fig. 24(c) and (f) indicate an effective oscillatory peak shift of the acoustic position of about ±25% the length of the array as the frequency is increased. I am not going to venture an opinion on whether or not this is audible.

Fig. 19. Polar magnitude responses for five-source unit-length Bessel array at fixed frequency of 10 Hz and different working distances. (a) 1.25 units. (b) 2.5 units. (c) 5 units. (d) 10 units. (e) 20 units. (f) 40 units. (g) 100 000 units. Note that, unlike previous arrays, polar ripple appears to get smaller and smaller the farther away you get from the array. Polar plots cover a range of 40 dB with +6 dB at the outer edge and −34 dB at the center. All polars are normalized so that on-axis level is 0 dB.

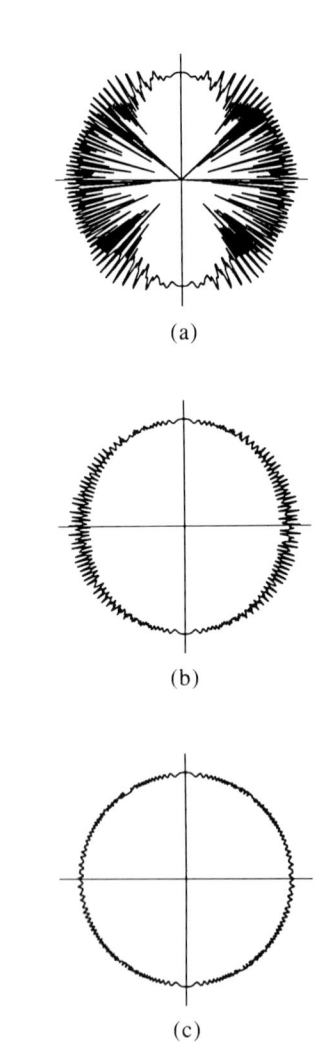

Fig. 20. Polar magnitude responses for five-source unit-length Bessel array, but at a much higher fixed frequency of 100 Hz and much farther working distances, covering the range of 10 to 1000 units in three steps of one decade each. (a) 10 units. (b) 100 units. (c) 1000 units. Note that even at this high frequency, where line is 100 wavelengths long, at large distances polar peak-to-peak ripple settles down to relatively small values. Polar plots cover a range of 40 dB with +6 dB at the outer edge and −34 dB at the center. All polars are normalized so that on-axis level is 0 dB.

4.4.3 Polar Peak-to-Peak Ripple versus Frequency

Fig. 25 exhibits a plot of polar peak-to-peak ripple, in decibels, versus frequency for the five-source Bessel array at a working distance of 20 units. Observe that the ripple increases much more gradually with increasing frequency as compared to the equal-level arrays. Note also the much extended bandwidth of operation as compared to the previous arrays. Also observe the plateau in the curve between 0.5 and 1.1 Hz, where the ripple is about 1.3 dB. At a peak-to-peak ripple of 6 dB, operation extends up to a frequency of 18 Hz (line length of 18 wavelengths).

To investigate the behavior of ripple with increasing distance, numerous plots of ripple versus frequency were done over the distance range of 2.5 to 100 000 units. These data are plotted in Fig. 26. It is quite evident that the operation of the Bessel array improves in direct proportion to the working distance away from the array. Note that at points very far from the array, the peak-to-peak ripple attains a constant value of about 1.3 dB; this is the source of the plateau noted previously.

Fig. 27 shows a plot of maximum operating frequency versus operating distance for the five-source Bessel array. Contours of equal peak-to-peak ripple at values of 3, 6, and 9 dB are plotted. The direct relationship between maximum frequency and operating distance is evident. The contours of constant peak-to-peak ripple form straight lines on the graph, except for slight deviations at small distances.

The following approximate equations relate the maximum operating frequency for omnidirectional radiation f_{max} to array length L and operating distance D, for the five-element Bessel array. Note the dependence on distance, which was absent in the previous array equations [Eqs. (9)–(11)].

$$f_{max} \approx 0.40 \frac{c}{L} D, \text{ for 3-dB peak-to-peak polar ripple}$$

$$\approx 0.55 \frac{c}{L} D, \text{ for 4-dB peak-to-peak polar ripple}$$

$$\approx 0.90 \frac{c}{L} D, \text{ for 6-dB peak-to-peak polar ripple}$$

$$\approx 1.4 \frac{c}{L} D, \text{ for 9-dB peak-to-peak polar ripple} \quad (12)$$

where

D = normalized operating distance, = d/L
c = velocity of sound
d = working distance away from center of array
L = length of array (center-to-center distance of outside sources).

For comparison with the equations for the previous arrays [Eqs. (9)–(11)], the following equations evaluate Eqs. (12) at a distance of 20 units:

$$f_{max} \approx 8 \frac{c}{L}, \text{ for 3-dB peak-to-peak polar ripple}$$

$$\approx 11 \frac{c}{L}, \text{ for 4-dB peak-to-peak polar ripple}$$

$$\approx 18 \frac{c}{L}, \text{ for 6-dB peak-to-peak polar ripple}$$

$$\approx 28 \frac{c}{L}, \text{ for 9-dB peak-to-peak polar ripple} \quad (13)$$

Note the large multipliers as compared to the previous array equations.

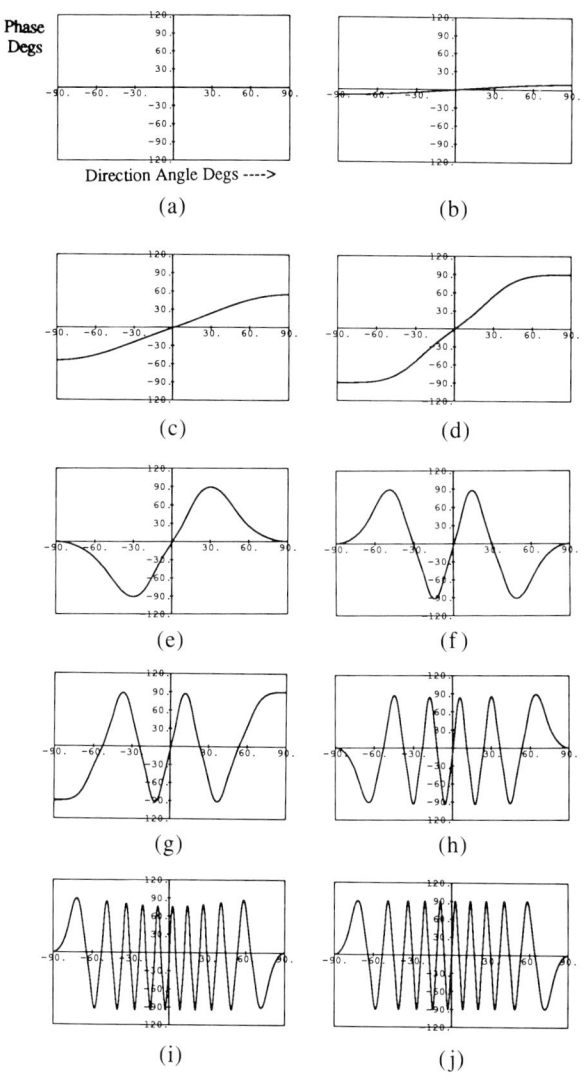

Fig. 21. Phase polar responses (phase versus direction angle) for five-source unit-length Bessel array at constant distance of 20 units and different frequencies. (a) 0 Hz. (b) 0.1 Hz. (c) 0.5 Hz. (d) 1 Hz. (e) 2 Hz. (f) 4 Hz. (g) 5 Hz. (h) 10 Hz. (i) 20 Hz. (j) 20 Hz at a distance of 100 000 units. Delay effects of working distance have been compensated for, thus making on-axis phase zero in every case. Phase curves exhibit highly nonlinear sinusoidal-like variation of phase with angle with a peak-to-peak amplitude of ±90°. For a fixed angular increment, number of oscillation cycles increases with frequency.

4.4.4 Discussion

The five-element Bessel array provides a very impressive increase in the bandwidth of operation when compared to equivalent two- and five-source equal-level equal-polarity equal-spaced arrays. The efficiency–bandwidth product, power–bandwidth product, and power–bandwidth product per unit are all very high in comparison to the previous arrays.

When compared to a two-source equal-level in-phase array, a five-source Bessel array is 2.4 dB less efficient, can handle 1.75 (+2.4 dB) more power, has the same maximum midband acoustic output power, and is usable for omnidirectional radiation 10 times higher in frequency. A working distance of 20 times the length of the Bessel array is assumed, with the length of the Bessel array (center-to-center measurement) being four times that of the two-source array.

The very nonlinear phase behavior with direction angle and frequency appears to be the single major problem with the Bessel array. Whereas a single point source has true omnidirectional radiation, it does not exhibit any variation of phase with angle or frequency (neglecting transport delay between source and sample point). The Bessel array's off-axis variation of phase with angle and frequency makes it very difficult to use it with any other sources. Computation of group delay versus frequency at an angle of 45° indicates an oscillatory movement of the acoustic position about ±25% of the array's length.

Eqs. (12) clearly show that the high-frequency limit of the Bessel array increases in direct proportion to the working distance from the array. This is in contrast to the behavior of the two- and five-source equal-level equal-polarity equal-spaced arrays, where the performance does not change beyond a fairly close distance measured in terms of the array length (about 10 times the array length). This means that the Bessel array is not like a conventional source that exhibits a typical near-field/far-field difference in its behavior. The Bessel array does not have a definite near-field/far-field boundary which defines its behavior.

Fig. 22. Magnitude versus frequency responses for five-source unit-length Bessel array at a constant working distance of 20 units. Responses cover the range from 0.1 to 10 Hz and are shown at angles ranging from 0 to +90°, with steps of 15°. (a) 0°. (b) 15°. (c) 30°. (d) 45°. (e) 60°. (f) 75°. (g) 90°. Unlike previous equal-level arrays, ripple does not increase continually with angle. Note that log of frequency is indicated (−1 = 0.1 Hz, 0 = Hz, etc.).

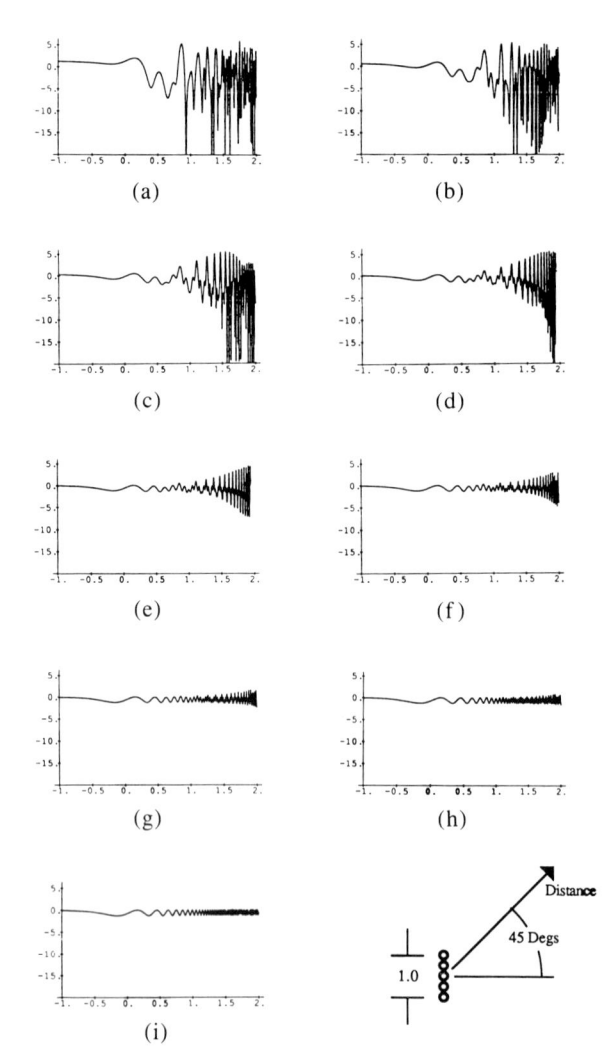

Fig. 23. Magnitude versus frequency responses for five-source unit-length Bessel array at a fixed angle of 45° and working distances of 1.25 to 160 units with 1:2 steps and at 100 000 units. Note wider frequency range of 0.1 to 100 Hz. (a) 1.25 units. (b) 2.5 units. (c) 5 units. (d) 10 units. (e) 20 units. (f) 40 units. (g) 80 units. (h) 160 units. (i) 100 000 units. Unlike previous arrays frequency response ripple decreases continually with distance until about a 2-dB peak-to-peak ripple is attained. Note that log of frequency is indicated (−1 = 0.1 Hz, 0 = 1 Hz, etc.).

4.5 Seven-Source Bessel Array with Overall Center-to-Center Length of 1.5 Units

As noted in Sec. 1, the seven-source Bessel array actually has six sources instead of seven, because the middle source has a drive level of zero, and thus does not have to be there physically. The space for the removed source must exist to preserve proper operation of the array, however. The length of the seven-source Bessel array was chosen to be 1.5 units (center-to-center spacing of outside sources). This specific length was selected because it is the length of the seven-source array when composed of the same-size units as the five-source array. The characteristics and calculated parameters for the seven-source Bessel array are shown in Table 6.

The efficiency of the seven-source Bessel array is actually about 11% (0.5 dB) less than that of one of the single sources that make up the array. The efficiency is also about 22% less than that of the five-source Bessel array. With the increased power handling of 4.5 (+6.5 dB), this generates a maximum output of 4 W (+6 dB), which is the same as the maximum output of the two-source equal-level equal-polarity array and the five-source Bessel array.

Because of the additional element required and the lower bandwidth, this array's power–bandwidth product per unit is less than half that of the five-source Bessel array. For this reason, very few response curves were generated for the seven-source Bessel array because of its relatively poor characteristics.

4.5.1 Polar Responses

Only one polar response was generated for the seven-source Bessel array. This is shown in Fig. 30 (Sec. 5.3), which compares the polars of all the arrays at a specific frequency and working distance.

4.5.2 Frequency Responses

Only one frequency response was calculated for the seven-source Bessel array and is shown in Fig. 31 (Sec. 5.4), where the response is compared with those of the other analyzed arrays.

4.5.3 Polar Peak-to-Peak Ripple versus Frequency

Fig. 28 exhibits a plot of the seven-source Bessel array's polar peak-to-peak ripple, in decibels, versus

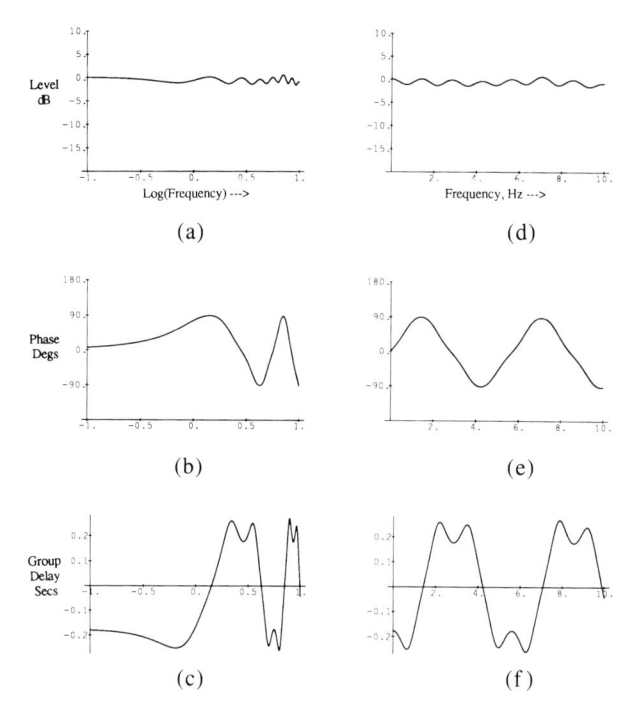

Fig. 24. Magnitude, phase, and group delay versus frequency responses for five-source unit-length Bessel array at 45° off axis and a working distance of 20 units. Both logarithmic and linear frequency scale plots are shown, up to a frequency of 10 Hz. (a) Magnitude, log scale. (b) Phase, log scale. (c) Group delay, log scale. (d) Magnitude, linear scale. (e) Phase, linear scale. (f) Group delay, linear scale. Magnitude response is mostly flat, with about a 2-dB peak-to-peak ripple. Phase varies nonlinearly in a somewhat sinusoidal manner with frequency, oscillating between ±90° which indicates a nonminimum phase response. Group delay plot indicates an effective oscillatory peak shift of acoustic position of about ±25% the length of the array, as the frequency is increased.

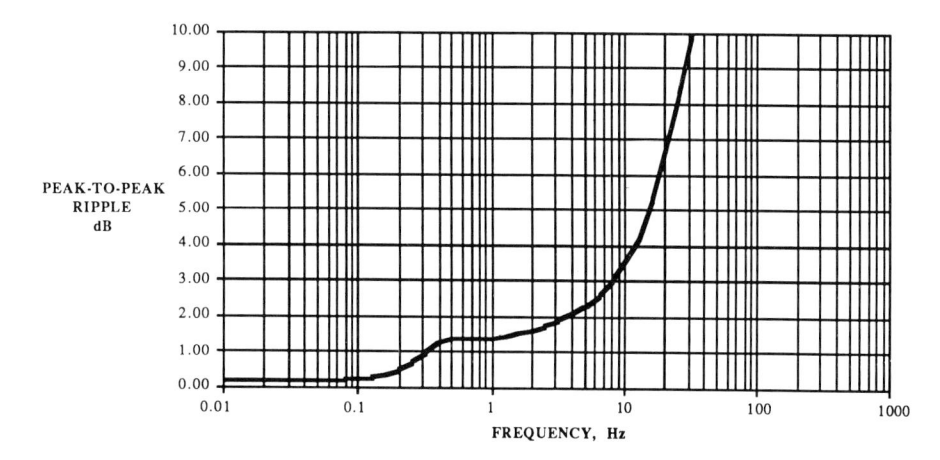

Fig. 25. Polar magnitude peak-to-peak ripple versus frequency for five-source unit-length Bessel array at a working distance of 20 units. Observe that ripple increases much more gradually with increasing frequency as compared to equal-level arrays. Note also much extended bandwidth of operation as compared to previous arrays. Velocity of propagation 1 unit/s.

frequency for working distances of 5, 10, 20, 100, and 100 000 units. At large distances, the ripple attains a minimum plateau value of about 1.0 dB. Close to the array ($D = 5$), the ripple does not go below about 2 dB. In general, the curves are shifted to the left, as compared to the five-source Bessel array, which indicates lower frequencies of operation.

4.5.4 Discussion

As stated in the introduction to this section, the disappointing performance of the seven-source Bessel array as compared to the five-source Bessel array makes it undesirable for practical use.

4.6 Nine-Source Bessel Array with Overall Center-to-Center Length of 2.0 Units

As noted in Sec. 1, the nine-source Bessel array actually has seven sources instead of nine, because two of the sources have drive levels of zero, and thus do not have to be in the array. The spaces for the removed sources must exist to preserve proper operation of the array, however. The overall length of the nine-source Bessel array is 2.0 units (center-to-center spacing of outside sources). This length was chosen because it results from using the same size sources as those used in the previous five-source arrays. The characteristics and the calculated parameters for the nine-source Bessel array are shown in Table 7.

The efficiency of the nine-source Bessel array is actually about 27% (1.4 dB) less than that of one of the single sources that make up the array. The efficiency is also about 36% less than that of the five-source Bessel array. With the increased power handling of 5.5 (+7.4 dB), this generates a maximum output of 4 W (+6 dB), which is the same as the maximum output of the two-source equal-level equal-polarity array and the five- and seven-source Bessel arrays. Also, the maximum upper frequency of the nine-source Bessel array (assuming 4-dB peak-to-peak ripple) is less than one-eighth that of the five-source Bessel array.

Because of the two additional elements required and drastically lower bandwidth, this array's power–bandwidth product per unit is less than one-tenth that of the five-source Bessel array. This very

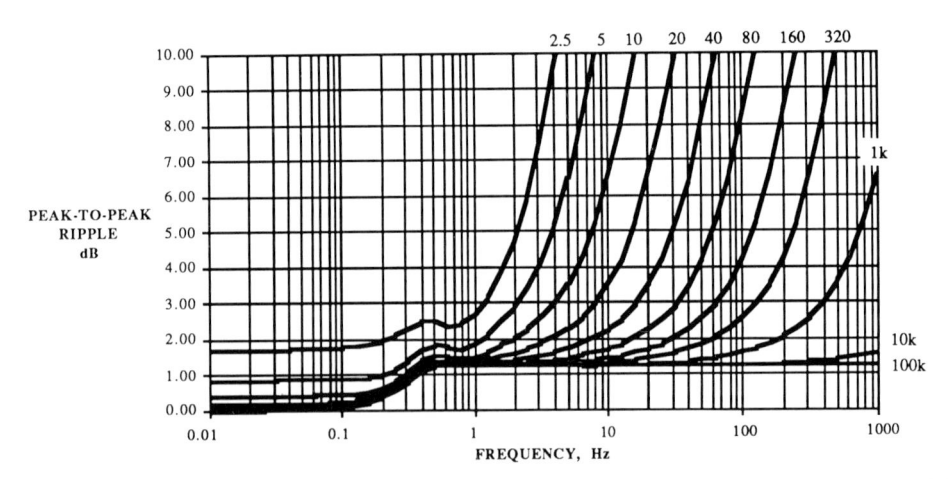

Fig. 26. Polar magnitude peak-to-peak ripple versus frequency five-source unit-length Bessel array at working distances of 2.5 to 100 000 units. It is quite evident that operation of the Bessel array improves in direct proportion to the working distance away from the array. Velocity of propagation 1 unit/s.

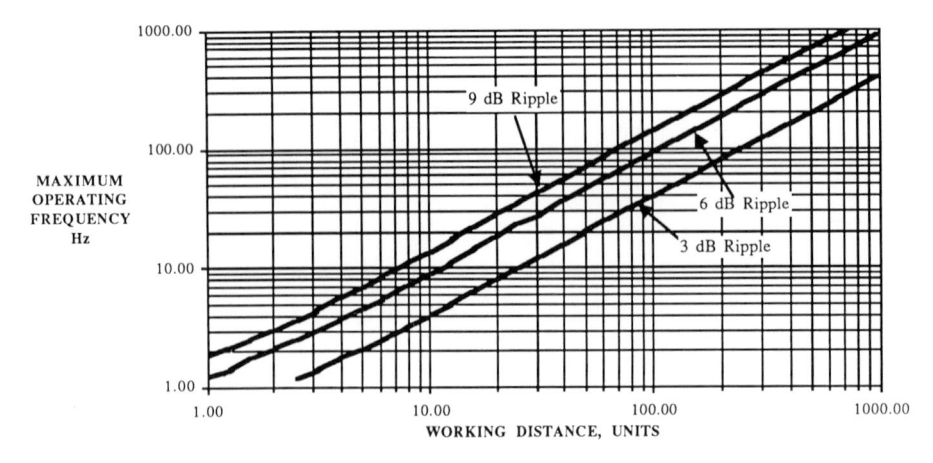

Fig. 27. Plot of maximum operating frequency versus working distance for five-source unit-length Bessel array. Contours of equal peak-to-peak ripple at values of 3, 6, and 9 dB are plotted. Direct relationship between maximum frequency and operating distance is clearly shown. Velocity of propagation 1 unit/s.

poor performance takes it out of the running for any practical application. For this reason, very few response curves were generated for the nine-source Bessel array.

4.6.1 Polar Responses

Only one polar response was generated for the nine-source Bessel array. This is shown in Fig. 30 (Sec. 5.3), which compares the polars of all the arrays at a specific frequency and working distance.

4.6.2 Frequency Responses

Only one frequency response was calculated for the nine-source Bessel array and is shown in Fig. 31 (Sec. 5.4), where the response is compared with those of the other analyzed arrays.

Table 6. Array type: seven- (six)-source Bessel array.

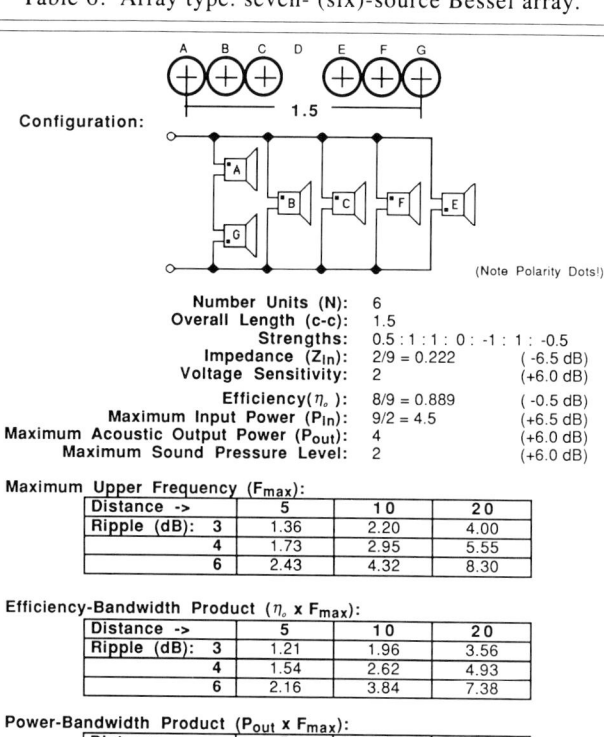

Configuration: A B C D E F G, length 1.5

(Note Polarity Dots!)

Number Units (N):	6
Overall Length (c-c):	1.5
Strengths:	0.5 : 1 : 1 : 0 : -1 : 1 : -0.5
Impedance (Z_{In}):	2/9 = 0.222 (-6.5 dB)
Voltage Sensitivity:	2 (+6.0 dB)
Efficiency (η_o):	8/9 = 0.889 (-0.5 dB)
Maximum Input Power (P_{In}):	9/2 = 4.5 (+6.5 dB)
Maximum Acoustic Output Power (P_{out}):	4 (+6.0 dB)
Maximum Sound Pressure Level:	2 (+6.0 dB)

Maximum Upper Frequency (F_{max}):

Distance →	5	10	20
Ripple (dB) 3	1.36	2.20	4.00
4	1.73	2.95	5.55
6	2.43	4.32	8.30

Efficiency-Bandwidth Product ($\eta_o \times F_{max}$):

Distance →	5	10	20
Ripple (dB) 3	1.21	1.96	3.56
4	1.54	2.62	4.93
6	2.16	3.84	7.38

Power-Bandwidth Product ($P_{out} \times F_{max}$):

Distance →	5	10	20
Ripple (dB) 3	5.44	8.80	16.00
4	6.92	11.80	22.20
6	9.72	17.28	33.20

Power-Bandwidth Product per Unit ($P_{out} \times F_{max} / N$):

Distance →	5	10	20
Ripple (dB) 3	0.91	1.47	2.67
4	1.15	1.97	3.70
6	1.62	2.88	5.53

Table 7. Array type: nine- (seven)-source Bessel array.

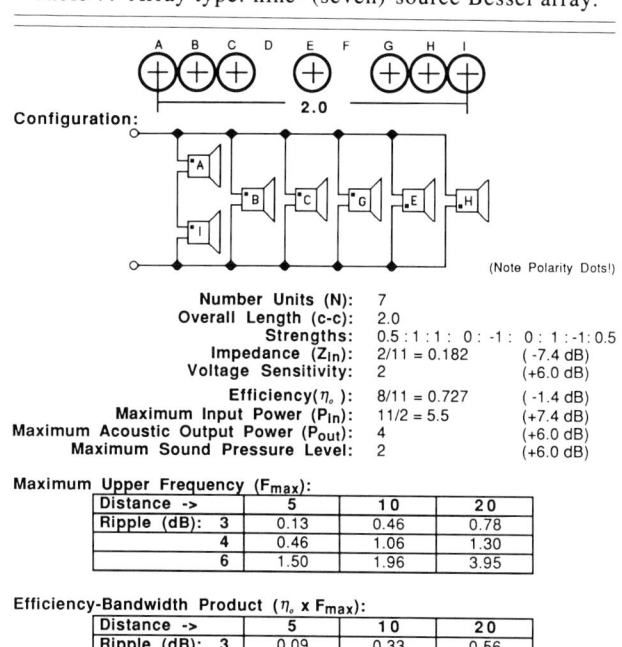

Configuration: A B C D E F G H I, length 2.0

(Note Polarity Dots!)

Number Units (N):	7
Overall Length (c-c):	2.0
Strengths:	0.5 : 1 : 1 : 0 : -1 : 0 : 1 : -1 : 0.5
Impedance (Z_{In}):	2/11 = 0.182 (-7.4 dB)
Voltage Sensitivity:	2 (+6.0 dB)
Efficiency (η_o):	8/11 = 0.727 (-1.4 dB)
Maximum Input Power (P_{In}):	11/2 = 5.5 (+7.4 dB)
Maximum Acoustic Output Power (P_{out}):	4 (+6.0 dB)
Maximum Sound Pressure Level:	2 (+6.0 dB)

Maximum Upper Frequency (F_{max}):

Distance →	5	10	20
Ripple (dB) 3	0.13	0.46	0.78
4	0.46	1.06	1.30
6	1.50	1.96	3.95

Efficiency-Bandwidth Product ($\eta_o \times F_{max}$):

Distance →	5	10	20
Ripple (dB) 3	0.09	0.33	0.56
4	0.33	0.77	0.95
6	1.09	1.42	2.87

Power-Bandwidth Product ($P_{out} \times F_{max}$):

Distance →	5	10	20
Ripple (dB) 3	0.52	1.84	3.12
4	1.84	4.24	5.20
6	6.00	7.84	15.80

Power-Bandwidth Product per Unit ($P_{out} \times F_{max} / N$):

Distance →	5	10	20
Ripple (dB) 3	0.07	0.26	0.44
4	0.26	0.61	0.74
6	0.86	1.12	2.26

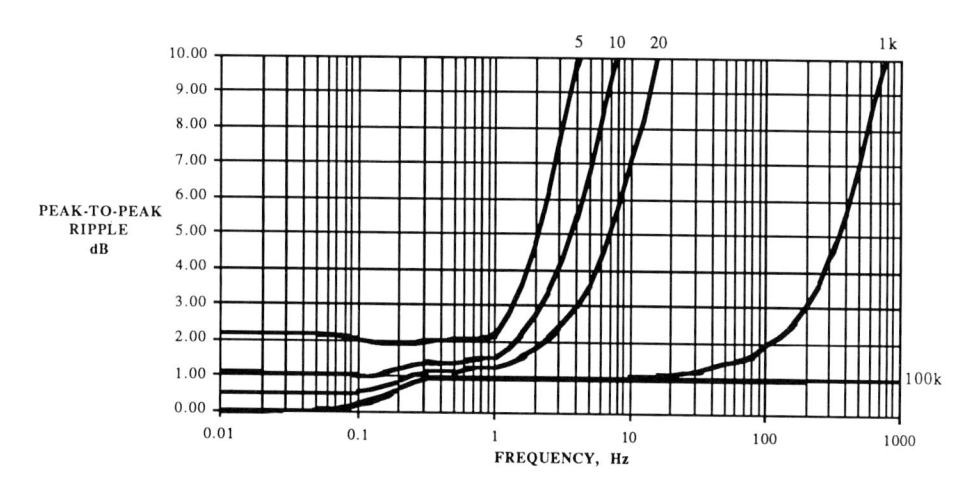

Fig. 28. Polar magnitude peak-to-peak ripple versus frequency for seven-source Bessel array of 1.5-unit length, at working distances of 5 to 100 000 units. At large distances, ripple attains a minimum plateau value of about 1.0 dB. In general, curves are shifted left as compared to five-source Bessel array, which indicates lower frequencies of operation.

4.6.3 Polar Peak-to-Peak Ripple versus Frequency

Fig. 29 shows a plot of the nine-source Bessel array's polar peak-to-peak ripple, in decibels, versus frequency for working distances of 5, 10, 20, 1000 and 100 000 units. At large distances, the ripple attains a minimum plateau value of about 3.6 dB, which is significantly higher than those of the previous Bessel arrays. As noted for the seven-source Bessel array, the curves are shifted even more to the left, as compared to the five-source Bessel array, which indicates an even lower bandwidth of operation.

4.6.4 Discussion

The performance of the nine-source Bessel array is significantly worse than even that of the seven-source Bessel array, which was previously judged undesirable for practical use. Its much lower efficiency, requirement of two more sources, and very much lower bandwidth definitely take it out of the running.

5 ARRAY COMPARATIVE ANALYSIS

A comparative analysis was done on all the analyzed arrays. This includes a master comparison table where all the array performance factors are shown, a series of performance ranking tables, a comparative display of polar responses and frequency responses, and a graph showing polar ripple versus frequency for all the arrays.

5.1 Tabular Comparison

Table 8 is a master tabular comparison of all the analyzed arrays, assuming a working distance of 20 units and a peak-to-peak polar ripple of 4 dB. The last four rows of the table indicate the clear superiority of

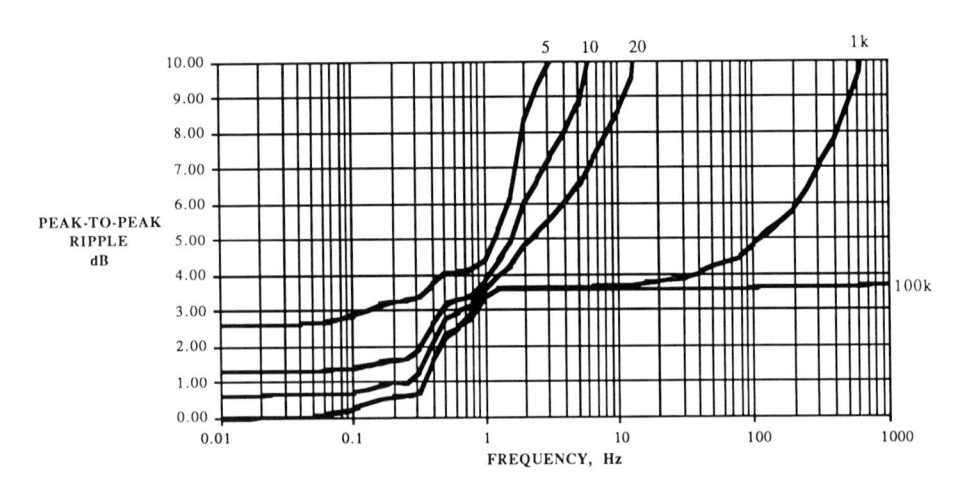

Fig. 29. Polar magnitude peak-to-peak ripple versus frequency for nine-source 2-unit-length Bessel array at working distances of 5 to 100 000 units. At large distances, ripple attains a minimum plateau value of about 3.6 dB, which is significantly higher than that of previous Bessel arrays. As noted for the seven-source Bessel array, curves are shifted even more left as compared to five-source Bessel array, which indicates an even lower bandwidth of operation.

Table 8. Comparison of array types.

ARRAY TYPE =	Single Source	2 Sources Equal Level (L=0.25)	2 Sources Equal Level (L=1.0)	5 Sources Equal Level & Spacing	5 Source Bessel	7(6) Source Bessel	9(7) Source Bessel
Configuration (to scale) =	o	∞	o o	ooooo	ooooo	ooo ooo	ooo o ooo
Number Units =	1	2	2	5	5	6	7
Overall Length (c-c) =	0	0.25	1.0	1.0	1.0	1.5	2.0
Impedance =	1.000	0.500	0.500	0.200	0.286	0.222	0.182
Voltage Sensitivity =	1	2	2	5	2	2	2
Efficiency =	1.000	2.000	2.000	5.000	1.143	0.889	0.727
Maximum Input Power =	1.0	2.0	2.0	5.0	3.5	4.5	5.5
Max. Output Power =	1.0	4.0	4.0	25.0	4.0	4.0	4.0
Maximum Sound Pressure Level =	1.0 (0 dB)	2.0 (+6 dB)	2.0 (+6 dB)	5.0 (+14 dB)	2.0 (+6 dB)	2.0 (+6 dB)	2.0 (+6 dB)
Maximum Upper Frequency = (Distance = 20, P-P Ripple = 4 dB)	Infinity	1.10	0.28	0.40	11.00	5.55	1.30
Efficiency-Bandwidth Product = (Distance = 20, P-P Ripple = 4 dB)	Infinity	2.20	0.56	2.00	12.57	4.93	0.95
Power Bandwidth Product = (Distance = 20, P-P Ripple = 4 dB)	Infinity	4.4	1.1	10.0	44.0	22.2	5.2
Power-Bandwidth Product per Unit = (Distance = 20, P-P Ripple = 4 dB)	Infinity	2.20	0.56	2.00	8.80	3.70	0.74

the five-source Bessel array as compared to the other analyzed arrays. The much higher bandwidth of operation is reflected in the high values of all the bandwidth products.

5.2 Performance Rankings

This section displays rankings for each of the analyzed arrays, for all the major array characteristics.

5.2.1 Efficiency

Table 9 shows the comparative rankings of the analyzed arrays for efficiency. As expected, the five-source equal-level equal-polarity equal-spaced array is at the top of the list. However, its high efficiency is mostly offset by its lower bandwidth of operation. The nine-source Bessel array is at the bottom of the list (27% less efficiency than a single source.).

5.2.2 Power Handling

The comparative rankings for input power handling are shown in Table 10. The nine-source Bessel array is at the top of this list. This is fortunate because it also has the lowest efficiency (Table 9). It would make a good heater.

5.2.3 Maximum Acoustic Output Power

Table 11 displays the ranking order for the array's maximum acoustic output power. The five-source equal-level array is seen to head the list. Even though this array provides high acoustic output power, its high frequency capabilities are limited. As can be seen, most of the analyzed arrays have maximum outputs of four times a single unit.

5.2.4 Maximum Operating Frequency

Table 12 ranks all the analyzed arrays for maximum operating frequency. Excluding the single source, the five-source Bessel array is seen to head the list with a large two-to-one margin. The widely separated two-source array is at the bottom of the list.

5.2.5 Efficiency–Bandwidth Product

The rankings for the efficiency–bandwidth product are shown in Table 13. Again, after excluding the single source, the five-source Bessel array's superiority is clearly shown, with a margin of greater than 2.5 over the second-place entry. The nine-source Bessel array is in next to last place.

5.2.6 Power–Bandwidth Product

Table 14 shows the rankings for the power–bandwidth product. The five-source Bessel array again heads the list, after excluding the single source. The seven- and nine-source Bessel arrays do a bit better in this comparison. The wide-separation two-source array is in last place.

5.2.7 Power–Bandwidth Product per Unit

Table 15 lists the rankings for the power–bandwidth product per unit. This parameter is a good figure of merit for comparing the arrays in that it shows how good the performance is on a per-unit basis. The five-source Bessel array is again on top, with the exception

Table 9. Ranking for efficiency.

RANK	VALUE	ARRAY TYPE
1	5.00	5 Sources, Equal Level and Spacing
2	2.00	2 Sources, Equal Level (L=0.25)
3	2.00	2 Sources, Equal Level (L=1.0)
4	1.14	5 Source Bessel
5	1.00	Single Source
6	0.89	7(6) Source Bessel
7	0.73	9(7) Source Bessel

Table 10. Ranking for maximum input power.

RANK	VALUE	ARRAY TYPE
1	5.5	9(7) Source Bessel
2	5.0	5 Sources, Equal Level and Spacing
3	4.5	7(6) Source Bessel
4	3.5	5 Source Bessel
5	2.0	2 Sources, Equal Level (L=0.25)
6	2.0	2 Sources, Equal Level (L=1.0)
7	1.0	Single Source

Table 11. Ranking for maximum output power.

RANK	VALUE	ARRAY TYPE
1	25	5 Sources, Equal Level and Spacing
2	4	2 Sources, Equal Level (L=0.25)
3	4	2 Sources, Equal Level (L=1.0)
4	4	5 Source Bessel
5	4	7(6) Source Bessel
6	4	9(7) Source Bessel
7	1	Single Source

Table 12. Ranking for maximum operating frequency. Distance = 20 units, peak-to-peak ripple = 4 dB.

RANK	VALUE	ARRAY TYPE
1	Infinity	Single Source
2	11.00	5 Source Bessel
3	5.55	7(6) Source Bessel
4	1.30	9(7) Source Bessel
5	1.10	2 Sources, Equal Level (L=0.25)
6	0.40	5 Sources, Equal Level and Spacing
7	0.28	2 Sources, Equal Level (L=1.0)

Table 13. Ranking for efficiency–bandwidth product. Distance = 20 units, peak-to-peak ripple = 4 dB.

RANK	VALUE	ARRAY TYPE
1	Infinity	Single Source
2	12.57	5 Source Bessel
3	4.93	7(6) Source Bessel
4	2.20	2 Sources, Equal Level (L=0.25)
5	2.00	5 Sources, Equal Level and Spacing
6	0.95	9(7) Source Bessel
7	0.56	2 Sources, Equal Level (L=1.0)

Table 14. Ranking for power–bandwidth product. Distance = 20 units, peak-to-peak ripple = 4 dB.

RANK	VALUE	ARRAY TYPE
1	Infinity	Single Source
2	44.0	5 Source Bessel
3	22.2	7(6) Source Bessel
4	10.0	5 Sources, Equal Level and Spacing
5	5.2	9(7) Source Bessel
6	4.4	2 Sources, Equal Level (L=0.25)
7	1.1	2 Sources, Equal Level (L=1.0)

of the single source. The seven-source Bessel array is in a fairly strong second-place position. The nine-source Bessel array is in next to last place with a power–bandwidth product of about one-twelfth that of the five-source Bessel array.

5.3 Polar Response Comparison

Fig. 30 shows a comparison of the polars for all the analyzed arrays. All the polars were run at the same frequency (10 Hz) and working distance (20 units). The peak-to-peak polar ripple is listed on each plot. The superiority of the five-source Bessel array [Fig. 30(d)] is clearly evident.

5.4 Frequency Response Comparison

Fig. 31 displays a comparison of off-axis frequency responses for all the analyzed arrays. The response curves were all run at the same off-axis angle (+45°) and working distance (20 units) and covered the same frequency range (0.1–20 Hz). Again, the five-source Bessel array has the smoothest and most extended response.

5.5 Ripple versus Frequency Comparison

Fig. 32 shows a comparison of the polar peak-to-peak ripple versus frequency for all the analyzed arrays at a working distance of 20 units. The superiority of the five-source Bessel array is again quite clear.

6 CONCLUSIONS

When compared to the other analyzed arrays, the five-source Bessel line array is the clear winner, considering 1) polar response, 2) off-axis frequency re-

Table 15. Ranking for power–bandwidth product per unit. Distance = 20 units, peak-to-peak ripple = 4 dB.

RANK	VALUE	ARRAY TYPE
1	Infinity	Single Source
2	8.80	5 Source Bessel
3	3.70	7(6) Source Bessel
4	2.20	2 Sources, Equal Level (L=0.25)
5	2.00	5 Sources, Equal Level and Spacing
6	0.74	9(7) Source Bessel
7	0.56	2 Sources, Equal Level (L=1.0)

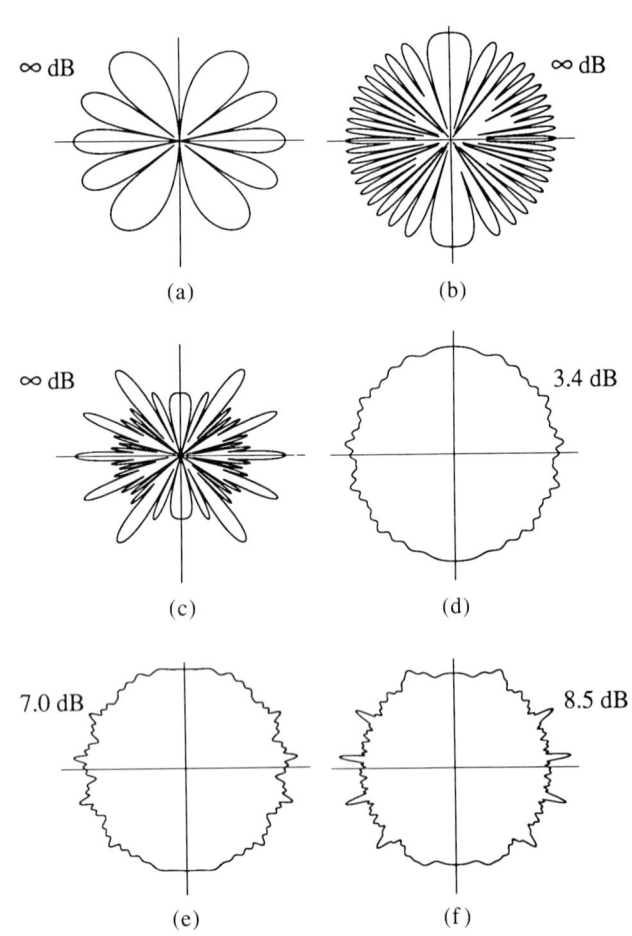

Fig. 30. Comparison of magnitude polars for all analyzed arrays. Polars were all run at a frequency of 10 Hz and a working distance of 20 units. Peak-to-peak polar ripple is listed on each plot. (a) Two-source equal-level equal-polarity equal-spaced array of 0.25-unit spacing. (b) Two-source equal-level equal-polarity equal-spaced array of 1.0-unit spacing. (c) Five-source equal-level equal-polarity equal-spaced array with 1.0-unit center-to-center length. (d) Five-source Bessel array with 1.0-unit center-to-center length. (e) Seven-source Bessel array with 1.5-unit center-to-center length. (f) Nine-source Bessel array with 2.0-unit center-to-center length. Superiority of five-source Bessel array (d) is very clear.

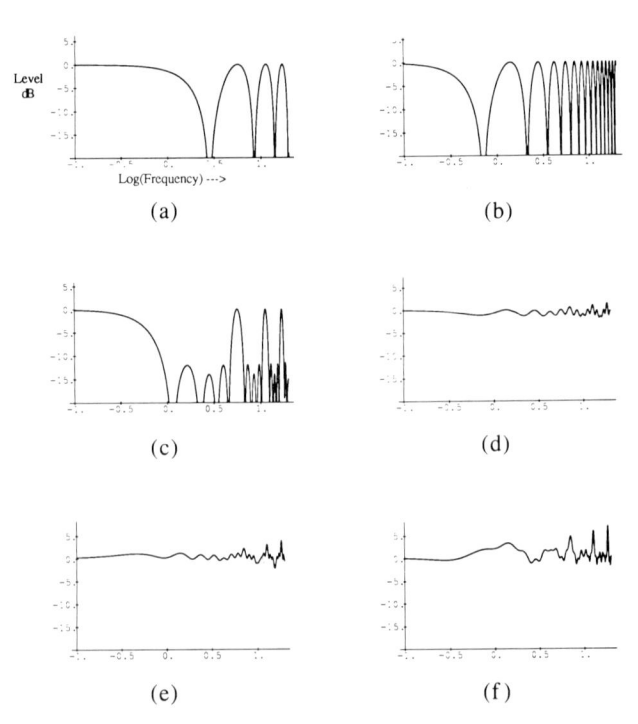

Fig. 31. Comparison of off-axis magnitude frequency responses for all analyzed arrays. Response curves were all run at +45° with a working distance of 20 units, and cover the same frequency range of 0.1 to 20 Hz. Note that log of frequency is indicated (−1 = 0.1 Hz, 0 = 1 Hz, etc.). (a) Two-source equal-level equal-polarity equal-spaced array of 0.25-unit spacing. (b) Two-source equal-level equal-polarity equal-spaced array of 1.0-unit spacing. (c) Five-source equal-level equal-polarity equal-spaced array with 1.0-unit center-to-center length. (d) Five-source Bessel array with 1.0-unit center-to-center length. (e) Seven-source Bessel array with 1.5-unit center-to-center length. (f) Nine-source Bessel array with 2.0-unit center-to-center length. Again, five-source Bessel array has smoothest and most extended response.

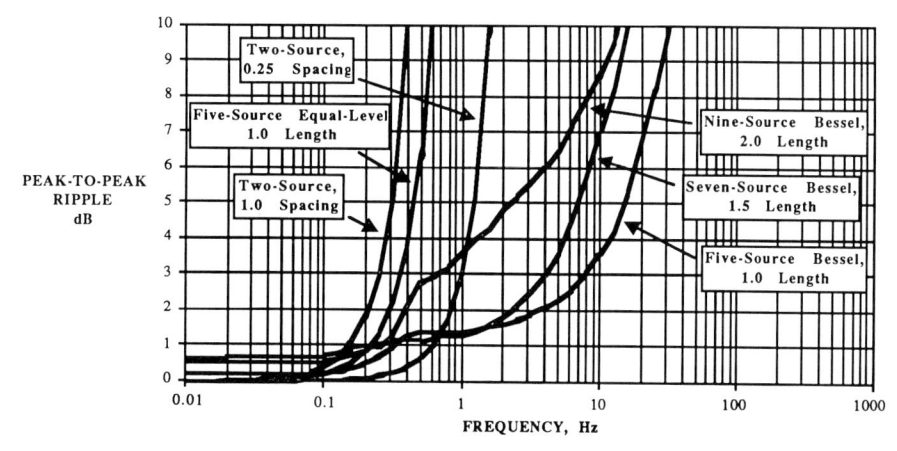

Fig. 32. Comparison of polar magnitude peak-to-peak ripple versus frequency for all analyzed arrays at a working distance of 20 units. Superiority of five-source Bessel is again quite clear. Velocity of propagation 1 unit/s.

sponse, 3) bandwidth of operation, 4) efficiency–bandwidth product, 5) power–bandwidth product, and 6) power–bandwidth product per unit.

Considering the maximum frequency of operation for omnidirectional radiation, at a typical working distance of 20 times the length of the array, the Bessel array outperforms the same-length five-source equal-level equal-polarity equal-spaced array by a factor of 28 and the one-quarter-length equal-level equal-polarity equal-spaced array by a factor of 10. Its power–bandwidth product exceeds that of its nearest competitor, a seven-source Bessel array, by a factor of 2.

The seven- and nine-source Bessel line arrays were found to be effectively unusable due to poor performance, as compared to the five-source Bessel array. Their much lower efficiency, requirement of additional sources, and much lower bandwidth placed them at a severe performance disadvantage.

The Bessel array's singular main problem is its nonlinear phase behavior with direction and frequency. This nonlinear behavior makes it difficult to use the array in conjunction with any other source. Crossing it over to a high-frequency device would be difficult and would require a high slope crossover to minimize off-axis lobing effects in the crossover region. The off-axis phase versus frequency response of the Bessel array is nonminimum phase and exhibits an oscillating phase characteristic. The Bessel array's 45° off-axis group delay versus frequency performance indicates that its time center ranges over a peak-to-peak shift of greater than 25% of the length of the array as the frequency increases.

The Bessel array does not exhibit normal near-field/far-field behavior. Its performance characteristics and high-frequency response get better and better the farther away you are from the array. This is in sharp contrast to the analyzed two- and five-source equal-level equal-polarity equal-spaced arrays, where there was a definite shift from near-field behavior, where the characteristics changed strongly with the working distance, to far-field behavior, where the characteristics changed very little with distance.

An analysis was not done on the 25 (5 × 5)-element planar (panel) source. Presumably the strong performance advantages of the five-source Bessel line array would carry over to this configuration.

7 ACKNOWLEDGMENT

The author would like to acknowledge the information he received and discussions he had with several people, including Mike Lamm, formerly of J. W. Davis & Company and currently with Atlas/Soundolier, Mark Gander of JBL Professional, and Marshall Buck of Cerwin-Vega (who suggested he look at phase). He would also like to thank Don Eger of Techron, a Division of Crown International, for allowing him the time and resources to do this study.

8 REFERENCES

[1] N. V. Franssen, "Direction and Frequency Independent Column of Electro-Acoustic Transducers," U.S. patent 4,399,328 (1983 Aug.), assigned to the U.S. Philips Corp.

[2] W. J. W. Kitzen, "Multiple Loudspeaker Arrays Using Bessel Coefficients," *Electron. Components & Appl.*, vol. 5 (1983 Sept.).

[3] "Bessel Panels—High-Power Speaker Systems with Radial Sound Distribution," literature given out by Philips at the 73rd Convention of the Audio Engineering Society, Eindhoven, The Netherlands, 1983 Mar. 15–18.

[4] V. H. Kuttruff and H. P. Quadt, "Elektroakustische Schallquellen mit ungebündelter Schallabstrahlung" (in German), *Acustica*, vol. 41, pp. 1–10 (1978).

[5] V. H. Kuttruff and H. P. Quadt, "Ebene Schallstrahlergruppen mit ungebündelter Abstrahlung" (in German), *Acustica*, vol. 50, pp. 273–279 (1982).

[6] E. Kreyszig, *Advanced Engineering Mathematics* (Wiley, New York, 1965).

[7] J. M. Kates, "Loudspeaker Cabinet Reflection Effects," *J. Audio Eng. Soc.*, vol. 27, pp. 338–350 (1979 May).

THE AUTHOR

D. B. (Don) Keele, Jr., was born in Los Angeles, CA, in 1940. After serving in the U.S. Air Force for four years as an aircraft electronics technician, he attended California State Polytechnic University at Pomona, from which he graduated with honors and B.S. degrees in both electrical engineering and physics. Mr. Keele worked as an audio systems engineer for Brigham Young University in Provo, Utah, where he received his M.S. degree in electrical engineering in 1975 with a minor in acoustics.

From 1972 to 1976, Mr. Keele worked at Electro-Voice, Inc. in Buchanan, MI, as a senior design engineer in loudspeakers, concentrating on high-frequency horns and low-frequency vented-box loudspeaker systems. He is the primary designer of their HR series of constant-directivity horns on which he holds the patent. For one year, starting in 1976, he worked for Klipsch and Associates in Hope, AR, as chief engineer involved in the company's commercial line of loudspeakers. From 1977 to 1984, he was with JBL, Inc, in Northridge, CA, as a senior transducer engineer working on horn and monitor loudspeaker system design. He also holds two patents on JBL's Bi-Radial series of constant-directivity horns.

Mr. Keele was employed by the Techron Division, Crown International, Elkhart, IN from 1984 to 1989, where he was manager of software development and responsible for the TEF System 12 time-delay spectrometry analyzer software. While at Techron, he was the programmer for two software packages for the TEF System: EasyTEF, a program for doing general-purpose TDS measurements; and TEF-STI, a program for measuring speech intelligibility.

Since October of 1989, he has been a self-employed independent consultant with his own company, DBK Associates, working primarily for *Audio* magazine, Diamandis Communications, as their Senior Editor in charge of loudspeaker reviews. He also is a consultant to Crown working with advanced TEF system development.

A member and fellow of the Audio Engineering Society, Mr. Keele has presented and published a number of papers on loudspeaker design and measurement methods, among them the paper for which he won the AES Publications Award, "Low-Frequency Loudspeaker Assessment by Nearfield Sound-Pressure Measurement" (*J. Audio Eng. Soc*, vol. 22, p. 154 (1974 Apr.). He is a frequent speaker at AES section meetings and workshops, is a member of several AES committees, and is on the AES *Journal* review board.

Microalignment of Drivers via Digital Technology*

JOHN A. MURRAY, *AES Member*

TOA Electronics, Inc., South San Francisco, CA 94080, USA

A suggested "in the field" method for aligning transducers within a loudspeaker system via a digital signal processor is presented. Cone drivers and horn-loaded compression drivers are addressed in adjacent passband (crossover) applications. Selection of crossover frequencies and filter slopes, driver alignment, and required digital signal delay resolution are discussed. The resulting frequency, phase, and polar response effects are illustrated in the examples of measurement data.

0 INTRODUCTION

Today's digital signal processing (DSP) technology makes available to the sound system contractor, audio production specialist, and common facility operator a choice of parameter control that is far more sophisticated than what was on the market just a few years ago. Any end user can choose from a large number of crossover frequency and filter slope combinations, adjust signal delay in high resolution, and fine-tune a variety of other parameters. The author has seen a great need in the field for information on how to employ properly the great amount of flexibility and control now at one's fingertips.

The information and the suggested methodology given in this engineering report are not rocket science. It is common knowledge to engineers employed by all the major audio manufacturers, and is a summation of what the author has learned from others. The aim here is not to break new ground in audio science. The intent of this engineering report is to give reasonable and, above all, practical information to those in the field who have a new and powerful tool and want to use it to its best advantage on site without the benefit of an anechoic chamber.

1 DEFINITION OF METHODOLOGY

Driver alignment involves two separate issues: 1) simultaneous impulse signal start times (call it signal alignment) and 2) phase alignment at the crossover frequency. In the case of popular analog crossovers that employ analog all-pass filters, phase alignment is the only option. Most current DSP devices do not provide the capability to simultaneously address both driver alignment issues either. This is because most horn/driver and woofer combinations will not allow both to be resolved at any one setting of acceptable signal parameters. One must generally choose either signal alignment with a phase misalignment at crossover, or phase alignment at crossover with a signal misalignment.

The method described in this engineering report provides phase alignment at crossover with minimal signal misalignment. When faced with choosing only one of the alignment choices, the author has found this method to be preferred by most. Listeners seem to be more sensitive to frequency response anomalies at crossover than to nonsynchronous arrival times of frequencies octaves apart with drastically different wavelengths and rise times.

2 MEASUREMENT CONDITIONS

2.1 Recommended Test Equipment

The measurement microphone ideally should be a calibrated, small-diaphragm, omnidirectional condenser. If one is not available, any small omnidirectional microphone can be used. Avoid directional microphones as they have a rougher frequency response and use cancellation via flanking-path ports for the directional characteristics. These ports can interfere with the boundary of "pressure zone" while making measurements.

The best type of spectrum analyzer for these measurements is a three-dimensional analyzer like a TEF-

---

* Presented but not preprinted at the 95th Convention of the Audio Engineering Society, New York, 1993 October 7–10; manuscript received 1993 November 22.

20 or other fast-Fourier-transform (FFT)-based device that shows not only amplitude and frequency but has time analysis as well. This resolves phase information, making the measurement and subsequent adjustments easier and much more accurate. If one is not available, a standard real-time analyzer can be used if it has a relatively fine dB sound-pressure-level scale with no more than 3 dB per step and at least a 20–30-dB window on display at one time. The step resolution will determine the relative accuracy of the measurement.

A piece of absorptive material approximately 1 m (3 ft) square is highly recommended for the measurements involving frequencies at or above 500 Hz. [See Fig. 1(b) and (c) for applications.] This should be similar in nature to a piece of Sonex, preferably the 4-in (101.6-mm) type. Avoid very thin absorptive materials [less than 4 in (101.6 mm) thick] as they will not attenuate reflections in the 500-Hz region adequately.

2.2 Measurement Microphone Positioning

The multiway loudspeaker system used in Fig. 1 was selected as a typical example utilizing a horn-loaded cone woofer, a compression driver with horn midrange, a compression-type tweeter, and a separate subwoofer. Its configuration and the methods employed permit adaptation to most devices one may encounter in the field.

The measurement-microphone positions suggested for these measurements are also applicable for the equalization process. Particularly when analyzing

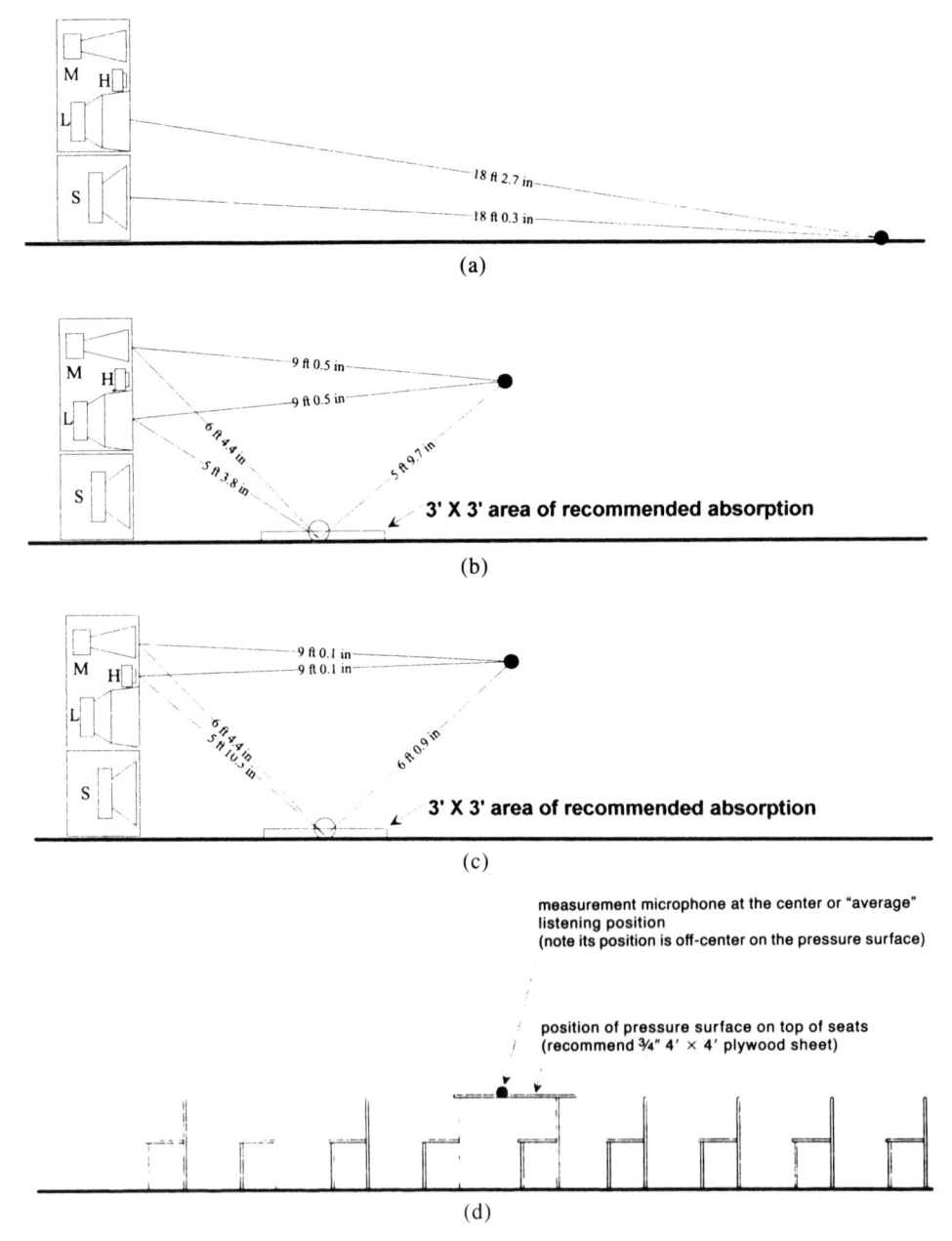

Fig. 1. Positions of measurment microphone. (a) For subwoofer-to-low-frequency driver alignment. (b) For low-frequency-to-mid-frequency driver alignment. (c) For mid-frequency-to-high-frequency driver alignment. (d) For driver alignment of a permanently positioned enclosure. H—high-frequency driver; M—mid-frequency driver; L—low-frequency driver; S—subwoofer driver; ●—measurement microphone position; ○—approximate floor reflection point.

loudspeaker arrays, one should take care that during these measurements only one device is connected per passband. Mutual coupling of multiple devices resulting in higher output at lower frequencies can be dealt with afterward via far-field real-time analysis and subsequent equalization adjustments.

2.2.1 Positioning for Subwoofer-to-Low-Frequency Alignment

Fig. 1(a) indicates the position for the measurement microphone when aligning the subwoofer to the low-frequency woofer. In many cases it is not necessary to align a subwoofer and woofer if they are both front mounted and stacked in line on top of one another, as the phase difference may be inconsequential. In cases of horn-loaded drivers, the setup illustrated in Fig. 1(a) is applicable and alignment may be beneficial.

If the subwoofer is placed on the floor and the low-frequency driver flown with the high-frequency section, a position on the floor in the center of the seating is recommended. If seats are fixed in the direct path between the loudspeakers and the microphone, use a pressure surface as illustrated in Fig. 1(d), but with a surface dimension of 8–16 ft (2.44–4.88 m) in the microphone-to-loudspeaker direction. Remember, walls and ceiling must be sufficiently far with respect to the wavelength of the frequency of interest.

The microphone is positioned at 18 ft (5.49 m) from the front of the loudspeaker enclosures in the example. This distance was chosen as being close to 5.5 m to satisfy both English and metric scales. A boundary or "pressure zone" miking technique is used by putting the microphone's grille directly on a hard, smooth-surfaced floor. This prevents any floor reflections from contaminating the measurement, which requires resolving long wavelengths for low-frequency accuracy. It is crucial to ensure that there are no other reflective surfaces within 18 ft (5.49 m) of the loudspeaker system or microphone, including the ceiling. This maintains at least 6 dB of isolation for the direct sound from the loudspeakers.

Generally a crossover frequency anywhere from 70 to about 200 Hz is used. In this example, the microphone is approximately 3° and 9° off axis from the subwoofer and low-frequency woofer, respectively. Considering the frequency range, this position's respective levels are approximately equal to being on axis with both drivers. The path lengths from each of the two drivers under measurement are 18 ft 0.3 in and 18 ft 2.7 in (5.50 and 5.56 m). This yields a path-length difference of 2.4 in (61 mm). For 200 Hz this is a phase shift of roughly 13°, for 70 Hz it is a phase shift of a mere 4.5°. Given the respective levels and phase shift at this measurement position, it is essentially equal to a free-field measurements centered between the drivers.

2.2.2 Positioning for Low-Frequency-to-Mid-Frequency Alignment

Fig. 1(b) indicates the position for the measurement microphone when aligning the low-frequency woofer to the mid-frequency driver. The position is equidistant from both driver positions at the enclosure's edge, and at 9 ft (3 m) this is about as close as the microphone should be to the loudspeaker system. The floor reflections at this distance are 2 and 3 ft (0.6 and 0.9 m) longer in path length than the direct paths of the woofer and horn, respectively. An FFT-based analyzer can isolate these reflections from the measurement. A real-time analyzer cannot isolate these reflections, but the off-axis angles of each, at 32.5° and 45°, respectively, combined with the absorption material placed at the reflection point, greatly attenuate their level and interference capabilities. At greater distances these reflections become much stronger due to having more on-axis level, and have less effective attenuation in the grazing path of reflection on the absorption material. They also become more of a problem to an FFT-based analyzer, being closer in path length to the direct sound, and they can no longer be sweep-filter passband isolated if frequency resolution is to be maintained.

This positioning is applicable for crossover frequencies ranging from 500 Hz to about 3 kHz and includes most common two- and three-way systems. In this example the microphone is only 5° off axis from each driver, well within each driver's coverage pattern, and within 1 dB of each driver's maximum on-axis level. This should yield tuning results very close to those made in a far-field anechoic environment.

If the loudspeaker system is to be installed permanently, particularly in a recording studio's control room environment, an average or "ideal" seating position, as indicated in Fig. 1(d), can be chosen.

2.2.3 Positioning for Mid-Frequency-to-High-Frequency Alignment

Fig. 1(c) indicates the position for the measurement microphone when aligning the mid-frequency driver to the high-frequency driver. This position has all the same parameters as were applicable to the low-to-mid-frequency positioning, with the exception of one. Extremely small changes in microphone positioning will change the phase relationship between the mid and high drivers due to the very short wavelengths involved. At these frequencies, generally around 8 kHz, a 180° phase shift requires only a 0.85-in (21.6-mm) shift in position. Turning one's head will produce a phase cancellation of the signal with respect to each ear's position. Once the drivers are synchronized at some point in front of the loudspeaker enclosure, it becomes academic how well aligned they are at any one position. Once again, in a permanent installation, such as a recording studio or postproduction facility, the optimum seating position illustrated in Fig. 1(d) can be used.

For the purpose of this engineering report the position used for all ETC and TDS files is that of Fig. 1(b). This position is close to being on axis of the high-frequency tweeter and should suffice for demonstration purposes since the mid-to-high measurement point is somewhat academic.

3 CHOOSING CROSSOVER TYPES AND SLOPES FOR EACH DRIVER

3.1 Custom versus Standard Symmetrical Crossover Networks

Using symmetrical crossover slopes and hinge frequencies as offered in standard active crossover networks can be a great compromise in potential system response uniformity. Using the right combination of crossover slopes, hinge frequencies, and delay times can produce much better system performance. The example loudspeaker system used for this engineering report is a three-way tri-amplified system using a 15-in (381-mm) woofer, a 2.8-in (11.1-mm) diaphragmed compression driver on a 90 by 40° horn for a midrange driver, and a compression-driven exponential ring radiator for a tweeter. This device's characteristics apply to most of the systems one might use in the field.

Figs. 2–4[1] represent symmetrical crossover networks used as the common frequency selections of 800 Hz and 8 kHz. They each employ 12-dB/oct Butterworth (BW), 18-dB/oct BW, and 24-dB/oct Linkwitz–Riley (L–R) filters, respectively. No signal alignment is used

[1] In frequency response plots, A is amplitude response, P is phase response.

in any of these as it is intended that they appear as if one purchased an active crossover and applied it to one's loudspeaker system without special consideration or analysis. The anomalies occurring in the crossover regions are of a non-minimum-phase nature and, therefore, cannot be corrected with equalization. Fig. 5 shows how much better the same loudspeaker system can perform with the right combination of crossover slopes, hinge frequencies, high-frequency lift, and signal alignment.

3.2 Driver Response Analysis for Crossover Filter Selection

Fig. 6 is the raw response of the system's 15-in (381-mm) woofer. All drivers are, in effect, bandpass filters, and they combine in series with the crossover filters used. If a driver's response is rolling off at 6 dB/oct and it is filtered with an 18-dB/oct low-pass filter, the resulting response will roll off at 24 dB/oct.

The rising response that peaks at 2 kHz seen in this on-axis measurement does not exist off axis as it is an artifact of the collapsing beamwidth at higher frequencies. However, it can be a problem in combination with the response of the midrange horn on axis. The woofer's response is smoothest up to 800 Hz, so that would be a good crossover point. Since the on-axis

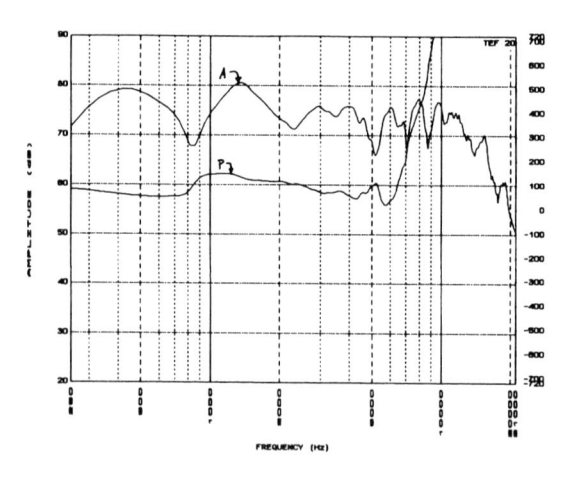

Fig. 2.

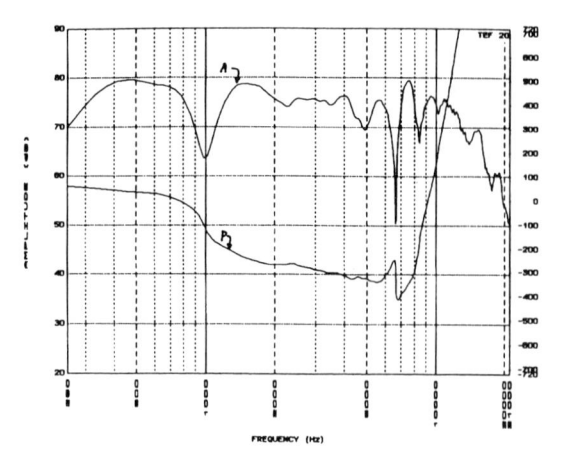

Fig. 3.

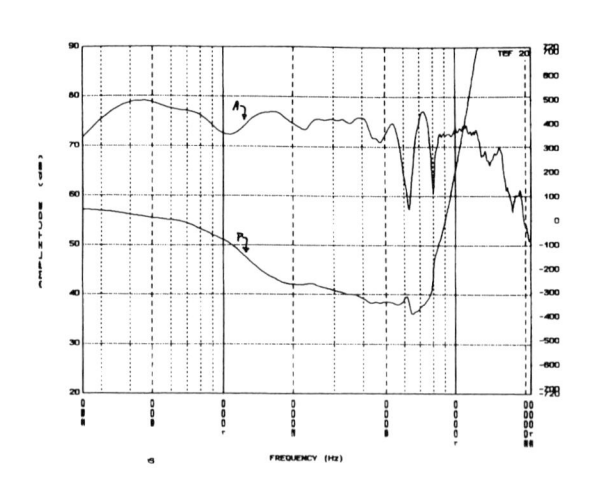

Fig. 4.

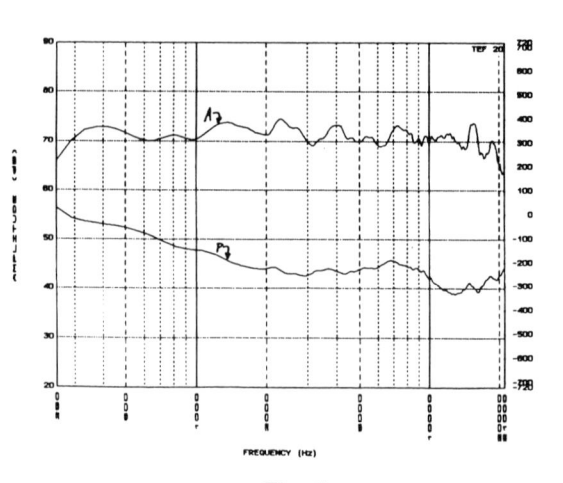

Fig. 5.

response above 800 Hz is rising, it would be wise to choose a steep low-pass filter for its crossover slope.

Some of the better DSP processors have a variety of crossover filter types, and choosing the best one for the application can be a trial-and-error process. This is due to the fact that each make and model driver series, as a bandpass filter, is unique. Each model's phase and frequency response, combined with the crossover filter at crossover, produces a different combination when united with the device to which it is crossed over. Brand A, model X may work best with 18-dB/oct BW filters, while brand B, model Y may work best with 24-dB/oct L–R filters. Yet another manufacturer may choose Bessel (BS) function filtering for loudspeaker systems for their low group-delay characteristics. Since the combintion of loudspeaker and electrical filtering often produces asymmetrical passband slopes, the acoustical crossover frequency is usually shifted from what was selected electronically.

Fig. 7 is the woofer with the chosen 800-Hz 24-dB/oct L–R low-pass filter. Figs. 8 and 9 are the midrange horn's and high-frequency tweeter's respective raw responses. Fig. 10 is the horn's response with the chosen 800-Hz 12-dB/oct L–R high-pass filter and the 9-kHz 12-dB/oct L–R low-pass filter. Fig. 11 is the tweeter's response with a 9-kHz 24-dB/oct L–R high-pass filter.

3.3 High-Frequency Lift/Horn Compensation Filtering

Another feature offered in some of the better DSP processors is high-frequency lift or CD-horn equalization. The processor used for this engineering report employs a 6-dB/oct high-frequency-boost filter with the amount of boost at 20 kHz indicated. Custom, device-specific boost filters are also selectable. For this particular system, the response of both the horn and

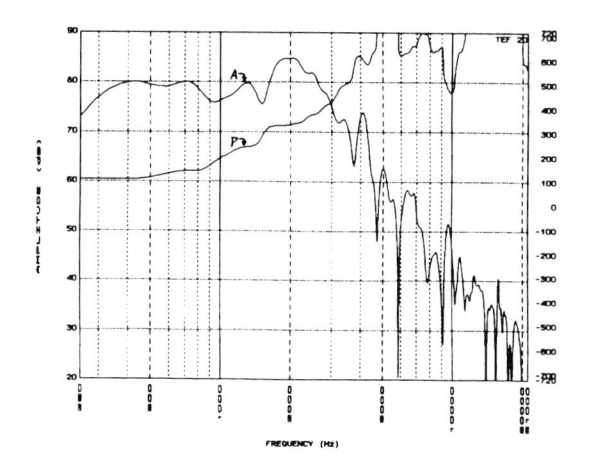

Fig. 6.

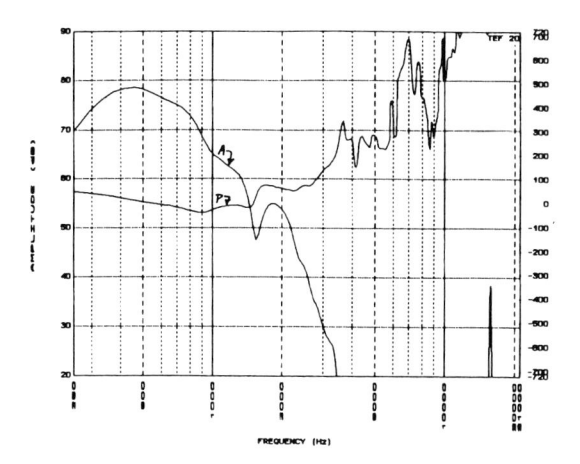

Fig. 7.

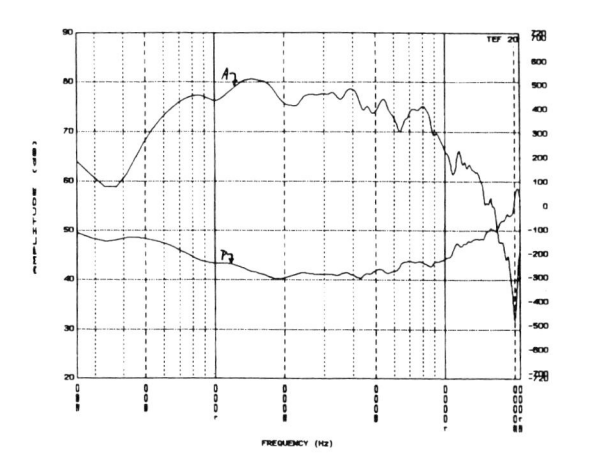

Fig. 8.

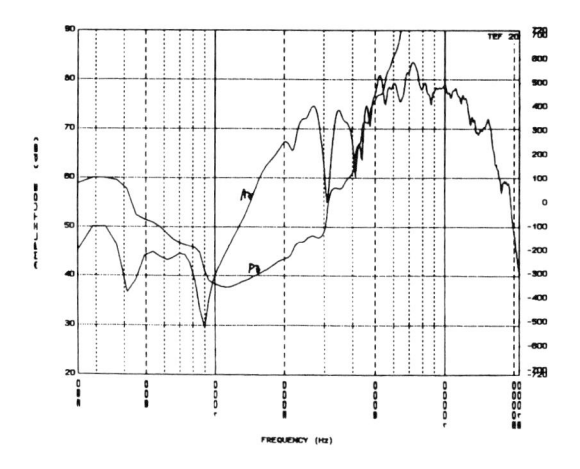

Fig. 9.

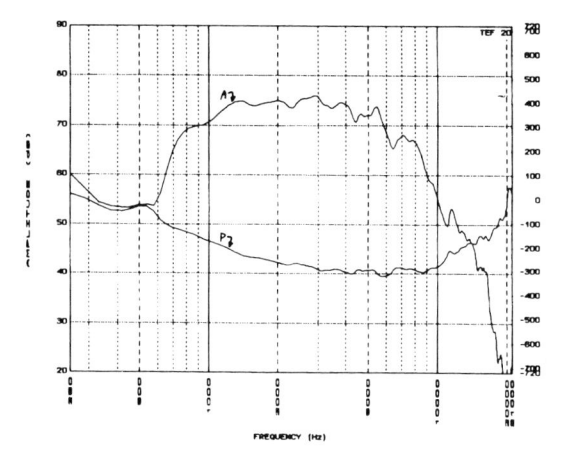

Fig. 10.

ENGINEERING REPORTS

the tweeter are flattened through use of these lift filters. Fig. 12 shows the horn's response with a 6-dB/oct boost filter yielding +15 dB at 20 kHz. Fig. 13 is the tweeter with a +8-dB setting at 20-kHz.

4 SIGNAL DELAY FOR DRIVER ALIGNMENT

4.1 Arrival Times of Different Drivers

Figs. 14–16 show the respective energy–time curves (ETC) for the mid-frequency horn, the high-frequency tweeter, and the low-frequency woofer. Using a linear sine-wave sweep, the measurement system is biased toward higher frequencies, and therefore the woofer's level has been raised for demonstration purposes. The arrival time of maximum energy for the tweeter is 7.45 ms, for both the horn and the woofer the time is 9.05 ms. This tells nothing about the time offset between unlike drivers at crossover frequencies.

The measurement signal is swept from 50 Hz to 21 kHz and shows the total energy output with respect to

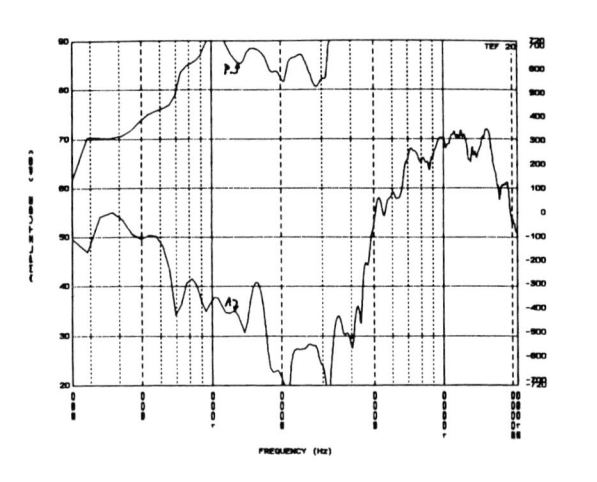

Fig. 11.

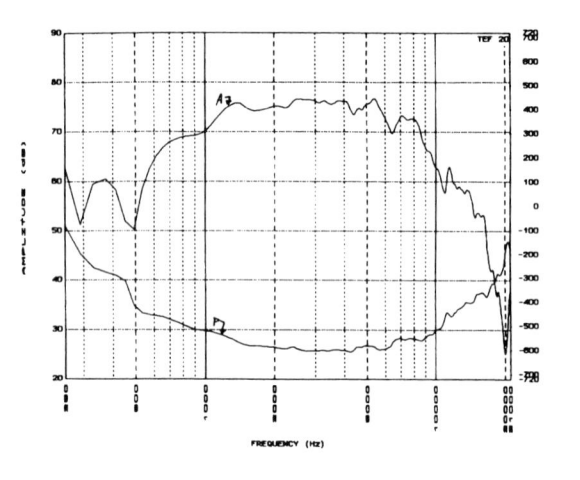

Fig. 12.

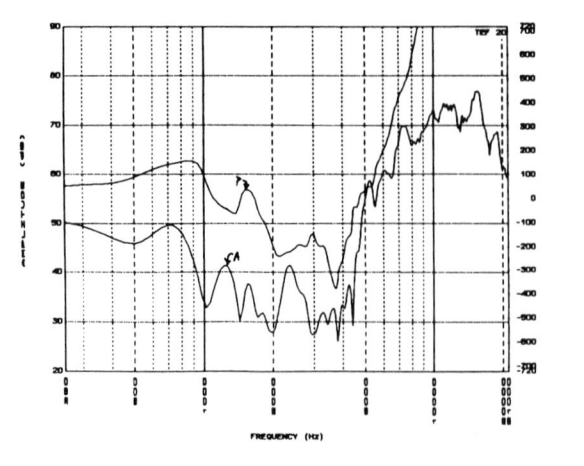

Fig. 13.

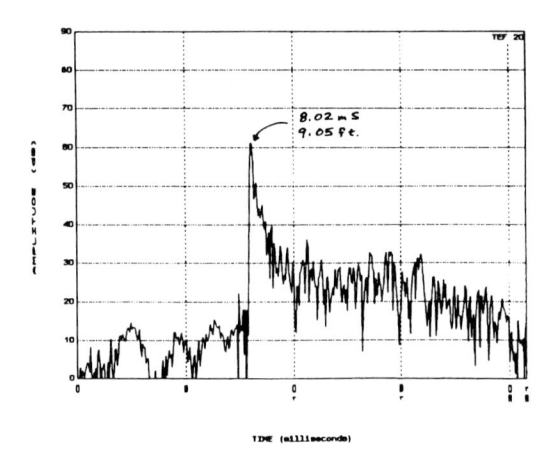

Fig. 14.

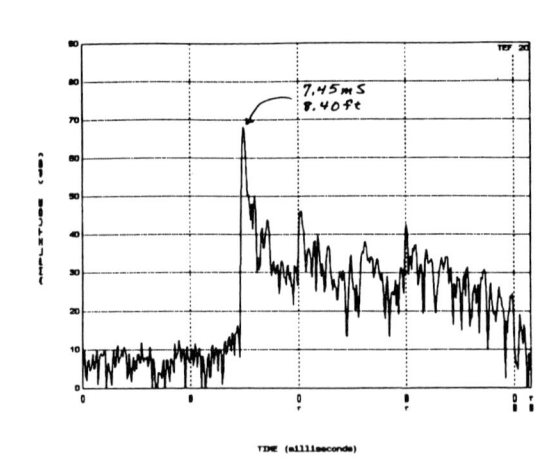

Fig. 15.

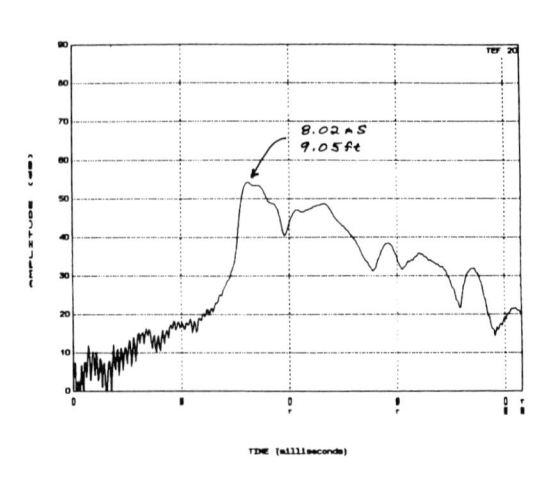

Fig. 16.

time for each device. It is therefore a frequency-blind measurement and cannot be used to determine signal-delay offsets between different types of drivers. While it can be used in cases of *identical* devices, the author has seen many in our industry attempting to determine signal arrival-time offset of different devices employing crossover networks via ETCs. This cannot be done. The phase shift of each driver's bandpass filtering characteristics and those of nonsymmetrical crossover filters change the phase relationship between drivers such that a frequency- and phase-blind measurement cannot resolve the relationship. This is not to say that the ETC is of no use for these measurements, however. It is useful in determining which phase "cycle," or 360° phase shift, to use in aligning unlike drivers whose overall energy outputs are several cycles apart with respect to the crossover frequency's wavelength.

4.2 Alignment Methodology

Fig. 17 shows the nondelayed response of the woofer and mid-frequency horn combination using the crossover parameters chosen earlier. Note that the response in the 800-Hz crossover region shows some phase cancellation due to nonsynchronization of the drivers. The slope of the phase in this region also shows a nonlinearity. Fig. 18 shows their relationship with the woofer's polarity reversed. At this point it looks better than the "in-polarity" response. Note that the woofer's rise time to maximum level is within a wavelength of the horn's total energy output in their ETCs. Also, assume that the higher frequencies passed by the woofer are both highest in level, due to the linear frequency sweep, and the first to reach maximum level, due to their shorter wavelengths. It would follow that the amount of delay to align the devices will be within one wavelength of 800 Hz.

Stepping the delay of the woofer in the available 21-μs steps upward from 0 ms, maximum notch depth is achieved at 0.521 ms (Fig. 19). This is the point at which the woofer and horn are 180° out of phase. Note that the phase response with this delay setting is a vertical line at the crossover frequency. Reversing the polarity of the woofer once again, the resulting response is Fig. 20. Though the phase response is not as flat overall as in Fig. 18, it is more linear through the crossover region, and the amplitude response is much flatter.

Using the same technique for the horn-to-tweeter relationship, Fig. 21 is the nondelayed response using the crossover parameters chosen in Section 2 for each. In Fig. 22 the polarity of the tweeter is reversed, and Fig. 23 yields a maximum notch depth with a 0.625-

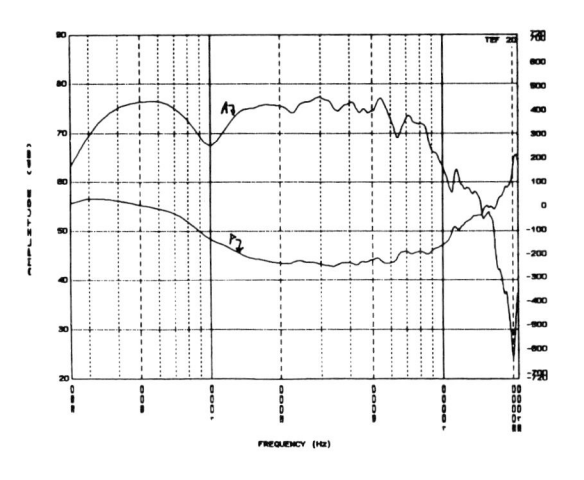

Fig. 17.

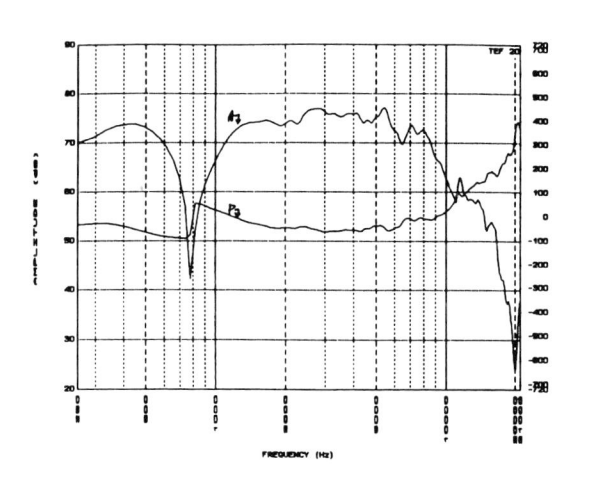

Fig. 19.

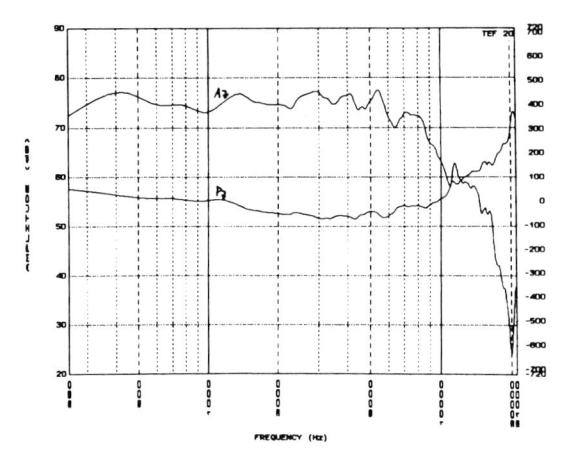

Fig. 18.

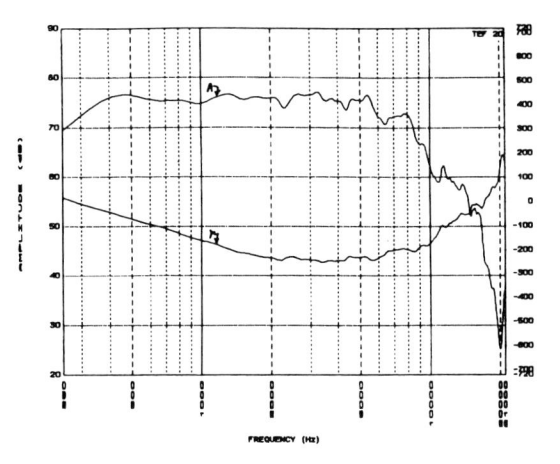

Fig. 20.

ms delay on the tweeter. Note that the notch depth is not as deep as measured with the woofer–horn combination. Absolutely equal levels, exactly out of phase, cancel completely with an infinite notch depth. Three things interfere with this: 1) the levels right at the crossover frequency may not be exactly equal, 2) exact phase cancellation may be between delay steps (only occurs at higher frequencies), and 3) ambient noise taints the measurement (usually the cause at lower frequencies). The author's experience has shown that under most reasonable conditions, a notch depth of 20 dB or so is usually an adequate indication of the proper delay for alignment. Finally, with the proper polarity restored, the resulting response can be seen in Fig. 24.

Putting the full-range response of the loudspeaker system together using all three drivers with proper alignment, the result is shown in Fig. 25. Using the graphic equalization available in the DSP processor sparingly, Fig. 26 displays an optimized response for this loudspeaker system.

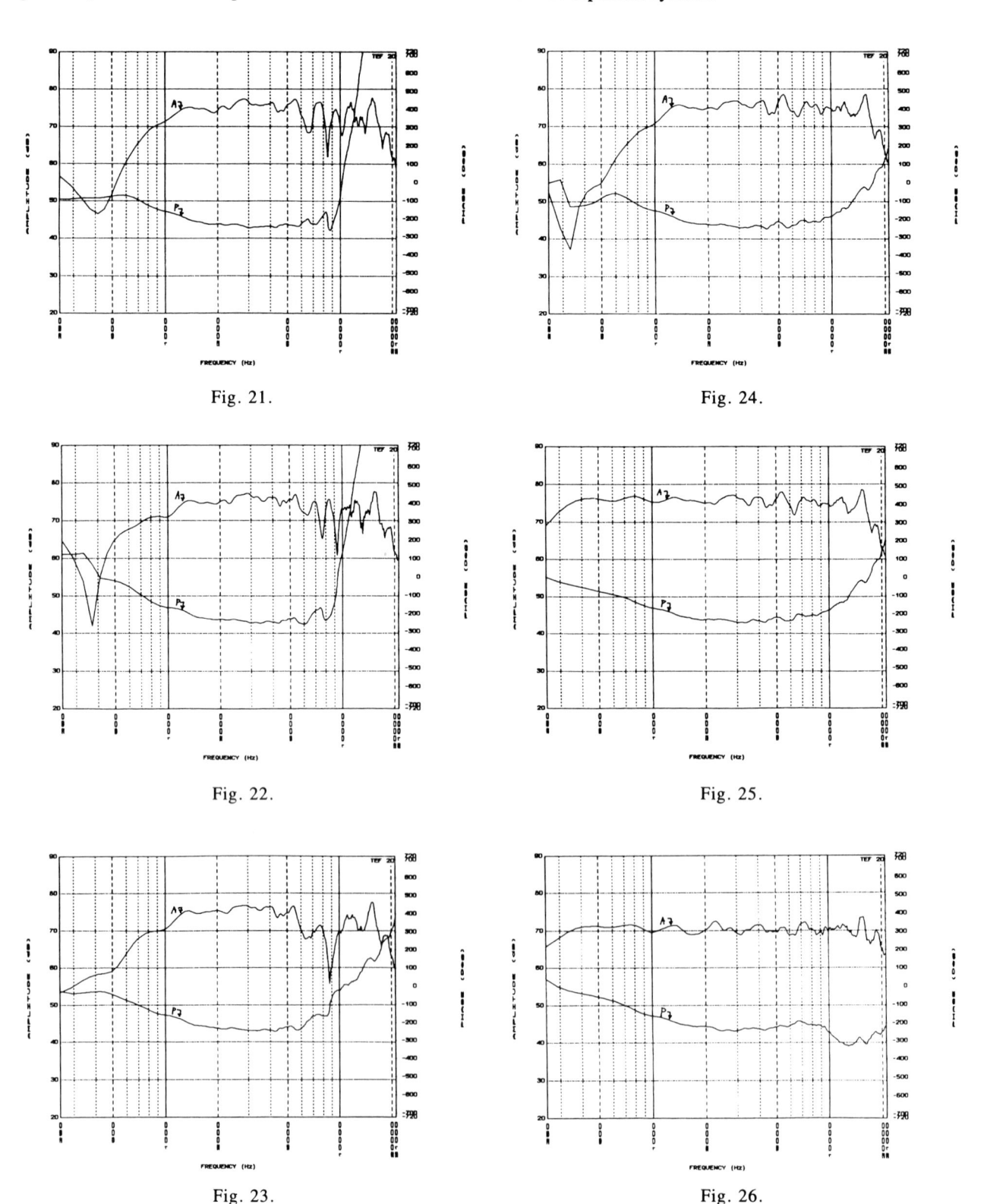

Fig. 21.

Fig. 22.

Fig. 23.

Fig. 24.

Fig. 25.

Fig. 26.

5 NECESSARY DELAY RESOLUTION FOR DRIVER ALIGNMENT

5.1 Frequency Response Considerations

As an employee of a manufacturer producing digital alignment delay devices, the author has been concerned with determining the needed resolution for signal delay increments. To that end, measurements were made using a digital delay unit with 10-µs delay steps. This yields a maximum error of 5µs, since an "absolute alignment" delay can be no more than 5 µs from a step increment either over or under that point in time.

Using the same loudspeaker system as in Sections 2 and 3, though measured in the anechoic chamber at the engineering facility, and employing slightly different setup parameters, the results are clear. Using the methodology outlined in Section 3, optimum delay settings were determined and measured. Subsequent measurements were made using increasing and decreasing delay offsets.

Fig. 27(a) shows the following three responses overlaid: 1) optimum delay; 2) +10 µs delay; and 3) +20 µs delay. Fig. 27(b) is an overlay of 1) optimum delay; 2) −10 µs delay; and 3) −20 µs. Obviously very little difference in frequency can be observed, even at very high frequencies. Once again, turning one's head can result in more drastic cancellations due to path-length differences to each ear. Therefore as long as the delay is ±10 µs of "absolute" alignment, the resolution is adequate for good frequency response.

Fig. 28 demonstrates the same situation for the reverse-polarity, delay-tuning setup and shows why this method is good for alignment delay setup. The out-of-polarity notches are more sensitive to phase changes than the in-polarity summations. At the 8-kHz crossover, the +20-µs setting in Fig. 28(a) changes the level at that point by about 12 dB. The −20-µs setting in Fig. 28(b) changes it by about 10 dB.

5.2 Polar Response Considerations

Fig. 29 shows ±10 and 20-µs offsets from the optimum delay for vertical polar responses at the acoustical crossover frequency of 944 Hz. At these wavelengths the amount of delay needed to produce significant variation would be measured in hundreds of microseconds, not 10 or 20.

Vertical polar plots for the acoustical crossover of 8.175 kHz are illustrated in Fig. 30. While the nulls do change significantly with increasing offsets, the general shape of the polar responses remains very consistent. If one sits up in a seat or moves much at all, more has been done to change what is heard than what is seen in these graphs.

6 CONCLUSIONS

It has been shown that a reasonable field method for the alignment of drivers within a multiway loudspeaker system can be achieved in the field without the need

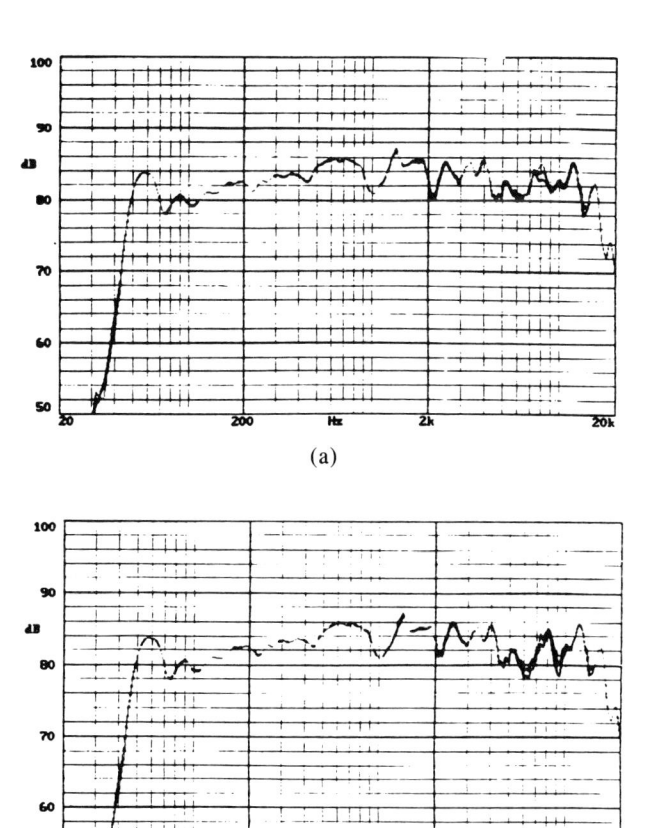

Fig. 27. Frequency response of 380SE drivers in polarity.

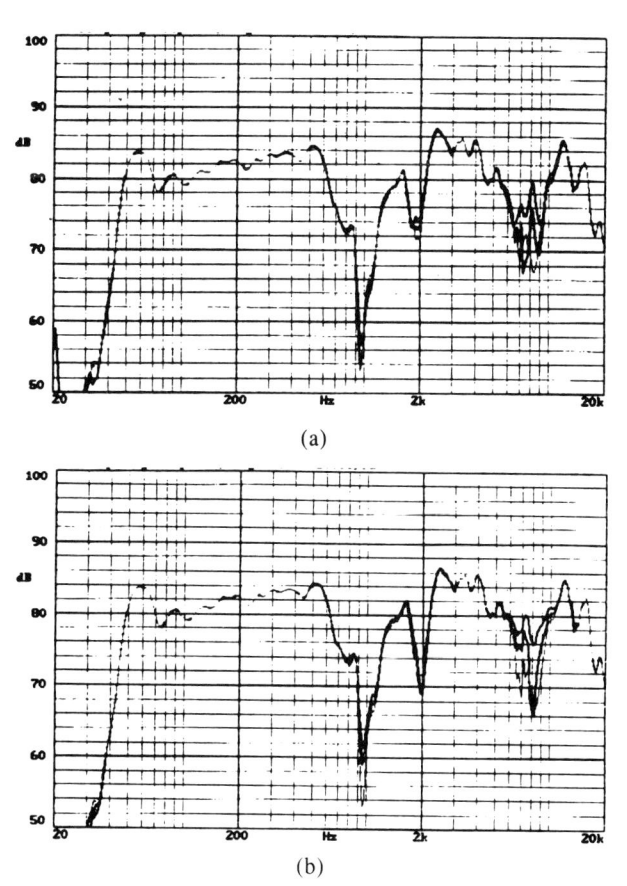

Fig. 28. Frequency response of 380SE woofer and tweeter in reverse polarity.

(a)

(a)

(b)

(b)

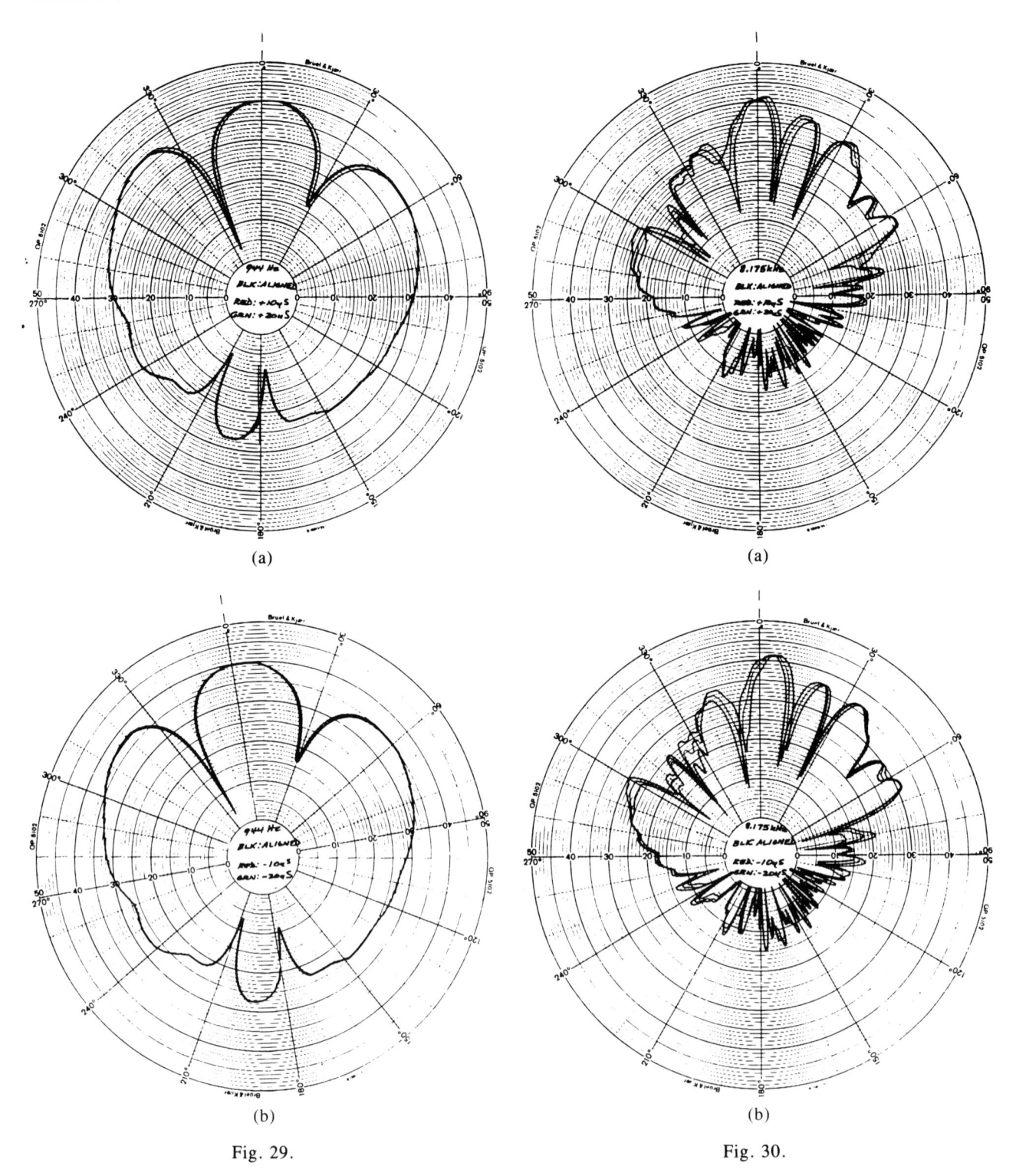

Fig. 29.

Fig. 30.

for an acoustical laboratory. This is not to declare the obsolescence of such facilities, but rather to show that current technology enables the "common" audio practitioner to use state-of-the-art equipment to its best without depending wholly on factory support.

It has also been shown that digital delay resolution finer than the common 21-μs step resolution is not a practical necessity. Response graphs can be made to look better, but the listening experience is not enhanced by such resolution.

The methods suggested here are just that, *suggested*, a place to start. If this author has learned anything about engineering, be it audio components, audio systems, or any other type, it is the fine art of compromise we learn. Keep in mind that these methods are not absolute. What seems very straightforward in print will run into difficulties requiring compromise in the field, just as it does in the engineering laboratories.

7 ACKNOWLEDGMENT

The author would like to thank Kurt Graffy of Paoletti Associates for suggesting he do this report; Tsutomu Yoshioka and staff of TOA Corporation, Japan, for making all the anechoic measurements; David Gunness of Electro-Voice, Naoki Fukumoto, and Chuck McGregor of TOA Electronics, Inc., for their help and tutorage; Don Davis of Syn-Aud-Con for his encouraging disagreement and educating most all of us; Dr. Gene Patronis for this patience, Tecron for the support, beta-test software, and free repairs, and everyone else from whom he has stolen all he knows.

THE AUTHOR

John A. Murray was born in New Rochelle, NY, and grew up in Springfield, OH. In 1975 he received a B.S. in radio-TV production and engineering from Ohio University. He worked as a sound contractor in Dayton, OH, and Indianapolis, IN, from 1976 to 1989, at which time he was the first TEF operator in Ohio and a beta tester for Acousta-CADD.

Mr. Murray was the professional sound marketing manager at Electro-Voice through 1991 and was involved in the development of Altec's Variable-Intensity horns and EV's MH-series of horns. In November of 1992 he organized and chaired a Syn-Aud-Con workshop on horn arrays. He is now manager of engineered systems for TOA Electronics and currently resides in Yellow Springs, OH. He has been a member of the AES off and on since 1977.

c.

reconciling speech and music

Sound Reinforcement Systems in Early Danish Churches*

DAN POPESCU

Danmarks Radio, Copenhagen, Denmark

The aspects of intelligibility and naturalness in churches with long reverberation time, large volume, considerable distance between talker and listeners, as well as blind areas produced by architectural pillars are discussed. The solution, which has been adopted in a considerable number of early Danish churches, uses a distributed loudspeaker system with special applications of delay equipment related to the various distances from the altar or pulpit to the listeners. Aesthetic and architectural considerations are also covered.

0 INTRODUCTION

Churches in Denmark, which became a Christian nation in the 10th century, can be divided into two categories: village churches and town churches.

Most Danish village churches were founded within a comparatively limited period between 1050 and 1250. Almost 2000 churches were built, and they remain today as monuments to perhaps the most intense building period in Denmark's history.

The second category, town churches, is related to the foundation of market towns and their economic growth and decline. The various functions of the church buildings were decisive for their size and appearance. Town churches can be placed into separate groups: cathedrals, churches founded in relation to the establishment and growth of market towns, and abbey churches.

The majority of churches built in the first centuries following the introduction of Christianity in Denmark are romanesque or early gothic. These have certain room characteristics which are of consequence in the use of these buildings for speech.

1 BASIC CHARACTERISTICS OF TOWN CHURCHES

The acoustic problems, which relate to the intelligibility in these churches, can be seen to result from the size and architectural form of the body of the church.

Quite a few of these churches are of considerable size; volumes of 10 000 m^3 or more are not uncommon. In connection with this we should note the considerable distance between the speaker and the listeners, just as it is typical that there are varied distances between the altar and pulpit and the listeners.

We ought to mention that, following the reformation, there was a greater emphasis put on the spoken word (the Danish church is also called "the church of the spoken word"), which resulted in the fact that in many churches the pulpit was moved toward the center of the nave to develop a better contact between the preacher and the congregation.

The development of the church form with nave and aisle, where large pillars create areas of sound shadows, is a factor which considerably reduces the intelligibility in the area of the aisles.

Finally there is the important fact that most of these larger churches have a reverberation time of between 4 and 6 seconds at 1000 Hz, with higher figures at low frequencies (Fig. 1). The churches are excellent rooms for the reproduction of music, especially choral and organ music, but without the use of electroacoustical systems intelligibility is limited to distances of a few meters from the speaker. Table 1 lists seven Danish churches which belong to the aforementioned group of town churches, describing the architectural and acoustic parameters.

2 DESIGN CRITERIA

During the past three years sound problems in many early Danish churches were solved through the installation of distributed loudspeaker systems, using delay equipment to synchronize the amplified sound with the direct live speech signal.

The prerequisites present in the design of those systems can be summarized as follows:

---

* Presented at the 63rd Convention of the Audio Engineering Society, Los Angeles, 1979 May 15–18; revised 1980 June 5.

1) Homogeneous sound coverage as regards the intelligibility and the natural reproduction of speech

2) The equipment should harmonize as much as possible with the churches' architectural styles

3) The equipment should be as free as possible from manual operation.

The basis for the calculation was the method by Peutz and Klein [1, 2] which takes into consideration the articulation loss for consonants. The basic formulas

$$\mathrm{AL}_{\mathrm{cons}} = \frac{200 D^2 RT_{60}^2}{VQ} + a/[\%] \quad (1)$$

for distances less than the critical distance and

$$\mathrm{AL}_{\mathrm{cons}} = 9 RT_{60} + a/[\%] \quad (2)$$

for distances greater than the critical distance where

RT_{60} = reverberation time (at 1400 Hz), in second
V = volume of room, in cubic meter
D = distance of source to listener, in meter
Q = directivity factor of sound source
D_c = critical distance ($D_c = 3.16 R_c$)

was modified by Staffeldt [3] into

$$\mathrm{AL}_{\mathrm{cons}} = \frac{287 D^2 RT_{60}^2}{VQ} + a/[\%] \quad (3)$$

so that, among other things, Eqs. (1) and (2) give the same result when $D = 3.16 R_c$.

In connection with the use of the calculation method of Peutz and Klein we must mention the following:

1) The criterion adopted for acceptable intelligibility in the case of sound reinforcement systems in churches was $\mathrm{AL}_{\mathrm{cons}}$ less than 10%. This takes into consideration the fact that many churchgoers are elderly with a tendency to deafness.

2) The value of RT_{60} is assessed as the value for an empty church, since the size of the congregation varies from Sunday to Sunday. In most instances the parish council insists that the intelligibility be acceptable even with very few people in the church.

3) The factor n in

$$\mathrm{AL}_{\mathrm{cons}} = \frac{287 D^2 RT_{60}^2 n}{VQ} + a/[\%] \quad (4)$$

in instances where a distributed loudspeaker system is used

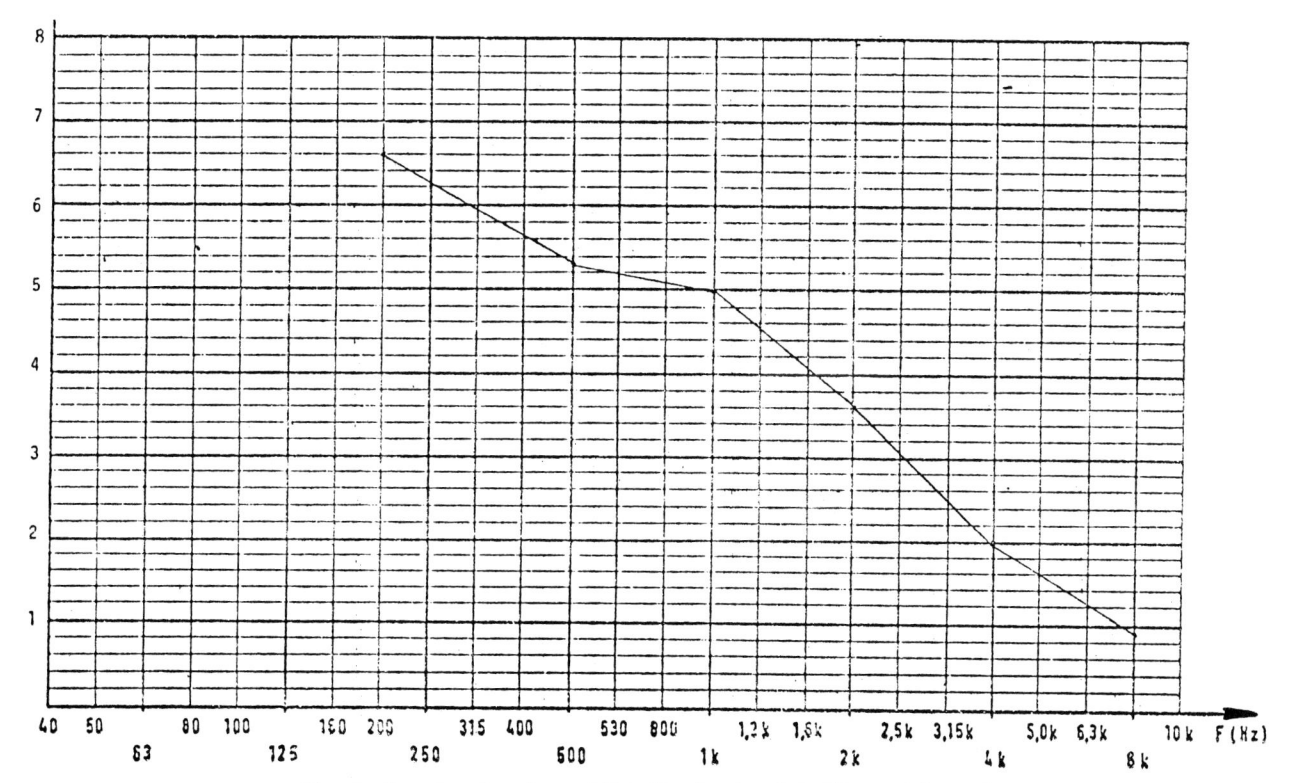

Fig. 1. Reverberation time of Sct. Bendts church in Ringsted (empty).

Table 1. Various parameters of some early Danish churches.

	RT_{60} at 1400 Hz, Empty /[second]	Volume /[m³]	Maximum clergy–listener distance /[m]	Pillar diameter /[m]	Distance between Pillars /[m]
Århus cathedral	3.4	24 000	60/20	3.3	6.5
Haderslev cathedral	5.0	13 000	40/18	2.4	7.5
Sorø church	4.3	15 000	55/27	2.2	5.0
Assens church	4.3	12 000	38/20	2.3	7.2
Sct. Bendts church	4.3	15 000	52/30	2.4	6.5
Sct. Mortens church	3.2	10 000	34/15	2.0	5.5
Løgumkloster church	3.3	10 000	34/14	3.0	5.0

is employed as defined by Davis [4, 5], n being the number of loudspeaker groups.

The measurement of R_c in previously installed distributed systems confirmed this definition, especially in instances where the maximum distance from source to listeners was not greater than $2 R_c$.

As one can deduce from Table 1, the distance from one pillar to another, which corresponds frequently to the distance between loudspeaker columns, is in most early Danish churches not greater than 6–7 m.

3 ELECTRICAL DESIGN CONSIDERATIONS

As mentioned already, it is a characteristic of protestant churches that the distances between the altar and pulpit and the listeners are different. An example of this is Århus cathedral, where the maximum distance of 60 m from the altar to the congregation is reduced to 20 m from the pulpit (Fig. 2). In such a case the installation of a delayed distributed loudspeaker system demands that the delay for the respective loudspeaker groups be changed according to where the minister is located.

Fig. 3 shows the three solutions. In Fig. 3(a) a supplementary delay is introduced with certain disadvantages in terms of reduced quality (double A/D–D/A conversion in the case of a digital delay system). In Fig. 3(b) relay switching is at the output of the delay modules. A change within the digital part of the time delay unit is shown in Fig. 3(c). This is the most flexible method as the two "sittings" can be adjusted independently of each other (Fig. 4).

In the cathedral at Århus the solution used was a combination of alternatives (b) and (c) (Fig. 5) because of the numerous address positions with correspondingly varied

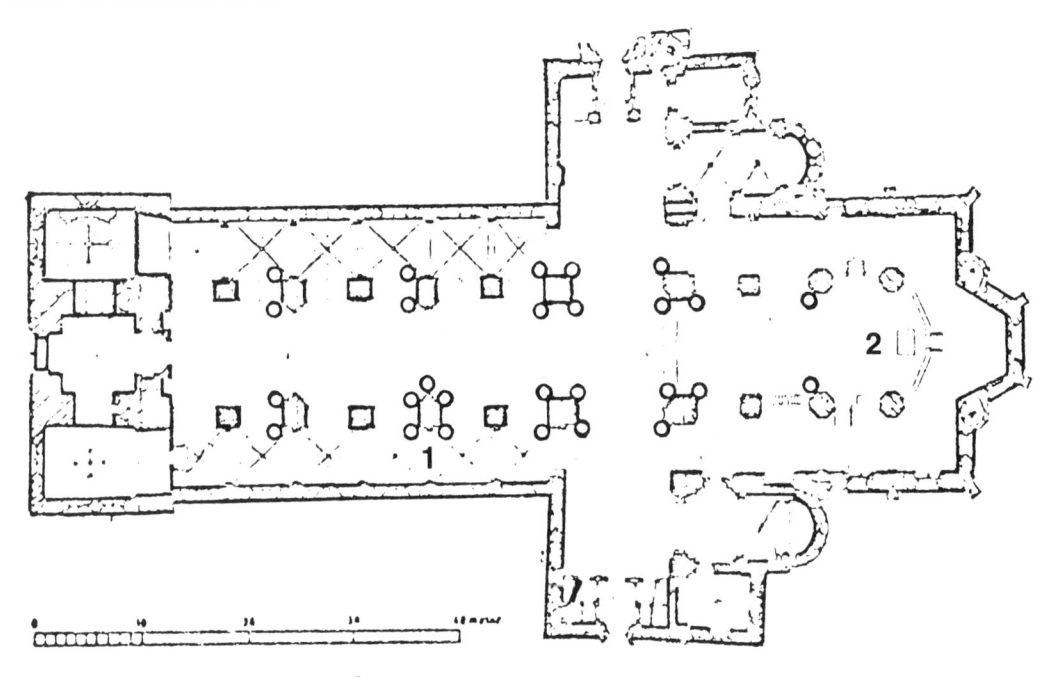

Fig. 2. Simplified plan of Århus cathedral. 1—pulpit; 2—altar; O—loudspeaker columns.

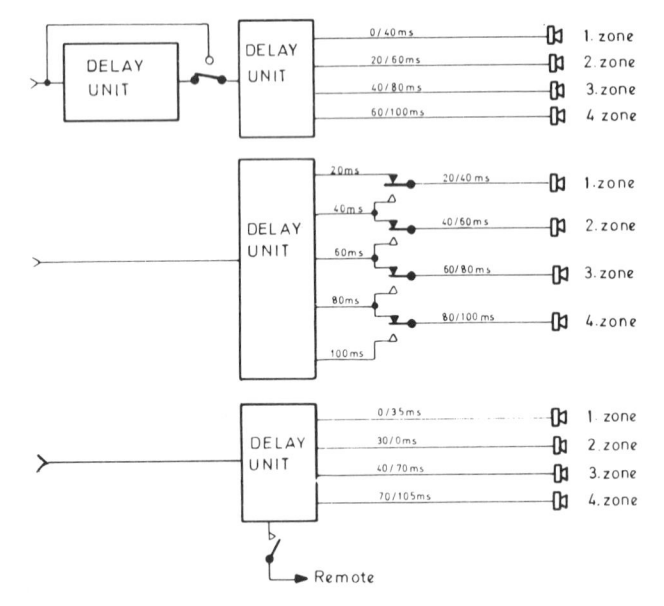

Fig. 3. Various solutions to change the delay time of the different loudspeakers' zones.

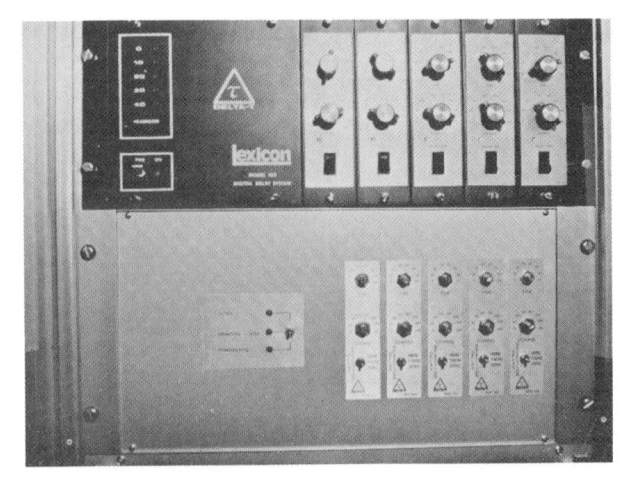

Fig. 4. Digital delay system + remote control units in Jesus church in Copenhagen.

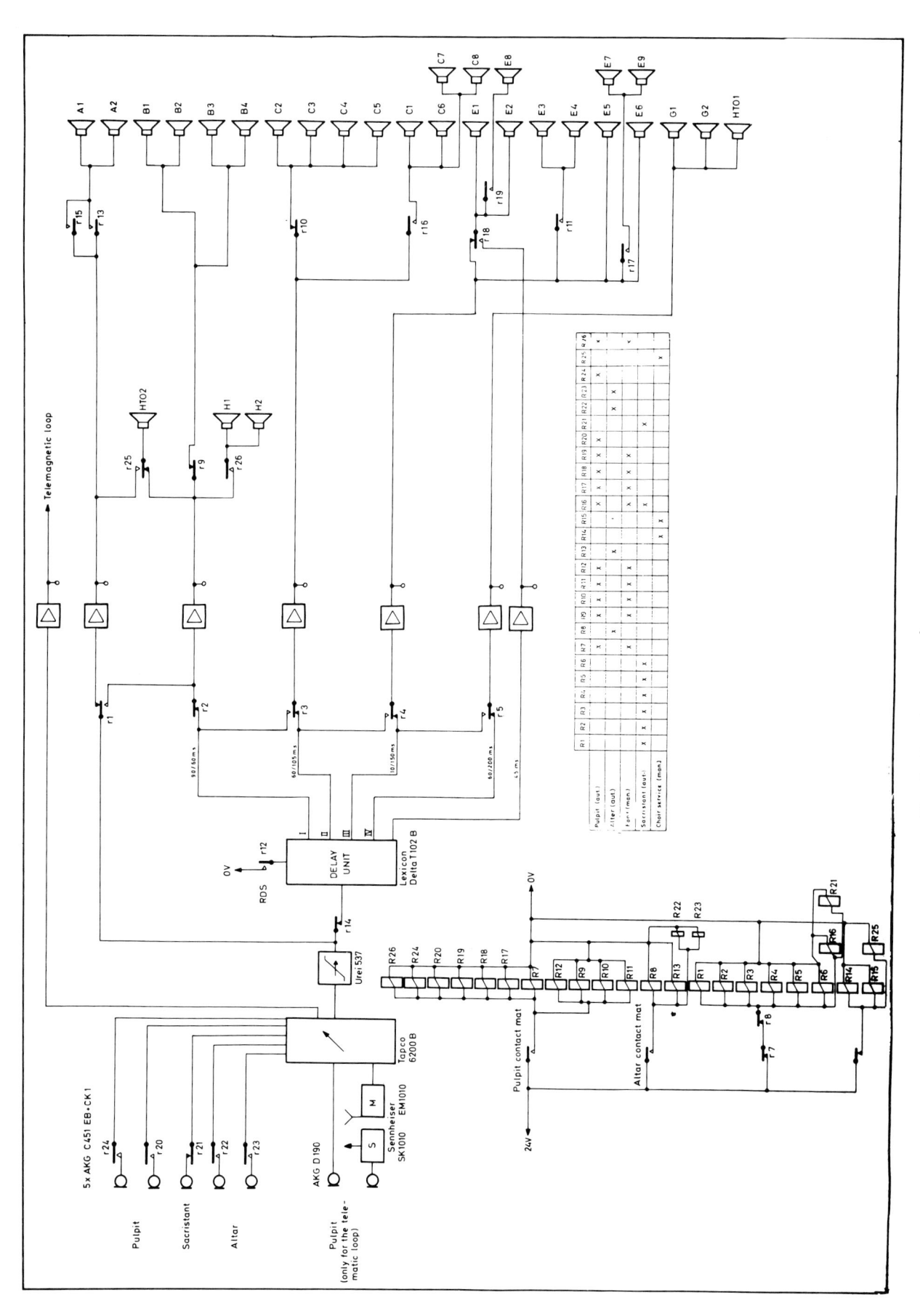

Fig. 5. Sound system of Århus cathedral.

speaker-to-audience distance characteristics for this large church.

One of the most important requirements of a modern sound reinforcement system in a church is that in spite of its complexity, the system should be able to function without manual operation.

Contact mats have been employed in many of the distributed loudspeaker systems for churches, which have been installed in Denmark during the past three years, to control time delay as well as the switching of the various microphones and loudspeaker columns. Those mats are usually placed near the altar and in the pulpit. In this manner an automatic control is established for at least three address places in the church—the altar, the pulpit, and the choir—with control of the appropriate time delays. The only operation necessary is the adjustment of levels to meet with the loudness of the different voices of the clergy.

In churches with long reverberation times, directional microphones are often used. Good results have recently been obtained with radio microphones using omnidirectional microphones, but the final choice is left to the clergyman's own preference as to whether or not he wishes to operate the technical equipment.

4 ARCHITECTURAL CONSIDERATIONS

One questionable point is the loudspeaker columns' matching the aesthetic and architectural demands in churches built 700–800 years ago.

Utilization of a great density of distribution by means of delay zones for every pillar zone means that the demand for a high Q factor to obtain the desired AL_{cons} is not so important, which permits the installation of columns with reduced physical dimensions.

An architect is always responsible for the integration of these foreign, but absolutely necessary, elements such as loudspeaker columns.

In many instances it has been sufficient to use white painted columns since most early Danish churches have lime-washed interiors. In a few cases the architect has been unwilling to accept the necessary inclination of the columns of approximately 10° when the columns are mounted at a height of 2 m. The solution here was to incorporate a standard column (for example, Philips LBC 3052) in an architect-designed box mounted vertically (Fig. 6).

5 ACKNOWLEDGMENT

The author wishes to acknowledge the excellent assistance provided by Mr. Jørgen Eriksen, Mr. E. Karing, and Mr. Erik Vikkelsoe from Oticon, Denmark, audio specialists responsible for the installation of distributed loudspeaker systems in various early Danish churches during the past three years.

Fig. 6. Architect-designed box including a standard column in Løgumkloster church.

6 REFERENCES

[1] V. M. A. Peutz, "Articulation Loss of Consonants as a Criterion for Speech Transmission in a Room," *J. Audio Eng. Soc.*, vol. 19, pp. 915–919 (1971 Dec.).

[2] W. Klein, "Articulation Loss of Consonants as a Basis for the Design and Judgment of Sound Reinforcement Systems," *J. Audio Eng. Soc.*, vol. 19, pp. 920–922 (1971 Dec.).

[3] H. Staffeldt, "Elektroakustik, Taletransmission i lukkede rum," unpublished.

[4] D. Davis and C. Davis, *Sound System Engineering* (Howard W. Sams, Indianapolis, IN, 1975).

[5] D. Davis, "Equivalent Acoustic Distance," *J. Audio Eng. Soc.*, vol. 21, pp. 646–649 (1973 Oct.).

[6] P. H. Parkin and J. H. Taylor, "Speech Reinforcement in St. Paul's Cathedral," *Wireless World*, vol. 58, p. 54 (1952 Feb.); vol. 58, p. 109 (1952 Mar.).

[7] D. L. Klepper, "Sound Systems in Reverberant Rooms for Worship," *J. Audio Eng. Soc.*, vol. 18, pp. 391–401 (1970 Aug.).

THE AUTHOR

Dan Popescu was born in Cluj, Rumania, in 1929. He received a degree in electronic engineering at The Polytechnic Institute in Bucharest in 1953. From 1952 to 1965 he worked with the Rumanian Broadcasting in Bucharest in the areas of broadcasting studio engineering.

In 1966 he joined The Danish Broadcasting in Copenhagen, Denmark. He is head of the design group for audio facilities in radio and television studios. Since 1976 he has been a consultant for The Danish Church Department, where he studies sound reinforcement systems in churches.

The Acoustical Design of a 4000-Seat Church*

A. H. MARSHALL, C. W. DAY, AND L. J. ELLIOTT

Marshall Day Associates, Auckland 1, New Zealand

The recently opened 4000-seat church auditorium for the Auckland Assembly of God presented several acoustical challenges to the consultants. Architectural acoustics, noise control, and electroacoustic systems are discussed.

0 INTRODUCTION

In 1979 the fast growing Auckland Assembly of God church decided that they needed to build their own place of worship. At that time they were regularly filling the Auckland Town Hall (2000 seats) twice each Sunday. An auditorium to seat 5000 people was initially investigated, with the final design reduced to 4000 seats due to town planning requirements. The site chosen for the church was adjacent to a major highway (7000 vehicles per hour) and a medium-flow access road. The auditorium was thus very close to a residential boundary and a major traffic route (Fig. 1).

The Assembly of God is a Pentecostal church with its services revolving around, and strongly dependent on, modern amplified music. The design of the church provided several challenging aspects for the acoustical consultant in room acoustics, noise control, and sound system design.

1 CONSULTANT'S BRIEF

Based on the client's experience with other facilities and a study by the consultant of buildings the client had found acceptable and unacceptable, an acoustical design brief was developed. In summary, the brief included the following:

1) *Room acoustics*: To provide acoustics appropriate for congregational singing, amplified music, and high levels of speech intelligibility.

2) *Air-conditioning noise control*: To control the level of air-conditioning noise (NR20).

3) *Sound insulation*: To control the intrusion of noise from external sources.

4) *Environmental noise control*: To control the level of noise from the church transmitted to the residential boundary.

5) *Sound system*: To provide high-quality music and speech amplification and recording systems.

To meet these objectives within stringent cost restraints, the following analysis was carried out.

2 ROOM ACOUSTICS

2.1 Concept

The acoustical properties of an auditorium to accommodate 4000 people comprise a major dimension of the design. An obvious requirement is the provision of excellent speech clarity throughout the space, and therefore the integration of the acoustic and sound reinforcement system design is essential. These factors were considered regularly throughout the design process, from the shaping of the room to the distribution of absorbing and diffusing areas to minimize echoes and the selection of suitable seating types.

However, a church is not just an auditorium. An essential aspect of its corporate worship is uplifting congregational singing, and special attention was given to ensuring that the acoustical space enhances this. Further, as a Pentecostal church, spontaneous congregational expressions of worship occur in most services. In these events it has been observed that the reverberant sound to which the members each contribute a tiny part enhances the sense of unity.

2.2 Reverberation Time

In order to provide an environment that would enhance the congregational worship, the protection of the reverberant field was one of our prime objectives, even though it made the design of a sound system for speech

---
* Presented at the 81st Convention of the Audio Engineering Society, Los Angeles, CA, 1986 November 12-16.

clarity rather more difficult.

The practical consequence is a larger volume (31 000 m^3, about 1 million ft^3) than the minimum required to accommodate so many people in the choice of materials for the church interior.

During the design process, as modifications occurred, their implication on the reverberation time was reviewed regularly. The desired reverberation time for the fully occupied condition was as followed:

Octave-band center frequency (Hz)	Reverberation time (seconds)
125	2.3
250	2.2
500	2.0
1000	1.9
2000	1.8
4000	1.6

To achieve this would require a considerable area of low-frequency absorbent, which would have significant cost implications. In view of this it was decided that initially no special low-frequency absorbers would be installed. The need to retrofit them was to be reviewed after subjective evaluation of the space. This resulted in the following predicted reverberation time, again for the fully occupied condition:

Octave-band center frequency (Hz)	Reverberation time (seconds)
125	3.1
250	2.6
500	2.2
1000	1.9
2000	1.8
4000	1.6

2.3 Model Study

Although careful consideration had been given at the conceptual design stage to the shape of the room in order to minimize potential echoes, it was deemed important to undertake a model study of the auditorium.

A 1:100 scale architectural model had been constructed, and this was sufficiently detailed to use for acoustic testing. The source was a spark generator, which was located both on the platform and at the main loudspeaker location. Several listening locations were tested with the impulse response being displayed on a storage oscilloscope.

Speech intelligibility and musical clarity rely primarily on the presence of sufficient early, integrable reflections. Reflections arriving within 50 ms (depending upon the nature of the sound) are integrated, by the hearing process, with the direct sound, thus contributing to the loudness, clarity, and intelligibility of the sound. Reflections arriving beyond 50 ms, if sufficiently strong, create echoes that are destructive to clarity and speech intelligibility [1], [2].

Despite the size of the space, the model tests generally exhibited a significant amount of useful early integrable sound and little troublesome late energy.

Usually in such spaces any surfaces that are the cause of late arriving energy, such as the rear wall, would be treated with a highly absorbent material. In this particular space it was decided to treat such problem surfaces with diffusing panels in order not to reduce the reverberation time significantly.

3 AIR-CONDITIONING NOISE CONTROL

The design criterion for background noise levels within the auditorium was established by the client based on their satisfactory experience at their existing facility, that is, NR20. This criterion was later relaxed to NR25 to allow broad-band air regenerated noise to provide some masking of traffic noise peaks (NR20).

3.1 Duct-Borne Fan Noise

The main supply and return air fans for the auditorium were located in a free-standing plant tower at the rear of the auditorium to isolate structural noise from the plant room. This location also permitted silencers to be installed with sufficient space between fan and auditorium. Axial flow fans were preferred by the mechanical engineer due to the low system resistance (low

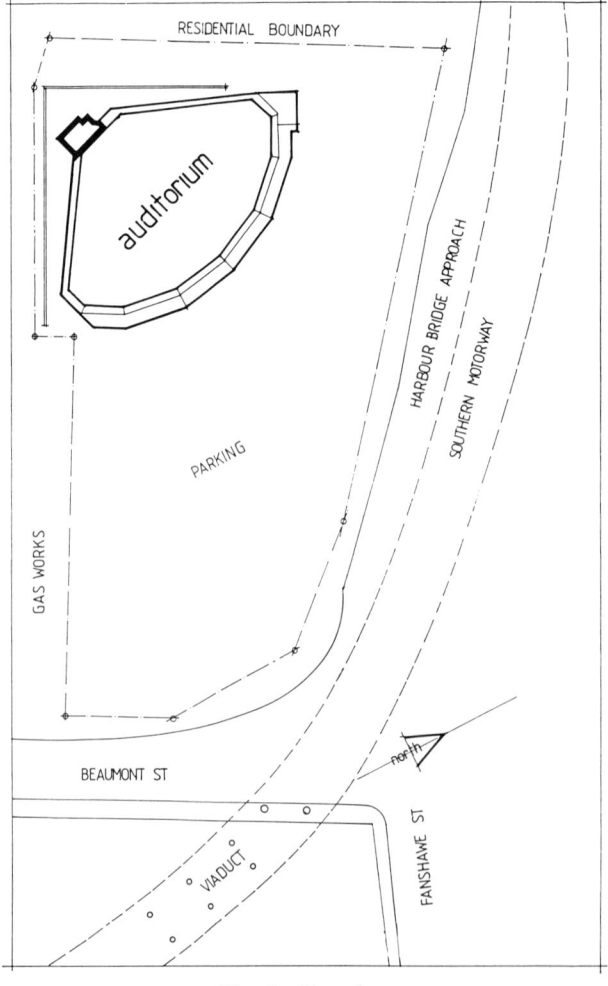

Fig. 1. Site plan.

velocities and simple ductwork) and also because of cost savings. After some investigation of the higher noise levels of these fans and thus additional attenuation requirements, the axial fan approach was adopted.

Calculation of the fan noise transmitted down duct to the auditorium was carried out using conventional techniques similar to those prescribed in ASHRAE and Woods and the consultant's experience in other auditoriums.

The most unusual aspect of the calculation was the need to predict the down-duct attenuation of plywood duct work. Exposed timber duct work was required by the architect for the predominantly timber auditorium. The supply air duct-work layout is shown in Fig. 2.

As with most low-noise situations (<NR25), the high attenuation required encouraged the use of primary and secondary silencers. This approach is used wherever possible since two separate silencers are generally more efficient than one single silencer of the same total length. It also enables one to select from the proprietary silencers available on the market and attenuate noise generated within the system. Unfortunately space limitations in the return air system dictated the use of a single silencer 3.6 m long, designed as a "one-off" for this project. The return air system is shown in Fig. 3.

3.2 Regenerated Air Noise

Noise generated by air turbulence at duct elements and diffusers is a significant problem for low-noise situations. The normal design approach is to ensure that the duct and discharge velocities are low enough to obtain low levels of regenerated noise. Unfortunately low air velocities at the discharge of supply air diffusers can cause dumping of cold air on the occupants. For this large hall with the ducting located around the perimeter, the mechanical engineer needed a diffuser with a throw of up to 50 m.

The consultants were made aware of a U.S. diffuser known as Punka louvre with particularly good throw characteristics and low noise levels. Unfortunately catalog test data are generally obtained under ideal laboratory conditions. In practice these conditions are very rarely realized on site. Noise generated by the diffuser is a result of the air flow conditions created by the combination of run-out duct, damper, and diffuser. For this reason it was decided to import a sample louvre and carry out a mock-up test based on the specific design for the church.

Each diffuser required individual balancing, so a terminal silencer/run-out duct was included to attenuate noise generated upstream of the diffuser and to straighten the airflow onto the diffuser. The test sample thus consisted of a balancing damper followed by an 1800-mm-long terminal silencer/lined run-out duct (1800 mm wide by 400 mm high), with the Punka louvre fitted neatly to the end.

This arrangement was tested with a specially designed "quiet" air flow system at the Acoustics Institute Laboratory, University of Auckland. Air was supplied to the test rig at the design volume, with the fan noise attenuated well below the air flow noise to be measured. The testing program both confirmed that the proposed arrangement would meet the design criterion and allowed the design of the terminal silencer to be refined, resulting in a reduction in the bulk of each unit and a subsequent cost saving.

4 SOUND INSULATION, TRAFFIC NOISE

The proximity of the church to the highway created potential noise intrusion from heavy vehicles. Several noise surveys were carried out on the site to determine the level of noise incident on the church at different times. The measured noise level was approximately 70 dBA (L10) during peak-hour traffic and 62 dBA on Sunday morning and evening (periods of predominant use).

From octave-band analysis of the on-site noise levels calculations were carried out to determine the required transmission loss of the building envelope for traffic noise.

To meet the required transmission loss using a concrete construction would have been relatively simple—75 mm of concrete would have provided sufficient safety margins for the consultant to sleep well at night. Structural considerations were paramount as the architect wanted to span the 80-m-wide and 50-m-long building without internal support, and alternative lightweight constructions had to be investigated.

Two different constructions were designed by the consultant team. The first utilized a 1-m-deep gang-nail truss with top and bottom plywood diaphragms (Fig. 4). The second comprised a 65-mm solid-timber ceiling with 200-mm-deep spacers and an upper surface of plywood, fiberglass, and metal roofing (Fig. 5). A certain degree of uncertainty crept into the prediction of sound transmission loss for these unusual constructions. With the threat of a town planning condition restricting operation of the church, two options were available:

1) Overdesign, with subsequent cost implications.
2) Test a section of the roof in the laboratory.

Both structures were tested for transmission loss at the Acoustics Institute Laboratory. The gang-nail truss system achieved the design criterion with a safety factor of 1 dB (at 500 Hz), as did the solid-timber alternative with 3 dB to spare at 500 Hz. The solid-timber alternative was chosen due to a significant cost advantage over the gang-nail truss. The cost savings achieved by the testing program were on the order of 10 times the consultant's total fee for the project. The building went ahead on this basis with regular site supervision to make sure that the normal degradation of laboratory performance did not occur on site.

5 ENVIRONMENTAL NOISE CONTROL

The proposed location of the church met with several objections from local residents, and the subsequent

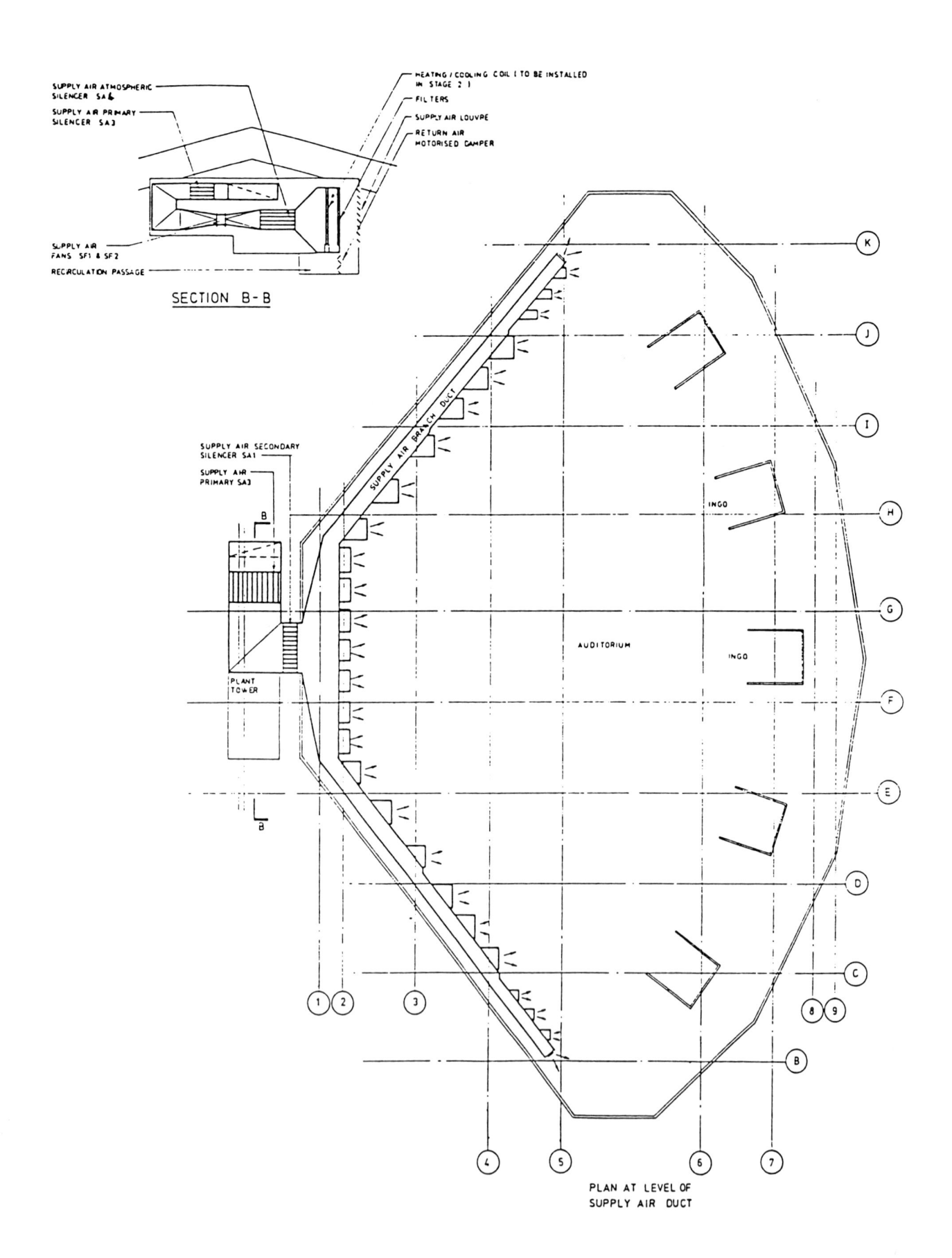

Fig. 2. Supply air system.

town planning hearing placed a condition, restricting noise produced by the church at the residential boundary. The noise control design thus involved protecting the residents from 1) the noise produced by 4000 people singing and a powerful high-quality sound system and 2) the air-conditioning plant located at the rear of the building close to the residential boundary.

5.1 Design Criteria

Following an environmental noise survey to establish the existing noise environment, the Auckland City Council's Planning Committee established a noise condition that in summary specified that noise emissions from the church should not exceed an L10 level of 50 dBA at the residential boundary (see Fig. 1).

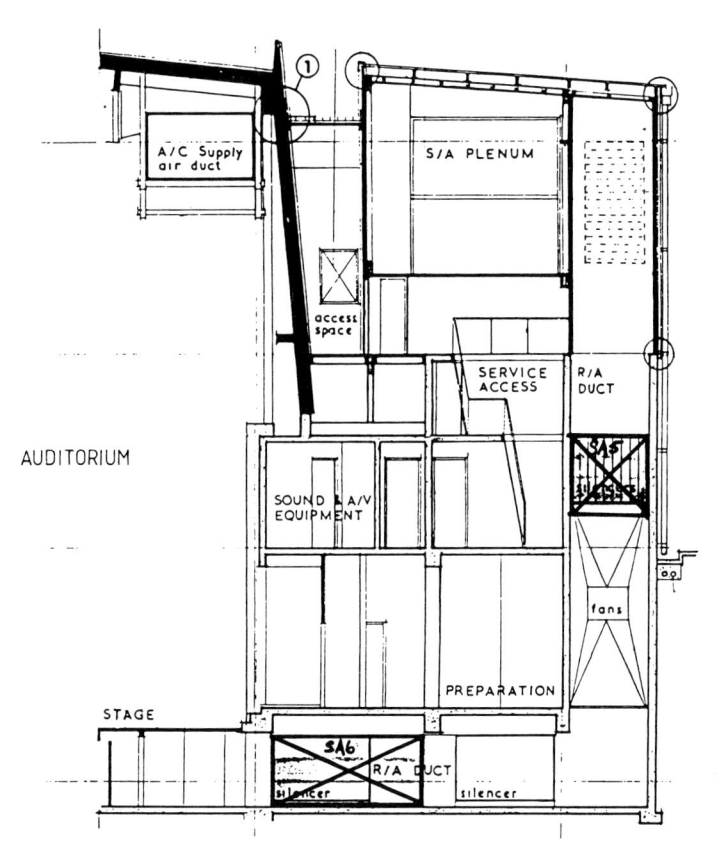

Fig. 3. Return air system.

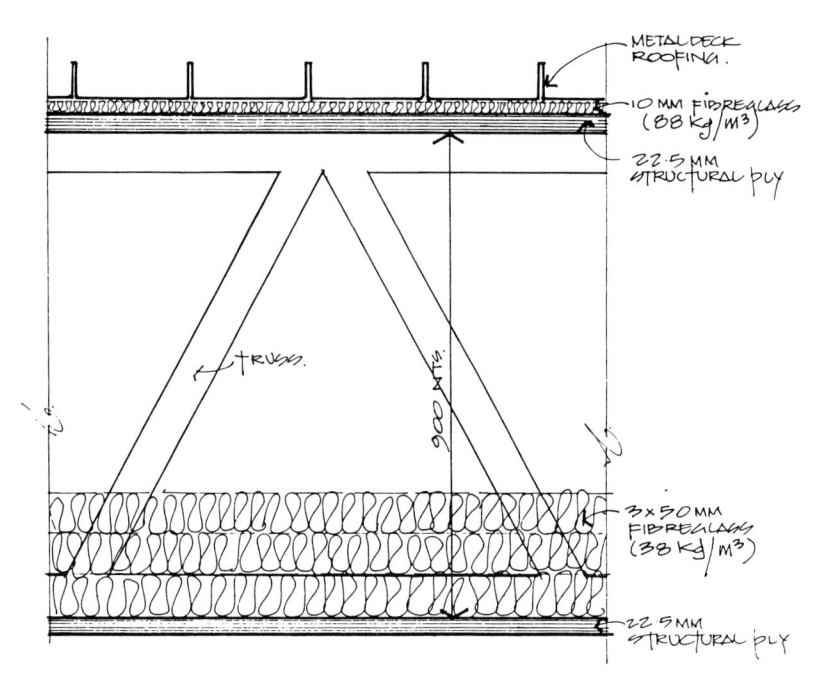

Fig. 4. Gang-nail truss system.

5.2 Congregation Noise

To calculate the level of congregation noise transmitted to the residential boundary, noise measurements were made in the Town Hall during a typical service. This survey showed that the combined level from singing, musicians, and the sound system was 90–95 dB, predominantly in the low and mid-frequencies. Using maximum levels of 95 dB in each octave, and the physical properties of the auditorium, the transmission loss required to meet the town planning requirements was calculated (Fig. 6).

5.3 Mechanical Services Noise

To meet the council's criterion, the following aspects of the mechanical design had to be analyzed:

1) Duct-borne fan noise to the atmosphere,
2) Fan noise transmitted through the plant tower cladding,
3) Noise from the free-standing air-cooled condensers.

5.4 Duct-Borne Fan Noise

The 17-m-high plant tower, located 20 m from the residential boundary, contains four large axial flow fans with sound power levels of almost 110 dB. Calculations including directivity and distance factors permitted selecting proprietary silencers and installing them on the atmospheric side of both the supply and the return air fans. Resultant noise levels are reported in Section 7.

5.5 Plant Room Breakout

Calculations of the sound transmission through the plant tower envelope showed that the outer framing had to be clad on both sides with two layers of 14.5-mm gypsum board. An outer skin of metal cladding was provided for weather protection. Attention to details and site supervision ensured that the required sound transmission loss was achieved.

5.6 Air-Cooled Condensers

The air-cooled condensers were originally to be located 5–10 m from the residential boundary. The bank of condensers was specifically designed to allow the installation of inlet and outlet silencers, selected to meet the boundary noise criteria. The condensers contained several fans, and the calculations thus included corrections for distance and directivity of a multiple-source configuration.

6 SOUND SYSTEMS

6.1 Concept

Given the size and seating layout of the auditorium, it was decided that although a stereophonic system was preferred for the musical aspects of the church, the most appropriate approach was to utilize a mono central system located above the platform. This would ensure

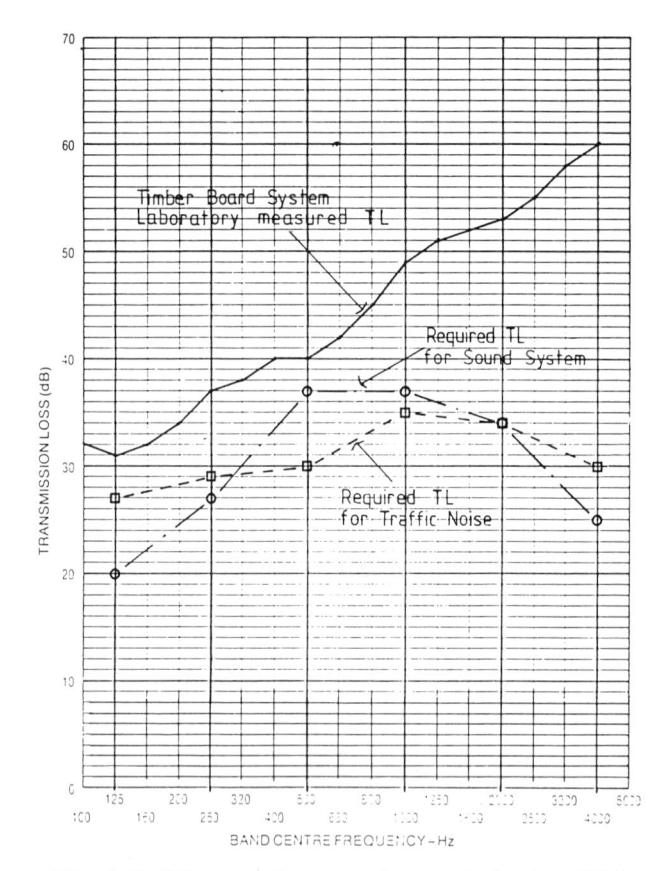

Fig. 6. Building envelope sound transmission loss (TL).

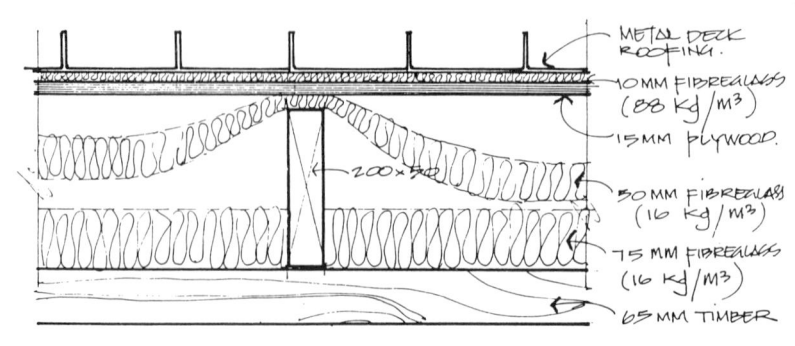

Fig. 5. Timber-board system.

good directional realism for the majority of the congregation and not involve the additional expense of trying to achieve a stereo image throughout the entire seating area.

This having been determined, the functional requirements of the systems were established.

6.2 Functional Requirements

The auditorium would be equipped with sound systems having the following features:

1) Reinforcement of both speech and music events taking place on the platform with coverage of the entire seating areas.

2) Reinforcement back to the performers using portable loudspeaker systems, with control for this system being available from either the main console or a supplementary console located on the platform.

3) Reproduction of single-channel motion picture sound.

4) Reproduction of single-channel audio-visual sound.

5) Playback of single-channel program material from tape.

6) Single- or dual-channel tape recording at the console location and multitrack recording in the sound control room.

7) Signal feeds for television and radio broadcasting.

8) Program relay to the backstage circulation areas and administration building.

9) Distribution of program to public circulation areas.

10) Distribution of emergency announcements.

11) Production communications between platform, sound and lighting control, and projection room.

6.3 Loudspeaker Systems

The auditorium sound system uses a central cluster of directional loudspeakers located above the platform. The cluster consists of five constant-directivity high-frequency horns and four dual-driver low-frequency horn-loaded cabinets. The cluster was designed using the Prohs–Harris system [4].

In view of the large distance to the rear seating areas, it was considered that the main loudspeaker system alone would not ensure uniform sound levels throughout the entire seating areas. Therefore four supplementary loudspeaker systems were provided to serve the rear area of seating. These loudspeakers operate with electronically delayed signals to preserve the directional realism and prevent artificial echoes.

Further supplementary loudspeakers were housed in the front of the platform to even out the coverage of direct sound to the front rows of seating and to lower the apparent acoustical location to one that correlates better with the visual location.

6.4 Microphone Receptacles

Microphone receptacles are provided throughout the platform area, in the front of the platform on the flat area of flooring, and on the lighting bridge for suspending microphones for recording.

6.5 Sound Control Positions

The permanent sound control position was provided at the rear of the main seating area. A 32-input 8-group console is located at this point.

A sound control room is provided for recording or broadcast of services. At this stage the control room is not equipped with specialized recording equipment, but this is hired when it is required.

On-stage monitoring signals are controlled from a console located on stage.

6.6 Control Equipment

The sound system control equipment is housed in an equipment room adjacent to the control room. For easy optimization of levels throughout the auditorium each loudspeaker device in the main cluster is served by its own power amplifier. Each of the four supplementary clusters has its own power amplifier. Levels for each power amplifier are not adjusted using the built-in volume control but using an independent system. This means that if required, the spare amplifier may be substituted in any circuit without requiring adjustment of gain.

Each of the signal lines to the amplifiers serving the main loudspeaker system, the stage monitor loudspeakers, and the supplemental loudspeakers is provided with one-third-octave band equalizers. Fig. 7 shows a simplified functional diagram for the auditorium sound system.

7 COMMISSIONING MEASUREMENTS

7.1 Reverberation Time

Reverberation time was measured using an impulse signal. The measured values for the unoccupied auditorium are as follows:

Octave-band center frequency (Hz)	Reverberation time (seconds)
125	3.5
250	2.8
500	2.2
1000	2.0
2000	1.6
4000	1.5

The higher than desired reverberation time at the low frequencies has not caused too many difficulties for the regular church services. However, when the venue has been used for some contemporary music concerts, there have been some problems. It is envisaged that the low-frequency absorbers originally proposed will eventually be installed as funding permits.

7.2 Air-conditioning Noise to Auditorium

The noise level within the auditorium was significantly higher than the design criterion when the system was initially commissioned. Investigation found that

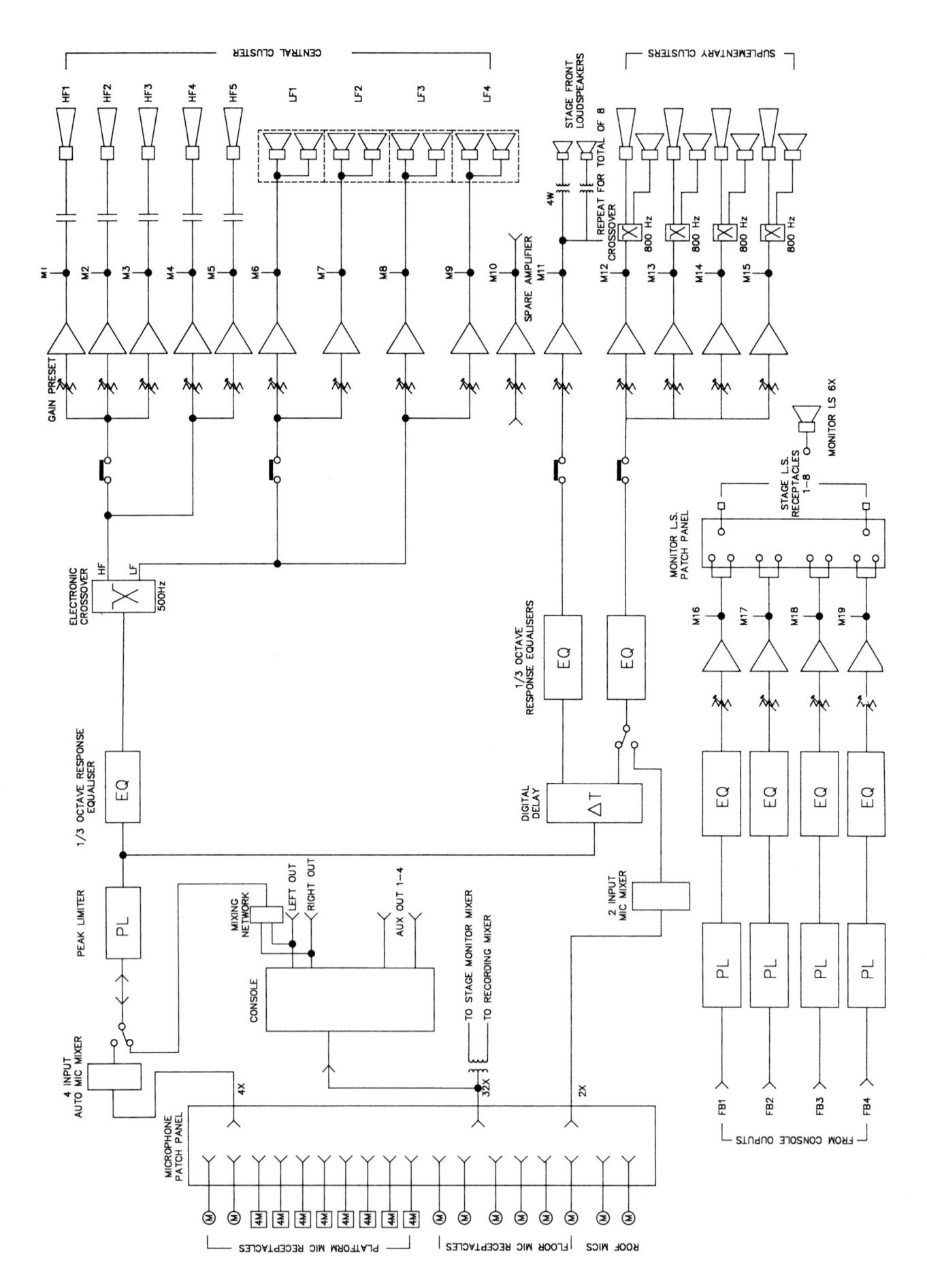

Fig. 7. Simplified functional diagram of Auckland Assembly of God sound system.

one of the "identical" supply air fans was producing an offensive tone in the 250-Hz octave band. Narrow-band analysis showed the frequency to be related to the blade passage frequency of the fan, and was probably caused by interaction between the fan blades and the motor supports.

Discussions with the fan supplier and subsequent minor modifications to the blade pitch angle have reduced the pure tone significantly.

Incorrect installation of the large return air silencer has allowed fan noise to leak around the silencer and contribute to the noise excess. The noise level in the auditorium is currently NR29 (4 dB over the design criterion), and with further rectification work the design objective should be achieved.

7.3 Air-Conditioning Noise to Residential Boundary

To ensure that the noise measurements made at the residential boundary were from the air-conditioning plant only, commissioning measurements were carried out at 3:00 a.m. At this time the background noise from traffic noise was sufficiently below the plant noise so as not to significantly affect the readings. With both the supply and the return air systems operating, the measured noise level was 46 dBA at the closest residential boundary—4 dB inside the 50-dBA criterion.

7.4 Congregation Noise to Residential Boundary

To measure the effective sound insulation of noise from the auditorium transmitted to the residential boundary, a controlled source was used in preference to the congregation noise. Pink noise amplified through the auditorium sound system produced a sound pressure level of 100 dB in each octave band, within the auditorium.

The noise level measured at the residential boundary was 51 dBA. With the actual maximum levels of congregation noise of 95 dB, the boundary noise level would be 46 dBA—again 4 dB within the town planning requirement.

7.5 Traffic Noise to Auditorium

Measurements of the traffic noise levels were made during the noisiest possible conditions (4:40 p.m. Friday) to ensure adequate signal-to-noise ratio within the auditorium. Noise levels both inside and outside the auditorium were measured.

The maximum (L10) level of traffic noise within the auditorium was NR30. At the same time the external noise level was 8 dB higher than the levels previously measured for a Sunday. On this basis the design criterion of NR20 would be achieved during services on Sunday.

7.6 Sound System

Uniformity of coverage was measured using an octave band of pink noise centered on 4 kHz. Variations throughout the seating area were within ±2 dB.

The overall system frequency response was measured in one-third-octave bands and equalized to a response that is essentially flat from 50 to 2500 Hz and slopes uniformly from 0 dB 2500 Hz to −8 dB at 10 000 Hz. This response appears to provide a good compromise for both speech and music signals.

The maximum sound pressure level was measured using a precision-grade sound level meter with a standard "fast" meter damping. Wide-band recorded music was used as a test signal, and levels of 105 dBC were measured throughout the seating areas. At this level the system peak limiters are just beginning to operate.

Acoustic gain was measured using a small loudspeaker fed with an A-weighted pink-noise signal. With a source-to-microphone distance of 500 mm, which relates to the "normal" talker-to-microphone distance at the pulpit, the measured acoustic gain at the rear row of the auditorium was 27 dB. This correlates within 2 dB with that predicted during the design of the system.

8 CONCLUSIONS

1) Low levels of fan noise (less than NR30) can be achieved with designs utilizing axial flow fans.

2) Pure-tone noise problems can occur with axial-flow fans due to fan blade interaction with the motor supports. These problems can be solved with relatively minor alterations to the impeller/support arrangement.

3) Low levels of air-regenerated noise can be achieved with diffusers giving a throw of up to 50 m. Laboratory testing of unusual installations is recommended.

4) Return air systems that return a large proportion of the air from predominantly one location must be designed carefully and stringently supervised to ensure that design noise levels are achieved.

5) Relatively inexpensive and lightweight building envelope constructions can be designed to meet a high degree of sound insulation. For this project both traffic noise and amplified music have been isolated.

6) Acoustical design combined with a program of laboratory testing can eliminate overdesign from predictive analysis, resulting in significant cost savings for the project.

7) Stringent acoustical criteria can be achieved with detailed acoustical design, diligent site supervision, and laboratory testing when required.

7 REFERENCES

[1] J. P. A. Lochner and J. F. Burger, "The Influence of Reflections on Auditorium Acoustics," *J. Sound Vibration*, vol. 1, pp. 426–454 (1964).

[2] V. M. A. Peutz, "Articulation Loss of Consonants as a Criterion for Speech Transmission in a Room," *J. Audio Eng. Soc.*, vol. 19, pp. 915–919 (1971 Dec.).

[3] I. Sharland, *Woods Practical Guide to Noise Control* (Woods Acoustics, Colchester, 1972).

[4] J. R. Prohs and D. E. Harris, "An Accurate and Easily Implemented Method of Modeling Loudspeaker Array Coverage," *J. Audio Eng. Soc.*, vol. 32, pp. 204–217 (1984 Apr.).

THE AUTHORS

H. Marshall    **C. Day**    **L. Elliott**

Harold Marshall comes to acoustical design from a multidisciplinary background. He graduated from Auckland University in 1956 with a science degree in mathematics and physics, and an honors degree in architecture. Professional work in Europe and the far East was followed by a year at the Munich Academy of Fine Arts. He earned his Ph.D. in engineering at the Institute of Sound and Vibration at Southampton University in 1967. He was associate professor in building science at the University of Western Australia from 1967 to 1972. He is now professor of architecture and head of the Acoustics Research Centre, University of Auckland, where he holds a personal chair. After 20 years of acoustical consultation he was joined in a new practice, Marshall Day Associates, by Chris Day and Larry Elliott in 1979.

His research interests have centered in concert hall acoustics, where he has been a major contributor to understanding the importance of lateral reflections. In recent years his work has revealed the necessary and sufficient conditions for instrumental and vocal ensemble.

Major concert halls in which he has contributed acoustical design include the Christchurch Town Hall (New Zealand), The Michael Fowler Centre (New Zealand), The Perth Concert Hall (Western Australia), Segerstrom Hall (Orange County Centre for the Performing Arts), and the Tsim Sha Tsui Cultural Centre in Hong Kong.

An active musician, Dr. Marshall sings regularly with the Auckland Bach Cantata Society and performs light opera, oratorio, and lieder.

Larry J. Elliott was born in Auckland, New Zealand, in 1947. From 1966 to 1982 he was involved as a sound engineer for both radio and television broadcasting as well as recording studios. In 1982 Mr. Elliott joined the acoustical consulting firm of Marshall Day Associates, where he has specialized in sound reinforcement systems design. His projects have included churches, concert halls, education facilities, and sports facilities both in New Zealand and Malaysia.

Mr. Elliott is a member of the Audio Engineering Society and the New Zealand Acoustical Society.

●

Christopher Day graduated from Monash University (Melbourne) in 1973 with a degree in mechanical engineering. Pursuing an interest in acoustics, he joined Sound Research Laboratories Ltd. of the U.K., gaining experience in laboratory testing, industrial and mechanical services noise control, and architectural acoustics.

In 1975 he returned to Australia to work for a noise control supply company. He moved to New Zealand in 1978, and after a year of studying background noise levels under sail in the Pacific Islands, he joined Harold Marshall in partnership to form Marshall Day Associates. His projects have included television studios in Malaysia and New Zealand, noise control for the Michael Fowler Centre, Wellington, and 1:10 model testing of the Orange County Performing Arts Centre, California. He is currently president of the New Zealand Acoustical Society.

ENGINEERING REPORTS

A Holographic Approach to Acoustic Control*

A. J. BERKHOUT

Delft University of Technology, Delft, The Netherlands

In the past the temporal aspects of sound fields have obtained considerably more attention than the spatial aspects. Therefore most electroacoustic arguments are based on temporal frequencies and temporal reflection sequences. It is shown that sound control should be based on "acoustic holography," featuring the spatial reconstruction of direct and reflected wave fields with desired wavefront properties at each moment of time. As holographically reconstructed sound fields cannot be distinguished from true sound fields, it is argued that holographic sound systems are the ultimate in sound control. A description is given of the holographic sound system ACS (patent pending) and measurements are shown.

0 INTRODUCTION

The total sound field within an enclosed space can be subdivided into three uncorrelated parts (Fig. 1):

$$\overline{p_{\text{tot}}^2} = \overline{p_{\text{dir}}^2} + \overline{p_{\text{rev}}^2} + \overline{p_{\text{noise}}^2} \tag{1}$$

where

$\overline{p_{\text{tot}}^2}$ = power of total field

$\overline{p_{\text{dir}}^2}$ = power of direct field

$\overline{p_{\text{rev}}^2}$ = power of reverberant field

$\overline{p_{\text{noise}}^2}$ = power of noise.

Generally, power levels are specified in frequency bands such as octave bands or one-third-octave bands.

A necessary condition for any good acoustics is a low noise level, say less than 30 dB(A). The reverberant field is determined by the enclosing surface and consists of first- and higher order reflections. Reflections define "the acoustics" of an enclosed space. The direct field leaves the source area and reaches the audience directly. Apart from internal reflections at the source and listening area, the direct field does not contain any energy that has been reflected against the enclosed surface. In sound control the direct field, the reverberant field, and the noise should all be addressed.

For a distribution of uncorrelated omnidirectional point sources we may write

$$\overline{p_{\text{dir}}^2} = \sum_i \frac{\rho c}{4\pi r_i^2} W_i \tag{2a}$$

and, assuming a diffuse reverberant field,

$$\overline{p_{\text{rev}}^2} = \sum_i \frac{4\rho c(1 - \overline{\alpha})}{\overline{\alpha} S} W_i . \tag{2b}$$

In expressions (2) W_i equals the emitted power of point source i and r_i is the distance from source i to a listener, $\overline{\alpha}$ is the average statistical absorption coefficient and S equals the total area of the enclosed surface, ρ and c are the density and the velocity for air, respectively.

Fig. 2 gives a sketch of the mean-square sound pressure of both fields as a function of the distance from one source ($i = 1$) to a listener. Close to the source the direct field dominates. Far away from the source the reverberant field dominates. From Eqs. (2) we may conclude that the direct field and the reverberant field have equal strength for the source–listener distance,

$$r_0 = \frac{1}{4} \sqrt{\frac{\overline{\alpha} S}{1 - \overline{\alpha}}} \tag{3}$$

the so-called reverberation distance. It is important to bear in mind that $\overline{\alpha}$, and therefore r_0, depends on the frequency band under consideration.

---

* Manuscript received 1988 May 23.

Eqs. (1) and (2) apply only under theoretical conditions. For instance, experience shows that due to a highly absorptive seating area expression (1) should be replaced by a frequency-dependent relation for receiver locations in the seating area. With respect to expression (2), experience also shows that for receiver locations in the seating area the reverberant field level decreases by about 1 dB per 10-m source–receiver distance. However, both idealized expressions are still very useful to derive some valuable field properties. We illustrate this with two simple examples.

In Table 1 the change in reverberation sound level is given as a function of the hall inner surface S for a constant absorption coefficient according to

$$\Delta \log \overline{p_{\text{rev}}^2} = -\Delta \log S . \qquad (4a)$$

Note that the reverberation level decreases by 3 dB if the area is doubled. From experience we know that a 3-dB change in reverberation level means perceptionally a major change.

In Table 2 the change in reverberant sound level is given as a function of the average absorption coefficient $\overline{\alpha}$ for a constant size and shape of the hall (S being constant) according to

$$\Delta \log \overline{p_{\text{rev}}^2} = -\Delta \log \frac{\overline{\alpha}}{1 - \overline{\alpha}} . \qquad (4b)$$

For small $\overline{\alpha}$ and $1 - \overline{\alpha}$ values (<0.1) the reverberation sound level decreases by about 3 dB for an increase of $\overline{\alpha}$, or a decrease of $1 - \overline{\alpha}$, by a factor of 2. However, for other $\overline{\alpha}$ values the decrease is about 2 dB if $\overline{\alpha}$ is raised by 0.1. According to Eq. (2b), $p_{\text{rev}}^2 = 0$ for $\overline{\alpha} = 1$ (full absorption).

0.1 Reverberation Decay

A reverberation decay curve for a diffuse reverberant field can be represented (according to Eyring) by

$$\overline{p_{\text{rev}}^2}(t) = \overline{p_{\text{rev}}^2}(0)(1 - \overline{\alpha})^{\beta t} \qquad t > r/c \qquad (5)$$

with $\beta = cS/4V$, where V represents the volume of the hall and r equals the source–listener distance. $\overline{p_{\text{rev}}^2}(0)$ is given by Eq. (2b) and gives the mean-square reverberation pressure in the stationary situation. At $t = 0$ the sound source is turned off. Note that for $\overline{\alpha} = 0$ (no absorption) the reverberation level stays constant after the source is turned off. However, for $\overline{\alpha} \neq 0$ the decay of the reverberation level increases for increasing $\overline{\alpha}$. For $\overline{\alpha} = 1$ there is no reverberation at all times. From Eq. (5) it follows that the reverberation time can be approximated by[1]

---

[1] At T_{60} the reverberant field has decayed by 60 dB.

Table 1. Decrease of reverberation level for increasing S, $\overline{\alpha}$ being kept constant.

S/S_0	1.1	1.2	1.5	2.0	3.0	4.0
ΔL_{rev}	-0.4 dB	-0.8 dB	-1.8 dB	-3 dB	-4.8 dB	-6 dB

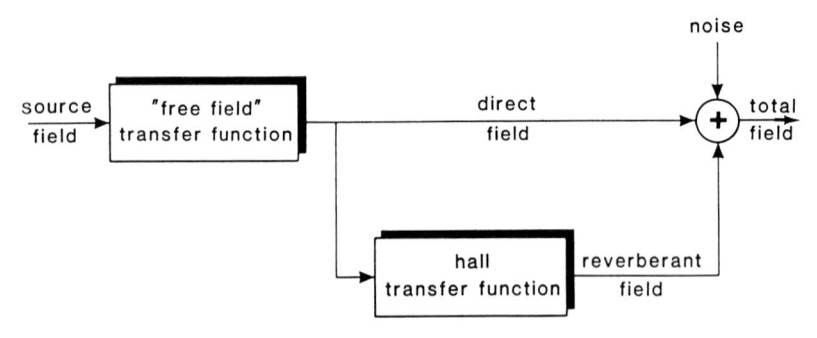

Fig. 1. Total sound field within an enclosed space can be subdivided into three (uncorrelated) parts. Often, the early reflections in a reverberant sound field are addressed separately.

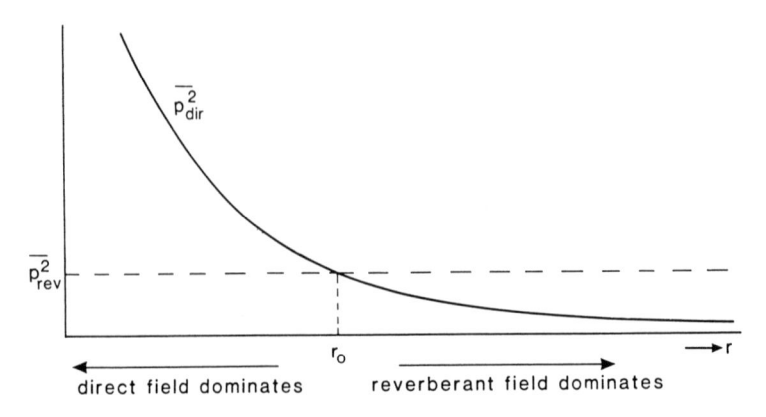

Fig. 2. Mean-square sound pressure for direct sound and reverberant sound as a function of source–listener distance r.

Table 2. Decrease of reverberation level for increasing $\bar{\alpha}$, S being kept constant.

α	0.1→0.2	0.2→0.3	0.3→0.4	0.4→0.5	0.5→0.6	0.6→0.7	0.7→0.8	0.8→0.9
ΔL_{rev}	−3.5 dB	−2.3 dB	−1.9 dB	1.7 dB	−1.7 dB	−1.9 dB	−2.3 dB	−3.5 dB

$$T_{60} \approx \frac{1}{6} \frac{V}{\overline{\alpha'} S} \tag{6}$$

with $\overline{\alpha'} = \underline{\ln} (1 - \bar{\alpha})^{-1}$. In halls with low absorption properties $\underline{\alpha'}$ is generally replaced by $\bar{\alpha}$ (Sabine).

Note that for a constant average absorption coefficient $\bar{\alpha}$, the reverberation time T_{60} will increase linearly with length parameter V/S. Hence if the hall is enlarged by a factor of 2, then T_{60} will double. Note also that for a constant size and shape of the hall T_{60} is *inversely* proportional to absorption coefficient α'. If α' is increased by a factor of 2, then T_{60} will be halved.

It is interesting to realize that for $r < r_0$ the decay curve may have a double slope: a fast decay of the early sound (direct field including the source area reflections) followed by a slower decay of the weaker reverberant field. For $r > r_0$ the decay properties of the reverberant field dominate (Fig. 3).

If a sound system contains a reverberation module, where reverberation levels and reverberation times can be chosen independently, then double decay curves can be realized throughout the hall. Low-level single decay curves and double decay curves are essential for multipurpose applications since the "definition" property of the sound field can be improved without shortening T_{60} or, alternatively, definition can be maintained while T_{60} is increased. Fig. 4 shows the type of decay curves that can be chosen with the Acoustic Control Systems ACS.[2]

0.2 Electroacoustics

In the past electroacoustics has been approached mainly from the principles of directional reinforcement and temporal filtering. On the one hand this means frequency-dependent enhancement of the direct field; on the other hand it means optimization of the reverberant field by manipulating the reflection sequence in the impulse response time function. With respect to the temporal aspects, the spatial aspects of sound fields have obtained much less attention, and, generally, arguments based on the properties of wave fronts are hardly utilized. In this paper the spatial properties of sound fields play a central role and it is argued that electroacoustics should be based on the principles of holography.

1 PRINCIPLES OF ACOUSTIC HOLOGRAPHY

Acoustic holography describes the reconstruction of acoustic wave fields from measurements. The principle is shown in Fig. 5. If an array of microphones measure the pressure of a propagating wave field in the plane $x = x_1$,

$$p = p(x_1, y, z, t)$$

at each position, and the microphone signals are fed to related loudspeaker positions, then the resulting wave field can be made identical to the original wave field for $x > x_1$. Mathematically (see, for example [2]),

$$P(x, y, z, \omega) = \iint_{S_1} P(x_1, \eta, \zeta, \omega)(1 + jkr) \times \cos \varphi \, \frac{e^{-jkr}}{r^2} \, d\eta \, d\zeta \tag{7a}$$

where $P(x, y, z, \omega)$ is the frequency domain presentation (Fourier transform) of pressure distribution $p(x, y, z, t)$ and

$$x > x_1$$

$$k = \omega/c$$

$$c = \text{sound velocity}$$

$$r = \sqrt{(x - x_1)^2 + (y - \eta)^2 + (z - \zeta)^2}$$

$$\cos \varphi = (x - x_1)/r .$$

Note from Eq. (7a) that for a perfect reconstruction the loudspeakers should have a dipole characteristic. In practical situations the integral (7a) should be replaced by a summation,

$$P(x, y, z, \omega) = \sum_m \sum_n P(x_1, y_m, z_n, \omega)(1 + jkr_{mn}) \times \cos \varphi_{mn} \, \frac{e^{-jkr_{mn}}}{r_{mn}^2} \, \Delta y \, \Delta z$$

or

$$P(x, y, z, \omega) = \sum_m \sum_n$$
$$\times W(x - x_1, y - y_m, z - z_n, \omega)$$
$$\times P(x_1, y_m, z_n, \omega) \tag{7b}$$

---

[2] A trademark of Birch Wood Acoustics B.V., The Netherlands.

where

$$W(x - x_1, y - y_m, z - z_n, \omega)$$

$$= (1 + jkr_{mn}) \cos \varphi_{mn} \frac{e^{-jkr_{mn}}}{r_{mn}^2} \Delta y \, \Delta z$$

with

$$r_{mn} = \sqrt{(x - x_1)^2 + (y - y_m)^2 + (z - z_n)^2}.$$

Eq. (7b) represents a spatial convolution along y and z.

According to sampling theory an exact reconstruction can be made if $\Delta x = \Delta y \leq \lambda/2$. Hence sampling should be denser for higher frequencies. Perception experiments have taught us that for reverberation reconstruction the sampling intervals can be coarser.

In the generalized version of acoustic holography, the measured microphone signals are fed into a processor and propagation ("extrapolation") to another plane, say $x = x_2$, is carried out numerically (Fig. 6),

$$P(x_2, y, z, \omega) = \sum_m \sum_n W(x_2 - x_1, y - y_m, z - z_n, \omega) P(x_1, y_m, z_n, \omega) \quad (8a)$$

where $x_2 > x_1$.

Next, reemission into space ("reconstruction") occurs by an array of dipole loudspeakers in the plane $x = x_2$,

$$P(x, y, z, \omega) = \sum_m \sum_n W(x - x_2, y - y_m, z - y_n, \omega) P(x_2, y_m, z_n, \omega) \quad (8b)$$

for $x > x_2$. Control of direct wave fields as well as

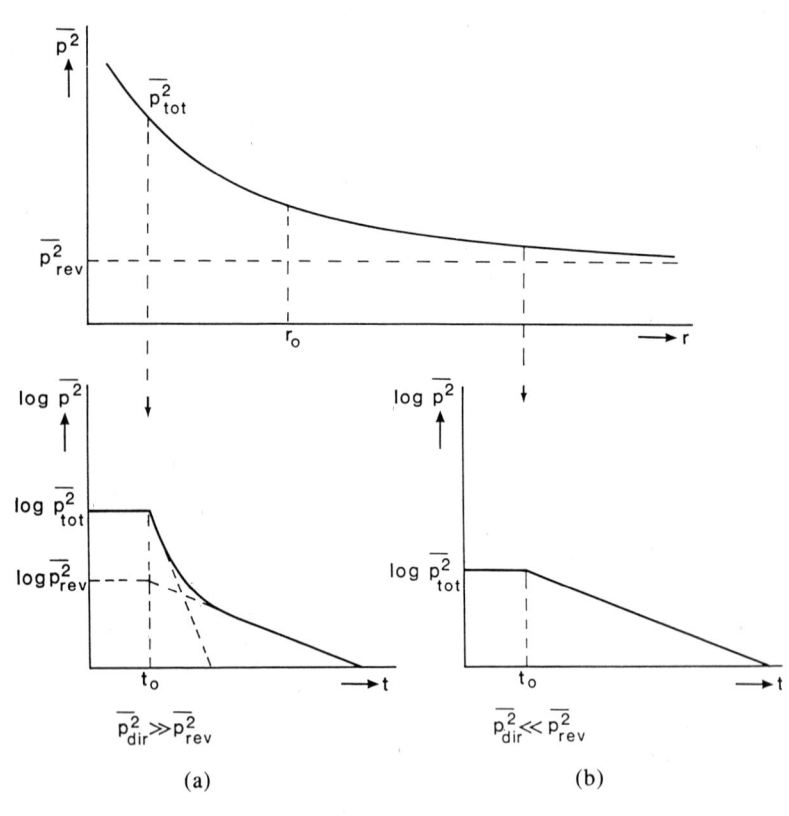

Fig. 3. Decay properties close to and far away from sound source.

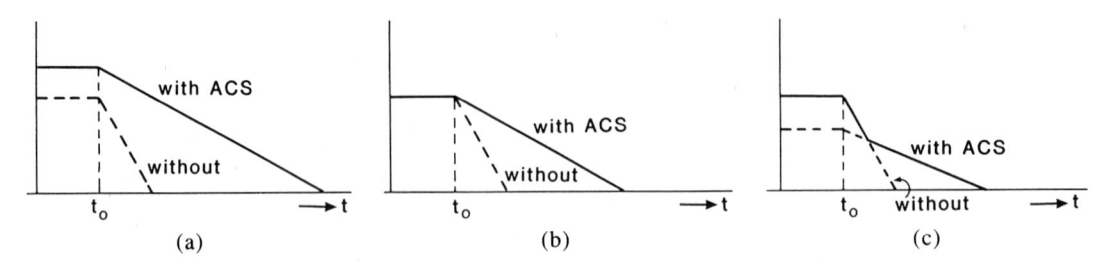

Fig. 4. Type of decay curves that can be generated with ACS. (a) Single decay, normal level. (b) Single decay, low level. (c) Double decay.

reflected wave fields should be based on holographic expressions (8a) and (8b).

In summary, by measuring the pressure of an incoming wave field with an array of microphones and applying to the microphone signals an extrapolation process, the wave field can be fully reconstructed everywhere in space by an array of loudspeakers. The listener would not notice the difference between the real field and the reconstructed field since recording, extrapolation, and reconstruction can be made as perfect as desired.

Figs. 7–9 illustrate the principle with an example. An impulsive sound field of a point source, traveling in the positive x direction, is shown at different points of time (Fig. 7). In Fig. 8 the recorded microphone signals are shown at $x = x_1$. Note that the pulses at positions farther away from the point source arrive later. Next the recorded signals are extrapolated numerically to $x = x_2$. The extrapolation result is shown in Fig. 8. Note the increased travel times. Finally the extrapolated signals are fed to the loudspeaker array at $x = x_2$. The resulting sound field is shown in Fig. 9 for two points of time. Note that the reconstructed wave field and the actual wave field are identical (compare Fig. 9 with Fig. 7 for t_3 and t_4).

In acoustic holography the measured microphone signals may be stored prior to extrapolation and reconstruction. In this way extrapolation and reconstruc-

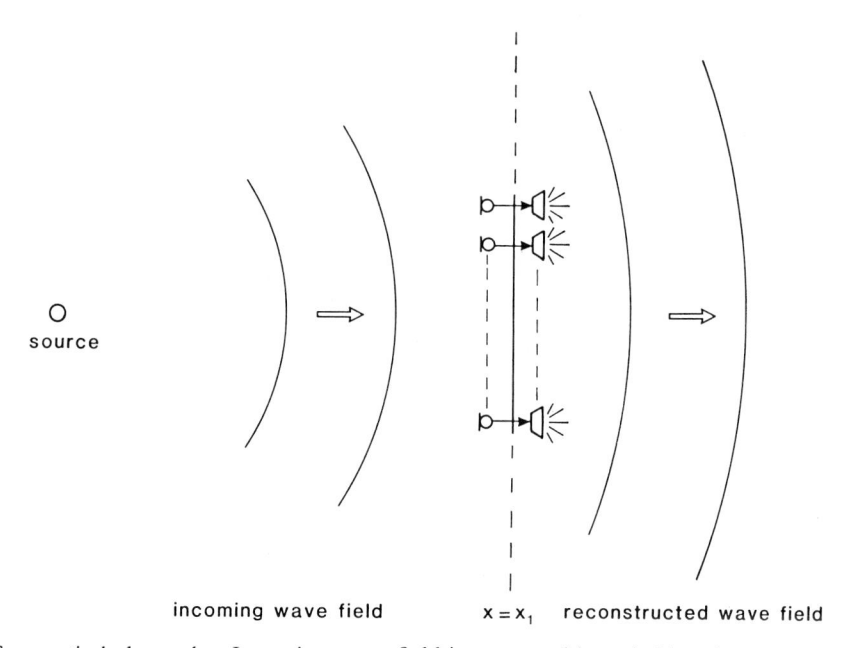

Fig. 5. Principle of acoustic holography. Incoming wave field is measured by suitable microphone array (acquisition) and reemitted by suitable loudspeaker array (reconstruction). Optionally, measured microphone signals may be stored before reemission occurs.

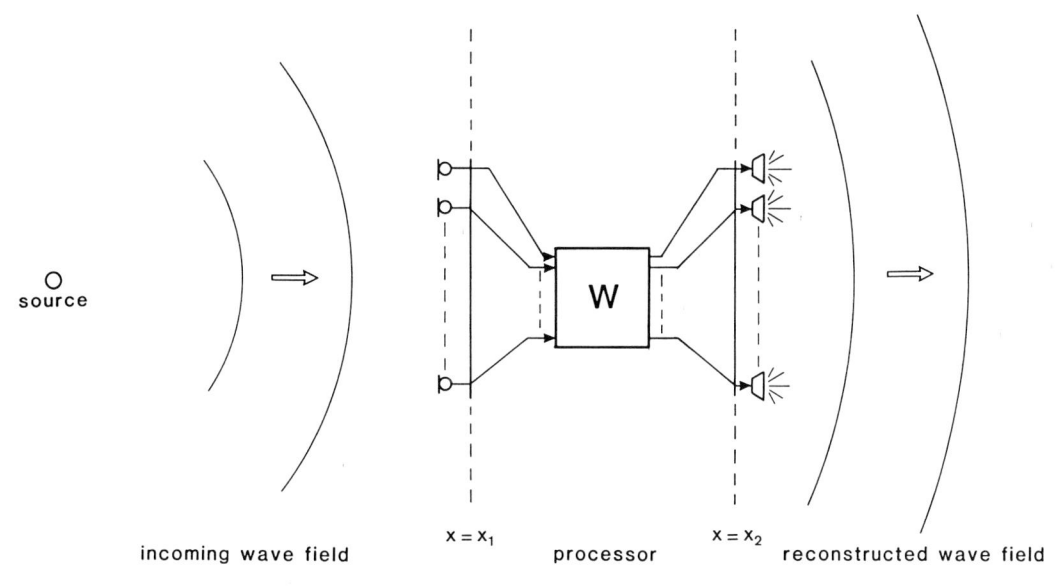

Fig. 6. In generalized version of acoustic holography, microphone array and loudspeaker array have different positions, and therefore measured wave field should be numerically extrapolated (by a processor) from location of acquisition to location of reconstruction.

tion may occur in another space at any other time (holographic sound reproduction).

If we want to reconstruct the reflected wave field from a wall S, then the reflected pressure P^+ should be used in Eqs. (8). If the wall is a plane surface, then P^+ can be represented elegantly by the wave field of an image source (Fig. 10). For a good understanding of reverberation control, image sources will be used later on.

In the foregoing we reconstructed the wave field from the (reflected) pressure distribution of *one* surface. However, for an enclosed space the reflected pressure of *all* surfaces should be included. For a rectangular hall this means

$$P(x, y, z, \omega) = \sum_{i=1}^{6} \iint_{S_i} W(x - \xi_i, y - \eta_i, z - \zeta_i, \omega) P^+(\xi_i, \eta_i, \zeta_i, \omega) \, dS_i \quad (9)$$

where P^+ represents the reflected pressure wave field leaving a wall. Note that for a fully absorptive wall $P^+ = 0$.

In summary, *direct* sound fields (Figs. 7–9) and *reflected* sound fields (Fig. 10) can be fully reconstructed from microphone signals by making use of the holographic principle based on expressions (7) and (8). This means that methods based on acoustic holography have full control over the direct and reverberant properties of a sound field.

2 DIRECT SOUND CONTROL

2.1 Introduction

Today most electroacoustic systems aim at direct sound enhancement. Basically, one or more directional microphones pick up the direct sound and, after some frequency-dependent amplification, reemission occurs by a cluster of loudspeakers which direct the amplified sound to the audience. Optionally a delay may be involved to compensate for the difference in position between microphones and loudspeakers. In many practical situations the source is perceived at the wrong location: intelligibility is improved at the cost of localization.

By using holographic arrays, intelligibility as well as localization can be handled properly: the wave front of the direct wave field is fully reconstructed at any moment of time with the correct phase spectrum

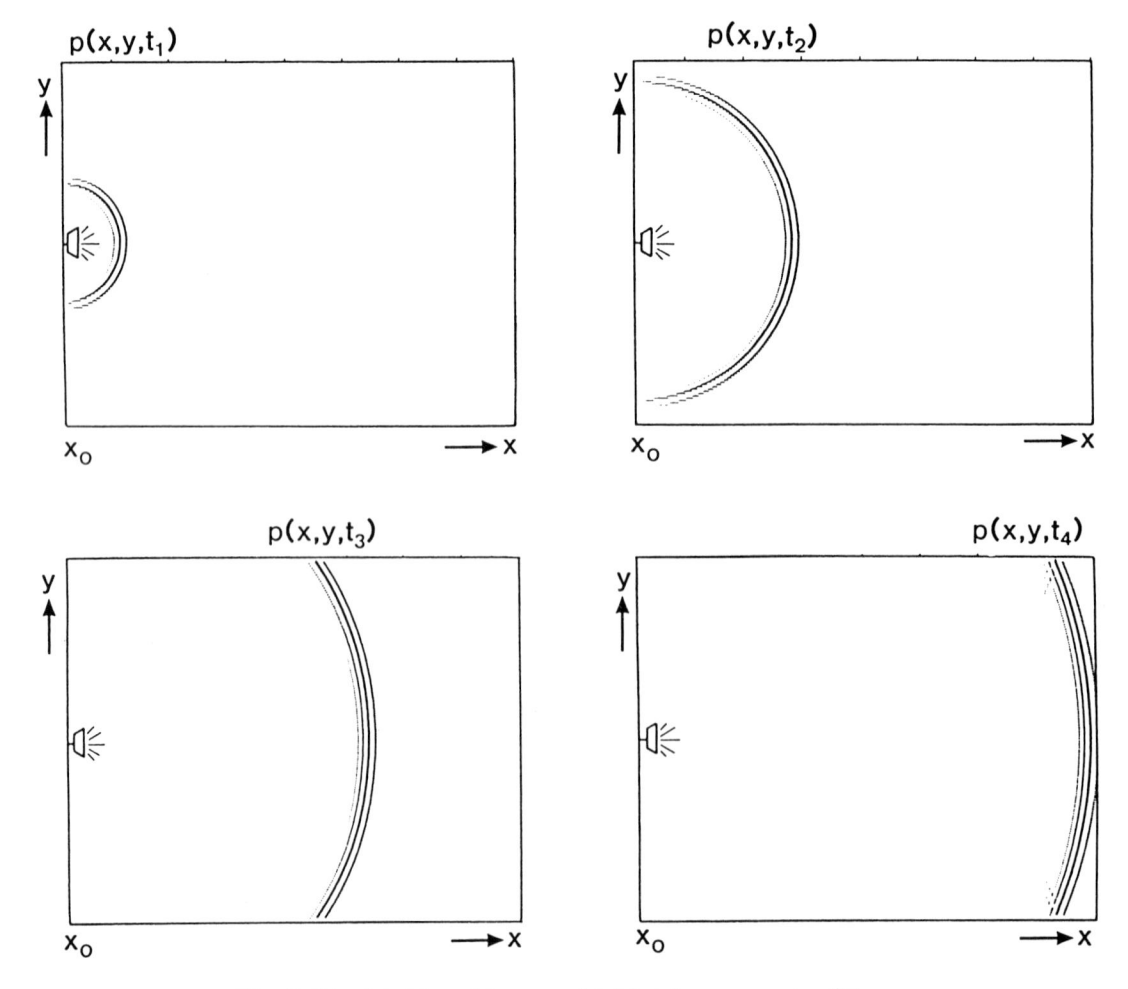

Fig. 7. Simulated impulsive sound field at four moments of time.

and with a desirable frequency-dependent enhancement.

2.2 Holographic Approach

To explain the holographic principle for direct sound control consider the two-dimensional situation in Fig. 11. In front of the source area a distribution of microphones is mounted to measure the direct wave field at $x = x_1$. Farther away, at $x = x_2$, we want to reconstruct the direct wave field by a distribution of loudspeakers. Using the two-dimensional version of Eq. (8a), extrapolation from $x = x_1$ to $x = x_2$ is given by

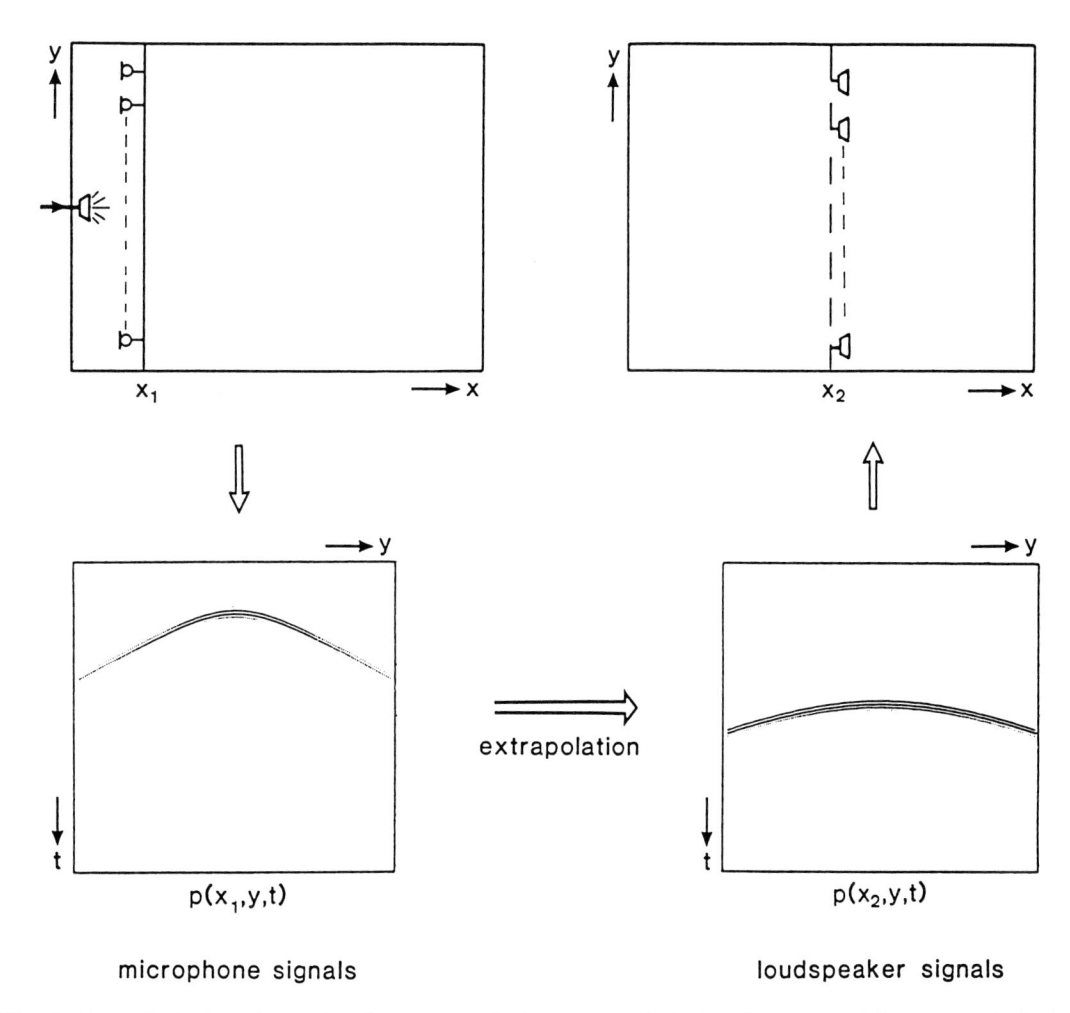

Fig. 8. Recorded microphone signals at $x = x_1$ (before extrapolation) and at $x = x_2$ (after extrapolation).

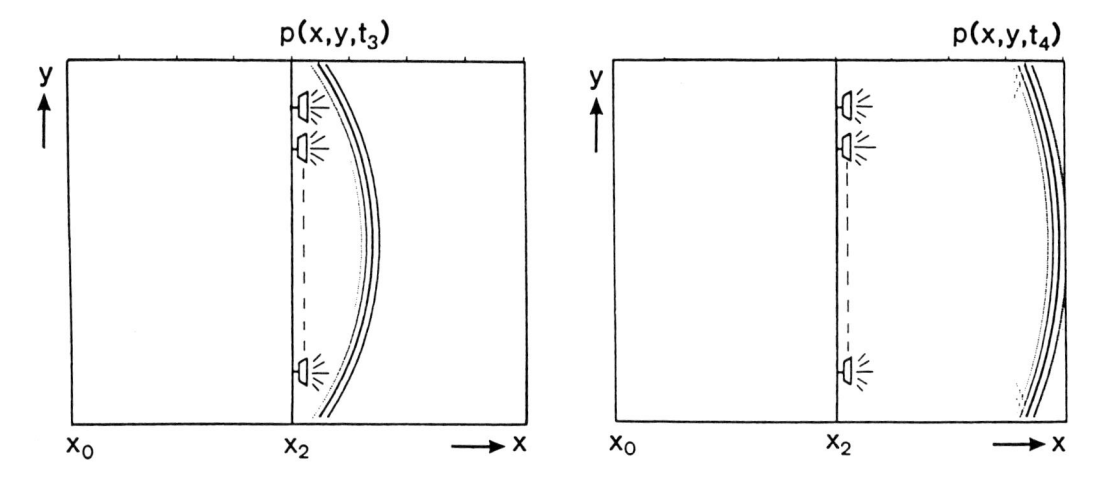

Fig. 9. Reconstructed wave field (for $x > x_2$) after acquisition (at $x = x_1$) and extrapolation (from x_1 to x_2) at two moments of time.

$$P(x_2, y, \omega) = \sum_m W(x_2 - x_1, y - y_m, \omega)$$
$$\times P(x_1, y_m, \omega) \quad (10a)$$

where

$$W(x_2 - x_1, y - y_m, \omega) = \sqrt{jk} \cos \varphi_m \frac{e^{-jkr_m}}{\sqrt{r_m}} \quad (10b)$$

and reconstruction by the dipole loudspeakers at $x = x_2$ is given by

$$P(x, y, \omega) = \sum_m W(x - x_2, y - y_m, \omega) P(x_2, y_m, \omega) \quad (10c)$$

for $x > x_2$.

In Eq. (10b) the factor $\sqrt{jk}$ gives a phase shift of $\pi/4$ and an enhancement of the high frequencies of 3 dB per octave, the term $\cos \varphi_m / \sqrt{r_m}$ causes attenuation, and the term $\exp(-jkr_m)$ represents travel time.

Fig. 12 shows the implementation. As mentioned before, extrapolation (10a) represents a spatial convolution and, therefore, one microphone is connected with different weighting factors and different delays to *all* loudspeakers. Simpler microphone–loudspeaker connections must lead to an incorrect reconstruction of the direct field, causing artifacts such as localization distortion.

Fig. 13 shows the reconstructed direct sound field for different positions of the sound source. Note the important property that the reconstructed wave fronts are consistent with the position of the source.

In practical situations the array of microphones may be mounted in front of the stage and the array of loudspeakers may be mounted everywhere in the hall, at the ceiling, at the edge of a balcony, and so on. In addition, similar to conventional techniques, it is advantageous to choose the loudspeakers directional *perpendicular* to the line array and directed to the audience.

3 REVERBERATION CONTROL

3.1 Introduction

Unfortunately many auditoriums have been built and are still being built with unsatisfactory acoustics for music performances. In particular, many auditoriums turn out to be "too dry," meaning that not enough reverberation is added to the music. Pronounced examples are modern churches and outdoor music pavilions. Many music halls also suffer from a severe lack of reverberation in the hall as well as on the stage. The reasons that auditoriums are still being built with unsatisfactory acoustics are twofold:

1) *Multipurpose function*. The hall is used for significantly differing purposes, such as lectures, pop concerts, musicals, and symphony concerts. These activities are acoustically difficult to combine in one hall.

2). *Creative architecture*. The hall is built by an architect with the ambition to give a personal touch to the architectural design. Unfortunately experience has taught that interesting designs do not always support good acoustics.

In conclusion, acoustic requirements may set a severe limit to the use of the hall and the creativity of the architect. Needless to say that any solution to circumvent both limitations would be wholeheartedly welcomed.

Traditional architectural acoustics recommend remodeling solutions which are often very costly or even prohibited for aesthetic reasons or because of historic preservation. An attractive alternative is the application of electroacoustic solutions, which show the following basic advantages:

1) *No physical changes*. The physical architecture is not touched except for the provision of microphones and loudspeakers (electroacoustic sound control).

2) *Flexibility*. The setting of the electroacoustic parameters can be optimally adjusted for each performance (variable acoustics).

3) *Low cost*. Looking at the present-day prices of modern electronic components, electroacoustic solu-

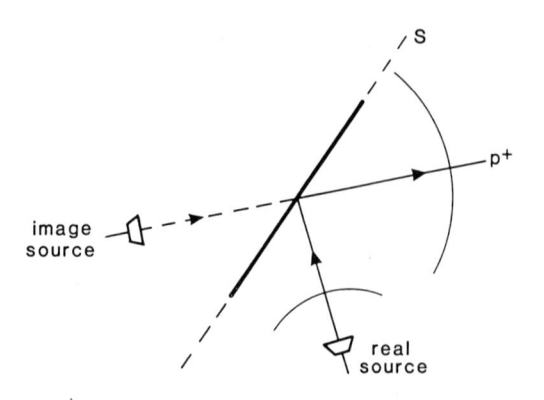

Fig. 10. Reflected pressure P^+ from planar surface S may be elegantly represented by wave field of image source.

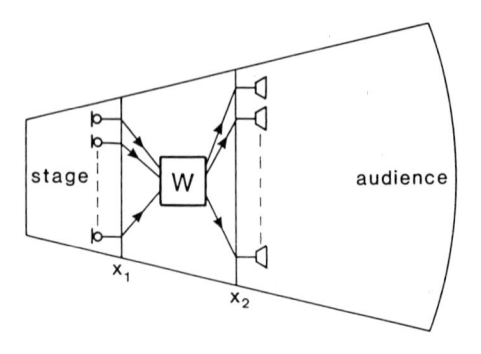

Fig. 11. Holographic direct sound control consists of three steps: acquisition by microphone array (near stage), extrapolation by signal processor, and reconstruction by loudspeaker array (near audience).

tions for reverberant field enhancement are generally considerably less costly than physical changes in the architecture.

Since the 1960s a number of electroacoustic solutions have been proposed aiming at the optimization of reverberation times and impulse responses in enclosures. On the one hand there are the systems designed especially for the enhancement of reverberation time and level in auditoriums (AIRO, Philips). Basically those systems are built up from a (usually large) number of narrow-band or broad-band channels, each consisting of a microphone, an amplifier/delay system, and a loudspeaker (Fig. 14). Reverberation enhancement is based on acoustic feedback so that the loop gain is determined not only by the electronic transfer function $A_3(\omega)$ but also by the acoustic transfer function $A_2(\omega)$ between loudspeaker and hall microphone. Unfortunately $A_2(\omega)$ is influenced by geometric variations in the hall (for example, due to the entry of an audience) as well as by temperature variations, so that the danger of coloration or even howlback becomes apparent for such systems. These undesired effects can only be avoided by reducing $A_3(\omega)$ to a "safe" level, which limits the effectiveness and flexibility of the system considerably. Franssen (1968) has calculated that for one broad-band loop the transfer function $A_3(\omega)$ should be kept below -17 dB, which means that for those systems many loops have to be included. On the other hand, there are the systems aiming at optimizing the impulse response at a listener position by adding synthetic reflection patterns radiated from loudspeakers placed at various positions. It is typically a time filter approach. Nowadays small-scale systems based on this principle are available for domestic use in order to overrule the natural acoustics of the listening room by acoustic conditions better fitting the music to be reproduced (such as Roland or Yamaha). Here the contribution of each loudspeaker channel is determined by an electronic delay line (Fig. 15). As long as the input of the system is electrical (such as a CD or record signal), coloration and howlback effects cannot occur. If the system is used to process the signal of a sound source within the room, using a directive microphone as the input device, acoustical feedback can be reduced to a safe level without limiting its performance since feedback is *not* the basis of the system.

In Sec. 3.2 reverberation control is approached from the holographic principle. Therefore it differs fundamentally from the above two concepts. Holographic reverberation control has been implemented in ACS.

3.2 Holographic Approach

According to expression (9), the reverberant wave field inside a hall is fully defined if the reflected wave fields at the six surfaces are given. Hence if we generate with dipole loudspeakers desired reflection patterns at

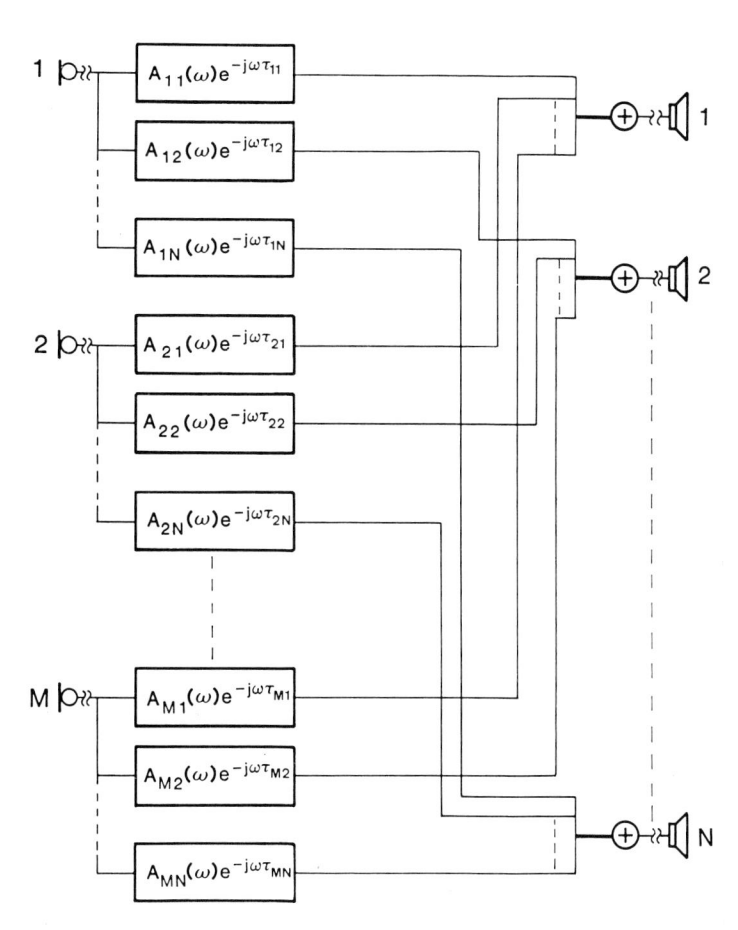

Fig. 12. Diagram of signal processor that extrapolates direct sound field from microphone array to loudspeaker array. Filter coefficients $A_{mn}(\omega)$ are based on Kirchhoff integral.

each position of the walls, then a desired wave field is created everywhere inside the hall. The desired reflection patterns are simulated in a processor for an acoustically desired hall. Figs. 16 and 17 show the principle for one microphone–loudspeaker pair. Here the desired hall is rectangular. The impulsive source in the desired hall has the same position as the microphone in the real hall. The receiver in the desired hall has the same position as the loudspeaker in the real hall. If $r_{mn}^+(t)$ represents the impulse response at receiver position n due to impulsive source m, the superscript + indicating that only waves leaving the wall are considered, then the desired reflection patterns at wall position n of the real hall is given by the temporal convolution

$$p_{mn}^+(t) = s_m^+(t) * r_{mn}^+(t) \tag{11a}$$

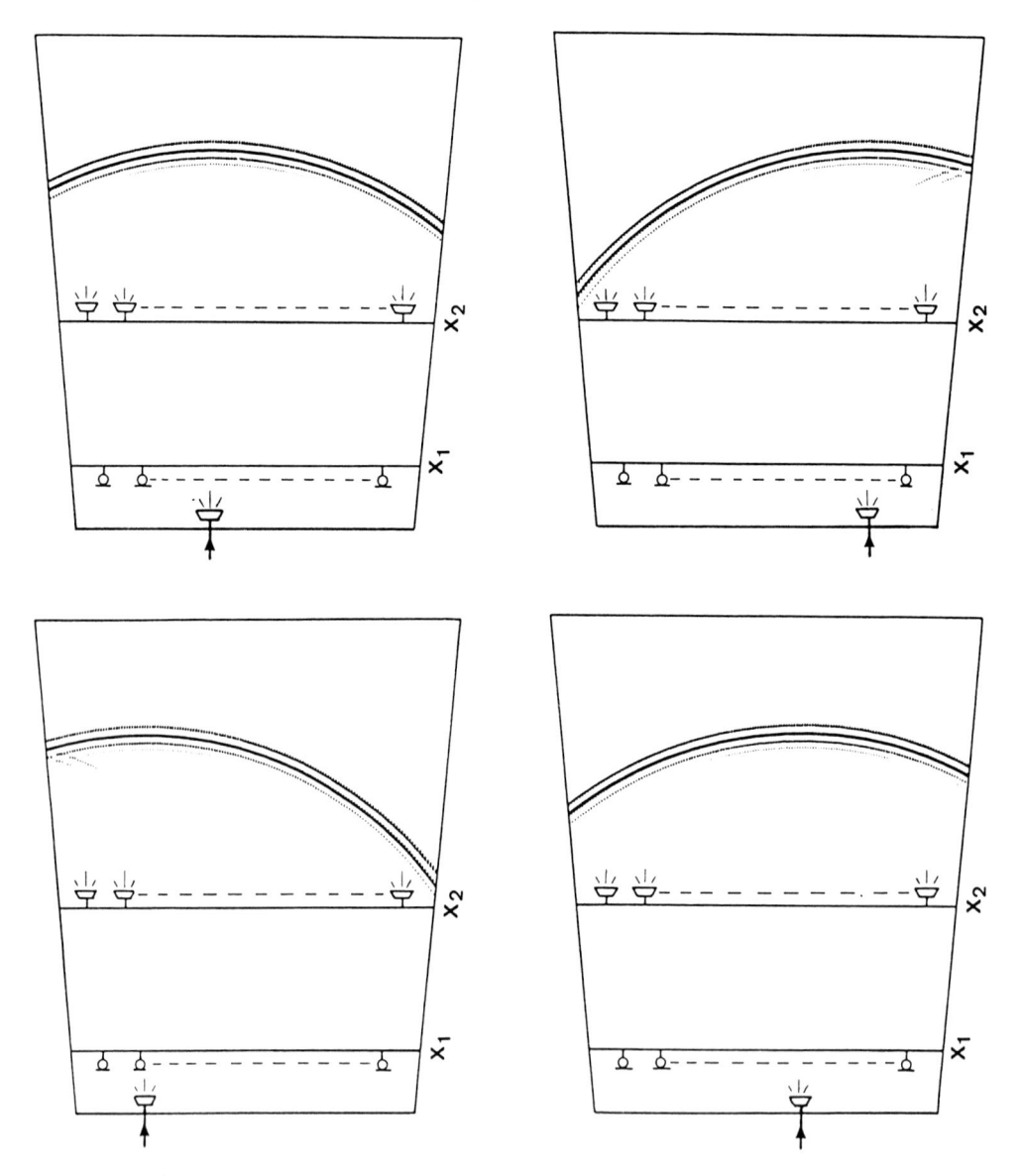

Fig. 13. Direct sound field after acquisition, extrapolation, and reconstruction for different positions of impulsive sound source.

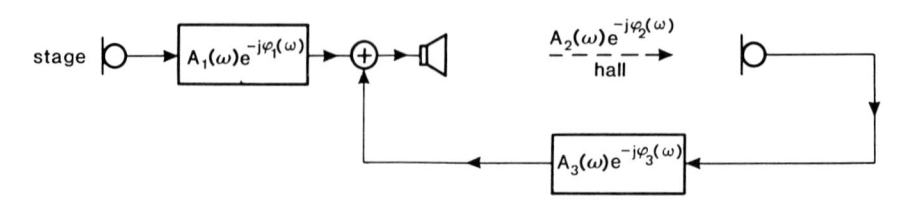

Fig. 14. Principle of electroacoustic reverberation enhancement systems based on acoustic feedback (single-channel version); stage microphone is optional. Note that for those systems the *acoustic* communication between hall microphones and hall loudspeakers is essential.

where $s_m^+(t)$ represents the microphone signal of the *direct* sound at position m. However, in the real hall part of the response p_{mn}^+ will be fed back to microphone m, and therefore impulse response r_{mn}^+ should be modified to f_{mn}^+ to compensate for the feedback phenomenon,

$$p_{mn}^+(t) = s_m(t) * f_{mn}^+(t)$$

$$= [s_m^+(t) + g_{nm}(t) * p_{mn}^+(t)] * f_{mn}^+(t) \quad (11b)$$

where $s_m(t)$ represents the microphone signal of the

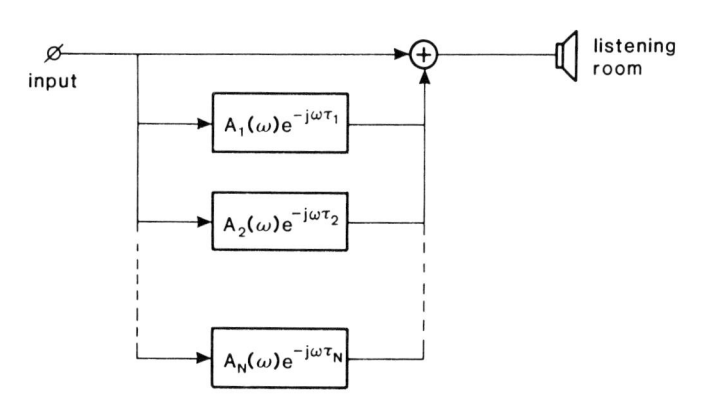

Fig. 15. Principle of electronic systems for simulation of "reverberation tails" (single-channel version). Transfer function $A_n(\omega) \exp(-j\omega\tau_n)$ defines property of one simulated reflection path.

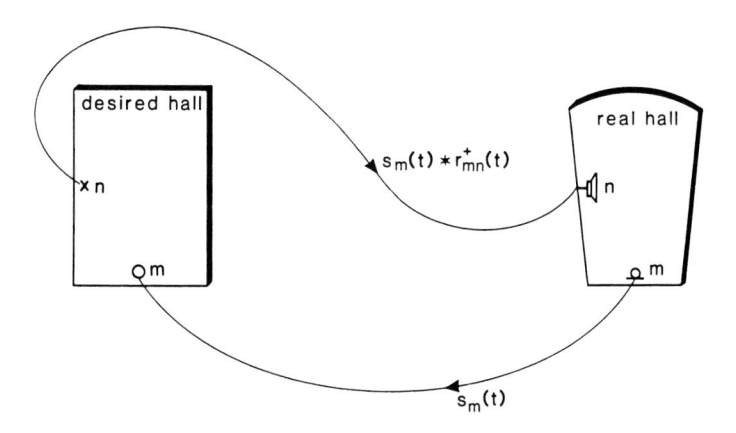

Fig. 16. Principle of holographic reverberation control, showing one microphone–loudspeaker pair only. Direct sound is measured by microphone near stage of real hall. Microphone signal is reemitted in fictive, acoustically desired hall. The resulting reverberant field is picked up and reemitted into real hall.

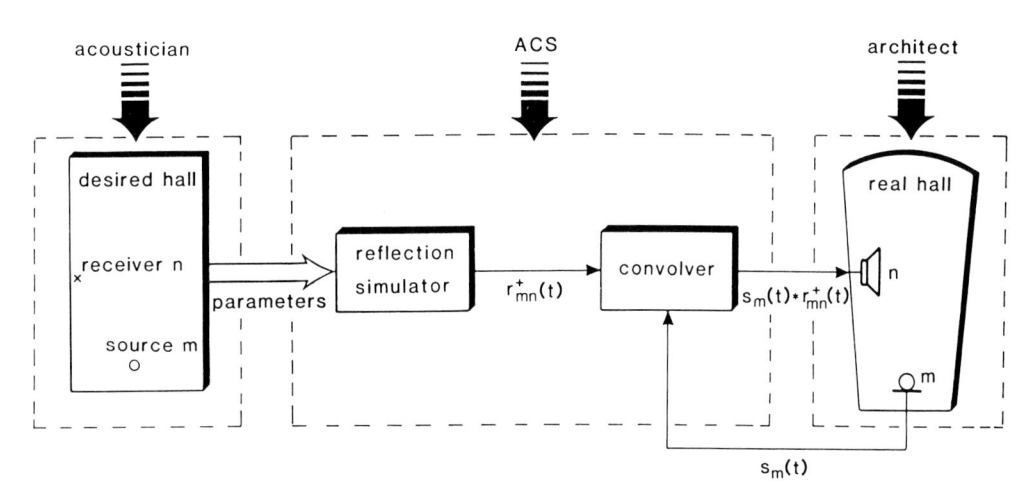

Fig. 17. Basic system diagram for holographic reverberation control, illustrated for one microphone–loudspeaker pair. Reverberation module of ACS is based on this diagram.

total sound at position m and $g_{nm}(t)$ is the real hall impulse response between loudspeaker n and microphone m.

Using the Fourier domain, the required impulse response with feedback compensation can be computed easily from Eqs. (11a) and (11b),

$$F_{mn}^+(\omega) = \frac{R_{mn}^+(\omega)}{1 + G_{nm}(\omega)R_{mn}^+(\omega)} . \quad (12)$$

In addition, the desired reflection pattern can be written as

$$P_{mn}^+(\omega) = S_m(\omega) \frac{R_{mn}^+(\omega)}{1 + G_{nm}(\omega)R_{mn}^+(\omega)} . \quad (13)$$

Note the fundamental difference between $R_{mn}^+(\omega)$ and $G_{nm}(\omega)$:

$R_{mn}^+(\omega)$ is a *simulated* transfer function in the *desired* hall

$G_{nm}(\omega)$ is a *measured* transfer function in the *real* hall.

In its simplest form $G_{nm}(\omega) = A_{nm}(\omega) \exp(-j\omega r_{nm}/c)$.

In ACS the feedback phenomenon (quantified by G_{nm}) is made very small, that is,

$$|G_{nm}(\omega)R_{mn}^+(\omega)| << 1 \quad (14)$$

for all m and n by taking the following measures:

1) The loudspeakers direct their energy to the absorptive audience area as much as possible.

2) The microphones have maximum sensitivity in the direction of the source area and no sensitivity in the opposite direction ($g_{nm} \to g_{nm}^+$).

3) The microphones are mounted near the source area where the direct sound level is significantly higher than the reverberant sound level.

4) The parameters of the desired impulse response $R_{mn}^+(\omega)$ are made time variance [see also expression (19)].

Hence in ACS we aim for

$$F_{mn}^+(\omega) \approx R_{mn}^+(\omega) . \quad (15)$$

Note that in practical situations $G_{nm}^+(\omega)$ can be evaluated easily by analyzing the signal of microphone m due to an impulse from loudspeaker n.

According to the holographic principle, a proper distribution of microphones should be used to define the source signal correctly. Each microphone position in the real hall is related to an impulsive source position in the desired hall with its own impulse response. Hence the complete desired reflection pattern at wall position n can be written as

$$P_n^+(\omega) = \sum_m S_m(\omega) F_{mn}^+(\omega) \quad (16a)$$

or, if feedback has been sufficiently minimized,

$$P_n^+(\omega) = \sum_m S_m^+(\omega) R_{mn}^+(\omega) . \quad (16b)$$

The total desired reverberant sound field in the real hall is given by the interference pattern of all loudspeaker responses,

$$P_{\text{rev}}(x, y, z, \omega) = \sum_n W(x - x_n, y - y_n, z - z_n, \omega) P^+(x_n, y_n, z_n, \omega) \quad (17a)$$

or, in the time domain,

$$p_{\text{rev}}(x, y, z, t) = \sum_n w(x - x_n, y - y_n, z - z_n, t) * p^+(x_n, y_n, z_n, t) \quad (17b)$$

where (x_n, y_n, z_n) defines the position of loudspeaker n, (x, y, z) defines a listener position, and $*$ means temporal convolution.

Finally, if ACS would be installed in a lively hall, then Eq. (16b) should be updated to

$$P_n^+(\omega) = \sum_m S_m^+(\omega) R_{mn}^+(\omega) + \sum_m S_{\text{rev},m}^+(\omega) R_{mn}^+(\omega) \quad (18)$$

where $S_{\text{rev},m}^+(\omega)$ represents the reverberant field as picked up by directive microphone m [expressions (11)].

In practical situations the influence of $S_{\text{rev},m}^+(\omega)$ can be easily evaluated by connecting the ACS input unit to the ACS processor some time (say 100 ms) *after* the source has been switched off.

In ACS the contribution of $S_{\text{rev},m}^+(\omega)$ may be optionally attenuated if

$$10 \log \left[\frac{|S_m^+(\omega)|^2}{|S_{\text{rev},m}^+(\omega)|^2} \right] < L_{\min}(\omega) \quad (19)$$

where $L_{\min}(\omega)$ depends on the ACS setting and the hall.

3.3 The Use of Image Sources

One of the advantages of the holographic approach is that the acoustically desired hall is fully under user control. Its shape may be chosen simple in order to minimize the user specifications and to simplify the simulation process. Although any simulation algorithm based on wave theory may be used, the image source approach is very instructive for desired halls with a simple geometry. Fig. 18 shows the principle for the well-known rectangular hall with one microphone. Note that the distribution of image sources is determined by the microphone positions and the geometry of the desired hall. The source signature $S_k(\omega)$ of the image source at (x_k, y_k, z_k) is determined by the related microphone signal and the absorption properties of the walls of the desired hall. For a given loudspeaker position (x_n, y_n, z_n), the loudspeaker signal can be written as

$$P^+(x_n, y_n, z_n, \omega) = \sum_k W(x_n - x_k, y_n - y_k, z_n - z_k, \omega) S_k(x_k, y_k, z_k, \omega). \quad (20)$$

In Eq. (20) k ranges over all image sources that are situated in the half-plane behind loudspeaker n.

4 DESCRIPTION OF ACS

4.1 The ACS Architecture

The ACS may be subdivided into three subsystems (Figs. 19 and 20):
1) Acquisition subsystem
2) Extrapolation subsystem
3) Reconstruction subsystem.

The acquisition subsystem measures the direct sound field with an array of high-quality broad-band microphones near the stage. The microphone signals are amplified, optionally equalized, and sent to the extrapolation subsystem. The latter consists of a number of

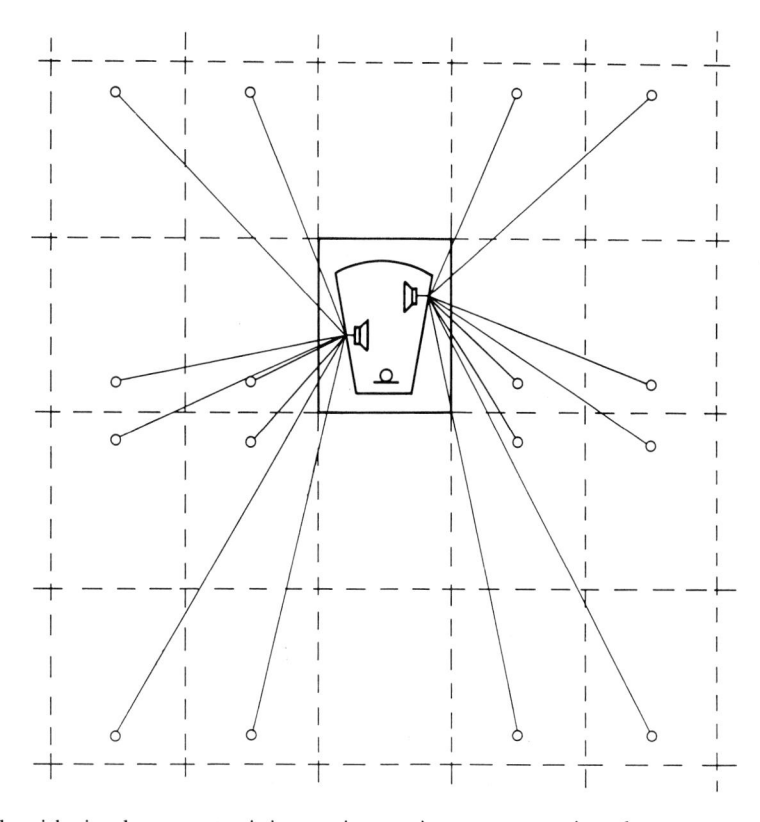

Fig. 18. For desired halls with simple geometry it is very instructive to generate impulse responses $r^+_{mn}(t)$ with image source approach. Here, a number of image sources are shown for one microphone position. Ray paths are shown for two loudspeaker positions.

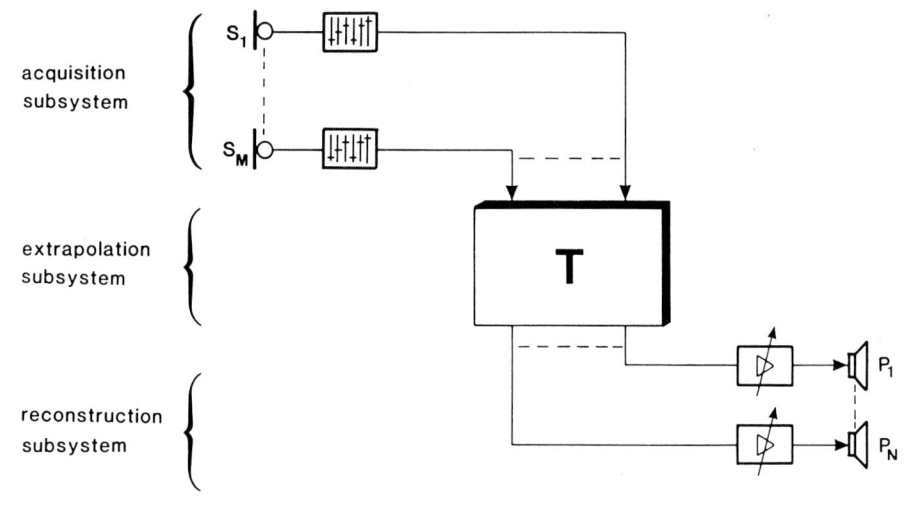

Fig. 19. ACS may be subdivided into three subsystems.

reflection simulation units. Depending on the maximum T_{60} required and the size of the hall, many reflection simulation units may be needed to include the necessary high-order reflections in $r_{mn}^{\pm}(t)$. The reconstruction subsystem transmits the simulated reflections back into the hall by an array of high-quality broad-band loudspeakers, distributed along the surfaces of the entire hall. It is important to realize that at a given position in the hall the reflection tail is not made by one loudspeaker but is synthesized by contributions of all loudspeakers: holography is *principally* mutichannel.

4.2 The Extrapolation Subsystem

If the M input signals of the extrapolation subsystem are represented by input vector S and the N output signals by output vector P, then input and output are related by transfer matrix T:

$$P = TS \qquad (21a)$$

where

$$T = \begin{pmatrix} t_{11} & t_{12} & \cdots & t_{1N} \\ t_{21} & t_{22} & \cdots & t_{2N} \\ & & \vdots & \\ t_{M1} & t_{M2} & \cdots & t_{MN} \end{pmatrix}^T . \qquad (21b)$$

Matrix element t_{mn} is specified by the transfer function between microphone m and loudspeaker n, to be written as

$$t_{mn}(\omega) = A_{mn}(\omega)e^{-j\varphi_{mn}(\omega)} . \qquad (22a)$$

Note that for one reflection simulation unit

$$\varphi_{mn}(\omega) = \omega\tau_{mn} . \qquad (22b)$$

Note also that multiplication by $t_{mn}(\omega)$ in the frequency domain means convolution in the time domain. In ACS, transfer matrix T is designed per octave band in the following way:

$$T = \sum_{ijk} T_{ijk}$$

where i indicates the number of reflections by the sidewalls, j the number of reflections by the front and back walls, and k the number of reflections by the ceiling and floor. Hence T_{ijk} represents the transfer function between the M sources in the image hall (i, j, k) and the relevant loudspeakers in the real hall (Figs. 21 and 22). If the floor is considered to be fully absorptive, then $k = 0$ or 1 only. If the back wall is considered to be fully absorptive, then $j = 0$ or 1 only. For direct sound control $i = 0$, $j = 0$, and $k = 0$.

It is important to realize that one reflection simulation unit T_{ijk} has the same architecture as the wave field extrapolation unit for direct sound control (Fig. 12).

4.3 The ACS Design Procedure

Before any ACS installation is planned in an existing hall, measurements should be taken to evaluate reverberation times T_{60} and sound pressure levels L_p at least in the one-third-octave bands of 125–8000 Hz. For

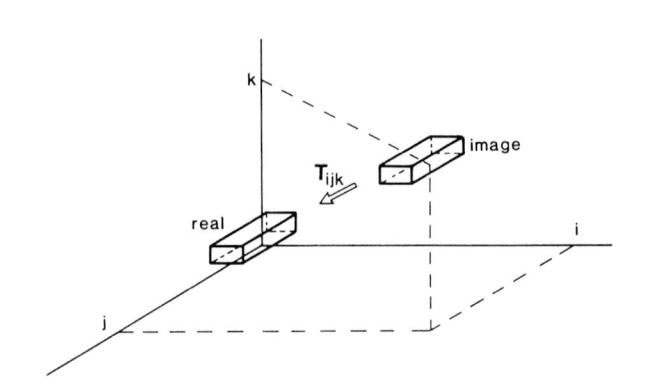

Fig. 21. Image sources in image hall (i, j, k) contribute to reverberant field in real hall $(0, 0, 0)$ via $M \times N$ transfer matrix T_{ijk}.

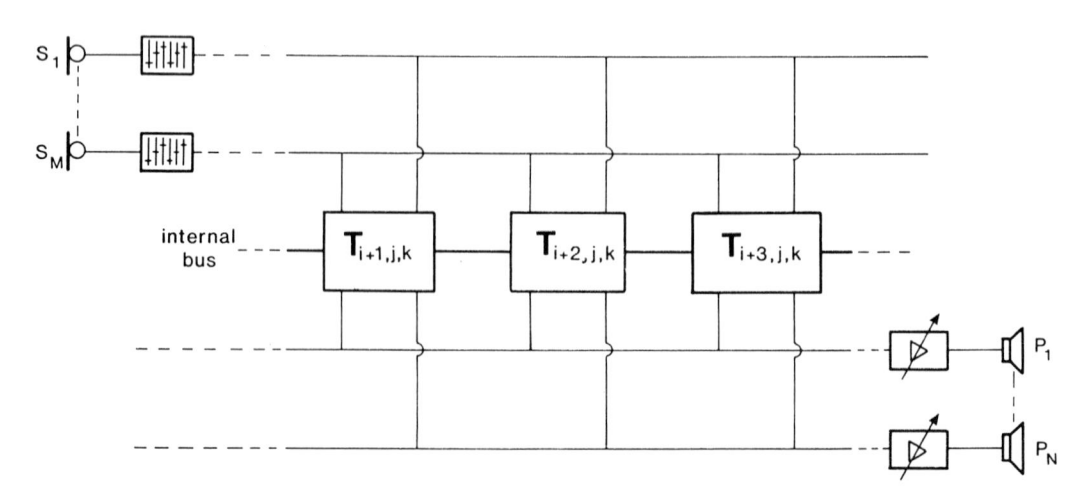

Fig. 20. Extrapolation subsystem may consist of many reflection simulation units.

those frequency bands where T_{60} values are undesirably high, a proper amount of suitable absorption material should be brought into the hall. Strong single echoes and flutter echoes should be cured with conventional means as well; weaker echoes may be masked effectively by the ACS wave fields. Therefore it is generally advantageous to evaluate undesirable echo phenomena *after* ACS has been installed.

A second action is to decide on the size and shape of the desired hall. Generally a rectangular hall is chosen just around the existing hall (Fig. 23). When the positions of the microphones and loudspeakers are determined, the ACS impulse responses $r^\pm_{mn}(t)$, can be computed for different absorption parameters. With this information the system will be built and preprogrammed at the manufacturer's site.

It is very important to realize that for the simulation of ACS impulse responses the real hall does *not* play a role! This is a major advantage in the design procedure.

4.4 The ACS Fine-Tuning Procedure

After ACS has been installed, the fine-tuning procedure can start. The principle is as follows. A reference setting is searched for by carrying out interactive measurements[3] such that T_{60} values and sound pressure levels are according to specification. The reference setting could be chosen such that, when switching the system on, the reverberation time values in octave bands measured in the hall correspond to those in the Amsterdam Concertgebouw [Fig. 24(a)], with reverberant sound pressure levels related to the reverberation times according to physical laws [Fig. 24(b)]. As mentioned before, appropriate ratios of early-to-late and lateral-to-frontal energy could be aimed for.

Starting from the reference setting that is stored in the ACS memory, the preference settings can be adjusted to "instantaneous multipurpose requirements" or "subjective alternatives" by varying 19 ACS fine-tuning parameters, labeled 1–19:

1–8	Individual reverberation time values in eight octave bands from 63 Hz up to 8 kHz
9–16	Individual pressure levels in same octave bands
17	Scaling factor for *all* reverberation times
18	Input amplification of *all* microphones
19	Output amplification of *all* loudspeakers.

The principle is illustrated in Fig. 25, where the characteristics of the Amsterdam Concertgebouw are chosen as the reference. Each fine-tuning parameter can be varied by means of a 16-step selection switch.

Note that these changes concern the performance of ACS. Resulting changes in the sound field within the hall also depend on the interference between the natural reverberant field and the nonperfect ACS-generated contribution (difference between theory and practice). However, the total result can be evaluated and adjusted by the aforementioned interactive measurement procedure.

Finally, as a result of the fine-tuning procedure, a small number of the most preferred settings are selected, largely subjectively. This has the advantage that in practical situations the operator simply chooses one of those preferred settings by a pushbutton selector.

4.5 ACS Modules

The ACS can be composed of a number of holographic modules with a variable number of input and output ports. (Each input port is connected with one microphone, each output port with one loudspeaker box.) In this way a large number of different systems can be designed to optimally suit a large variety of different purposes.

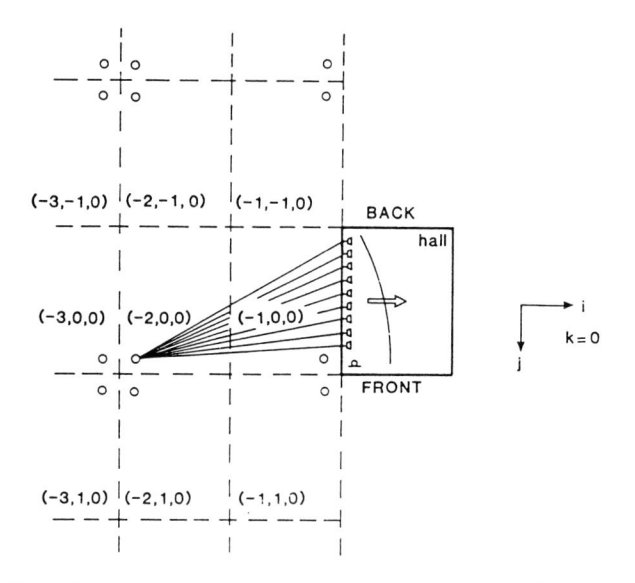

Fig. 22. Ray paths from one image source to a distribution of loudspeakers, defining the wavefront of a laterally reflected wave field.

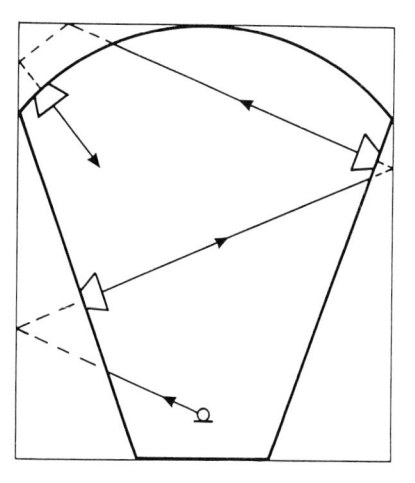

Fig. 23. In ACS, acoustically desired hall is generally chosen around real hall.

---

[3] For this purpose a special measurement device is used for fast interactive evaluation of reverberation times and sound pressure levels.

There exist four ACS modules:

1) The stage early reflection module generates a variable amount of early reflections on the stage for optimum communication between musicians ("ensemble").

2) The hall early reflection module generates a variable amount of early reflections in the hall for music intelligibility and spaciousness (early *lateral* reflections).

3) The hall reverberation module for small, simple halls generates variable reverberation in a small hall with a simple shape (rectangular, fan).

4) The hall reverberation module for any type of hall generates variable reverberation in any type of hall.

The four ACS modules can be combined into different systems in any desired way. For halls with large balconies more than one hall module may be required. By using presettings variable acoustics can be simply realized by a pushbutton selector.

4.6 ACS Models

The ACS 1000–5000 are six examples from the many possibilities of integrating ACS modules (Fig. 26):

1) ACS 1000 S/H. It consists of one or more early reflection modules for stage (S) or hall (H). Changes in reflection level can be set per octave band.

2) ACS 2000. It consists of one reverberation module for small, simple halls. Levels and reverberation times can be changed per octave band. A choice can be made out of four presettings.

3) ACS 3000. It consists of one or more reverberation modules for any type of hall. Levels and reverberation times can be changed per octave band. A choice can be made out of eight presettings.

4) The hall reverberation module for any type of hall generates variable reverberation in any type of hall.

5) ACS 5000. It combines the 1000 S, 1000 H, and 3000.

5 ACS MEASUREMENTS

Recently an ACS was installed at the auditorium of Delft University in The Netherlands and at the Decoustics/ACS Centre for Acoustical Research at York University (DACARY), Toronto, Ont., Canada. Fig. 27 gives the natural reverberation times together with two preferred ACS settings. Figs. 28 and 29 show the type of decay curves that can be generated with ACS. It is seen that very small decay rates can be generated without the slightest indication of coloration. In addition, experience shows that settings with relatively

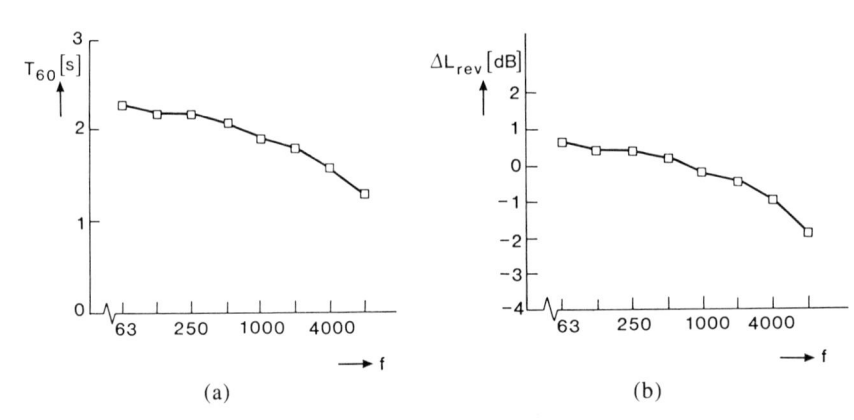

Fig. 24. (a) Reverberant times and (b) reverberant pressure levels in octave bands, corresponding to those in Amsterdam Concertgebouw.

FROM REFERENCE TO PREFERENCE

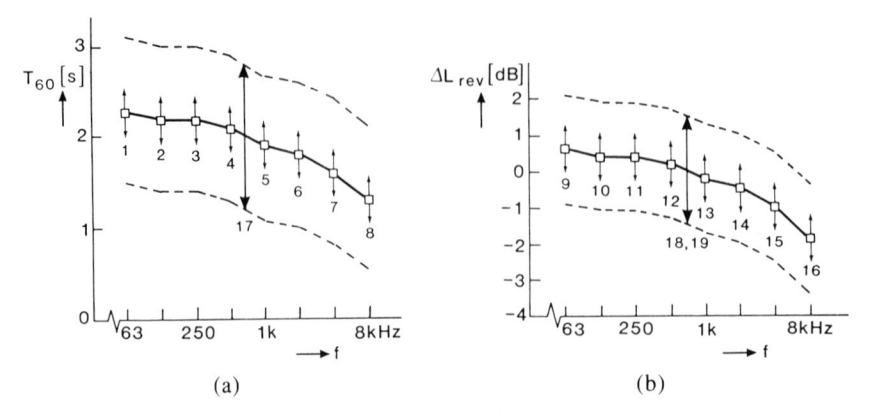

Fig. 25. From reference to preference. Effect of changing ACS fine-tuning parameters 1–19.

strong early reflections (or relatively weak late reflections) create excellent intelligibility, even with reverberation times as large as 4 s. For further details the reader is referred to [3].

6 CONCLUSIONS

In this paper it is argued that sound control should be based on wave-field extrapolation (holography). The underlying theory has been presented in terms of the Kirchhoff integral. It has been shown that wave-field extrapolation involves "spatial convolution." Wave fields, constructed in a holographic way, are in principle indistinguishable from genuine wave fields. It has been shown that generalized holography for sound control consists of three basic steps:

1) Wave field *acquisition* by a microphone array
2) Wave field *extrapolation* by a multichannel processor
3) Wave field *reconstruction* by a loudspeaker array.

The microphone array should be located near the source area to acquire the direct wave field; each microphone signal should be measured separately (no mixing). The processor applies a complex-valued spatial convolution process. This process can be elegantly formulated in terms of a matrix multiplication. The loudspeaker array reemits the desired wavefronts into the hall toward the audience. The resulting sound field is not optimum for a few points only, but it is optimum for the entire enclosed space (a holographic solution is a spatial solution).

For direct sound control the acquired direct wave field is extrapolated *under free-field conditions* toward the positions of the loudspeakers. The contributions of all loudspeakers reconstruct a truly natural direct sound field with any desired strength.

For reverberation control the acquired direct wave field is extrapolated in a fictive *acoustically desired hall*. The resulting reverberant field is picked up at positions in the fictive hall that correspond to the loudspeaker positions in the real hall. The contributions of all loudspeakers reconstruct the desired, truly natural, reverberant sound field. By changing the parameters of the acoustically desired hall (volume, shape, absorption), variable acoustics can be created in the real hall in a truly natural way. If relatively low-energy reverberant fields are reconstructed, then large reverberation times can be realized with an acceptable loudness (particularly important for small concert halls) and with an acceptable intelligibility.

If in practical situations one or more microphone–loudspeaker pairs contain an unacceptable large acoustic feedback, then it is shown that a filter can be derived to compensate for the feedback effect.

ACS is based on these holographic principles. ACS systems control the early and late reflections in an en-

	early reflection module for stage	early reflection module for hall	reverberation module for small simple halls	reverberation module for any type of hall
model 1000S	X			
model 1000H		X		
model 2000			X	
model 3000				X
model 4000S	X			X
model 4000H		X		X
model 5000	X	X		X

Fig. 26. Standard models of ACS.

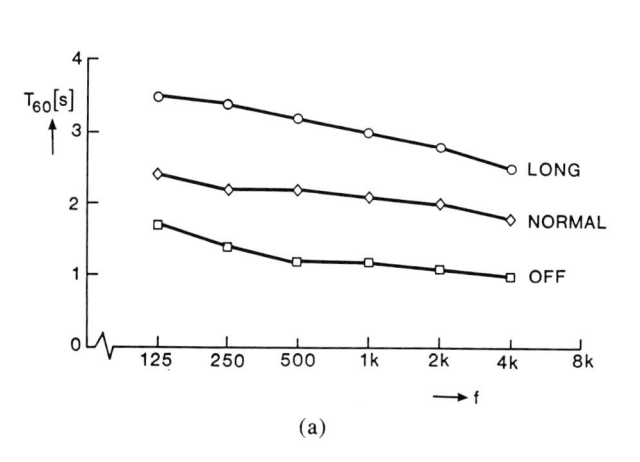

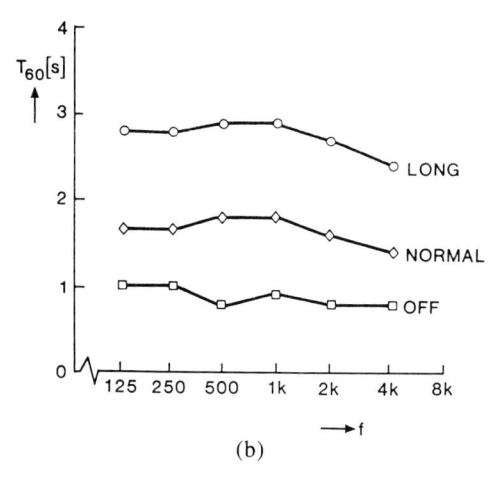

Fig. 27. Natural reverberation times with two preferred ACS settings. (a) At Delft auditorium. (b) At DACARY.

closed space, that is, ACS controls the acoustics in an enclosed space. Experience with ACS shows that the system guarantees the realization of *preferred*, *natural* sound-field characteristics as well as the realization of *instantaneous*, *variable* acoustics. If the source signals of the ACS microphones are recorded individually, then ACS can be used as a holographic reproduction system for the direct sound field together with any desired reverberant field. According to the author this is the ultimate in audio recording and reproduction technology.

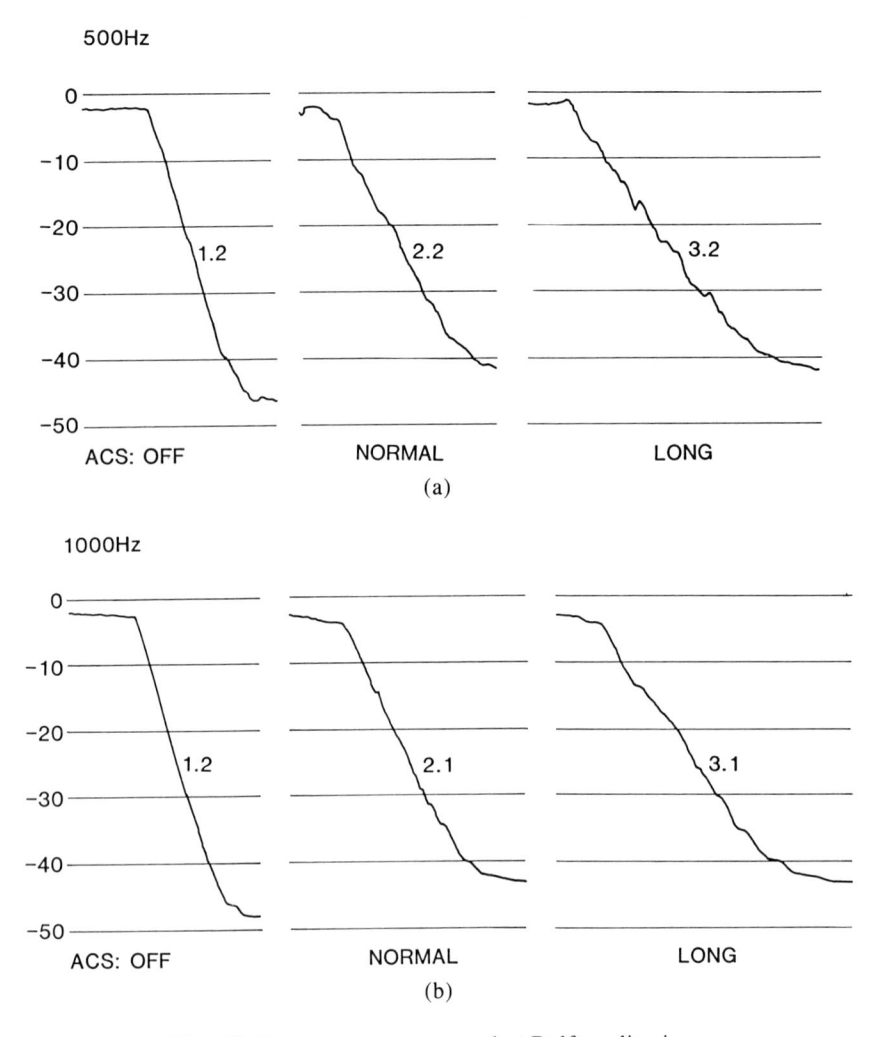

Fig. 28. Decay curves measured at Delft auditorium.

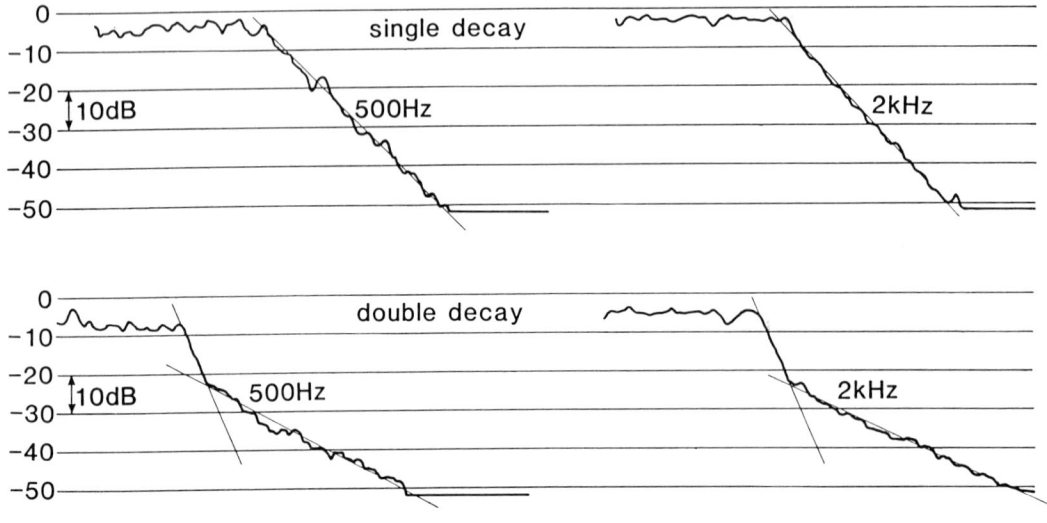

Fig. 29. Decay curves measured at DACARY.

7 ACKNOWLEDGMENT

The ACS measurements were made by Dr. D. de Vries in the auditorium of Delft University of Technology, The Netherlands, and by J. Hemingway in the Decoustics/ACS Centre for Acoustical Research at York University, Toronto, Ont., Canada. The ACS systems were manufactured by Electronica-Griffioen B.V., The Netherlands.

8 REFERENCES

[1] Ahnert, W., "The Complex Simulation of Acoustical Sound Fields by the Delta Stereophony System (DSS)," presented at the 81st Convention of the Audio Engineering Society, *J. Audio Eng. Soc. (Abstracts)*, vol. 34, p. 1035 (1986 Dec.), preprint 2418.

[2] Berkhout, A. J., *Applied Seismic Wave Theory* (Elsevier, Amsterdam, 1987).

[3] Berkhout, A. J., D. de Vries, J. R. Hemingway, and A. Griffioen, "Experience with the Acoustical Control System 'ACS'," presented at the 1988 Sound Reinforcement Conference (Nashville, TN, May 5-8, 1988).

[4] X. Franssen, "Sur l'Amplification des Champs Acoustiques," *Acustica*, vol. 18, pp. 315-223 (1968).

[5] Jaffe, C., "Sound System Design for the Eugene Performing Arts Centre, Oregon," *Record. and Eng.*, pp. 86-93 (1982 Dec.).

[6] Pösselt, Ch., Ch. Jaffe, K. Genuit, and J. Blauert, "Application of Physical—and Computer—Modeling Tools in the Process of Planning Room Acoustics," presented at the Spring Conference, Acoustics '88 (Cambridge, Apr. 1988).

THE AUTHOR

A. J. Berkhout received a master's degree in electrical engineering, and a doctorate in physics from the Delft University of Technology, The Netherlands, in 1963 and 1970 respectively. He was a lecturer in physics at the Royal Dutch Navy Academy in 1963 and 1964. From 1965-1971 Dr. Berkhout was a research geophysicist with Shell International Research, mainly engaged with analysis and processing of seismic wave fields. From 1971-1974 he was employed by Brunei Shell Petroleum Co., developing and applying methods for direct gas and oil detection from seismic data. In 1975 and 1976 he served as adviser for Shell International Petroleum Co., The Hague, on data analysis and processing in seismic exploration.

Since 1976 Dr. Berkhout is professor of seismics and acoustics at the Delft University of Technology. His primary field of interest includes seismic inversion ultrasonic imaging and architectural acoustics.

FORUM

ACOUSTICAL NEGATIVE FEEDBACK FOR GAIN CONTROL

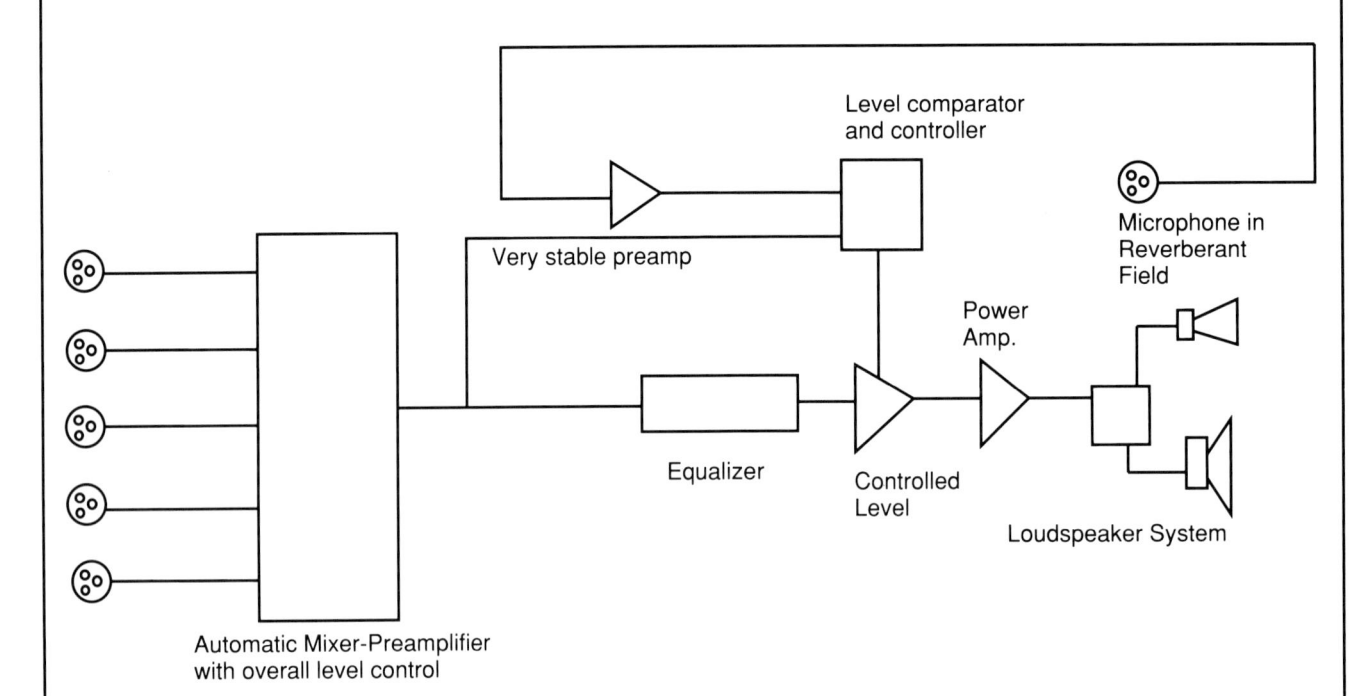

Figure 1: Block diagram for acoustical negative feedback for level adjustment with varying occupancy of reverberant spaces

Great progress has been made in recent years in the invention of "smart electronics" that can assist in providing sound systems with unattended operation, including automatic microphone mixers, in particular, as first developed by Daniel Dugan. These mixers recognize the microphone input while suppressing others, and reduce overall gain by an appropriate factor when the number of open microphones is greater than one.

Further development by Industrial Research Products, Ivie Electronics, and Innovative Electronic Designs (among others) has resulted in units that also provide overall gain control, similar to conventional AGC action, but without gain going to maximum when no signal is present. This can result in a very natural-sounding speech reinforcement system, with the overall operation very similar to, and possibly better than, that provided by a skilled human operator.

A further refinement is suggested to compensate for changes in room acoustic gain with differences in occupancy. This further development should result in completely stable speech reinforcement systems in many reverberant worship spaces with hard seating, such as the typical Roman Catholic of Eastern Orthodox church, many Protestant churches, and possibly some Jewish and Moslem houses of worship also.

1. ACOUSTICAL NEGATIVE FEEDBACK

The technique employs one or more controller microphones, usually omnidirectional, in the reverberant field, but not too far from the loudspeaker system(s) for signal delay to become a great factor. This microphone would feed a very stable preamplifier. A comparison circuit would be employed, comparing the output of the control console or automatic mixer-preamplifier, with the output of the controller microphone's preamplifier and the level difference would provide negative feedback control of an output level controller to keep the level difference (mixer output vs. controller microphone output) as constant as possible. Generally, the typical Roman Catholic or Eastern Orthodox church experiences changes in reverberation times of the order of three-to-one to two-to-one, meaning that about 5 or 6 dB of total level adjustment is all that is required for this additional control system.

Fig. 1 is a simplified block diagram of the proposed system. The comparison circuit and level control can be implemented by either digital or analog means. In certain cases, only changes in programming to existing hardware (NOALA-type circuits, for example), or simple modifications and the addition of the controller microphone are all that will be required to supplement existing automatic microphone mixers with this concept.

In most cases, the controlling microphone and circuitry should be as wide-band as the complete reinforcement system. The integration period of the comparison circuit should be long enough to fit the delay inherent in the controller microphone's distance from the loudspeaker system. However, experience with other portions of an automatic level control for speech reinforcement systems indicates that 0.1s is usually the longest integrating period that can be tolerated. Remember that loud coughs, loud music not requiring amplification such as organ music, and other nonreinforced sound sources, may be sensed by the system as an increase in room gain, and cause the system to effect gain reduction. An integration period of 0.1s or less should minimize any possible harmful effects of this momentary gain reduction.

2. BENEFITS

Addition of this concept to exiting systems employing automatic microphone mixers can result in the reduced need to provide additional manual level control, stability of equalization settings to control feedback and possibly some degree of additional feedback prevention. Using hard-pew seating in a variety of worship spaces, improves congregational singing. Sound is also more efficiently distributed by the sound-reflecting pew surfaces.

3. AVAILABILITY

Our firm is not currently in the electronic design and manufacturing business and has not developed the necessary circuitry to make this concept a marketable item. A patentability investigation conducted by our attorneys, and completed August 20, 1990, indicated an opinion that the concept is patentable. However, since the need for this concept, particularly in houses of worship, is considered great, we are not applying for a patent but are making this concept available to any interested manufacturer. We will be pleased to provide consultation upon receipt of an application.

4. ACKNOWLEDGMENTS

The author wishes to acknowledge the great contribution of Daniel Dugan to electronic control of sound systems. This particular idea was, however, inspired by John Meyer's development of "Source Independent Measurement" for equalization. The present idea is, of course, intended for level control, not equalization, but a combination of the two concepts is certainly possible.

DAVID KLEPPER
KMK Associates
White PLains, New York

ENGINEERING REPORTS

The Acoustics of St. Thomas Church, Fifth Avenue*

DAVID LLOYD KLEPPER, AES Fellow

Klepper Marshall King Associates, Ltd., White Plains, NY 10603, USA

The acoustical improvement program for St. Thomas Episcopal Church in New York City is described, including ceiling treatment and a pew-backed loudspeaker system. Reverberation time measurements, energy–time curves, and early-to-reverberant-sound ratios are given.

0 INTRODUCTION

The St. Thomas Episcopal Church in New York City held its first service in its Fifth Avenue and 53rd Street building in 1913. The church is 100 ft (30 m) wide by 214 ft (65 m) long. The nave is 95 ft (29 m) high and 43 ft (13 m) wide between columns. The volume of the church is approximately 2,200,000 ft^3 (62,304 m^3).

As originally built, the ceiling of the church was constructed with Guastovino tile. The history of this material is well covered in the paper "Gothic Sound for the Neo-Gothic Chapel of Duke University" by James G. Ferguson, an architect in Chapel Hill, NC, and the late Robert B. Newman, a founding partner of Bolt Beranek and Newman, Inc. [1]. The material is an artificial stone, produced by firing clay with embedded carbon particles and having sound-absorbing properties. It was typically applied to neo-Gothic churches and cathedrals to produce a dry acoustical environment where speech would be intelligible—prior to the advent of sound systems capable of high intelligibility within highly reverberant rooms.

1 ACOUSTICAL TREATMENT

The acoustical improvement program was begun when scaffolding was erected in 1962 for cleaning and repair of the raredos. William Self and vestry member Medley G. B. Whelpley argued for sealing the ceiling area that could be reached from that scaffolding, and the author recommended the Borden Chemical Kyanize product that had been formulated to Bolt Beranek and Newman's requirements for their acoustical consulting work at Riverside Church. The results of this very minimal treatment were a noticeable improvement.

In 1970 and 1971 a pew-back speech loudspeaker system designed by Larry S. King and the author was installed by Commercial Radio Sound under the direction of Seymore Gerber. The concept of a pew-back system for St. Thomas Church had been discussed by G. Donald Harrison, the tonal director of the Aeolian Skinner Organ Company in 1955, at the time the organ was installed, preparing the way for the design concept of Mr. King and the author. It is the very first system to use both pew-back loudspeakers and solid-state digital memory signal delay for synchronizing local loudspeaker sound with live sound from the front of the church, a design concept that has proven valuable for systems in many cathedrals, churches, and synagogues throughout North America. As a cost-saving measure, certain of the best of the old column loudspeakers from the previous system were reused for coverage of the south balcony, and these 30-year-old loudspeakers are still performing well in that location, where their characteristics are an optimum fit to the architectural and acoustical requirements. Although most pew-back systems require three loudspeakers on the back of each normal-size church pew, St. Thomas Church obtains good results from only two, which cover 3.2 linear meters. The system electronics and certain microphones have been upgraded over the years to keep the system "state of the art," but the original pew-back KLH 12.5 loudspeakers, a loudspeaker designed by Henry Kloss for the public-address system at Washington's Dulles Airport and the 1970 KLH compact FM radio, continue to perform well. Both Robert Lin of Comco Systems, Inc., and the late John Klanatsky of Information Transmission Systems were involved in the later electronic improvements.

Also in 1971, the sealing treatment of the ceiling that had proved effective in the raredos area was extended over the entire ceiling. The increase in reverberation time was not as great as expected. Inspection several years later indicated that the sealing process was still incomplete because of an insufficient number of coats, and subsequently, with support raised by the rector, Dr. John Andrew, and the church's treasurer, Mr. Edward

---

* Manuscript received 1995 February 12; revised April 26.

Weist, plans were made to provide up to 10 additional coats of the Kyanize sealer, with much lower cost scaffolding arranged by use of the supports for the suspended chandeliers. Also, the chantry and south balcony ceilings were treated. The author and Mr. King inspected sample ceiling treatments after the second and the fourth coats and suggested that four coats were enough. The late Dr. Robert B. Newman of Bolt Beranek and Newman also was involved in this phase.

2 RESULTS

Subjective reactions from the church musicians, the clergy, and the congregants indicated that this judgment was correct for the St. Thomas Church ceiling. This is supported by the subsequent measurement of the early-to-late-sound ratio and the reverberation time, as well as by the time–energy curves (Figs. 1–3).

With the initial suggestion of church member Will Carter, glazing was applied to the chancel side openings, further improving communication among choir members as well as reflected energy to the congregation.

Five of the products that are being used for this type of treatment are:

- Coronado 70-200, aquaplastic urethane, Coronado Paint Company. P.O. Box 308, Edgewater, FL 32032, (800) 874-4193, attention Cynthia Black.
- Kyanize L-0560 sealer with L-0561 flat finish, Hudson-Shatz Painting Company, 429 West 53rd Street, New York, NY 10019.
- M. A. B. Hydro-Clear (acrylic latex), M. A. B. Paints anc Coatings, 600 Reed Road, Broomall, PA 15008, (215) 353-5100.
- Okon four-times-solid-density sealer, Okon, Inc., 6000 West 13th Avenue, Lakewood, CO 80214, (303) 232-3571, attention Frank Livingston.

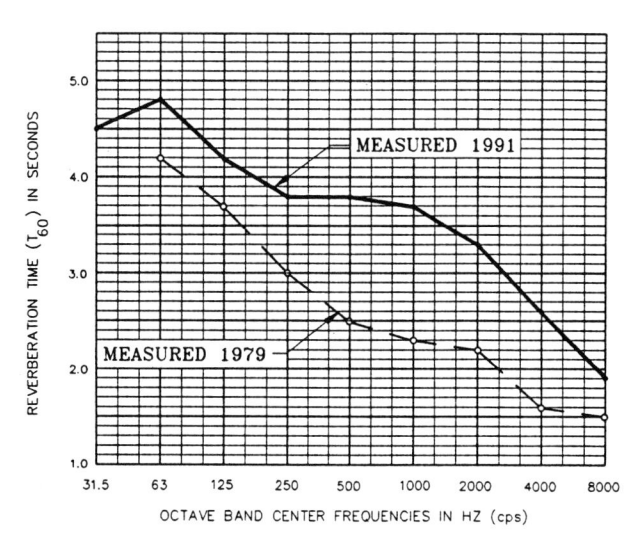

Fig. 2. Reverberation time of St. Thomas Church. Measured using Bruel & Kjaer graphic level recorder with organ tone clusters recorded by the author.

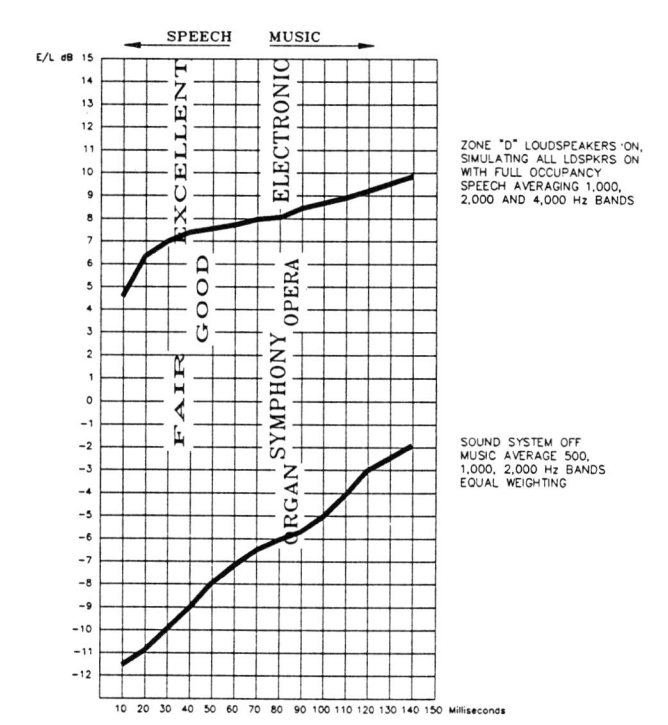

Fig. 1. Early-to-late-sound ratio versus time, St. Thomas Church unoccupied. Measured by L. Gerald Marshall using his ELR Technique [2].

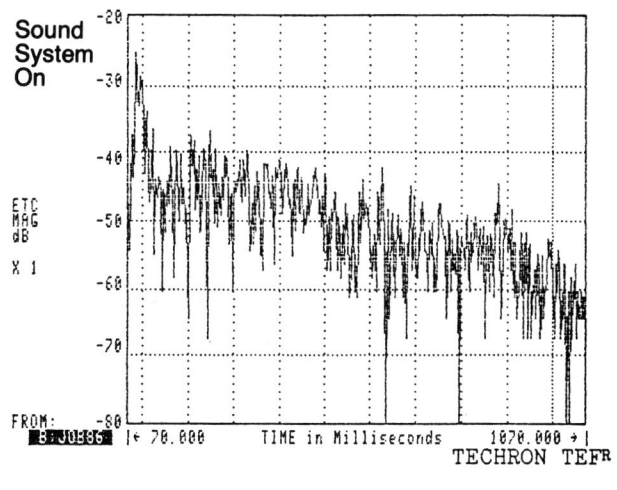

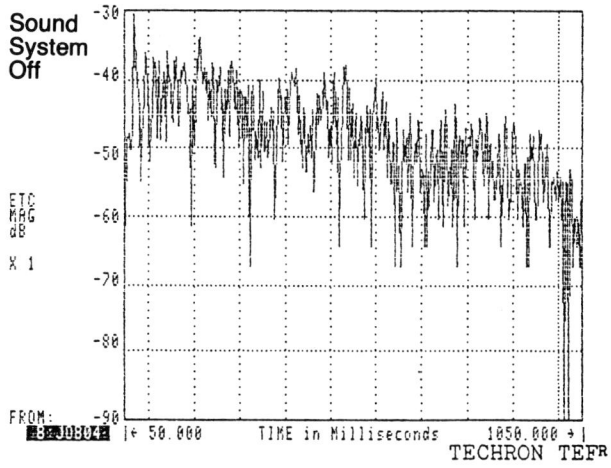

Fig. 3. St. Thomas Church energy–time curves for 1000-Hz band with pew-back sound system off and on. Data obtained by L. Gerald Marshall using Techron TEF-12+ system.

- Pratt and Lambert acrylic latex varnish dull Z-39, Pratt and Lambert Regional Office, 25 Truman Drive South, Edison, NJ 08817, (908) 985-0770.
- UVS sealer, ProSoCo, Inc., P.O. Box 171677, Kansas City, KS 66117, (913) 281 2700.

Quoting from former music director William Self's book [3]:

> St. Thomas Church, with its now correct ceiling, its happily missing tapestries, and its complete lack of carpets is at last a building where music sounds as it should. The spoken word can be heard in every pew and the services are heightened by an appropriately spiritual and mystical atmosphere. A group of 16 or 18 boys has the volume of 40, and the tone of the complete choir takes on a silken quality that was impossible to obtain before. Equally enhanced is the tonal depth and color of the organ; yet the words of the speaker are heard with satisfying clarity. This atmosphere encourages congregational singing, and what was once an aching void in the service is now a thrilling component of it.

3 REFERENCES

[1] J. G. Ferguson and R. B. Newman, "Gothic Sound for the Neo-Gothic Chapel of Duke University." Available from Acentec, Cambridge, MA.

[2] L. Marshall, "An Acoustics Measurement Program for Evaluating Auditoriums Based on the Early/Late Sound Energy Ratio," *J. Acoust. Soc. Am.*, vol. 96, pp. 2251–2261 (1994 Oct.).

[3] W. Self, *For Mine Eyes Have Seen* (Worcester Chapter of the American Guild of Organists, Worcester, MA, 1990), pp. 179–180.

THE AUTHOR

David Lloyd Klepper was born in New York City in 1932. He received his B.S.E.E. in 1953 and M.S.E.E. in 1957, both from MIT. His M.S. thesis was the development of a binaural microphone system for concert hall evaluation. From 1957 to 1971, he worked in the Cambridge, Downers Grove, and New York offices of Bolt Beranek and Newman Inc. on numerous architectural acoustics projects involving sound system design and specification, room acoustics design, mechanical equipment noise and vibration control, and sound isolation. Since 1971 he continues his architectural acoustics work with L. Gerald Marshall and Joanna Stachowicz at Klepper Marshall King Associates, Ltd., which was founded in 1971. He also teaches architectural acoustics at New York's City College.

Mr. Klepper is a fellow of the Audio Engineering Society, past chairman of the Midwest and New York Sections, past member of the Board of Governors, past Eastern Vice President, a recipient of its Silver Medal, and a member of the *Journal*'s review board. He is also a member of the Institute of Electrical and Electronics Engineers, Acoustical Society of America, Institute of Noise Control Engineering, American Guild of Organists, and the Electric Railroaders' Association. His interests include classical music, singing in synagogue choirs, and serving as a cantoral soloist. He has a broad range of theological interests, as well as an interest in worship as an art form.

D.
intelligibility evaluation

Acoustic Feedback—Its Influence on Speech Intelligibility*

SVERRE STENSBY, ASBJØRN KROKSTAD, AND SVEIN SØRSDAL

Acoustics Laboratory/ELAB, The Norwegian Institute of Technology, N-7034 Trondheim-NTH, Norway

Available acoustic gain in speech reinforcement systems will often be limited by acoustic feedback. The changes in intelligibility caused by ringing effects in systems working near the point of instability are used as a basis for determining the optimum gain. The results show no significant influence of acoustic feedback in the domain from 1.5 dB below instability to no feedback.

0 INTRODUCTION

It is important that a sound system give sufficient sound level over the entire listening area. The usable sound level may be too low because the available gain is limited by acoustic feedback from loudspeaker to microphone. When the amplifier gain is slowly increased, the system will turn into oscillations at the frequency where the conditions of unity gain and zero phase angle in the open loop are simultaneously fulfilled. The margin of stability M is defined as $M = -20 \log A$, where A is the open-loop gain at the frequency where howlback occurs.

The subjective influence of a small stability margin is audible ringing. When the ringing is pronounced, it is considered annoying and also detrimental to speech intelligibility. The purpose of the present investigation was to measure intelligibility at various stability margins and find the lowest margin which did not reduce the intelligibility significantly.

1 INSTRUMENTAL SETUP AND TEST ROOM CHARACTERISTICS

Speech intelligibility tests were carried out under controlled conditions in seven rooms. Geometric and acoustic data for the different rooms are shown in Table 1.

The sound system used in all the rooms is shown in Fig. 1. The test persons were seated in a small area. The test vocabulary was recorded on magnetic tape and played through a laboratory loudspeaker with power response within ± 3 dB in the frequency range of 100 Hz to 5 kHz. The feedback signal was picked up with an omnidirectional dynamic microphone and mixed with the signal from the tape recorder after amplification in the controlled amplifier. This amplifier can be used as an automatic volume control to establish unity gain and nondestructive howlback in the system. When the point of unity gain is found at the frequency where howlback occurs, the margin of stability may be selected simply by calibrated attenuation.

2 VOCABULARY FOR THE INTELLIGIBILITY TEST

In normal speech under reverberant conditions a word will be partly masked by the reverberation from the previous words. Single isolated syllables will not be masked in this manner and would probably not be suited for intelligibility tests under reverberant conditions. The use of a limited number of carrier sentences has the disadvantage that the number of masking spectra is small compared with that in normal speech.

This lack of randomness is avoided in the test described here by the use of "nonsense sentences." Each test sentence consists of a sequence of consonant sounds (C) and vowel sounds (V) in this order:

CVCVCVC   CV*CVC* VC.

---

* Presented at the 47th Convention of the Audio Engineering Society, Copenhagen, Denmark, 1974 March 25–29.

Table 1. Some properties of the rooms.

Room	Volume [m³]	Average Reverberation Time for Octave Bands 500, 1000, and 2000 Hz [second]	Distance from Loudspeaker to Feedback Microphone [meter]	Distance from Loudspeaker to Listening Area [meter]
Lecture room	800	0.46	7.7	7.5
Assembly room	500	1.2	7.3	18
Assembly hall	2000	1.6	4.3	14
Concert hall	3500	1.6	3.4	18
Church	6000	3	8.5	17.5
Cathedral	50000	3.8	18	25
Reverberation Chamber	270	4.1	8.5*	8.4*

* Loudspeaker facing a corner.

The sounds are drawn randomly with replacement among the allowed sounds. Examples of test sentences are:

PYPOJAK   FIN*AI*SEG

REILIBUD   ROK*JEG*OP.

In the Norwegian language almost all such combinations of sounds may be pronounced without difficulty.

Groups of 25 test sentences were compiled in sentence lists and presented as a unit in the intelligibility tests. The listeners were given a special version of the sentence lists. Here the middle part of the second word in each sentence (italics above) was omitted and was to be filled in when listening to the test tape. The percentage of correctly identified phonemes in each list was used as a measure of intelligibility.

The test sentences were recorded with 10-second time intervals. They were spoken distinctly by one male speaker with "normalized" Norwegian linguistic background.

3 TEST PROCEDURE

The tests were carried out with 1.5-dB margin of stability, 3-dB margin of stability, and without feedback. The speech signal was presented at two levels with maximum sound pressure levels of 60 dB(A) and 70 dB(A), respectively, measured with fast response in the listening area. The background noise in the rooms varied between N25 and N35. The frequency of howlback oscillation was different in different rooms, but it was made sure that oscillation occurred at frequencies in the central part of the speech frequency domain.

For each combination of room, level, and margin, 3–4 lists were presented to 6–10 persons. The listeners consisted of students from the university. The maximum deviation from the normal hearing threshold for the listeners measured with a pure-tone audiometer was ± 10 dB in the frequency range of 125–8000 Hz. The listening test in each room lasted 2 hours, including a short break, and care was taken to avoid weariness.

4 RESULTS AND EVALUATION

The intelligibility results covered the range from 100 to 60%. Median and upper/lower 25% intelligibility values are shown in Fig. 2 for each room as a function of speech level and margin. The data had a skew distribution. For this reason the nonparametric "median test" [1] was used on the null hypothesis: "The intelligibility result at the three margins of stability is drawn from identically distributed populations." The intelligibility values are arranged in a table showing the values for each stability margin falling above and below the median of the combined set from one room and one speech level in turn. The table is analyzed as a contingency table obtaining a χ^2 statistic with 2 degrees of freedom. The hypothesis is rejected if the observed χ^2 is larger than the critical value. The test did not reject the null hypothesis at the 10% level for any of the two sound levels and seven rooms.

5 DISCUSSION

In [2] it is assumed that "the permissible gain (in sound reinforcement systems) is about 2 dB below the gain at which instability (singing) occurs," while [3] says "it has been found that if the loop gain is 3 dB below the point of instability, the system has no noticeable ringing effects." Working on the sound system in our setup we found that the ringing was still noticeable with a 3-dB margin of stability. However, the data obtained show no significant influence on speech intelligibility from acoustic feedback when the margin is 1.5 dB or higher.

To the extent that the test procedure is representative of real speech reinforcement systems, speech intelligibility may be improved in such systems in noisy environments by increased amplification beyond the point where noticeable ringing effects occur. However, if quality aspects other than intelligibility are taken into account, such as fidelity and naturalness, it is evident that the margin should be high enough to prevent audible ringing.

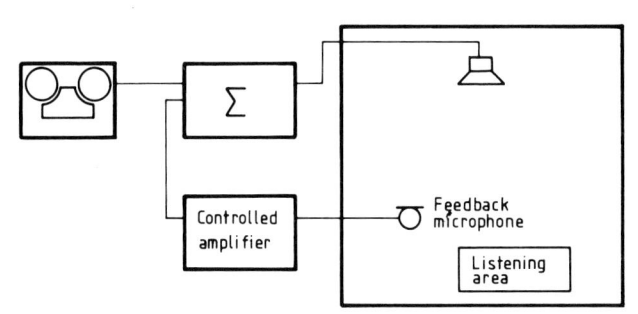

Fig. 1. Sound system used in the intelligibility tests.

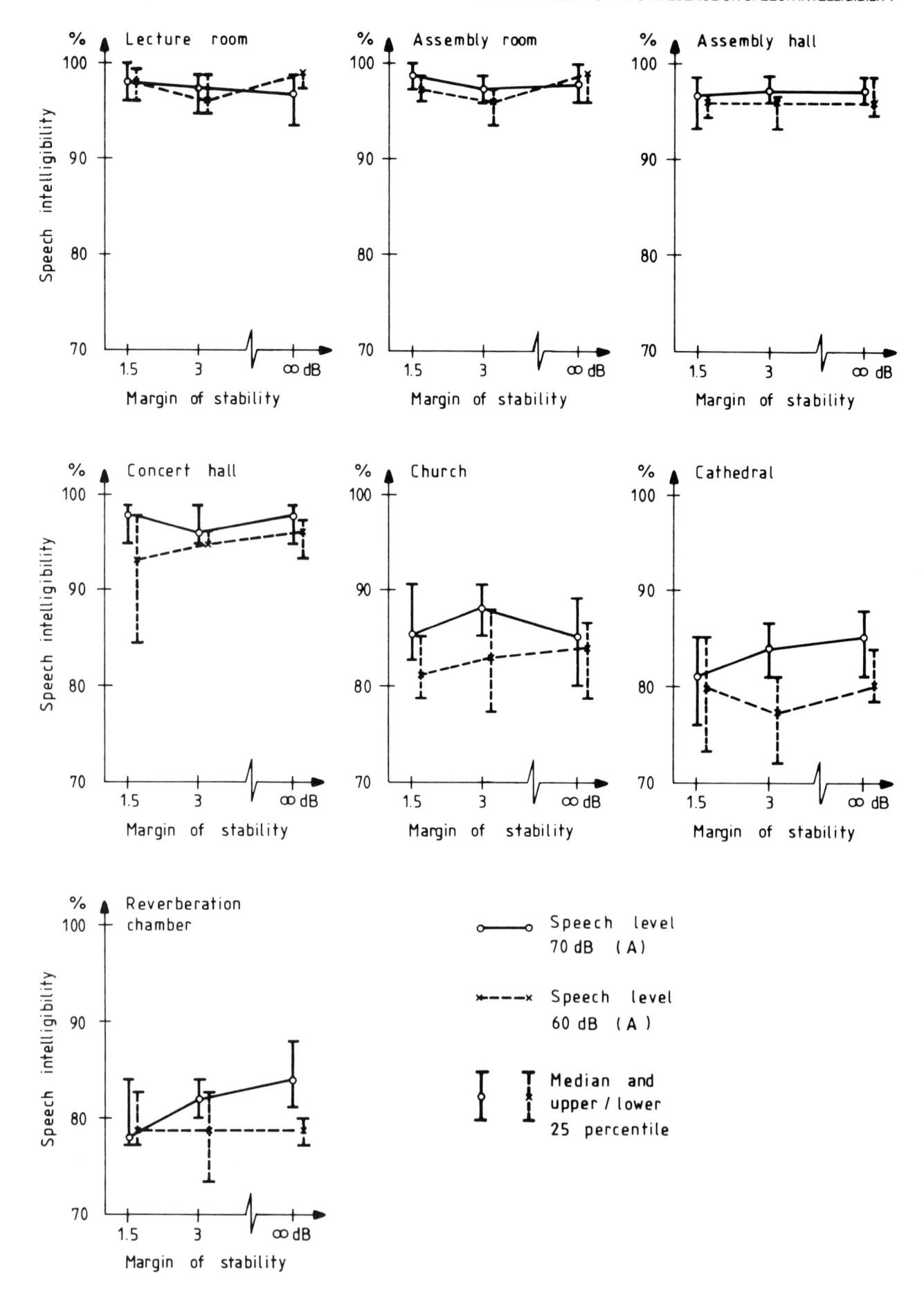

Fig. 2. Test results. Speech intelligibility versus margin of stability in seven rooms.

6 ACKNOWLEDGMENT

This work was supported by the Royal Norwegian Council for Scientific and Industrial Research.

7 REFERENCES

[1] W. J. Dixon and F. J. Massey, Jr., *Introduction to Statistical Analysis* (McGraw-Hill, Tokyo, 1957), p. 295.

[2] M. R. Scroeder, "Improvement of Acoustic-Feedback Stability by Frequency Shifting," *J. Acoust. Soc. Am.*, vol. 36, pp. 1718–1724 (1964 Sept.).

[3] R. W. Guelke and A. D. Broadhurst, "Reverberation Time Control by Direct Feedback," *Acustica*, vol. 24, no. 1, pp. 33–41 (1971).

THE AUTHORS

Sverre Stensby was born in 1946. He studied at The Norwegian Institute of Technology, graduating in 1971. Since then he has worked at Acoustics Laboratory/ELAB, concentrating on speech intelligibility and sound systems.

•

Asbjørn Krokstad was born in 1931. He graduated from the Norwegian Institute of Technology in 1956, and later received the degree of Dr.ing. in acoustics in 1963. He has been a professor in acoustics at the Division of Telecommunications, The Norwegian Institute of Technology, since 1970. He is a member of the Audio Engineering Society.

•

Svein Sørsdal was born in 1942. He graduated from The Norwegian Institute of Technology in 1967, where he is now a lecturer.

Those Early Late Arrivals! Mr. Haas, What Would You Do?*

CECIL CABLE

Cecil R. Cable and Associates, Edmonton, Alberta, Canada (with Hilliard & Bricken, Inc., Santa Ana, CA 92700, USA)

AND

R. CURTIS ENERSON

Alberta Government Telephones, Edmonton, Alberta, Canada

An in-depth time delay spectrometry study of a multidesktop distributed sound system in a 131 000-ft³ (3668-m³) room has been made. The validity of the fusion principle for arrivals from nearby desktop loudspeakers is questioned. The differences were neutralized by electronic delays of approximately 5 and 10 ms with improved articulation and an apparent reduction in RT_{60}. Subsequent research supports this argument. Hard data and spectrograms are presented.

0 INTRODUCTION

Mr. Haas, what would you do about those early late arrivals (ELA)? You have said that echoes arriving within 30 ms, more or less, are integrated with the first arrival. You have shown this to be true. When a second source of the same intensity follows the first source by not more than 30 ms it raises the apparent sound pressure level (SPL) to the listener 3 dB more than either source heard alone. The total sound appears to come from the first source only. From this and other experiments you concluded that all these ELAs were fused or integrated with the first arrivals and indeed could be considered as first arrivals in terms of transfer of information from sender to receiver [1].

But wait a moment, Mr. Haas. What about this transfer of information? You said that you could not identify the source of the ELAs. The sound became louder when the energy of the two sources fused. The energy was integrated, but somehow a portion of the information identifying the source was lost. Would not the location of the second source be important information?

Mr. Haas, you speak of severe distortion "when primary sound and echo are emitted from one loudspeaker.... According to the particular delay difference of the echo, certain frequency ranges are very much reinforced whereas others are completely wiped out." We are familiar with this and now refer to it as the comb filter effect. However, you say that these "interferences are not perceived in binaural hearing of sound and echo from two loudspeakers.... Only consequence of small delay differences in acoustic superposition is the described directional impression." We know that the comb filter effect of swept-frequency tones heard from primary and delayed acoustic sources is clearly audible in even moderately reverberant rooms. This important information is transmitted to us through the acoustic medium. However, for acoustic coupling you say that for speech (and perhaps music) only total energy additions from the echo are perceived. Now we have not only lost the directional information from the echo but also the information regarding the comb filter effect as well.

Mr. Haas, we are confused! How much information are we going to lose from this echo which we are led to believe is perceived in every respect as a first arrival?

---

* Presented at the 63rd Convention of the Audio Engineering Society, Los Angeles, 1979 May 15–18.

You undertook to qualify the effect of early echoes on the audibility of speech. This proved to be difficult. You finally resorted " . . . to give a continuous text true to practical conditions and with various echo disturbances, then leave to the observers the decision whether they felt disturbed or not." You did not ask if the early echo contributed to the transfer of information in the hearing of connected speech. We wonder if it does.

After thirty years why are we asking these questions?

1 DISCOVERY

The case history which prompts the above questions concerns the Legislature of the Province of Alberta in Canada. The Legislature is the meeting place of the elected officials of the Provincial Government, which is analagous to a State Government in the United States. The area of interest is the main meeting room, referred to as the House. A floor plan and two sections of this room are shown in Fig. 1.

Alberta Government Telephones (AGT) became involved in providing sound equipment and an associated recording system for this room in the mid-1960s, the Assembly having functioned without the benefits of modern sound technology prior to that time. Acoustically the room is adequate for listening to oration properly executed. However, not all present-day politicians can orate properly. Hence a sound system is required. The first system was installed on a ten-year lease contract, and when this expired, a new system was installed for the fall of 1976 (on another ten-year contract). Unfortunately the new system suffered some problems. The bulk of these were corrected during the 1976 Christmas break, and the system performed quite well for the spring session of 1977. The Honourable Members were still not completely satisfied. One of the authors (cc) was engaged to perform an exhaustive study and make recommendations concerning possible improvements.

After a comprehensive battery of "normal" tests, including things such as system hum, noise, distortion, and frequency response, along with room concerns (reverberation times, background noise, etc.), exploration of time effects within the room commenced. A diligent search for reflections and other anomalies was carried out. The primary instrumentation used for this purpose was a high-quality spectrum analyzer especially modified for time delay spectrometry (TDS)

Significant results of the testing are as follows: The reverberation times averaged 1.4 seconds. The room noise measured NC 28. The system had more than enough output capability. The acoustic gain was definitely adequate, although lower than desired. (The customer's aesthetic requirements prevent the hanging of a central array and/or shortening the talker-to-microphone distance.) With all these things going for the system, what was the problem?

Let us take a quick look at the system layout. Each member's desk has a loudspeaker and a microphone (Fig. 2). This forms the bulk of the sound system on the floor. Other loudspeakers are located in the dais desk, in Mr. Speaker's chair, on the Assembly table, and in the galleries. Other microphones are located on the Assembly table and on the dais desk. Microphones are manually selected by an operator who also controls the system overload and gain settings.

An analysis of the customer's complaints indicated an articulation problem, primarily in the back rows of the desks. But time and level analysis tells us that from the point of view of a back-row auditor,

1) the wall behind him causes a reflection 12 ms late which is lower in level than a number of loudspeakers acoustically nearest him, and

2) of the 21 loudspeakers in a typical worst case quarter of the Assembly (Fig. 3) the "latest" loudspeaker is only 12 ms late, and it is 17 dB down relative to the nearest

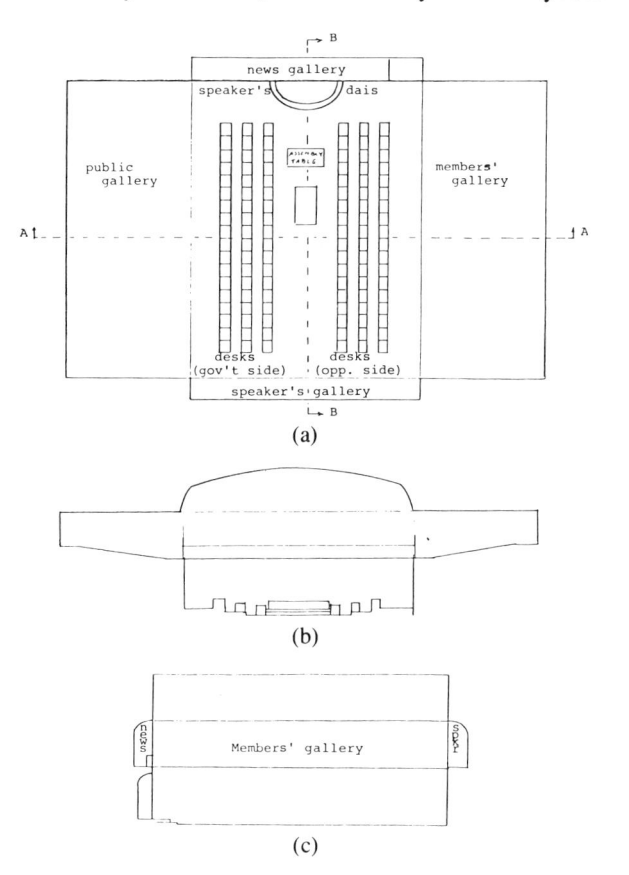

Fig. 1. Legislature of Alberta. (a) Floor plan. (b) Section A–A. (c) Section B–B.

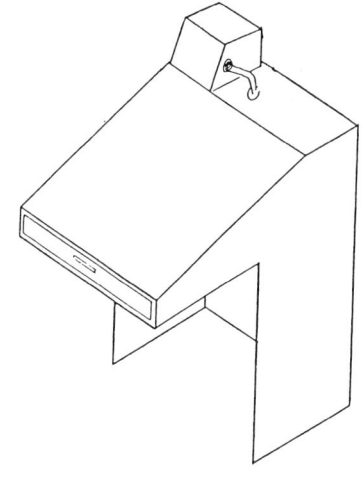

Fig. 2. Member's desk (not to scale).

loudspeaker.

According to the 20-ms rule of thumb this should present no problems. But it was suggested that time delay be attempted. Concern entered the designer's mind. Reviewing the information gleaned during testing confirmed that there was indeed a significant amount of late energy arriving at the third-row auditor's position, and that it had to be coming primarily from other loudspeakers. A review of current literature revealed that the people involved in the study of hearing who talk about echo had not done anything with delays shorter than approximately 15–20 ms, except for making the assumption that because echoes received within that time limit are not consciously perceived, the echoes must be making a contribution to the hearing process. Such an assumption, of course, does not constitute proof that early echoes do not interfere with hearing.

It was decided to try delay. A fine-resolution (20-μs) digital delay line with three delay channels was installed along with the necessary extra power amplification and equalization required. The "zero-delay" channel was connected to the innermost (front) rows of desks. One channel of the delay was applied to the second rows, another to the third rows and the last delayed channel was applied to the gallery feeds. The system was tuned and the delay set to bring the front and second rows in time with the third rows, as observed (objectively) from a typical third-row listening position. The timing is estimated to be within 0.5 ms.

Setting the time closer than this is not practical considering the path length variations due to lateral movement of the listeners. The gallery time delay was set for a nominal listening position.

The system was tested and found to have the same acoustic gain. Listening tests were conducted in the back rows of the floor, and a definite positive effect of the delay was observed. The inclusion of a bypass switch made possible instantaneous A–B comparisons. (Tapes of the difference of sound have been made.)

However, the acid test was the customer usage of the system. It was heartening to notice that the negative feedback from the Honourable Members ceased. We have concluded that an enhancement must have been realized.

The effect is not particularly striking, but it is definitely perceptible. While the full spectrum of sound is audible without the delay operating, when delay is injected, the clarity increases. The tapes show this as an apparent decrease in reverberation. The impression on site is subtly different. There is something psychoacoustical going on here which definitely requires deeper research.

The most significant thing is that, subtle or not, a change was realized which removed a source of annoyance to listeners *uneducated* in hearing tasks (and sometimes inattentive). Hearing must have been made easier for them.

2 CONTINUING RESEARCH

It is believed that the energy of the ELAs in the Legislative Assembly, that energy arriving within the fusion period, does not wholly contribute to the transfer of information from the talker to the listener. On the contrary, ELAs may even provide destructive interference. Is it possible that somehow our processor looks at those arrivals as reverberant energy not to be properly categorized with the direct sound field? The Legislative Assembly certainly sounds more reverberant without the digital delay lines in the line.

A significant time delay spectrogram of the room sound can be seen in Fig. 4. The microphone was placed in a typical auditor's position as shown in Fig. 3. Each trace is offset 10 dB for clarity. The first arrival, shown as trace 1, is the energy directly from the near loudspeaker. Trace 2, approximately 4 dB higher, (2.5 times more energy) is the sum of all arrivals within the first 6 ms, or from all those primary and secondary sources within 6.78 ft (2 m) of the near loudspeaker. A further opening of the time/space window to 20 ms, or 22.6 ft (6.9 m), gave an additional

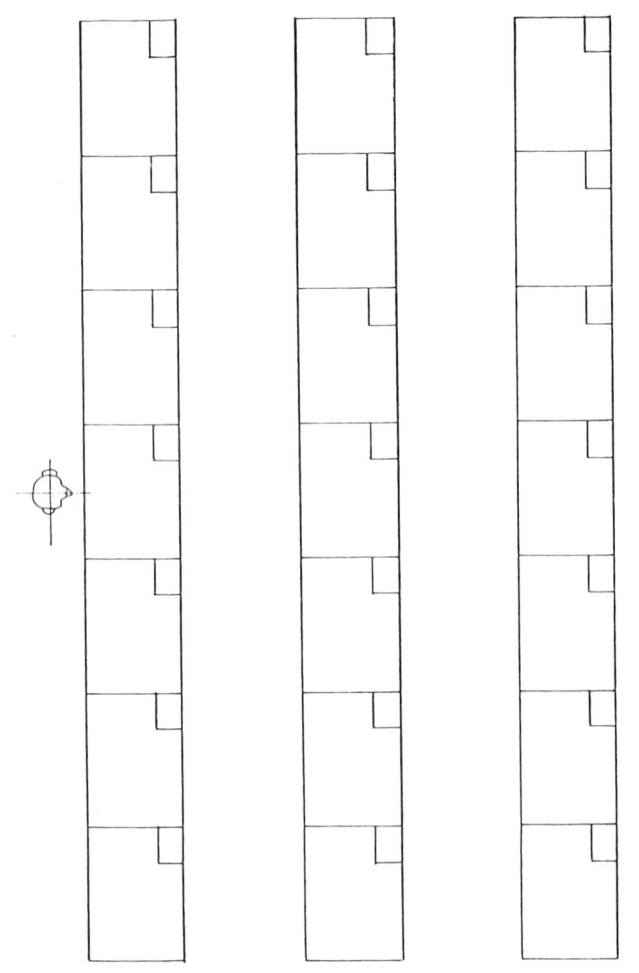

Fig. 3. Plan view of an auditor in the back row.

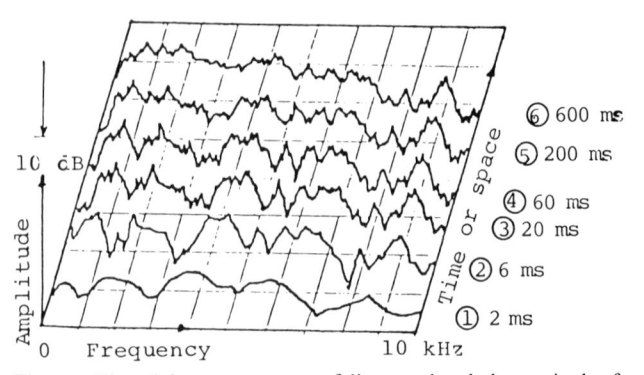

Fig. 4. Time delay spectrogram of direct and early late arrivals of integrated energy from loudspeakers and room.

increase of about 2 dB (1.58 times more energy) for trace 3. A further expansion of the window caused acceptance of relatively little additional energy. This suggests that much of the first 10 dB of decay after the first arrival is from the multisource origin rather than the reverberant field.

With the same setup a 5-ms pulse of pink noise was fed to the system. Fig. 5 is a spectrogram of the signal received at the typical listener's position of Fig. 3. The pulse from the near loudspeaker is clearly seen at zero time. About 5 ms later the sound arrives from the second-row loudspeakers and is down about 4 dB. About 12 ms after the first arrival the sound from the front row is perceived down about 6 dB. Following this is the beginning of a semireverberant field. The slight rise appearing from 35 to 50 ms may be the reflection from the far wall. This would be the early sound from the three rows of loudspeakers facing the other way.

A crude simulation of the early field was attempted. Setup was as shown in Fig. 6. The locale was an out-of-doors asphalt-covered parking lot with the nearest possible reflection (other than ground) 50 ft (15 m) distant.

For test 1 tape-recorded connected speech was fed to loudspeaker 1 and picked up by a cardioid microphone very close to the hard asphalt 12.9 ft (3.9 m) from the loudspeaker. This was then recorded on the second track of the tape recorder. The upper trace in Fig. 7 shows the transfer function of the setup. The frequency response on the linear scale is from 0 to 5000 Hz, or 500 Hz per major division. This TDS response includes all early echoes arriving at the microphone within 22.5 ms. The response of the upper curve of Fig. 7 changed very little when the TDS window was narrowed to 2 ms, indicating that no significant reflections were included in the response.

For test 2 (lower trace of Fig. 7) the microphone was raised to 5.75 ft (1.75 m) above ground. Three identical loudspeakers were energized by connecting them in parallel with loudspeaker 1. The microphone now received the first sound followed by seven reflections arriving 3.55, 10.25, 10.69, 11.69, 11.96, 12.02, 12.84, and 14.00 ms late. This included the direct sound from each loudspeaker to the microphone plus the ground reflections, all of which were recorded on track 3 of the tape recorder.

Last, segments of tracks 2 and 3 were alternately recorded on track 4. Besides the alteration in the tonal balance not too much imagination is required to feel that the program which included the seven ELAs was taken in a small reverberant room, whereas the program without the echoes could have been recorded in a less reverberant space.

We would again draw your attention to the spectrogram of Fig. 7. The upper trace is the response of the loudspeaker without echoes. The lower trace with seven echoes added is not too dissimilar to a display of ordinary room modes. But in this instance there was no room. We were outside. However, it sounds as if there were a room. We wonder how often time misalign anomalies (TMA) of early reflections are really considered as room modes? The two phenomena are entirely different in their construction and behavior. When looking at the lower trace of Fig. 7 it is not surprising that the sound reproduction system sounds like a room. The trace looks like a room response.

The fusion theory, precedent or Haas effect, says that because the ELAs all arrive at the listener (microphone) within 14 ms, they all fall within the fusion period. They should integrate with the direct sound and subjectively carry the same information as the direct sound. This appears not to be true. It may be argued that monophonic microphone pickup in some way destroys the effectiveness of the fusion of information. When a person takes the place of the microphone position, one hears the same coloration and room effect as is heard on the tape. All sound seems to come from the near loudspeaker, but the other effects are clearly audible in the perceived sound. This suggests that although the energy of the echoes is integrated with the first arrival, our processor is cognizant of what is really going on. It processes the first arrival for information, then perhaps casts the ELAs into a different slot to be processed in a different way. It does relate the ELAs to the first arrival when it acknowledges sound coloration through the comb filter effect. (This may say something about the hard rear wall of a studio control room "disappearing" because of the "Haas effect.") It may or may not process all or part of the ELA's speech information and place it with the first arrival. Considerable parts of the ELA energy seems to process in another way, such as providing the sensation of a

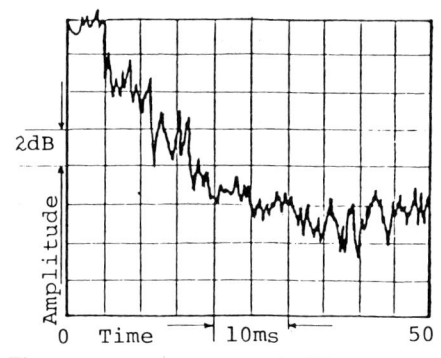

Fig. 5. Time–energy response at typical listener's position (Fig. 3) to a 5-ms pulse of pink noise over the multisource system.

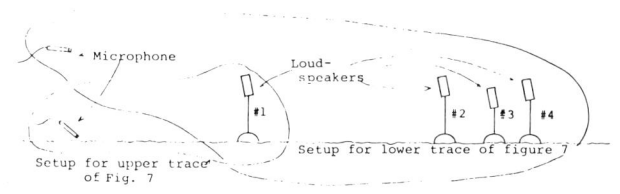

Fig. 6. Setup for simulation of early field.

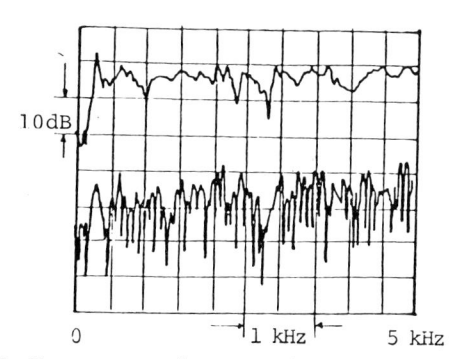

Fig. 7. Upper trace—direct out-of-doors response of loudspeaker with microphone close to ground; lower trace—microphone 5.75 ft (1.75 m) above ground with three additional loudspeakers energized (see Fig. 6).

quasi-reverberant room.

Our experience with the Legislative Assembly suggests that the ELAs carry little speech information to the listener. The more likely case is that most of the ELA energy contributes to the reverberant field. This is clearly heard on the tape and is subjectively apparent by personal observation in the real room.

The effect of ELAs on human perception was looked at another way. The circuit of Fig. 8 was employed. The potentiometer P_1 was calibrated in ratio of signal from the two digital delay lines DDL_1 and DDL_2 as well as in the depth of the TMAs caused by a difference in the arrival time of signals from DDL_1 and DDL_2. Recorded connnected speech was fed to the two digital delay lines. With equal delay in each DDL, movement of the calibrated potentiometer P_1 caused only a slight level change.

When DDL_2 was delayed a few milliseconds more than DDL_1, a movement of P_1 away from either terminal caused a coloration of tone quality and added a roughness to the quality of speech. The roughness is thought to be a modulation of the speech by the comb filter effect. The roughness was perceived differently for different delay times and relative levels of DDL_2 referenced to DDL_1. Maximum roughness did not appear at the center position of P_1 but rather when it was positioned somewhat more toward DDL_2.

Many observations of this phenomenon were made and the averages plotted in Fig. 9. The graph displays the perception of roughness as a parameter with the relative levels and delay times of the primary and echo sources as variables. The left-hand margin indicates the level of the echo channel referenced to the primary channel. At zero decibels they are equal in level. The right-hand margin indicates the depth of the TMAs referenced to the relative levels of the primary and echo signals. When they are equal in level, the anomalies are at a maximum amplitude. When the primary and echo signals are 20 dB different in level, the TMA becomes a mere ripple with a maximum and a minimum of only 1.74 dB.

Maximally perceived roughness throughout most of the delay range falls a few decibels above the equal loudness, or maximum anomaly 0.0-dB level line. This suggests that when the level or the echo signal is equal to the primary signal, the Haas or precedence effect inhibits the ELA by the few decibels shown by the offset maximum perception line of Fig. 9. Raising the late arrival by this much may cancel out the inhibition.

It is more interesting to note a wider deviation between the barely perceived roughness as a function of the primary to echo signal level. When the primary sound is louder than the echo, the roughness is just perceived when the echo level is brought up to within 10 dB of the primary sound. Here the TMAs are about 5.69 dB in amplitude. When the echo level is predominating, the primary sound need only be brought up to within 20 dB of the echo level for the roughness to be perceived. Here the TMA amplitude is only 1.74 dB. This appears as an overall shift in the threshold level for a perception of roughness of 10 dB when the Haas effect is negated by raising the echo level well above the primary sound, compared to when the primary sound level is predominant. This suggests that the ELA, when lower in level than the primary sound, may be inhibited by the Haas effect. Indirectly this lowers the roughness as perceived by our processor because of the lowered TMA caused by the inhibition of the ELA.

These observations are fragile. The degree of the perception of the roughness of speech is only presumed to be caused by the modulation of speech by the comb filter effect or the TMAs — changes from moment to moment, depending on the immediate previous listening history of the subject. A sudden onset of anomalies is perceived to a greater degree than the same level of disturbance after a few moments of continuous exposure. When after continuous listening to maximally perceived roughness DDL_1 and DDL_2 are suddenly brought into the same time frame, removing the anomalies, the perceived roughness increases. Then after fractions of a second, the roughness dies away and clear speech tones are heard. The roughness is perceived throughout any portion of the spectrum but appears to vary directly with the received spectrum bandwidth.

These data are tentative. The fragility of the subjective listening response to the phenomenon in question makes it impossible to derive any absolute numbers on the perception of roughness as related to an inhibition of the ELA by the Haas effect. The data do appear to show a shift, or offset, in perception, which may be due to the Haas effect.

Although reduced in apparent level, the roughness is perceived through a two-channel loudspeaker system in much the same way as in single-channel headphone listening.

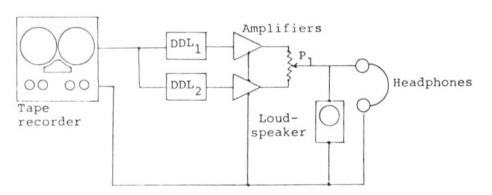

Fig. 8. Setup for perception of roughness when listening to the first and early late arrivals of speech.

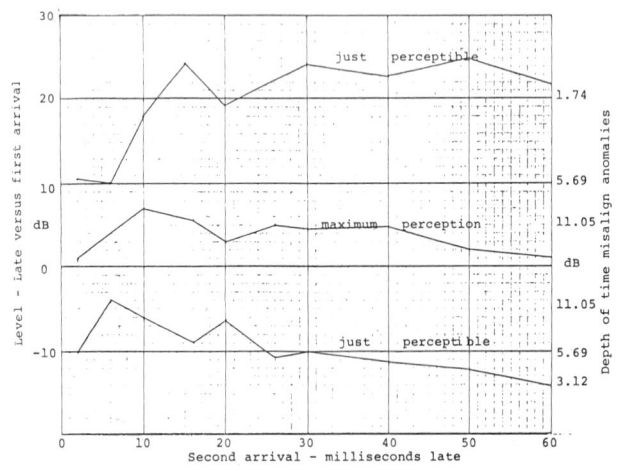

Fig. 9. Perception of roughness in listening to speech with one early late arrival.

3 CONCLUSION

Our experience with the ELAs at the Legislative Assembly, supported by preliminary research, has convinced us that the ELAs within the fusion period under some circumstances do not contribute to the transfer of information from talker to listener. On the contrary, they may even provide destructive interference by contributing energy to a subjectively perceived quasi-reverberant field.

A search of the literature discloses little direct evidence to disturb this premise. Some investigators have reasoned that usually more level provides a greater transfer of information from talker to listener. They assume that because the integration of the energy of the ELAs provides more level, there is a greater transfer of information. Or as quoted earlier from Haas, the delay of the ELAs has been extended until auditors were disturbed by the ELA. It was then presumed that all earlier ELAs contributed to the transfer of information from talker to listener.

Continued research may provide more insight into the way our processor accepts and operates on the early sound arrivals. A better understanding of these processes may remove some of the black magic from the design of listening rooms such as auditoriums, lecture rooms, and recording control rooms.

4 ACKNOWLEDGMENTS

The authors wish to acknowledge the extended loan of equipment without which the current investigation and future research on this subject would not be possible. Their thanks go to John Hilliard and Bill Putnam.

5 REFERENCES

[1] H. Haas, "The Influence of a Single Echo on the Audibility of Speech," *J. Audio Eng. Soc.*, vol. 20, pp. 146–159 (1972 Mar.). [Original publication in German in 1949.]

[2] J. P. A. Lochner and J. F. Burger, "The Influence of Reflections on Auditorium Acoustics," *J. Sound Vib.*, vol. 1, pp. 426–454 (1964).

[3] M. Wallach, E. G. Newman, and M. R. Rosenzweig, "The Precedence Effect in Sound Localization," *Am. J. Psychol.*, vol. 6, pp. 315–336 (1949).

[4] A. K. Nabelek and L. N. Robinette, "Influence of the Precedence Effect on Word Identification by Normally Hearing and Hearing-Impaired Subjects," *J. Acoust. Soc. Am.*, vol. 63, pp. 187–194 (1978).

[5] A. K. Nabelek and L. N. Robinette, "Reverberation as a Parameter in Clinical Testing," *Audiology*, vol. 17, pp. 239–259 (1978).

[6] T. Somerville, C. L. S. Gilford, N. F. Spring, and R. D. M. Negus, "Recent Work of the Effects of Reflectors in Concert Halls and Music Studios," *J. Sound Vib.*, vol. 3, pp. 127–134 (1966).

THE AUTHORS

C. R. Cable

R. C. Enerson

Cecil R. Cable was born in the Province of Saskatchewan, Canada, in 1917, and received his formal education there.

Since 1939 Mr. Cable has been involved with design and application of sound reinforcement systems. During World War II he served as instructor in electronics to Air Force classes. Soon after, he was engaged in field research with Gulf Research and Development in the application of the airborne magnetometer in mineral prospecting. For several years he was principal examiner for tradesmen's qualifications for the Province of Alberta. During 1977–78 he traveled with Synergetic Audio Concepts, demonstrating his work with time-delay spectrometry.

Mr. Cable now operates a private consulting practice in electroacoustics. He is currently associated with Hilliard & Bricken, Acoustical Engineers, and continues to do independent research.

•

R. C. Enerson was born in the province of Saskatchewan, Canada, in 1944. He later moved to Alberta.

He has a B.Sc. degree in electrical engineering and an M.Sc. in computing science, both from the University of Alberta. He is a licensed Professional Engineer.

In 1973 he joined Alberta Government Telephones, the provincially-owned telephone operating company which serves most of the province. His personal involvement with sound system engineering developed during his undergraduate years and led to his involvement in designing systems for AGT. Until his recent transfer into a planning group, Mr. Enerson specialized in complex sound system design for AGT's Special Products group. Presently he is involved in VF equipment standardization.

The Practical Application of Time-Delay Spectrometry in the Field*

CECIL R. CABLE AND JOHN K. HILLIARD

Hilliard and Bricken, Santa Ana, CA 92701, USA

Time-delay spectrometry moves the laboratory into the world of real rooms and situations. Subtleties of sound behavior not written into equations or included in laboratory simulations are automatically included in measurement results. These results, which display the way it is, frequently depart significantly from the accepted norm. Energy distributions through the frequency and time domains are readily quantified in situ. Time-delay spectrometry techniques and applications, as practiced in the field, are discussed by the authors.

0 INTRODUCTION

As the name implies, time-delay spectrometry (TDS) [1] is a process of spectral analysis delayed in time. This procedure can be used to determine many acoustical parameters, parameters difficult or impossible to determine otherwise.

A spectrum analyzer is a tunable receiver which in this case covers the audio range. Associated with the tunable audio receiver is a variable audio generator, which is usually arranged to accurately track along with the receiver as the frequency is varied.

For normal spectral analysis (energy versus frequency) the output of the generator is fed to the input of a device to be tested. The output of the device is then coupled to the input of the variable receiver. The output of the receiver is displayed on the vertical axis of an oscilloscope (commonly built into the spectrum analyzer which may or may not have storage capability for the retention of images).

The horizontal axis of the oscilloscope is coupled to the tuning control of the receiver and tracking generator. Thus a graph or transfer function showing the frequency response of the device is displayed on the oscilloscope screen as the frequency of the spectrum analyzer is varied. It is assumed that the time delay through the tested device is short in respect to the rate at which the spectrum analyzer tuning is varied, that is, if the receiver has a bandwidth of 10 Hz, it must not shift more than 5 Hz in frequency during the time it takes the signal to pass through the device from the generator to the receiver. If it does, the signal transmitted by the generator cannot be accepted by the receiver, which has now moved to a new frequency.

Most passive and active electronic devices have short transfer times in respect to the rate of change of the frequency scan. When electroacoustic systems are to be tested, the sound receptor (microphone connected to the input of the spectrum analyzer) may be at some distance from the sound transmitter (loudspeaker driven by the tracking generator). A rapid frequency scan will cause the signal, delayed in time by its passage through air, to arrive at

---

* Presented at the 60th Convention of the Audio Engineering Society, Los Angeles, 1978 May 2–5; revised 1979 September 12.

the receiver "off tune." Therefore proper response cannot be displayed on the spectrum analyzer oscilloscope.

It is this very phenomenon, the mistracking of the generator and receiver through excessive signal delay in respect to frequency scan time through the test specimen, which forms the basis for TDS.

1 PROCEDURE

The tracking generator[1] may be coupled to an audio amplifier which in turn drives a loudspeaker. If a microphone, connected to the input of the spectrum analyzer, is placed directly in front of and very close to the loudspeaker, the time delay will be short, so the receiver will still pass the signal from the tracking generator, and the graph will now display the transfer function, or frequency response, of the amplifier and loudspeaker.

The receiver may be adjusted so that it will only tune in signals which are ± 5 Hz of the frequency to which it is tuned. The audio generator must now track the receiver very accurately, or it will get "off tune" and the receiver will not respond to it. The time-delay spectrometer has this inherent accuracy built into it. However, as soon as the signal from the tracking generator is reproduced by a loudspeaker with a microphone placed any finite distance away from it, a problem is introduced. While the electrical signals travel through the system at near 186 000 mi/s (300 000 km/s), sound waves travel through air slowly, at approximately 1130 ft/s (344 m/s). If the microphone were placed 10 ft (3 m) away from the loudspeaker, it would take the sound 10/1130 seconds or 8.85 ms to travel from the loudspeaker to the microphone. When the frequency scan of the analyzer is set to scan the audio range from 0 to 10 000 Hz in 1s, or 10 000 Hz/s, the receiver has moved 88.5 Hz in 8.85 ms. While it was *exactly* in tune with the tracking generator when the sound left the loudspeaker, it was 88.5 Hz out of tune with the received signal by the time it reached the microphone, resulting in the receiver not responding to the signal.

There is a solution to the problem. The tracking generator may be readjusted so that it is tuned to any given frequency 8.85 ms before the receiver is tuned to it as the frequency is swept. The sound now does not reach the receiver 88.5 Hz "off tune" but is exactly "in tune," because the tracking generator had a head start. This exactly "makes up" for the length of time it took the sound to travel 10 ft (3 m) or 8.85 ms from the loudspeaker to the microphone.

Now if the microphone is moved closer or farther away from the loudspeaker, the sound will again be "off tune," and the receiver will not respond to it. This would imply, and it is indeed true, that the microphone may be placed at any reasonable distance from the loudspeaker. The audio signal may then be kept "on tune" by giving the tracking generator exactly the right amount of head start to make up for the length of time it takes for the sound to travel from the loudspeaker to the microphone. Once this adjustment is made, any other sound paths longer or shorter than the one specifically adjusted for will be "off tune" and rejected by the receiver.

This capability opens the way for a completely new domain of study in the behavior of sound not heretofore available. Already, old principles have been clarified and new principles have evolved in the field of acoustics [2]–[5]. Every large room offers the realization of selective anechoic measurements.

2 PRACTICE

1) For the first time ever it is possible to make anechoic response measurements of loudspeakers in situ, while they are loaded by a real reverberant room.

2) Interference anomalies of multisource radiators such as monitor and high-fidelity loudspeakers and loudspeaker clusters in commercial installations may be observed, in situ, free from interference from the reverberant field. Loudspeaker design may be facilitated by examining time misalign anomalies caused by time-displaced primary and secondary sources.

3) Low-order reflections may be examined while the direct and later reverberant field is rejected.

4) Direct versus direct plus reverberant sound may be observed at any position in any room.

5) Loudspeaker directivity is easily observed in situ by comparing the direct and reverberant fields at various locations within a room.

6) The early sound field, the direct sound plus the early reflections of auditoriums and other rooms, may be qualified and quantified.

7) Direct sound distribution from sound reproduction or reinforcement loudspeakers may be checked for uniformity. Hot spots and voids may be examined without interference from reverberant field masking.

8) Room boundaries, or any boundary, may be analyzed in situ, providing absorptive and reflective coefficients.

9) Sound transmission coefficients of walls, windows, roofs, etc., may be defined on a per unit basis, yielding information difficult to quantify by other means.

---

[1] Modern spectrum analyzers used for TDS are basically superheterodyne receivers which convert the audio spectra to a relatively high frequency, pass the signal through variable-bandwidth fixed-frequency filters, and then down-convert to the audio spectrum. The frequency conversions are conventional where a local variable high-frequency oscillator beats against the incoming signal, which is then passed through a modulator to produce a fixed intermediate frequency (IF) signal to be passed through the variable-bandwidth filters.

The local high-frequency oscillator, which controls the frequency to which the receiver is tuned, is also down-converted, beats against a second local oscillator, and is passed through a modulator to produce the audio frequency output of the tracking generator. Since the receiver IF signal and the tracking generator are derived from the same variable high-frequency oscillator, the tracking generator must accurately track the receiver as the tuning is varied. Further, if the second local oscillator which generates the tracking generator signal is of the same frequency as the receiver IF, this generator may be mixed with the IF signal to produce a zero frequency beat note as the spectrum is swept. However, if there are time perturbations between the tracking generator output and the receiver input, the receiver IF frequency will vary directly with the perturbations. The output of the mixed IF frequency and the local oscillator will vary from zero. When these frequency deviations are identified by an integrating real-time frequency analyzer (400-line FFT), time perturbations of the medium (including the transducers) will be read out directly on the real-time analyzer.

3 DISCUSSION

3.1 Anechoic Measurements

It has long been debated that the response of a loudspeaker is altered when it is subjected to ordinary room loading, as compared with the frequency response curves taken in an anechoic chamber. This should not be confused with 4π, 2π, 1π, and $\pi/2$ loading, which occurs when a loudspeaker is moved from the center of a room to various positions near the room boundaries. Fig. 1 shows the frequency response of a small single-cone loudspeaker taken near the center of a 5000-ft^3 (140-m^3) room and a second trace taken under identical conditions out of doors. The two traces appear nearly coincident.

To observe loudspeaker anechoic response, place the loudspeaker away from nearby reflective surfaces. Place the measurement microphone connected to the time-delay spectrometer at a designated distance from the loudspeaker. Drive the loudspeaker with a TDS tracking generator through the frequency span of interest. "Tune" the time-delay spectrometer to the distance between loudspeaker and microphone. Observe the loudspeaker frequency response on the spectrometer storage oscilloscope. Discrimination against nearby reflective surfaces will be inversely proportional to the bandwidth of the receiver and directly proportional to the cycles per second of the frequency scan. Conversely, the resolution (display of closely spaced anomalies in the loudspeaker response) is directly proportional to the bandwidth and inversely proportional to the cycles per second of the frequency scan (see Fig. 1). The window of observation expressed in terms of time or space W_s is

$$\text{space window } W_s = \frac{\text{filter bandwidth} \times \text{speed of sound}}{\text{sweep rate}}.$$

For example,

space window (or bandwidth)

$$W_s = \frac{10 \text{ Hz} \times 1130 \text{ ft } (344 \text{ m})}{5000 \text{ Hz/s}} = 2.26 \text{ ft } (0.68 \text{ m}).$$

$$\text{time window } W_t = \frac{2.26}{1130} = 0.002 \text{ second, or 2 ms}.$$

The frequency scale is linear from 0 to 5000 Hz, showing a valid response curve from 500 to 5000 Hz. The vertical scale is 2 dB per major division. As can be seen, the maximum variation between the two curves, which is hardly visible, is ± 0.15 dB. In this case the room loading did not significantly alter the mid- and high-frequency response of the loudspeaker.

3.2 Interference Anomalies

As is well known, multisource reproduction of a spectrum may cause frequency response anomalies in the perceived sound. Few sounds we hear arrive at our ears as single sources. Out of doors we have ground reflections as a secondary source mixing with the prime source to cause interference anomalies. Indoors we have many early reflections mixing with the prime source and with each other to cause a ragged frequency response in the perceived early field. Fig. 2 displays the direct sound frequency response of a small loudspeaker out of doors as well as the direct sound plus an early reflection.

The time-delay spectrometer setup was similar to that for an anechoic response of a loudspeaker. The "free air" response curve was taken with the loudspeaker out of doors away from the ground surface. The "near flat surface" response curve was taken with a reflective surface placed under and forward of the loudspeaker. In both cases the microphone was 3 ft (0.9 m) forward and 30° above on axis of the loudspeaker. The time misalign anomaly caused by the early reflection is clearly visible. The anomaly peaks are about 1 kHz apart. This signifies that the reflection arrives $1130/1000 = 1.13$ ft (0.34 m) or 1 ms after the direct sound. The anomalies in the "free air" trace at about 2750 and 3600 Hz could suggest an early reflection from the rear of the cabinet 0.66 ft (0.2 m) away, passing through the cone to mix with the direct sound, causing the frequency response anomalies shown. To observe interference patterns of early sound, we "tune" to a mean between the distance of direct and/or early reflections with the window broad enough to accept signals of interest. Then we observe

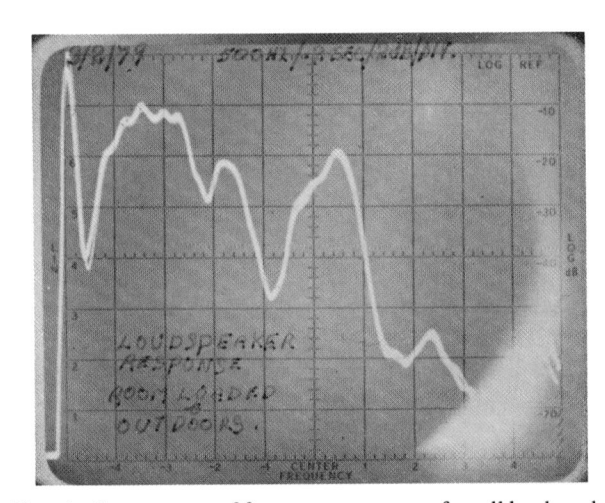

Fig. 1. Spectrogram of frequency response of small loudspeaker measured in center of 5000-ft^3 (140-m^3) room and again in free air out of doors. Frequency scale—linear, 500 Hz/div; vertical scale—2 dB/div. Two traces do not deviate more than 0.15 dB in respect to each other with and without room loading.

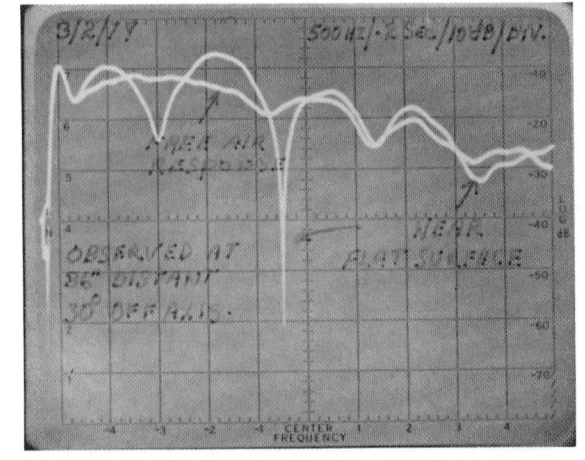

Fig. 2. Spectrogram of frequency response of small loudspeaker, at 3-ft (0.9-m) distance 30° off axis. Trace 1—free air; trace 2—near reflective surface. Frequency scale—linear, 500 Hz/div; vertical scale—10 dB/div.

an interference pattern on the time-delay spectrometer oscilloscope.

3.3 Low-Order Reflections

It has been shown that the window related to time and space is a selectable parameter. Its position in time and space is also selectable. The window may be kept closed until after the direct sound has passed the microphone, then opened only long enough to admit those early reflections of interest. This is the technique used to examine the reflectance (1 − absorption) of room boundaries and walls (see Figs. 4–12).

TDS has been used to determine the echo energy return from a bluff 1500 ft (457 m) away. A city park was planned next to the bluff. Neighbors feared that the reflections of the recreational activities would be annoying. The time-delay spectrometer measured the maximum reflected energy and found the reflected energy to be 30 dB below the source, which was located in the center of the activity area. It is interesting to note that such out-of-doors measurements are best made on cloudy, cool, and windless days. Local thermals developed on hot breezy days cause rapid random changes in the air temperature. This introduces changes in the velocity of sound in air (considered to be constant in the TDS equations) and severe refractive and time-arrival perturbations, displayed these effects in a manner not previously observed by the authors. This effect has not been seen as a problem within rooms of less than a few hundred feet in linear dimension.

3.4 Direct versus Reverberant Sound

A narrow window may be opened to coincide with the passage of the direct sound, rejecting all later sound. Then by slowing the frequency sweep rate and opening the filter bandwidth, the open window time may be expanded to equal or exceed the reverberation time of the room. The solid lines of Fig. 3 are the direct sound. The broad fuzzy lines showing the room modes are the total sound field, including the direct field. The direct sound trace window was 4.5 ft (1.37 m) or 4.0 ms wide. The total sound trace window was 1356 ft (413 m) or 1200 ms wide. The reverberation time of the 5000-ft³ (140-m³) room was 0.7 s.

3.5 Source Directivity

The top traces of Fig. 3 were taken with the microphone 6 ft (1.8 m) on axis from an 8-in (200-mm) cone-type loudspeaker mounted in a small cabinet. The lower traces were taken under identical conditions, except that the loudspeaker was turned 45° off axis in respect to the microphone. The solid line shows the direct sound only. The broad line is composed of the total sound and displays many room modes. In general the rms average power of the total sound is just below the room mode peak envelope. When the total sound trace is 3 dB above the direct sound, the direct and total sounds are equal in level. When the direct sound solid line rides near the top of the total field envelope as at the higher frequencies of the upper set of traces, the direct sound is predominant. When the direct sound line is more than 3 dB below the upper edge of the total sound trace, the reverberant field becomes dominant. The ratio of direct to reverberant sound at a given distance within a given room is an indication of the directivity factor Q of a source. This is clearly visible in Fig. 3.

The on-axis response (upper set) indicates a Q rising with frequency. At the high end of the spectrum the direct sound is predominant. Near 1 kHz the lower on-axis Q of the loudspeaker contributes less direct sound to the on-axis microphone position. At the 45° off-axis position, the ratio of direct to reverberant sound is low at all frequencies. This indicates a low loudspeaker Q 45° off axis. Further, at this angle the Q varies greatly with frequency, as indicated by wide amplitude variations in the direct sound as the frequency is varied.

3.6 Early Sound Field

The preceding can be employed to fully characterize the sound fields within a large or small room or auditorium. The space and time domains can be explored in respect to frequency. Arrival times of early reflections from any point to any point in the room may be determined. The spectra of the early arrivals may be compared to the direct sound spectrum. This delineates the absorption, diffusion, or reflectance of the specific surfaces involved. The effects of diffusion or focusing can be observed. Time interference anomalies caused by early late arrivals may be read directly from the time-delay spectrometer display. Not only are the anomalies immediately identifiable, the frequency spacing between anomalies defines the difference in the travel distance of the signals involved. The difference distance equals c/f_s, where c is the speed of sound and f_s is anomaly spacing in frequency. By slowly tuning a relatively narrow window out into space, the entire field may be examined throughout time and space. When the time-delay spectrometer has appropriate oscilloscope storage available, actual waterfall (three-dimensional) displays may be constructed, where the X axis displays frequency, the Y axis

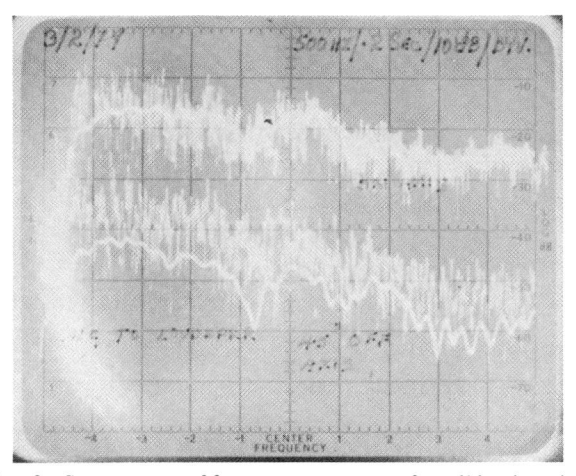

Fig. 3. Spectrogram of frequency response of small loudspeaker in center of 5000-ft³ (140-m³) room. Solid line, top set—on-axis direct sound; fuzzy line—total loudspeaker response including room reflections; lower set—as top set, but 45° off loudspeaker axis. Frequency scale—500 Hz/div; vertical scale—10 dB/div.

displays amplitude, and the Z axis displays time and space. Any unusual bumps or voids in the response usually indicate an aberration in the room response. It will be identified in the frequency and space domains and can be isolated for further study.

This technique readily discloses inappropriate boundary construction of rooms and auditoriums. Early late arrivals (ELA) [5], reaching the listener after the direct sound, have a measurable effect on the audibility of speech and a more obscure effect on the quality of music as perceived by an observer. Time-honored architectural and acoustical procedures rely upon boundary construction to control the reflection and diffusion of sound to enhance the observer's response to the directly received sound along with the ELAs. Unfortunately, until the advent of TDS and energy time curves (ETC) there was no simple way for the designer, other than through subjective listening response, to evaluate the effectiveness of his work.

Through observance and TDS measurements taken in many real rooms and auditoriums, the authors have found large discrepancies between the avowed design criteria and real relationships between the directly heard sound and the ELAs. These discrepancies appear to have arisen, in part, from the difference in the textbook versions defining reflective and diffusing surfaces and the in situ measurement of such surfaces.

The three-dimensional displays mentioned above are in effect composite TDS and energy time curves (ETC). The individual TDS curve of the waterfall gives a full energy versus frequency display of the signal at that point in time, which the TDS curve represents. A three-dimensional family of TDS curves not only displays variances in energy with frequency, but also displays energy variances with time throughout the entire spectrum of interest. ETCs, as practiced with time-delay spectrometers interfaced with a fast Fourier transform (FFT) instrument, integrate each frequency spectrum's energy displaced in time by a perturbation in the transmission path and display signal energy on a single time scale. Frequency information is lost.

The displays provide information in a way that is similar to the pulse method of room analysis as practiced by the authors for over 30 years. The significant difference is that pulse measurements were usually made with single frequencies displayed in the time domain, while ETCs integrate all frequencies of interest and display total spectrum energy in the time domain. Once anomalies in the system (source, room, and receiver) are identified in time, the time-delay spectrometer must be tuned to that time to display the energy versus frequency distribution of energy. As the time-delay spectrometer provides instant readout of energy versus frequency, ETCs provide instant readout of energy versus time. TDS waterfalls combine all three parameters into one display. Hence there is considerable intrinsic value in using a time-delay spectrometer with high storage capability, which is required for waterfall displays.

3.7 Sound Distribution

Regardless of the various objectives of sound reinforcement and/or sound reproduction systems, there are usually one or more listeners to whom the system is dedicated. This can range from a single listener in a recording studio control room to thousands of listeners in auditoriums, arenas, and stadiums. The task of the system designer is to lay down a sound field to each listener to meet the design objective. In large rooms where people gather to hear, this often calls for a uniform distribution of sound throughout the frequency spectrum over a broad area. The distribution of the total sound throughout the space is easily determined. However, it is usually the reverberant sound which is predominant, as in the lower set of traces in Fig. 3. So it is really the reverberant field which is being checked, not the direct distribution of the sound radiators. Frequently the distribution is checked at some high frequency, perhaps 4 kHz, where the direct field may be predominant, as in the upper set of traces in Fig. 3. Section 5 indicates that the distribution of any sound radiator may vary widely with frequency. The readings may then reflect the distribution of the sound radiators at 4 kHz, but says little about the distribution at lower frequencies.

The distribution as measured with a narrow TDS window displays the energy directly from the radiators on a spectrum-wide basis for any position in the house, without interference from the reverberant field. This is represented by the solid line traces in Fig. 3.

3.8 Boundary Absorption

The absorption and reflectance of room boundaries or any boundary or material is easily measured with a time-delay spectrometer. A sound source driven by the TDS tracking generator output is placed several feet away and facing toward the specimen or boundary to be analyzed. A microphone connected to the TDS receiver input is placed between the source and the subject to be tested. The time-delay spectrometer is then tuned to the total distance from source to subject to microphone. The direct sound from the source to the microphone will be off tune, as will all other unwanted sound paths, and thus it will be rejected by the time-delay spectrometer. Discrimination against the direct source to microphone response can be further enhanced by the use of a directional microphone oriented to reject the direct source sound.

In common with most acoustical test facilities, this TDS method of measuring subject reflectance/absorption has a low-frequency cutoff related to the size of the specimen and the proximity of other reflective surfaces in relation to wavelength. Generally a minimum sample dimension should not be less than 1 wavelength, and the minimum distance from nearby reflective surfaces should not be less than 2 wavelengths of the lowest test frequency of interest.

Fig. 4 graphically illustrates on a 10-dB/div scale the absorption of homespun drapery material mounted in front of a plywood reflecting panel. The upper trace of each set shows the reflectance of the plywood. The lower trace of the upper set shows the absorption when the drapery material is close to the plywood. The absorption is only effective above 2000 Hz. The lower trace of the middle set shows the absorption with the curtain hanging loosely in pleats so that double the material is exposed. The absorption is effective at lower frequencies. The lower trace of the lower set is similar to the upper set except that the spacing between drape and plywood is 3 in (76 mm). In this case the average absorption is greater than the center or upper set at frequencies above 500 Hz and is much more effective above 2000 Hz.

Fig. 5 shows on a 2-dB/div scale the reflections of lightweight carpet in contact with the plywood board. Only a small amount of low-frequency attenuation is obtained.

Fig. 6 shows on a 2-dB/div scale reflections of a heavy carpet with underlay in contact with the plywood board. The increased overall thickness and spacing provide an increased low-frequency attenuation.

Fig. 7 shows the reflectance of 2-in (50.8-mm) thick fiberglass, having a density of 0.8 lb/ft³ (12.8 kg/m³), spaced 6 in (152 mm) in front of the plywood panel. The absorption starts at about 500 Hz and averages 10 dB in the upper frequencies.

Fig. 8 shows the reflectance with 6-in (152-mm) fiberglass spaced at 4 in (101 mm) and 1 in (25 mm). Absorption is effective beginning at 500 Hz and averages 15 dB above 1000 Hz.

Fig. 9 shows the reflectance of ¾-in (19-mm) thick fiberglass board having a density of 6 lb/ft³ (96.1 kg/m³), in contact with the hard board. The average attenuation above 250 Hz is a minimum of 8 dB.

Fig. 10 shows the fiberglass board spaced 4 in (101 mm) from the hard plywood board. This measurement shows the increased attenuation due to air space between the hard surface and the absorbing material. As much as 20 dB of attenuation is obtained at 800–1000 Hz, indicating an absorption coefficient of 0.99.

Examples of the effect of absorbing materials at various spacings from a hard surface are shown in Figs. 11 and 12. The upper curve in Fig. 11 is the reflectance from a 4 ft (1.2 m) by 4 ft (1.2 m) by ⅜ in (9.5 mm) plywood board. A 4 ft (1.2 m) by 4 ft (1.2 m) by 2 in (50.8 mm) fiberglass blanket, having a density of 0.8 lb/ft³ (12.8 kg/m³), was placed in contact with the hard surface. The reflected energy through the fiberglass was attenuated only above 1000 Hz.

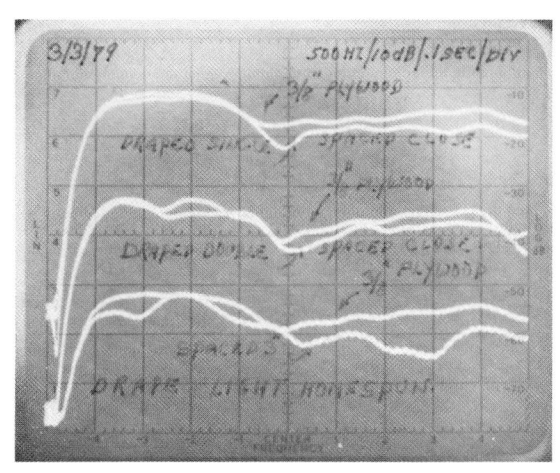

Fig. 4. Top set, top trace—spectrogram of frequency response of first reflection sheet of ⅜-in (9.5-mm) plywood; top set, bottom trace—drapery material against and in front of plywood; middle set, top trace—reflection from plywood; bottom trace—drapery material pleated to ½ width; lower set, top trace—reflection from plywood; bottom trace—drapery material spaced from plywood 3 in (76 mm). Frequency scale—500 Hz/div; vertical scale—10 dB/div.

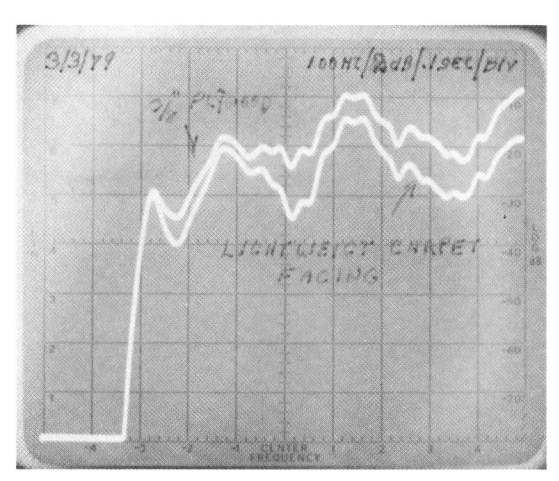

Fig. 5. Top trace—spectrogram of frequency response of first reflection off ⅜-in (9.5-mm) plywood; lower trace—reflection with light carpet in contact with plywood face. Frequency scale—100 Hz/div; vertical scale—2 dB/div.

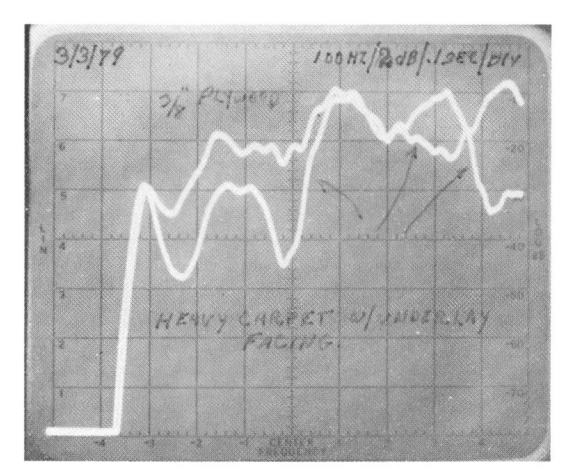

Fig. 6. As Fig. 5, but heavy carpet with underlay in contact with plywood.

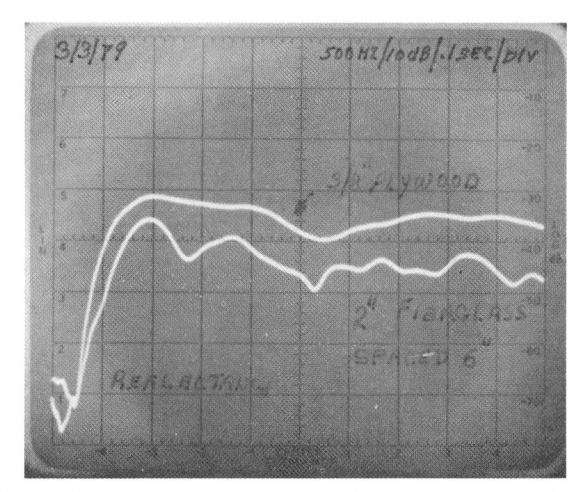

Fig. 7. Top trace—spectrogram of frequency response of first reflection off ⅜-in (9.5-mm) plywood. Lower trace—as top trace with 2-in (50-mm) 0.8-lb/ft³ (12.8-kg/m³) fiberglass placed 6 in (152 mm) in front of plywood. Frequency scale—500 Hz/div; vertical scale—10 dB/div.

Fig. 12 shows the reflectance when the fiberglass, having a density of 0.8 lb/ft^3 (12.8 kg/m^3), was spaced 4 in (101 mm) from the hard surface. The reflected energy was attenuated by more than 10 dB above 500 Hz.

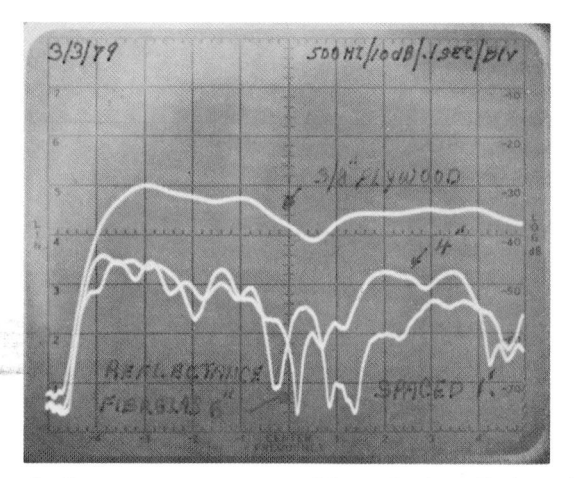

Fig. 8. Top trace—spectrogram of first reflection off plywood; lower traces—6 in (152 mm) fiberglass spaced 4 in (101 mm) and 1 in (25 mm) from plywood. Frequency scale—500 Hz/div; vertical scale—10 dB/div.

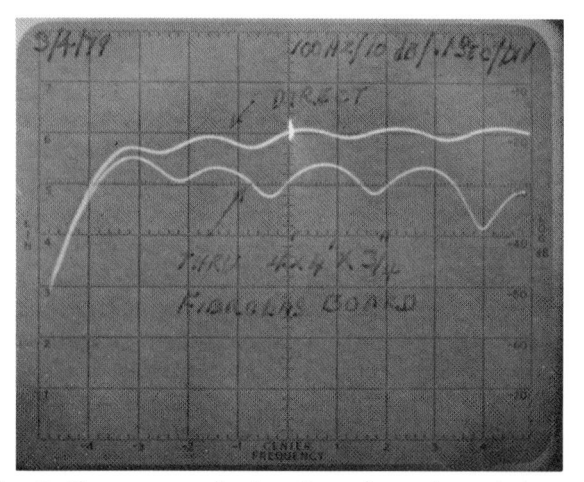

Fig. 9. Top trace—reflection direct from plywood; bottom trace—reflection through ¾-in (19-mm) fiberglass board. Density 6 lb/ft^3 (96.1 kg/m^3), placed against plywood. Frequency scale—100 Hz/div; vertical scale—10 dB/div.

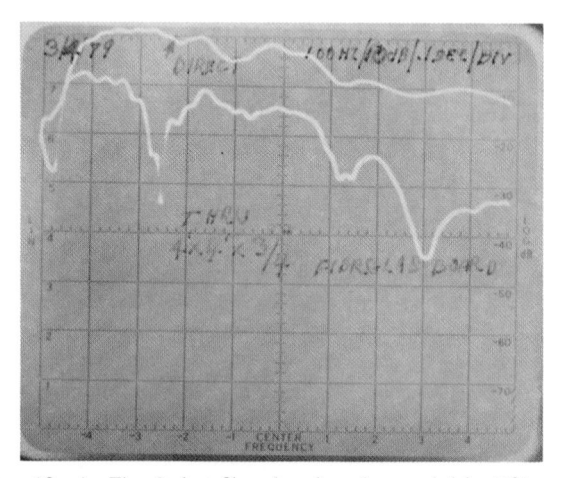

Fig. 10. As Fig. 9, but fiberglass board spaced 4 in (101 mm) from plywood.

These measurements of reflectivity by the TDS method were completed in an open parking space with buildings as close as 20 ft (6 m) away. Residential traffic and background noise were present. Sound pressure levels from the loudspeaker system at the measurement point were approximately 75–80 dB. Discrimination against interfering broadband noise can be as much as

$$10 \log \frac{\text{analyzer bandwidth}}{\text{noise bandwidth}}.$$

For example,

$$10 \log \frac{10}{10\,000} = -30 \text{ dB}.$$

Background noise down 20 dB from the measurement signal will cause less than 0.05 dB error in the measurement. These reflectance measurements were taken with the sound propagation normal to the face of the test specimens. The absorption/reflectance of sound propagated at any angle of incidence to the subject is easily accomplished by proper positioning of the source, subject, and microphone.

3.9 Transmission Loss

The use of TDS in measuring the transmission loss of a partition has experimentally verified its value in detecting flanking paths or leaks in a common partition or "party wall." Sound from one side of a partition has many path lengths. Some energy is propagated through the wall, while other flanking paths at the sides, ceiling, or floor combine with that coming through the wall to constitute an overall component of energy which has a multitude of phase and amplitude relationships. Under these conditions the apparent source from the generation side is not at the wall but several feet away into the room as shown on the time-delay spectrometer oscilloscope. TDS allows a more quantitative method of locating these flanking paths so that proper correction can be made.

In situ noise reduction through closed windows and doors is easily measured. The response of the sound source loud-

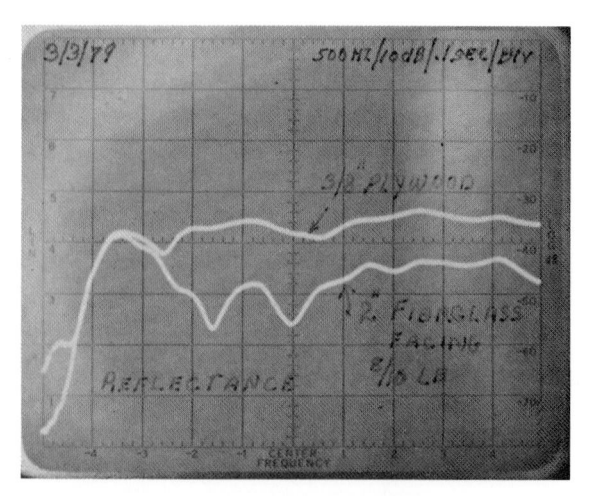

Fig. 11. Top trace—spectrogram of frequency response of first reflection off ⅜-in (9.5-mm) plywood; bottom trace—2-in (50-mm) fiberglass blanket in contact with plywood. Frequency scale—500 Hz/div; vertical scale—10 dB/div.

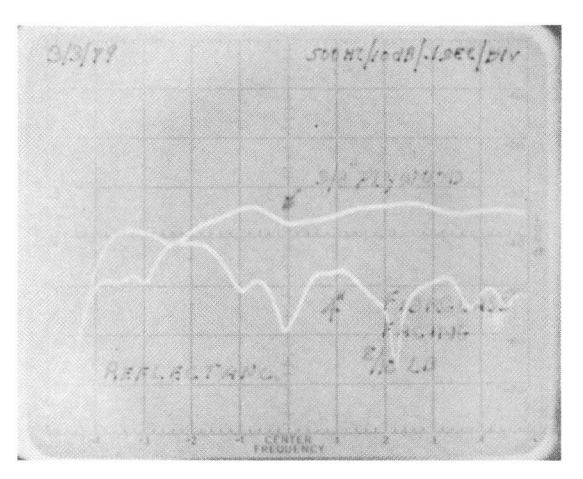

Fig. 12. As Fig. 11, but fiberglass blanket spaced 4 in (101 mm) from plywood.

speaker is characterized by taking a time-delay spectrometer on-axis measurement at a distance of, say, 10 ft (3 m), from the source. This trace is "stored" in the time-delay spectrometer. The source is then set up normal to and 5 ft (1.52 m) from the subject door or window. The microphone connected to the time-delay spectrometer is set up normal to and 5 ft (1.52 m) from the other side of the subject. The time-delay spectrometer bandwdith is set broad enough to pass all signal paths through and around the perimeter of the subject, but narrow enough to reject all other flanking paths. A second trace is run. The difference between the first and second traces is a spectrum-wide measurement of the noise reduction of the subject, including its mounting in the structure. It should be noted that this method is similar to a sound transmission loss measurement in free space and results will differ from a chamber method of measurement.

4 CONCLUSION

Time-delay spectrometry has placed us upon a new plateau in the measurement and our understanding of the behavior of sound within rooms and out of doors. Many phenomena such as feedback of regenerative sound systems within rooms have been solved in general terms by an expansion of theory and computer studies. In this one instance, however, it remained for the in situ studies in real rooms by time-delay spectrometry to *explicitly* define the feedback mechanism for these rooms. These mechanisms differed in detail and principle from previous studies.

It is reasonable to believe, as TDS becomes the tool of many investigators, that the essence of good auditorium design may be qualified.

TDS analysis of sound behavior relating to objects and barriers may increase our ability to control unwanted sound and noise intrusion.

5 REFERENCES

[1] R. C. Heyser, "Acoustical Measurements by Time-Delay Spectrometry," *J. Audio Eng. Soc.*, vol. 15, pp. 370–382 (1967 Oct.).

[2] R. C. Heyser, "Loudspeaker Phase Characteristics and Time-Delay Distortion, Pts. I and II," *J. Audio Eng. Soc.*, vol. 17, pp. 30–40 (1969 Jan.); pp. 130–137 (1969 Apr.).

[3] R. C. Heyser, "Determination of Loudspeaker Signal Arrival Times, Pts. I–III," *J. Audio Eng. Soc.*, vol. 19, pp. 734–743 (1971 Oct.); pp. 829–834 (1971 Nov.); pp. 902–905 (1971 Dec.).

[4] C. R. Cable, "Time-Delay Spectrometry Investigation of Regenerative Sound Systems," *J. Audio Eng. Soc.*, vol. 26, pp. 114–119 (1978 Mar.).

[5] C. R. Cable and R. C. Enerson, "Those Early Late Arrivals! Mr. Haas, What Would You Do?" *J. Audio Eng. Soc.*, vol. 28, pp. 40–45 (1980 Jan./Feb.).

THE AUTHORS

J. K. Hilliard

John K. Hilliard was born in Wyndmere, North Dakota, in 1901. He received a B.S. degree in physics at Hamlin University, St. Paul, Minnesota in 1925 and did graduate work in electrical engineering at the University of Minnesota. He received an honorary doctorate in 1951.

Dr. Hilliard spent 14 years at MGM in the development of recording and reproducing film and tape equipment and the design of microphones and loudspeakers for theaters. He also worked for many years on high-intensity environmental sound equipment. From 1932 to 1960, Dr. Hilliard was with Altec Lansing as vice president of the advanced engineering department, where he was responsible for transducers and communication equipment.

Dr. Hilliard is a fellow of the Acoustical Society of America, the Audio Engineering Society and the Society of Motion Picture and Television Engineers. He has received the John H. Potts Award of the AES, and is a member of the American Physical Society, the Armed Forces Committee on Hearing, Bioacoustics and Biomechanics, Eta Kappa' Nu, the Institute of Environmental Engineers, the Institute of Noise Control Engineers and the National Council of Noise Consultants. He was director of the LTV Western Research Center between 1960 and 1970. At the present time he is president of Hilliard and Bricken, Inc., acoustical and energy consultants.

●

Mr. Cable's biography was published in the January/February issue.

LETTERS TO THE EDITOR

COMMENTS ON "THE PRACTICAL APPLICATION OF TIME-DELAY SPECTROMETRY IN THE FIELD"

In the above interesting paper,[1] an application to the measurement of transmission loss of in-situ partitions is described. I would be grateful if the authors could comment upon the following observations.

When a loudspeaker is used to irradiate one wall of a room, the propagating wave fronts fall on the surface at varying angles of incidence, and with varying intensities, depending upon the distance of the source from the wall and the directivity characteristics of the source. Some portion of the sound is reflected from the other source room surfaces before falling upon the wall concerned. The resulting incident pressure field is therefore extremely complex. The wall is set into bending vibrations from which sound is radiated in many directions, dependent upon its stiffness and mass, and upon the space–time structure of the incident field. There is no certainty that the resulting "directly" transmitted field bears any resemblance to the incident field. In addition, bending vibrations of any particular frequency continue for some time, depending upon the damping of the structure. Hence the transmitted field in any frequency band will not fall upon the receiving microphone within any easily specified time interval. Some of the "directly" transmitted disturbances will only reach the microphone after reflection from a number of receiving room surfaces. Hence the resulting "transmission loss" will depend upon the chosen frequency time window of the system.

It is quite conceivable that some flanking sound, possibly involving dispersive wave paths in the structure, will arrive at the microphone during the decay time of the primary wall. Since it is known that the true frequency response characteristics of the transmission system can only be evaluated from the total impulse response time history, there would seem to be some question about the validity of the use of time-delay spectrometry in the evaluation of transmission paths involving vibration energy storage elements, particularly lightly damped ones such as windows.

The authors quite correctly state that the "transmission loss" evaluated by the time-delay spectrometry technique will differ from that measured using a double room transmission suite. However, I would suggest that the reason is not simply that the incident sound field does not correspond to the conventionally assumed diffuse field, but that the factors mentioned above are also of considerable significance.

F. J. FAHY
Institute of Sound and Vibration Research
The University
Southampton SO9 5NH, England

AUTHORS' REPLY

The authors, in compliance with the title of their paper, endeavored to show actual field practices without too much involvement with the more intricate aspects of the behavior of sound which are encountered in any measurement. We welcome this opportunity to discuss some of the complications which can arise in specific applications of time-delay spectrometry (TDS) in the field.

Mr. Fahy perceives correctly that "when a loudspeaker is used to irradiate one wall of a room, the propagating wave fronts fall on the surface at varying angles of incidence, and with varying intensities, depending on the distance of the source from the wall and the directivity characteristics of the source. Some portion of the sound is reflected from other source room surfaces before falling upon the wall concerned. The resulting incident pressure field is therefore extremely complex."

The "standard" field sound transmission class (FSTC) measurement attempts to sum and average all sound energy within a reverberant room impinging upon the subject and then determine what portion of that energy passes through the structure to the space beyond. Further, when the subject is a portion of a sound impervious partition in a "double room transmission suite," all sound passing through the subject is collected by the nonabsorptive walls of the receive room and returned to the microphone as a reverberant field for measurement. After referencing against a "standard curve," a single number FSTC is obtained. Through experience we have learned to relate these single numbers to real-life experience. When the test subject (a solid partition between fairly live rooms) meets the standard's criteria, the authors employ the standard method using computerized equipment to yield FSTC numbers, repeatable to better than ±1 dB.

---

[1] C. R. Cable and J. K. Hilliard, *J. Audio Eng. Soc.*, vol. 28, pp. 302–309 (1980 May).

LETTERS TO THE EDITOR

However, when the subject is the window of a heavily furnished living room or bedroom facing a freeway, most of the sound energy impinging upon the subject is nondiffuse and may be near normal to the subject. No real reverberant field is developed in the receive room as most scattered sound is absorbed by the room boundaries and/or contents and is not returned to the measurement microphone.

The microphone may therefore measure, for the most part, only the incident sound as it passes from the subject to the receive space. This in no way meets the requirement of the standard, and the measurement leads to the uncertainties of measurement Mr. Fahy talks about.

This situation may be carried to the extreme. The wall containing the window may be suspended in free space with the sound source on one side and the measurement microphone on the other, with some means to qualify or eliminate flanking paths around the wall. This is what a TDS measurement simulates. Mr. Fahy appears to have overlooked this when the authors stated, "It should be noted that this method is similar to a sound transmission loss measurement in free space. . . ." They intended to imply just that. That is, the subject and the associated wall only, not the room, are effectively in free space.

In this case we are only concerned with that sound from the loudspeaker which impinges upon the sample. That which goes elsewhere is not included within the measurement. Also we are concerned only with that energy which passes through the sample to the measurement microphone. That energy which passes through the sample in directions other than to the microphone, through scattering or for other reasons, is lost and gone forever.

Since the essence of TDS is measurement of frequency in the time domain, this method enhances the classical free space measurement when it determines which portion of the time domain it shall include within the measurement. This causes Mr. Fahy concern, and rightly so. Imagine an infinite pane of 3/16-in (4.8-mm) thick glass in free space with a loudspeaker and microphone spaced 10 acoustical feet (3 m) apart on either side of the glass and normal to it. When the TDS window is set to 1 ms [1.13 ft (0.35 m)] all sound passing through the glass within a circle with a radius of 2.443 ft (0.75 m) centered on a line between source and microphone and reaching the microphone, will be within the TDS time frame (3 dB down points). What happens outside this 4.886-ft (1.5-m) circle is not included within the measurement because it arrives too late to pass the TDS filter (see Fig. 1). Here the directivity of the loudspeaker is of some importance. However, by scanning the area of the 4.886-ft (1.5-m) circle in the same plane the subject will occupy, the directional characteristics of the loudspeaker over the 52° angle of radiation we are concerned with, may be determined.

It is the directional characteristics on the send side of the subject that really concerns us. If the subject has no scattering and transmits all sound energy in the same direction it receives it, only that energy flowing along the line connecting source and receiver will be registered. All other sound will be lost in space. This gives a "specific loss" for the sample similar to the "specific electrical resistance" of a substance. If there is scattering during the passage of the sound through the sample, energy may reach the measurement position by paths other than on the line between the loudspeaker and the microphone. If scattering is over a broad angle, there will be a greater contribution of energy from the outer edges of the "time pass" 4.886-ft (1.5-m) circle. Whichever way the sound goes, the measured loss relates to the real loss under this circumstance between the source and the receiver and often relates more closely to some real-world situations than the double room transmission suite.

When this idealized case becomes a real-life window, the same principles apply. The variance of time arrivals, and the TDS capability to pick and choose which arrivals it will look at results in enhancement of the measurement facility instead of confusion. When the glass is bounded by a frame set in a wall, TDS may look at only the transmission through the glass or may include leakage through the surround. Or in special cases it may "tune in" on a specific time arrival which is identified with an unwanted flanking path. Indeed, as Mr. Fahy states, ". . . the resulting transmission loss will depend upon the chosen frequency time window."

When a TDS signal is used to excite a portion of the boundary of a real send room such as a door or window, only that energy accepted by the TDS time window enters the measurement. Normally all reflections from other boundaries and the reverberant field arrive too late to be passed by the TDS filter. However, all this late

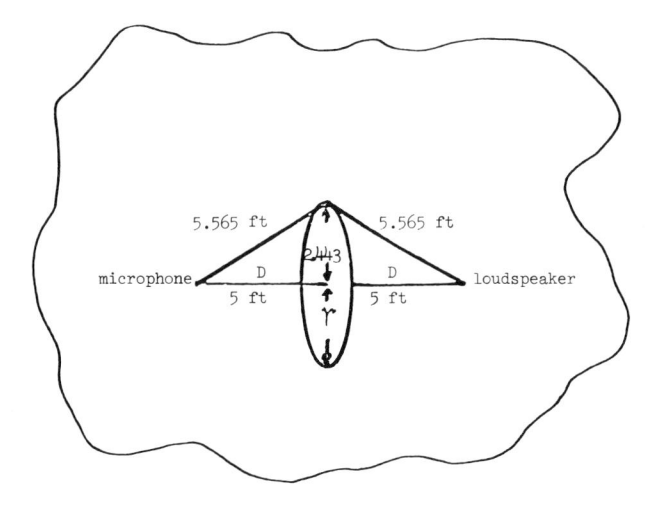

Fig. 1. For a 1.0-ms window (±0.5 ms) the maximum time elapse from the center of the window is 0.5 ms at the −3-dB points. For the example cited this is exceeded when sound through the sample deviates from the normal path by more than 2.443 ft (0.75 m) at the surface of the sample. For this example

$$r = \left[\sin \cos^{-1} \left(\frac{D}{\frac{t \times 1.13}{2} + D} \right) \right] \left(D + \frac{t \times 1.13}{2} \right)$$

where D and r are in feet and t is the time window in milliseconds.

energy does impinge upon the subject causing "bending vibrations" and may in some way modify the transmission of energy which is observed by the measurement system. We believe that this may, at least partially, account for Mr. Fahy's statement, "Since it is known that the true frequency response characteristics of the transmission system can only be evaluated from the total impulse response time history, there would seem to be some question about the validity of the use of TDS in the evaluation of transmission paths involving vibration storage elements, particularly lightly damped ones such as windows."

Although perhaps not sufficiently emphasized in the paper, the footnote on page 303 describes the means to fully qualify and quantify the energy time arrivals by displaying the frequency difference between the send (the driving signal for the loudspeaker) and the receive signal (at the microphone) on a 400-line FFT analyzer. The analyzer readout plotted by an *XY* recorder may be calibrated in feet, meters, milliseconds, or whatever units desired. This measurement is known as the energy time curve (ETC).

Fig. 2 is such a measurement. The subject was a 4-ft by 6.5-ft (1.2-m by 1.95-m) sliding glass door with the microphone and loudspeaker positioned 4 ft (1.2 m) above the floor, centered on opposite sides of the glass door, 19 in (0.48 m) apart. The dashed trace indicates the energy arrival times from the loudspeaker to the microphone with the glass door open. The first energy arrival, somewhat smeared over most of a millisecond, is direct from the loudspeaker. The sharp rise at 3.2 ms is a reflection from the side door frames. A small floor reflection is seen at 7.8 ms. The solid trace, offset upward 30 dB, is with the door closed—nothing else changed. The direct sound passing through the glass is down approximately 43 dB from the open condition. Early late arrivals are seen starting at 3.8, 5.8, and 7.7 ms. These are flanking paths around the door, poorly fitted to the frame. Each late arrival is smeared over considerable time since the side, top, and bottom edges of the rectangular door are at variable distances from the line normal to the loudspeaker and microphone.

It is interesting to note that the first arrival times with the door open and with the door closed fall within very few microseconds of each other. (Each minor division is 100 μs.) The "energy storage element" in this case did not appear to significantly delay the passage of energy from loudspeaker to microphone. It is a simple matter with most FFT analyzers to integrate the total power in each arrival time bundle. Relative transmission loss through the subject and through the flanking paths is then easily established in the time domain.

The authors did not intend to imply that the TDS method of measuring transmission loss should replace the two-room suite. It does offer additional insights, not previously available, into the behavior of sound through and around boundaries, barriers, and impediments. In special cases TDS studies in transmission loss of a subject may place the consultant in a better position to diligently apply the use of the material to the real-life case than do the numbers provided by the two-room suite measurement.

It should be clear that this type of measurement will be of little value to the novice. The capability of those who are most knowledgeable will be dramatically expanded by its use.

We again thank Mr. Fahy for his thoughtful comments.

C. R. CABLE AND J. K. HILLIARD
Hilliard and Bricken
Santa Ana, CA 92701, USA

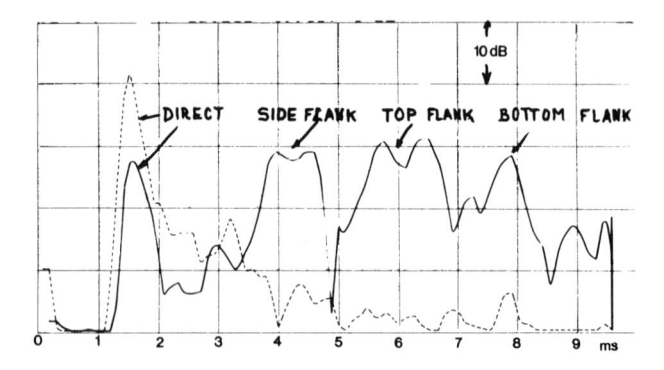

Fig. 2. Energy arrival times. Dashed line—with door open, offset −30 dB; solid line—with door closed, offset 0 dB.

The Modified Hopkins–Stryker Equation*

DON DAVIS

Synergetic Audio Concepts, San Juan Capistrano, CA 92693, USA

From the original work of Hopkins and Stryker in 1948 through its development by Beranek in 1949 and its use by Davis in 1968 and Boner in 1969 in acoustic-gain calculations, the Hopkins–Stryker equation has proved highly useful to a myriad of users. In recent years this versatile tool has been modified in light of measurements by Peutz and Davis to account for multiple sources, semireverberant spaces, modifiers of critical distance, and various electroacoustic modifiers of the ratio of direct sound to reverberant sound. A thorough discussion of these modifications and their proper application in acoustic calculations is given.

0 INTRODUCTION

The Hopkins–Stryker paper published in 1948 [1] provided Beranek in 1949 [2] with a key equation for use in calculating acoustic level changes. Davis in 1968 [3] and Boner in 1969 [4] employed this equation in the calculation of acoustic gain. In recent years this versatile tool has been modified in light of measurements by Peutz [5] and Davis [6] to account for multiple sources, semireverberant spaces, modifiers of critical distance, and various other electroacoustic modifiers of the ratio of direct sound to reverberant sound.

The Hopkins–Stryker equation and its derivatives are based on the same assumptions that were used by Sabine [7] in his classical study of reverberation. Sabine predicated a stochastic process in an ergodic enclosure (that is, randomly mixing homogeneous space). It is this description that qualifies a "reverberant" sound field as an entity distinct from a discrete reflection or a "limited train" of discrete reflections.

It was in this context that the equation called Hopkins–Stryker came into being with separate terms to account for the direct sound field and the reverberant sound field.

Measurements and empirical calculations have provided an additional term, when required, for semireverberant situations. With these caveats let us proceed to the equation and its variations.

---

* Presented at the 72nd Convention of the Audio Engineering Society, Anaheim, CA, 1982 October 23–27; revised 1984 July 2.

1 MODIFYING THE HOPKINS–STRYKER EQUATION

1) Basic equation:

$$L_T = L_W + 10 \log \left(\frac{QM_e}{4\pi(D_\chi)^2} + \frac{4N}{S\bar{a}M_a} \right) + K .$$

Use when $\Delta dB \leqq 1.0$ or less.

2) Direct sound level:

$$L_D = L_W + 10 \log \left(\frac{QM_e}{4\pi(D_\chi)^2} \right) + K .$$

Use when ΔdB is $\geqq$ than 5.0. $K = 10.5$ in the English system; $K = 0.2$ in the metric (SI) system, that is, $10 \log (pc/400)$.

3) Reverberant sound level:

$$L_R = L_W + 10 \log \left(\frac{4N}{S\bar{a}M_a} \right) + K .$$

Use when $\Delta \ll 0.5$.

4) Actual sound level:

$$L_{act} = L_W + 10 \log \left(\frac{Q}{4\pi(D_c)^2} \right)$$

$$+ \left(0.734* \left(\frac{\sqrt{V}}{h \cdot RT_{60}} \right) \left(\log \frac{D_c}{D_\chi > D_c} \right) \right) + K .$$

[* Metric (SI) = 1.329.]
Use when ΔdB falls between 1.0 and 5.0 dB.

5) Glossary:

L_T = total sound pressure level in decibels at D_χ (ref. 20 µPa)

L_D = direct sound pressure level in decibels at D_χ (ref. 20 µPa)

L_R = reverberant sound pressure level in decibels at D_χ (ref. 20 µPa)

L_{act} = actual total sound pressure level in decibels that occurs in semireverberant sound fields at $D_\chi > D_c$ (ref. 20 µPa)

L_W = sound power level in decibels for the device providing L_D at D_χ (ref. 10^{-12} W)

ΔdB = number of decibels L_{act} is below the L_T predicted by the basic Hopkins–Stryker equation at $2D_c$

$$\Delta dB = 0.221* \left(\frac{\sqrt{V}}{h \cdot RT_{60}} \right)$$

[* Metric (SI) = 0.4.] (ΔdB > 6 dB = 6 dB).

Q = directivity factor (dimensionless) for the device providing L_D at D_χ

D_χ = distance in feet or meters from the source to where L_χ is established

M_e = any electroacoustic modifier that changes L_D but not L_R (that is, useful directivity effects, such as microphone to loudspeaker random incidence efficiency)

N = total acoustic power radiated by system divided by acoustic power radiated by device or devices producing L_D at D_χ

$S\bar{a}$ = total absorption in sabins per square foot or square meter

M_a = architectural modifier

$$M_a = \left(\frac{1 - \bar{a}}{1 - a_c} \right) \left(\frac{Q_{THEOR}}{Q_{ACT}} \right)$$

V = internal volume of enclosed space in cubic feet or cubic meters

h = height of ceiling in feet or meters

RT_{60} = "apparent" reverberation time in seconds for 60 dB of decay

D_c = critical distance (that is, distance at which the Hopkins–Stryker equation makes $L_D = L_R$) in feet or meters

$$D_c = 0.141 \sqrt{\frac{Q S\bar{a} M_e M_a}{N}}$$

$$= 0.03121* \sqrt{\frac{QV M_a M_e}{RT_{60}(N)}}$$

[* Metric (SI) = 0.057.] The first equation is dimensionless.

0.734* = constant obtained by multiplying 0.221 by 3.322 [6, p. 23].

[* Metric (SI) = 1.329.] This constant allows calculating the log multiplier:

$$\text{log multiplier} = 3.322 \Delta dB$$

$$= 0.734 \left(\frac{\sqrt{V}}{h \cdot RT_{60}} \right) .$$

pc = ambient pressure times volocity of sound.

2 THE EIGHT KEY PARAMETERS AND THEIR ROLE IN THE ESTABLISHMENT OF THE TWO SOUND FIELDS

L_D is determined by L_W, Q, D_χ, and M_e.
L_R is determined by L_W, $S\bar{a}$, M_a, and N.

In the limiting case of total output of the sound source into a totally absorptive surface with no leakage to other surfaces, Q can be said to affect L_R (that is, $M_a = \infty$).

3 NOTES ON ABOVE DATA

1) The asymptotic limit for D_χ in the Hopkins–Stryker equation is

$$10 \log \left(\frac{4N}{S\bar{a}M_a} \right) .$$

This is significant only when $D_\chi >>> D_c$ (that is, $D_\chi = 10 D_c$).

2) In the English system sabins are per square foot; in metric (SI) sabins are per square meter.

3) When English system dimensions are employed, the constant 10.5 in the Hopkins–Stryker equation is derived as follows:

$$\frac{0.282 \text{ ft}}{1} \cdot \frac{12 \text{ in}}{1 \text{ ft}} \cdot \frac{2.54 \text{ cm}}{1 \text{ in}} \cdot \frac{1 \text{ m}}{100 \text{ cm}} = 0.08595 \text{ m}$$

$$20 \log \left(\frac{0.282 \text{ m}}{0.08595 \text{ m}} \right) = 10.3 \text{ dB} .$$

The temperature and barometric pressure correction factor is

$$dB_{corr} = -10 \log \left[\frac{\sqrt{°F + 460}}{527} \left(\frac{30}{B} \right) \right]$$

where B is the barometric pressure in inches of mercury and °F is the temperature in degrees Fahrenheit. 67°F and 30 in Hg result in a correction factor of 0 decibels. The additional 0.2 dB in the constant 10.5 allows for variation in standard temperature and pressure (STP).

4) Originally L_W was defined to allow 1 acoustic watt from a source with a $Q = 1.0$ to produce a sound pressure level at 0.282 ft (0.08595 m) of 130 dB (ref. 10^{-13} W).

The current L_W allows 1 acoustic watt from a source with a $Q = 1.0$ to produce a sound pressure level at

0.282 m (0.925 ft) of 120 dB (ref. 10^{-12} watt).

Thus the reference power was raised and the reference distance was increased:

$$10 \log \left(\frac{10^{-12} \text{ W}}{10^{-13} \text{ W}} \right) = 10 \text{ dB}$$

$$20 \log \left(\frac{0.282 \text{ m}}{0.08595 \text{ m}} \right) = 10.3 \text{ dB}.$$

5) Further qualification of the M_a factor:

$$M_a = \left(\frac{1 - \bar{a}}{1 - a_c} \right) \left(\frac{Q_{act}}{Q_{theor}} \right)$$

where:
- Q_{act} = measured Q
- Q_{theor} = theoretical Q which the C_L suggests [6, p. 44]
- $\bar{a}$ = average absorption coefficient in the space
- a_c = absorption coefficient of area where the first reflection occurs; $a_c > \bar{a}$ must occur.

6) Delta levels ΔD_χ: If L_W is removed from the equations, they become ΔD_χ equations yielding *relative* levels:

$$\Delta D_\chi = 20 \log \left(\frac{0.282}{D_\chi < D_c} \right) + 10 \log Q$$

remembering that

$$10 \log \left(\frac{1.0}{4\pi(0.282)^2} + \frac{4}{S\bar{a}} \right) \simeq 0.$$

7) Inverse functions:

  a) $D_\chi < D_c$:

$$D_\chi = \sqrt{\frac{Q}{4\pi(10^{\Delta D_\chi/10})}}$$

  b) $D_\chi \geqq D_c$:

$$D_\chi = \frac{\Delta D_\chi}{[10^{0.734(\sqrt{V/h} \cdot RT_{60})}]^{D_c}}$$

  c) Basic equation:

$$D_\chi = \sqrt{\frac{Q}{4\pi(10^{\Delta D_\chi/10} - 4/S\bar{a})}}$$

8) Other useful variations:

$$S\bar{a} = \frac{(D_c)^2 N}{0.019881 Q M_a M_e}$$

$$Q = \frac{(D_c)^2 N}{0.019881 S\bar{a} M_a M_e}.$$

See also [6, app. VIII, pp. 261–270].

When a D_χ value of 0.282 is used in the Hopkins–Stryker equation ($Q = 1$), it yields one square unit of area for the surface of a sphere of that radius in whatever dimension D_χ is expressed. If D_χ is in feet, then the area becomes 1 ft^2. If D_χ is in meters, then the area becomes 1 m^2 (0.282 ft = 0.08595 m, 0.282 m = 0.925 ft).

$$20 \log \left(\frac{0.282}{0.08595} \right) = 10.3 \text{ dB}.$$

When the old standard reference for L_W was 10^{-13} W, 1 W was $L_W = 130$ dB. The new standard reference for L_W is 10^{-12} W (1 pW), and 1 W is $L_W = 120$ dB.

Since in both cases L_W is a given *fact* when available, no power adjustment of the level ΔD_χ is required. What is required is an adjustment in ΔD_χ for the dimensional units feet or meters. If meters are used, it is correct as written plus 0.2 dB for L_W referenced to 10^{-12} W. If feet are used, then 10.5 dB must be added to the equation as written because the distance is shorter (10.3 + 0.2 = 10.5 dB).

When the equation is used to obtain ΔD_χ numbers without L_W, no correction is required as the ΔD_χ numbers are *relative* numbers. They become absolute levels only when used with an L_W.

9) An alternative view of sound fields. In our depiction of sound level versus distance (Fig. 1), the failure of the curve to follow the inverse square law rate of level change with increasing distance tells us that an L_R is present. If looked at as sound level versus time, L_D arrives first, followed, after an interval known as the initial time delay gap, by the early reflected sound levels. After a sufficient interval has elapsed to allow mixing of the total reflected sound energy (enough time for hundreds of mean free path distances to occur), the exponentially growing and decaying L_R appears. Thus in order to obtain accurate information about the presence or lack of a reverberant sound field, both sufficient distance from the source must be established and sufficient time must be provided for it to develop (see Fig. 2).

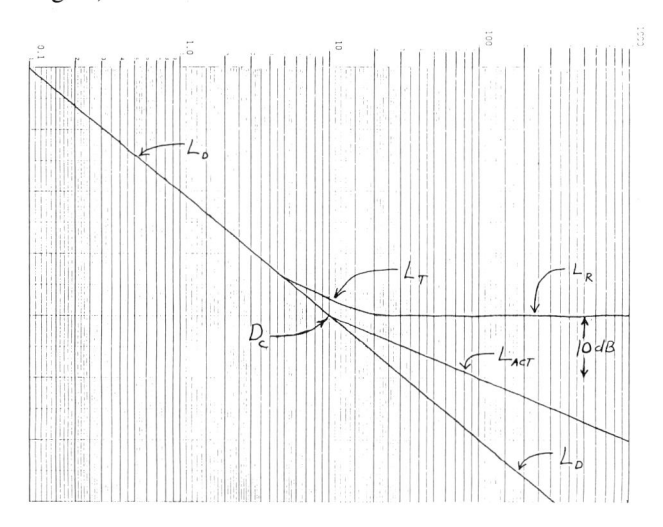

Fig. 1. Acoustic level versus distance.

4 PEUTZ MODIFICATION OF HOPKINS–STRYKER EQUATION

1) For D_χ less than *apparent* D_c:
$$\Delta D_\chi = 10 \log \left[\frac{Q}{4\pi(D_\chi)^2}\right].$$

2) For D_χ equal to or greater than *apparent* D_c:
$$\Delta D_\chi = 10 \log \left[\frac{Q}{4\pi(D_c)^2}\right]$$
$$+ 0.734\left(\frac{\sqrt{V}}{h \cdot RT_{60}}\right)\left[\log\left(\frac{D_c}{D_\chi \lesseqgtr D_c}\right)\right].$$

3) Inverse calculations:
a) $D_\chi < D_c$:
$$D_\chi = \sqrt{\frac{Q}{4\pi(10^{D_\chi/10})}}.$$

b) $D_\chi \lesseqgtr D_c$:
$$D_\chi = \frac{\Delta D_\chi}{[10^{0.734(\sqrt{V}/h \cdot RT_{60})}] D_c}.$$

Note that the reference distance for $20 \log (\text{ref}/D_\chi)$ is 0.282. Apparent $D_c = 0.03121* \sqrt{QV/RT_{60}}$. [* Metric (SI) = 0.057.*]

5 DESCRIBING Q MORE ACCURATELY

The measurement of the directivity factor *Q is always at a point*. There can be a series of points within an area that have the same Q, thus allowing the concept of an "average of Qs" within an area. It is a normal practice to measure Q on axis (the zero-angle axis usually being the highest output as well). Let us call this measurement Q_{axis}.

The value Q is both frequency dependent, $Q_{\text{axis}}(f)$, and, for real-life devices, angularly dependent. Q_{axis} specifies the angle relative to the transducer. For angles other than the on-axis position we could specify a Q_{rel}:

$$Q_{\text{rel}} = Q_{\text{axis}}\left(10^{\pm C_\angle \text{dB}/10}\right)$$

where $\pm C_\angle \text{dB}$ indicates the level in decibels of the particular angle *relative* to the level in decibels on axis.

A complete descriptive may be specified by

$$Q_{\text{rel}} = Q_{\text{axis}}\left(10^{\pm C_\angle \text{dB}/10}\right)(f)$$

where f is the frequency at which the measurement is made.

A further useful convention would be to agree that where no f is specified, the one-third-octave band at 2000 Hz is indicated.

In the design of a sound system we use $Q_{\text{min(ss)}}$, where ss stands for single source and which usually is synonymous with Q_{axis} but may, on occasion, actually be a Q_{rel}. The term "min" indicates that it is the minimum value that will allow the $\%AL_{\text{cons}}$ required at that *point*.

If more than one source is used, we encounter the term NQ_{min}, where we increase the Q of the first device proportionately to the number N of additional devices (of equal acoustic power output).

We also employ the term Q_{avail} whereby we can calculate the N required for a multiple-source system:

$$N = \left(\frac{Q_{\text{min}}}{Q_{\text{avail}}}\right).$$

A further refinement is the direct calculation of a distance D_2 at which Q_{avail} results in the same ratio of direct sound to reverberant sound as NQ_{min} would have provided:

$$D_{2\text{max}} = \frac{D_{2\text{ss}}}{N}.$$

At the current time we utilize the following Q de-

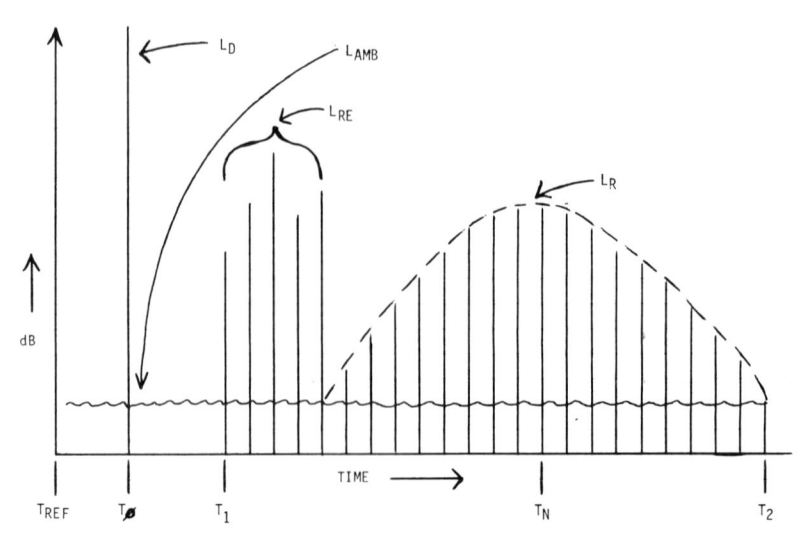

Fig. 2. Defining sound field levels versus time.

scriptives: Q_{axis}, Q_{min}, Q_{rel}, and Q_{avail}, along with the descriptive modifiers ss, $\pm C_\angle$ dB, N, and f.

6 A SUBTLETY REGARDING Q BY PLACEMENT

An often misinterpreted point with regard to establishing a directivity factor Q by placement of the source near a reflecting surface (mirror images) is that the source must be at, *not in*, the surface.

Loudspeakers mounted in the wall will, at lower frequencies, exhibit "mutual coupling" (see Fig. 3).

When a single loudspeaker is mounted *in* a wall, half the power goes into *another* space. When mounted near the wall, half the power is reflected back into the space.

7 ΔD_χ AND THE USE OF Q

Users of the Hopkins–Stryker equation often question whether different values of Q_s should be employed for D_s, D_1, D_2, and D_0 when obtaining $\Delta D\chi$. Normally the answer is "use the loudspeaker Q for all distances." Differing Q (for example, the talker with $Q = 2.5$ and the loudspeaker with $Q = 50$) will establish quite different values of D_c.

In the normalized gain equations D_0 drops out and is replaced by EAD. This means that both the ΔD_s and the ΔEAD are *normally* in the direct sound field where inverse square law level change is the rule, and the Q chosen can be arbitrary so long as it is the same for both distances. In any case, the remaining two distances D_1 and D_2 are dependent upon the loudspeaker's Q.

8 HOW TO USE DIFFERING VALUES

If it is desired to use separate Q values for the talker and the loudspeaker (and perhaps to assign a Q to the microphone as well), you are free to obtain an absolute level change rather than a relative one. This is accomplished by using a reference point and a D_χ point for each value and taking the ΔD_χ of both points, followed by using the difference between them as the ΔD_χ in the gain equations.

What you may not do legitimately is to use differing Q in the Hopkins–Stryker equation when obtaining *relative* ΔD_χ values. The reference point chosen should usually be less than 0.5 ft.

9 USING THE HOPKINS–STRYKER EQUATION

The sound power level L_W is referenced to 10^{-12} W. In using the Hopkins–Stryker equation to obtain an expected sound pressure level L_D at some distance D_χ, it is important not to use the L_W of the array but only the L_W of that part of the array that supplies L_D and D_χ. To do otherwise is to miscalculate the L_D, which is dependent *only* upon the L_W of the devices also producing the L_D at the point of observation (measurement). The N factor inserted into the equation in opposition to the total absorption $S\bar{a}$ correctly adjusts for the ratio of total L_W to the L_W producing L_D because this portion of the Hopkins–Stryker equation affects only the reverberant sound field level L_R.

Always bear in mind that L_D is affected by that part of L_W that produces L_D at the point of measurement, a distance D_χ from the array, the Q of the device that produces L_D at D_χ (not the Q of the array), and the distance from the array D_χ. Thus we avoid the difficulties of overestimating the level of L_D at D_χ.

L_R is affected by the *total* L_W of the array (which may be properly accounted for by the ratio N, which scales the level appropriately to the L_W of the single device producing L_D) and the total absorption present. Here it is important to note the sometimes significant role of the architectural acoustic modifier M_a. A substantial M_a can lower L_R (but RT_{60} or the decay rate remains the same). The M_a factor is invariably lower than would be expected because of the difference between the Q of real-life devices and the coverage angles employed. It is important to note that this effect cannot operate unless $a_c > \bar{a}$.

10 SUMMARY

These instructive equations reveal the interaction of each of the primary parameters controlling the various sound fields. We have discussed additional parameters that have direct bearing on the modified behavior of the primary parameters. In its modified form, the Hopkins–Stryker equation has kept pace with measurements in the sense that our prediction accuracy has kept pace with our measurement capability.

11 REFERENCES

[1] H. F. Hopkins and N. R. Stryker, "A Proposed Loudness-Efficiency Rating for Loudspeakers and the Determination of System Power Requirements for Enclosures," *Proc. IRE* (1948 Mar.).

[2] L. L. Beranek, *Acoustics* (McGraw-Hill, New

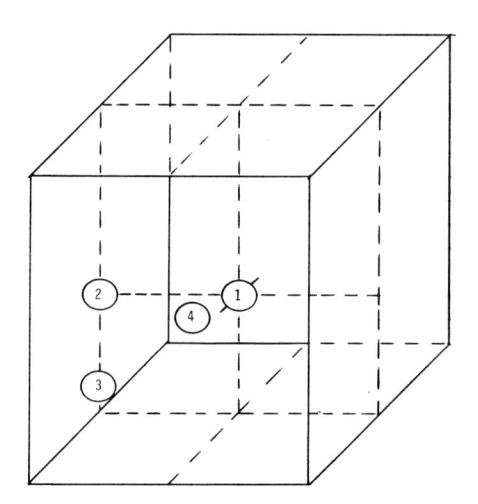

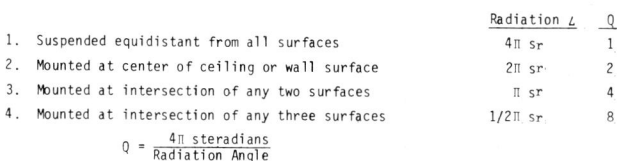

	Radiation $\angle$	Q
1. Suspended equidistant from all surfaces	4π sr	1
2. Mounted at center of ceiling or wall surface	2π sr	2
3. Mounted at intersection of any two surfaces	π sr	4
4. Mounted at intersection of any three surfaces	$1/2\pi$ sr	8

$$Q = \frac{4\pi \text{ steradians}}{\text{Radiation Angle}}$$

Fig. 3. Determining minimum Q by loudspeaker placement.

York, 1949), pp. 314–315.

[3] D. Davis, internal Altec memorandum to A. A. Ward, John Hilliard, Jim Noble, and Arthur C. Davis with outline of computer program supporting Altec Acousta Voicing contractors, 1968 Jan.

[4] C. P. Boner and R. E. Boner, "The Gain of a Sound System," *J. Audio Eng. Soc.*, vol. 17, pp. 147–150 (1969 Apr.).

[5] V. M. A. Peutz, *Publikaties*. (Dutch Acoustical Society, 1954–1982).

[6] D. Davis and C. Davis, *Sound System Engineering* (Howard W. Sams, Indianapolis, IN, 1973).

[7] W. C. Sabine, *Collected Papers on Acoustics*. (Howard University Press, Washington, DC, 1922).

THE AUTHOR

Don Davis is president of Synergetic Audio Concepts and is active in teaching seminars in acoustics and electroacoustics. He is also an international consultant on acoustics.

A former vice president of Altec and of Klipsch and Associates, Davis has authored three books, *Acoustical Tests and Measurements*, *How to Build Loudspeaker Enclosures*—written with the late Alexis Badmaieff, and *Sound System Engineering*, coauthored with his wife, Carolyn. He is the author of over a hundred technical articles on audio subjects, and the inventor of the acousta-voice method of room-sound system equalization as well as the live-end-dead-end (LEDE™) control room design technique. Mr. Davis' firm, Synergetic Audio Concepts, is the appointed agent of the California Research Institute Foundation for the licensing of time-energy-frequency (TEF™) measurements under the Heyser-Cal Tech patents.

The recipient of numerous awards in the field of acoustics and electroacoustics for his work in the design of sound systems and their acoustic environments, Mr. Davis has in the past three years concentrated on the interface of electroacoustic transducers to acoustic environments requiring nontraditional analysis techniques (such as control rooms). He is a fellow of the AES, a senior member of the IEEE, a member of the ASA, a member of the SMPTE, and a member of the National Council of Acoustical Consultants.

Subjective and Predictive Measures of Speech Intelligibility—The Role of Loudspeaker Directivity*

KENNETH D. JACOB

Bose Corporation, Framingham, MA 01701, USA

An experiment has been conducted to determine whether the speech intelligibility in rooms is related in a simple way to the loudspeaker directivity Q. Three loudspeakers of widely differing Q were used to subjectively test intelligibility in five auditoria. These results indicate that intelligibility and Q are not directly related. In addition, impulse response measurements were made so that several methods of predicting intelligibility could be compared with subjective scores. One method, which assumes a linear relationship between Q and intelligibility, was shown to be the least accurate predictor. Two other methods, one based on the psychophysics of the auditory system and the other based on the modulation transfer function, proved to be better predictors of intelligibility.

0 INTRODUCTION

Installed sound systems for speech and music are judged primarily by the answers to four questions: Is the system loud enough? Is the sound coverage even? Is the frequency response acceptable? Is speech intelligible?

In many situations, an installed system is used exclusively, or at least primarily, for speech reinforcement and reproduction. In these cases system intelligibility becomes the primary criterion upon which a system's performance is judged.

Loudspeaker directivity Q is generally thought of as a parameter affecting speech intelligibility. Subjective experiments designed to determine this relationship have not, to this author's knowledge, been the subject of published research. Based only on theoretical considerations, a commonly used technique for predicting speech intelligibility (after Klein [1]) assumes a linear relationship between Q and intelligibility;

$$\%\text{AL}_{\text{cons}} = \frac{200D^2T^2}{QV} \quad (1)$$

where

$\%\text{AL}_{\text{cons}}$ = percentage articulation loss of consonants
D = source-to-listener distance
T = reverberation time of room
V = room volume
Q = loudspeaker directivity

In order to obtain a better understanding of sound system performance, an experiment was designed to answer two questions: 1) Is there a clear relationship between loudspeaker directivity and speech intelligibility? 2) Which of several techniques for predicting speech intelligibility are most accurate?

These will include Klein's [1] formula [Eq. (1)]; Lochner and Burger's procedure [2], which is based on the degree to which the auditory system integrates room reflections with direct sound; and the modulation transfer function (MTF), which quantifies the blurring effect reverberation has on speech [3].

Five auditoria were chosen for their broad range of reverberation times (0.9–3.5 s) and intended applications (cinema, theater, and meeting hall). In each room, two listening locations were chosen, roughly in the middle and at the rear of the auditorium floor. Using 14 trained listeners and three loudspeakers having dif-

---

* Manuscript received 1985 May 27, revised 1985 September 25.

ferent Q values, physical and subjective measurements were made.

The subjective results indicate that there is no simple relationship between Q and intelligibility, as measured by actual intelligibility tests. Even in reverberant rooms, where it might be assumed that a very directional loudspeaker would be required, medium and high Q loudspeakers performed nearly identically. Analysis of physical data revealed the formula which assumes a linear relationship between Q and intelligibility [Eq. (1)] to be the least accurate of the three predictive techniques. Furthermore, the two other techniques are better at predicting intelligibility over a wider range of situations.

1 EXPERIMENTAL DESIGN

The goal of the experimental design was to investigate loudspeaker directivity and its effect on intelligibility. Other variables known to affect intelligibility were held constant.

Three loudspeakers of differing directivities were used[1]:

High Q ($Q = 17$)	constant-directivity horn
Medium Q ($Q = 7.5$)[2]	array of identical drivers
Low Q ($Q = 1.0$)	spherical source

Five auditoria with various acoustical qualities, including two rooms known for their intelligibility problems, were chosen. Room parameters relating to this study are presented in Table 1.

Subjects were chosen from the general public. Each was screened for normal hearing as defined by American National Standards Institute (ANSI) Std S3.2-1960 [4]. Qualified subjects were trained using the same standard's guidelines. In anticipation of testing in rooms with known intelligibility problems, additional training was conducted in which background noise was used intentionally in order to create difficult conditions for intelligibility. The same 14 subjects were used throughout the experiment.

Intelligibility test material was in the form of monosyllabic English words embedded in a carrier sentence. Twenty lists of 50 words each, as defined by the ANSI standard, were used. These lists have the attribute of being phonetically balanced—in other words, individual speech sounds are represented with about the same frequency as in normal speech. Furthermore, the lists are approximately equivalent in difficulty.

Word lists were read by two different talkers at a rate of about 15 words per minute. Recordings of the word lists were made in an anechoic chamber using an instrumentation-grade omnidirectional microphone and a talker-to-microphone distance of 0.5 m. Thus a spectrally accurate on-axis speech recording was made.

A portable pneumatic tower was used to configure and elevate the three loudspeakers. The medium and high Q loudspeakers were individually aimed to provide the best coverage over the listening positions. Fig. 1 shows a sketch of the loudspeakers and tower in a typical location. In each room this location could be described approximately as the middle top of the stage proscenium. The configuration of the three loudspeakers was held constant from day to day.

Special care was taken to eliminate any hum or distortion in the test loudspeaker systems. In addition, because it was necessary to eliminate differences in loudspeaker-to-loudspeaker frequency responses as a variable in this experiment, a one-third-octave real-time analyzer was used at the listener locations to equalize the energy responses of each of the three sources. This was accomplished to within ±2 dB. In all cases the subjective quality of the speech being reproduced was very similar.

Finally, in order to minimize background noise as a variable affecting speech intelligibility, a speech signal-to-noise ratio of greater than 30 dBA was maintained. This was ascertained by measuring the background noise and adjusting the speech system gain so as to guarantee the signal-to-noise ratio. This is the accepted signal-to-noise ratio beyond which background noise has no significant effect on intelligibility [5].

In each room, two listener locations were chosen to coincide roughly with 1) the critical distance of the

Table 1. Room parameters.

Room	T (seconds)	V (m³)
Berklee Performance Center (primarily music)	0.9	5 450
Coolidge Corner Cinema (cinema)	1.0	4 590
Huntington Theater (speech)	1.1	3 190
Saint Bridget's Church (primarily speech)	2.0	3 810
Nevins Hall (primarily speech)	3.5	10 620

---

[1] The terminology of high, medium, and low Q is used here to reflect the importance given Q in the %AL$_{cons}$ formula. It is understood that higher Q sources exist.

[2] Array-type loudspeakers do not in general exhibit Q values as a function of frequency which are as stable as those of constant-directivity horns. The array loudspeaker used here has the following octave-band Q values: $Q = 7.9$ at 1 kHz, $Q = 4.5$ at 2 kHz, and $Q = 9.5$ at 4 kHz. The value given and used here for computations is an average of these three octave bands.

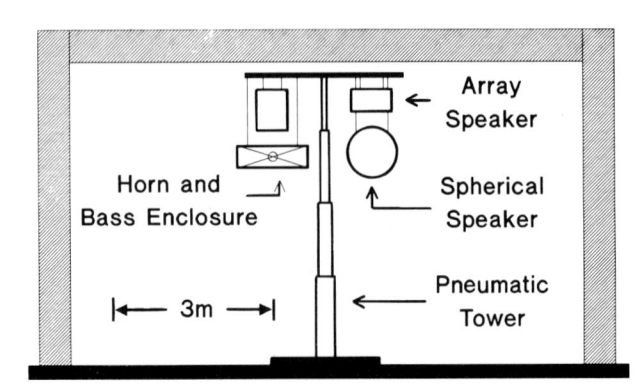

Fig. 1. Configuration of three loudspeakers in typical location in room.

high Q source, and 2) the "intelligibility distance" of the high Q source (Fig. 2). Thus the listeners would would be within the distance range where intelligibility is predicted by Eq. (1) to vary directly with loudspeaker directivity.

The testing was carried out over five successive days. In each room, in each position, and with each source, four word lists were played. Or, in other words, there were 2800 data points for each loudspeaker in each position ($4 \times 50 \times 14 = 2800$). There was never more than one list in a row played successively over a given loudspeaker. In addition, the order of the 24 lists was changed from day to day. Logos were omitted from loudspeaker products, and subjects were not told the purpose or methodology of the experiment.

For each loudspeaker, in each position, and in each room, impulse response measurements were taken and stored digitally. (Once a system's impulse response has been captured, a system's frequency response, reflection arrival times, and reverberation time can be found.)

2 RESULT OF SUBJECTIVE TESTING

Word lists were scored by percentage of phonetically correct words. (Words could be spelled incorrectly and still be scored correctly, so long as they were phonetically correct, as specified by ANSI Std S3.2-1960 [4].)

In order to determine the accuracy of average subjective scores, statistical analysis was performed on the data using the Student's t test. Subjective scores are shown in Table 2. Using the t test and a confidence level of 95%, average subjective intelligibility scores were generally within an accuracy of 1–2%.

Fig. 3 is a bar chart of the average intelligibility scores. From Table 2 and Fig. 3 it can be seen that there is little or no statistical difference between the scores of the high and medium Q loudspeakers, while the low Q source is at times significantly less intelligible. These data indicate that the theory which assumes intelligibility to be directly proportional to loudspeaker directivity [Eq. (1)] may not be correct.

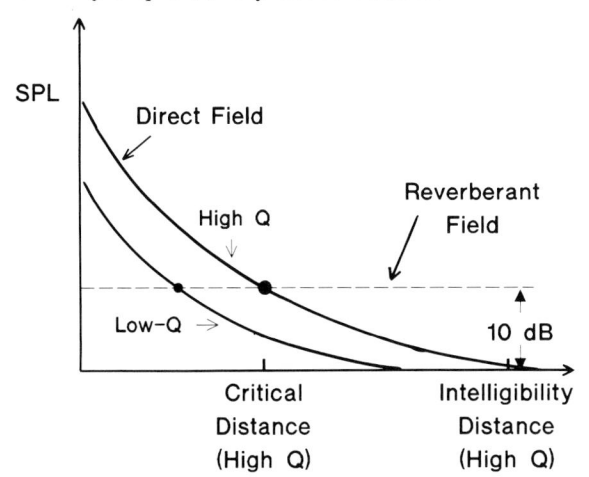

Fig. 2. Critical and "intelligibility" distances for two different Q loudspeakers as defined by classical statistical reverberation theory.

3 PREDICTIVE TECHNIQUES

Three well-known techniques are used for predicting speech intelligibility in rooms:

3.1 Articulation Loss of Consonants

Peutz [6] used an omnidirectional loudspeaker in a variety of rooms in order to investigate intelligibility empirically. He found that subjective intelligibility could be based on the percentage of correctly understood consonants in special monosyllabic nonsense words. Peutz showed that articulation loss varied with the square of the source-to-listener distance. This relationship held until a certain distance was reached, beyond which the articulation loss remained constant; he called this the critical distance. (In order to avoid confusion with the more traditional definition of critical distance, namely, the distance at which the direct and reverberant fields from a source in a room are equal, we shall call Peutz's critical distance the *intelligibility distance*.) The intelligibility distance is about 3.2 times greater than the critical distance. Peutz's formula for the articulation loss of consonants is

$$\%\text{AL}_{\text{cons}} = \frac{200 D^2 T^2}{V} \qquad (2)$$

where

$\%\text{AL}_{\text{cons}}$ = percentage articulation loss of consonants
D = source-to-listener distance
T = reverberation time of room
V = room volume

Klein [1] modified Peutz's formula for articulation loss by including loudspeaker directivity. He utilized the fact that critical distance is related to loudspeaker directivity by

$$C_d = \left[\frac{QV}{T}\right]^{1/2} \qquad (3)$$

where

C_d = critical distance
Q = loudspeaker directivity

Table 2. Subjective intelligibility scores.

Room/position	High Q	Medium Q	Low Q
1 1	98 ± 0.7	96 ± 1.0	96 ± 1.0
1 2	96 ± 1.0	96 ± 1.0	93 ± 1.0
2 1	97 ± 0.6	97 ± 0.6	97 ± 0.6
2 2	91 ± 1.2	94 ± 1.1	90 ± 1.6
3 1	94 ± 1.1	95 ± 1.0	94 ± 1.3
3 2	92 ± 1.1	89 ± 1.4	86 ± 1.7
4 1	93 ± 1.2	92 ± 1.2	92 ± 1.1
4 2	86 ± 1.5	88 ± 1.5	82 ± 2.1
5 1	89 ± 1.6	87 ± 3.0	78 ± 3.0
5 2	90 ± 1.6	89 ± 1.1	89 ± 1.6

Klein assumed that Peutz's formula could be modified to account for the dependence of the critical distance on loudspeaker directivity. The resulting formula is

$$\%\text{AL}_{\text{cons}} = \frac{200 D^2 T^2}{QV}.$$

3.2 Signal-to-Noise Procedure

Lochner and Burger [2] concentrated upon the fact that speech intelligibility was dependent on, among other things, the ratio of speech signal to background noise. In early work they established this relationship through subjective testing. They later hypothesized that this basic relationship between speech signal and masking noise could be adapted to include reverberation. They reasoned that for a certain period of time the hearing system integrates energy in the form of room reflections with the sound energy arriving directly from the source. (This is a well-known phenomenon in psychoacoustics [7].) They designed an experiment to determine the degree to which reverberant energy was integrated by the hearing system. They postulated that the portion of energy integrated could be considered signal and that the remaining reverberant energy could be considered noise. This led to a formula for the effective signal-to-noise ratio,

$$S/N_{\text{eff}} = \frac{\int_0^{95\ \text{ms}} p^2(t) a(t)\ dt}{\int_{95\ \text{ms}}^{\infty} p^2(t)\ dt} \qquad (5)$$

where

S/N_{eff} = effective signal-to-noise ratio
$p(t)$ = impulse response of system
$a(t)$ = weighting function for integration properties of the hearing system

The effective signal-to-noise ratio was then used in conjunction with their subjective data to predict intelligibility.

Lochner and Burger made both subjective and physical measurements in a variety of rooms and found excellent agreement between predicted and measured values of speech intelligibility. Other investigators [8] have also found excellent correlation.

3.3 Modulation Transfer Function

The modulation transfer function (MTF) technique for predicting intelligibility relies on the fact that reverberation and background noise have the effect, at the output of a system, of smearing, or blurring, an input waveform. The modulation transfer function was a technique first used to measure the accuracy and clarity of optical systems. Houtgast and Steeneken [3] adapted the modulation transfer function in order to predict intelligibility in speech transmission channels.

In the modulation transfer function technique, speech-band noise (the carrier) is modulated by frequencies which coincide with the modulating frequencies of natural speech. As the modulated noise passes through a speech transmission system, the smearing effect can be measured by the change from input to output of the modulation depth.

Once the modulation transfer function has been generated, it is weighted and summed to yield a single number, the speech transmission index (STI), which is an indicator of speech intelligibility. Houtgast and Steeneken have found very good correlation between predicted and measured values using a variety of systems.

Schroeder [9] derived that connection between the impulse response of a system and its modulation transfer function,

$$m(F) = \left| \frac{\int_0^{\infty} p^2(t)\ e^{-j\omega t}\ dt}{\int_0^{\infty} p^2(t)\ dt} \right| \qquad (6)$$

where

$p(t)$ = system impulse response
$m(F)$ = modulation transfer function

In words, the modulation transfer function is proportional to the magnitude of the Fourier transform of the squared impulse response.

4 ANALYSIS

Klein's formula for predicting speech intelligibility was calculated by computing room volume, reverberation time, source-to-listener distance, and source directivity Q. It should be noted that reverberation time was taken from the impulse response measurements and was not calculated using predictive methods, which can cause significant errors.

Lochner and Burger's signal-to-noise procedure was calculated by processing the digitally stored impulse responses according to Eq. (5). The modulation transfer function was generated according to Eq. (6).

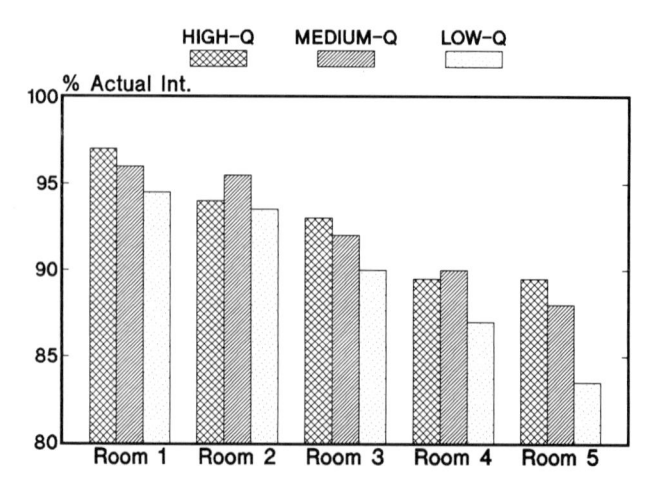

Fig. 3. Actual subjective intelligibility scores for three different Q loudspeakers in five rooms, averaged over two positions.

5 SUBJECTIVE VERSUS PREDICTIVE INTELLIGIBILITY SCORES

Results of the three predictive methods were translated using mapping graphs provided by Houtgast and Steeneken [3] and Lochner and Burger [2] into percentage phonetically balanced (PB) word intelligibility scores so that they could be compared with the subjective data.

Fig. 4 shows such a comparison for each of the three predictive methods in the five rooms. (If the predictive methods were ideal, all points would fall on the line indicated in Fig. 4.) These plots show immediately that the predictive method that assumes a linear relationship between Q and intelligibility [Eq. (1)] has the most scattering and is therefore the least accurate. These deviations can be examined more closely. Fig. 5 shows the variance of predicted versus actual values for each predictive technique in each room. In almost all cases, the variance produced by using Eq. (1) is greater than that due to the other two techniques.[3] It is also important to note that the variance in points predicted by Klein's formula is highest in rooms where the reverberation time is high (rooms 4 and 5). In comparison, the variance of the other two techniques also increases, but the levels are much lower. Another way to interpret the results is to compute the mean differences between predicted and actual intelligibility scores. Fig. 6 shows these differences. It is clear from these data that the signal-to-noise method consistently predicts intelligibility scores that are 2-5% too high, whereas the modulation transfer function method predicts scores that are 3-6% too low.

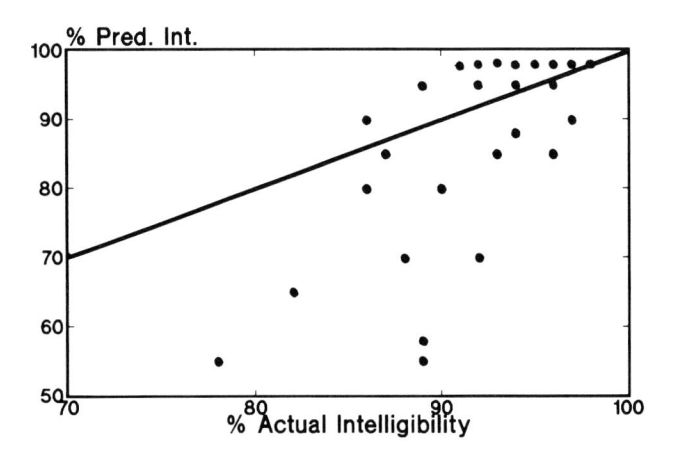

6 DISCUSSION

The data show that Klein's method of predicting intelligibility can be inaccurate, especially in highly reverberant rooms. In these cases the formula predicts intelligibility scores too high for the high Q source and too low for the low and medium Q loudspeakers. These are rooms where the prediction of intelligibility is most crucial since they are most likely to have intelligibility problems.

In general, as measured by actual intelligibility tests, high and medium Q loudspeakers performed equally well. The low Q loudspeaker performed significantly less well in some cases, especially in the two reverberant rooms. These results can be explained in an intuitive way by considering the impulse responses typical of the three systems.

The high Q source provides the highest ratio of direct to reverberant energy. The medium Q device has a lower ratio of direct to reverberant energy, but provides, in most rooms, much more energy in the form of early reflections than the high Q source. So long as these early reflections are early enough to be integrated by the hearing system, they can be considered signal, and therefore contribute to intelligibility by improving the

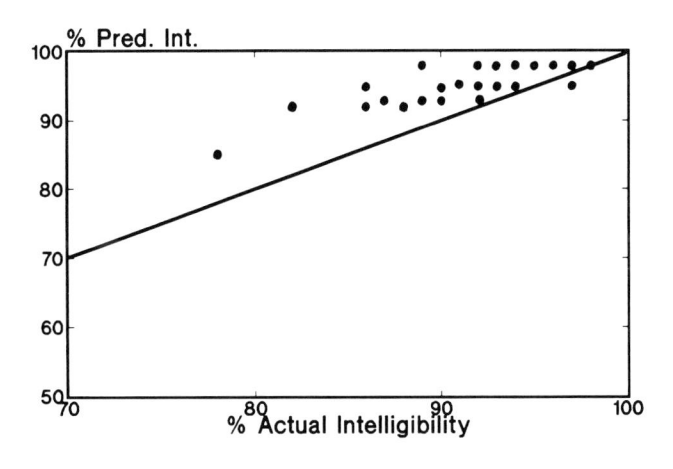

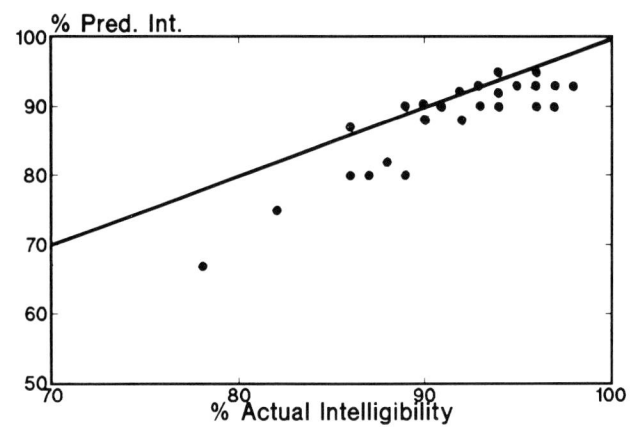

Fig. 4. Scattergrams of predicted versus actual intelligibility scores.

---

[3] Houtgast and Steeneken [3] have shown that for the theoretical case of a room whose impulse response is perfectly exponential, speech intelligibility as predicted by the modulation transfer function and Eq. (1) will be the same. However, real rooms usually deviate substantially from this ideal. Intelligibility as predicted by the modulation transfer function method takes these deviations into account while the %AL$_{cons}$ method Eq. (1) does not; thus predicted scores can be significantly different.

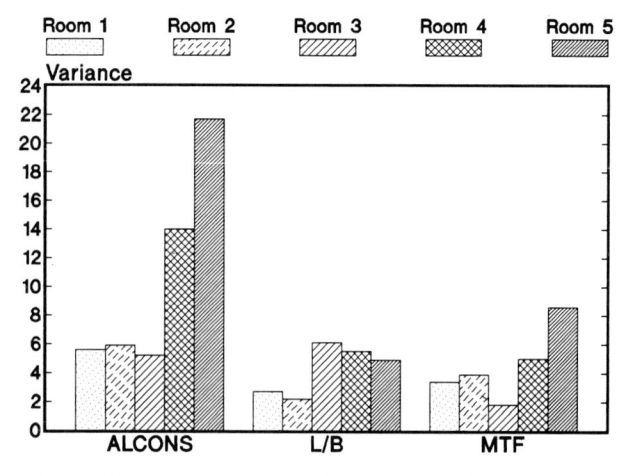

Fig. 5. Variance of predicted versus actual intelligibility scores.

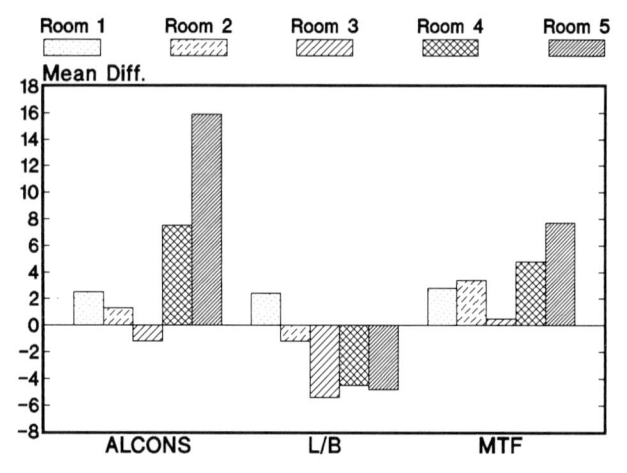

Fig. 6. Mean differences of predicted versus actual intelligibility scores.

effective signal-to-noise ratio. The positioning of the low Q source in the rooms used in this study was unfavorable for intelligibility by the same argument; the loudspeakers were in all cases positioned at the front and middle elevation of the stage, and hence were not particularly near surfaces that would be necessary to enhance early reflections.

Although the signal-to-noise and modulation transfer functions were better in predicting intelligibility, they have significant disadvantages. Foremost is that in their present form they require an impulse response as input. While methods exist for predicting an electroacoustic system's impulse response (for example ray-tracing and image-model techniques), they are difficult to implement and are computationally intensive. This presents a paradoxical situation for the sound system designer—Klein's formula is simple, but can be highly inaccurate; the signal-to-noise and modulation transfer function techniques are more accurate but are much more difficult to implement. Clearly this points the way toward future research. First, more rooms need to be characterized, particularly those with potential or real intelligibility problems. A more complete data base must be established. Second, the widespread access to computers means that predictive techniques need not be restricted to simple algebraic expressions.

7 REFERENCES

[1] W. Klein, "Articulation Loss of Consonants as a Basis for the Design and Judgment of Sound Reinforcement Systems," *J. Audio Eng. Soc.*, vol. 19, pp. 920–922 (1971 Dec.).

[2] J. P. A. Lochner and J. F. Burger, "The Influence of Reflections on Auditorium Acoustics," *J. Sound Vibration*, vol. 1, pp. 426–454 (1964).

[3] T. Houtgast and J. M. Steeneken, "A Review of the MTF Concept in Room Acoustics and Its Use for Estimating Speech Intelligibility in Auditoria," *J. Acoust. Soc. Am.*, vol. 77 (1985 Mar.).

[4] ANSI Std S3.2-1960 (R 1971), "Method for Measurement of Monosyllabic Word Intelligibility."

[5] K. D. Kryter, "Methods for the Calculation and Use of the Articulation Index," *J. Acoust. Soc. Am.*, vol. 34 (1962 Nov.).

[6] V. M. A. Peutz, "Articulation Loss of Consonants as a Criterion for Speech Transmission in a Room," *J. Audio Eng. Soc.*, vol. 19, pp. 915–919 (1971 Dec.).

[7] D. Green, *An Introduction to Hearing* (Wiley, New York, 1976).

[8] H. G. Latham, "The Signal-to-Noise Ratio for Speech Intelligibility—An Auditorium Design Index," *Appl. Acoust.*, vol. 12 (1979 July).

[9] M. R. Schroeder, "Modulation Transfer Functions: Definition and Measurement," *Acustica*, vol. 49 (1981).

THE AUTHOR

Ken Jacob is a member of the engineering staff of Bose Corporation, Framingham, MA. He received a bachelor's degree in acoustics from the University of Minnesota in 1981 and a master's degree in acoustics from the Massachusetts Institute of Technology in 1984, where his thesis topic was acoustic emission from superconducting magnets.

His interest in acoustics grew out of work as a sound designer and instructor in professional theater. He is also involved in music recording, and has produced two record albums.

LETTERS TO THE EDITOR

COMMENTS ON "SUBJECTIVE AND PREDICTIVE MEASUREMENTS OF SPEECH INTELLIGIBILITY—THE ROLE OF LOUDSPEAKER DIRECTIVITY"[†]

In the above paper[3] Mr. Jacob apparently misunderstands the concept of loudspeaker Q and the way in which it affects speech intelligibility in reverberant spaces. The generally specified value for the Q of a loudspeaker is an *axial* Q, that is, the value of Q with

---

[†] Manuscript received 1986 February 3.
[3] K. D. Jacob, *J. Audio Eng. Soc.*, vol. 33, pp. 950–955 (1985 Dec.).

reference to the loudspeaker's axis. The Q as perceived by a listener off axis is given by

$$Q_o = Q_a \times 10^{dB/10}$$

where Q_o is the perceived Q off axis, Q_a is the axial Q, and dB is the sensitivity of the loudspeaker at the off-axis angle relative to on axis (e.g., -6 dB for 20° off axis of a 40° horn). For most loudspeakers, those whose sensitivity falls off axis, the perceived Q off axis is lower than that on axis. Therefore as a result of this lowered perceived Q, and all other things being equal, the intelligibility is reduced off axis.

The effects on intelligibility of this lowered Q may be compensated for by the inverse square law. This is evident from the variable D^2, the source-to-listener distance, in the equation for $\%AL_{cons}$ [Eq. (1) in Mr. Jacob's paper]. Therefore if a listener were on the -6-dB angle of the loudspeaker *and* twice as close as a listener on axis, the effects of lowered Q and the inverse square law would cancel each other, and both listeners would enjoy the same intelligibility. This is the reason for all of the emphasis on achieving even direct sound coverage in reverberant spaces.

It is assumed that Mr. Jacob used *axial Q* to predict the $\%AL_{cons}$ at each listening location. (We are not told where the Q values for the three loudspeakers come from.) If the direct sound coverage was not even, and this appears to be the case, then errors of the type experienced by Mr. Jacob are to be expected. It has been my experience that when uniform direct sound coverage has been achieved, the intelligibility will be uniform and in good agreement with that predicted by the $\%AL_{cons}$ method. Had Mr. Jacob determined the Q at each listener position and used these individual values when predicting the $\%AL_{cons}$ at the individual locations, his predictions would have matched the intelligibility tests much more closely.

Also, on page 950, Mr. Jacob states that ". . . two listening locations were chosen, roughly in the middle and at the rear of the auditorium floor." He then states on page 951 that ". . . two listener locations were chosen to coincide with 1) the critical distance of the high Q source, and 2) the 'intelligibility distance' of the high Q source. . . ." It is unlikely that both of these statements will be satisfied at the same time in one room, let alone *five*. As the critical and limiting (intelligibility) distances are Q dependent, they will vary off axis in the same manner as the Q. I doubt that *any* listener was at the limiting distance, even for the low Q loudspeaker, in any of the five rooms.

In addition, I would like to point out that by deriving the intelligibility from the measured impulse response of the room in the cases of signal-to-noise procedure and modulation transfer function, Mr. Jacob has in effect measured the intelligibility. His Fig. 4 therefore compares not *predicted* but *measured* intelligibility with actual, the actual being the intelligibility measured in yet another way. The $\%AL_{cons}$ is the only value actually predicted, and then with erroneous input. Quick and easy methods do exist to measure $\%AL_{cons}$ directly—TEF, for example.

There seem to have been a great many variables in Mr. Jacob's experiments: Q varying with frequency, listeners at various off-axis angles from the loudspeakers, and so on. In order to limit better the variables to Q alone, I suggest, as an experimental setup, a sound system consisting of a bass cabinet and a high-frequency horn at one end of the test room. Aim the loudspeakers with their axes parallel to the floor and at ear height such that a listener may move fore and aft along the axis to any desired distance. Several manufacturers offer "families" of constant-directivity high-frequency horns. By substituting low, medium, and high Q horns from the same family into the sound system, the Q may be varied in a known way. Under these conditions, the axial Q is the Q perceived by the listeners, and a more accurate test of the effects of Q on intelligibility may be made.

FARREL M. BECKER
Audio Artistry
Kensington, MD 20795, USA

ADDITIONAL COMMENTS[‡]

It was with regret that we noted the failure of the *Journal* reviewers to detect and correct the many errors in the above paper.[3] For example, Eq. (3) leaves out the dimensional constant 0.4 normally associated with that equation when SI parameters are used. There is a nondimensional form of the same equation that is much more widely used:

$$D_c = 0.141 \sqrt{QS\bar{a}}.$$

(See Klein and Davis.[4]). The use of C_d for critical distance when *Journal* usage for the past 15 years has been D_c and the use of "intelligibility distance" in place of the agreed-upon "limiting distance" D_L would seem to indicate unfamiliarity with the literature by both the author of this paper and the *Journal* reviewers. A further irritation is the failure to provide dimensional labels in Eq. (2).

Neither Jacob nor the reviewers recognize that the audible difference to be expected from going from $Q = 1.0$ to $Q = 7.5$ is

$$10 \log \left(\frac{7.5}{1}\right) = 8.7 \text{ dB}$$

[‡] Manuscript received 1986 February 10.
[4] W. Klein and D. Davis, "Formulas for Distributed Loudspeaker Systems," *J. Audio Eng. Soc.* (*Letters to the Editor*), vol. 20, pp. 401–402 (1972 June).

whereas the difference in level between $Q = 7.5$ and $Q = 17$ is only

$$10 \log \left(\frac{17}{7.5}\right) = 3.5 \text{ dB} .$$

Perhaps the reason Jacob heard little difference between his "medium Q" array and his "high Q" loudspeaker is that it was the difference between a "medium Q" device and a "medium Q" device.

To have been consistent, he should have gone from $Q = 7.5$ to $Q = 7.5 \times 7.5 = 56.3$ for another

$$10 \log \left(\frac{56.3}{7.5}\right) = 8.7 \text{ dB} .$$

Since loudspeakers listed in today's catalogs have Q values up to 80+, it is a pity that Jacob did not try a true "high Q" device.

This writer feels that arrays do not have a Q, but rather an N (see Davis[5]). My personal experience has been that because of "lobing" in arrays, the calculation of their Q values becomes a highly subjective matter. Therefore it would have been useful to have seen the polar data used in Jacob's calculations.

Q is always at a point and not an area. Therefore any listener not sitting on the axis of the device is not experiencing the indicated Q value. The relative Q at other positions can be calculated from the polar response by

$$Q_{\text{rel}} = Q_{\text{axis}} (10^{\pm \text{dB}_{\text{CL}}/10}) .$$

In order to evaluate the accuracy of Jacob's data, we would require:
1) The seating plan of the listeners relative to the on-axis line of the loudspeakers
2) The distance to the listener
3) The variation in level with angle to each listener.

Other authors, notably Houtgast and Steeneken, have shown the close correlation between V. M. A. Peutz's %AL$_{\text{cons}}$ and intelligibility calculated from the MTF technique as validation of MTF (see Jacob's Ref. [3]). Jacob uses the same techniques to do the reverse. The truly remarkable work of Peutz during the past 14 years was completely overlooked by both Jacob and the reviewer.

As a user of all of the techniques named and a number not referred to in the measurement of speech intelligibility in real systems in real installations, may I be permitted the comment that rarely are signal-to-noise ratios or direct-to-reverberant sound ratios the culprit when poor intelligibility is encountered. Rather, poor coverage, misalignment of alike devices, improper equalization, and discrete high-level late reflections are the causes found and corrected.

The question raised, namely, "What is the relationship of the Q rating of a loudspeaker to the intelligibility experienced?" is a worthy one, but it must be explored with the other variables under control and documented.

DON DAVIS
Synergetic Audio Concepts
San Juan Capistrano, CA 92693, USA

AUTHOR'S REPLY[6]

In Response to F. M. Becker's Concerns

The author regrets not having stated more clearly the relationship between the listeners and the major axes of the loudspeakers. In all cases, listeners were within 10° of being on axis of the loudspeakers. (See Fig. A and Fig. 2 in the paper.) This being the case, axial Q was used as input to the %AL$_{\text{cons}}$ predictive formula. This clarification should also relieve concerns about the listening locations and their relation to "critical distance" and "intelligibility distance."

Readers should note that every axis is a "major" axis in the case of the omnidirectional loudspeaker. Note too that the largest errors in the %AL$_{\text{cons}}$ formula occurred not with the $Q = 17$ device, but rather with the $Q = 7.5$ and $Q = 1$ devices.

Speech intelligibility is a subjective parameter. By definition, any test paradigm for measuring intelligibility must include listeners. The measurement of a system's impulse response—or any other objective measurement for that matter—cannot, therefore, be used to measure speech intelligibility.

In Response to D. Davis's Concerns

It is not clear which audible differences are to be expected from changes in the directivity index ($= 10 \log Q$). If, by audible differences, differences in actual subjective speech intelligibility is meant, it is unknown whether the writer is referring to published or private data relating the two. The predictive formula shown in the paper to be least accurate uses as an input parameter loudspeaker directivity Q, rather than the directivity index ($= 10 \log Q$),

$$\%\text{AL}_{\text{cons}} = \frac{200 D^2 T^2}{QV} \tag{1}$$

(units are SI metric). Therefore the loudspeaker Q values in this experiment were chosen in part to reflect the importance placed on them by this predictive formula, as was stated in the body of the paper.

It is not true that "Houtgast and Steeneken have

---

[5] D. Davis, "The Modified Hopkins–Stryker Equation," J. Audio Eng. Soc., vol. 32, pp. 862–867 (1984 Nov.).

[6] Manuscript received 1986 April 28.

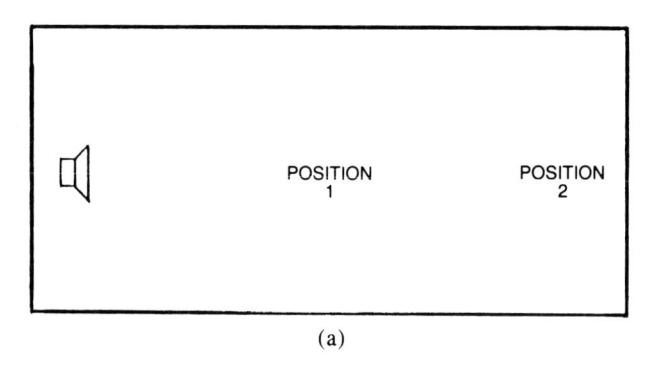

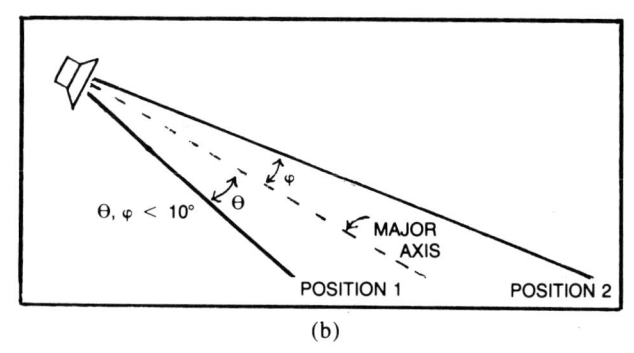

Fig. A. Relationship between listening positions and loudspeaker major axis. (a) Plan view. (b) Elevation.

shown the close correlation between . . . %AL$_{cons}$ and intelligibility calculated from the MTF technique." What has been stated by Mr. Steeneken (private communication and [3]), and what was stated in the body of the paper, is that intelligibility as predicted by the two techniques will agree only in the hypothetical case of an exponentially decaying squared room impulse response.

While Q is unambiguously defined for an array loudspeaker, it is true that these loudspeakers tend to exhibit lobes in certain frequency bands due to complex pressure summation from individual array elements. The effect of lobing on intelligibility, if any, should be treated in a separate experiment.

Readers should be aware that other algorithms have been developed by Peutz for predicting %AL$_{cons}$ since the one used here. While preliminary results indicate that Eq. (1) can be inaccurate, this experiment was in no way intended to discredit the work of Dr. Peutz or his associates.

KENNETH D. JACOB
Bose Corporation
Framingham, MA 01701, USA

Measurement of %AL$_{cons}$*

CAROLYN P. DAVIS

San Juan Capistrano, CA 92693, USA

Measurement of the %AL$_{cons}$ and RASTI scores for a sound system in a reverberant church are made and compared. A conversion chart is provided. The measurable intelligibility differences between a signal-aligned and a signal-unaligned array are presented. It is suggested that the time has arrived to use such tools in sound system specifications.

We recently had the opportunity to measure the effect on %AL$_{cons}$ by changing the loudspeaker directivity factor Q in a reverberant church (3.85 s for the one-third-octave band at 62 Hz and 2.36 s for the 2-kHz octave band. See Fig. 1 and Table 1.

The articulation measurements were made with a Techron TEF analyzer and double checked using a Bruel & Kjaer RASTI meter operated by R. Green and A. Perman of Bruel & Kjaer. We wished to compare directly the modulation transfer function method as exemplified in the Bruel & Kjaer instrument with the Peutz %AL$_{cons}$ method as used in the Techron TEF 10 and 12 analyzers The two methods were found to confirm the relationship shown in Table 2.

The church undergoing testing had a volume V = 230 000 ft^3 (6513 m^3). The measurement distance was 124 ft (37.8 m). The directivity factor of the lower device Q_L was 15 and that of the higher device Q_H was 45. Using the original Peutz equation, but with Q added, we can calculated for Q_L,

$$\%AL_{cons} = \frac{200(D_2)^2 (RT_{60})^2}{VQ_L}$$

$$= \frac{200(37.8 \text{ m})^2 (2.36 \text{ s})^2}{(6513 \text{ m}^3) (15)} = 16.3\%$$

and for Q_H,

$$\%AL_{cons} = \frac{200(D_2)^2 (RT_{60})^2}{VQ_H}$$

$$= \frac{200(37.8 \text{ m})^2 (2.36 \text{ s})^2}{(6513 \text{ m}^3) (45)} = 5.4\%$$

where the factor 200 is replaced by 656 if the U.S. customary measuring system is used and where

D_2 = distance from loudspeaker to listener
RT_{60} = reverberation time in seconds for a 60-dB change in level once the signal is turned off
%AL$_{cons}$ = percentage of articulation loss of consonants.

We measured for Q_L (Q = 15),

%AL$_{cons}$ = 18.2%   (see Fig. 2)
RASTI = 0.41   (see Table 2)

and for Q_H (Q = 45.9),

%AL$_{cons}$ = 7.4%   (see Fig. 3)
RASTI = 0.58   (see Table 2)

The conversions for %AL$_{cons}$ to RASTI are courtesy of F. Becker, who also made the Techron TEF measurements during this comparative measurement session.

---

* Manuscript received 1986 July 2; revised 1986 September 20.

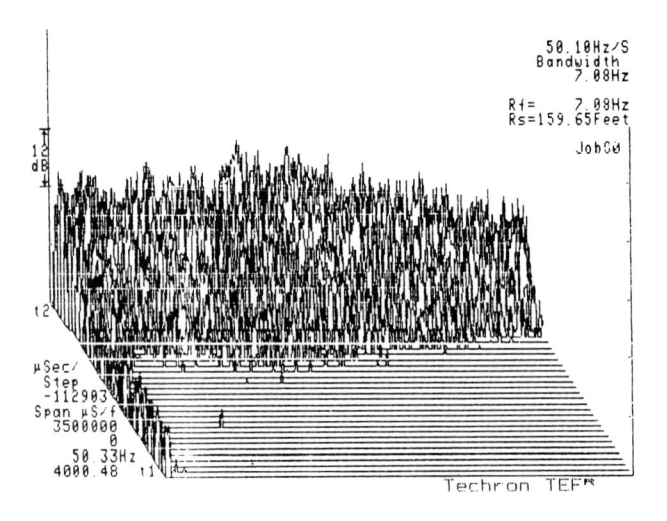

Fig. 1. Three-dimensional display of reverberation at Saint Boniface Church, Anaheim, CA.
Vertical scale—12 dB/div, with base of display at 73.4 dB; 0 dB is located at 0.00002 Pa. Horizontal scale—50.33 Hz to 4000.48 Hz; 1080.01 Hz/in or 425.20 Hz/cm. Resolution—1.5965E+02 ft and 7.0778E+00 Hz. Time of test—3 500 000 μs 3.9550E+03 ft (*front* to 0 μs 0.0000E+00 ft (*back*); −112 903 μs/step or −127.5806451613 ft. Sweep rate and bandwidth—50.10 Hz/s and 7.0778E+00 Hz. Input configuration—noninverting with 18 dB of input gain and 6 dB of IF gain.

The measurement techniques described here also accurately show what happens when the system is misaligned by the manufacturer or the installer. The measurements in Figs. 4–6 were taken in the overlap zone between a two-cell and a 10-cell horn. A listener standing sideways with one ear toward the array and one ear toward the rear wall heard sound from both directions at almost equal levels in the unaligned state, while in the aligned state the sound was heard only from the array. A 300-μs misalignment caused spurious lobes to excite the ceiling and sidewalls, whereas application of electronic alignment resulted in the actual polar patterns behaving as they would if each horn had been used alone.

In order to measure the polar effect of such misalignments in detail, we later measured a pair of small monitor loudspeakers set up first in alignment and then in approximately 300-μs misalignment. The polar response for a chosen frequency is shown in Figs. 7 and 8.

To summarize, the obtained data demonstrate the following.

1) %AL$_{cons}$ and RASTI measurements correlate to a remarkable degree of accuracy in this test.

2) The directivity factor Q directly affects speech intelligibility in a manner accurately predicted by the

Table 1. Reverberation time analysis at Saint Boniface Church, Anaheim, CA.

| Ln | Time | \multicolumn{7}{c}{f_c (Hz)} |
		62	125	250	500	1000	2000	4000
0	0	47.0	46.0	41.0	43.6	46.7	50.2	49.5
1	−113	41.6	46.0	42.5	43.1	46.3	49.6	48.6
2	−226	43.5	41.5	40.0	43.4	46.6	48.5	47.2
3	−339	44.7	47.8	42.3	44.1	44.1	49.3	48.6
4	−452	44.7	46.1	42.2	43.0	47.5	50.7	50.1
5	−565	43.1	43.2	42.4	44.7	47.1	48.9	48.4
6	−677	42.9	40.7	42.4	44.5	46.5	50.7	50.1
7	−790	31.0	39.6	41.6	44.0	47.3	49.7	49.5
8	−903	45.9	47.3	41.3	44.3	47.4	50.0	49.5
9	−1016	43.4	43.9	40.8	43.5	46.1	48.5	47.7
10	−1129	42.1	42.2	40.1	44.3	46.4	48.7	47.9
11	−1242	39.0	40.8	44.7	42.6	46.2	48.3	48.6
12	−1355	47.8	49.2	42.2	43.9	47.4	49.3	48.9
13	−1468	39.9	45.4	42.8	43.0	46.7	50.0	49.8
14	−1581	36.8	42.2	42.6	44.9	47.5	50.4	50.4
15	−1694	41.8	43.8	40.5	43.3	47.4	49.9	49.9
16	−1806	40.5	44.6	43.5	42.8	47.4	50.1	49.3
17	−1919	44.4	45.4	43.1	43.6	48.2	50.7	50.0
18	−2032	43.3	47.1	44.5	43.5	47.7	50.8	50.1
19	−2145	43.4	49.0	47.0	44.1	47.9	50.5	48.9
20	−2258	44.6	51.7	49.2	45.9	48.8	51.8	48.8
21	−2371	47.3	49.3	51.8	48.4	51.4	53.8	50.0
22	−2484	48.9	53.6	52.1	49.8	54.8	56.0	48.8
23	−2597	41.5	55.1	56.5	52.9	57.1	59.0	50.4
24	−2710	48.6	58.2	57.9	55.6	59.7	62.1	53.0
25	−2823	48.5	57.5	59.5	57.4	63.2	66.0	56.7
26	−2935	57.0	62.2	63.7	60.1	66.2	69.5	60.8
27	−3048	57.1	63.1	63.9	64.4	69.6	72.5	64.7
28	−3161	54.5	64.6	69.1	67.8	72.8	76.8	70.2
29	−3274	64.0	65.8	70.2	71.4	76.2	80.2	73.1
30	−3387	62.9	63.7	70.2	71.5	76.4	82.2	74.5
31	−3500	57.2	55.3	61.9	63.2	69.6	76.0	67.4
\multicolumn{9}{c}{Reverberation times}
RT(21/31)		3.85	6.18	3.89	3.00	2.79	2.36	2.48

original Peutz equations when such equations are used correctly.

3) The listening group of approximately 20 persons concurred that the perceived changes in intelligibility matched the measurement figures. (The 20 observers were members of a special workshop on the application of TEF analysis plus the Bruel and Kjaer personnel present. The group consisted of persons experienced in the design and installation of sound reinforcement systems.)

4) Misalignment measurably affects intelligibility.

Since %AL$_{cons}$ can be accurately measured with the Techron TEF analyzer and with the Bruel and Kjaer RASTI meter, the measurement should become part of every sound system specification, as should the measurement of signal alignment now that several manufacturers have announced the availability of precision alignment devices.

Table 2. Converting RASTI to %AL$_{cons}$.

	RASTI	%AL$_{cons}$
Bad	0.20	57.7
	0.22	51.8
	0.24	46.5
	0.26	41.7
	0.28	37.4
Poor	0.30	33.6
	0.32	30.1
	0.34	27.0
	0.36	24.2
	0.38	21.8
	0.40	19.5
	0.42	17.5
	0.44	15.7
Fair	0.46	14.1
	0.48	12.7
	0.50	11.4
	0.52	10.2
	0.54	9.1
	0.56	8.2
	0.58	7.4
Good	0.60	6.6
	0.62	5.9
	0.64	5.3
	0.66	4.8
	0.68	4.3
	0.70	3.8
	0.72	3.4
	0.74	3.1
	0.76	2.8
	0.78	2.5
	0.80	2.2
	0.82	2.0
	0.84	1.8
	0.86	1.6
Excellent	0.88	1.4
	0.90	1.3
	0.92	1.2
	0.94	1.0
	0.96	0.9
	0.98	0.8
	1.00	0.0

AL$_{cons}$ = 170.5405*EXP(−5.419*STI)
STI = −0.1845*Ln(%AL$_{cons}$) + 0.9482

Source: Courtesy Farrel Becker.

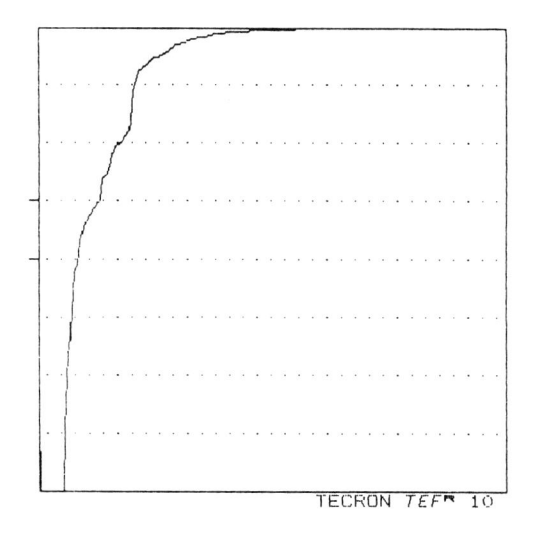

Fig. 2. Integrated ETC of house system at Saint Boniface Church, Anaheim, CA. Rear of church; $Q = 15$, %AL$_{cons}$ = 18.2%.

Vertical scale—linear relative. Horizontal scale—0 to 2 003 340 μs. Sweep range—20 000 μs. Reverberation time—2.24 s. Time interval used for calculation—$T_1 = 180\ 753$ μs; $T_2 = 1\ 220\ 080$ μs. Energy ratio $E(T_A - T_B)/E(T_B - T_C) = -23.36$ dB. Time settings used for calculation—$T_A = 95\ 397$ μs; $T_B = 110\ 460$ μs; $T_C = 2\ 003\ 340$ μs. Stationary SPL—64 dB. AL$_{cons}$ without noise = 18.2%.

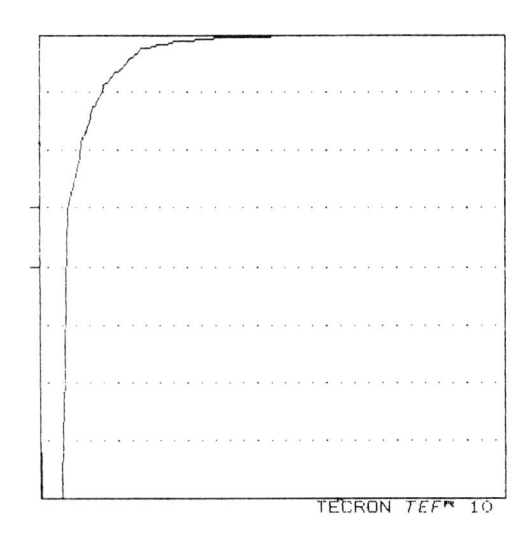

Fig. 3. Integrated ETC at Saint Boniface Church, Anaheim, CA. Rear of church; $Q = 45.9$; %AL$_{cons}$ = 7.4%.

Vertical scale—linear relative. Horizontal scale—0 to 2 003 340 μs. Sweep range—1900 Hz to 2099 Hz. Time constant—20 000 μs. Reverberation time—2.18 s. Time interval used for calculation—$T_1 = 175\ 732$ μs; $T_2 = 1\ 139\ 750$ μs. Energy ratio $E(T_A - T_B)/(T_B - T_C) = -0.70$ dB. Time settings used for calculation—$T_A = 80\ 335$ μs; $T_B = 115\ 481$ μs; $T_C = 2\ 003\ 340$ μs. Stationary SPL—68 dB. AL$_{cons}$ without noise = 7.4%. Minimum needed signal-to-noise ratio for AL$_{cons}$ = 10%—14 dB.

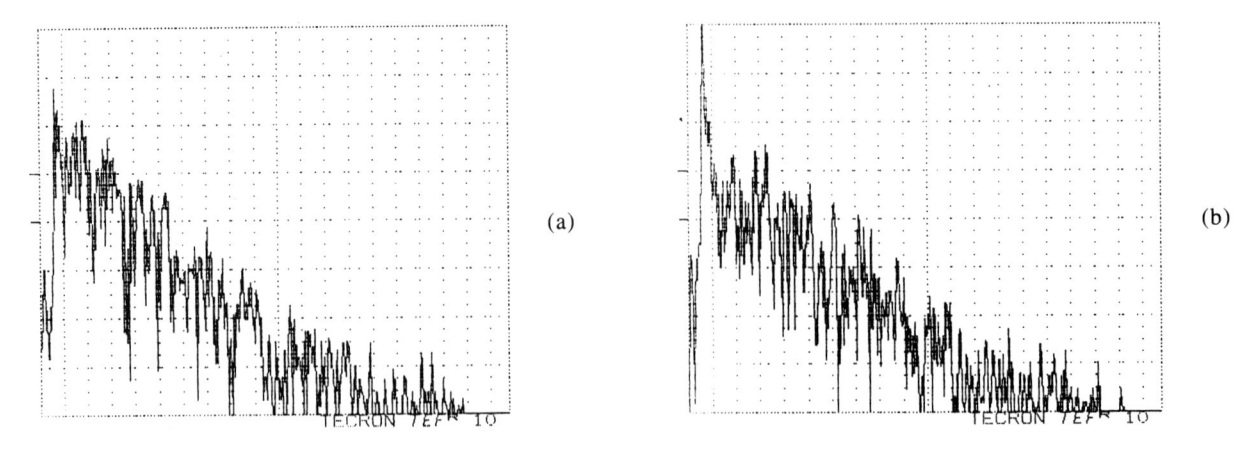

Fig. 4. ETC of house system alignment at Saint Boniface Church, Anaheim, CA.
Vertical scale—6 dB/div with base of display at 13.4 dB; 0 dB is located at 0.00002 Pa. Horizontal scale—0 µs or 0 ft to 1 971 547 µs or 2227.84 ft; 6.0912E+02 ft/in or 2.3981E+02 ft/cm; 5.39039E+5 µs/in or 2.1222E+5 µs/cm. Line spacing—4941.21 µs or 5.58357 ft. Line width—6720.05 µs or 7.59366 ft. Sweep rate—20.04 Hz/s. Sweep range—1399.52 Hz to 1601.90 Hz. Input configuration—Noninverting with 18 dB of input gain and 6 dB of IF gain. (a) Middle of misalignment area; no delay. (b) Same as (a) but with 300-µs signal delay on near-throw horns.

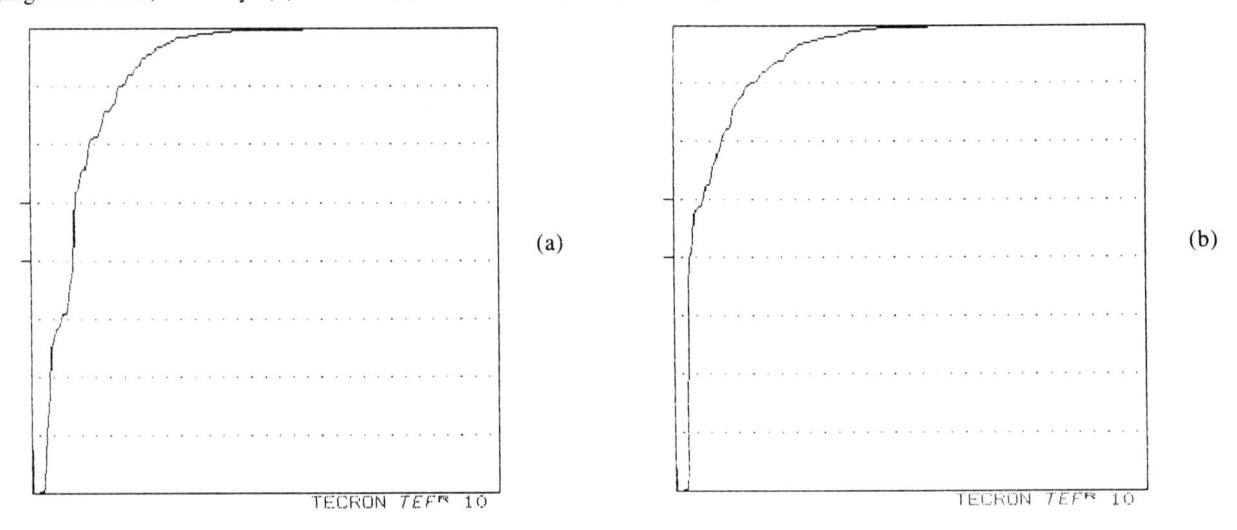

Fig. 5. Integrated ETC of house system alignment at Saint Boniface Church, Anaheim, CA.
Vertical scale—linear relative. Horizontal scale—0 to 2 003 340 µs. Sweep range—1900 Hz to 2099 Hz. Time constant—20 000 µs. (a) 10.5 %AL$_{cons}$ in overlap area when unaligned. Reverberation time—2.16 s. Time interval used for calculation—T_1 = 200 836 µs; T_2 = 1 079 500 µs. Energy ratio $E(T_A - T_B)/E(T_B - T_C)$ = -5.76 dB. Time settings used for calculation—T_A = 40 167 µs; T_B = 85 356 µs; T_C = 2 003 340 µs. Stationary SPL—65 dB. (b) 6.9 %AL$_{cons}$ in overlap area when aligned. Reverberation time—2.35 s. Time interval used for calculation—T_1 = 251 046 µs; T_2 = 1 024 270 µs. Energy ratio $E(T_A - T_B)/E(T_B - T_C)$ = 0.75 dB. Time settings used for calculation—T_A = 45 188 µs; T_B = 85 356 µs; T_C = 2 003 340 µs. Stationary SPL—68 dB. Minimum needed signal-to-noise ratio for AL$_{cons}$ = 10%—12 dB.

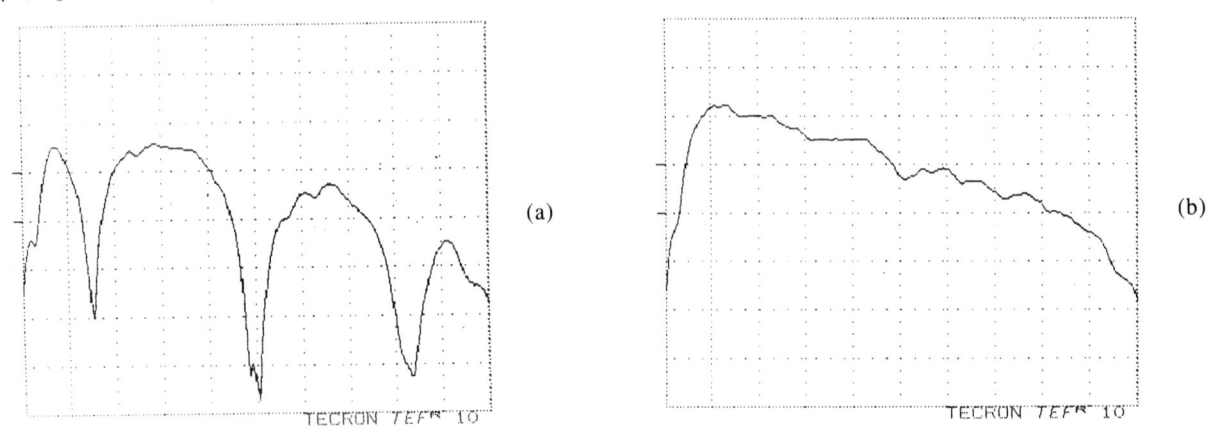

Fig. 6. Mag versus frequency (EFC) of house system alignment at Saint Boniface Church, Anaheim, CA.
Vertical scale—6 dB/div with base of display at 34.4 dB; 0 dB is located at 0.00002 Pa. Horizontal scale—50.33 Hz to 10001.20 Hz; 2720.67 Hz/in or 1071.13 Hz/cm. Resolution—2.1053E+00 ft and 5.3674E+02 Hz. Time of test—61 198 µs, 6.9153E+01 ft. Sweep rate and bandwidth—10 734.80 Hz/and 2.0000E+01 Hz. Input configuration—noninverting with 18 dB of input gain and 9 dB of IF gain. (a) Linear frequency response in overlap area when unaligned. (b) Linear frequency response in overlap area when aligned.

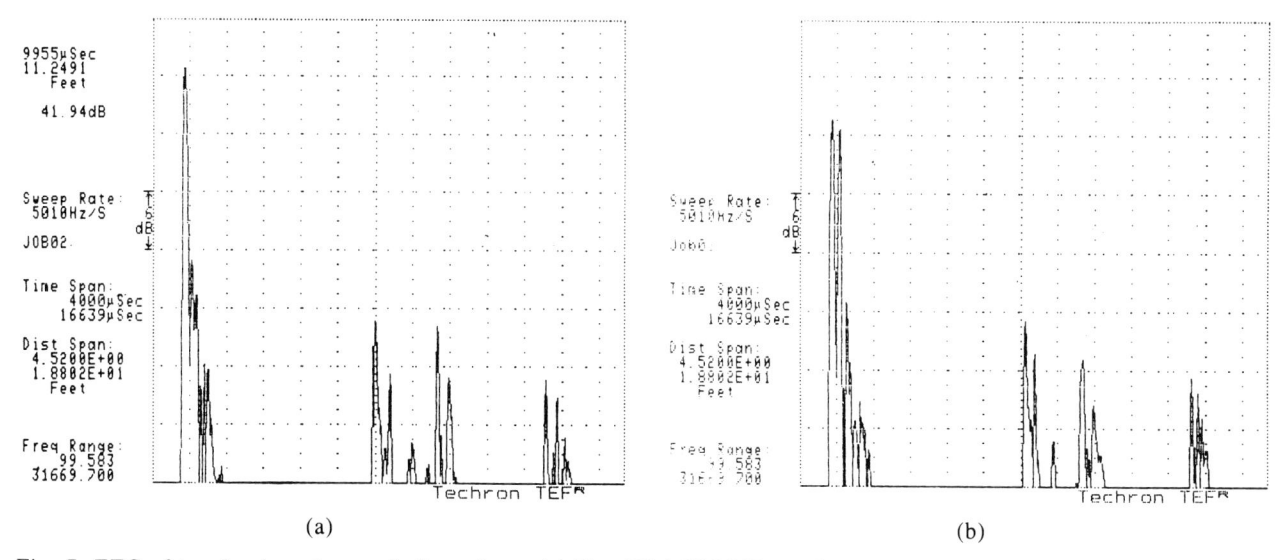

Fig. 7. ETC of two loudspeakers at Indiana farm. (a) Two TOA 280 ME monitors stacked in alignment. (b) Two TOA 280 ME monitors stacked out of alignment.

Vertical scale—6 dB/div with base of display at 25.4 dB; 0 dB is located at 0.00002 Pa. Horizontal scale—4000 μs or 4.52 ft to 16 639 μs or 18.8015 ft; 3.9047E+00 ft/in or 1.5373E+00 ft/cm; 3455 μs/in or 1360 μs/cm. Line spacing—31.6755 μs or 3.57934E−2 ft. Line width—43.0787 μs or 0.048679 ft. Sweep rate—5009.55 Hz/s. Sweep range—99.58 Hz to 31669.70 Hz. Input configuration—noninverting with 18 dB of input gain and 12 dB of IF gain.

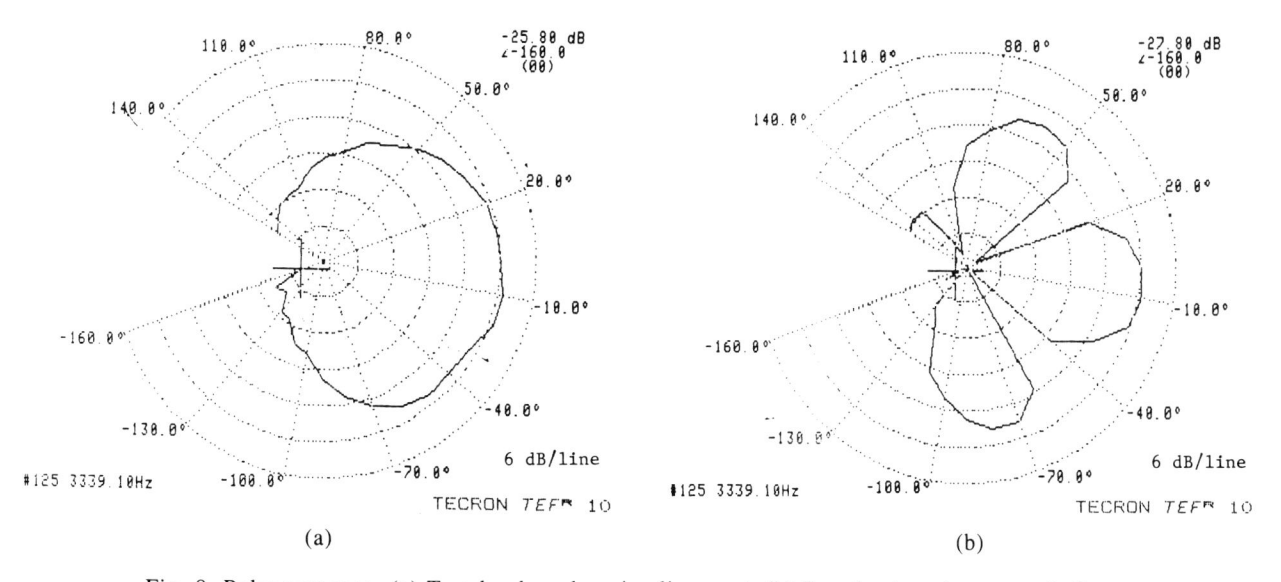

Fig. 8. Polar response. (a) Two loudspeakers in alignment. (b) Two loudspeakers out of alignment.

THE AUTHOR

Carolyn Davis attended Purdue University, class of 1951, and has been actively engaged in audio for over 30 years. She worked with her husband, Don Davis, in 1967 on the development of one-third-octave equalization in their home laboratory, and was the first user of one-third-octave real-time analyzers for sound system equalization. She co-authored, with her husband, *Sound System Engineering*, which was published in 1975 by Howard W. Sams. For the past 13 years, Mrs. Davis has co-authored the Syn-Aud-Con Newsletter, a publication of approximately 50 pages per issue. For the last 14 years she has assisted in conducting audio engineering seminars and workshops for over 6000 audio personnel. Mrs. Davis has been a member of the Audio Engineering Society since 1968, serving as a session chairman in 1977 and convention chairman in 1980. She and her husband received the annual achievement award in 1978 from the Upper Midwest Chapter of the Acoustical Society of America. She was a recipient of the AES Board of Governors Award.

Application of Speech Intelligibility to Sound Reinforcement*

DON DAVIS AND CAROLYN DAVIS

Synergetic Audio Concepts, Norman, IN 47264, USA

Sound reinforcement and reproduction systems have grown to a scale where church committees and auditorium owners spend up to $1 000 000 for them. In the past, such systems were purchased without any assurance that acceptable speech intelligibility would be achieved. Today speech intelligibility can be specified in advance, designed for, and objectively measured with an accuracy as good as that achieved using a panel of "live" listeners. The competing techniques are described and evaluated, and some of the remaining problem areas encountered in analyzing nonstandard systems are outlined. The Peutz percent articulation loss of consonants technique ($\%AL_{cons}$), the speech transmission index (STI), and the rapid speech transmission index (RASTI) are all defined, compared to live listener tests made with a large listener sample, and illustrated using currently available analyzers.

0 INTRODUCTION

Speech intelligibility should be a primary goal of a sound reinforcement system. Until recently the only reliable way to measure the ability of a sound reinforcement system to improve speech intelligibility over that produced by an unassisted live talker was to conduct lengthy and expensive tests with a significant number of trained listeners.

Thus measured, the speech intelligibility score is that score made by such live listeners in a given test situation. However, such scores can also be predicted using calculations at the drawing board during the design stage of a project by means of algorithms intended to correlate with speech intelligibility scores. The correctness of such predictions can also be measured in an actual space, with or without a sound system, and these measurements may, if necessary, be verified by live listeners.

Speech intelligibility tests, including the percentage of articulation loss of consonants $\%AL_{cons}$, are made using standardized lists of words, phrases, or sentences, or combinations of these. On the other hand, the predictions are made with techniques such as the information index (II), the articulation index (AI), predicted $\%AL_{cons}$, and modulation transfer functions such as the speech transmission index (STI) and the rapid speech transmission index (RASTI).

In 1987, four different measurement techniques came into practical use:
1) Techron TEF$\%AL_{cons}$
2) Techron TEF full STI
3) Techron TEF RASTI
4) Bruel & Kjaer RASTI.

The measurement techniques relate to the II, the AI, and the STI, all of which attempt to quantify the elements that relate to speech intelligibility.

Two analyzers have been verified in the presence of qualified commercially disinterested witnesses, the TEF analyzer and the Bruel & Kjaer analyzer. The Techron system 12 utilizes the Heyser transform for the gathering of data. Fig. 1 shows the Heyser transform.

The more familiar Fourier transform (Fig. 2) is a special degenerate case of the more general Heyser transform. The Fourier transform is a zero curvature hyper*plane*, which restricts its use to linear systems only, whereas the Heyser transform is a hyper*surface*, which can accommodate nonlinear systems.

The Techron system 12 measures:
1) $\%AL_{cons}$

---

* Presented at the 85th Convention of the Audio Engineering Society, Los Angeles, 1988 November 3–6; revised 1989 June 1 and September 29.

2) Full STI (about 1½-min measurement)
3) RASTI (about ½-min measurement).

The Techron system 12 utilizes the Schroeder modulation transfer function (MTF) equation for STI and RASTI measurements (Fig. 3), which is the mathematical expression for "the magnitude of the Fourier transform of the squared impulse response" (Fig. 4). Fig. 5 shows one form MTF displays can take. The second analyzer, the Bruel & Kjaer RASTI analyzer, reduces the seven octave bands of STI to two (the 500- and 2000-Hz octave bands) and utilizes the modulation transfer function technique proposed by Houtgast and Steeneken. This particular embodiment of the MTF has been found by Peutz [1] of The Netherlands and Humes [2] of Indiana University, in separate research, to be a masking paradigm rather than a more general measure of speech intelligibility.

Speech intelligibility algorithms include 1) the Peutz algorithm as exemplified in the TEF analyzer; 2) the MTF (Schroeder transform), again as exemplified in the full STI program in the TEF analyzer; and 3) the Bruel & Kjaer RASTI version of the MTF. Partial algorithms are 1) the Lochner–Burger signal-to-noise algorithm and 2) the Bradley signal-to-noise algorithm.

Such analyzers must be shown to have good correlation to accepted standard intelligibility tests using live listeners. The analyzers should also provide safeguards against misleading reading, which can result with techniques such as impulse measurements utilizing personal computers, where indications of nonlinearity, noncoherence, or poor signal-to-noise ratios are not provided in the programming.

To test the two analyzers described, and to compare the viability of their respective predictors to other predictors such as the Lochner–Burger and the Bradley signal-to-noise algorithms, the 1986 Speech Intelligibility Workshop was organized.

1 1986 SPEECH INTELLIGIBILITY WORKSHOP

Under the auspices of Syn-Aud-Con a special speech intelligibility workshop was convened in the Chicago area in order to compare the two analyzers and four techniques against conventional intelligibility tests using phonetically balanced modified rhyme tests from Dynastat in Austin, TX, where the test word is imbedded in a carrier sentence. A group of 90 audio professionals participated in the evaluation of three speech-reinforcement systems. Three different acoustic environments were employed.

1) An excellent space—a classic restored motion picture theater of the early 1930s, where an unaided voice easily provided excellent intelligibility to all 2000 seats
2) A difficult space—a concert hall of about 800 seats that exhibited severe speech intelligibility problems 20 ft (6 m) from a talker on stage
3) A highly reverberant 1 000 000-ft^3 (764 600-m^3) Catholic Church.

The 90 listeners were divided into three groups of 30 each so that during a three-day testing period all three groups experienced all three environments. The group of 30 were divided again into two groups of 15 each, and while one group was being scored in the near field of the loudspeakers under test, the second group was in the far reverberant field. We are only concerned here with the measurements and tests made in the far field. The members of each group of 15 sat in close proximity to each other (Fig. 6). Each group had the opportunity to hear all three systems in all three test sites from identical listening locations. Three sound-reinforcement systems on electric hoists were in each environment.

1) A very low Q (directivity factor), nearly omnidirectional

Fig. 1. Heyser transform illustrated as a hypersurface. This is the general case transform of which the Fourier and Hilbert transforms are degenerate special cases. The relevance of this transform to intelligibility measurements is its ability to allow processing of nonlinear systems. (*Courtesy R. C. Heyser.*)

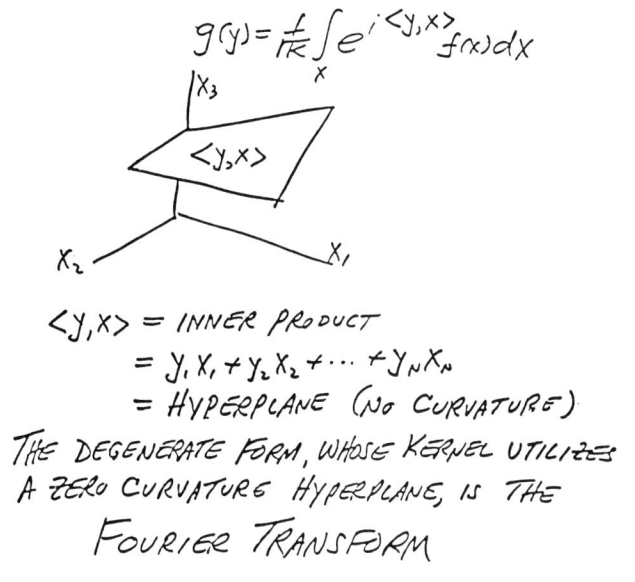

Fig. 2. Fourier transform illustrated in the same manner as Fig. 1, but as a hyperplane. Fourier transform is only valid for linear system analysis. (*Courtesy R. C. Heyser.*)

2) A medium Q, approximately 4–5
3) A high Q, approximately 50.

The tests were conducted under the supervision of experienced professional design consultants, David Klepper, Rolly Brook, and V. M. A. Peutz. Instrumentation was supervised by Bruel & Kjaer, Techron, and Bose personnel. Loudspeakers were provided by BES, Bose, Community, EV, and JBL [3], [4].

$$\text{MTF} = \frac{\int_0^\infty P^2(t)e^{-j\omega t}\,dt}{\int_0^\infty P^2(t)\,dt}$$

Fig. 3. Three key transforms for use in the development of impulse measurements for speech intelligibility analysis: Heyser transform, Fourier transform, and Schroeder transform. This Schroeder MTF transform is the underlying transform function. The technique used in the Bruel & Kjaer RASTI is an approximation of this transform, not an implementation of it.

1.1 Loudspeaker Measurements

Each loudspeaker used in the speech tests had its polar response, its frequency response (both amplitude and phase), and its energy–time curve measured prior to use in the test system. Each loudspeaker was again measured in the system, with the measuring microphone placed in the seating area of the live listening group. Each system was equalized under the supervision of the engineers from the manufacturer of the loudspeakers used. High-frequency horns were used with associated low-frequency units and crossover to ensure full-range loudspeakers for each test. Levels were chosen to ensure at least 25-dB signal-to-noise ratio in the 1000- and 2000-Hz one-octave band.

The authors cannot stress too vigorously that all tests were conducted under the supervision of recognized authorities and with full agreement at the time of the tests from the manufacturers' engineers, who were present and participating in the tests.

In the attempt to predict the behavior of given loudspeakers in advance of the actual measurements, it was

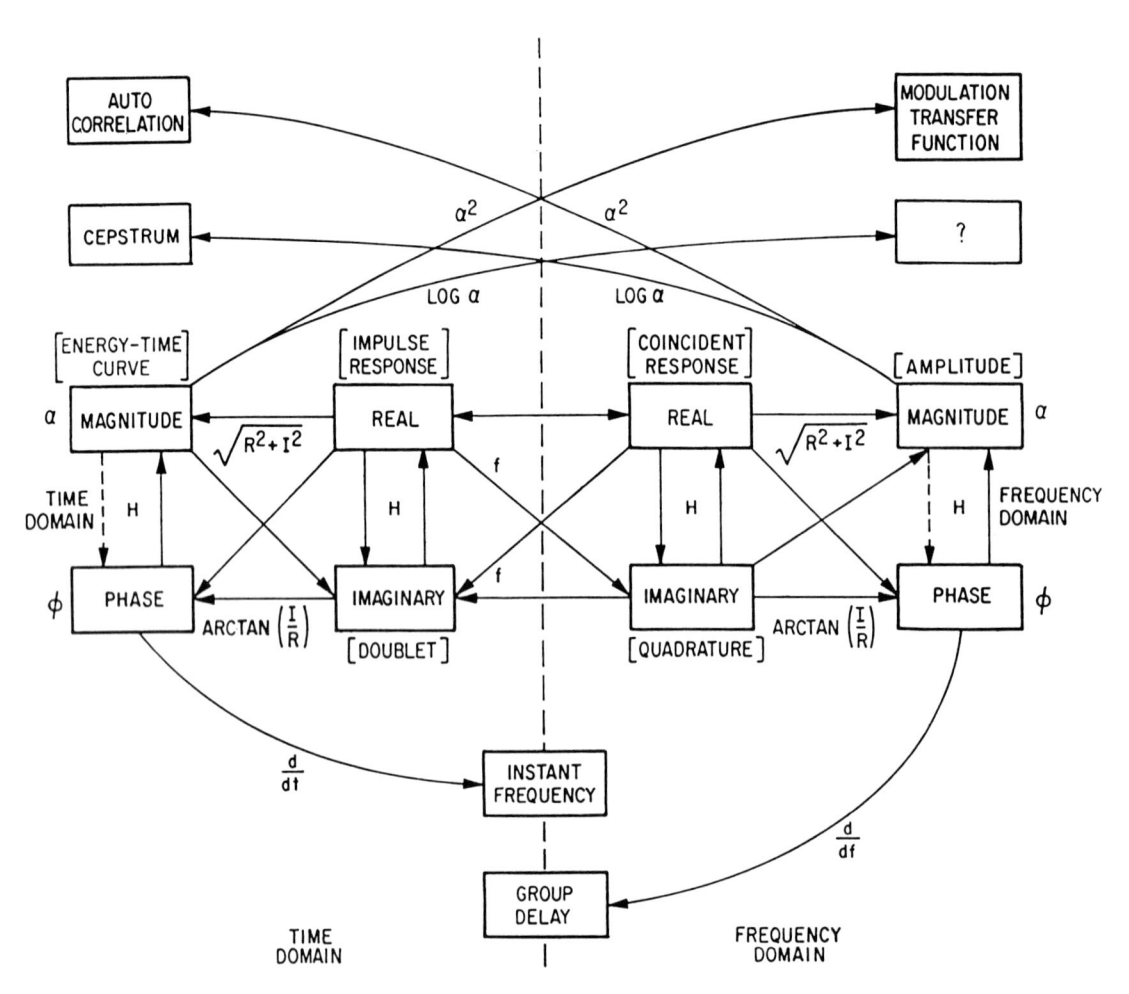

Fig. 4. Conventional measurement domains and their offspring. F—Fourier transformations; H—Hilbert operations. Note particularly that the relationship between magnitude and phase is a Hilbert operator only if the system under test is a minimum phase system. There is also a modulation domain and a delay-plane domain, originally identified by Heyser, and which promise many future insights. Those conversant with FFT technology often make fundamental errors of judgment when first introduced to TEF analysis, with its frequency-domain acquisition of signals and its nonreal-time postprocessing of impulse and ETC responses, such as not realizing where "windowing" can be applied in the processing or what type would be optimum or appropriate.

found that, for devices with accurate parameter specifications, accurate predictions could be made, but that for others the frequency-dependent variations, in polar response, for example, would not allow either accurate specification or prediction (Fig. 7). If the polar response of a loudspeaker is too uneven to derive a single number Q from it, as in Fig. 7(a), it is impossible to use the Peutz %AL$_{cons}$ formula to predict an accurate performance of a loudspeaker in a given space,

$$\%AL_{cons} = \frac{656*(D_2)^2(RT_{60})^2}{VQ} \quad *Si = 200$$

where

V = internal volume of room, cubic feet or cubic meters
D_2 = distance from source to listener, feet or meters
Q = directivity factor of source, dimensionless
RT_{60} = time for reverberant sound field to decay 60 dB, seconds.

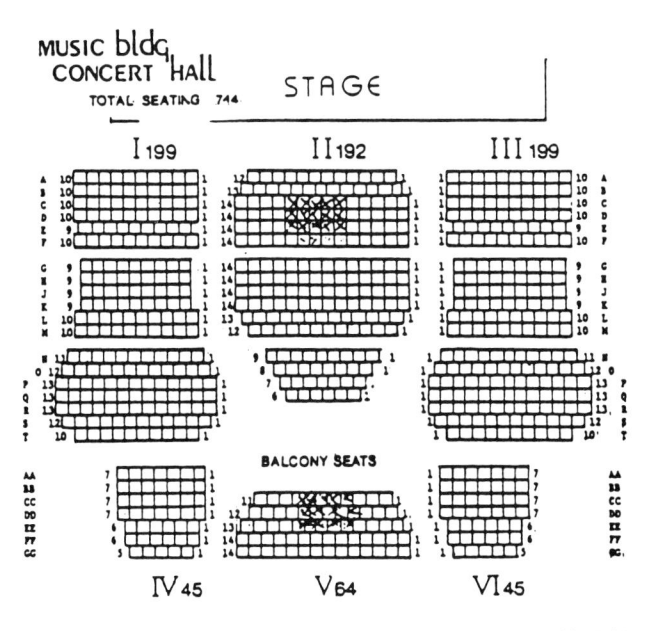

Fig. 6. Seating plan of Music Building Concert Hall, 744 seats, showing the location of the two groups of 15 live listeners at front and back of auditorium.

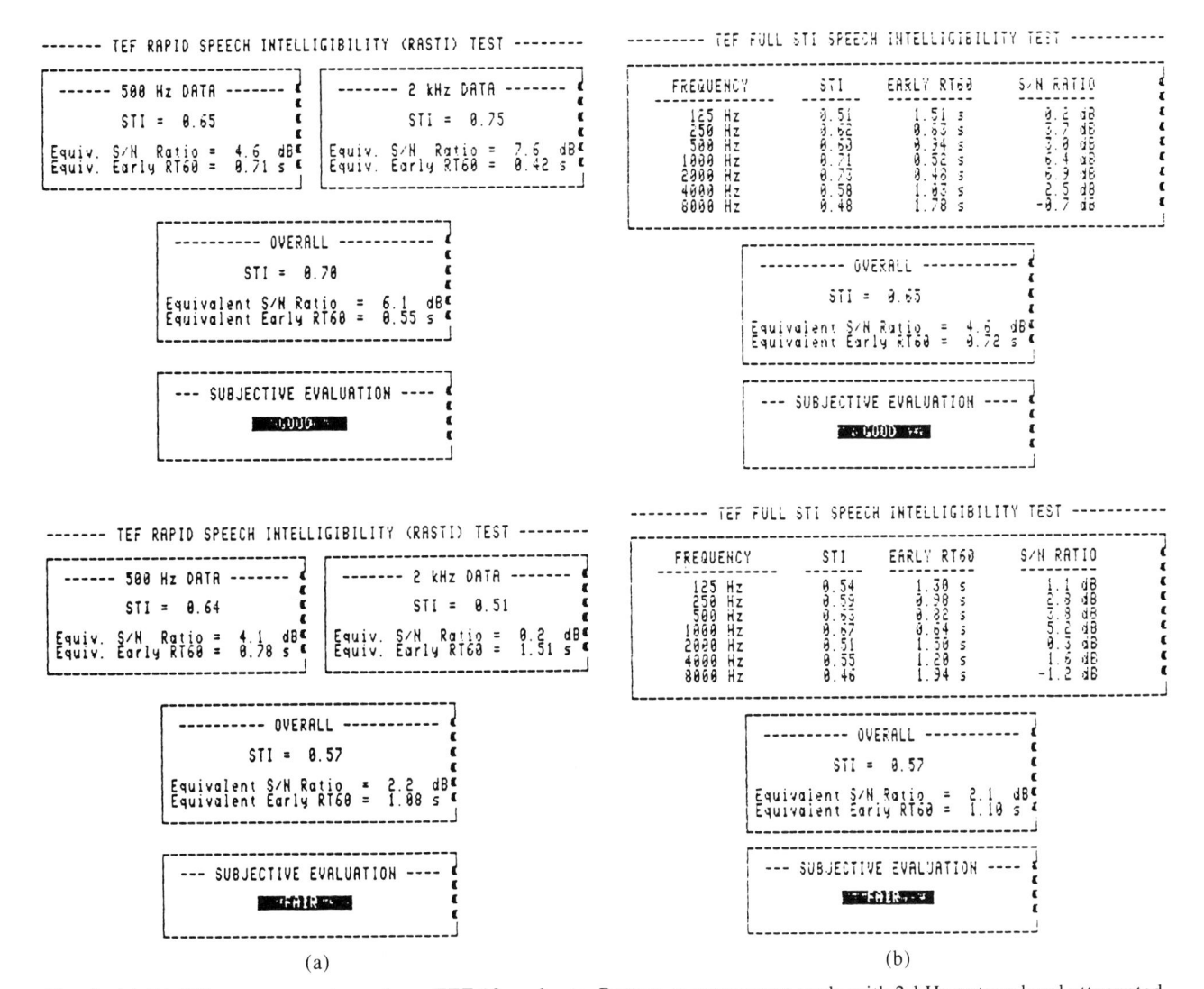

Fig. 5. (a) RASTI measurements made on TEF 12 analyzer. Bottom measurement made with 2-kHz octave band attenuated. The degradation was introduced by generating a comb filter that attenuated the 2-kHz frequency. (b) Full STI of same tests as 5(a).

(a) (b) (c)

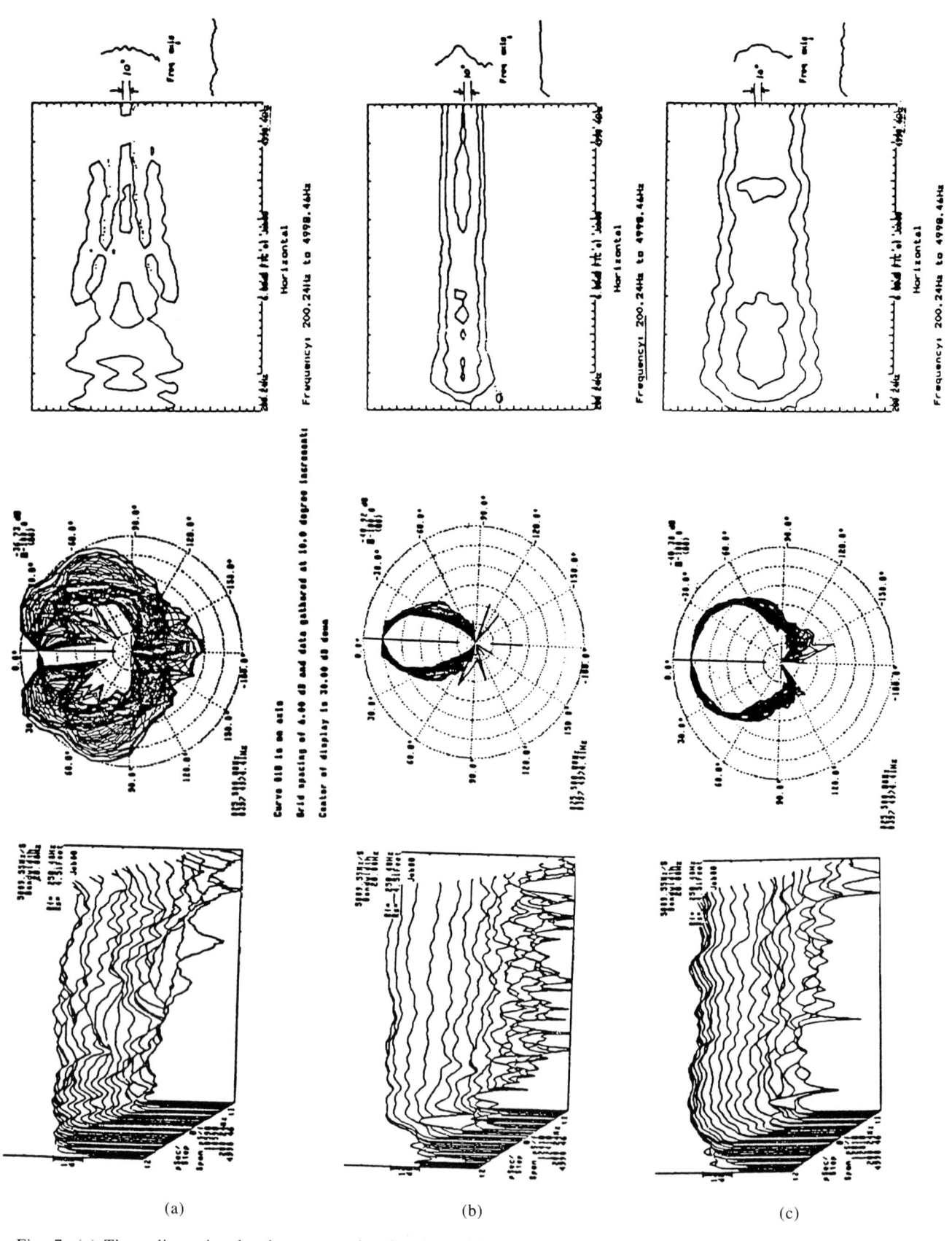

Fig. 7. (a) Three-dimensional polar, conventional and overlaid polars, and frequency versus polar plots of a Bose 802 loudspeaker system. A system that varies this dramatically between 500 and 3000 Hz (shown here) makes predictions of its performance with regard to speech intelligibility impossible. By prediction we mean its performance predicted by an assigned Q and coverage angle for 1000 and 2000-Hz octave bands actually representative of the frequency range covered. The original Peutz equation has proved its accuracy for over two decades when controlled polar pattern devices are used. (b) Same plots as in (a), but for EV 40 × 20 constant-directivity horn. As can be seen, this device lends itself to precise predictions of performance. (c) JBL loudspeaker system showing that good polar control is possible for both wide and narrow coverage patterns. Again, a perfectly predictable device in terms of speech intelligibility.

Using this equation with the data for the EV horn used in the tests, excellent correlation between measurement and prediction was obtained. While the performance of the Bose units could be "measured," it could not be predicted using this equation of useful architectural parameters. We believe that this does not constitute a fault of the equation, but a problem in the device.

The polar pattern of one cone or horn within a loudspeaker system may be excellent, but if missynchronized with the other cones or horns within the loudspeaker system, the polar pattern becomes completely different than the polar response of the individual cone or horn. The sound of missynchronization in music may be pleasant, but often it degrades the quality of speech, especially in a reverberant space, since the lobing of the loudspeaker often is directed toward the hard surfaces of the floor, ceiling, and sidewalls, and not where the loudspeaker is aimed. The reader should keep in mind that the polar responses of microphones can become quite erratic, especially when using more than one at a time. Individual devices with well-behaved polar responses and smooth frequency responses can become quite erratic when missynchronized with similar devices (Fig. 8 [5]).

1.2 Results of the Tests

The directivity factor Q was found to be a fundamental parameter in speech intelligibility in difficult acoustic environments. Table 1 and Fig. 9 summarize the results of the test conducted during this workshop. As expected, in the well-controlled Paramount Theater the loudspeakers performed well. It was in the reverberant Cathedral of St. Raymond that the live listener and the

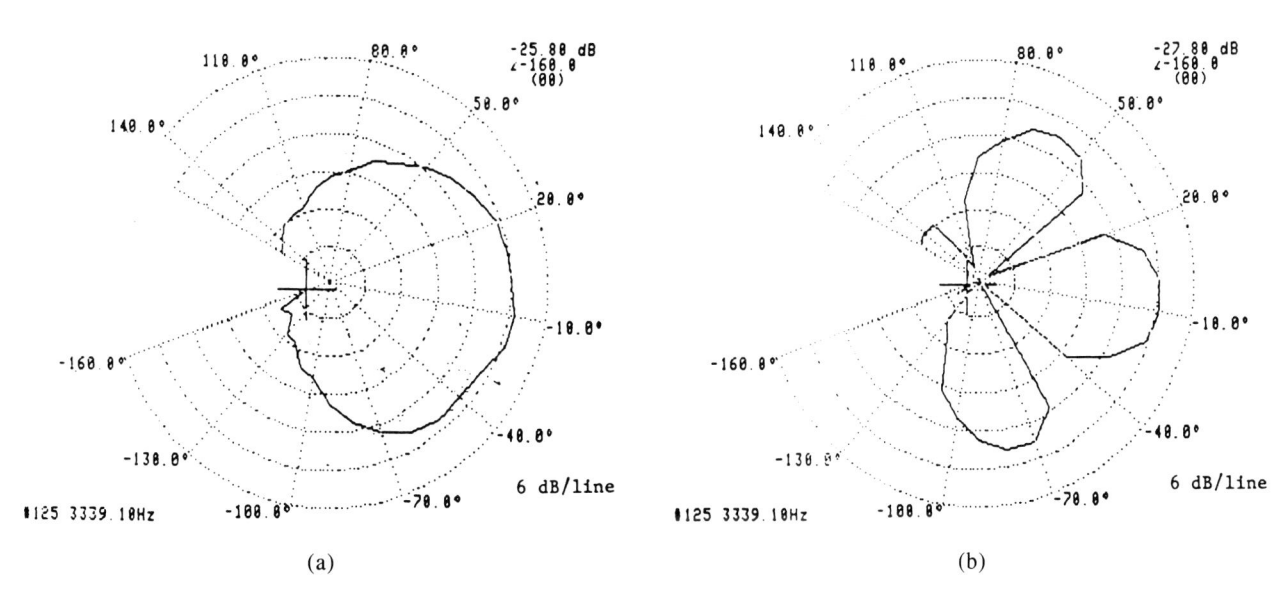

Fig. 8. (a) Polar plot of two identical loudspeakers aimed at same location. Here loudspeakers are in perfect synchronization. (b) Same two loudspeakers as in (a), but 4 in (0.1 m) out of synchronization. This kind of unexpected "lobing" can excite reverberant spaces in highly detrimental ways.

Table 1. Summary of average live listener %AL$_{cons}$ scores compared to B&K RASTI and TEF %AL$_{cons}$.*

Loudspeaker	Live listener %AL$_{cons}$	Standard deviation	B&K RASTI	Equivalent %AL$_{cons}$ †	TEF %AL$_{cons}$
1. Paramount Theater (RT_{60} = 1.5 s at 2 kHz; 2000 seats)					
BES SM200	6.6	2.8	0.62	5.9	7.87
Bose 802	7.05	3.1	0.62	5.9	7.87
EV HP 40 × 20	4.0	3.1	0.78	2.5	4.01
2. Northern Illinois University Concert Hall (RT_{60} = 2.0 s at 2 kHz; 744 seats)					
BES SM200	11.6	4.3	0.41	18.8	12.53
Bose 802	13.1	5.3	0.50	11.4	12.88
EV HP 40 × 20	4.5	2.5	0.67	4.5	4.14
3. Cathedral of St. Raymond (RT_{60} = 3.0 s at 2 kHz; 1500 seats)					
BES SM200	15.4	6.3	0.42	17.5	15.05
Bose 802	15.2	5.3	0.51	10.5	14.11
EV HP 40 × 20	6.4	3.3	0.66	4.8	4.71

* Results of a study of 90 participants in three different environments—a well-controlled 2000-seat theater, a very difficult acoustic 744-seat concert hall, and a 3-s cathedral, using a low-Q (BES 200), a medium-Q (Bose 802), and a high-Q loudspeaker (EV HP 40 × 20 horn with DH-1A driver).
† B&K RASTI scores converted to an equivalent %AL$_{cons}$ by means of the chart in Fig. 9a.

BES and Bose had the most difficulty.

A mass of data was acquired and their analysis is still ongoing. We felt that there were three outstanding results immediately evident.

1) The Techron TEF analyzer and the Bruel & Kjaer RASTI analyzer both worked reliably in conditions where no bizarre singularities such as severe nonlinearity (that is, speech compression) existed. Their scores correlated with a group of live listeners.

2) The ear–brain system responded to what, in measurement, was a single source, either direct sound or, in special cases, a high-quality specular reflection, and that integration of both direct and early reflected sound did not allow the score to be correlated with live listeners.

3) The TEF RASTI and STI, using the Schroeder integral to obtain the MTF, correlated with the Bruel & Kjaer RASTI scores under identical conditions.

4) The workshop demonstrated that higher Q loudspeakers give better intelligibility scores than lower Q devices. We occasionally hear people say that sound fields mask missynchronized sources. The results from the workshop demonstrated that this is not the case. Some feel that the only penalty paid for missynchronization is amplitude variations (comb filters). (Fig. 5 shows that the introduction of a comb filter in the 2kHz region lowers intelligibility from RASTI excellent to RASTI good.) Erratic polar responses often cause excitation of boundary surfaces in highly reverberant spaces with disastrous consequences to the signal-to-noise ratio.

It is important to note that this was a large sample supervised by competent personnel, and at the present writing only two instruments tested have been verified independently. A significant number of those who attended and participated in the tests have found that they can now judge the likely intelligibility score just by listening to speech in a room, having had the experience of hearing a range of conditions during the tests.

To interpret and understand these results, a review of some theory of intelligibility and the hearing process in large spaces is desired.

2 FUNDAMENTALS REVIEWED

The physical acoustic factors that influence speech intelligibility by causing the signal-to-noise ratio (S/N) to deteriorate are:

1) Background noise (S/N)
2) Too early reflections
3) Too late reflections.

	RASTI	%ALcons
	0.20	57.7
	0.22	51.8
BAD	0.24	46.5
	0.26	41.7
	0.28	37.4
	0.30	33.6
	0.32	30.1
	0.34	27.0
POOR	0.36	24.2
	0.38	21.8
	0.40	19.5
	0.42	17.5
	0.44	15.7
	0.46	14.1
	0.48	12.7
	0.50	11.4
FAIR	0.52	10.2
	0.54	9.1
	0.56	8.2
	0.58	7.4
	0.60	6.6
	0.62	5.9
	0.64	5.3
	0.66	4.8
GOOD	0.68	4.3
	0.70	3.8
	0.72	3.4
	0.74	3.1
	0.76	2.8
	0.78	2.5
	0.80	2.2
	0.82	2.0
	0.84	1.8
	0.86	1.6
EXCELLENT	0.88	1.4
	0.90	1.3
	0.92	1.2
	0.94	1.0
	0.96	0.9
	0.98	0.8
	1.00	0.0

$$ALcons = 170.5405 * EXP(-5.419 * STI)$$

$$STI = -0.1845 * Ln\ (\%AL_{cons}) + 0.9482$$

(b)

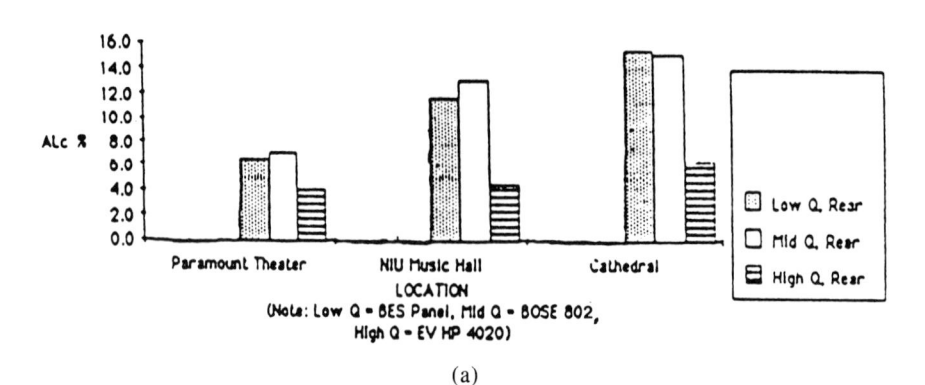

Fig. 9. (a) Graph of articulation loss of consonants versus system Q, rear position, for data in Table 1. (b) Converting RASTI to %AL$_{cons}$. (*Courtesy Farrel Becker.*)

Too early reflections are partially a function of the early decay time and hence the early ratio, $L_D - L_R$, as established by the loudspeaker directivity factor Q. At the listening position (or point of observation), L_D is the level of the direct sound, L_R the level of the reverberant sound.

Too late reflections may be individual high-level specular reflections arriving at the listener more than 90 ms after the direct sound. They may also be the ensemble of the multiple late reflections that make up reverberation.

These S/N factors must be considered together with the signal bandwidth—that is, the amplitude—frequency response of the speech system—before the more sophisticated analyses are attempted.

The frequency range and signal-to-noise ratios for speech intelligibility are fundamental parameters. The frequency range from approximately 300 to 3000 Hz [6] has been demonstrated to suffice for telephone intelligibility. Figs. 10 and 11 show the speech power and one-third-octave-band contribution to articulation and reveal the key role the 2000-Hz one-octave band plays. (It is this region that contains the consonants in speech.) Fig. 12 shows the temporal distribution of energy into useful and harmful divisions. (This chart was originally intended to be used with environments where music was to be the primary source.) The area entitled "range of useful reflections" is a good depiction of the delay interval and relative level of useful reflections that will not interfere with speech intelligibility or aid it.

Fig. 13 illustrates the range of signal-to-noise ratios desirable for various levels produced by talkers under various conditions.

2.1 Directivity Factor Q

The role of the loudspeaker Q can best be seen by its effect on the early decay time as determined by

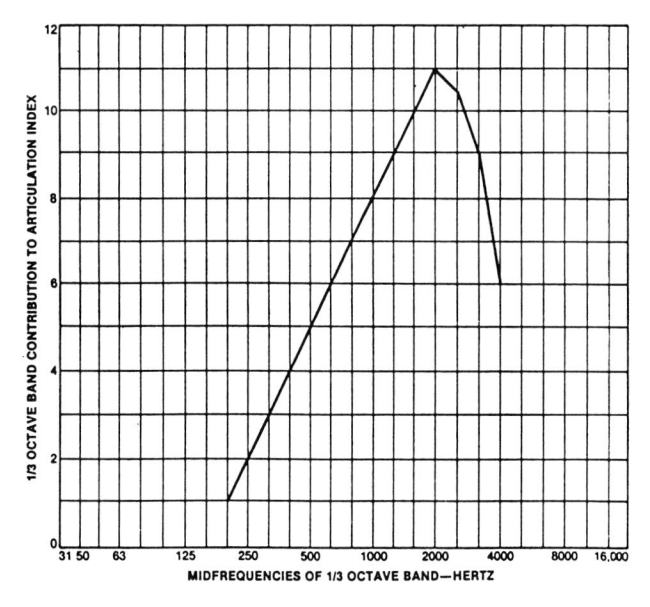

Fig. 11. One-third-octave-band contribution of articulation. (*Courtesy Michael Rettinger.*)

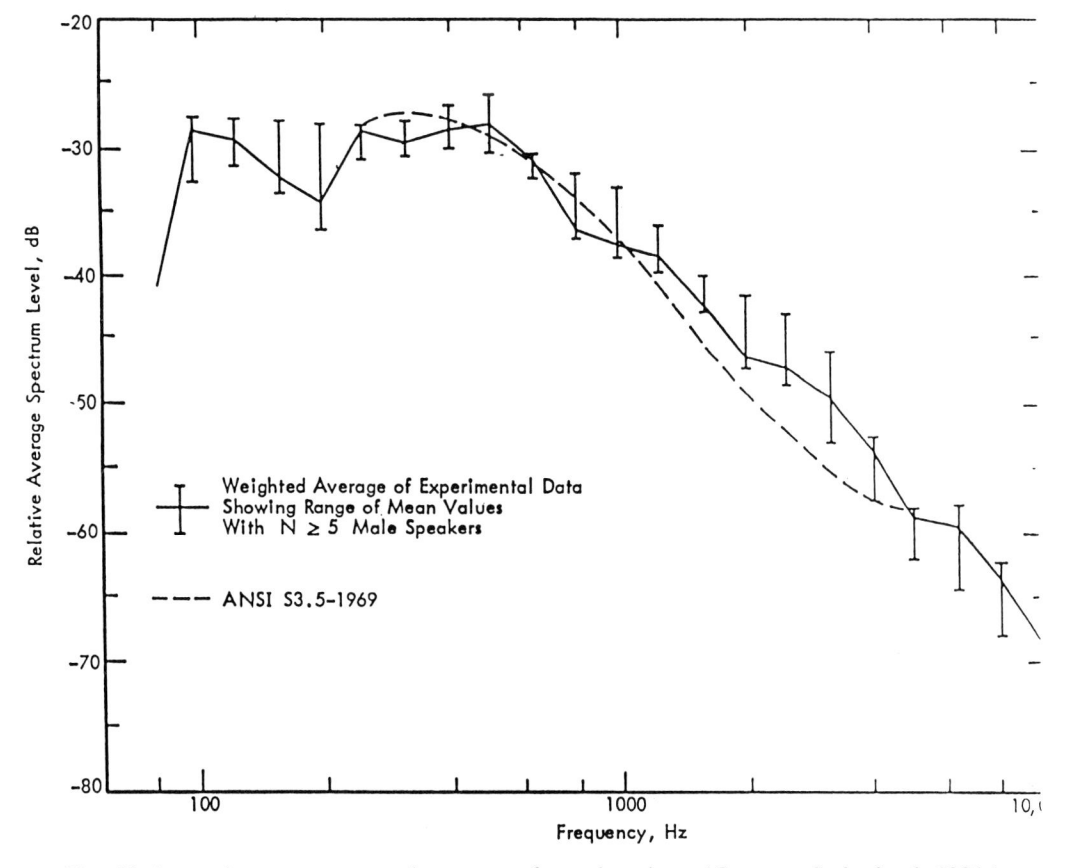

Fig. 10. Long-time average speech spectrum for male voices. (*Courtesy Sutherland, 1981.*)

Schroeder's method (see Heyser [23]). Even in a reverberation chamber that has been modified with a single surface area of high absorption, we can observe this change in decay time for the initial arrivals. In such chambers the Hopkins–Stryker equation is a reliable predictor,

$$L_P = L_W + 10 \log \left[\frac{Q_{ME}}{4\pi(D_x)^2} + \frac{4N}{S\bar{a}M_a} \right] + 0.5$$

where

- L_P = sound pressure level at a given observation point, dB (re 20 μPa)
- L_W = sound power level of sources providing L_D at the observation point, dB (re 10^{-12} W)
- Q = directivity factor of source providing L_D at the observation point, dimensionless
- D_x = distance from source providing L_D to the observation point, feet or meters
- N = ratio of other sources providing L_W to space at the observation point, but not providing L_D to the observation point
- $S\bar{a}$ = number of sabins in space, square feet or square meters
- M_a = architectural modifier that results from L_D encountering an area of high absorption before it is reflected to the space as a whole,

$$M_a = \left(\frac{1 - \bar{a}}{1 - a_c} \right) \left(\frac{Q_{act}}{Q_{theor}} \right)$$

with

- $\bar{a}$ = average absorption coefficient
- a_c = absorption coefficient for area of high absorption
- Q_{theor} = theoretical Q for a given area's angular projection
- Q_{act} = actual Q of device aimed at that area of special absorption.

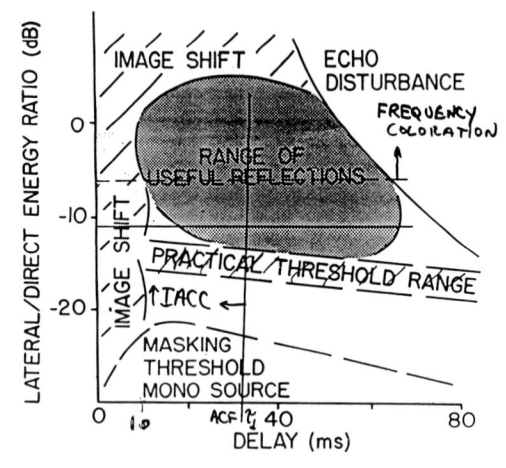

Fig. 12. Useful reflections for music. The shaded area basically embodies the Haas effect. Classical music is more tolerant of delays than contemporary music, and speech is the least tolerant. In spite of the fact that much of this chart was derived from music, it has been shown by Bilello, D'Antonio, Davis, and others to have remarkable correlation to the judgment of imaging in small rooms and speech intelligibility in large rooms. (*Courtesy Barron, modified by Marshall and Hyde and D'Antonio.*)

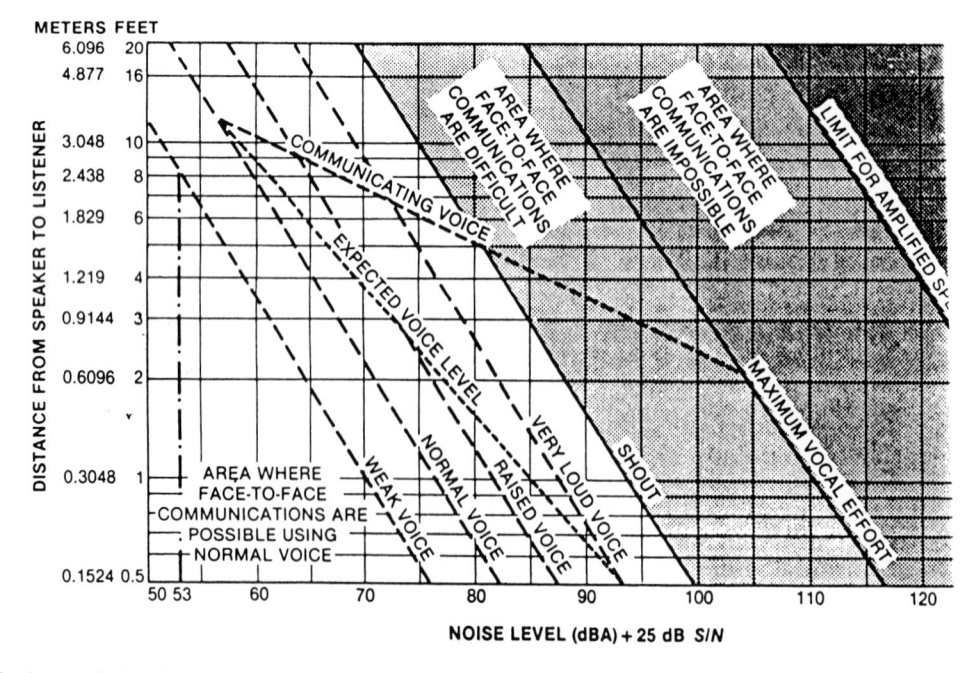

Fig. 13. Typical speech levels and talker types showing level to be expected at a given distance from talker. By adding 25 dB to the existing ambient noise level and entering the chart at the bottom scale, then rising vertically to intercept the type of talker, followed by going horizontally to the left scale you can read the maximum distance at which the proper signal-to-noise ratio can be maintained. This signal-to-noise algorithm does in chart form what the Lochner–Burger equation does in a mathematical form.

3 HOW DO WE HEAR?

For steady-state signals in an enclosed space, the distance at which $L_D = L_R$, the critical distance, is

$$D_c = 0.141\sqrt{\frac{Q S \bar{a} M_a}{N}}.$$

However, speech is a transient signal. It supplies energy to the reverberant sound field intermittently. Thus as the reverberant field fills, its level L_R increases. Thus the critical distance only approaches its limit at the steady-state value.

Similarly, some observers feel that L_D can also increase due to the fusing of early reflections into the direct sound, with the result that the effective critical distance would increase. This turns out not to be true.

3.1 What Is Meant by Fusion?

One way to approach it is to regard integration time as the time the hearing system can still use something from the past. Thus it relates to the growth of loudness. In contrast, fusion time is really a period during which all inputs still provide the listener with a "unified auditory image" [7]. It is usually a difference of arrival times between the ears, but sometimes it is associated with arrival times between reflections.

If what we mean is, "can the direction from which it comes be detected, or is it perceived as a separate source?" then the ear indeed integrates early reflection energy into the direct sound energy over a fairly long interval (50–80 ms under some favorable circumstances).

If, however, we mean, "which energy arrival contains 'information' that our ear–brain system will accept?" then the answer appears to be different. We know that the first energy arrival starts a mental clock which runs until it hears the first reflection of energy arrive, after which our brain ignores direction and interval. We know that this latter energy is accepted as slightly increased loudness. We also know that, depending on the delay interval and level, it can be detrimental to speech intelligibility. We have at hand no objective data saying that it can be helpful to speech intelligibility.

The ear–brain fusion of the early reflections with the direct sound, in terms of perceived loudness versus the integration of measured early reflected energy in establishing the ratio of direct-to-reverberant energy, has been demonstrated not to be related so far as intelligibility is concerned. The ear–brain system chooses a single energy return (the highest one possessing a suitable speech spectrum, that is, 300–3000 Hz).

3.2 Should We Integrate Received Energy in Measurements?

It would appear from the evidence gathered with the Techron TEF that speech intelligibility measurements are dependent on a single energy spectrum, not a series of spectra integrated. The evidence is that only the first arrival is used (defined as energy that has not undergone reflection of any kind) in the case where the first arrival is of sufficient level. When a single reflection is the choice, because the first arrival is too low in level, then only that reflection and no others provides the information received. Fig. 14 shows typical examples of where it is necessary to separate the direct sound from the reflected sound in order to match the subjective scores with the objective %AL$_{cons}$ measurement. Implicit in these data is the fact that for the energy received,

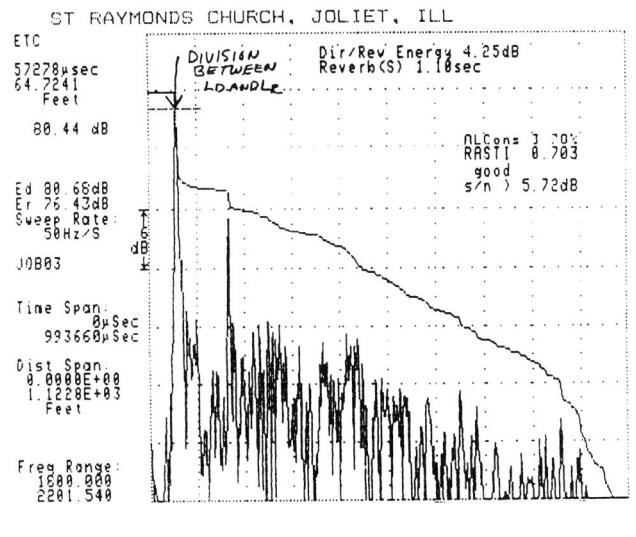

Fig. 14. Energy–time curves (a) Two loudspeakers, showing where to place the cursor on the TEF. The ETC is of a high-Q device showing the division between L_D and L_R that allowed correlation with live listener data (that is, no integration of early reflections). (b) Medium-Q device under identical conditions as in (a). Note the increase in reverberant energy, increased early decay time, and poorer intelligibility score.

the $1/f$ interval for the lowest frequency is approximately 10–20 ms. It is conceivable that past workers may have mistakenly interpreted the $1/f$ interval as an integration interval.

3.2.1 Theories

The subjective aspects of reflections were investigated by Haas [8], with the precedence effect. Lochner and Burger [9] made assumptions on integration and masking effects, and Nickson [10] on the difference between speech and music. Barron and Marshall [11], [12] proposed an objective index for clarity of music, a ratio of early to late energy. It should be borne in mind that much work reported in the literature was not from measurement but from theoretical considerations, usually primitive ray tracing.

3.2.2 Guesses and Measurements

Atal et al. [14] were the first to point out the significance of Schroeder's early decay time to speech intelligibility. Yegnanarazana and Ramakrishna [15] verified the subjective significance for speech intelligibility of Atal's suggestions.

A remarkably large number of workers have written on what constitutes a detrimental reflection, with Niese [16] proposing 33 ms, Meyer and Thiele [17] 50 ms, and Lochner and Burger 95 ms. In contrast, and most interesting because the next two had access to impulse squared measurements compared to subjective scores, Mankovsky [18] and Kuttruff and Jusofie [19] imply that all reflections may be detrimental to speech intelligibility and that the integration time should be 0 ms. We believe that there exists genuine confusion among researchers between the ear–brain's behavior with running speech studies and the proper way to evalaute the impulse response as an energy distribution over time relevant to speech intelligibility. These are perhaps quite separate problems, but we are prepared to support with accurate data the correlation between first arrivals being treated as the total direct sound and *all* reflections as reverberant sound in the sense that $L_D - L_R$ affects speech intelligibility.

3.3 How Do We Listen to Speech?

To answer that question properly we need to ask what we mean by "listen to." Do we mean speech level, the direction speech came from, speech intelligibility, or the tonal quality of speech? If we mean speech level, we do indeed "integrate" early reflections into a fusion with first arrivals. Spatially and in terms of intelligibility the law of the first wavefront (a truly descriptive term) would seem to predominate. If tonal quality were the primary concern, then we would need to account for too early reflections as well as for those that fall at the outer edge (in a time sense) of the area known as useful reflections.

An excellent description of some parameters needing consideration is given in Blauert [20, p. 179];

The primary sound arrives first at the position of the subject, generating a primary auditory event. The primary sound also elicits an inhibitory effect, in conforming with the law of the first wavefront. For a certain length of time the forming of further auditory events is suppressed. After a time interval corresponding to the echo threshold, either a strong reflection leads to the forming of an echo and to further inhibition, or else the intervening reverberation has been strong enough that a precisely located auditory event is no longer formed. The higher the level of the primary sound in comparison with that of the diffuse field, the more precisely located is the primary auditory event. If the level of the diffuse field is overwhelmingly higher than that of the primary sound, there is no primary event but only a diffusively located reverberant auditory event. (Emphasis added.)

If, instead of a diffuse reverberant sound field, the late arriving auditory event is a specular reflection with a fully developed spectrum, it will become the primary auditory event and the first wavefront will become simply an interfering noise (that is, reduce the signal-to-noise ratio).

A fully developed spectrum has a $1/f$ time associated with it (that is, the primitive period of the lowest frequency). The full-spectrum specular reflection is one in the frequency region of four times the Schroeder frequency (large room or critical frequency) with a reasonably smooth spectrum from 300 to 3000 Hz. Actual measurement reveals that from the front edge of the first arrival to its rear edge (in time) we obtain the $1/f$ time, or roughly 10–20 ms. Integration beyond that point leads to overly optimistic estimates of the ratio of direct to reflected sound.

4 SYSTEM ERRORS THAT REDUCE SPEECH INTELLIGIBILITY

4.1 Reflections on Reflections

Reflections can be either useful or detrimental. Their level relative to the direct sound (here defined as the first arrival) and their spectrum are the parameters that decide their utility. For example, as we show in Fig. 11, the 2-kHz octave band is key to articulation. If we have a reflection with a level equal to the direct sound we can calculate the difference in travel distance for the reflection versus the direct sound that would generate an initial null frequency (INF) of 2-kHz by

$$D = 0.5c/\text{INF}$$

$$= 0.5 (1130)/2000$$

$$= 0.28 \text{ ft or } 3.39 \text{ in}$$

$$= 0.09 \text{ m or } 8.61 \text{ cm}.$$

This means that reflections are separated from the direct sound by this short distance (whether due to poorly synchronized sources or to other delays) and cause a deep notch to fall into the 2-kHz region. In some cases the second, third, or other notch in a series of comb filters falls right on 2-kHz,

$$\text{INF} = 0.5c/D = 0.5(1130)/0.84 = 672.62$$

$$\text{2nd frequency} = (\text{INF} + \text{NFI}) = 672.62 + 1345.24 = 2017.86$$

where

$$\text{NFI} = c/D = 1130/0.84 = 1345.$$

Because Q dramatically affects the early delay time (EDT), it also in the same proportion affects speech intelligibility whenever EDT exceeds the 1.5-s lower limit. When EDT is less than 1.5 s, reverberation will not be the primary parameter affecting speech intelligibility, as Peutz's original AES paper clearly demonstrated [21].

A further detrimental effect is the change in polar response of the source device when interfered with in this manner (Fig. 8). Thus, high-level signals within the first 3 ms can cause loss of sound pressure level (not energy—energy is redistributed and conserved) in the speech spectrum because of comb filtering, 1 ms being particularly noticeable because it causes a deep notch near 2 kHz, thus reducing the level of consonants, often below the noise floor.

4.2 Detrimental High-Level Reflections

A reflection with a spectrum that provides the range and smoothness required for useful reflections and is at a higher level than the direct sound will become the reflection the ear–brain uses to determine intelligibility. (Fig. 12). Reflections that fall within the "range of useful reflections" and do not predominate in level are useful because they are perceived as increased loudness and are not harmful to speech intelligibility in either a polar or a frequency sense. Fig. 12 shows how investigators into entirely different situations site the polar pattern variations (image shift), the loss of frequency information (coloration), and interference (echoes). This same chart, originally plotted for musical purposes, contains a wealth of clues with regard to the temporal behavior of any signal.

Lochner and Burger equated loudness with intelligibility, but there is no justification for this in their work—mere loudness level over masking was not shown to be equivalent to the reflection-free stimuli used to investigate masking in headphone-based studies. In Fig. 12, Barron et al. [10], [11] were concerned with the enjoyment of music and not with information transfer, still their work contains information valuable to research into speech intelligibility.

4.3 Signal-to-Noise Ratio

Background noise deteriorates intelligibility according to the spectrum of the noise. This frequency dependence is often overlooked. Low-frequency signals tend to mask high frequencies, and narrow-band noises have specific effects.

4.4 Loudspeaker Directivity

Loudspeaker Q dramatically affects EDT. Similarly, EDT affects speech intelligibility. As pointed out by Peutz, if EDT is less than 1.5 s, reverberation will not be the primary parameter affecting speech intelligibility [21].

5 MEASURING SOUND SYSTEMS

5.1 Interpreting Impulse Responses

The term impulse is applied to a number of quite different measurement systems. What many engineers mean when they use the term is a sufficiently narrow (in time) spike of signal which has a more or less uniform frequency spectrum over the range of interest for the test.

Serious problems arise whenever a narrow time spike of high voltage is used to obtain the "impulse response" of a real-world acoustic transducer such as a loudspeaker or microphone. In order to obtain the frequency response, as computed by the Fourier transform, it is necessary to show that all transducers are truly operating in a linear mode because the Fourier transform is only true for a linear system.

It is the usual practice to extract the pressure waveform for display (that is, voltage amplitude versus time elapsed). This results in a confusing pattern of decaying oscillatory components. Extracting useful information from such a display is subject to substantial error (somewhat akin to forecasting weather from the fur on wooly worms). In the pressure waveform of an impulse response there are many instants when the pressure display passes through zero, that is, the local equilibrium pressure in the measurement environment. Does this mean that there is no "sound" at these moments? The answer, of course, is no. It only means that there is no pressure excess at those moments. The fact that there are pressure rates of change at these times of zero pressure excess indicates that there is energy density available in perhaps another form, and indeed there is.

5.2 The Analytic Signal

From energy balance considerations it is evident why Gabor [22] was forced to develop the analytic signal. When we apply the analytic signal to the measurement of electroacoustic transducers, we find a most useful tool available, called the Hilbert transform. (Perhaps it would be more accurate to call it the Hilbert operator, as it is a local-to-local operation rather than a global-to-local transform.)

5.3 Acoustic Impulse Response

In acoustic measurements the pressure is proportional to the square root of the potential energy density. If the pressure impulse response is one term of the analytic signal, then the complete analytic signal represents the magnitude and partitioning of the energy density components. The significance of the analytic signal to electroacoustic measurements now becomes evident. When an electric impulse is fed to a loudspeaker, the sound does not instantly emerge. In fact, the total acoustic signal will arrive in discrete bundles for those components that are the result of diffraction and reflection in both the loudspeaker enclosure and the acoustic environment itself.

The magnitude of the analytic signal (the square root of the sum of the real part squared plus the imaginary part squared) will show a peak for each discrete component of energy arrival. The rate of change of the phase of the analytic signal bears a relation to the spectral distribution of the separate energy arrivals. When the analytic signal is used, the myriad variations in the scalar impulse response are quickly revealed as discrete arrivals in signal energy and the subsequent decrement in energy for each arrival.

This means that an energy calculation based solely on either a measured pressure response or a measured particle velocity response will be correct to within a factor of 2, even if the analytic signal is not used [23]. Put another way, an arrival at a discrete time of pure kinetic energy can dramatically affect, in time, a nearby signal of pure potential energy in terms of total energy by producing comb filters or lobing of the polar responses of two sources containing the partitioned energy. It is true that theoretically the scalar impulse response contains all the necessary information, but in a form that makes it difficult to see the interaction of time-offset energy arrivals.

5.4 Frequency-Domain Signal Acquisition

When we acquire our real and imaginary parts of the analytic signal by means of the Heyser transform and we do so in the frequency domain, we can use the Fourier transform to obtain the energy–time curve or envelope of the impulse response. When we do this, we are presented with a 20 000:1 signal-to-noise advantage through the use of a swept tone and a tracking filter. An energy–time curve displays the amount of total energy returned and how it is distributed over time. Energy–time curves reveal that dividing the energy associated with the first arrival in time that has a full enough spectrum (250–5000 Hz), and classifying all subsequent energy arrivals as "reverberant," yields the most accurate estimate of speech intelligibility. The single exception is the case where a subsequent specular reflection has a sufficiently full spectrum *and* is significantly higher in level than the first arrival. When this occurs, the reflection is treated as the direct sound level, and the first arrival is included in the reverberant sound field's level. The Haas effect, the brain's processing time, and the $1/f$ associated with full spectra are all remarkably similar in a time sense. The way in which the ear–brain processes running speech is a continuing research project for many workers.

5.5 Integration Considerations

Integrating measured impulse energy data for 50 ms, 80 ms, or more and calling the result the direct sound level is, we believe, an error, as it gives too optimistic a score in large spaces. In small rooms, integrating over 90 ms results in all devices measuring the same. One manufacturer has tried to turn this to its advantage by incorporating such a long integration

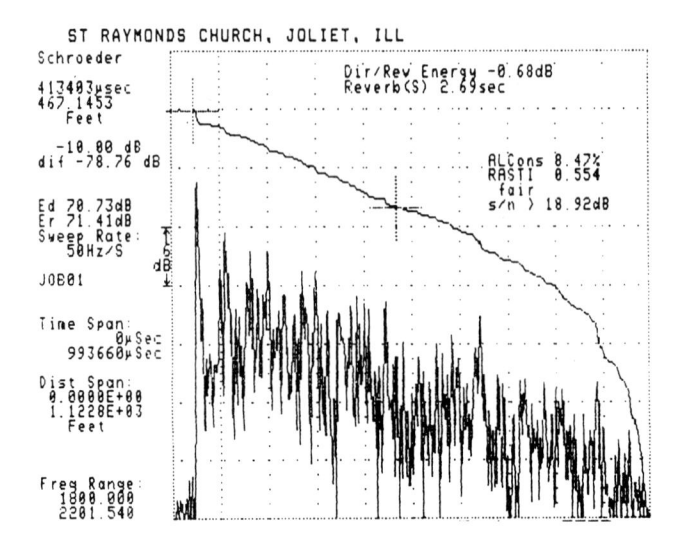

Actual score as measured by three groups of 30 people in three different spaces:

%AL$_{cons}$ 15.2   0.44 RASTI equiv.

TEF measurement:
%AL$_{cons}$ 14.1   0.46 RASTI equiv.

90-ms integration:
%AL$_{cons}$ 8.5   0.55 RASTI equiv.

(a)

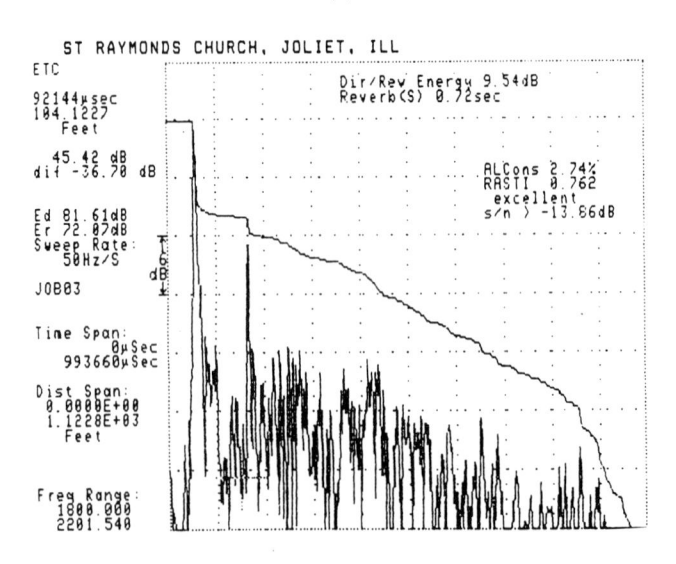

Actual score:
%AL$_{cons}$ 6.4   0.60 RASTI equiv.

TEF measurement:
%AL$_{cons}$ 4.7   0.66 RASTI equiv.

90-ms integration:
%AL$_{cons}$ 2.7   0.76 RASTI equiv.

(b)

Fig. 15. (a) Medium-Q device measured using 90-ms of integration as direct sound. (b) Same data for high-Q loudspeaker.

(Fig. 15). How our brain processes running speech is a different question than how our analyzers analyze the sound system and the environment's effect on that running speech, but we believe that they correlate with each other.

We can report with assurance that accurate, easy to use tools are now available for the estimation of speech intelligibility over sound systems in all kinds of environments. We believe that such a measurement will serve to force meaningful improvements in such systems.

It is fortuitous that we can predict with meaningful accuracy the speech intelligibility score using such simple tools. Our measurement system may indeed mimic our ear–brain process. We do a full frequency sweep in the frequency domain, much as our ear–brain listens to full spectrum returns. We acquire the real and imaginary parts via the Heyser transform (coincident and quadrature responses). We transform them into the time-domain equivalent real and imaginary parts (impulse and doublet responses; see domain chart in Fig. 4) and from there to the magnitude of the energy–time response. Here we are left to choose which energy return arrival contains sufficient information (broad and smooth enough spectra) to seize our brain's attention. The Haas effect, the precedence effect, Fay Hall's experiments, and clear back to Joseph Henry's work on this subject in 1847, all attest to the fact that our ear–brain system does select some energy returns spread out in time at the expense of others. Indeed when our ear–brain system prefers the higher in level reflection (higher in level than L_D) and that reflection has a smooth enough spectrum to contain the necessary information, then the L_D actually constitutes interferences. Delayed signals in the 20–30-ms range are easily handled by the ear–brain system, as they occur after each $1/f$ spectrum has had a chance to be heard.

5.6 Step-by-Step Measurement Technique

Forgetting for the moment what the ear–brain system might actually be doing in a temporal way, let us address the question from the original direction. In analyzing the test results for 90 listeners listening to three different sound systems in three different acoustic environments over which Dynastat modified rhyme test word lists were played from cassette recordings, we declared the resulting scores as "the truth." We then took the Techron energy–time curves and examined by means of postprocessing which approach came nearest to "the truth" and was it unique to each test or could we use the same technique every time. What we found was that the postprocessing of the energy–time curve (ETC) if done in the following manner unequivocally predicted the live subject scores:

1) Do a Schroeder "backward" integration of the ETC by pressing key I. Find the RT_{60} for the very first 10 dB of decay by placing the cursor and then pressing key R.

2) Place the cursor on the ETC by pressing E and move it to include the peak of the direct sound level (defined as the first arrival) by moving it one step past the peak level and separating the direct from the reverberant level by pressing key 3.

3) Call up %AL$_{cons}$, RASTI, and the required S/N for a %AL$_{cons}$ not to exceed 10% by pressing "shift, &." What we can conclude from these tests is that you do not integrate early reflections into your data for speech intelligibility measurements. They may indeed add to the speech loudness and perceived quality, and provide clues about spaciousness, and so on, but the evidence is that they do not contribute to speech intelligibility measurements as part of the direct sound.

The literature clearly indicates that this is a controversial subject, but it is interesting to note that those who have participated in precision acoustic measurements of energy arrivals have advocated exactly what we are finding in our measurements (namely, zero integration).

With a full STI available in just a 1½-min measurement there would seem to be less need of a RAPID STI employing compromises. While we have, at this writing, only been able to test both systems in a dozen environments, we have found that the full STI nearly always is different (lower score) than the RASTI (Figs. 16 and 17).

6 MEASURING SPEECH PROCESSORS

Allen [24], at the Navy Ocean System Center in San Diego, CA, developed for the U.S. Navy a clever speech processor that uses single-sideband modulation to shift audio frequencies to radio frequencies, severely peak clip them, then use a low-pass filter (still at radio frequencies), and finally demodulate the signal back down to the audio range. This processor has shown remarkable ability to raise the level of consonants in speech up to levels equal to vowels without raising the level of the vowels. The result is an improvement in speech intelligibility in noisy environments that is the equivalent of nearly 10 dB in signal-to-noise ratio while maintaining the same peak power.

When we measured this system with the Bruel & Kjaer RASTI analyzer, it indicated the presence of excessive noise and was unable to provide an estimation of intelligibility. The Techron TEF 12 measured the system as if the device were not present, and there was no improvement in intelligibility. Using the Techron TEF 12, we were able to establish a measurement criterion that merely required an 8-dB offset of the data to supply a reliable prediction of the speech intelligibility score of the various areas covered by the shipboard system employing this unique processor (an aircraft carrier).

7 SUMMARY

Of even greater interest than the fact that we had at long last achieved a useful and accurate measure of speech intelligibility was the finding, at least in terms of the measurements, that including early reflections as part of the so-called direct sound was *not* valid. As discussed in this paper, early reflections may indeed have other legitimate uses, but this is not one of them. The role of increased Q in shortening the early decay times was directly correlated to improved intelligibility scores.

The intelligibility workshop revealed that, in terms of intelligibility as measured on a TEF analyzer, only the direct sound level should be used as L_D. Integration of early reflections is *not* valid. Very early reflections (under 5 ms) cause not only frequency-domain anomalies, but polar response anomalies as well. This in turn affects intelligibility by moving energy off the listener and onto the walls and ceiling. $\%AL_{cons}$, as exemplified by the Peutz algorithm in the TEF analyzer, proved to be the most accurate predictor. The key measurement parameters became the signal-to-noise ratio, the early decay time, and the ratio of direct sound level to *all* other reflected levels. It was determined that all algorithms can be "tricked" and that $\%AL_{cons}$ was found to have the least error.

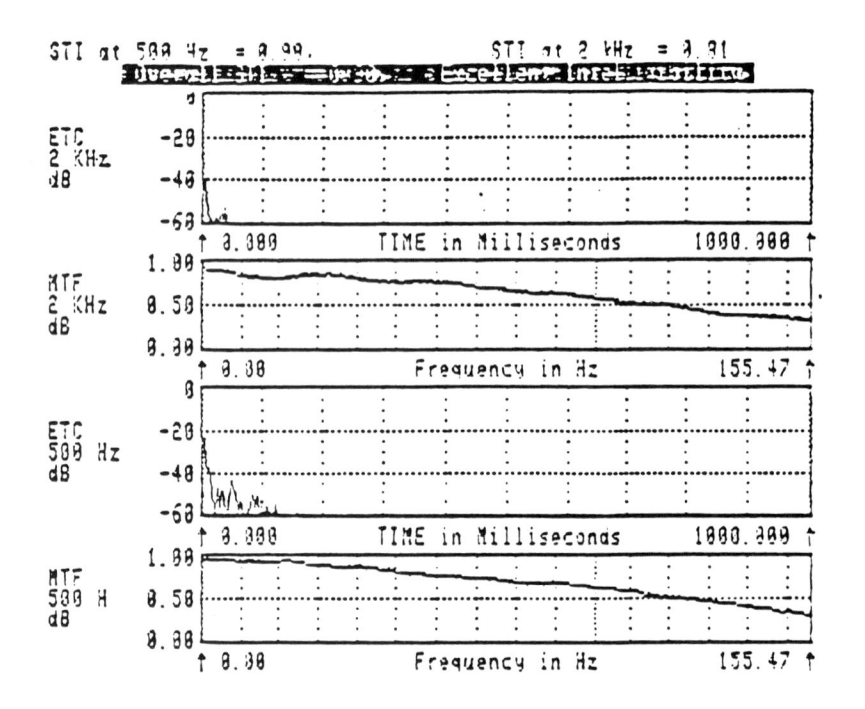

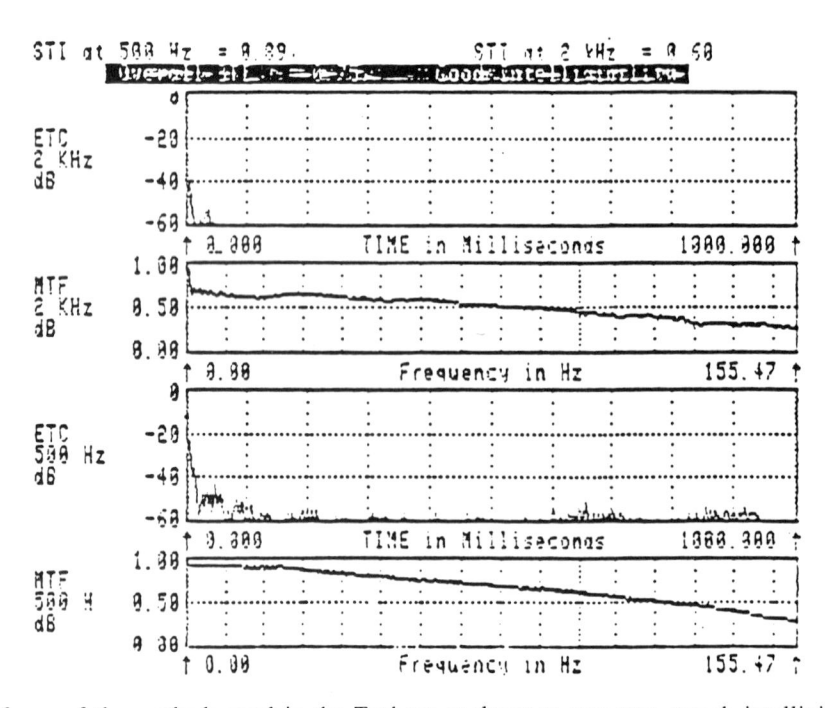

Fig. 16. Example of one of the methods used in the Techron analyzer to measure speech intelligibility via the RASTI technique.

8 ACKNOWLEDGMENT

The remarkable work of V. M. A. Peutz, Richard C. Heyser [25],[1] and Gerald Stanley [26] in producing, programming, and demonstrating how to successfully measure %AL$_{cons}$ with accuracy has resulted in reliable objective speech intelligibility measurements. Because the Heyser transform allows extreme accuracy in the presence of noise; because the Stanley TEF analyzer is small, rugged, inexpensive, and reliable; because of V. M. A. Peutz's 30 years of intelligent work in the development of parameters affecting %AL$_{cons}$; and because of the work of Dr. Peter D'Antonio and John Konnert [27] in programming the phase of the energy–time curve, now we all have research opportunities undreamed of even a decade ago. This paper is dedicated to these men and to those readers who will use their work as a starting point for equally dedicated endeavors reported with equally undeviating integrity.

---

[1] Any engineer or scientist in acoustic measurements must read all of the late Richard C. Heyser's papers (many available in the *AES Loudspeaker Anthology*, vol. 1-25). This is even more important when what you are trying to measure is in the time domain because before Heyser no one had ever properly accounted for the total acoustic energy in an impulse measurement.

9 REFERENCES

The authors have made no attempt to be exhaustive in terms of references. The important points in this paper are based on original material from the 1986 Intelligibility Workshop. The references are examples of fruitful areas of collaborative research

[1] V. M. A. Peutz, personal communication (1987).

[2] L. Humes, personal communication (1987).

[3] C. Davis and D. Davis, "Speech Intelligibility Workshop," *Syn-Aud-Con Tech Topic*, vol. 14, no. 1, Fall 1986; no. 8, Summer 1987.

[4] C. P. Davis, "Measurement of %AL$_{cons}$," *J. Audio Eng. Soc.*, vol. 34, pp. 905–909 (1986 Nov.).

[5] D. Davis and C. Davis, *Sound System Engineering*, 2nd ed. (Howard W. Sams, Indianapolis, IN, 1986).

[6] H. Fletcher, *Speech and Hearing in Communication* (Van Nostrand, New York, 1953).

[7] Y. Tobias, *Foundations of Modern Auditory Theory* (Academic Press, 1970).

[8] H. Haas, "Uber den Einfluss eines Einfachechos auf die Hörsamkeit von Sprache (transl.)," *J. Audio Eng. Soc.*, vol. 20, pp. 146–159 (1972 Mar.).

[9] J. P. A. Lochner and J. F. Burger, "The Intelligibility of Reinforced Speech," *Acustica*, vol. 9 (1959).

[10] A. F. B. Nickson, R. W. Muncey, and P. DuBout, "The Acceptability of Artificial Echoes with Reverberant Speech and Music," *Acustica*, vol. 4 (1954).

[11] M. F. E. Barron, "The Effects of Early Reflections on Subjective Acoustical Quality in Concert Halls," Ph.D. dissertation, Southampton University, UK, 1974.

[12] A. H. Marshall, "Acoustical Determinants for the Architectural Design of Concert Halls," *Archit. Sci. Rev.*, vol. 11, pp. 81–87 (1968).

[13] W. Reichardt, O. A. Alim, and W. Schmidt, "Definition und Messgrundlage eines objektiven Masses zur Ermittlung der Grenze zwischen brauchbarer und unbrauchbarer Durchsichtigkeit bei Musikdarbietung," *Acustica*, vol. 32 (1975).

[14] B. S. Atal, M. R. Schroeder, and G. M. Sessler, "Subjective Reverberation Time and Its Relation to Sound Decay," presented at the 5th Int. Cong. on Acoustics, Liege, Belgium 1965.

[15] B. Yegnanarazana and B. S. Ramakishina, "Intelligibility of Speech under Nonexponential Decay Conditions," *J. Acoust. Soc. Am.*, vol. 58 (1975).

[16] H. Niese, "Untersuchung über die Knallform bei o raumakustischen Impulsmessungen," *Hochfre-*

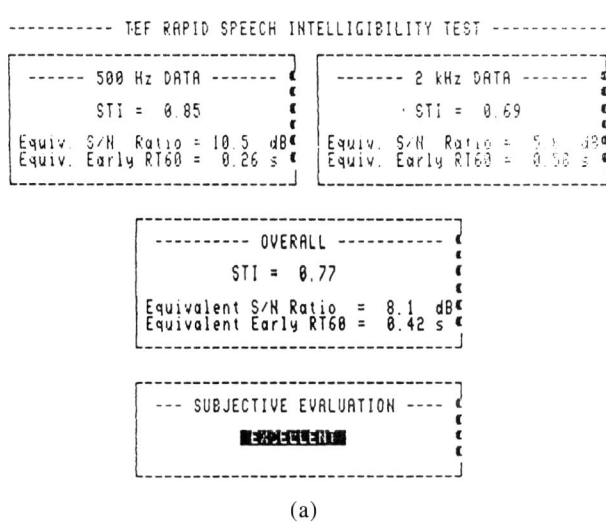

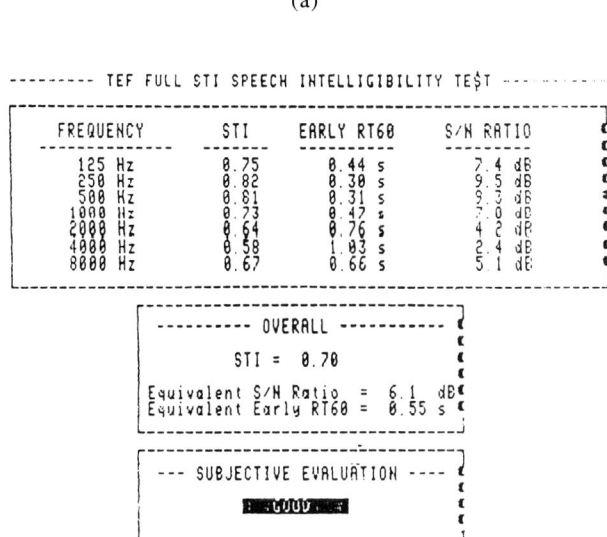

Fig. 17. Displays for Techron 12 analyzer. (a) RASTI. (b) Full seven-frequency STI. Both measurements were made under identical circumstances, thus revealing the optimistic tendency of RASTI.

quenztech. *Electroakust.*, vol. 66 (1957).

[17] E. Meyer and R. Thiele, "Raumakustische Undersuchungen in zahlreichen Konzertsälen und Rundfunkstudios unter Anwendung Neuerer Messverfahren," *Acustica*, vol. 6 (1956).

[18] V. S. Mankovsky, *Acoustics of Studios and Auditoria* (Focal Press, London, 1971).

[19] H. Kuttruff and M. F. Jusofie, "Nachhallmessungen nach dem Verfahren der integrierten Impulsantworte," *Acustica*, vol. 19 (1967).

[20] J. Blauert, *Spatial Hearing* (Hirzel Verlag, Stuttgart, West Germany, 1974; new transl., M.I.T. Press, Cambridge, MA, 1983).

[21] V. M. A. Peutz, "Speech Reception and Information," *Proc. 9th Int. Cong. of Acoustics* (Madrid, Spain, 1977); "Articulation Loss of Consonants as a Criterion for Speech Transmission in Room," *J. Audio Eng. Soc.*, vol. 19, pp. 915–919 (1971 Dec.); both also in *Publiikaties*, a publication of the Netherlands Acoustical Society honoring V.M.A. Peutz, 1983.

[22] D. Gabor, "Theory of Communication," *Proc. IEE*, vol. 93 (1946). (London).

[23] R. C. Heyser, "Concepts in the Frequency and Time Domain Response of Loudspeakers," *Monitor—Proc. IEE* (1967 Mar.). (London).

[24] C. Allen, "Speech Processor," *Syn-Aud-Con Tech Topic*, vol. 13, no. 7, 1986.

[25] R. C. Heyser, "Acoustical Measurements by Time Delay Spectrometry," *J. Audio Eng. Soc.*, vol. 15, p. 370 (1967).

[26] G. R. Stanley, "A Micro-Processor-Based TEF Analyzer," in D. Davis and C. Davis, *Sound System Engineering*, 2nd ed. (Howard W. Sams, Indianapolis, IN, 1986), app. IX.

[27] P. D'Antonio and J. Konnert, "Complex Time-Response Measurements Using Time-Delay Spectrometry," *J. Audio Eng. Soc.*, vol. 37, pp. 674–690 (1989 Sept.).

APPENDIX

1. STI (noise predominant):

$$STI = \frac{S/N + 15}{30}$$

where S/N is in decibels (that is, $L_S - L_N$).

2. Peutz equation (architectural):

$$\%AL_{cons} = \frac{656*(D_2)^2(RT_{60})^2 N}{VQM} \qquad *SI = 200$$

3. Peutz equation (measurement algorithm):

$$A = -0.32 \log\left(\frac{L_R + L_N}{10L_D + L_R + L_N}\right).$$

For $A \geq 1$, $A = 1$.

$$B = -0.32 \log\left(\frac{L_N}{10L_R + L_N}\right).$$

For $B \geq 1$, $B = 1$.

$$C = -0.5 \log\left(\frac{RT_{60}}{12}\right)$$

$$AL_C = 100(10^{-2[(A+BC)-ABC]}) + 0.015$$

4. Lochner–Burger algorithm:

$$S/N_{effective} = \frac{\int_0^{95\ ms} P^2(t)a(t)\,dt}{\int_0^{95\ ms} P^2(t)\,dt}$$

THE AUTHORS

C. and D. Davis

Don Davis has been president of Synergetic Audio Concepts since 1973 and active in teaching seminars in acoustics and electroacoustics. He is also an international consultant in audio and acoustics.

Mr. Davis has authored three books: *Acoustical Tests and Measurements*; *How to Build Loudspeaker Enclo-*

sures, written with the late Alexis Badmaieff; and *Sound System Engineering*, co-authored with his wife, Carolyn. He is the author of over a hundred technical articles on audio subjects. The Davises have written and published the *Syn-Aud-Con Newsletters and Tech Topics* for the past 17 years. Today a full set of these has become a collector's item among the more than 7000 people who have attended Syn-Aud-Con seminars and workshops. Mr. Davis is co-inventor of the Acousta-Voicing method of sound system equalization and is the developer of the Live End-Dead End[1] (LEDE[1]) control room design technique.

The Davises are the recipient of numerous awards in the field of acoustics and electroacoustics for their work in the design of sound systems and their acoustic environments.

Mr. Davis is a member of ASA, a senior member of the IEEE, and a fellow of the AES. Mrs. Davis is a member of the AES.

---

[1] Trademarks of Synergetic Audio Concepts.

Correlation of Speech Intelligibility Tests in Reverberant Rooms with Three Predictive Algorithms*

KENNETH D. JACOB

Bose Corporation, Framingham, MA 01701, USA

The data base used in an earlier speech intelligibility study has been extended. The rooms added in the present study involve higher reverberation times and represent more difficult environments for sound reinforcement. The results of subjective intelligibility tests are compared to the predictions of three published algorithms. One of these predictors, based on the modulation transfer function, is being considered as an international standard (IEC Report 268-16, 1988), and its accuracy is confirmed in this study. A second predictor, based on the Lochner–Burger signal-to-noise ratio, is shown to be equally accurate when data are analyzed in a manner similar to that of two other intelligibility studies. The third algorithm, which assumes that all reflections degrade intelligibility, is shown to be inferior. The study confirms the beneficial effect of early reflections and the detrimental effect of late reflections on speech intelligibility assumed by the two more accurate algorithms.

0 INTRODUCTION

The clarity with which a sound system transmits speech is one of the fundamental parameters governing its quality. Acceptable speech intelligibility is therefore a basic objective of sound system designers. It is well known that frequency response, distortion, and background noise affect intelligibility, and the relationship between these factors and intelligibility has been studied extensively [1]–[3]. As a result, techniques have been developed to aid designers in predicting their effects. Another parameter known to affect speech intelligibility is reverberation, and studies to quantify its effect have also led to predictive algorithms. Two of these algorithms assume that early reflections contribute to intelligibility. A third assumes that all reflections degrade intelligibility and that intelligibility is monotonically related to loudspeaker directivity. Because of their different assumptions, it can be expected that these algorithms will produce different results when applied to reverberant environments.

In a preliminary study by this author [4], an experimental data base consisting of five rooms, three different loudspeakers, and two listener positions was used to test the accuracy of these same three predictive algorithms. The rooms were moderately reverberant, and the conclusions of the first study therefore were limited in scope. Because it is also important that sound system designers be capable of accurately predicting intelligibility in more reverberant environments, five rooms of higher reverberation time were added to the original data base. Conclusions reached in the present study can be applied to the broad range of reverberant environments in which sound systems typically operate.

In the present study, for each combination of room, loudspeaker type, and listener position, subjective intelligibility tests have been conducted to obtain estimates of the actual intelligibility. Against these actual intelligibility scores, the accuracy of each of the three predictive algorithms has been tested. The results are presented in side-by-side comparisons. In addition, the experiment has been documented to facilitate reproduction by other investigators. Methodology, experimental technique, and intermediate results are described explicitly.

1 SOURCE, ROOM, AND RECEIVER PARAMETERS

1.1 Loudspeakers

The three different loudspeaker sources used are listed in Table 1.

1.2 Room Parameters

A total of 10 rooms located in the Boston metropolitan area were studied, as given in Table 2.

---

* Presented at the 85th Convention of the Audio Engineering Society, Los Angeles, 1988 November 3–6; revised 1989 July 24 and September 25.

1.3 Source Positions

Sound sources were positioned in the part of the room normally used for speaking. In the rooms that had proscenium-type stage areas, loudspeakers were always placed in the middle and on the audience side of the proscenium. In some rooms, the loudspeakers were raised to approximately the middle top of the proscenium opening using a pneumatic tower, and in others, loudspeakers were placed upon 6-ft (2-m) stands, as shown in Fig. 1. In all cases, efforts were made to place loudspeakers in locations suitable for either permanent or temporary sound systems.

1.4 Listener Positions

In each room, two listener locations were selected, one corresponding approximately to one-third the distance from the loudspeaker to the rear wall, and the other at the rear of the auditorium. These positions were chosen in the expectation of yielding a wide range of intelligibility scores. In all cases, the relationship between loudspeakers and listeners was such that listeners were within 7.5° of the major axes of the loudspeakers.

2 SOUND-SYSTEM ADJUSTMENTS

To study the effect of reverberation on intelligibility, efforts were made to minimize the effects of other factors known to affect intelligibility. The loudspeakers were adjusted individually to have equivalent frequency response to within ±2 dB as measured on a one-third-octave analyzer. Bandwidth was limited to from 100 Hz to 10 kHz. Any audible noise, hum, or distortion from the loudspeakers was eliminated before tests were conducted.

Background noise was minimized as a factor by ensuring a ratio of speech level to background noise of at least 30 dB as measured on a instrumentation-grade sound pressure level meter set on A-weighting. In general, a ratio of speech signal to background noise of at least 15 dB is required to eliminate noise as a factor [2].

3 SUBJECTIVE TESTING

To establish an accurate estimate of the true speech intelligibility, subjective tests were conducted for every combination of room, sound source, and listener location. The subjective tests were administered according to ANSI S3.2-1960(R1982) [5].

In the ANSI test, subjects are presented with a combination of 20 lists of 50 phonetically balanced (PB) monosyllabic English words. When presented to the listener, the words are embedded in a carrier phrase, for example, "you will write the word *test* now." The carrier phrase is necessary to simulate the interfering effect of continuously spoken words. Words are scored according to phonetics rather than spelling; for example, "payed" is equivalent to "paid." If any part of the word is phonetically incorrect, the entire word is scored incorrect. The percent intelligibility is defined as the percentage of phonetically correct words, and is denoted by %PB_{act}.

Word lists reproduced through the loudspeakers were originally recorded in an anechoic chamber using an omnidirectional instrumentation-grade microphone at a distance of 0.5 m from the talker in order to create an on-axis speech recording containing negligible reverberation.

Subjects were chosen from the general public. At least 3 years of college and normal hearing as checked by an audiometer were required. Subjects were not told the nature of the experiment.

In the first five rooms (see Table 2) subjective testing was conducted in the rooms. For the second group of five rooms, a Sennheiser dummy head and dual microphone combination was used at the two listener positions to record the word lists.[1] These binaural recordings were played back over headphones to subjects at a later date. Thus testing was binaural in all cases.

Testing was conducted over two periods of five successive days each. Different subjects were used for the two periods. Word lists were scrambled to avoid any long-term training of the listeners, and the same word lists were never played on the same sources or in the

---

[1] A preliminary experiment was conducted to quantify the accuracy of the binaural recording and playback system. Results showed negligible change in intelligibility scores when they were compared to scores from the same test taken with subjects in the room.

Table 1. Loudspeakers

Name	Type	Axial Directivity
Soundsphere 2212-1	Omni radiator	1.1
Bose 802-II	Eight-driver array	7.3
Electro Voice HR6040A (with TL806AX)	Constant-directivity horn	17.7

Table 2. Room parameters.

Name	T_{60} (s)	Volume (m^3)
Berklee Performance Center, Boston	0.9	5 450
Coolidge Corner Movie House, Brookline	1.0	4 590
Huntington Theater, Boston	1.1	3 190
Saint Bridget's Church, Framingham	2.0	3 810
Nevins Hall, Framingham	3.5	10 620
Jordan Hall, Boston	2.2	4 530
Mechanics Hall, Worcester	2.2	11 582
South End Cathedral, Boston	3.3	59 152
The Cyclorama, Boston	3.5	11 610
MIT Indoor Track, Cambridge	4.6	42 475

same listener positions from day to day. In addition, two lists were never played in succession for any one condition. The total number of words presented for each room, source, and listener position combination ranged from 2000 to 2800 words.

4 OBJECTIVE MEASUREMENTS

For each room, source, and listener position combination, system impulse responses representing the transfer functions from loudspeaker input terminals to listener were recorded. A microcomputer-based data-acquisition system was used to measure, store, and analyze the impulses. The sampling rate was 9615 Hz, representing a usable bandwidth of approximately 4 kHz. 8192-point buffers were used for each impulse.

The directivity of the constant-directivity horn was taken from the manufacturer's data sheet. For the omni radiator and the array loudspeaker, polar responses were taken in an anechoic chamber using multiple microphone locations in order to measure full-sphere radiation. The directivity of both of these devices is not constant in the frequency range above 1000 Hz due to the interaction of multiple drivers and diffraction effects. The directivities quoted in Table 1 represent averages of the 1-, 2-, and 4-kHz octave bands.

Reverberation time was calculated in each room as follows. For each impulse response, 1) the impulse was squared, 2) the squared impulse was integrated from maximum time (about 800 ms) to zero time (so-called reverse integration), 3) the resulting curve was converted to decibels to form the curve that would be measured if the room were filled with steady-state noise and then shut off, 4) a linear regression was applied to the curve from 100 to 600 ms after the onset of decay, 5) the reverberation time for the impulse was estimated as 60 divided by the decay rate of the regression line. Finally, the times for several impulses were averaged. However, reverberation times calculated for impulses within a given room were consistent within ±3%.

The room volume was calculated by analyzing blueprints. The distance from source to listener was computed from the impulse response by multiplying the delay time from source to measurement microphone by the speed of sound.

5 PREDICTIVE ALGORITHMS

5.1 Modulation Transfer Function (STI)

The modulation transfer function can be used to quantify the amount of degradation an input signal undergoes as it passes through a system [6]. The modulation method has been used as a quantitative measure of how organisms process visual signals, and in general for quantification of the performance of optical systems [7]. In room acoustics, signal degradation occurs in the time domain due to a variety of factors, including reverberation, distortion, and background noise. Because these factors are also known to degrade speech intelligibility, the modulation method was proposed by Houtgast and Steeneken as early as 1971 [8] as a quantitative predictor of intelligibility.

The algorithm for applying the modulation transfer function to the problem of predicting speech intelligibility uses a range of modulating frequencies (0.63–12.5 Hz) corresponding to the modulating frequencies of speech, and broadband noise corresponding to the average spectrum of speech. The modulation transfer function is calculated at discrete frequencies, weighted, summed, and normalized to yield a single index of speech intelligibility, called the speech transmission index (STI). Steeneken and Houtgast [9] showed that the technique was useful for a wide variety of different conditions, including reverberation, distortion, and background noise. Subsequently, the technique was simplified for the purpose of computational efficiency and is called the rapid speech transmission index (RASTI) [6].

Schroeder [10] derived the relationship between the impulse response of a system and its modulation transfer function. Schroeder's formula makes it practical to implement the modulation method on microcomputer-based data acquisition and analysis systems. This formula has been implemented in the present study:

$$m(F) = \left| \frac{\int_0^\infty p^2(t) \, e^{-j2\pi F t} \, dt}{\int_0^\infty p^2(t) \, dt} \right| \quad (1)$$

where

$m(F)$ = modulation transfer function
$p(t)$ = system impulse response
F = modulation frequency, hertz.

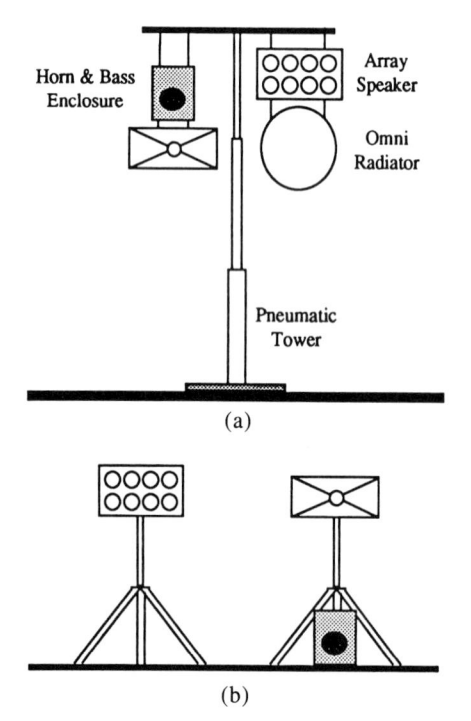

Fig. 1. Two different loudspeaker configurations used.

From Eq. (1) it can be seen that the modulation transfer function is proportional to the magnitude of the Fourier transform of the squared impulse response. More simply, the modulation transfer function is the very low frequency response of the squared impulse response. [It should be noted that direct implementation of Eq. (1) does not properly account for the effect of background noise on the STI. The STI algorithm specifies an input signal spectrum which is not flat, but rather approximates the average power spectrum of the human voice. Thus the signal-to-background-noise ratio (a factor directly affecting the STI) would be different than the ratio found by simply measuring the system's impulse response. In this experiment, however, a speech signal-to-background-noise ratio of more than 30 dB was ensured in all cases in order to minimize background noise as a factor. This allows Eq. (1) to be used with minimal error.]

From the fact that the STI is calculated for modulation frequencies $\leq$ 12.5 Hz, it can be shown that early reflections arriving within a certain time limit can never lower the STI; in fact, in the presence of late reverberation, early reflections increase the STI. Houtgast et al. [11] calculated the STI in a range of exponentially decaying systems. They used a variety of hypothetical cutoff times between useful and detrimental reflections and found that a cutoff of approximately 75 ms corresponded best to actual intelligibility. This agrees in general with the work of Lochner and Burger [12]. It is central to the modulation method that early reflections have a beneficial, and at worst no detrimental, effect on intelligibility.

5.2 Lochner–Burger Signal-to-Noise Ratio

Lochner and Burger [12] performed fundamental research on how speech energy, particularly in the form of room reflections, is processed by the hearing system in order to understand speech. They found that a percentage of the energy in a reflection is integrated, or summed, with the direct sound depending on both its level and the arrival time. For example, a reflection whose level is 5 dB less than the first arrival will be totally integrated by the ear if it arrives within 40 ms of delay, according to their findings.

Lochner and Burger showed that the integrated energy from early reflections could be lumped into a signal part, and energy from late arriving reflections combined with background noise lumped into a noise part, to form a signal-to-noise ratio for speech intelligibility. Several other investigators have used the same basic idea of a psychoacoustic signal-to-noise ratio to form objective measures of subjective impressions [13]. Lochner and Burger's formula for the signal-to-noise ratio is

$$R_e = 10 \log \frac{\int_0^{95 \text{ ms}} p^2(t)\alpha(t, l) \, dt}{\int_{95 \text{ ms}}^{\infty} p^2(t) \, dt} \quad (2)$$

where

R_e = Lochner–Burger signal-to-noise ratio, decibels
$p(t)$ = impulse response of system
$\alpha(t, l)$ = weighting curve applied to reflections according to delay t and level l.

Lochner and Burger did not provide conclusive experimental data showing the exact relationship between their signal-to-noise ratio and speech intelligibility. Other investigators, most notably Bradley [14] and Latham [15], have found that the Lochner–Burger signal-to-noise ratio contains the information necessary to predict speech intelligibility. However, they found it necessary to develop their own correlation functions between objective and subjective measures.

5.3 %AL$_{cons}$ Formula for Directional Sources

Peutz [16] developed a subjective intelligibility test method which scored word lists according to the mean percentage of misunderstood consonants. The resulting score, the percent articulation loss of consonants (denoted here by %AL$_{cons \, act}$) can be shown (see Sec. 6.2) to be simply related to %PB$_{act}$ as measured and scored using the ANSI standard [5].

Peutz conducted an experiment using an omnidirectional loudspeaker in a variety of reverberant rooms. His data indicated that %AL$_{cons}$ as measured using the subjective test increased linearly as a function of the square of the source-to-listener distance up to a certain limiting distance, beyond which there was no further increase. The data also indicated that this limiting distance was simply related to the critical distance of the room–loudspeaker combination, defined as the source-to-listener distance at which the direct field and the steady-state reverberant field are equal in intensity [17]. From these trends, a simple formula was written to predict %AL$_{cons}$:

$$\%\text{AL}_{\text{cons pred}} = \begin{cases} 200 D^2 T^2 / V, & D < D_L \approx 3.2 D_C \\ 9T, & D > D_L \end{cases} \quad (3)$$

where

D = source-to-listener distance, meters
T = reverberation time, seconds
V = room volume, cubic meters
D_L = limiting distance, meters
D_C = is critical distance, meters

No statistical analysis of the data was offered to show the accuracy with which the articulation loss could be predicted, except for the stated accuracy of $\pm 10\%$. A 10% error in %AL$_{cons}$ corresponds to a 17% error in %PB as measured by the ANSI standard (see Fig. 3).

Klein [18] subsequently hypothesized that because the critical distance can be shown to be theoretically related to loudspeaker directivity through the use of a statistical reverberation formula, Peutz's limiting distance D_L could be extended by increasing directivity.

The statistical reverberation formula is

$$D_C = K\left(\frac{Q_{AXIAL}V}{T}\right)^{1/2} \quad (4)$$

where

- K = a constant
- Q_{AXIAL} = source directivity on the axis from source to listener.

The modified Peutz formula introducing loudspeaker directivity Q was published by Klein [18], but is attributed to Davis [19],

$$\%AL_{cons\ pred} = \begin{cases} 200D^2T^2/Q_{AXIAL}V, & D < D_L \approx 3.2D_C \\ 9T, & D > D_L. \end{cases} \quad (5)$$

No data were offered by Klein to substantiate his hypothesis. To this author's knowledge, no attempt has been made to correlate Eq. (5) with actual speech intelligibility scores except for this author's previous study, which indicated poor correlation [4]. This is significant in light of the fact that the equation has been implemented in many manuals, design guides, and computer programs for designing sound systems.

It should be stressed that $\%AL_{cons\ act}$ is a parameter specified by a particular subjective intelligibility test, analogous to the ANSI test. The $\%AL_{cons\ act}$ parameter must be separated from any formulas that have been written to predict $\%AL_{cons}$. In this study, one of these formulas, Eq. (5), has been chosen for examination. However, the subjective intelligibility test yielding $\%AL_{cons\ act}$ values is not under examination.

6 RESULTS

6.1 Results of Subjective Testing

Intelligibility was scored for each subject in each room, source, and listener position combination, and a mean value was computed. The resulting average is denoted by $\%PB_{act}$. The accuracy of these mean values was estimated to within a 95% confidence interval using Student's t test [20]. Results of the subjective testing are shown in the Appendix, Table A.1.

6.2 Converting Objective Data to %PB$_{pred}$ Scores

The aim of this study is to compare the accuracy of three speech intelligibility predictors. The modulation and the signal-to-noise methods both predict the same parameter as used in this study's subjective testing, %PB, allowing a direct comparison of predicted versus actual scores.

STI values were converted to $\%PB_{pred}$ scores using the best-fit third-order regression curve established by Steeneken and Houtgast [9] for a wide variety of speech transmission conditions. The formula for this regression curve is

$$\%PB_{pred} = -43.0 + 279.8 STI - 31.2 STI^2 - 124 STI^3. \quad (6)$$

The technique used to convert the Lochner–Burger signal-to-noise ratios to $\%PB_{pred}$ for each room, source, and listener position combination is similar to that used by Bradley [14] and Latham [15]. A second-order polynomial regression curve was calculated using $\%PB_{act}$ versus signal-to-noise ratio data, as shown in Fig. 2. The equation for the regression curve can then be used to convert signal-to-noise ratios to $\%PB_{pred}$ scores.

The regression line of Fig. 2 is represented by

$$\%PB_{pred} = 79.458 + 2.99R_e - 0.1368R_e^2. \quad (7)$$

The percent articulation loss of consonants $\%AL_{cons}$ is a method of subjectively evaluating speech articulation which differs from methods used to obtain %PB scores. Similarly, a $\%AL_{cons}$ value predicted by a formula cannot be directly compared to a predicted %PB value. A relationship is required to convert $\%AL_{cons}$ scores to %PB scores.

To establish the relationship between $\%AL_{cons}$ and %PB, the word lists collected in this study were rescored according to the $\%AL_{cons}$ procedure as outlined by Peutz [16]. An excellent relationship was established between the two methods of scoring, as shown in Fig. 3, and

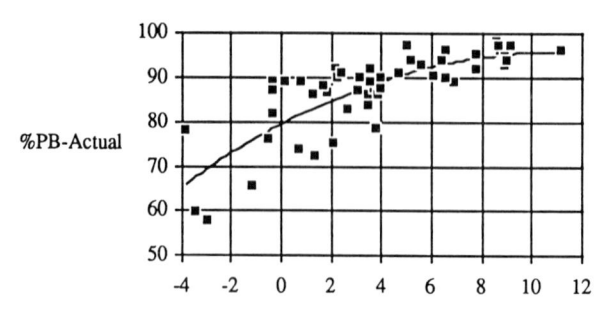

Fig. 2. Lochner–Burger signal-to-noise ratios versus $\%PB_{act}$ and regression line.

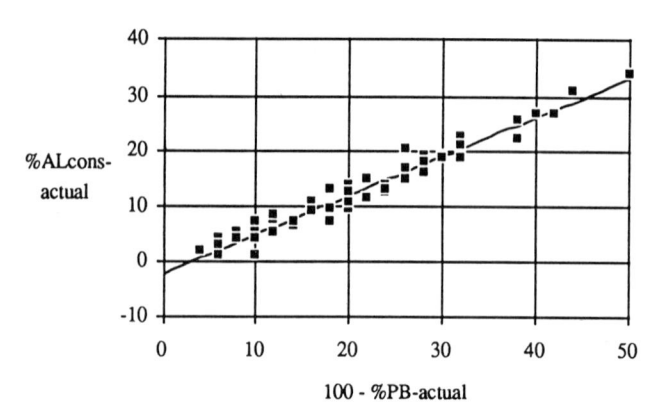

Fig. 3. Scattergram showing relationship between $\%AL_{cons\ act}$ and $\%PB_{act}$, both scored from the same subjective tests.

is used to convert %AL$_{cons\ pred}$ to %PB$_{pred}$ scores.

The regression line of Fig. 3 is represented by

$$\%PB_{act} = 100 - \frac{\%AL_{cons\ act} + 2.12}{0.70}. \quad (8)$$

To summarize, the STI scores were converted to %PB$_{pred}$ scores using a previously published formula; Lochner–Burger signal-to-noise ratios were converted by calculating a best curve fit to the data collected in this study; and the predicted %AL$_{cons}$ scores were converted by establishing a relationship between the %PB and %AL$_{cons}$ methods of scoring intelligibility word lists. Using these conversions, the three predictive algorithms can be directly compared on %PB$_{act}$ versus %PB$_{pred}$ graphs.

6.3 Correlation of Predicted Scores with Subjective Scores

Scatter plots of %PB$_{act}$ versus %PB$_{pred}$ for each of the three predictive methods are shown in Fig. 4. The graphs can be interpreted visually by realizing that perfect predictions fall on the straight line. Points below the straight line represent predicted scores that are too high when compared to actual scores, and points above the line represent predicted scores that are too low.

The data show a relatively even distribution of points about the straight line in Fig. 4(a) and (b). Significantly more scattering and some bias toward points above the line can be seen in Fig. 4(c).

6.4 Overall Standard Deviations

The data presented in the scattergrams can be examined more closely, and the degree of accuracy of the predictor quantified. The most straightforward measure of overall accuracy is the standard deviation, which is formed by squaring and adding the errors between actual and predicted scores, dividing by the total number of points, and taking the square root of the result. The overall standard deviation of the three predictive methods is shown in Fig. 5. The chart shows that the signal-to-noise method has the lowest overall standard deviation, followed closely by the modulation method; the standard deviation for the %AL$_{cons}$ method is significantly higher. It is interesting to note that the standard deviation of 7% found for the modulation method is almost exactly the same as that quoted by Steeneken and Houtgast [9]. Furthermore there is some agreement between the standard deviation of 11.5% found for the %AL$_{cons}$ method compared to the 17% accuracy (10%AL$_{cons}$) quoted by Peutz [16]. Different data bases, however, do not make these overall deviations exactly comparable.

6.5 Mean Errors

Another way to examine the data is to compute the mean differences between predicted and actual scores. This analysis yields information about whether the data contain any biases. For example, if most of the points in a scattergram fall below the straight line, meaning that the predictions were consistently too high, the mean difference would be positive. Zero mean difference signifies no biases in the predictor. The overall mean differences for the three predictors are shown in Fig.

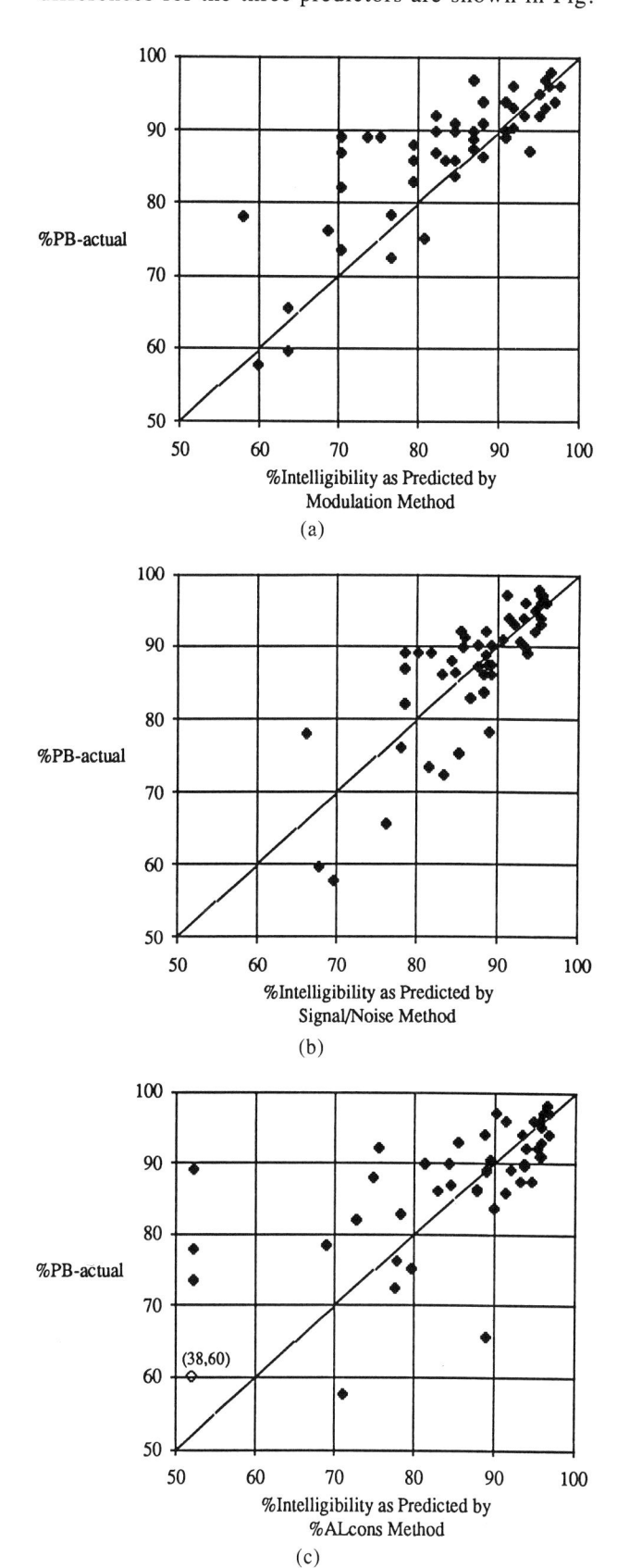

Fig. 4. Scattergrams of %PB$_{act}$ versus %PB$_{pred}$ for three predictive algorithms. (a) Modulation method. (b) Signal-to-noise method. (c) %AL$_{cons}$ method.

6. The analysis shows no significant bias in the signal-to-noise procedure, an expected outcome of the second-order curve fitting (see Fig. 2). Both the modulation and the %AL$_{cons}$ methods show a bias toward predictions, which are too low compared to the actual scores.

It must be stressed that the method used to convert Lochner–Burger signal-to-noise ratios to %PB$_{pred}$ scores was developed from the data in this study only, and cannot be generalized without further study. The regression analysis was required because no universal method has been established by previous studies.

6.6 Correlation of Scores in Areas of Most Interest to Designers

Another way to examine the data is to ask whether there are any regions of special interest to sound system designers. In the context of designing sound systems, there are at least four. What is the accuracy of the three predictors when

1) Actual intelligibility is less than 85%?
2) Actual intelligibility is greater than 85%?
3) Predicted intelligibility is less than 85%?
4) Predicted intelligibility is greater than 85%?

The value of 85%PB intelligibility has been chosen because it is generally accepted as the transition between fair-to-good and good-to-excellent intelligibility. The first question addresses the accuracy of the predictors in cases where it is already known that the intelligibility is unacceptable, the second where actual intelligibility is known to be good. The third question is relevant for cases where predictions show unacceptable intelligibility, and the fourth where predictions show acceptable intelligibility. The results are shown in Fig. 7.

Fig. 7 shows that in each of the four regions of special interest to the sound system designer, the modulation and signal-to-noise methods have lower standard errors than the %AL$_{cons}$ method. The data contained in the charts also indicate that prediction is poorer overall in the region below 85% intelligibility.

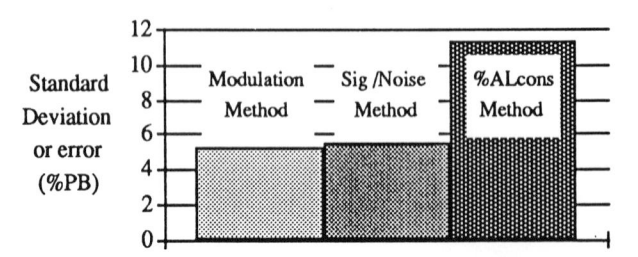

Fig. 5. Overall standard deviation of three predictors.

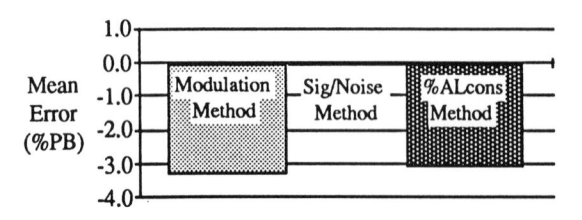

Fig. 6. Mean difference between predictive algorithms and actual intelligibility. Negative mean difference corresponds to predictions too low when compared to actual scores.

6.7 Illustrative Examples

Examination of the actual intelligibility data (see Appendix, Table A.1) reveals an unexpected case where the array loudspeaker was more intelligible than the horn loudspeaker of higher directivity (the rear position in the cathedral). In this case the modulation and signal-to-noise methods predict the same trend, while the %AL$_{cons}$ method predicts that the horn should have been significantly more intelligible, as shown in Fig. 8.

Another unexpected result was observed when intelligibility was measured to be higher in the rear position than the front position; the horn in Mechanics Hall is an example. Again, the modulation and signal-to-noise methods predict the same trend, while the %AL$_{cons}$ method predicts the opposite, as shown in Fig. 9.

These individual conditions suggest the need to in-

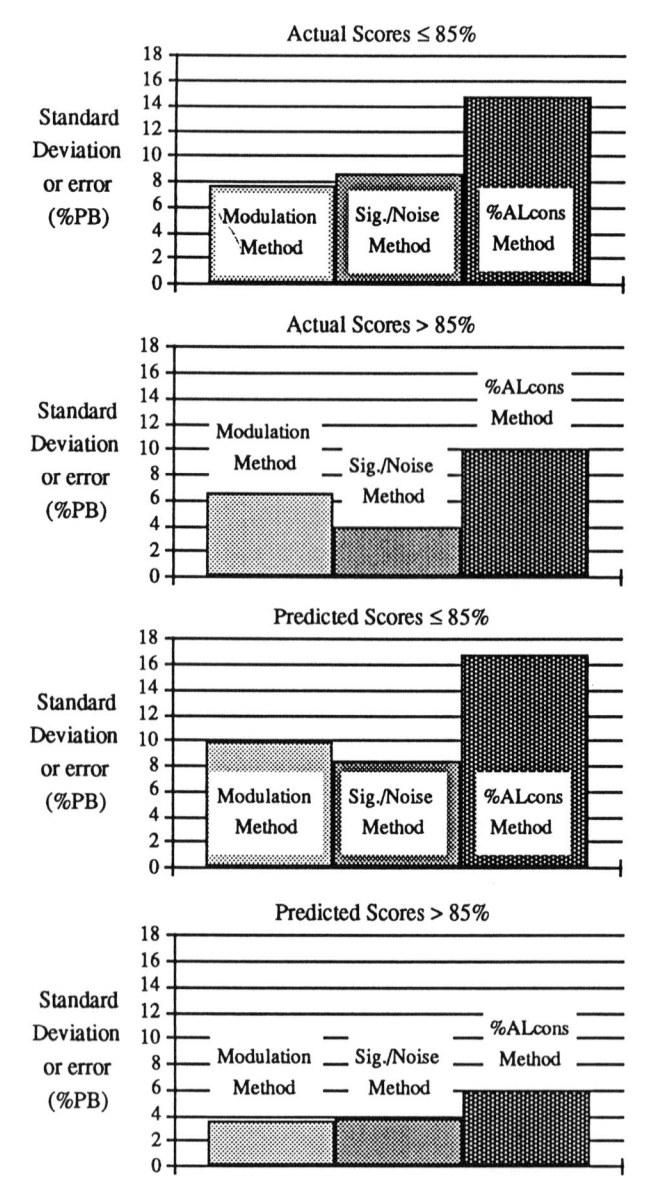

Fig. 7. Standard deviations for four special cases of interest to the sound system designer.

clude early reflections in the prediction of intelligibility. Furthermore, these cases show that while the additional direct sound provided by an increase in source directivity is useful for improving intelligibility, neglect of other factors affecting the ratio of early to late sound can lead to serious errors.

7 CONCLUSION

The data presented here support the conclusion that the speech transmission index, computed using the modulation method, is an accurate predictor of speech intelligibility in reverberant rooms. Correlation between actual and predicted intelligibility scores was very good overall, in both the under 85% and the over 85% regions. These results provide additional evidence of the suitability of the STI as an international standard for predicting speech intelligibility.

The Lochner–Burger signal-to-noise method was shown to be accurate as a predictor when a function unique to the data base of this study was used to translate signal-to-noise ratios to %PB$_{pred}$ scores. Two other investigations using the method required different functions [14], [15] for this purpose. Thus while the signal-to-noise ratio has been shown to contain the information necessary to predict intelligibility within this data base accurately, no universal method of translating signal-to-noise ratios to %PB intelligibility has been established. Additional investigation is required to establish a general relationship between the Lochner–Burger signal-to-noise ratio and intelligibility.

The formula for predicting %AL$_{cons}$, as defined in Eq. (5), has been shown to be the least accurate. From the results of this extended data base it must be concluded that intelligibility as predicted by the formula can lead to serious errors. In addition, the hypothesis that intelligibility can be increased monotonically simply by increasing loudspeaker directivity should be rejected. It is important to note that under the same speech conditions, the modulation and signal-to-noise methods predict intelligibility accurately, independent of loudspeaker directivity. The data therefore support the conclusion that loudspeaker directivity is important only in so far as it affects the ratio of early to late sound.

Finally, this experiment has been conducted and presented in such a manner as to allow repetition by other investigators. The hope is that the data base of rooms, sound sources, and listener positions can be

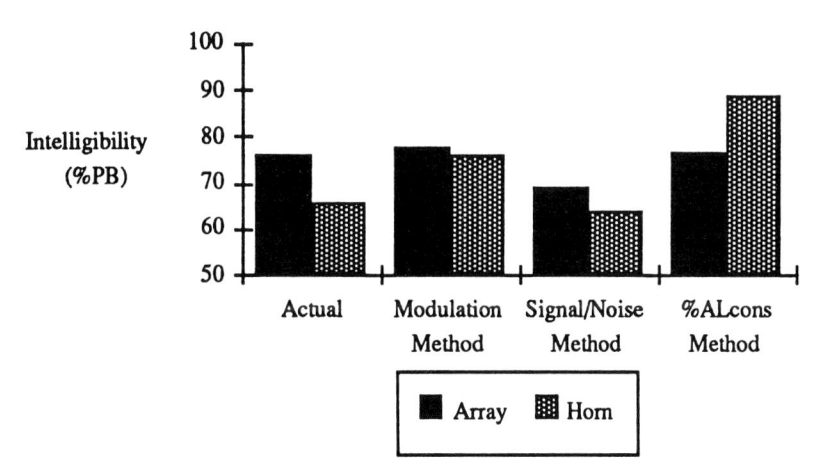

Fig. 8. Actual and predicted intelligibility scores in the rear of the South End Cathedral. Notice that the modulation and signal-to-noise methods predict the same trend as actual scores, while the %AL$_{cons}$ method predicts the opposite trend.

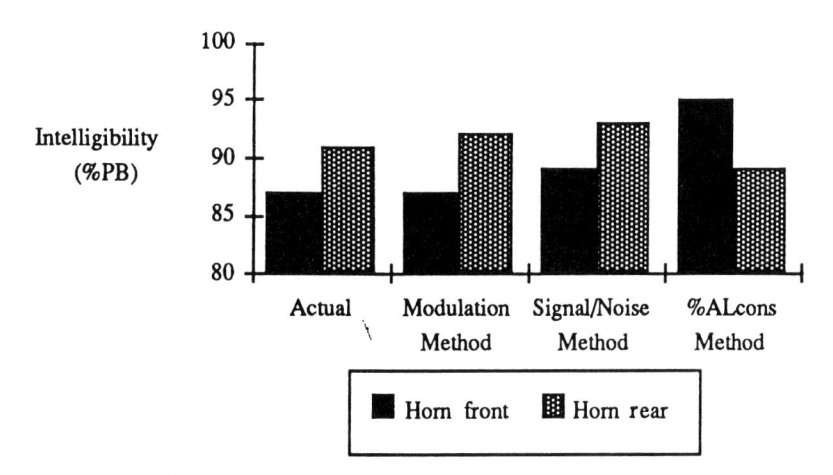

Fig. 9. Actual and predicted intelligibility scores in the front and rear of Mechanics Hall using the horn loudspeaker (see Appendix, Table A.1). Notice that the modulation and signal-to-noise methods predict the same trend as actual scores, while the %AL$_{cons}$ method predicts the opposite.

further extended in an effort to establish increasingly more accurate predictors of intelligibility.

8 ACKNOWLEDGMENT

The author gratefully acknowledges the help of Dr. Amar Bose in the preparation of this manuscript, and Chris Ickler for discussions on the analysis and interpretation of data.

9 REFERENCES

[1] N. R. French and J. C. Steinberg, "Factors Governing the Intelligibility of Speech Sounds," *J. Acoust. Soc. Am.*, vol. 19, pp. 90–119 (1947).

[2] K. D. Kryter, "Methods for the Calculation and Use of the Articulation Index," *J. Acoust. Soc. Am.*, vol. 34, pp. 1689–1697 (1962).

[3] C. V. Paclovic, "Derivation of Primary Parameters and Procedures for Use in Speech Intelligibility Predictions," *J. Acoust. Soc. Am.*, vol. 82, no. 2 (1987).

[4] K. D. Jacob, "Subjective and Predictive Measures of Speech Intelligibility—The Role of Loudspeaker Directivity," *J. Audio Eng. Soc.*, vol. 33, pp. 950–955 (1985 Dec.).

[5] ANSI Std. S3.2-1960 (R1982), "American National Standard Method for Measurement of Monosyllabic Word Intelligibility," Am. Nat. Stand. Inst., New York.

[6] T. Houtgast and H. J. M. Steeneken, "A Review of the MTF Concept in Room Acoustics and Its Use for Estimating Speech Intelligibility in Auditoria," *J. Acoust. Soc. Am.*, vol. 77, no. 3 (1985).

[7] *Opt. Acta*, vol. 2, no. 18 (1971).

[8] T. Houtgast and H. J. M. Steeneken, "Evaluation of Speech Transmission Channels by Using Artificial Signals," *Acustica*, vol. 25 (1971).

[9] H. J. M. Steeneken and T. Houtgast, "A Physical Method for Measuring Speech Transmission Quality," *J. Acoust. Soc. Am.*, vol. 67, no. 1 (1980).

[10] M. R. Schroeder, "Modulation Transfer Functions: Definition and Measurement," *Acustica*, vol. 49 (1981).

[11] T. Houtgast, H. J. M. Steeneken, and R. Plomp, "Predicting Speech Intelligibility in Rooms from the Modulation Transfer Function; Part 1: General Room Acoustics," *Acustica*, vol. 46, no. 1 (1971).

[12] J. P. A. Lochner and J. F. Burger, "The Influence of Reflections on Auditorium Acoustics," *J. Sound Vib.*, vol. 1, no. 4 (1964).

[13] H. Kuttruff, *Room Acoustics*, 2nd ed. (Appl. Sci. Publ., Ltd., London, 1979), p. 180.

[14] J. S. Bradley, "Predictors of Speech Intelligibility in Rooms," *J. Acoust. Soc. Am.*, vol. 80, no. 3 (1986).

[15] H. G. Latham, "The Signal-to-Noise Ratio for Speech Intelligibility—An Auditorium Acoustics Design Index," *Appl. Acoust.*, vol. 12, pp. 253–320 (1979).

[16] V. M. A. Peutz, "Articulation Loss of Consonants as a Criterion for Speech Transmission in a Room," *J. Audio Eng. Soc.*, vol. 19, pp. 915–919 (1971 Dec.).

[17] L. L. Beranek, *Acoustics* (Am. Inst. of Physics, New York, 1986).

[18] W. Klein, "Articulation Loss of Consonants as a Basis for the Design and Judgment of Sound Reinforcement Systems," *J. Audio Eng. Soc.*, vol. 19, pp. 920–922 (1971 Dec.).

[19] D. Davis, private communication, 1985.

[20] See, for example, H. Kohler, *Statistics for Business and Economics* (Scott, Foresman and Co., Glenview, IL, 1985).

APPENDIX

Table A.1. Subjective scores.*

Room	Source	Position	Score
Berklee	Sphere	Near	96 ± 1.0
		Far	93 ± 1.0
	Array	Near	96 ± 1.0
		Far	96 ± 1.0
	Horn	Near	98 ± 0.7
		Far	96 ± 1.0
Coolidge	Sphere	Near	97 ± 0.6
		Far	90 ± 1.6
	Array	Near	97 ± 0.6
		Far	94 ± 1.1
	Horn	Near	97 ± 0.6
		Far	91 ± 1.2
Huntington	Sphere	Near	94 ± 1.3
		Far	86 ± 1.7
	Array	Near	95 ± 1.0
		Far	89 ± 1.4
	Horn	Near	94 ± 1.1
		Far	92 ± 1.1
Bridget's	Sphere	Near	92 ± 1.1
		Far	82 ± 2.1
	Array	Near	92 ± 1.2
		Far	88 ± 1.5
	Horn	Near	93 ± 1.2
		Far	86 ± 1.5
Nevins	Sphere	Near	78 ± 3.0
		Far	89 ± 1.6
	Array	Near	87 ± 3.0
		Far	89 ± 1.1
	Horn	Near	89 ± 1.6
		Far	90 ± 1.6
Jordan	Array	Near	89 ± 1.4
		Far	78 ± 2.4
	Horn	Near	90 ± 1.9
		Far	87 ± 1.6
Mechanic's	Array	Near	86 ± 2.0
		Far	83 ± 2.0
	Horn	Near	87 ± 2.0
		Far	91 ± 1.6
Cathedral	Array	Near	90 ± 1.8
		Far	76 ± 3.7
	Horn	Near	91 ± 1.7
		Far	66 ± 2.6
Cyclorama	Array	Near	86 ± 2.6
		Far	73 ± 2.3
	Horn	Near	87 ± 2.3
		Far	72 ± 2.9
MIT Track	Array	Near	75 ± 2.5
		Far	60 ± 3.2
	Horn	Near	84 ± 2.9
		Far	58 ± 3.4

* Room—auditoriums listed in Table 2; Source—loudspeakers given in Table 1; Position—listener positions defined in Section 1.4; Score—mean over all subjects with the accuracy relative to the true intelligibility computed using Student's t test.

Table A.2. Speech transmission indexes, signal-to-noise ratios, and predicted %AL$_{cons}$ for all conditions.*

Room	Source	Position	STI	L/B ratio	%AL$_{cons\ pred}$
Berklee	Sphere	Near	0.65	6.6	8
		Far	0.71	8.9	8
	Array	Near	0.72	8.8	1
		Far	0.72	8.9	5
	Horn	Near	0.73	8.6	1
		Far	0.78	11.2	2
Coolidge	Sphere	Near	0.60	5.0	5
		Far	0.56	3.2	9
	Array	Near	0.71	9.2	1
		Far	0.64	6.4	3
	Horn	Near	0.71	8.7	0
		Far	0.61	4.7	1
Huntington	Sphere	Near	0.61	5.2	6
		Far	0.57	3.5	10
	Array	Near	0.70	7.8	1
		Far	0.64	6.9	4
	Horn	Near	0.74	9.0	0
		Far	0.67	7.8	1
Bridget's	Sphere	Near	0.56	2.2	17
		Far	0.48	−0.3	17
	Array	Near	0.70	3.6	3
		Far	0.54	1.7	17
	Horn	Near	0.65	5.6	1
		Far	0.54	1.3	7
Nevins	Sphere	Near	0.41	−3.8	32
		Far	0.48	−0.3	32
	Array	Near	0.48	−0.3	11
		Far	0.51	0.8	32
	Horn	Near	0.50	0.2	4
		Far	0.60	4.0	12
Jordan	Array	Near	0.60	3.6	7
		Far	0.52	3.8	20
	Horn	Near	0.64	6.6	2
		Far	0.56	3.1	9
Mechanic's	Array	Near	0.58	3.9	5
		Far	0.54	2.7	16
	Horn	Near	0.60	4.0	2
		Far	0.65	6.1	6
Cathedral	Array	Near	0.58	2.3	3
		Far	0.47	−0.5	17
	Horn	Near	0.58	2.4	1
		Far	0.44	−1.1	6
Cyclorama	Array	Near	0.61	1.9	8
		Far	0.48	0.7	32
	Horn	Near	0.68	3.7	3
		Far	0.52	1.4	14
MIT Track	Array	Near	0.55	2.1	15
		Far	0.44	−3.4	41
	Horn	Near	0.58	3.5	5
		Far	0.42	−2.9	19

* STI—speech transmission index generated from modulation tranfer function, as described in Section 5.1; L/B Ratio—measured Lochner–Burger signal-to-noise ratio as discussed in Section 5.2; %AL$_{cons\ pred}$—predicted percent articulation loss of consonants as discussed in Section 5.3.

THE AUTHOR

Ken Jacob is a staff engineer of Bose Corporation, Framingham, Massachusetts. He received his master's degree from the Massachusetts Institute of Technology and his bachelor's degree from the University of Minnesota, both with specializations in acoustics.

In addition to research in speech intelligibility, Mr. Jacob is the acoustic engineer of the Bose 102[1] commercial sound system and the Acoustic Wave[1] Cannon system, and is the project engineer of the Sound System[1] software family including the Modeler[1] design program.

[1] Trademarks of Bose Corporation.

Editor's note: *The previous two papers[1,2] appear to use similar data to arrive at very different conclusions. I am sure that Don, Carolyn, and Ken have modified their overall views of intelligibility modeling over the decade, as have we all. Both papers enjoyed Journal status and as such deserve anthology status. It would be difficult to pick one over the other for inclusion so we are making the best decision, to include both of them.*

David Lloyd Klepper

---

[1] "Application of Speech Intelligibility to Sound Reinforcement," Don Davis and Carolyn Davis (Vol. 37, pp. 1002–1019 [1989 December])

[2] "Correlation of Speech Intelligibility Tests in Reverberant Rooms with Three Predictive Algorithms," Kenneth D. Jacob (Vol. 37, pp. 1020–1030 [1989 December])

Speech Intelligibility in Some German Sports Stadiums*

K. RIJK, F. BREUER, AND V. M. A. PEUTZ

Akustikberatung Peutz, Düsseldorf, Germany, and Nijmegen, The Netherlands

The Landesministerium für Umwelt, Raumordnung und Landwirtschaft of the state of Nordrhein-Westfalen has commissioned a study on optimizing speech intelligibility and reducing noise pollution caused by sound systems in sports stadiums. A detailed analysis of eight facilities, involving time-delay spectrometry and subjective AL_{cons} tests, shows that sound system performance is generally poor when compared with the maximum attainable speech intelligibility under given circumstances, that is, acoustics and crowd noise. Possibilities for improvement are outlined.

0 INTRODUCTION

In sports stadiums, leisure parks, and other public places the audience is provided with information by means of loudspeakers. The announcements (lineup of teams, commercials, emergency calls, and sometimes extensive live coverage of the event) may also be audible in the surrounding neighborhood and can be annoying. Aggravating factors are the high attention-drawing value of the signals and their occurrence at quiet times.

The stadiums have to comply with the German law that states that "harmful effects on the neighborhood must be avoided if technically feasible." When assessing existing situations or planning new facilities, local authorities find this law to be lacking in accuracy and liable to misinterpretation.

Therefore the present study is concerned with assessing the technical possibilities of reducing sound emission without seriously affecting speech intelligibility. As these two requirements are naturally conflicting, this will lead to compromises and, in some cases, to a reduction of the sound levels to the absolute minimum required.

At the outset a general survey of (mainly soccer) stadiums was made by means of a questionnaire to gather data on activities, location, and sound systems. From this overview, some representative stadiums from each category were selected for detailed analysis.

1 ANALYSIS

Each stadium was investigated with regard to speech intelligibility and noise emission.

1.1 Speech Intelligibility

Speech intelligibility can be considered as a statistical phenomenon: a message is either understood (recognized) or not. At this moment, direct "measurement" of speech intelligibility, taking into account all detrimental influences, is only available and valid for some exceptional and relatively simple situations [1].

Therefore a subjective judgment of the speech intelligibility in the sports stadiums was obtained by means of listening tests. Lists of phonetically balanced nonsense syllables embedded in carrier sentences (Fig. 1) were played over the sound system, and were recorded binaurally at several locations in the stadium, namely, in the main seating area, opposite the seating area, and in curved areas. These head-related calibrated recordings were used to carry out listening tests in the laboratory with groups of test subjects—with and without crowd noise added—to obtain the articulation loss of consonants AL_{cons}. Three hundred syllables were evaluated by at least four listeners per listener position and location.

---

* Presented at the 86th Convention of the Audio Engineering Society, Hamburg, Germany, 1989 March 7–10; revised 1990 September 12.

The AL_{cons} without crowd noise provides the optimum intelligibility that can be attained for given circumstances, namely, stadium acoustics and sound system. In order to obtain an indication of the speech intelligibility in practice, crowd noise was mixed into the recorded word lists at a level corresponding to the level measured in the stadium. The statistical descriptor L_{50}, the level that is exceeded 50% of the time, was used as an indicator for the nonconstant crowd noise level. A typical sequence during a match was chosen and attenuated or amplified so that L_{50} equaled the L_{50} value measured at that location. Thus an "average" speech intelligibility was obtained—the intelligibility will be worse during noisy periods and better during quiet periods.

In addition, during two matches per stadium, binaural stereo recordings were made simultaneously at the different locations for sound level indication and for reference. During each match the intelligibility at some locations was also assessed subjectively by means of a questionnaire.

In unoccupied facilities, the sound system and stadium acoustics were analyzed using time-delay spectrometry. For each measurement location the energy–time curve (ETC) around 500 Hz and 2000 Hz, with a time span of 0.4 s, was measured, as well as the delay spectrum of the direct sound. Important parameters derived from the measurements are the quality of the direct sound, the ratio of direct to reverberant sound, the reverberation time, and occasional echoes.

1.2 Noise Pollution

During two matches per stadium, sometimes with different events (American football, soccer, different numbers of spectators), sound level measurements were taken both inside and outside the confines of the facility. The sound levels were analyzed statistically. Of special interest are the average level L_m (approximately corresponding to the ISO L_{eq}) and the maximum level L_{max}, both with respect to the distance from the stadium. In addition, the reduction in sound level from inside to outside the confines was measured at several locations using pink noise.

2 RESULTS

2.1 Speech Intelligibility Tests

Fig. 2 shows the AL_{cons} score for all measured listener positions, with and without crowd noise added. Without noise, the loss of consonants varies between 12 and 35% (in the unoccupied ice stadium it is even higher), confirming our suspicion during the matches.

With crowd noise added, the loss of consonants is even greater. Fig. 3 shows AL_{cons} as a function of the signal-to-noise ratios in the stadiums. As expected, there is a tendency toward better intelligibility with higher signal-to-noise ratios. However, even among locations with the same signal-to-crowd-noise ratio considerable differences in AL_{cons} were found.

Fig. 4 summarizes the announcement sound levels outside the facilities (time averaged to obtain L_m, averaged at a distance of 100 m around the stadium). The loudest stadium measures about 75 dBA, and the most quiet one about 55 dBA (both with a residential area well within 100 m). The reduction in sound level (from inside to outside) varies due to different layouts (shielding) and loudspeaker directivities.

Fig. 5 illustrates the eight facilities tested.

2.1.1 Stadium 1 [Fig. 5(a)]

Stadium 1 has a distributed loudspeaker system, that is, compression horn loudspeakers are distributed throughout the stadium near the spectators. Speech intelligibility is insufficient due to a number of factors: the sound level is too low, high frequencies are absent, a shrill sound quality, and a reverberant effect due to inhomogeneous sound distribution. Emission sound levels are low due to the extremely low level inside, the "closed" layout and shielding, and the distributed sound system. The rather complex sound system is not reliable. Defects are not detected.

2.1.2 Stadium 2 [Fig. 5(b)]

Intelligibility is insufficient in stadium 2 despite the high sound levels (about 12 dB above crowd noise L_{50}). The covered main seating area is reverberant; the frequency response of the direct sound shows dips and lacks high frequencies. Compression loudspeakers distributed over the roof provide the sound for the opposite zones and the curved areas, where the sound is harsh or shrill. Noise emission is high, due to the high sound level inside, and to the low inside-to-outside reduction (caused by the direct sound from the long-throw horns, the open structure, and by adjacent buildings reflecting the sound). For this stadium, surrounded by a residential area, a distributed sound system, as in stadium 1, would be more suited.

Zuerst kommt	NAN	als eins.
Jetst kommt	NED	als zwei.
Darauf folgt	KUF	als drei.
Achten Sie auf	SCHEL	als vier.
Jetzt kommt	SOS	als fünf.
Und hiernach	MES	als sechs.
Dann wieder	REZ	als sieben.
Es folgt jetzt	FEIS	als acht.
Hören Sie auf	FITZ	als neun.
Und auf	LEHN	als zehn.
Dann kommt	CEZ	als elf.
Nun folgt	SAAR	als zwölf.
Darauf folgt	CHES	als dreizehn.
Achten Sie auf	GAK	als vierzehn.
Nun kommt	NEP	als fünfzehn.
Und hierauf	TIS	als sechzehn.
Dann wieder	ZIG	als siebzehn.
Es folgt dann	REF	als achtzehn.
Hören Sie auf	NIDS	als neunzehn.
Und jetzt auf	SUHN	als zwanzig.
Dann kommt	GIM	als einundzwanzig.
Nun folgt	MUS	als zweiundzwanzig.
Darauf folgt	SAUG	als dreiundzwanzig.
Achten Sie auf	MAUN	als vierundzwanzig.
Und zum Schluß	FIR	als fünfundzwanzig.

Fig. 1. Example of word list with nonsense syllables for listening tests.

2.1.3 Stadium 3 [Fig. 5(c)]

Intelligibility is insufficient in stadium 3, mainly because of low sound levels. The covered zone is lacking high frequencies, has ragged frequency response of the direct sound, and loud vowels are distorted. In the open-air zones the loudspeakers on the opposite roof sound harsh or distorted, and there is an echo from loudspeakers placed in opposite corners of the stadium (Fig. 6). Emission is low due to the low sound level inside the stadium.

2.1.4 Stadium 4 [Fig. 5(d)]

In stadium 4 one of the seating areas is covered. Here intelligibility is poor due to excess low frequencies. In the curves the sound is reverberant. The sound quality of the compression loudspeakers is relatively good. Sound emission is high due to direct radiation from the central loudspeakers.

2.1.5 Stadium 5 [Fig. 5(e)]

One of the seating areas in stadium 5 is covered. Intelligibility is insufficient due to a lack of high frequencies, and in the open area it is also due to the echo from the main stand. Emission is not so high because the central loudspeakers are quiet.

2.1.6 Stadium 6 [Fig. 5(f)]

Stadium 6 is an open-air stadium with two loudspeakers located behind the main zone. Intelligibility is fair due to a good signal-to-noise ratio. In some parts of this area, intelligibility is worse, being too far off the loudspeaker axes. On the other side and around the field intelligibility is insufficient because of a lack of high frequencies and an echo from the two loudspeakers at different distances. Emission is low due to the low emission level, despite the open architecture.

2.1.7 Ice Stadium [Fig. 5(g)]

In this ice stadium, which is seating 11 000, a central loudspeaker cluster and two delayed clusters are suspended from the ceiling. The results of the listening tests with word lists only have a limited value, because the recordings were made in a (reverberant) unoccupied location. Intelligibility is sufficient, despite the low level of the central cluster with respect to the delayed clusters, introducing an unnecessary reverberant effect in most areas. The delayed clusters have different frequency responses and equalization settings for no apparent reason. Emission is high due to the high levels reached inside and to the openings in the stadium confines. The emission spectrum is relevant up to 2500 Hz, in contrast with the other stadiums.

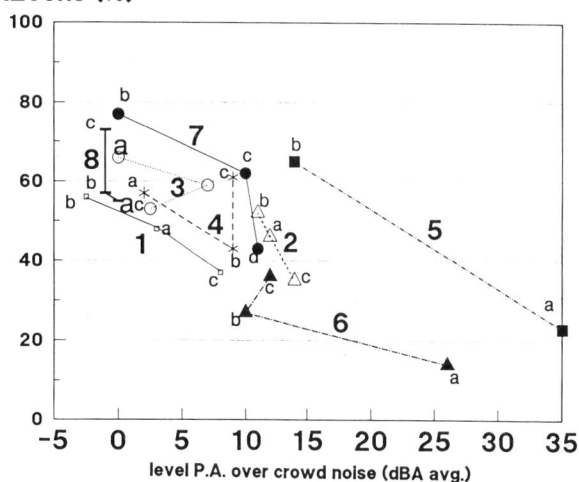

Fig. 3. Articulation loss of consonants AL_{cons} as a function of signal-to-noise ratio (announcement level versus crowd noise level) for eight facilities. a—main seating area; b—opposite seating area; c—curved area.

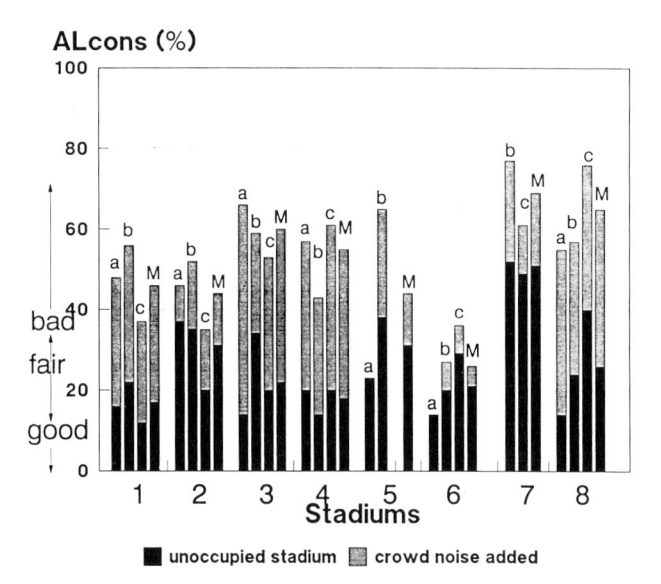

Fig. 2. Articulation loss of consonants AL_{cons} from listening tests for all measured locations in eight facilities, with and without crowd noise added. a—main seating area; b—opposite seating area; c—curved area; M—average.

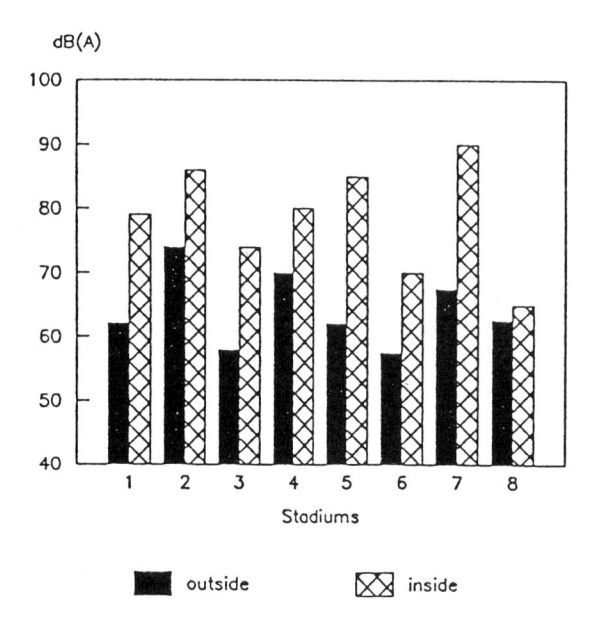

Fig. 4. Summary of sound system noise levels in dBA measured outside facilities (time-averaged L_m, averaged at a distance of 100 m beyond facilities).

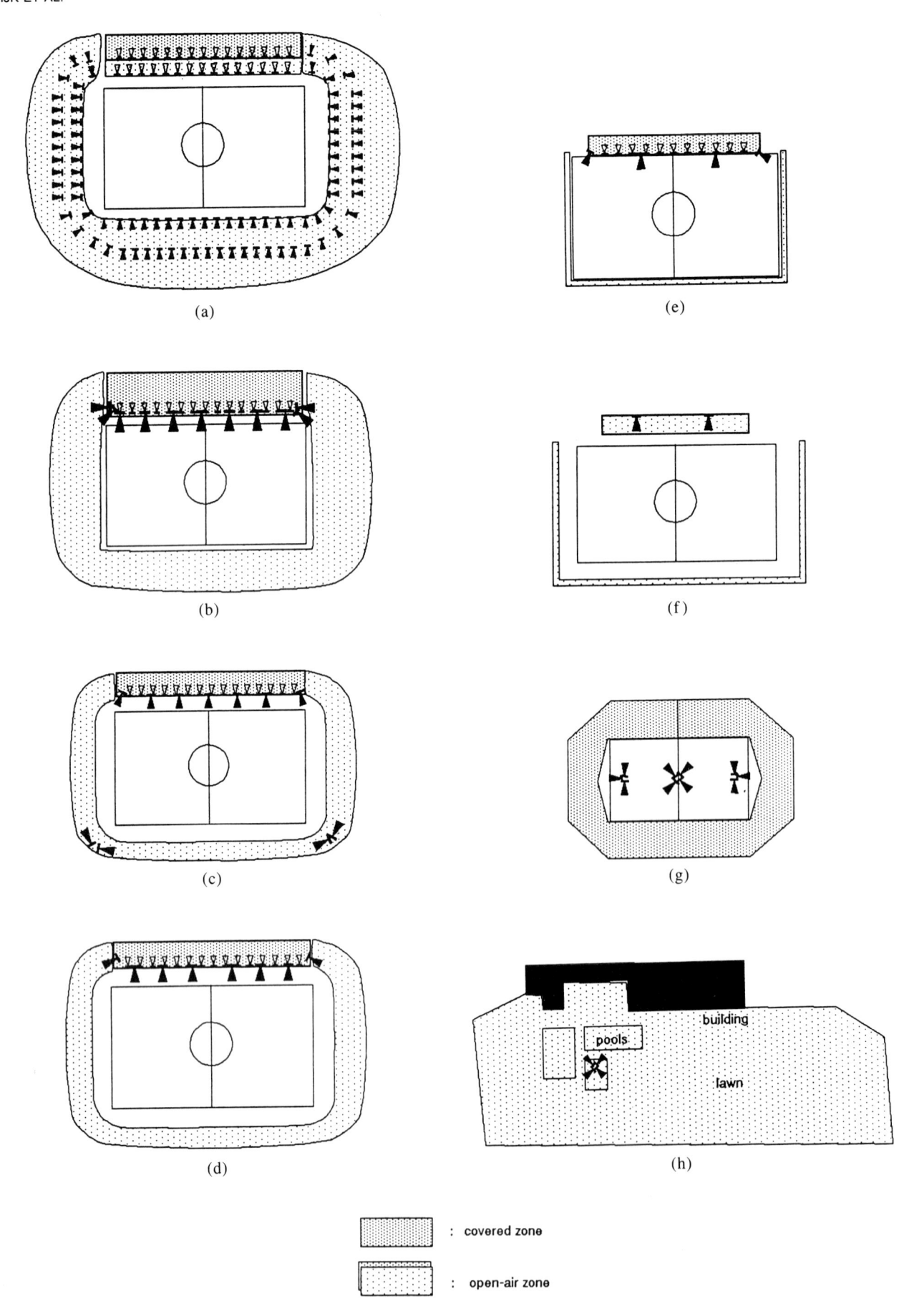

Fig. 5. Layouts of eight facilities studied. (a) Stadium 1. (b) Stadium 2. (c) Stadium 3. (d) Stadium 4. (e) Stadium 5. (f) Stadium 6. (g) Ice stadium. (h) Leisure park.

2.1.8 Leisure Park [Fig. 5(h)]

The leisure park has a central omnidirectional loudspeaker cluster. Intelligibility is insufficient due to a lack of high frequencies and because the sound levels are too low. In some areas the sound is reverberant and echoic, which is caused by reflections and obstacles. Emission is low due to low sound levels.

Summarizing the results, we can state that, apart from crowd noise, the following factors cause the unsatisfactory performance of the sound systems.

1) The lack of high frequencies is due to the fact that in all stadiums, except the ice stadium, the frequency response is cut off at 2500 Hz, in some cases already at 2000 Hz.

2) Different arrival times are due to the number of loudspeakers and the directional characteristics as well as the use of unsuitable loudspeaker columns placed horizontally.

3) In most situations reverberation is an important factor, especially in the unoccupied ice stadium and the covered stands of stadiums 1 and 2. However, it appears that most time-delay spectrometry measurements show the ratio of early to reverberant sound to be such that excellent intelligibility is, in principle, still possible. Some measurements apparently showing too much "reverberation" are in fact caused by incorrect loudspeaker placement or directivity.

4) Harsh, distorted, and shrill sound is due to the inherent distortion, comb-filtering, and resonant character of funnel-shaped compression loudspeakers.

5) The random positioning of control settings and an imbalance in different groups of loudspeakers result in a faulty use of the sound system.

2.2 Speech Intelligibility Computation Model

The effect of reverberation (reverberation time and ratio of direct to reverberant sound), signal-to-noise level, and frequency range can be determined the method of Peutz [1], [2]. The curves in Fig. 7 indicate the predicted AL_{cons} as a function of the signal-to-noise level. Curve A indicates the best intelligibility that can be obtained (advanced signal processing and dynamic compression not considered). For curves B and C the frequency range is limited to 4000 Hz, and for curve C the ratio of direct to reverberant sound is 0 dB (no information in the reverberant sound). Most spectator positions in sports stadiums are between curves B and C.

The attainable intelligibility is much better than the measured values of Fig. 3, which were obtained through listening tests. Even when considering that in reality the masking signal is crowd noise of varying levels, it is clear that other factors not considered in the computation model are significant. It must be stated here that the AL_{cons} model is used mostly for situations with relatively good intelligibility, such as conference halls, theaters, and concert halls with public-address systems. Thus the model may be somewhat inaccurate for situations with low signal-to-noise ratios as well as considerable reverberation.

To assess the difference, a simulation was done, bandpassing the original word lists and adding the same echo and reverberation characteristics as in reality. The speech intelligibility of the simulated word lists was surprisingly good, and the AL_{cons} appeared lower than in reality, confirming the computation model. The audible differences (a harsh and distorted sound) are due to the tonal quality of the stadium loudspeakers, caused by resonances and reflections inherent in the type of loudspeaker used (mostly funnel-shaped horns).

3 RECOMMENDATIONS

3.1 Sound Systems

Sound system concepts or recommendations have been developed for minimum emission and maximum intelligibility. Even though the diversity of layouts, the acoustics, and the demands on the sound systems

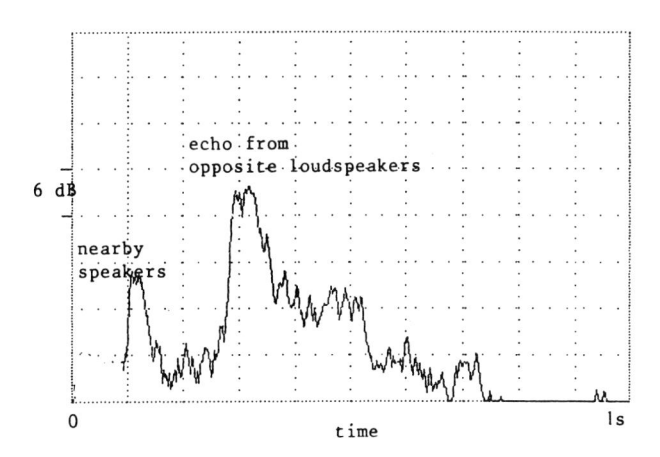

Fig. 6. Energy–time curve of stadium 3 sound system. A serious echo occurs due to (undelayed) loudspeakers opposite main stand.

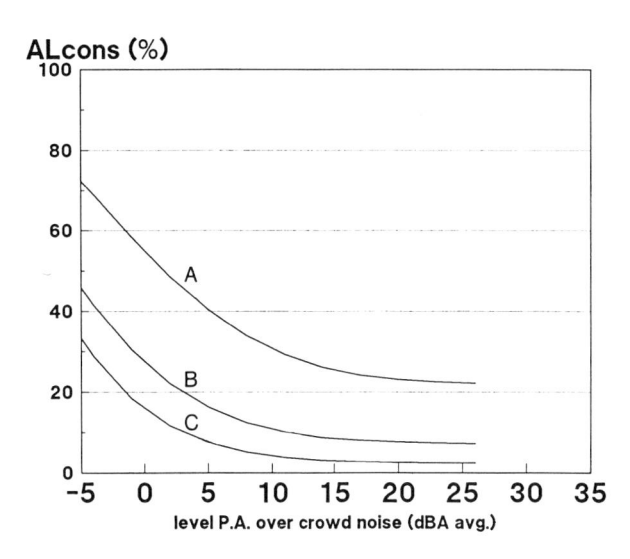

Fig. 7. Attainable speech intelligibility as a function of signal-to-noise ratio. Dynamic compression or more advanced signal processing are not considered. A—optimum (theoretical); B—frequency limited to a 4 kHz; C—as B, plus a 0-dB early-to-reverberant sound ratio with no information in reverberant sound.

are considerable, some general guidelines are proposed (Table 1).

The most important choice to be made is the circumstances under which the announcements must still be intelligible (Table 2). It must be mentioned here that "good" intelligibility is generally considered equivalent to an AL_{cons} score of 15% or less. With an experienced talker and uncomplicated messages we can be more lenient.

Commercials (sometimes fairly complicated messages) occur at quiet times, and announcements relevant to the match may occur during noisy periods. Excluding complicated alarm messages in noisy situations, we arrive at the recommended minimum sound levels given in Table 3.

3.2 Loudspeakers

A choice can be made between a central cluster and a distributed loudspeaker system. As an example, an environmental emission computation according to VDI guidelines shows that a central system on the roof is up to 15 dBA noisier than a distributed (more expensive) system, with the same sound level inside the stadium (Fig. 8), even for a fairly "closed" facility.

In most sound systems presented, the loudspeakers limit the frequency range and the quality of the high frequencies, essential to consonants. A frequency range up to 4000 Hz is recommended, which poses no problem for the noise emission since frequencies above 2000 Hz are hardly found to be dBA relevant. Playing frequencies higher than 4000 Hz is, in most cases, not necessary and is impractical because of attenuation in the air over long distances.

For loudspeaker positioning and directing, well-known and universally accepted design methods apply, which obtain uniform coverage up to the required 4000 Hz. Loudspeakers with a directivity matching the geometry of the stadium and varying little with frequency should be chosen. The sound should preferably be directed at the audience, coming from in front, such as by making use of the roof edge or loudspeaker stands at the edge of the field.

4 CONCLUSION

From this research, and especially from the comparison of measured intelligibility (Fig. 3) attainable intelligibility (Fig. 7), it can be concluded that either better intelligibility can be achieved at the same sound level, or the same intelligibility can be obtained at a lower sound level. Where low sound emission is required, a distributed loudspeaker system is effective. Recommendations include organizational aspects (reliability, operating ergonomics, protection), sound levels and sound level dynamics control, and directional and frequency characteristics of loudspeakers [3].

Table 1. Two categories of sports stadiums.

	Large stadium	Small stadium
Audience	Larger than 10 000	Mostly smaller than 2000
Crowd noise	L_{50} = 75–80 dBA	L_{50} = 70–75 dBA
		L_{50} = 60–65 dBA in open areas
Layout	Closed	Open
Areas to be covered	Large and wide	Line shaped
Acoustics	Reverberant, echoic	Occasional echo
Loudspeakers	Positions constrained	Positions dictated by physical possibilities
	Must withstand violence	

Table 2. Proposed AL_{cons} criteria for intelligibility in sports stadiums.

	Good	Fair
Complicated announcement	<10%	<15%
Simple announcement	<20%	<40%*

* At AL_{cons} = 40% one can still grasp the essentials, such as names, team lineup, and scores.

Table 3. Recommended minimum sound levels.

	Crowd noise level	Recommended sound level L_{eq}, dB(A)	
Facility	L_{50}, dB(A)	AL_{cons} = 20%	AL_{cons} = 40%
Large stadium	75*–80†	80–90	75–85
Small stadium:			
Covered area	70–75	75–85	70–80
Open area	60–65	65–75	60–70

* Quiet audience.
† Excited audience.

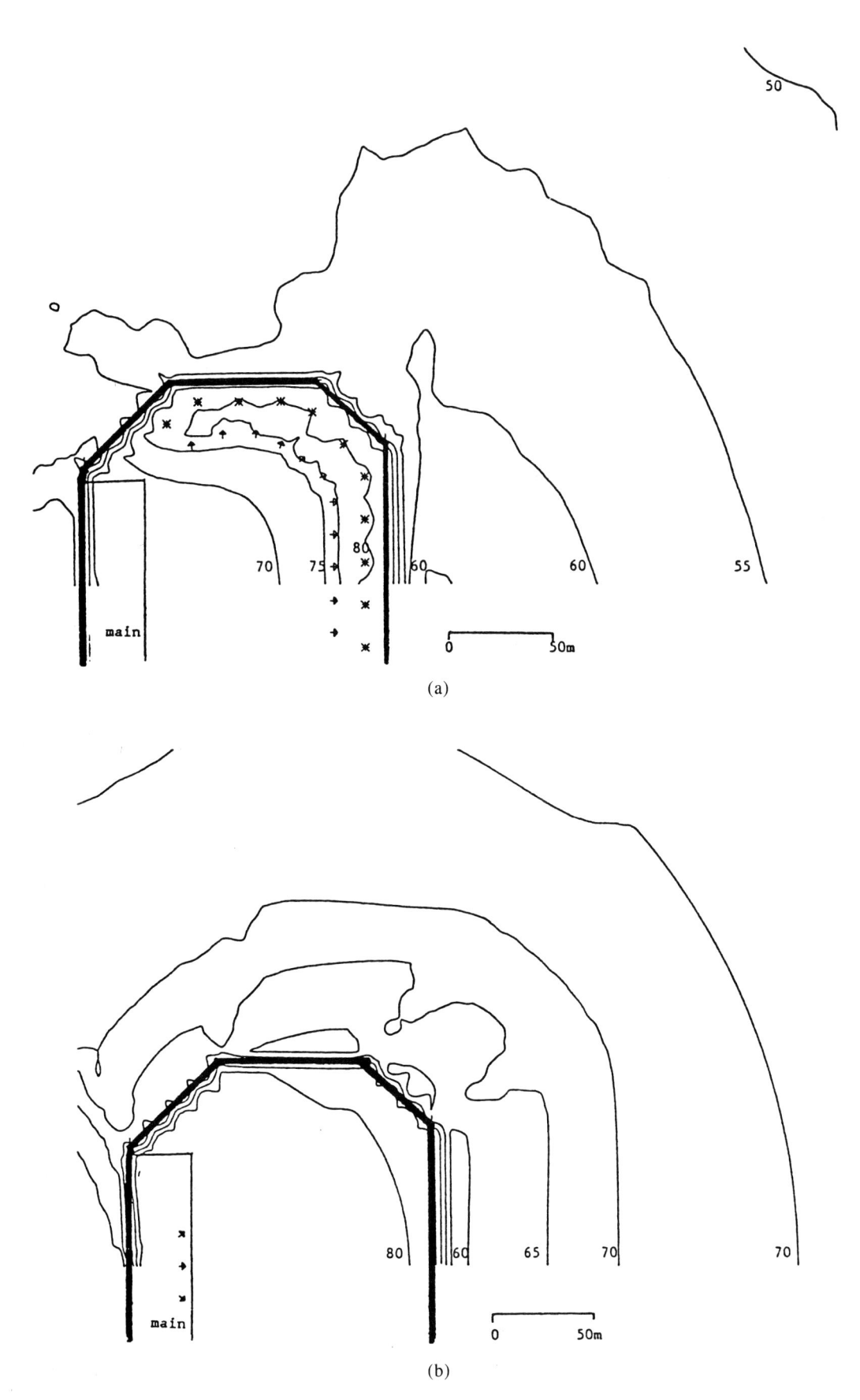

Fig. 8. Announcement sound levels (dBA isophone map) computed according to VDI 2720/2714, taking into account shielding and a speechlike spectrum. (a) Distributed loudspeaker system. (b) Central loudspeaker cluster.

5 REFERENCES

[1] V. M. A. Peutz, and F. Breuer, "Sprachverständlichkeit—Zum Einfluß von Frequenzbeschränkungen," Deutsche Arbeits gemeinschaft für Akustik, 1987.

[2] V. M. A. Peutz, "Designing Sound Systems for Speech Intelligibility," presented at the 48th Convention of the Audio Engineering Society, *J. Audio Eng. Soc. (Abstracts)*, vol. 22, p. 360 (1974 June).

[3] F. Breuer, K. Rijk, and V. M. A. Peutz, "Untersuchung zur Informationsoptimierung bei gleichzeitiger Verminderung der Störwirkung bei Außenlautsprecheranlagen," Rep. VB2, Ministerium für Umwelt, Raumordnung und Landwirtschaft, Düsseldorf, Germany, 1988.

THE AUTHORS

K. Rijk  F. J. Breuer  V. M. A. Peutz

Kees Rijk studied physics at the Eindhoven University of Technology, from which he graduated in 1985. Subsequently, he joined Peutz & Associates, one of the major European consultancy groups in acoustics and building physics, where he is currently involved in the design and construction of auditoriums, conference centers, and theaters, etc., as well as in research. Mr. Rijk has been a member of the Audio Engineering Society since 1986.

•

Franz J. Breuer, studied electrical engineering at the University FH Düsseldorf and received his degree of Diplom Ingenieur in 1978. He has been active in the consulting group Peutz & Associates for more than 10 years as a chief consulting engineer at Akustikberaung Peutz GmbH in Düsseldorf. His recent work includes noise control consultancy in town planning and area development, complex acoustical and vibration measurements, and developing computer programs for acoustics and noise control.

•

Victor M.A. Peutz graduated in physics from the Delft University of Technology following a scholarship at the University of Paris. After a period as assistant professor (acoustics) at Delft University, he worked at the University of Nijmegen in the field of audiology.

Since 1954, Mr. Peutz has been active as an independent acoustical consultant. After chairing the board of Peutz & Associates for a very long period, he is now a senior consultant with the company. In 1981, he was general chairman of the 10th International Conference on Noise Control Engineering (InterNoise) in Amsterdam, and now is a director of the International I.N.C.E. He is an honorary member of the Dutch Acoustical Society, Member of the Royal Dutch Institute of Engineers, Fellow of the Audio Engineering Society, the N.C.A.C., and of many other acoustical societies. In 1990, Mr. Peutz was awarded the AES Silver Medal.

His most interesting projects in the acoustics field include: development and verification in practice of a theory for measuring, evaluating, and predicting speech intelligibility in rooms through electroacoustic systems; the validation of reverberation formulas and determination of sound level distribution in rooms; and developing optimal acoustical conditions in music rooms, concert halls, etc., in which the acoustical properties of halls are compared and correlated to their subjective quality for music. The latter research has resulted in preliminary design rules for music rooms and concert halls, but it is still far from being completed.

CORRECTIONS

CORRECTION TO "SPEECH INTELLIGIBILITY IN SOME GERMAN SPORTS STADIUMS"[1]

We regret that labels A and C in Fig. 7 of the above paper were transposed. The correct figure is shown below.

The authors wish to thank John Woodgate for bringing this to their attention.

K. RIJK, F. BREUER, AND V.M.A. PEUTZ
*Akustikberatung Peutz
Dusseldorf, Germany and
Nijmegen, The Netherlands*

---

[1] K. Rijk, F. Breuer, and V.M.A. Peutz, *J. Audio Eng. Soc.*, vol. 39, p. 39–46 (1991 Jan./Feb.).

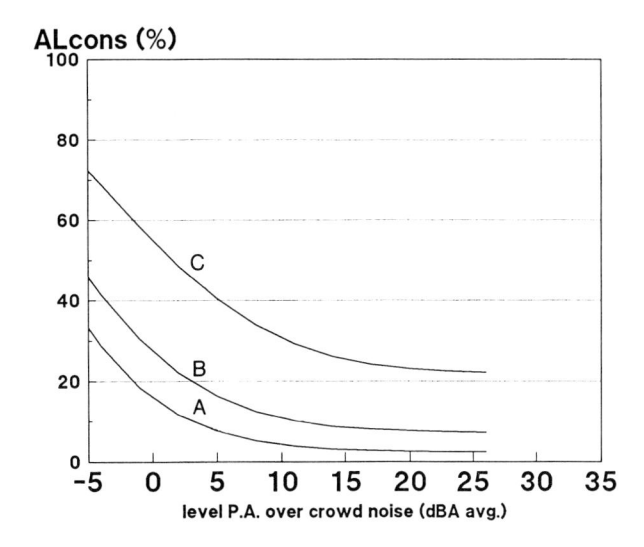

Fig. 7. Attainable speech intelligibility as a function of signal-to-noise ratio. Dynamic compression or more advanced signal processing are not considered. A—optimum (theoretical); B—frequency limited to a 4 kHz; C—as B, plus a 0-dB early-to-reverberant sound ratio with no information in reverberant sound.

Use of the L. G. Marshall–Crown–Techron ELR Program for Adjusting Digital Units in Sound Reinforcement Systems*

DAVID LLOYD KLEPPER, *AES Fellow*

Klepper Marshall King Associates, Ltd., White Plains, NY 10603, USA, and Jerusalem, Israel

The history of the use of signal delay in sound reinforcement systems is reviewed. The application of the Crown-Techron TEF-ELR (early-to-late ratio) program is discussed in this historical setting.

1 BACKGROUND

The original commercial digital audio signal delay equipment was produced by two manufacturers in the winter of 1971–1972. Shortly afterward Larry S. King and the author completed the design of the first pew-back worship-space sound system employing digital delay at St. Thomas Church, Fifth Avenue, New York [1], and the first theater sound system employing digital delay for underbalcony and balcony loudspeakers at the Walnut Street Threatre, Philadelphia, PA. Today both sound systems remain in operation substantially unaltered.

From the beginning, digital delay units were adjustable, and early instructions were to set delays in accordance with the measured distances between the live source, the central system, and the delayed loudspeaker, the speed of sound, and the Haas or precedence effect. In this respect, these instructions followed the practice used with earlier signal delay equipment, the acoustically terminated tube with compression driver and spaced microphones furnished by Ancha Electronics, the magnetic recording tape loop equipment furnished by Philips and Acoustics Research, and a British magnetic disc unit. The Ancha unit had to be custom ordered to fit delay requirements established by the architectural drawings or measured distances for signal delays. The magnetic equipment was adjustable, and instructions (when they were provided) also suggested basing delay settings on measured distances derived from architectural drawings.

The Philips unit was unique on the American market in that it usually arrived without English language instructions. The delay settings were engraved on the face plate in meters (not milliseconds), backward from the longest delay. Needless to say, some consultants and contractors were confused. In contrast Lexicon's and IRP's instructions for the first digital units were clear and covered different applications (100% distributed systems and distributed systems supplementing main central systems, with an implication for distributed systems with precedence loudspeakers). The author had some involvement with both companies' instruction manuals.

2 REASONS FOR DEPARTURE FROM BASIC DELAY THEORY

Once digital delay with simpler adjustability led to a more frequent use of signal delay in many sound-reinforcement applications, contractors and consultants sometimes found that their clients were best satisfied and happiest with delay settings that departed from what considerations of intelligibility (direct sound and delayed local sound arriving as close to the same moment as possible) and naturalness (the precedence or Haas effect—direct sound arriving a few milliseconds before the delayed local sound) would indicate. These in-the-field departures from theory appear to come in two categories:

1) In certain cases, using the theoretically correct delay appears to increase apparent room reverberation or to lengthen a noticeable echo from a sound-reflecting surface (such as a concave rear wall), reducing intelligibility in part of the seating area, possibly in a portion considerably forward of the area covered by the delayed loudspeakers. In these compromise situations, a shorter delay, sometimes deemphasizing directional realism, provides acceptable intelligibility and overall sound-system performance. Of course, the obvious answer is that the offending sound-reflecting surfaces should have

---

* Manuscript received 1995 May 11; revised July 22.

been made sound diffusing or sound absorbing (or sometimes both), or the sound system should have been redesigned, but this is not always possible.

2) In other cases, lengthening the delay beyond that which is theoretically correct can provide best results because the major component of energy from the front of the hall (live or from a central loudspeaker system) is reflected sound energy, with a delay beyond the time direct sound transmission would involve, and the delay setting for the local loudspeakers must be set to match the time of arrival of the reflected-sound-energy path for highest intelligibility and realism, not the direct-sound-energy path.

3 APPLICATION OF ENERGY–TIME CURVES TO DELAY SETTING

The two departures from theory discussed in Section 2 were usually arrived at in the field, in the final adjustment of the sound system, by ear (careful listening). Richard Heyser's development of time-delay spectroscopy provided the first practical instrumentation to allow precise adjustment other than by listening.

The author believes the first time this technique was demonstrated to consultants and contractors was at a Syn-Aud-Con Workshop at the Wally Heider Recording Studio in 1979 May. Basically, sweep tones and computer processor controlled tracking windows can provide data on the sound-energy pattern (energy–time curves) in a room location for a variety of bandwidths, time durations, and band center frequencies [2]. Fig. 1 illustrates the energy–time curve with delay for the sound system in the Holy Cross Cathedral, Boston, MA. The early-to-late-sound ratio analysis is given in Fig. 2. The energy–time curve for the system without delay is shown in Fig. 3 and the early-to-late-sound ratio analysis in Fig. 4. These data could just as easily represent correctly adjusted delay and incorrectly adjusted delay [3]. We recommend using the 2000-Hz-band energy–time curve for delay setting because the 2000-Hz octave band contributes more to intelligibility than other bands and screen displays are clearer and easier to understand than displays at lower frequencies. On the screen displays the relative strength of direct energy and each early reflection can be seen as well as the proper delay chosen to maximize the pre-50-ms delay energy as compared with the post-50-ms delay energy. This is not always easy on the basis of energy–time curves because of the need to evaluate the contributions of many reflections of different intensities, and often a combination of careful listening and energy–time curve evaluation was necessary.

Proper use of the available cursors on energy–time curves in the earlier STC program can provide the information needed to calculate the overall C-50 before-to-after-50-ms ratios. The early-to-late-sound ratio (ELR) program is a time saving and a great convenience in performing the operation.

We note that the energy–time curves shown are not full octave bands because of the nature of the instrumentation. But they are believed to be representative, and the center frequency is 2000 Hz in both cases. Other bands octaves apart (not shown in this case) provide a full set of energy–time data.

4 EVALUATION WITH ELR PROGRAM

The ELR program introduces additional computer processing which gives a continuous plot of the early-to-late-sound ratio with varying time. This plot can be furnished for band center frequencies an octave apart from 125 to 4000 Hz and for an appropriate music average and a weighted speech average.

With the ELR program the changes in intelligibility (using speech weighting for C-50, supplemented possibly by a look at the 2000-Hz-band C-50 by itself) at specific seat locations with minor changes in delay settings can be evaluated quickly, with numerical ratios in decibels quickly compared. The time saved over a combination of scrutiny of the energy–time curve only and careful listening can allow checking many more seat locations during a system checkout and adjustment, and thus a closer to optimum final adjustment. The 2000-Hz-band energy–time curve is recommended for the first rough adjustments, with the early-to-late-sound ratio used for final fine-tuning, by maximizing C-50 for many typical seats. Be certain that the level balance between the central cluster (or any precedence loudspeakers) and the delayed loudspeakers is the same as the balance you wish to use in actual system operation. If there are questions about this balance, the ELR program may be helpful in balancing levels as well as in setting delays. Note that in the case of setting delays shorter than strict theory, the delayed loudspeaker sound energy is considered the initial or "direct" signal when setting the direct cursor for the ELR program's energy–time curve display because it arrives before the actual direct signal. Similarly, when experimenting with delays longer than strict theory, set the "direct" cursor on the first strong reflection and not on the earlier weak direct signal. Of course, delay settings in accordance with classical delay theory should be tried first.

5 REFERENCES

[1] D. L. Klepper, "The Acoustics of St. Thoms Church, Fifth Avenue," *Audio Eng. Soc. (Engineering Reports)*, vol. 43, pp. 599–601 (1995 July/Aug.).

[2] L. G. Marshall, "An Acoustics Measurement Program for Evaluating Auditoriums Based on the Early/Late Sound Energy Ratio," *J. Acoust. Soc. Am.*, vol. 96, pp. 2251–2261 (1994 Oct.).

[3] D. L. Klepper, "The Sound System at Holy Cross Cathedral, Boston—The Realization of a Dream," *Tech Topics, Supplement to Syn-Aud-Con Newsletter*, vol. 22, no. 2 (1994). Syn-Aud-Con, Norman, IN (Winter 1994/1995).

The biography for David Lloyd Klepper was published in the 1995 July/August issue of the *Journal*.

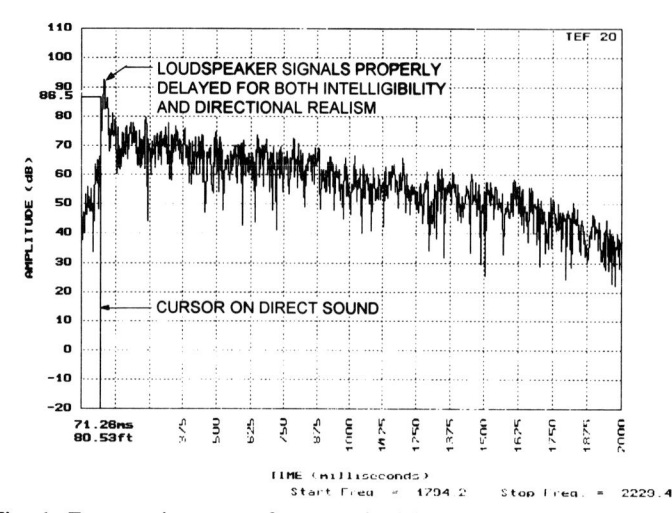

Fig. 1. Energy–time curve for properly delayed system, 2000-Hz band.

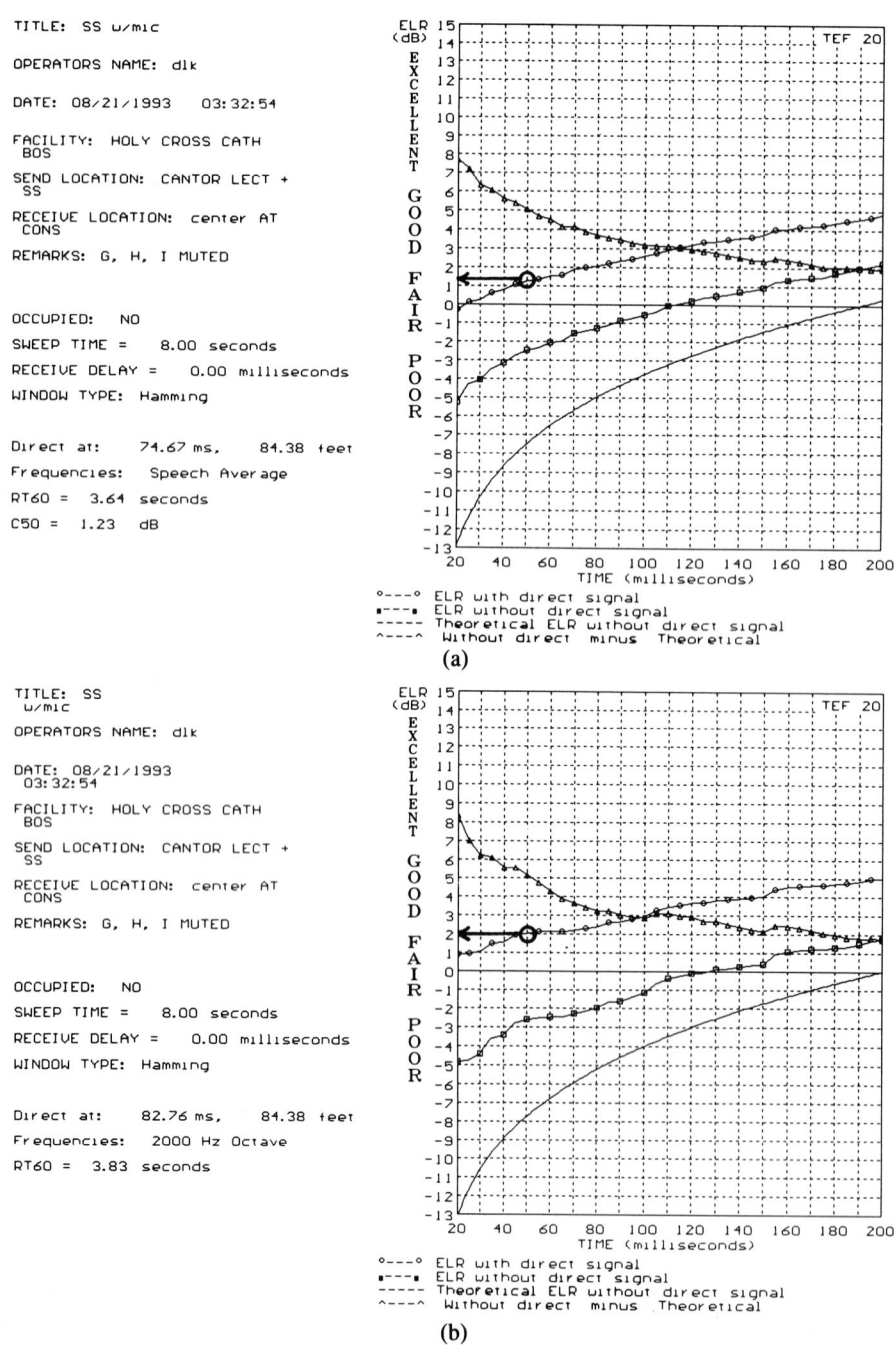

Fig. 2. Early-to-late ratio analysis of data from Fig. 1. (a) Speech average. (b) 2000-Hz band.

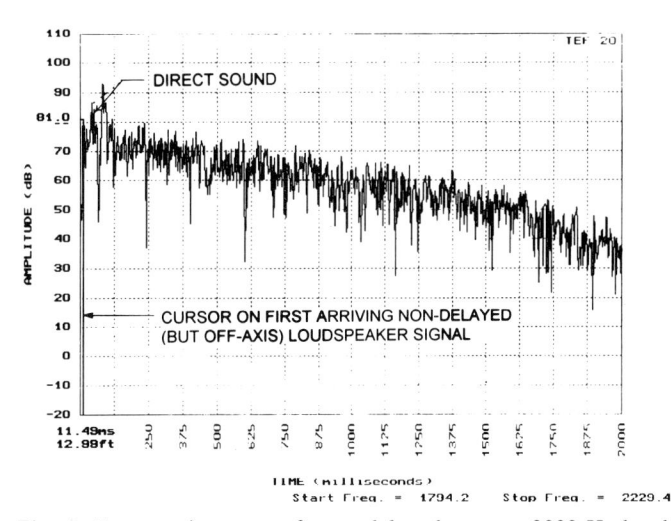

Fig. 3. Energy–time curve for nondelayed system, 2000-Hz band.

(a)

(b)

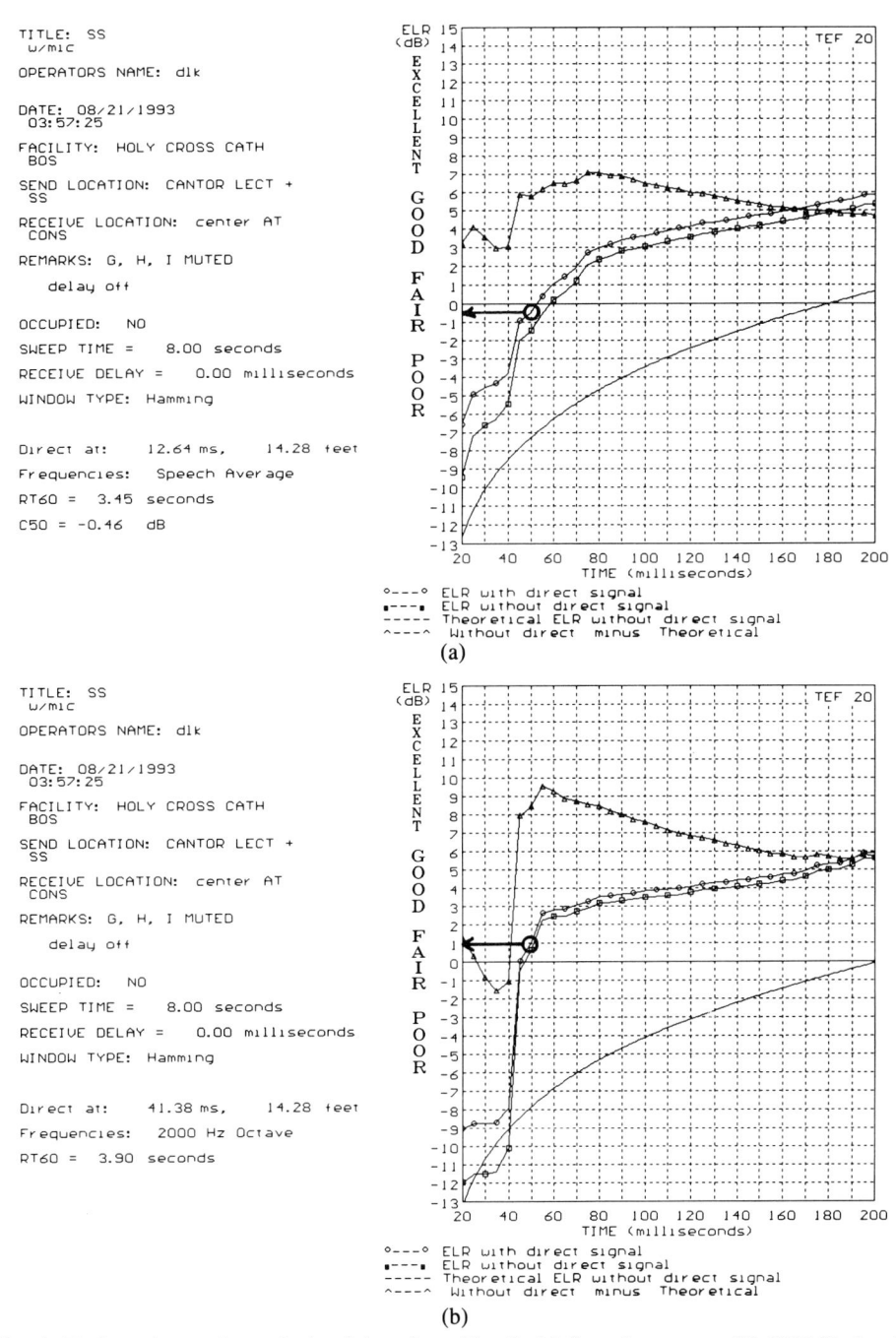

Fig. 4. Early-to-late ratio analysis of data from Fig. 3. (a) Speech average. (b) 2000-Hz band.

An Analysis Procedure for Room Acoustics and Sound Amplification Systems Based on the Early-to-Late Sound Energy Ratio*

L. GERALD MARSHALL

KMK Associates, White Plains, NY 10603, USA

The early-reflection portion of the sound decay process is of particular importance since it is most responsible for subjective impression. Curves showing early-to-late sound energy ratios (ELR) in the early-reflection period can be helpful in analyzing energy–time characteristics within that period. A measurement and analysis process is described which employs energy–time curves (ETC) and corresponding displays of early-to-late sound energy ratio values in the early-reflection period between 20 and 200 ms. Speech-weighted C_{50} values for quantifying speech intelligibility and frequency-averaged C_{80} values for quantifying music clarity are among the objective measures provided by the ELR procedure.

0 INTRODUCTION

Sound energy ratios in various forms have long been used to quantify and predict both clarity and reverberance in room acoustics [1, chaps. 3 and 7, app. B and D]. The 50-ms ratio C_{50} experiences use in speech-related analysis, and the 80-ms ratio C_{80} is commonly used in music-related analysis. Beyond those two well-known indices, however, a full display of ELR versus time in the early portion of a reflection sequence (say, the first 200 ms) may be viewed as a "signature" associated with a particular source–receive relationship in an auditorium, and can be helpful in examining reflection sequences.

Clarity, reverberance, and spaciousness are acoustic qualities determined by a sequence of events beginning with the arrival of the initial signal, followed by a continuing series of reflections that dissipate with time. The strength, time, and direction-of-arrival characteristics of these reflections in connection with the relative strength of the direct signal determine subjective impression. Two of the characteristics—strength and time—can be examined on energy–time curves (ETCs), as illustrated in Fig. 1. The details of the early, highlighted portion (consisting of discrete early reflections) are especially important because an auditorium's sound quality is largely established in this period. The ELR procedure

was created to assist in analyzing this critical early portion of an ETC [2], [3]. The effect of sound decay beyond the early period is mainly limited to the audible persistence of reverberation during rests, separations between notes, and so on, and the reverberation time RT effectively describes the later portion of a normal decay process.

1 ELR DISPLAY

From an energy–time response, early-to-late ratios are calculated for the first 200 ms, then displayed in several forms, as shown in Fig. 2. The upper measured curve shows ELR values containing the energy of the direct signal C_t, and the lower measured curve shows ELR values with the energy of the direct signal omitted C_{to}. The observed difference between the two measured curves C_t and C_{to} gives an indication of the relative strength of the direct signal. Another curve, C_x, shows theoretical ELR values for a pure exponential decay based on a decay time equal to that of the measured curves,

$$C_x = 10 \log(e^{13.82t/RT} - 1) \qquad (1)$$

where t is the ratio dividing time in seconds and RT is the reverberation time. This is the theoretical equivalent of the measured without-direct-signal curve C_{to} and serves as a useful comparison curve. To better illustrate

---

* Manuscript received 1996 January 2; revised 1996 March 6.

the comparison between C_{to} and C_x, a fourth curve is produced by subtracting C_x from C_{to}. Since C_{to} is solely a consequence of the room effect, the difference curve illustrates reflective behavior in a manner that is straightforward and easy to comprehend.

The remaining curve in this figure, which is the dashed curve labeled C_B, is not produced by the ELR measurement software used by the author and was calculated separately, then added to the figure as a useful supplement. C_B is a theoretical curve that includes the energy of the direct signal and is the theoretical equivalent of C_t. The equation for this curve comes from Barron's revised room acoustics theory [1, App. B],

$$C_B = 10 \log \left[\frac{V e^{(0.04r + 13.82t)/RT}}{312 RT r^2} + e^{13.82t/RT} - 1 \right] \quad (2)$$

where V is the room volume in cubic meters and r is the source–receive distance in meters.

As an aside, the author remembers deriving early-to-late ratios during the slide-rule era using Polaroid photos of oscilloscope traces of tape-recorded balloon-burst

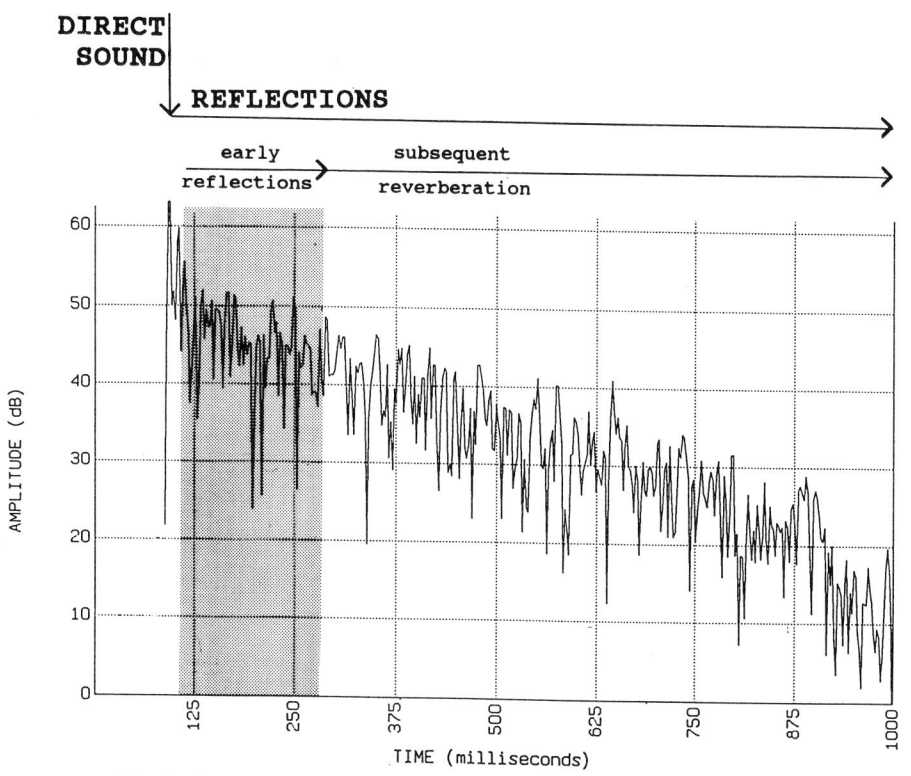

Fig. 1. Energy–time curve with ELR calculation period highlighted.

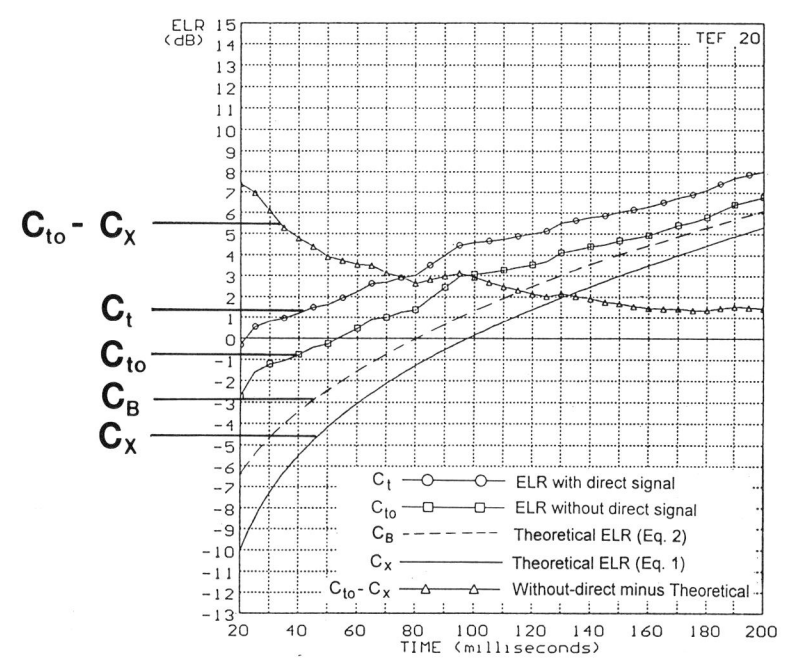

Fig. 2. ELR data curves.

decays—all hand-calculated with tedium and without great precision, as one would imagine. With today's affordable, portable digital instrumentation, the concern is no longer about ease and speed, but now only for degree of accuracy within a generally small range of variations between measuring systems due to differences in such things as calibration, resolution, filtering, windowing, and triggering [4], [5].

2 SPEECH C_{50} AND MUSIC C_{80} RATING SCALES

Good correlations exist between early-to-late ratios and speech intelligibility [1, chap. 7 and app. D], [6]. A speech-weighted C_{50} is put forth by the author as an objective measure of speech intelligibility. For weighting, ELR values at 0.5-, 1-, 2-, and 4-kHz octave bands are assigned respective contributions of 15, 25, 35, and 25%, then summed to obtain the speech-weighted values. Fig. 3 illustrates the speech intelligibility rating scheme for C_{50} proposed by the author. This scale may well be adjusted somewhat over time as data of C_{50} versus actual intelligibility are accumulated.

Music clarity cannot be so clearly defined as a function of frequency as speech because of the vastly greater number of variables in music. In the ELR procedure, composite "music" ELR values are produced by summing the valus (in decibels) of the 0.5-, 1-, and 2-kHz bands, then dividing by 3. A C_{80} of 0 dB ± 2 dB has been pretty well established as a good range for symphonic music in concert halls [1, chap. 3.9], [7]. Fig. 3 also shows a rating scale developed by the author in which "symphony" is centered on a three-frequency-averaged C_{80} of 0 dB. This scale is intended only for full-size performance spaces.

3 MEASUREMENT PROCESS

For room acoustics tests an omnidirectional microphone and either an omnidirectional or a speech-directivity loudspeaker is needed. When music acoustics is the specific interest, omnidirectional loudspeakers (dodecahedron, for example) are generally desired. For speech acoustics, compact loudspeakers with dimensions roughly corresponding to those of a human head are usually employed since these can have appropriate directivity factor characteristics ($Q \approx 2\frac{1}{2}$) in the speech-intelligibility frequency range.

One very practical use for the ELR procedure is to judge sound-system performance objectively. "Speech" curves obtained with a room's sound system both on and off readily show the change in early energy and C_{50} values affected by the reinforcement system (see Fig. 5). ELR measurements may be obtained through sound-reinforcement systems in two ways. First, and most directly related to actual sound system usage, is to place the ELR test loudspeaker at the talker's position in front of a live reinforcement system microphone. With this approach one must make certain that the reinforcement system's level is predominant, and this can be determined either by listening or by checking the signal strength of the reinforced sound and its time of arrival on an ETC. A variation would be appropriate in rooms where sound from the person talking and sound from a distant reinforcement loudspeaker combine to give degraded intelligibility because of poorly synchronized live and amplified sound components. In such cases somewhat equal signal levels at the receive position from the reinforcement and test loudspeakers could be used. The alternative method for measuring a sound-reinforcement system is to send the test signal directly to the system, omitting the ELR test loudspeaker and bypassing reinforcement system microphone pickup.

Another use for the procedure in sound-system analysis is to optimize delay settings by optimizing C_{50} values. After establishing the initial delay settings on a theoretical basis, one may fine-tune the settings by optimizing the speech C_{50} values [8].

4 ELR ANALYSIS

The ELR procedure comprises the following:

1) The clarity measures C_{50} and C_{80} in individual octave bands and in frequency-averaged and weighted forms

2) The behavior and relative strength of reflected energy in comparison with idealized exponential decay in the period between 20 and 200 ms after the first arrival, as indicated by the difference between C_{to} and C_x

3) The relative strength of the direct signal in relation to subsequent reflected energy, as indicated by the degree of separation between C_t and C_{to}, and how this separation compares with theory, as indicated by the separation between C_B and C_x

4) The relative strength of the combined direct signal

Fig. 3. Speech and music (C_{50} and C_{80}) rating scales.

and reflected energy in comparison with that predicted by theory, as indicated by the separation between C_t and C_B.

Fig. 4 shows before-and-after natural room-acoustics measurements using a voice-directivity test loudspeaker in a small theater that underwent a fairly minor renovation. Acoustic changes consisted only of reorienting some existing suspended sound-reflecting panels and portions of the sidewalls for improved ray diagramming. A comparison of the difference curves on the two displays of Fig. 4 clearly shows the resulting strong improvement in early sound reinforcement, with a C_{50} increase from $-\frac{1}{2}$ dB (fair) to $3\frac{1}{2}$ dB (good).

Fig. 5 shows sound-system on–off measurements in a church equipped with column loudspeakers distributed down side-aisle columns. For the sound system "on" measurements, the test signal was fed directly into the sound system, so there was no supplementary live signal from the lectern as there would be for speech reinforcement. Even though the distributed loudspeakers were not delay synchronized, a C_{50} increase from $-2\frac{1}{2}$ dB (poor/fair) to 3 dB (good) was effected by the sound system.

Fig. 6 shows two measurements performed at a commercial recording session involving a small chamber group in a 1300-seat auditorium. One of the two measurements indicates the natural acoustic response of the room at a mid-audience location using a dodecahedron source loudspeaker placed at the concertmaster position. For the other measurement a mono-mixdown signal was delivered directly from the recording console to the instrumentation. Thus the two displays give a comparison of live room-acoustic characteristics and recorded-signal characteristics. The differences in the displays are dramatic and worthy of some discussion.

The recordings took place in theater A at the State University of New York in Purchase. (Coincidentally, this room was used in an earlier paper by the author to illustrate the ELR procedure [3].) An undertaking of Delos Records with John Eargle as balance engineer, Bach's 6 Brandenburg Concerti were recorded in three days of sessions with the Chamber Music Society of the Lincoln Center. The ensemble (the concerto grosso) was generally arranged in a typical semicircular recital configuration on stage within an orchestra enclosure with soloists (the concertino) standing toward the front, not much different than if playing to an audience. A simple microphone arrangement consisted of three positions— right, left, and center—immediately in front of the ensemble, about 3 to $3\frac{1}{2}$ m (10 to 12 ft) off the floor [9]. (Separate pickup was used for soloists and was not included in the measurements.) Omnidirectional microphones were placed about $1\frac{3}{4}$ m (6 ft) directly to the right and left of the coincident cardioids used in the center. For the ELR measurements, a dodecahedron loudspeaker was placed at the concertmaster's position, and the source–receive distance (slant distance) was $2\frac{3}{4}$ m (9 ft) to the adjacent omni and $3\frac{1}{2}$ m (12 ft) to the center cardioids.

On the natural room acoustics "music" ELR of Fig. 6, the measured curves stay close to the theoretical curves except in the very early portion before 55 ms where there is a deficiency of strong reflections. The deficiency in early energy is to be expected at a middle-audience

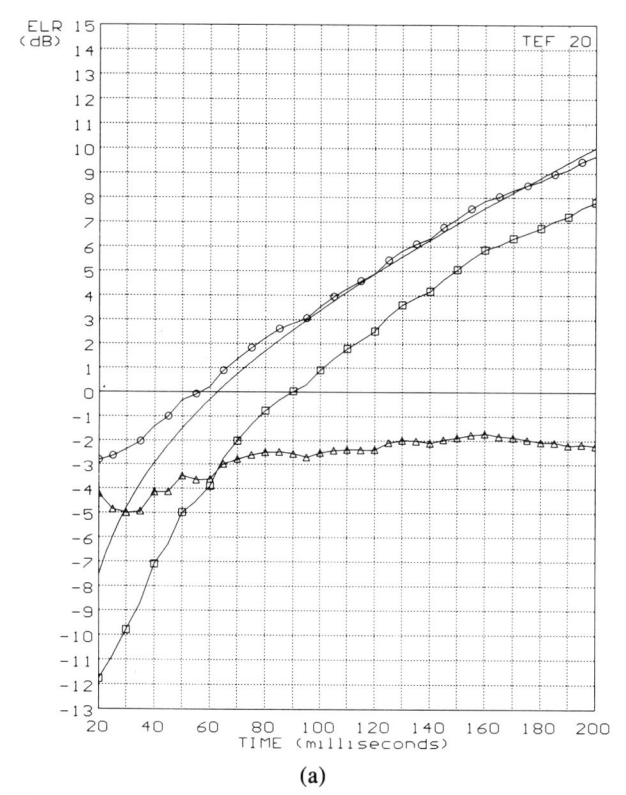

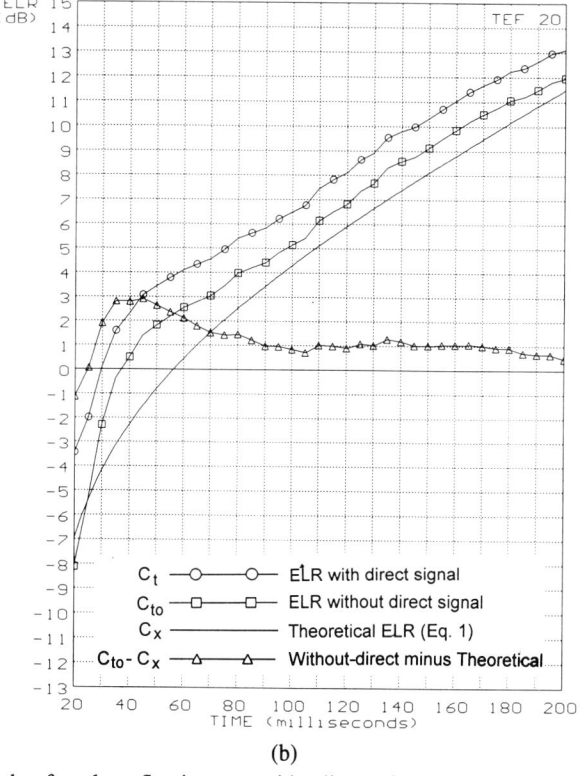

Fig. 4. ELR in renovated theater showing improvement in strength of early reflections at midaudience location. (a) Before renovation. (b) After renovation.

location well removed from sound-reflecting boundaries.

In contrast, the ELR for the recording system shows a C_{80} of 7 dB and a very strong pattern of early reflections beginning around 35 ms. Fig. 7 is the 2-kHz ETC associated with the recording system ELR, and it shows the strong initial signals from the three microphone positions occurring within the first 17 ms, after which there is a delay before strong reflections from the orchestra enclosure begin arriving. The energy significance of the signals is more clearly perceived on the ELR of Fig. 6, and this serves as a good illustration of the usefulness of the ELR procedure in evaluating ETCs.

The auditorium has good natural acoustics for orchestral performance, and this is reflected in the appropriateness of the measured midfrequency RT of 1.8 s and a

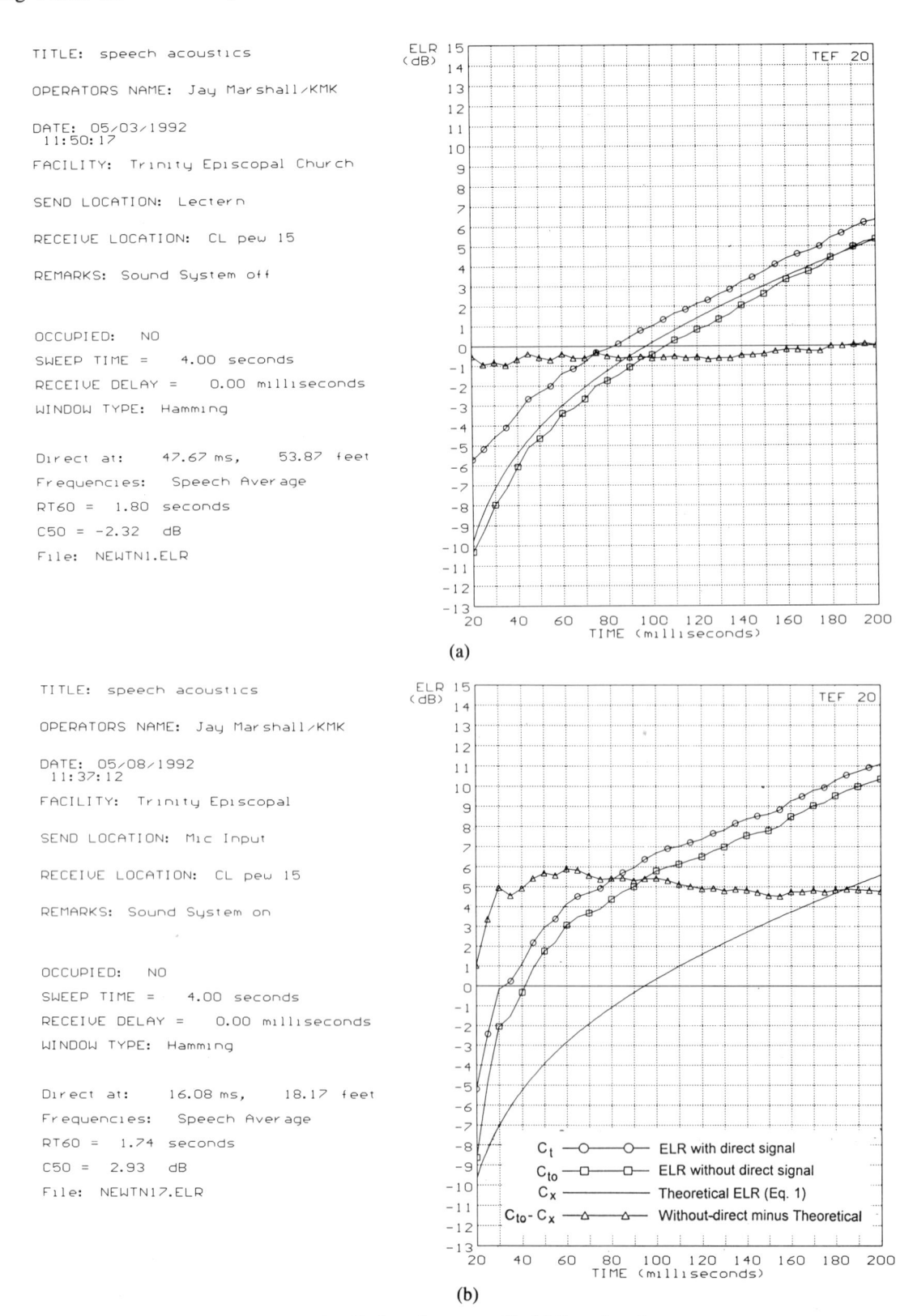

Fig. 5. ELR. (a) Sound system off. (b) Sound system on.

C_{80} near 0 dB. The sound on the recording is substantially better, however, and that poses a number of questions. Since it is well known that desirable C_{80} values in concert halls center around 0 dB [1, chap. 3.9], [7], how could the recorded sound not be terribly dry with a C_{80} of 7 dB? This is a value associated with excellent speech intelligibility, and in fact, musicians' comments

through the recording system were perfectly clear to engineers in the control room.

The answer is found in the differences between natural binaural listening and stereophonic (or monophonic) reproduction, and the merging of three "listening" positions (microphone positions) into one. As an illustration, if a listener were located as close to the musicians as

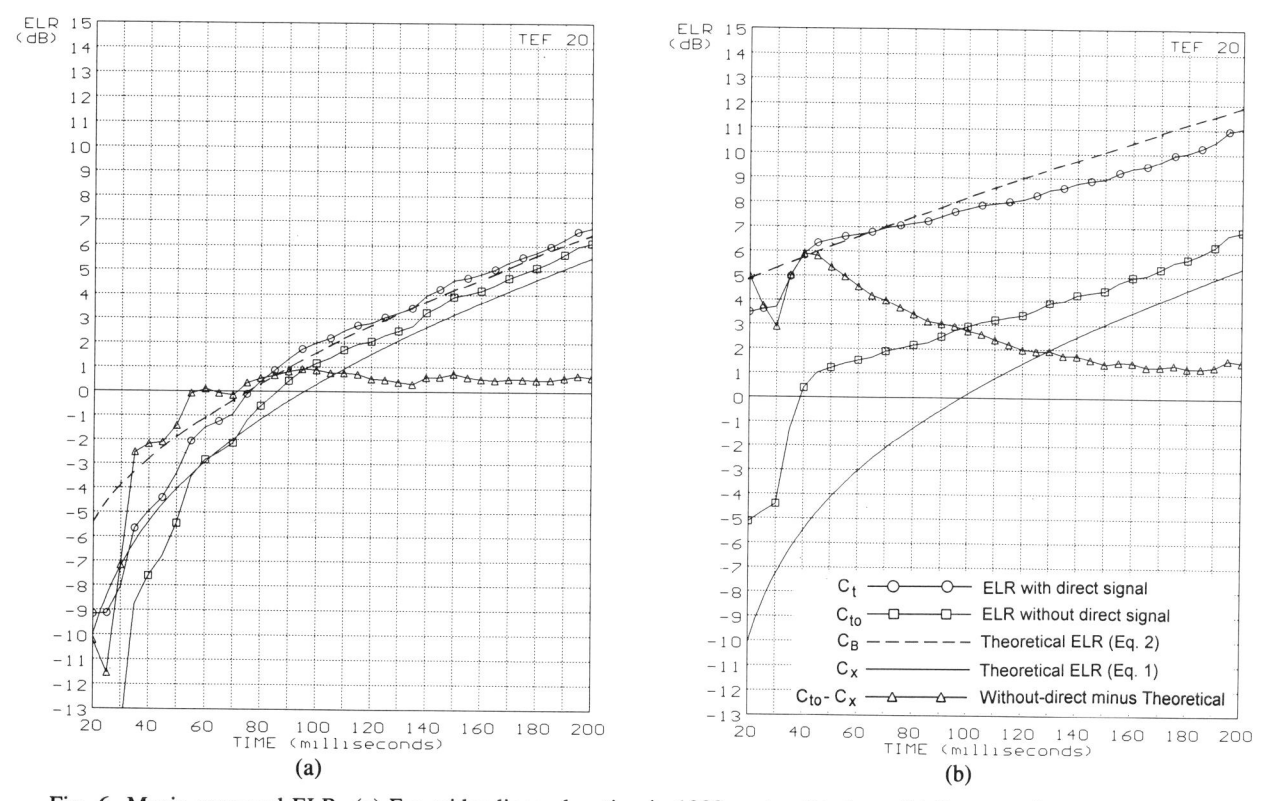

Fig. 6. Music-averaged ELR. (a) For midaudience location in 1300-seat auditorium. (b) For recording system output.

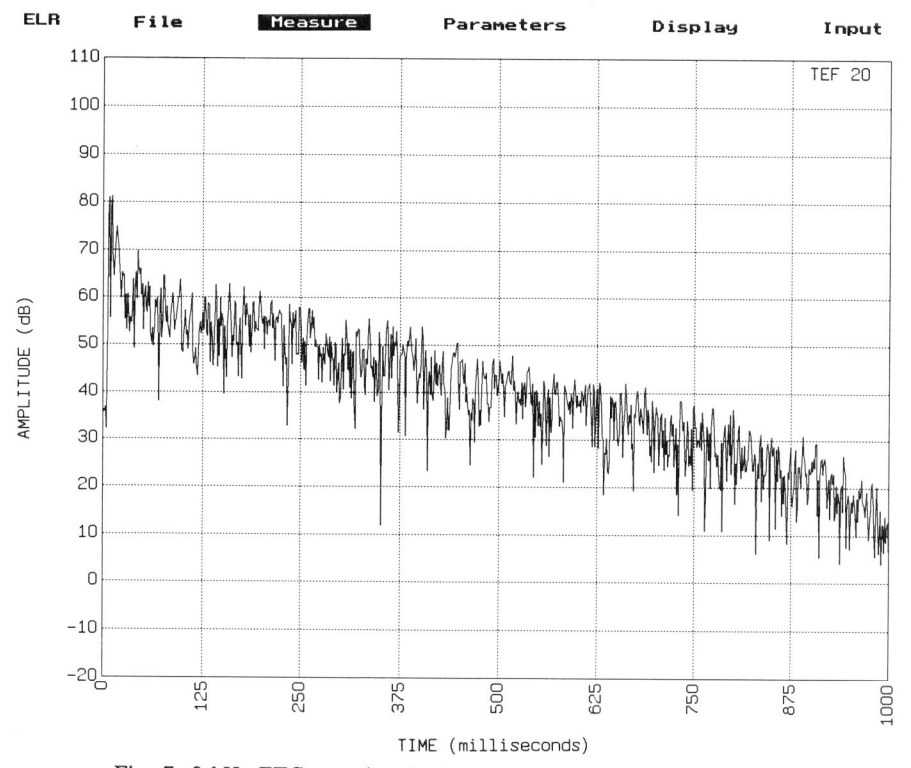

Fig. 7. 2-kHz ETC associated with recording system ELR of Fig. 6.

the main coincident pair of microphones, the acoustic impression would be very different from the one obtained from listening to the recordings. The former is a near-field, on-stage acoustic condition dominated by the direct sound from the instrumentalists, with little sense of hall sound other than an awareness of some late reverberation heard at rests. Once the signal from the spaced omnis is added in, however, there is no longer an analogy with real-life listening, and the spaced omnis add reverberance and spaciousness in a way that does not occur in the nonelectronic world. Fig. 8 shows the change with distance of theoretical speech C_{50} and music C_{80} values, using a directivity factor Q of 2½ for the former and of unity for the latter. [The use of Q with Eq. (2) is discussed in the following section.] Fig. 8 illustrates well that the microphones are in a near-field region of very high clarity.

Although not included as figures, measurements were also made with the omni pair alone and with the cardioid pair alone. The C_{80} of 9½ dB obtained with the coincident cardioids alone is understandably higher, because of directivity, than for all the microphones on. For the pair of omnis alone, C_{80} is essentially the same as the value obtained with all the microphones on.

The sound delivered to the recorder was unaltered within the four-octave frequency range covered by these measurements. The only tailoring by the engineers involved a little low-frequency deemphasis, and a small amount of high-frequency reverberation was to be added to the final product. Both alterations are minor and were prompted by the acoustic qualities of the hall. At the affected frequencies, a small reduction in C_{80} should result from the addition of high-frequency reverberation.

As an incidental observation, the great differences seen between natural room acoustic response and recording system response underscore the possible pitfalls in using synthesized sound fields and loudspeaker reproduction for research relating to room-acoustics criteria. Judgments made from listening tests with virtual sound fields may mislead when deviations from the physics of real-world listening are present.

Fig. 9 is for two different acoustic configurations in a 550-seat multipurpose hall. For these measurements the only physical changes occurred on stage, where one setup was for drama and included the normal complement of cloth legs, borders, and upstage traveler. For the second setup a full orchestra enclosure replaced the stage cloth. The only differences between the two sets of data, therefore, relate to stage conditions: 1) sound-absorptive drama configuration on stage, using a compact loudspeaker with a Q of about 2½ to simulate speech, and 2) sound-reflecting music configuration on stage with orchestra enclosure, stands, and chairs with upholstered seats, using a dodecahedron loudspeaker with omnidirectional characteristics. The receive location was in midaudience and the test-source locations were about 2½ m (8 ft) upstage of the proscenium on centerline.

A speech ELR and a music ELR are shown, and the differences are striking when one realizes that no physical changes occurred in the audience chamber. The speech measurement difference curve clearly seems to show more early reinforcement than the music measurement, even though that reinforcement is present in both, of course, since no sound-reflecting surfaces were removed for the music measurement. On the contrary, sound-reflecting boundaries were created on stage for the music configuration. In the speech ELR there is good agreement between C_{to} and the theoretical curve C_x, whereas there is a relative deficiency of early energy in the music ELR. Most of the difference in this instance is due to the altered balance of early-to-late energy produced by substituting sound-reflecting surfaces for sound-absorbing materials, rather than to a loss of early reflected energy. As a result, the auditorium produces reasonable speech C_{50} values with the drama stage and appropriate music C_{80} values in the full music configuration. Though not used during these two comparative measurements, the auditorium is equipped with adjustable sound-absorbing curtains, which can be extended to reduce RT and further increase C_{50} for drama.

Since the auditorium's music acoustics are quite good, the difference-curve shape for the music configuration is obviously an appropriate one. The author would not expect this to be an appropriate characteristic for a full-size concert hall, however. As the strength of response decreases with increasing room volume, the need for a strong early-reflection sequence (indicated by a positive difference curve) presumably increases with increasing volume.

5 DIRECTIVITY FACTOR Q FROM ELR

The first term within the bracketed portion of Eq. (2) represents the ratio of direct-to-late energy, and the

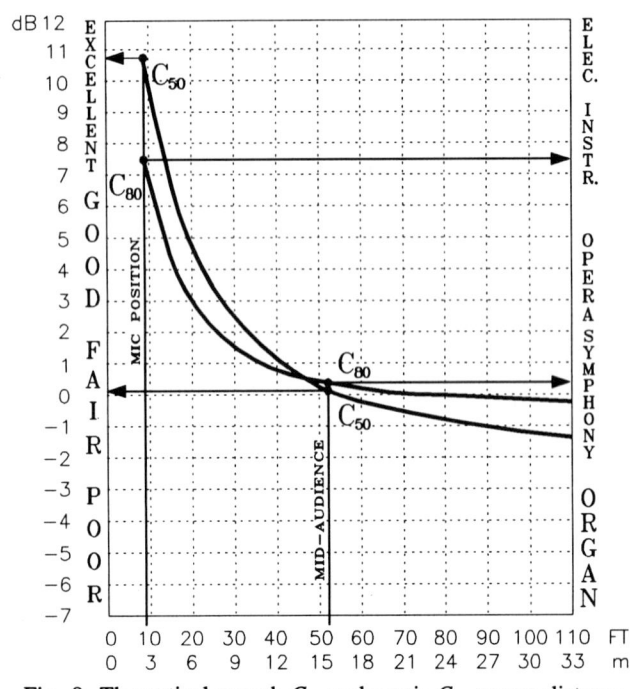

Fig. 8. Theoretical speech C_{50} and music C_{80} versus distance in 1300-seat auditorium with source–receive distances of Fig. 6 and rating scales of Fig. 3 indicated.

remainder represents the ratio of early-to-late, excluding the direct component [Eq. (1)]. Directivity factor Q can be applied to the direct component to obtain theoretical C_t values corresponding to sources having directional properties [10]. If solved for 50 ms (that is, C_{50}), the equation can be used to predict speech intelligibility in enclosed spaces for sound amplification systems and for unamplified voice,

$$C_{50} = 10 \log \left[\frac{QVe^{(0.04r + 0.691)/RT}}{312 RT r^2} + e^{0.691/RT} - 1 \right]. \quad (3)$$

In transposed form the equation may be used to calculate the minimum effective Q required to obtain a desired C_{50} intelligibility value,

$$Q_{MIN} = \frac{312 RT r^2}{Ve^{(0.04r + 0.691)/RT}} \left[\log^{-1} \left(\frac{C_{50}}{10} \right) - e^{0.691/RT} + 1 \right]. \quad (4)$$

Still another use of Eq. (3) is to derive the effective (or apparent) Q of an installed sound system from 50-ms C_t and C_{to} values, along with 50-ms C_B [from Eq. (3) with $Q = 1$] and C_x values [10],

$$Q_{DS} = \frac{\log^{-1}(C_t/10) - \log^{-1}(C_{to}/10)}{\log^{-1}(C_B/10) - \log^{-1}(C_x/10)}. \quad (5)$$

The assumption underlying Eq. (5) is that, except for the effect of Q on the direct signal (where the direct signal comes from the sound system and is picked up by an omnidirectional microphone), the differences between with- and without-direct-signal ratio values would be similar for measured and theoretical ELRs. As an example, if the 50-ms values in Fig. 2 are used with Eq. (5) (or values at any other time, for that matter), a Q_{DS} of 2½ is obtained for the compact loudspeaker used in the measurement.

Automatic calculation of Q_{DS} would be a desirable feature in the measurement routine. When initiating a set of sound-system measurements, the user would toggle the C_B option on, enter room volume, and Q_{DS} would be provided with each measurement.

6 SUMMARY

The ELR procedure was created for analyzing auditorium acoustics and sound-reinforcement system performance by employing 1) energy–time curves by octaves and 2) the corresponding displays of early-to-late sound energy ratio values obtained in 5-ms increments between 20 and 200 ms. That brief period of early reflections is of particular interest since it is the portion of the sound-decay process most responsible for subjective impression. Strength–time characteristics of the reflection sequence are illustrated with ETCs, and ELR curves can be helpful in examining the early reflection portion of an ETC.

More specifically, the ELR procedure provides the following.

1) The clarity measures C_{50} and C_{80} in individual octave bands and in frequency-averaged and weighted forms.

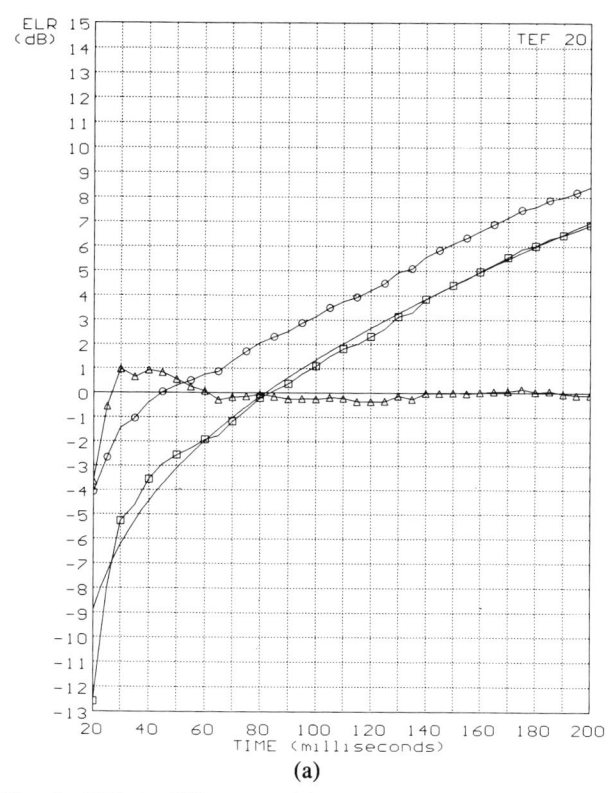

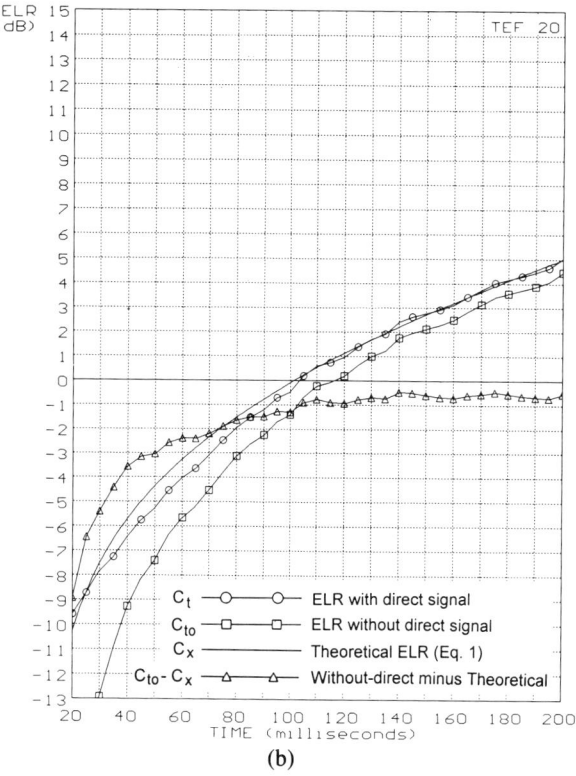

Fig. 9. ELR in 550-seat multipurpose auditorium. (a) Speech-weighted ELR for drama stage. (b) Music-averaged ELR for concert stage.

2) The behavior and relative strength of reflected energy in comparison with idealized exponential decay in the period between 20 and 200 ms after the first arrival, as indicated by the difference between C_{to} and C_x.

3) The relative strength of the direct signal in relation to subsequent reflected energy, as indicated by the degree of separation between C_t and C_{to}, and how this separation compares with theory, as indicated by the separation between C_B and C_x.

4) The relative strength of the combined direct signal and reflected energy in comparison with that predicted by theory, as indicated by the separation between C_t and C_B.

The effective Q of a sound system may also be derived using 50-ms ELR values along with Eqs. (3) and (5).

7 REFERENCES

[1] M. Barron, *Auditorium Acoustics and Architectural Design* (E and FN Spon, London, 1993).

[2] L. L. Beranek, "Concert Hall Acoustics—1992," *J. Acoust. Soc. Am.*, vol. 92, pp. 1–39 (1992 July).

[3] L. G. Marshall, "An Acoustics Measurement Program for Evaluating Auditoriums Based on the Early/Late Sound Energy Ratio," *J. Acoust. Soc. Am.*, vol. 96, pp. 2251–2261 (1994 Oct.).

[4] J. S. Bradley, A. C. Gade, and G. W. Siebein, "Comparisons of Auditorium Acoustics Measurements as a Function of Location in Halls," presented at the 125th ASA Meeting, Ottawa, Ont., Canada, 1993 May.

[5] M. Barron, "Impulse Testing Techniques for Auditoria," *Appl. Acoust.*, vol. 17, pp. 165–181 (1984).

[6] J. S. Bradley, "Predictors of Speech Intelligibility in Rooms," *J. Acoust. Soc. Am.*, vol. 80, pp. 837–854 (1986 Sept.).

[7] J. R. Hyde, in *Proc. Sabine Centennial Symp.*, Acoustical Society of America (New York, 1994), pp. 199–202.

[8] D. L. Klepper, "Use of the L. G. Marshall–Crown–Techron ELR Program for Adjusting Digital Units in Sound Reinforcement Systems," *J. Audio Eng. Soc. (Engineering Reports)*, vol. 43, pp. 942–945 (1995 Nov.).

[9] J. M. Eargle, *Music, Sound, Technology*, 2nd ed. (Van Nostrand Reinhold, New York, 1995), chap. 13.3.1.

[10] L. G. Marshall, "Speech Intelligibility Prediction from Calculated C_{50} Values," *J. Acoust. Soc. Am.*, vol. 98, pp. 2845–2847 (1995 Nov.).

Note: As a footnote, mention should be made of an ELR user's group established by Neil Shade to facilitate interaction among users and to produce occasional newsletters containing items of interest submitted by users. Neil can be reached at 703-533-0717 and fax 703-533-0739. Software for the ELR measurement and analysis routine described in this paper is produced by TEF Products for TEF instrumentation.

THE AUTHOR

L. Gerald Marshall has been a principal at the architectural acoustics consulting firm of Klepper Marshall King Associates, Ltd in White Plains, NY, since its formation in 1971. He holds a master's degree in music from the University of Oklahoma, a bachelor's degree in architectural engineering from the University of Colorado, and was a special graduate student in acoustics at MIT under Robert Newman. He began his career in acoustics with BBN in 1965. Prior to that was a professional musician, having played trumpet with the Denver Symphony, the Buffalo Philharmonic, the Oklahoma Symphony, and the USMA Band at West Point. Mr. Marshall is a fellow of the Acoustical Society of America.

E.
special systems

Conference Systems Using Infrared-Light Techniques*

H.-J. GRIESE

Sennheiser Electronic KG, Wedemark, Germany

0. INTRODUCTION

Wire-bound discussion and interpreter systems are difficult to handle at larger conferences, especially when the participants have no fixed positions. In these cases induction-loop systems are often used. These systems have the disadvantage that the signal cannot be confined to the room where it is supposed to be transmitted. It is possible to pick up the signal outside the conference room. Thus confidential matters are hard to safeguard. Furthermore, loop systems in adjacent rooms may interfere with each other. Because of the limited frequency band inherent to the principle, the quality of the transmission of loop systems is inferior and leaves much to be desired.

A new technique of using light as the transmission medium is closely related to modern semiconductor technology. Light-emitting diodes have made it possible to produce modulated light easily. For reception, efficient photodiodes are available.

1. THE PRINCIPLE

The basic principle of light-emitting diodes is as follows. If a p-n semiconductor diode is flown through by a forward current, electrons are traveling to the p-field and mobile holes to the n-2 field. Recombinations between charge carriers are taking place. With a so-called "shining recombination" the electron jumps (according to the band model) from the energetic higher placed conduction band to the energetic lower valence band and turns over the exceeding energy as electromagnetic radiation. The share of "shining" as compared to not shining recombination depends on the material of the semiconductors. Particularly suitable are the so-called III-V combination semiconductors from the periodic system of elements, that is, Ga, As, GaAsP, and GaP.

The wavelength of the radiated light is given by the material and doping.

For the transmission of sound, the light of gallium arsenide diodes is used because of their high efficiency. Its spectrum lies in the near infrared range. The power efficiency is about 7%.

The radiated light of a light-emitting diode is of a relatively small range of wavelength, but not single waved as laser light. This is an advantage because problems due to interference are missing.

For receiving modulated light silicon p-i-n diodes are used preferably. These diodes work in the reverse direction. They operate through pairs of charge carriers produced by the incoming light inside the barrier layer, by which the diode becomes more or less conductive. The designation p-i-n means that there is an intrinsic junction between the p and n junctions, which is a high-resistance layer. Thus the capacity of the diode is decreased and the efficiency increased. This leads to short switching times with high sensibility. In the ideal case each photon produces one pair of charge carriers. Fig. 1 shows different

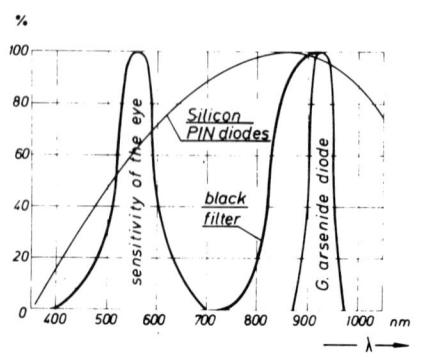

Fig. 1. Spectral curves.

---

* Presented at the 59th Convention of the Audio Engineering Society, Hamburg, Germany, 1978 Feb. 28–Mar. 3.

spectral curves. The wavelength range of the human eye is about 400 to 700 nm; the emission spectrum of gallium arsenide diodes is relatively small with a maximum at about 925 nm. This light is completely invisible and of course not dangerous. The silicon p-i-n diode shows a wide spectral sensitivity range. To keep away visible light, a so-called black light filter is used, which is practically an optical low pass.

The infrared light acts in a similar way as the visible light. There are light and shadow zones, but in the latter sufficient diffused light can be found to guarantee an unobjectionable transmission. To obtain diffusity, it would naturally be favorable to illuminate the room from different positions.

With regard to light power, one has to consider that the small receiving diode receives only an extraordinarily small part of the transmitted infrared light because the "receiving surface" is small. Because of the high cost of large-surface diodes, an increase of the surface has been attempted by the addition of optics. Unfortunately this leads to a more or less strong directional reception, which is unwanted if as much diffused light as possible has to be captured. The problem is to increase the receiving surface by optics, without decreasing the angle from which the light is received. Some people maintain that this is going against natural laws. However, this is not the case, as the following example shows. If the receiving diode is placed in the center of a glass globe, each incoming beam will be reflected to the inside, and the diode receives more light than before. This is valid for all directions because of spherical geometry. If the refraction coefficient of glass were unlimited, every light would concentrate on the center of the globe. Unfortunately there is no material with such a high refraction coefficient.

Fig. 2 shows the arrangement of the receiving diode used with a black light filter and a spherical lens made from plastic. The result can be seen in Fig. 3. The gain of light is limited, but a real gain does exist.

It is possible to modulate only the amplitude of the infrared light emission. In order to suppress the noise, the light stream is not modulated directly by the audio frequencies. Multiple modulations are used. For an interpreter system with a multitude of channels there are two possibilities, either frequency multiplexing or time multiplexing. For our interpreter system we decided upon the frequency-multiplex version because we felt that this leads to a rather simple receiver design. To obtain high quality transmission, frequency modulation is used. In a first step the different carrier frequencies are frequency modulated by the audio frequencies, and in a second step the amplitude of the light is modulated by this signal. Fig. 4 shows very roughly the modulation of the amplitude of the light by a frequency-modulated carrier frequency.

2. SYSTEM DESIGN

An important step in the design of an interpreter system is to determine the necessary number of channels. The specifications vary between four and twelve channels. The technical expenditure, naturally, grows with the number of channels. We feel that even in the future there will be no conferences requiring translation into twelve languages. This would lead to too many interpreter booths and even more interpreters. Therefore, the system was designed to have a maximum of nine channels. As will be explained later, whenever fewer channels are used, the light power per channel increases, so that the necessary investment for equipment will become less.

The frequency diagram of the entire system is shown in Fig. 5. The channel frequencies are equidistant with a channel spacing of 40 kHz, starting at 55 kHz und ending up at 375 kHz. The maximum frequency deviation is 7 kHz.

The receiver is a very simply designed superhet with an intermediate frequency that lies above the channel frequencies. In this case preselection is not necessary. The commonly used intermediate frequency of 455 kHz is chosen. Therefore the oscillator frequencies are higher than the intermediate frequency. They are also shown in Fig. 5.

A block diagram of the receiver is shown in Fig. 6. The

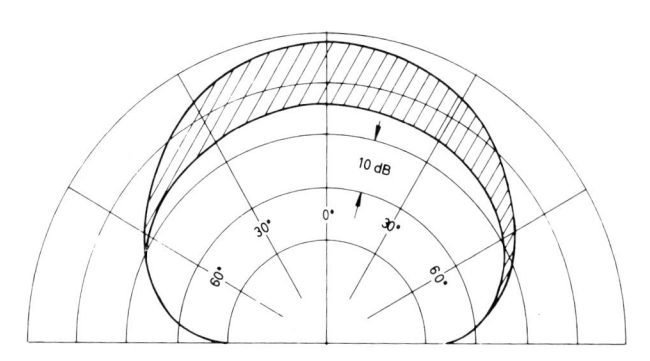

Fig. 3. Gain diagram of arrangement according to Fig. 2.

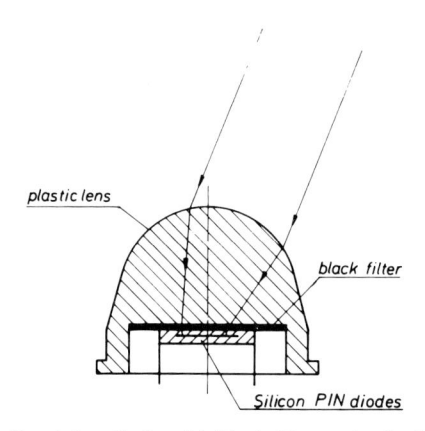

Fig. 2. Receiving diode with black filter and spherical lens.

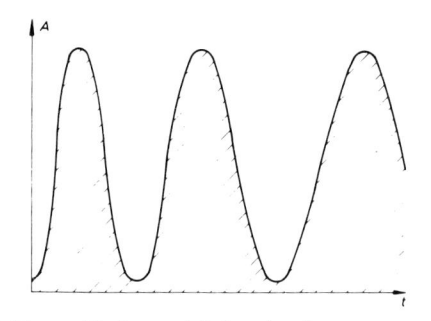

Fig. 4. Light amplitude modulation by frequency-modulated carrier.

receiver diode picks up the complex infrared signal containing a mixture of all different carrier frequencies. This signal is amplified and then passed on to the mixer. The filter for the intermediate frequency is a commercial high-quality ceramic filter.

The receiver is shown in Fig. 7. The only control elements in this receiver are the channel selector knob and the volume control. The unit is switched on by the insertion of the headphone plug into the bottom of the housing. At the lower corners, contacts are located for charging the internal battery. The receiver can be worn comfortably on the jacket by means of the attached clip or by means of a cord around the neck.

The transmitter consists essentially of a 9-channel modulating and driving unit and an appropriate number of power radiators, depending on the size of the room where the equipment is to be used. The block-diagram of the transmitter is shown in Fig. 8. Each channel is equipped with a symmetrical input, followed by a limiter amplifier to keep the modulation within the permitted range. Thus each channel delivers a sinusoidal signal with very low distortion. Low distortion is essential in order to have low crosstalk from one channel to the other. By means of summing resistors and a channel switch, each signal can be switched through to the power amplifier.

Fig. 9 shows the outside of the transmitter. Each channel is equipped with a modulation indicator. The power supply has been designed to feed four power radiators. The mixture of the different carrier frequencies is fed to the power radiators via coaxial cable. Fig. 10 shows a block diagram of a power radiator. At the input there is an amplifier with automatic gain control. The output voltage is independent of the number of channels used and of the loss in the cable carrying the modulated signal. Thus the maximum power capability is automatically emitted, regardless of the number of channels chosen. If, for instance, only three channels are used simultaneously, the modulation of the infrared light amplitude is three times the value obtained when all nine channels are used. The mixture of all carrier frequencies amplified as described is

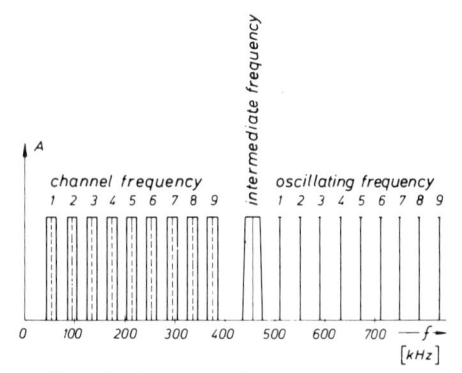

Fig. 5. Frequency diagram of system.

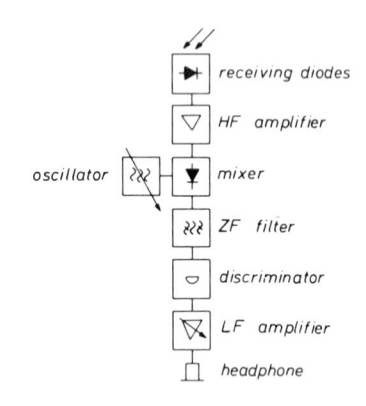

Fig. 6. Block diagram of receiver.

Fig. 7. Receiver.

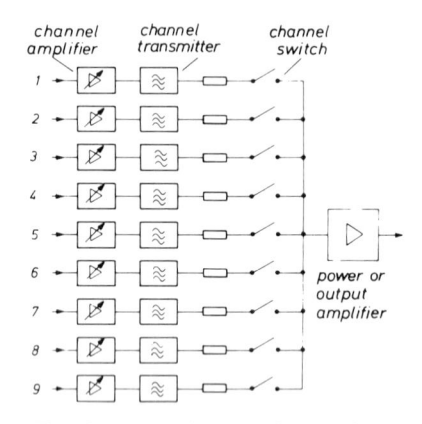

Fig. 8. Block diagram of transmitter.

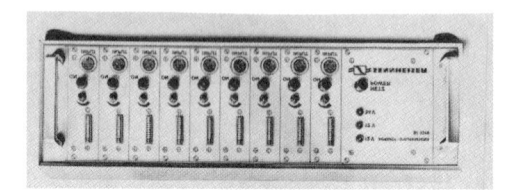

Fig. 9. Transmitter.

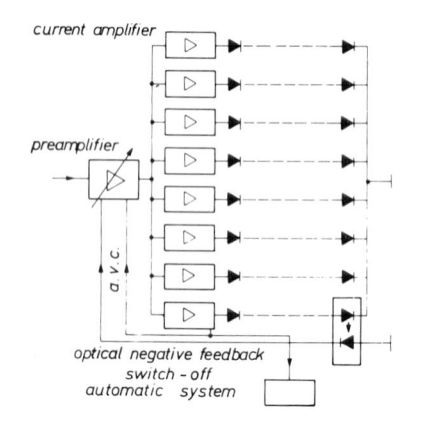

Fig. 10. Block diagram of power radiator.

passed into eight current amplifiers, each feeding fifteen light-emitting diodes in series.

From the design of optocouplers it is well known that p-i-n diodes show a very strict proportionality between the incoming light and the current passing through the diode. However, in this special application the normal linearity was not quite sufficient, leading to considerable crosstalk between the channels. To improve linearity an optical feedback technique had to be used. One light-emitting-diode is coupled directly onto a silicon p-i-n diode, and the current through the diode is used in the feedback loop of the input amplifier. Thus a very high signal-to-noise ratio is obtained.

Furthermore, each power radiator is equipped with an automatic switch which turns off the entire power radiator should a defect occur within the optical feedback loop. Failure of single diodes, or of up to three groups of diodes which are not incorporated in the optical feedback, results only in a reduction of the emitted light.

For use in large rooms almost any number of power radiators may be connected in parallel. This leads to a very simple adaptation of the complete conference system to the varying needs. This is of special importance when infrared conference systems are used in different rooms on various occasions. Fig. 11 shows a power radiator.

Complete conference systems need a certain number of powering units to recharge the large number of receivers. Fig. 12 shows such a charging unit for 100 receivers. The charging time can be preselected by a built-in clock. After completion of the selected charging cycle, the built-in circuitry switches automatically to a small current to keep the batteries fully powered. In addition, the charging unit contains five spare boxes and one more for testing purposes.

Fig. 11. Power radiator.

Fig. 12. Charging unit.

THE AUTHOR

Hans-Joachim Griese was born in 1915 in Wismar, Germany, and received a doctorate in communications theory from the Hannover Institute of Technology in 1942. From his graduation until 1944 he was with Feuerstein Laboratories as department head in charge of vocoder and long-range wireless communication systems development, and in 1945 he joined Sennheiser Electronic, a company specializing in the manufacture of microphones. Between 1945 and 1949, and again since 1956, Dr. Griese has been chief physicist of this company. In the intervening years he was associated with the Grundig Corporation and the Broadcasting Technology Research Institute of Nuernberg, working primarily on high-frequency techniques. He is the author of numerous papers in his field.